Principles of Terrestrial Ecosystem Ecology

Springer
New York
Berlin
Heidelberg
Hong Kong
London
Milan
Paris
Tokyo

F. Stuart Chapin III
Pamela A. Matson
Harold A. Mooney

Principles of Terrestrial Ecosystem Ecology

Illustrated by Melissa C. Chapin

With 199 Illustrations

Springer

F. Stuart Chapin III
Institute of Arctic Biology
University of Alaska
Fairbanks, AK 99775
USA
terry.chapin@uaf.edu

Pamela A. Matson
School of Earth Sciences
Mitchell Hall 101
Stanford University
397 Panama Mall
Stanford, CA 94305-2210
USA
matson@pangea.stanford.edu

Harold A. Mooney
Department of Biological Sciences
Herrin Hall, MC 5020
Stanford University
Stanford, CA 94305-5020
USA
hmooney@jasper.stanford.edu

Cover illustration: Waterfall and forests on Valean Poas in Costa Rica. Photograph by Peter Vitousek.

Library of Congress Cataloging-in-Publication Data
Chapin, F. Stuart (Francis Stuart), III.
Principles of terrestrial ecosystem ecology / F. Stuart Chapin III, Pamela A. Matson, Harold A. Mooney.
p. cm.
Includes bibliographical references (p.)
ISBN 0-387-95439-2 (hc :alk. paper)—ISBN 0-387-95443-0 (sc :alk. paper)
1. Ecology. 2. Biogeochemical cycles. 3. Biological systems. I. Matson, P.A. (Pamela A.) II. Mooney, Harold A. III. Title.
QH541 .C3595 2002
577′.14—dc21 2002017654

ISBN 0-387-95439-2 (hardcover)
ISBN 0-387-95443-0 (softcover)

Printed on acid-free paper.

Printed in the United States of America. (BS/EB)

9 8 7 6 5 4 3

SPIN 11007890 (softcover)
SPIN 10866301 (hardcover)

Springer-Verlag is a part of *Springer Science+Business Media*

springeronline.com

Preface

Human activities are affecting the global environment in myriad ways, with numerous direct and indirect effects on ecosystems. The climate and atmospheric composition of Earth are changing rapidly. Humans have directly modified half of the ice-free terrestrial surface and use 40% of terrestrial production. Our actions are causing the sixth major extinction event in the history of life on Earth and are radically modifying the interactions among forests, fields, streams, and oceans. This book was written to provide a conceptual basis for understanding terrestrial ecosystem processes and their sensitivity to environmental and biotic changes. We believe that an understanding of how ecosystems operate and change must underlie our analysis of both the consequences and the mitigation of human-caused changes.

This book is intended to introduce the science of ecosystem ecology to advanced undergraduate students, beginning graduate students, and practicing scientists from a wide array of disciplines. We also provide access to some of the rapidly expanding literature in the many disciplines that contribute to ecosystem understanding.

The first part of the book provides the context for understanding ecosystem ecology. We introduce the science of ecosystem ecology and place it in the context of other components of the Earth System—the atmosphere, ocean, climate and geological systems. We show how these components affect ecosystem processes and contribute to the global variation in terrestrial ecosystem structure and processes. In the second part of the book, we consider the mechanisms by which terrestrial ecosystems function and focus on the flow of water and energy and the cycling of carbon and nutrients. We then compare and contrast these cycles between terrestrial and aquatic ecosystems. We also consider the important role that organisms have on ecosystem processes through trophic interactions (feeding relationships), environmental effects, and disturbance. The third part of the book addresses temporal and spatial patterns in ecosystem processes. We finish by considering the integrated effects of these processes at the global scale and their consequences for sustainable use by human societies. Powerpoint lecture notes developed by one of the authors are available online (www.faculty.uaf.edu/fffsc/) as supplementary material.

Many people have contributed to the development of this book. We particularly thank our families, whose patience has made the book possible, and our students from whom we have learned many of the important ideas that are presented. In addition, we thank the following individuals for their constructively critical review of chapters in this book: Kevin Arrigo, Teri Balser, Perry Barboza, Jason Beringer, Kim Bonine, Rich Boone, Syndonia Bret-Harte, John Bryant, Inde Burke, Zoe Cardon, Oliver Chadwick, Scott Chambers, Melissa Chapin, Kathy Cottingham, Joe Craine, Wolfgang Cramer, Steve Davis, Sandra Diaz, Bill Dietrich, Rob Dunbar, Jim Ehleringer, Howie Epstein, Werner Eugster, Valerie Eviner, Scott Fendorf, Jon Foley, David Foster, Tom Gower, Peter Groffman, Paul Grogan, Diego Gurvich, Bill Heal, Sarah Hobbie, Dave Hooper, Shuijin Hu, Pilar Huante, Bruce Hungate, Jill Johnstone, Jay Jones, Jürg Luterbacher, Frank Kelliher, Jennifer King, Dave Kline, Christian Körner, Hans Lambers, Amanda Lynch, Michelle Mack, Steve MacLean, Joe McFadden, Dave McGuire, Sam McNaughton, Knute Nadelhoffer, Jason Neff, Mark Oswood, Bob Paine, Bill Parton, Natalia Perez, Steward Pickett, Stephen Parder, Mary Power, Jim Randerson, Bill Reeburgh, Peter Reich, Jim Reynolds, Roger Ruess, Steve Running, Scott Rupp, Dave Schimel, Josh Schimel, Bill Schlesinger, Guthrie Schrengohst, Ted Schuur, Stephen Parder Mark Serreze, Gus Shaver, Nigel Tapper, Monica Turner, Dave Valentine, Peter Vitousek, Lars Walker, and Katey Walter. We particularly thank Phil Camil, Valerie Eviner, Jon Foley, and Paul Grogan for comments on the entire book; Mark Chapin, Patrick Endres, and Rose Meier for comments on illustrations; Phil Camil for comments on educational approaches; and Jon Foley and Nick Olejniczak for providing global maps.

F. Stuart Chapin III
Pamela A. Matson
Harold A. Mooney

Contents

Part I
Context

1
The Ecosystem Concept

Ecosystem ecology studies the links between organisms and their physical environment within an Earth System context. This chapter provides background on the conceptual framework and history of ecosystem ecology.

Introduction

Ecosystem ecology addresses the interactions between organisms and their environment as an integrated system. The ecosystem approach is fundamental in managing Earth's resources because it addresses the interactions that link biotic systems, of which humans are an integral part, with the physical systems on which they depend. This applies at the scale of Earth as a whole, a continent, or a farmer's field. An ecosystem approach is critical to resource management, as we grapple with the sustainable use of resources in an era of increasing human population and consumption and large, rapid changes in the global environment.

Our growing dependence on ecosystem concepts can be seen in many areas. The United Nations Convention on Biodiversity of 1992, for example, promoted an ecosystem approach, including humans, to conserving biodiversity rather than the more species-based approaches that predominated previously. There is a growing appreciation of the role that individual species, or groups of species, play in the functioning of ecosystems and how these functions provide services that are vital to human welfare. An important, and belated, shift in thinking has occurred about managing ecosystems on which we depend for food and fiber. The supply of fish from the sea is now declining because fisheries management depended on species-based approaches that did not adequately consider the resources on which commercial fish depend. A more holistic view of managed systems can account for the complex interactions that prevail in even the simplest ecosystems. There is also an increasing appreciation that a thorough understanding of ecosystems is critical to managing the quality and quantity of our water supplies and in regulating the composition of the atmosphere that determines Earth's climate.

Overview of Ecosystem Ecology

The flow of energy and materials through organisms and the physical environment provides a framework for understanding the diversity of form and functioning of Earth's physical and biological processes. Why do tropical forests have large trees but accumulate only a thin layer of dead leaves on the soil surface, whereas tundra supports small plants but an abundance of soil organic matter? Why does the concentration of carbon dioxide in the atmosphere decrease in summer and increase in winter? What happens to that portion of the nitrogen that is added to farmers' fields but is not harvested with the crop? Why has the introduction of exotic species so strongly affected the productivity and fire frequency of grasslands and forests? Why does the number of people on Earth correlate so strongly with the concentration of methane in the Antarctic ice cap or with the quantity of nitrogen entering Earth's oceans? These are representative questions addressed by ecosystem ecology. Answers to these questions require an understanding of the interactions between organisms and their physical environments—both the response of organisms to environment and the effects of organisms on their environment. Addressing these questions also requires that we think of integrated ecological systems rather than individual organisms or physical components.

Ecosystem analysis seeks to understand the factors that regulate the **pools** (quantities) and **fluxes** (flows) of materials and energy through ecological systems. These materials include carbon, water, nitrogen, rock-derived minerals such as phosphorus, and novel chemicals such as pesticides or radionuclides that people have added to the environment. These materials are found in **abiotic** (nonbiological) pools such as soils, rocks, water, and the atmosphere and in biotic pools such as plants, animals, and soil microorganisms.

An **ecosystem** consists of all the organisms and the abiotic pools with which they interact. **Ecosystem processes** are the transfers of energy and materials from one pool to another. Energy enters an ecosystem when light energy drives the reduction of carbon dioxide (CO_2) to form sugars during photosynthesis. Organic matter and energy are tightly linked as they move through ecosystems. The energy is lost from the ecosystem when organic matter is oxidized back to CO_2 by combustion or by the respiration of plants, animals, and microbes. Materials move among abiotic components of the system through a variety of processes, including the weathering of rocks, the evaporation of water, and the dissolution of materials in water. Fluxes involving biotic components include the absorption of minerals by plants, the death of plants and animals, the decomposition of dead organic matter by soil microbes, the consumption of plants by herbivores, and the consumption of herbivores by predators. Most of these fluxes are sensitive to environmental factors, such as temperature and moisture, and to biological factors that regulate the population dynamics and species interactions in communities. The unique contribution of ecosystem ecology is its focus on biotic and abiotic factors as interacting components of a single integrated system.

Ecosystem processes can be studied at many spatial scales. How big is an ecosystem? The appropriate scale of study depends on the question being asked (Fig. 1.1). The impact of zooplankton on the algae that they eat might be studied in the laboratory in small bottles. Other questions such as the controls over productivity might be studied in relatively homogeneous patches of a lake, forest, or agricultural field. Still other questions are best addressed at the global scale. The concentration of atmospheric CO_2, for example, depends on global patterns of biotic exchanges of CO_2 and the burning of fossil fuels, which are spatially variable across the globe. The rapid mixing of CO_2 in the atmosphere averages across this variability, facilitating estimates of long-term changes in the total global flux of carbon between Earth and the atmosphere.

Some questions require careful measurements of lateral transfers of materials. A watershed is a logical unit in which to study the effects of forests on the quantity and quality of the water that supplies a town reservoir. A **watershed**, or catchment, consists of a stream and all the terrestrial surfaces that drain into it. By studying a watershed we can compare the quantities of materials that enter from the air and rocks with the amounts that leave in stream water, just as you balance your checkbook. Studies of input–output budgets of watersheds have improved our understanding of the interactions between rock weathering, which supplies nutrients, and plant and microbial growth, which retains nutrients in ecosystems (Vitousek and Reiners 1975, Bormann and Likens 1979).

The upper and lower boundaries of an ecosystem also depend on the question being asked and the scale that is appropriate to the

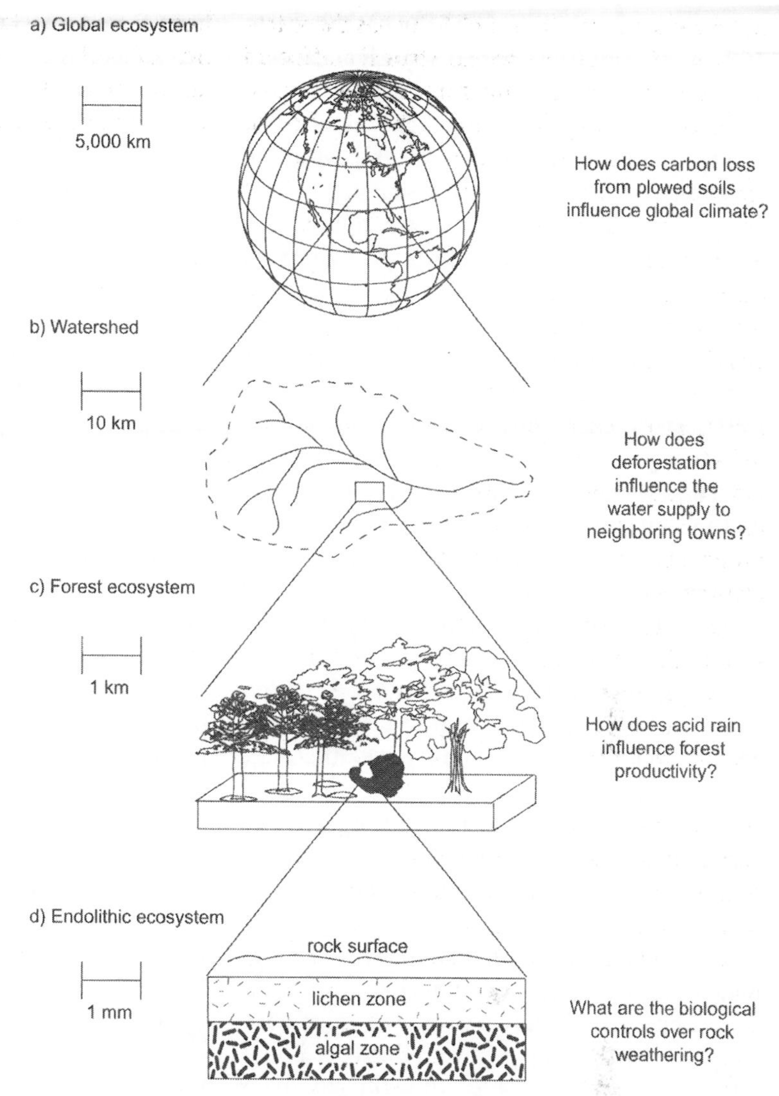

FIGURE 1.1. Examples of ecosystems that range in size by 10 orders of magnitude: an endolithic ecosystem in the surface layers of rocks, 1×10^{-3} m in height (d); a forest, 1×10^{3} m in diameter (c); a watershed, 1×10^{5} m in length (b); and Earth, 4×10^{7} m in circumference (a). Also shown are examples of questions appropriate to each scale.

question. The atmosphere, for example, extends from the gases between soil particles all the way to outer space. The exchange of CO_2 between a forest and the atmosphere might be measured a few meters above the top of the canopy because, above this height, variations in CO_2 content of the atmosphere are also strongly influenced by other upwind ecosystems. The *regional* impact of grasslands on the moisture content of the atmosphere might, however, be measured at a height of several kilometers above the ground surface, where the moisture released by the ecosystem condenses and returns as precipitation (see Chapter 2). For questions that address plant effects on water and nutrient cycling, the bottom of the ecosystem might be the maximum depth to which roots extend because soil water or nutrients below this depth are inaccessible to the vegetation. Studies of long-term soil development, in contrast, must also consider rocks deep in the soil, which constitute the long-term reservoir of many nutrients that gradually become incorporated into surface soils (see Chapter 3).

Ecosystem dynamics are a product of many temporal scales. The rates of ecosystem processes are constantly changing due to fluctuations in environment and activities of organisms on time scales ranging from microseconds to millions of years (see Chapter 13). Light capture during photosynthesis responds almost instantaneously to fluctuations in light availability to a leaf. At the opposite extreme, the evolution of photosynthesis 2 billion years ago added oxygen to the atmosphere over millions of years, causing the prevailing geochemistry of Earth's surface to change from chemical reduction to chemical oxidation (Schlesinger 1997). Microorganisms in the group Archaea evolved in the early reducing atmosphere of Earth. These microbes are still the only organisms that produce methane. They now function in anaerobic environments such as wetland soils and the interiors of soil aggregates or animal intestines. Episodes of mountain building and erosion strongly influence the availability of minerals to support plant growth. Vegetation is still migrating in response to the retreat of Pleistocene glaciers 10,000 to 20,000 years ago. After disturbances such as fire or tree fall, there are gradual changes in plant, animal, and microbial communities over years to centuries. Rates of carbon input to an ecosystem through photosynthesis change over time scales of seconds to decades due to variations in light, temperature, and leaf area.

Many early studies in ecosystem ecology made the simplifying assumption that some ecosystems are in **equilibrium** with their environment. In this perspective, relatively undisturbed ecosystems were thought to have properties that reflected (1) largely closed systems dominated by internal recycling of elements, (2) self-regulation and deterministic dynamics, (3) stable end points or cycles, and (4) absence of disturbance and human influence (Pickett et al. 1994, Turner et al. 2001). One of the most important conceptual advances in ecosystem ecology has been the increasing recognition of the importance of past events and external forces in shaping the functioning of ecosystems. In this nonequilibrium perspective, we recognize that most ecosystems exhibit inputs and losses, their dynamics are influenced by both external and internal factors, they exhibit no single stable equilibrium, disturbance is a natural component of their dynamics, and human activities have a pervasive influence. The complications associated with the current nonequilibrium view require a more dynamic and stochastic view of controls over ecosystem processes.

Ecosystems are considered to be at **steady state** if the balance between inputs and outputs to the system shows no trend with time (Johnson 1971, Bormann and Likens 1979). Steady state assumptions differ from equilibrium assumptions because they accept temporal and spatial variation as a normal aspect of ecosystem dynamics. Even at steady state, for example, plant growth changes from summer to winter and between wet and dry years (see Chapter 6). At a stand scale, some plants may die from old age or pathogen attack and be replaced by younger individuals. At a landscape scale, some patches may be altered by fire or other disturbances, and other patches will be in various stages of recovery. These ecosystems or landscapes are in steady state if there is no long-term directional trend in their properties or in the balance between inputs and outputs.

Not all ecosystems and landscapes are in steady state. In fact, directional changes in climate and environment caused by human activities are quite likely to cause directional changes in ecosystem properties. Nonetheless, it is often easier to understand the relationship of ecosystem processes to the current environment in situations in which they are not also recovering from large recent perturbations. Once we understand the behavior of a system in the absence of recent disturbances, we can add the complexities associated with time lags and rates of ecosystem change.

Ecosystem ecology uses concepts developed at finer levels of resolution to build an understanding of the mechanisms that govern the entire Earth System. The biologically mediated movement of carbon and nitrogen through ecosystems depends on the physiological properties of plants, animals, and soil microorganisms. The traits of these organisms are the products of their evolutionary histories and the competitive interactions that sort species into communities where they successfully grow, survive, and reproduce (Vrba and Gould 1986). Ecosystem fluxes also depend

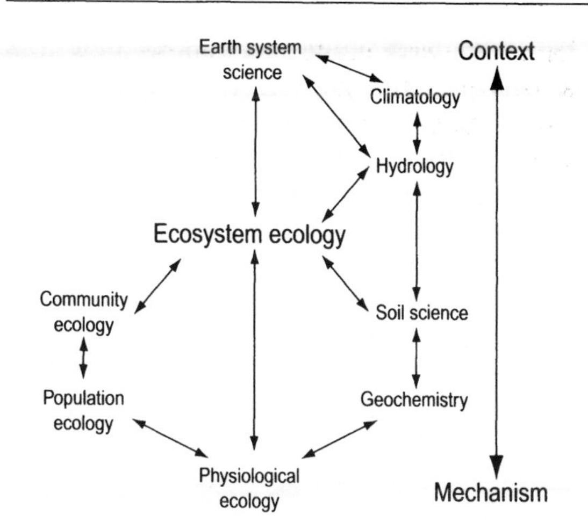

FIGURE 1.2. Relationships between ecosystem ecology and other disciplines. Ecosystem ecology integrates the principles of several biological and physical disciplines and provides the mechanistic basis for Earth System Science.

on the population processes that govern plant, animal, and microbial densities and age structures as well as on community processes, such as competition and predation, that determine which species are present and their rates of resource consumption. Ecosystem ecology therefore depends on information and principles developed in physiological, evolutionary, population, and community ecology (Fig. 1.2).

The supply of water and minerals from soils to plants depends not only on the activities of soil microorganisms but also on physical and chemical interactions among rocks, soils, and the atmosphere. The low availability of phosphorus due to the extensive weathering and erosional loss of nutrients in the ancient soils of western Australia, for example, strongly constrains plant growth and the quantity and types of plants and animals that can be supported. Principles of ecosystem ecology must therefore also incorporate the concepts and understanding of disciplines such as geochemistry, hydrology, and climatology that focus on the physical environment (Fig. 1.2).

Ecosystem ecology provides the mechanistic basis for understanding processes that occur at global scales. Study of Earth as a physical system relies on information provided by ecosystem ecologists about the rates at which the land or water surface interacts with the atmosphere, rocks, and waters of the planet (Fig. 1.2). Conversely, the global budgets of materials that cycle between the atmosphere, land, and oceans provide a context for understanding the broader significance of processes studied in a particular ecosystem. Latitudinal and seasonal patterns of atmospheric CO_2 concentration, for example, help define the locations where carbon is absorbed or released from the land and oceans (see Chapter 15).

History of Ecosystem Ecology

Many early discoveries of biology were motivated by questions about the integrated nature of ecological systems. In the seventeenth century, European scientists were still uncertain about the source of materials found in plants. Plattes, Hooke, and others advanced the novel idea that plants derive nourishment from both air and water (Gorham 1991). Priestley extended this idea in the eighteenth century by showing that plants produce a substance that is essential to support the breathing of animals. At about the same time MacBride and Priestley showed that breakdown of organic matter caused the production of "fixed air" (carbon dioxide), which did not support animal life. De Saussure, Liebig, and others clarified the explicit roles of carbon dioxide, oxygen, and mineral nutrients in these cycles in the nineteenth century. Much of the biological research during the nineteenth and twentieth centuries went on to explore the detailed mechanisms of biochemistry, physiology, behavior, and evolution that explain how life functions. Only in recent decades have we returned to the question that originally motivated this research: How are biogeochemical processes integrated in the functioning of natural ecosystems?

Many threads of ecological thought have contributed to the development of ecosystem ecology (Hagen 1992), including ideas relating to **trophic interactions** (the feeding relationships among organisms) and **biogeochemistry** (biological influences on the chemical processes in ecosystems). Early research on trophic interactions emphasized the transfer of energy among organisms. Elton (1927), an English zoologist interested in natural history, described the role that an animal plays in a community (its **niche**) in terms of what it eats and is eaten by. He viewed each animal species as a link in a **food chain**, which described the movement of matter from one organism to another. Elton's concepts of trophic structure provide a framework for understanding the flow of materials through ecosystems (see Chapter 11).

Hutchinson, an American limnologist, was strongly influenced by the ideas of Elton and those of Russian geochemist Vernadsky, who described the movement of minerals from soil into vegetation and back to soil. Hutchinson suggested that the resources available in a lake must limit the productivity of algae and that algal productivity, in turn, must limit the abundance of animals that eat algae. Meanwhile, Tansley (1935), a British terrestrial plant ecologist, was also concerned that ecologists focused their studies so strongly on organisms that they failed to recognize the importance of exchange of materials between organisms and their abiotic environment. He coined the term ***ecosystem*** to emphasize the importance of interchanges of materials between inorganic and organic components as well as among organisms.

Lindeman, another limnologist, was strongly influenced by all these threads of ecological theory. He suggested that energy flow through an ecosystem could be used as a currency to quantify the roles of organisms in trophic dynamics. Green plants (**primary producers**) capture energy and transfer it to animals (**consumers**) and **decomposers**. At each transfer, some energy is lost from the ecosystem through respiration. Therefore, the productivity of plants constrains the quantity of consumers that an ecosystem can support. The energy flow through an ecosystem maps closely to carbon flow in the processes of photosynthesis, trophic transfers, and respiratory release of carbon. Lindeman's dissertation research on the trophic-dynamic aspect of ecology was initially rejected for publication. Reviewers felt that there were insufficient data to draw such broad conclusions and that it was inappropriate to use mathematical models to infer general relationships based on observations from a single lake. Hutchinson, Lindeman's postdoctoral adviser, finally (after Lindeman's death) persuaded the editor to publish this paper, which has been the springboard for many of the basic concepts in ecosystem theory (Lindeman 1942).

H. T. Odum, also trained by Hutchinson, and his brother E. P. Odum further developed the **systems approach** to studying ecosystems, which emphasizes the general properties of ecosystems without documenting all the underlying mechanisms and interactions. The Odum brothers used radioactive tracers to measure the movement of energy and materials through a coral reef. These studies enabled them to document the patterns of energy flow and metabolism of whole ecosystems and to suggest generalizations about how ecosystems function (Odum 1969). Ecosystem budgets of energy and materials have since been developed for many fresh-water and terrestrial ecosystems (Lindeman 1942, Ovington 1962, Golley 1993), providing information that is essential for generalizing about global patterns of processes such as productivity. Some of the questions addressed by systems ecology include information transfer (Margalef 1968), the structure of food webs (Polis 1991), the hierarchical changes in ecosystem controls at different temporal and spatial scales (O'Neill et al. 1986), and the resilience of ecosystem properties after disturbance (Holling 1986).

We now recognize that element cycles interact in important ways and cannot be understood in isolation. The availability of water and nitrogen are important determinants of the rate at which carbon cycles through the ecosystem. Conversely, the productivity of vegetation strongly influences the cycling rates of nitrogen and water.

Recent global changes in the environment have made ecologists increasingly aware of the changes in ecosystem processes that occur in response to disturbance or other environmental changes. **Succession**, the directional change in ecosystem structure and functioning resulting from biotically driven changes in resource supply, is an important framework for understanding these transient dynamics of ecosystems. Early American ecologists such as Cowles and Clements were struck by the relatively predictable patterns of vegetation development after exposure of unvegetated land surfaces. Sand dunes on Lake Michigan, for example, are initially colonized by drought-resistant herbaceous plants that give way to shrubs, then small trees, and eventually forests (Cowles 1899). Clements (1916) advanced a theory of community development, suggesting that this vegetation succession is a predictable process that eventually leads, in the absence of disturbance, to a stable community type characteristic of a particular climate (the **climatic climax**). He suggested that a community is like an organism made of interacting parts (species) and that successional development toward a climax community is analogous to the development of an organism to adulthood. This analogy between an ecological community and an organism laid the groundwork for concepts of ecosystem physiology (for example, the net ecosystem exchange of CO_2 and water vapor between the ecosystem and the atmosphere). The measurements of net ecosystem exchange are still an active area of research in ecosystem ecology, although they are now motivated by different questions than those posed by Clements. His ideas were controversial from the outset. Other ecologists, such as Gleason (1926), felt that vegetation change was not as predictable as Clements had implied. Instead, chance dispersal events explained much of the vegetation patterns on the landscape. This debate led to a century of research on the mechanisms responsible for vegetation change (see Chapter 13).

Another general approach to ecosystem ecology has emphasized the controls over ecosystem processes through comparative studies of ecosystem components. This interest originated in studies by plant geographers and soil scientists who described general patterns of variation with respect to climate and geological substrate (Schimper 1898). These studies showed that many of the global patterns of plant production and soil development vary predictably with climate (Jenny 1941, Rodin and Bazilevich 1967, Lieth 1975). The studies also showed that, in a given climatic regime, the properties of vegetation depended strongly on soils and vice versa (Dokuchaev 1879, Jenny 1941, Ellenberg 1978). Process-based studies of organisms and soils provided insight into many of the mechanisms underlying the distributions of organisms and soils along these gradients (Billings and Mooney 1968, Mooney 1972, Larcher 1995, Paul and Clark 1996). These studies also formed the basis for extrapolation of processes across complex landscapes to characterize large regions (Matson and Vitousek 1987, Turner et al. 2001). These studies often relied on field or laboratory experiments that manipulated some ecosystem property or process or on comparative studies across environmental gradients. This approach was later expanded to studies of intact ecosystems, using whole-ecosystem manipulations (Likens et al. 1977, Schindler 1985, Chapin et al. 1995) and carefully designed gradient studies (Vitousek et al. 1988).

Ecosystem experiments have provided both basic understanding and information that are critical in management decisions. The clear-cutting of an experimental watershed at Hubbard Brook in the northeastern United States, for example, caused a fourfold increase in streamflow and stream nitrate concentration—to levels exceeding health standards for drinking water (Likens et al. 1977). These dramatic results demonstrate the key role of vegetation in regulating the cycling of water and nutrients in forests. The results halted plans for large-scale deforestation that had been planned to increase supplies of drinking water during a long-term drought. Nutrient addition experiments in the Experimental Lakes Area of southern Canada showed that phosphorus limits the productivity of many lakes (Schindler 1985) and that pollution was responsible for algal blooms and fish kills that were common in lakes near densely populated areas in the 1960s. This research provided the basis for regulations that removed phosphorus from detergents.

Changes in the Earth System have led to studies of the interactions among terrestrial

ecosystems, the atmosphere, and the oceans. The dramatic impact of human activities on the Earth System (Vitousek 1994a) has led to the urgent necessity to understand how terrestrial ecosystem processes affect the atmosphere and oceans. The scale at which these ecosystem effects are occurring is so large that the traditional tools of ecologists are insufficient. Satellite-based remote sensing of ecosystem properties, global networks of atmospheric sampling sites, and the development of global models are important new tools that address global issues. Information on global patterns of CO_2 and pollutants in the atmosphere, for example, provide telltale evidence of the major locations and causes of global problems (Tans et al. 1990). This gives hints about which ecosystems and processes have the greatest impact on the Earth System and therefore where research and management should focus efforts to understand and solve these problems (Zimov et al. 1999).

The intersection of systems approaches, process understanding, and global analysis is an exciting frontier of ecosystem ecology. How do changes in the global environment alter the controls over ecosystem processes? What are the integrated system consequences of these changes? How do these changes in ecosystem properties influence the Earth System? The rapid changes that are occurring in ecosystems have blurred any previous distinction between basic and applied research. There is an urgent need to understand how and why the ecosystems of Earth are changing.

Ecosystem Structure

Most ecosystems gain energy from the sun and materials from the air or rocks, transfer these among components within the ecosystem, then release energy and materials to the environment. The essential biological components of ecosystems are plants, animals, and decomposers. **Plants** capture solar energy in the process of bringing carbon into the ecosystem. A few ecosystems, such as deep-sea hydrothermal vents, have no plants but instead have bacteria that derive energy from the oxidation of hydrogen sulfide (H_2S) to produce organic matter. **Decomposer** microorganisms (microbes) break down dead organic material, releasing CO_2 to the atmosphere and nutrients in forms that are available to other microbes and plants. If there were no decomposition, large accumulations of dead organic matter would sequester the nutrients required to support plant growth. **Animals** are critical components of ecosystems because they transfer energy and materials and strongly influence the quantity and activities of plants and soil microbes. The essential abiotic components of an ecosystem are **water**; the **atmosphere**, which supplies carbon and nitrogen; and **soil minerals**, which supply other nutrients required by organisms.

An **ecosystem model** describes the major pools and fluxes in an ecosystem and the factors that regulate these fluxes. Nutrients, water, and energy differ from one another in the relative importance of ecosystem inputs and outputs vs. internal recycling (see Chapters 4 to 10). Plants, for example, acquire carbon primarily from the atmosphere, and most carbon released by respiration returns to the atmosphere. Carbon cycling through ecosystems is therefore quite open, with large inputs to, and losses from, the system. There are, however, relatively large pools of carbon stored in ecosystems, so the activities of animals and microbes are somewhat buffered from variations in carbon uptake by plants. The water cycle of ecosystems is also relatively open, with water entering primarily by precipitation and leaving by evaporation, transpiration, and drainage to groundwater and streams. In contrast to carbon, most ecosystems have a limited capacity to store water in plants and soil, so the activity of organisms is closely linked to water inputs. In contrast to carbon and water, mineral elements such as nitrogen and phosphorus are recycled rather tightly within ecosystems, with annual inputs and losses that are small relative to the quantities that annually recycle within the ecosystem. These differences in the "openness" and "buffering" of the cycles fundamentally influence the controls over rates and patterns of the cycling of materials through ecosystems.

The pool sizes and rates of cycling differ substantially among ecosystems (see Chapter 6). Tropical forests have much larger pools of carbon and nutrients in plants than do deserts or tundra. Peat bogs, in contrast, have large pools of soil carbon rather than plant carbon. Ecosystems also differ substantially in annual fluxes of materials among pools, for reasons that will be explored in later chapters.

Controls over Ecosystem Processes

Ecosystem structure and functioning are governed by at least five independent control variables. These **state factors**, as Jenny and co-workers called them, are **climate, parent material** (i.e., the rocks that give rise to soils), **topography, potential biota** (i.e., the organisms present in the region that could potentially occupy a site), and **time** (Fig. 1.3) (Jenny 1941, Amundson and Jenny 1997). Together these five factors set the bounds for the characteristics of an ecosystem.

On broad geographic scales, climate is the state factor that most strongly determines ecosystem processes and structure. Global variations in climate explain the distribution of **biomes** (types of ecosystems) such as wet tropical forests, temperate grasslands, and arctic tundra (see Chapter 2). Within each biome, parent material strongly influences the types of soils that develop and explains much of the regional variation in ecosystem processes (see Chapter 3). Topographic relief influences both microclimate and soil development at a local scale. The potential biota governs the types and diversity of organisms that actually occupy a site. Island ecosystems, for example, are frequently less diverse than climatically similar mainland ecosystems because new species reach islands less frequently and are more likely to go extinct than in mainland locations (MacArthur and Wilson 1967). Time influences the development of soil and the evolution of organisms over long time scales. Time also incorporates the influences on ecosystem processes of past disturbances and environmental changes over a wide range of time scales. These state factors are described in more detail in Chapter 3 in the context of soil development.

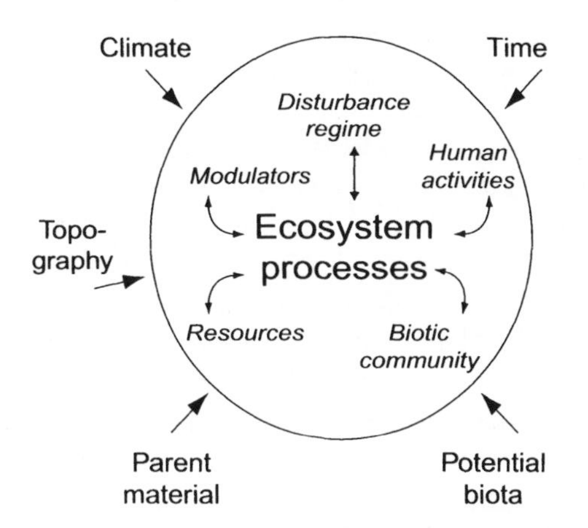

FIGURE 1.3. The relationship between state factors (outside the circle), interactive controls (inside the circle), and ecosystem processes. The circle represents the boundary of the ecosystem. (Modified with permission from *American Naturalist*, Vol. 148 © 1996 University of Chicago Press, Chapin et al. 1996.)

Jenny's state factor approach was a major conceptual contribution to ecosystem ecology. First, it emphasized the controls over processes rather than simply descriptions of patterns. Second, it suggested an experimental approach to test the importance and mode of action of each control. A logical way to study the role of each state factor is to compare sites that are as similar as possible with respect to all but one factor. For example, a **chronosequence** is a series of sites of different ages with similar climate, parent material, topography, and potential to be colonized by the same organisms (see Chapter 13). In a **toposequence**, ecosystems differ mainly in their topographic position (Shaver et al. 1991). Sites that differ primarily with respect to climate or parent material allow us to study the impact of these state factors on ecosystem processes (Vitousek et al. 1988, Walker et al. 1998). Finally, a comparison of ecosystems that differ primarily in potential biota, such as the mediterranean shrublands that have developed on west coasts of California, Chile, Portugal, South Africa, and Australia, illustrates the importance of evolutionary history in shaping ecosystem processes (Mooney and Dunn 1970).

Ecosystem processes both respond to and control the factors that directly govern their activity. For example, plants both respond to and influence their light, temperature, and moisture environment (Billings 1952). **Interactive controls** are factors that both *control* and *are controlled by* ecosystem characteristics (Fig. 1.3) (Chapin et al. 1996). Important interactive controls include the **supply of resources** to support the growth and maintenance of organisms, **modulators** that influence the rates of ecosystem processes, **disturbance regime**, the **biotic community**, and **human activities**.

Resources are the energy and materials in the environment that are used by organisms to support their growth and maintenance (Field et al. 1992). The acquisition of resources by organisms depletes their abundance in the environment. In terrestrial ecosystems these resources are spatially separated, being available primarily either aboveground (light and CO_2) or belowground (water and nutrients). Resource supply is governed by state factors such as climate, parent material, and topography. It is also sensitive to processes occurring within the ecosystem. Light availability, for example, depends on climatic elements such as cloudiness and on topographic position, but is also sensitive to the quantity of shading by vegetation. Similarly, soil fertility depends on parent material and climate but is also sensitive to ecosystem processes such as erosional loss of soils after overgrazing and inputs of nitrogen from invading nitrogen-fixing species. Soil water availability strongly influences species composition in dry climates. Soil water availability also depends on other interactive controls, such as disturbance regime (e.g., compaction by animals) and the types of organisms that are present (e.g., the presence or absence of deep-rooted trees such as mesquite that tap the water table). In aquatic ecosystems, water seldom directly limits the activity of organisms, but light and nutrients are just as important as on land. Oxygen is a particularly critical resource in aquatic ecosystems because of its slow rate of diffusion through water.

Modulators are physical and chemical properties that affect the activity of organisms but, unlike resources, are neither consumed nor depleted by organisms (Field et al. 1992). Modulators include temperature, pH, redox state of the soil, pollutants, UV radiation, etc. Modulators like temperature are constrained by climate (a state factor) but are sensitive to ecosystem processes, such as shading and evaporation. Soil pH likewise depends on parent material and time but also responds to vegetation composition.

Landscape-scale **disturbance** by fire, wind, floods, insect outbreaks, and hurricanes is a critical determinant of the natural structure and process rates in ecosystems (Pickett and White 1985, Sousa 1985). Like other interactive controls, disturbance regime depends on both state factors and ecosystem processes. Climate, for example, directly affects fire probability and spread but also influences the types and quantity of plants present in an ecosystem and therefore the fuel load and flammability of vegetation. Deposition and erosion during floods shape river channels and influence the probability of future floods. Change in either the intensity or frequency of disturbance can cause long-term ecosystem change. Woody plants, for example, often invade grasslands when fire suppression reduces fire frequency.

The nature of the **biotic community** (i.e., the types of species present, their relative abundances, and the nature of their interactions) can influence ecosystem processes just as strongly as do large differences in climate or parent material (see Chapter 12). These species effects can often be generalized at the level of **functional types**, which are groups of species that are similar in their role in community or ecosystem processes. Most evergreen trees, for example, produce leaves that have low rates of photosynthesis and a chemical composition that deters herbivores. These species make up a functional type because of their ecological similarity to one another. A gain or loss of key functional types—for example, through introduction or removal of species with important ecosystem effects—can permanently change the character of an ecosystem through changes in resource supply or disturbance regime. Introduction of nitrogen-fixing trees onto British mine wastes, for example, substantially increases nitrogen supply and productivity and alters patterns of vegetation development (Bradshaw 1983). Invasion by exotic grasses can alter fire frequency, resource supply, trophic interactions, and rates of most ecosystem processes (D'Antonio and Vitousek 1992). Elimination of predators by hunting can cause an outbreak of deer that overbrowse their food supply. The types of species present in an ecosystem depend strongly on other interactive controls (see Chapter 12), so functional types respond to and affect most interactive controls and ecosystem processes.

Human activities have an increasing impact on virtually all the processes that govern ecosystem properties (Vitousek 1994a). Our actions influence interactive controls such as water availability, disturbance regime, and biotic diversity. Humans have been a natural component of many ecosystems for thousands of years. Since the Industrial Revolution, however, the magnitude of human impact has been so great and so distinct from that of other organisms that the modern effects of human activities warrant particular attention. The cumulative impact of human activities extend well beyond an individual ecosystem and affect state factors such as climate, through changes in atmospheric composition, and potential biota, through the introduction and extinction of species. The large magnitude of these effects blurs the distinction between "independent" state factors and interactive controls at regional and global scales. Human activities are causing major changes in the structure and functioning of all ecosystems, resulting in novel conditions that lead to new types of ecosystems. The major human effects are summarized in the next section.

Feedbacks analogous to those in simple physical systems regulate the internal dynamics of ecosystems. A thermostat is an example of a simple physical feedback. It causes a furnace to switch on when a house gets cold. The house then warms until the thermostat switches the furnace off. Natural ecosystems are complex networks of interacting feedbacks (DeAngelis and Post 1991). **Negative feedbacks** occur when two components of a system have opposite effects on one another. Consumption of prey by a predator, for example, has a positive effect on the consumer but a negative effect on the prey. The negative effect of predators on prey prevents an uncontrolled growth of a predator's population, thereby stabilizing the population sizes of both predator and prey. There are also **positive feedbacks** in ecosystems in which both components of a system have a positive effect on the other, or both have a negative effect on one another. Plants, for example, provide their mycorrhizal fungi with carbohydrates in return for nutrients. This exchange of growth-limiting resources between plants and fungi promotes the growth of both components of the symbiosis until they become constrained by other factors.

Negative feedbacks are the key to sustaining ecosystems because strong negative feedbacks provide resistance to changes in interactive controls and maintain the characteristics of ecosystems in their current state. The acquisition of water, nutrients, and light to support growth of one plant, for example, reduces availability of these resources to other plants, thereby constraining community productivity (Fig. 1.4). Similarly, animal populations cannot sustain exponential population growth indefinitely, because declining food supply and increasing predation reduce the rate of population increase. If these negative feedbacks are weak or absent (a low predation rate due to predator control, for example), population cycles can amplify and lead to extinction of one or both of the interacting species. Community dynamics, which operate within a single ecosystem patch, primarily involve feedbacks among soil resources and functional types of organisms. Landscape dynamics, which govern changes in ecosystems through cycles of disturbance and recovery, involve additional feedbacks with microclimate and disturbance regime (see Chapter 14).

Human-Caused Changes in Earth's Ecosystems

Human activities transform the land surface, add or remove species, and alter biogeochemical cycles. Some human activities directly affect ecosystems through activities such as resource harvest, land use change, and management; other effects are indirect, as a result of changes in atmospheric chemistry, hydrology, and

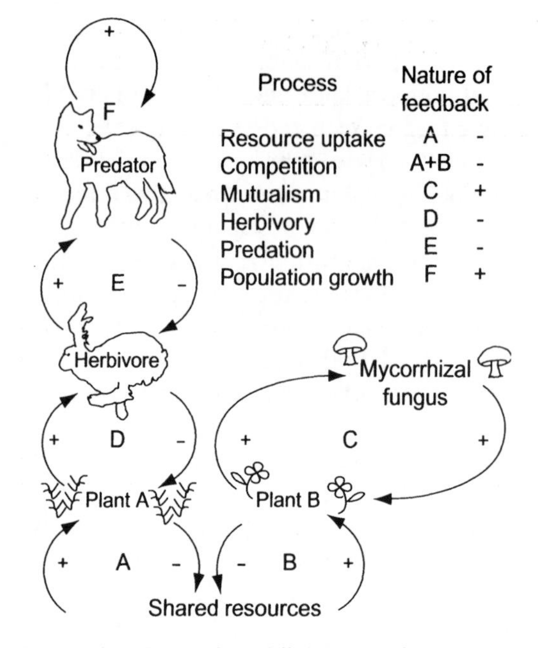

FIGURE 1.4. Examples of linked positive and negative feedbacks in ecosystems. The effect of each organism (or resource) on other organisms can be positive (+) or negative (–). Feedbacks are positive when the reciprocal effects of each organism (or resource) have the same sign (both positive or both negative). Feedbacks are negative when reciprocal effects differ in sign. Negative feedbacks resist the tendencies for ecosystems to change, whereas positive feedbacks tend to push ecosystems toward a new state. (Modified with permission from *American Naturalist*, Vol. 148 © 1996 University of Chicago Press, Chapin et al. 1996.)

climate (Fig. 1.5) (Vitousek et al. 1997c). At least some of these **anthropogenic** (i.e., human-caused) effects influence all ecosystems on Earth.

The most direct and substantial human alteration of ecosystems is through the transformation of land for production of food, fiber, and other goods used by people. About 50% of Earth's ice-free land surface has been directly altered by human activities (Kates et al. 1990). Agricultural fields and urban areas cover 10 to 15%, and pastures cover 6 to 8% of the land. Even more land is used for forestry and grazing systems. All except the most extreme environments of Earth experience some form of direct human impact.

Human activities have also altered freshwater and marine ecosystems. We use about half of the world's accessible runoff (see Chapter 15), and humans use about 8% of the primary production of the oceans (Pauly and Christensen 1995). Commercial fishing reduces the size and abundance of target species and alters the population characteristics of species that are incidentally caught in the fishery. In the mid-1990s, about 22% of marine fisheries were overexploited or already depleted, and an additional 44% were at their limit of exploitation (Vitousek et al. 1997c). About 60% of the human population resides within 100 km of a coast, so the coastal margins of oceans are strongly influenced by many human activities. Nutrient enrichment of many coastal waters, for example, has increased algal production and created anaerobic conditions that kill fish and other animals, due largely to transport of nutrients derived from agricultural fertilizers and from human and livestock sewage.

Land use change, and the resulting loss of habitat, is the primary driving force causing species extinctions and loss of biological diversity (Sala et al. 2000a) (see Chapter 12). The time lag between ecosystem change and species loss makes it likely that species will continue to be driven to extinction even where rates of land use change have stabilized. Transport of species around the world is homogenizing Earth's biota. The frequency of biological invasions is increasing, due to the globalization of the economy and increased international transport of products. Nonindigenous species now account for 20% or more of the plant species in many continental areas and 50% or more of the plant species on many islands (Vitousek et al. 1997c). International commerce breaks down biogeographic barriers, through both purposeful trade in live organisms and inadvertent introductions. Purposeful introductions deliberately select species that are likely to grow and reproduce effectively in their new environment. Many biological invasions are irreversible because it is difficult or prohibitively expensive to remove invasive species. Some species invasions degrade human health or cause large economic losses. Others alter the structure and functioning of ecosystems, leading to further loss of species diversity.

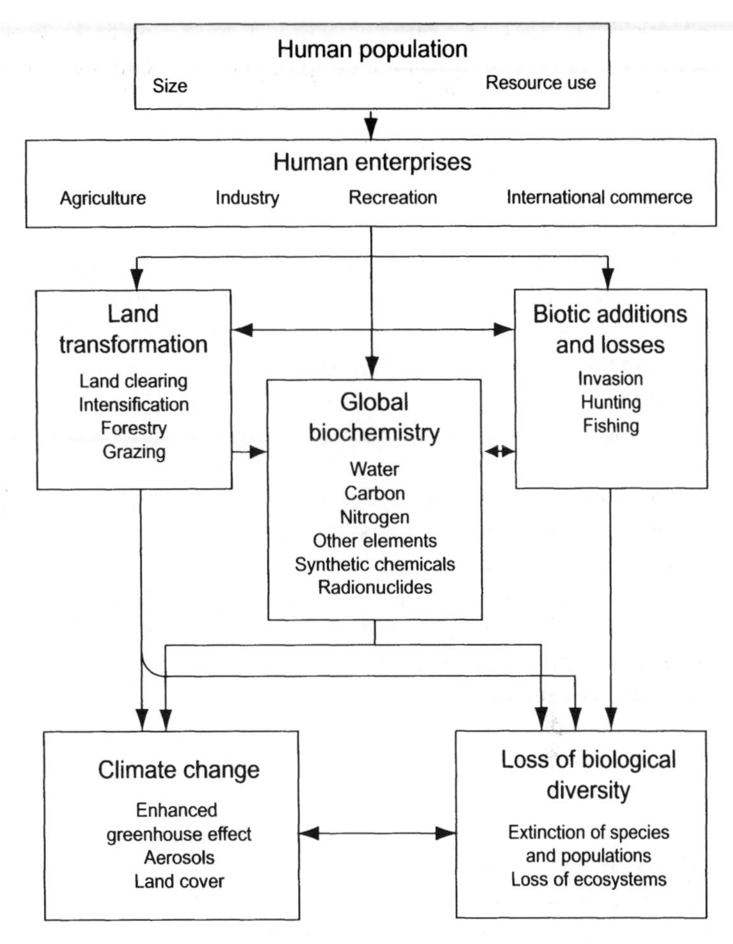

FIGURE 1.5. Direct and indirect effects of human activities on Earth's ecosystems. (Redrawn with permission from *Science*, Vol. 277 © 1997 American Association for the Advancement of Science; Vitousek et al. 1997c.)

Human activities have influenced biogeochemical cycles in many ways. Use of fossil fuels and the expansion and intensification of agriculture have altered the cycles of carbon, nitrogen, phosphorus, sulfur, and water on a global scale (see Chapter 15). These changes in biogeochemical cycles not only alter the ecosystems in which they occur but also influence unmanaged ecosystems through changes in lateral fluxes of nutrients and other materials through the atmosphere and surface waters (see Chapter 14). Land use changes, including deforestation and intensive use of fertilizers and irrigation, have increased the concentrations of atmospheric gases that influence climate (see Chapter 2). Land transformations also cause runoff and erosion of sediments and nutrients that lead to substantial changes in lakes, rivers, and coastal oceans.

Human activities introduce novel chemicals into the environment. Some apparently harmless anthropogenic gases have had drastic effects on the atmosphere and ecosystems. Chlorofluorocarbons (CFCs), for example, were first produced in the 1950s as refrigerants, propellants, and solvents. They were heralded for their nonreactivity in the lower atmosphere. In the upper atmosphere, however, where there is greater UV radiation, CFCs react with ozone. The resulting ozone destruction, which occurs primarily over the poles, creates a hole in the protective blanket of ozone that shields Earth's surface from UV radiation. This **ozone hole** was initially observed near the South Pole. It has expanded to lower latitudes in the Southern Hemisphere and now also occurs at high northern latitudes. As a result of the Montreal Protocol, the production of many CFCs has ceased. Due to their low reactivity, however, their concentrations in the atmosphere are only now beginning to decline, so their ecological effects will persist for decades. Persistent novel chemicals, such as CFCs, often have long-lasting ecological effects than cannot be predicted at the time they are first produced and which extend far beyond their region and duration of use.

Other synthetic organic chemicals include DDT (an insecticide) and polychlorinated biphenyls (PCBs; industrial compounds), which were used extensively in the developed world in the 1960s before their ecological consequences were widely recognized. Many of these compounds continue to be used in some developing nations. They are mobile and degrade slowly, causing them to persist and to be transported to all ecosystems of the globe. Many of these compounds are fat soluble, so they accumulate in organisms and become increasingly concentrated as they move through food chains (see Chapter 11). When these compounds reach critical concentrations, they can cause reproductive failure. This occurs most frequently in higher trophic levels and in animals that feed on fat-rich species. Some processes, such as eggshell formation in birds, are particularly sensitive to pesticide accumulations, and population declines in predatory birds like the perigrine falcon have been noted in regions far removed from the locations of pesticide use.

Atmospheric testing of atomic weapons in the 1950s and 1960s increased the concentrations of radioactive forms of many elements. Explosions and leaks in nuclear reactors used to generate electricity continue to be regional or global sources of radioactivity. The explosion of a power-generating plant in 1986 at Chernobyl in Ukraine, for example, released substantial radioactivity that directly affected human health in the region and increased the atmospheric deposition of radioactive materials over eastern Europe and Scandinavia. Some radioactive isotopes of atoms, such as strontium (which is chemically similar to calcium) and cesium (which is chemically similar to potassium) are actively accumulated and retained by organisms. Lichens, for example, acquire their minerals primarily from the atmosphere rather than from the soil and actively accumulate cesium and strontium. Reindeer, which feed on lichens, further concentrate cesium and strontium, as do people who feed on reindeer. For this reason, the input of radioisotopes into the atmosphere or water from nuclear power plants, submarines, and weapons has had impacts that extend far beyond the regions where they were used.

The growing scale and extent of human activities suggest that all ecosystems are being influenced, directly or indirectly, by our activities. No ecosystem functions in isolation, and all are influenced by human activities that take place in adjacent communities and around the world. Human activities are leading to global changes in most major ecosystem controls: climate (global warming), soil and water resources (nitrogen deposition, erosion, diversions), disturbance regime (land use change, fire control), and functional types of organisms (species introductions and extinctions). Many of these global changes interact with each other at regional and local scales. Therefore, all ecosystems are experiencing directional changes in ecosystem controls, creating novel conditions and, in many cases, positive feedbacks that lead to new types of ecosystems. These changes in interactive controls will inevitably change the properties of ecosystems and may lead to unpredictable losses of ecosystem functions on which human communities depend. In the following chapters we point out many of the ecosystem processes that have been affected.

Summary

Ecosystem ecology addresses the interactions among organisms and their environment as an integrated system through study of the factors that regulate the pools and fluxes of materials and energy through ecological systems. The spatial scale at which we study ecosystems is chosen to facilitate the measurement of important fluxes into, within, and out of the ecosystem. The functioning of ecosystems depends not only on their current structure and environment but also on past events and disturbances and the rate at which ecosystems respond to past events. The study of ecosystem ecology is highly interdisciplinary and builds on many aspects of ecology, hydrology, climatology, and geology and contributes to current efforts to understand Earth as an integrated system. Many unresolved problems in ecosystem ecology require an integration of systems approaches, process understanding, and global analysis.

Most ecosystems ultimately acquire their energy from the sun and their materials from the atmosphere and rock minerals. The energy and materials are transferred among components within the ecosystem and are then released to the environment. The essential biotic components of ecosystems include plants, which bring carbon and energy into the ecosystem; decomposers, which break down dead organic matter and release CO_2 and nutrients; and animals, which transfer energy and materials within ecosystems and modulate the activity of plants and decomposers. The essential abiotic components of ecosystems are the atmosphere, water, and rock minerals. Ecosystem processes are controlled by a set of relatively independent state factors (climate, parent material, topography, potential biota, and time) and by a group of interactive controls (including resource supply, modulators, disturbance regime, functional types of organisms, and human activities) that are the immediate controls over ecosystem processes. The interactive controls both respond to and affect ecosystem processes. The stability and resilience of ecosystems depend on the strength of negative feedbacks that maintain the characteristics of ecosystems in their current state.

Review Questions

1. What is an ecosystem? How does it differ from a community? What kinds of environmental questions can be addressed by ecosystem ecology that are not readily addressed by population or community ecology?
2. What is the difference between a pool and a flux? Which of the following are pools and which are fluxes: plants, plant respiration, rainfall, soil carbon, consumption of plants by animals?
3. What are the state factors that control the structure and rates of processes in ecosystems? What are the strengths and limitations of the state factor approach to answering this question.
4. What is the difference between state factors and interactive controls? If you were asked to write a management plan for a region, why would you treat a state factor and an interactive control differently in your plan?
5. Using a forest or a lake as an example, explain how climatic warming or the harvest of trees or fish by people might change the major interactive controls. How might these changes in controls alter the structure of or processes in these ecosystems?
6. Use examples to show how positive and negative feedbacks might affect the responses of an ecosystem to climatic change.

Additional Reading

Chapin, F.S. III, M.S. Torn, and M. Tateno. 1996. Principles of ecosystem sustainability. *American Naturalist* 148:1016–1037.

Golley, F.B. 1993. *A History of the Ecosystem Concept in Ecology: More Than the Sum of the Parts*. Yale University Press, New Haven, CT.

Gorham, E. 1991. Biogeochemistry: Its origins and development. *Biogeochemistry* 13:199–239.

Hagen, J.B. 1992. *An Entangled Bank: The Origins of Ecosystem Ecology*. Rutgers University Press, New Brunswick, NJ.

Jenny, H. 1980. *The Soil Resources: Origin and Behavior*. Springer-Verlag, New York.

Lindeman, R.L. 1942. The trophic-dynamic aspects of ecology. *Ecology* 23:399–418.

Schlesinger, W.H. 1997. *Biogeochemistry: An Analysis of Global Change*. Academic Press, San Diego.

Sousa, W.P. 1985. The role of disturbance in natural communities. *Annual Review of Ecology and Systematics* 15:353–391.

Tansley, A.G. 1935. The use and abuse of vegetational concepts and terms. *Ecology* 16:284–307.

Vitousek, P.M. 1994. Beyond global warming: Ecology and global change. *Ecology* 75:1861–1876.

2
Earth's Climate System

Climate is the state factor that most strongly governs the global distribution of terrestrial biomes. This chapter provides a general background on the functioning of the climate system and its interactions with atmospheric chemistry, oceans, and land.

Introduction

Climate exerts a key control over the distribution of Earth's ecosystems. Temperature and water availability determine the rates at which many biological and chemical reactions can occur. These reaction rates control critical ecosystem processes, such as the production of organic matter by plants and its decomposition by microbes. Climate also controls the weathering of rocks and the development of soils, which in turn influence ecosystem processes (see Chapter 3). Understanding the causes of temporal and spatial variation in climate is therefore critical to understanding the global pattern of ecosystem processes.

Climate and climate variability are determined by the amount of incoming solar radiation, the chemical composition and dynamics of the atmosphere, and the surface characteristics of Earth. The circulation of the atmosphere and oceans influences the transfer of heat and moisture around the planet and thus strongly influences climate patterns and their variability in space and time. This chapter describes the global energy budget and outlines the roles that the atmosphere, oceans, and land surface play in the redistribution of energy to produce observed patterns of climate and ecosystem distribution.

Earth's Energy Budget

The balance between incoming and outgoing radiation determines the energy available to drive Earth's climate system. An understanding of the components of Earth's energy budget provides a basis for determining the causes of recent and long-term changes in climate. The sun is the source of virtually all of Earth's energy. The temperature of a body determines the wavelengths of energy emitted. The high temperature of the sun (6000 K) results in emissions of high-energy **shortwave radiation** with wavelengths of 300 to 3000 nm (Fig. 2.1). These include visible (39% of the total), near-infrared (53%), and ultraviolet (UV) radiation (8%). On average, about 31% of the incoming shortwave radiation is reflected back to space, due to **backscatter** (reflection) from clouds (16%); air molecules, dust, and haze (7%); and Earth's surface (8%) (Fig. 2.2). Another 20% of the incoming shortwave radiation is absorbed by the atmosphere, especially by ozone in the upper atmosphere and by clouds and water vapor in the lower atmosphere. The remaining

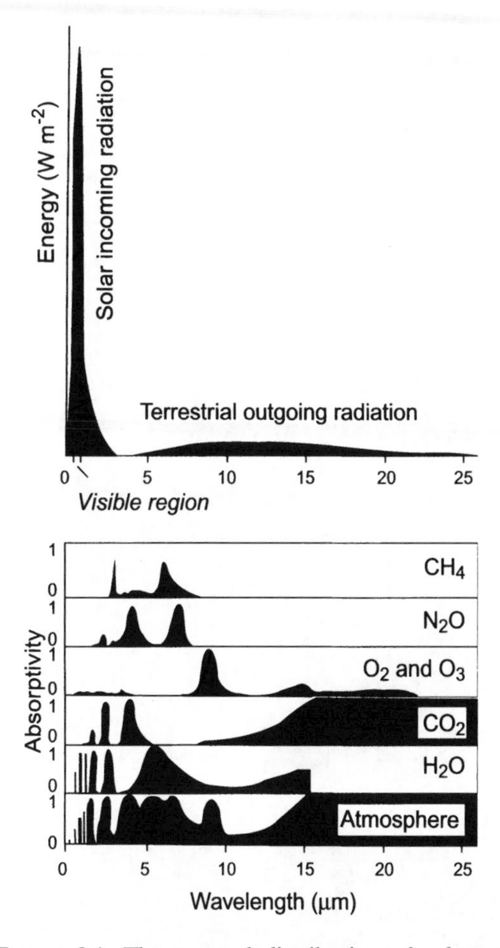

FIGURE 2.1. The spectral distribution of solar and terrestrial radiation and the absorption spectra of the major radiatively active gases and of the total atmosphere. These spectra show that the atmosphere absorbs terrestrial radiation more effectively than solar radiation, explaining why the atmosphere is heated from below. (Sturman and Tapper 1996, Barry and Chorley 1970.)

49% reaches Earth's surface as direct or diffuse radiation and is absorbed.

Over time scales of a year or more, Earth is in a state of radiative equilibrium, meaning that it releases as much energy as it absorbs. On average, Earth emits 79% of the absorbed energy as low-energy **longwave radiation** (3000 to 30,000 nm), due to its relatively low surface temperature (288 K). The remaining energy is transferred from Earth's surface to the atmosphere by the evaporation of water (**latent heat flux**) (16% of terrestrial energy loss) or by the transfer of heat to the air from the warm surface to the cooler overlying atmosphere (**sensible heat flux**) (5% of terrestrial energy loss) (Fig. 2.2). Heat absorbed from the surface when water evaporates is subsequently released to the atmosphere when water vapor condenses, resulting in formation of clouds and precipitation.

Although the atmosphere transmits about half of the incoming shortwave radiation to Earth's surface, it absorbs 90% of the longwave (infrared) radiation emitted by the surface (Fig. 2.2). Water vapor, carbon dioxide (CO_2), methane (CH_4), nitrous oxide (N_2O), and industrial products like chlorofluorocarbons (CFCs) effectively absorb longwave radiation (Fig. 2.1). The energy absorbed by these **radiatively active gases** is reradiated in all directions as longwave radiation (Fig. 2.2). The portion that is directed back toward the surface contributes to the warming of the planet, a phenomenon know as the **greenhouse effect**. Without a longwave-absorbing atmosphere, the mean temperature at Earth's surface would be about 33°C lower than it is today and would probably not support life. Radiation absorbed by clouds and radiatively active gases is also emitted back to space, balancing the incoming shortwave radiation (Fig. 2.2).

Long-term records of atmospheric gases, obtained from atmospheric measurements made since the 1950s and from air bubbles trapped in glacial ice, demonstrate large increases in the major radiatively active gases (CO_2, CH_4, N_2O, and CFCs) since the beginning of the Industrial Revolution 150 years ago (see Fig. 15.3). Human activities such as fossil fuel burning, industrial activities, animal husbandry, and fertilized and irrigated agriculture contribute to these increases. As concentrations of these gases rise, more of the longwave radiation emitted by Earth is trapped by the atmosphere, enhancing the greenhouse effect and causing the surface temperature of Earth to increase.

The globally averaged energy budget outlined above gives us a sense of the critical factors controlling the global climate system. Regional climates, however, reflect spatial

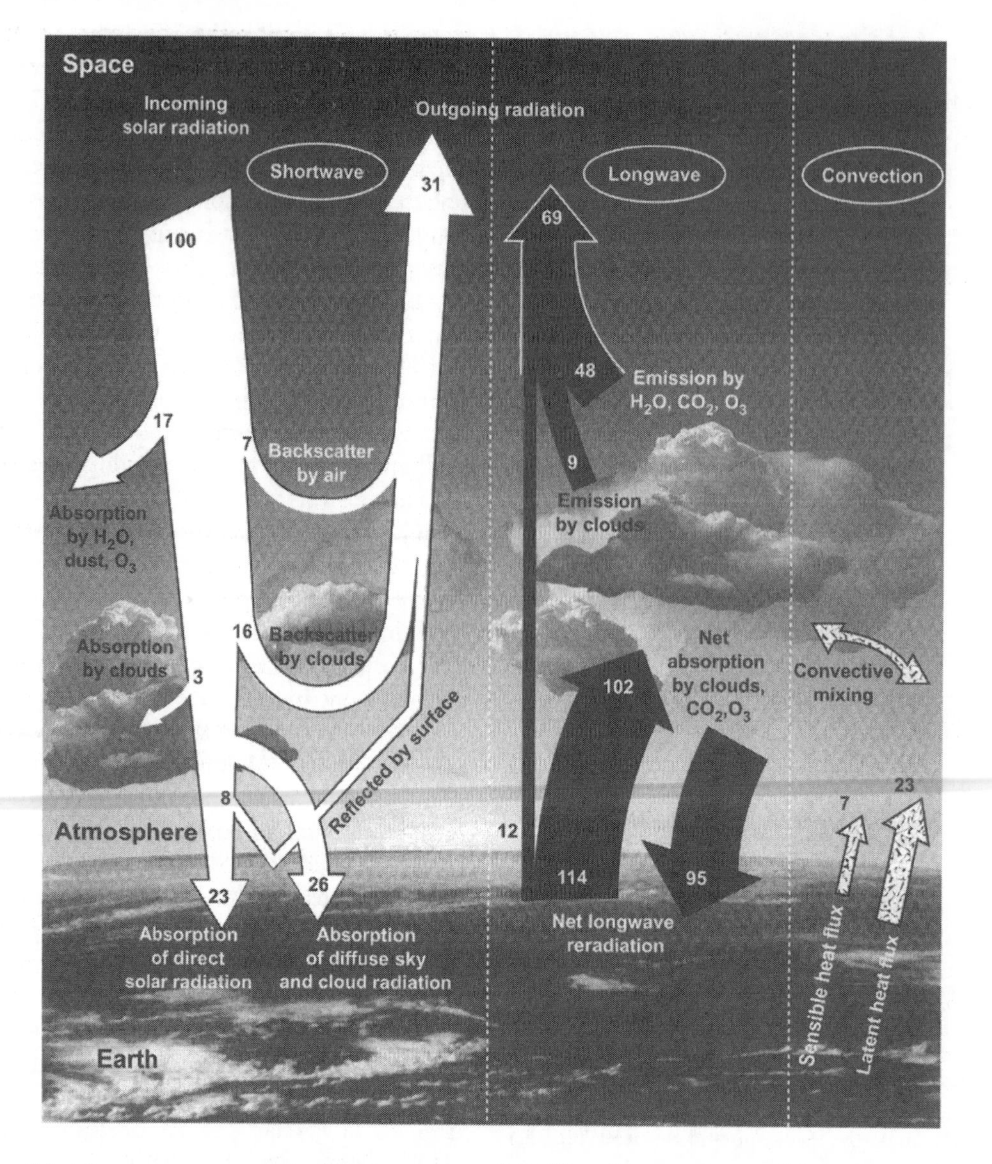

FIGURE 2.2. The average annual global energy balance for the Earth–atmosphere system. The numbers are percentages of the energy received as incoming solar radiation. At the top of the atmosphere, the incoming solar radiation (100% or 342 W m^{-2}) is balanced by reflected shortwave radiation (31%) and emitted longwave radiation (69%). Within the atmosphere, the absorbed shortwave radiation (20%) and absorbed longwave radiation (102%) and latent plus sensible heat flux (30%) are balanced by longwave emission to space (57%) and longwave emission to Earth's surface (95%). At Earth's surface the incoming shortwave radiation (49%) and incoming longwave radiation (95%) are balanced by outgoing longwave radiation (114%) and latent plus sensible heat flux (30%) (Graedel and Crutzen 1995, Sturman and Tapper 1996, Baede et al. 2001).

variability in energy exchange and in heat transport by the atmosphere and oceans. Earth experiences greater heating at the equator than at the poles, and it rotates on a tilted axis. Its continents are spread unevenly over the surface, and its atmospheric and oceanic chemistry and physics are dynamic and spatially variable. A more thorough understanding of the atmosphere and oceans is therefore needed to understand the fate and processing of energy and its consequences for the ecosystems of the planet.

The Atmospheric System

Atmospheric Composition and Chemistry

The chemical composition of the atmosphere determines its role in Earth's energy budget. Think of the atmosphere as a giant reaction flask, containing thousands of different chemical compounds in gas and particulate forms, undergoing slow and fast reactions, dissolutions and precipitations. These reactions control the composition of the atmosphere and many of its physical processes, such as cloud formation. These physical processes, in turn, generate dynamical motions crucial for energy redistribution.

More than 99.9% by volume of Earth's atmosphere is composed of nitrogen, oxygen, and argon. Carbon dioxide, the next most abundant gas, accounts for only 0.0367% of the atmosphere (Table 2.1). These percentages are quite constant around the world and up to 80 km in height above the surface. That homogeneity reflects the fact that these gases have long **mean residence times** (MRTs) in the atmosphere. MRT is calculated as the total mass divided by the flux into or out of the atmosphere over a given time period. Nitrogen has an MRT of 13 million years; O_2, 10,000 years; and CO_2, 4 years. In contrast, the MRT for water vapor is only about 10 days, so its concentration in the atmosphere is highly variable, depending on regional variations in surface evaporation, precipitation, and horizontal transport of water vapor. Some of the most important radiatively active gases, such as CO_2, N_2O, CH_4, and CFCs, react relatively slowly in the atmosphere and have residence times of years to decades. Other gases are much more reactive and have residence times of days to months. Reactive species occur in trace amounts and make up less than 0.001% of the volume of the atmosphere. Because of their great reactivity, they are quite variable in time and place. Some of the consequences of reactions among these trace species, such as smog, acid rain, and ozone depletion, threaten the sustainability of ecological systems (Graedel and Crutzen 1995).

TABLE 2.1. Major chemical constituents of the atmosphere.

Compound	Formula	Concentration (%)
Nitrogen	N_2	78.084
Oxygen	O_2	20.946
Argon	Ar	0.934
Carbon dioxide	CO_2	0.037

Data from Schlesinger (1997) and Prentice et al. (2001).

Some atmospheric gases are critical for life. Photosynthetic organisms use CO_2 in the presence of light to produce organic matter that eventually becomes the basic food source for all animals and microbes (see Chapters 5 to 7). Most organisms also require oxygen for metabolic respiration. Dinitrogen (N_2) makes up 78% of the atmosphere. It is unavailable to most organisms, but nitrogen-fixing bacteria convert it to biologically available nitrogen that is ultimately used by all organisms in building proteins (see Chapter 8). Other gases, such as carbon monoxide (CO), nitric oxide (NO), N_2O, CH_4, and volatile organic carbon compounds like terpenes and isoprene, are the products of plant and microbial activity. Some, like tropospheric ozone (O_3), are produced in the atmosphere as products of chemical reactions involving both **biogenic** (biologically produced) and anthropogenic gases and can, at high concentrations, damage plants, microbes, and humans.

The atmosphere also contains **aerosols**, which are small particles suspended in air. Some aerosol particles arise from volcanic eruptions and from blowing dust and sea salt. Others are produced by reactions with gases from pollution sources and biomass burning. Some aerosols are hydroscopic—that is, they have an affinity for water. Aerosols are involved in reactions with gases and act as **cloud condensation nuclei** around which water vapor condenses to form cloud droplets. Together with gases and clouds, aerosols determine the reflectivity (**albedo**) of the atmosphere and therefore exert major control over the energy budget of the atmosphere. The scattering (reflection) of incoming shortwave radiation by aerosols reduces the radiation reaching Earth's surface, which tends to cool the climate. The sulfur released to the atmosphere by the volcanic eruption of Mount Pinatubo in the Philippines in 1991, for example, caused a temporary atmospheric cooling throughout the globe.

Clouds have complex effects on Earth's radiation budget. All clouds have a relatively high albedo and reflect more incoming shortwave radiation than does the darker Earth surface. Clouds, however, are composed of water vapor, which is a very efficient absorber of longwave radiation. All clouds absorb and re-emit much of the longwave radiation impinging on them from Earth's surface. The first process (reflecting shortwave radiation) has a cooling effect by reflecting incoming energy back to space. The second effect (absorbing longwave radiation) has a warming effect, by keeping more energy in the Earth System from escaping to space. The balance of these two effects depends on the height of the cloud. The reflection of shortwave radiation usually dominates the balance in high clouds, causing cooling; whereas the absorption and re-emission of longwave radiation generally dominates in low clouds, producing a net warming effect.

Atmospheric Structure

Atmospheric pressure and density decline with height above Earth's surface. The average vertical structure of the atmosphere defines four relatively distinct layers characterized by their temperature profiles. The atmosphere is highly compressible, and gravity keeps most of the mass of the atmosphere close to Earth's surface. Pressure, which is determined by the mass of the overlying atmosphere, decreases exponentially with height. The vertical decline in air density tends to follow closely that of pressure. The relationships between pressure, density, and height can be described in terms of the hydrostatic equation

$$\frac{dP}{dh} = -\rho g \qquad (2.1)$$

where P is pressure, h is height, ρ is density, and g is gravitational acceleration. The hydrostatic equation states that the vertical change in pressure is balanced by the product of density and gravitational acceleration (a "constant" that varies with latitude). As one moves above the surface toward lower pressure and density, the vertical pressure gradient also decreases. Furthermore, because warm air is less dense than cold air, pressure falls off with height more slowly for warm than for cold air.

The **troposphere** is the lowest atmospheric layer and contains most of the mass of the atmosphere (Fig. 2.3). The troposphere is heated primarily from the bottom by sensible and latent heat fluxes and by longwave radiation from Earth's surface. Temperature therefore decreases with height in the troposphere.

Above the troposphere is the **stratosphere**, which, unlike the troposphere, is heated from the top. Absorption of UV radiation by O_3 in the upper stratosphere warms the air. Ozone is concentrated in the stratosphere because of a balance between the availability of shortwave UV necessary to split molecules of molecular oxygen (O_2) into atomic oxygen (O) and a high enough density of molecules to bring about the required collisions between atomic O and mol-

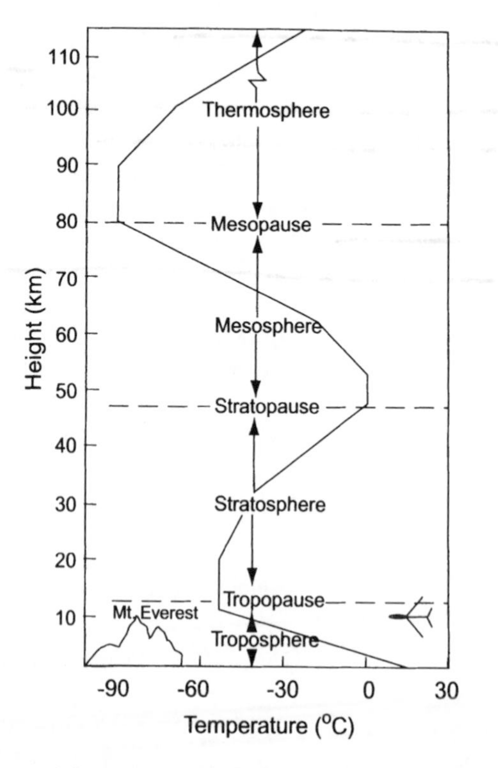

FIGURE 2.3. Average thermal structure of the atmosphere showing the vertical gradients in Earth's major atmospheric layers. (Redrawn with permission from Academic Press; Schlesinger 1997.)

ecular O_2 to form O_3. The absorption of UV radiation by stratospheric ozone results in an increase in temperature with height. The ozone layer also protects the biota at Earth's surface from damaging UV radiation. Biological systems are sensitive to UV radiation because it can damage DNA, which contains the information needed to drive cellular processes. The concentration of ozone in the stratosphere has been declining due to the production and emission of CFCs, which destroy stratospheric ozone, particularly at the poles. This results in an ozone "hole," an area where the transmission of UV radiation to Earth's surface is increased. Slow mixing between the troposphere and the stratosphere allows CFCs and other compounds to reach and accumulate in the ozone-rich stratosphere, where they have long residence times.

Above the stratosphere is the **mesosphere**, where temperature again decreases with height. The uppermost layer of the atmosphere, the **thermosphere**, begins at approximately 80 km and extends into space. The thermosphere has a small fraction of the atmosphere's total mass, composed primarily of O and nitrogen (N) atoms that can absorb very shortwave energy, again causing an increase in heating with height (Fig. 2.3). The mesosphere and thermosphere have relatively little impact on the biosphere.

The troposphere is the atmospheric layer in which most weather occurs, including thunderstorms, snowstorms, hurricanes, and high and low pressure systems. The troposphere is thus the portion of the atmosphere that directly responds to and affects ecosystem processes. The **tropopause** is the boundary between the troposphere and the stratosphere. It occurs at a height of about 16 km in the tropics, where tropospheric temperatures are highest and hence where pressure falls off most slowly with height (Eq. 2.1), and at about 9 km in polar regions, where tropospheric temperatures are lowest. The height of the tropopause also varies seasonally, being lower in winter than in summer.

The **planetary boundary layer** (PBL) is the lower portion of the troposphere, which is influenced by mixing between the atmosphere and Earth's surface. Air within the PBL is mixed by surface heating, which creates convective turbulence, and by mechanical turbulence, which is associated with the friction of air moving across Earth's surface. The PBL increases in height during the day largely due to convective turbulence. The PBL mixes more rapidly with the free troposphere when the atmosphere is disturbed by storms. The boundary layer over the Amazon Basin, for example, generally grows in height until midday, when it is disrupted by convective activity (Fig. 2.4). The PBL becomes shallower at night when there is no solar energy to drive convective mixing. Air in the PBL is relatively isolated from the free troposphere and therefore functions like a chamber over Earth's surface. The changes in water vapor, CO_2, and other chemical constituents in the PBL thus serve as an indicator of the biological and physiochemical processes occurring at the surface (Matson and Harriss 1988). The PBL in urban regions, for example, often has higher concentrations of pollutants than the cleaner, more stable air above. At night, gases emitted by the surface, such as CO_2 in natural ecosystems or pollutants in urban environments, often reach high concentrations because they are concentrated in a shallow boundary layer.

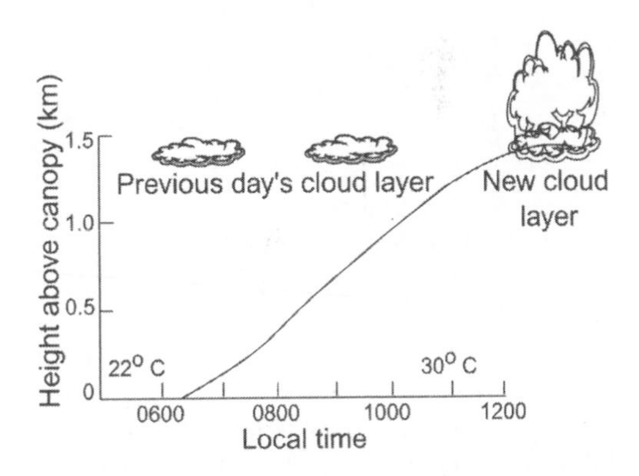

FIGURE 2.4. Increase in the height of the planetary boundary layer between 6:00 A.M. and noon in the Amazon Basin on a day without thunderstorms. The increase in surface temperature drives evapotranspiration and convective mixing, which causes the boundary layer to increase in height until the rising air becomes cool enough that water vapor condenses to form clouds. (Redrawn with permission from *Ecology*; Matson and Harriss 1988.)

Atmospheric Circulation

The fundamental cause of atmospheric circulation is the uneven heating of Earth's surface. The equator receives more incoming solar radiation than the poles because Earth is spherical. At the equator, the sun's rays are almost perpendicular to the surface at solar noon. At the lower sun angles experienced at high latitudes, the sun's rays are spread over a larger surface area (Fig. 2.5), resulting in less radiation received per unit ground area. In addition, the sun's rays have a longer path through the atmosphere, so more of the incoming solar radiation is absorbed, reflected, or scattered before it reaches the surface. This unequal heating of Earth results in higher tropospheric temperatures in the tropics than at the poles, which in turn drives atmospheric circulation.

Atmospheric circulation has both vertical and horizontal components (Fig. 2.6). The transfer of energy from Earth's surface to the atmosphere by latent and sensible heat fluxes and longwave radiation generates strong heating at the surface. This warming causes the surface air to expand and become less dense than surrounding air, so it rises. As air rises, the decrease in atmospheric pressure with height causes continued expansion (Eq. 2.1), which decreases the average kinetic energy of air molecules, causing the rising air to cool. The **dry adiabatic lapse rate** is the change in temperature experienced by a parcel of air as it moves vertically in the atmosphere without exchanging energy with the surrounding air and is about 9.8°C km^{-1}. Cooling also causes condensation and precipitation because cool air has a lower capacity to hold water vapor than warm air. Condensation in turn releases latent heat, which reduces the rate at which rising air cools by expansion. This release of latent heat can cause the rising air to be warmer than surrounding air, so it continues to rise. The resulting **moist adiabatic lapse rate** is about 4°C km^{-1} near the surface, rising to 6 or 7°C km^{-1} in the middle troposphere. The greater the moisture content of rising air, the more latent heat is released to drive convective uplift, which contributes to the intense thunderstorms and deep boundary layer in the wet tropics. The average lapse rate varies regionally, depending on the strength of surface heating but averages about 6.5°C km^{-1}.

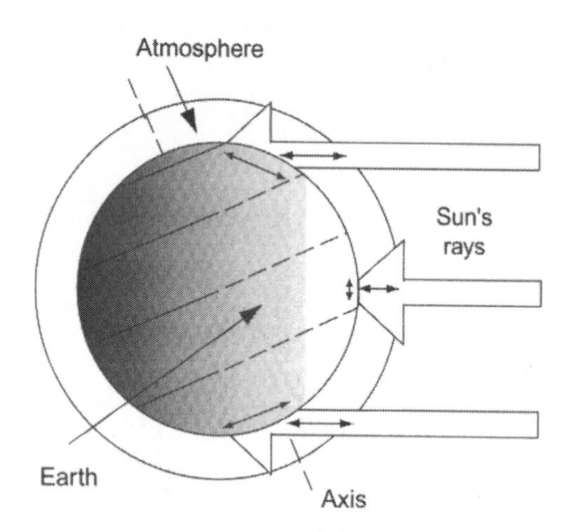

FIGURE 2.5. Atmospheric and angle effects on solar input at different latitudes. The arrows parallel to the sun's rays show the depth of the atmosphere that solar radiation must penetrate. The arrows parallel to Earth's surface show the surface area over which a given quantity of solar radiation is distributed. High-latitude ecosystems receive less radiation than those at the equator because radiation at high latitudes has a longer path through the atmosphere and is spread over a larger ground area.

Surface air rises most strongly at the equator because of the intense equatorial heating and the large amount of latent heat released as this moist air rises and condenses. This air rises until it reaches the tropopause. The expansion of equatorial air also creates a horizontal pressure gradient that causes the equatorial air aloft to flow horizontally from the equator along the tropopause toward the poles (Fig. 2.6). This poleward-moving air cools due to emission of longwave radiation to space. In addition, the air converges into a smaller volume as it moves poleward because Earth's radius and surface area decrease from the equator toward the poles. Due to the cooling of the air and its convergence into a smaller volume, the density of air increases, creating a high pressure that causes upper air to subside, which forces surface air back toward the equator to replace the rising equatorial air. Hadley proposed this model of atmospheric circulation in 1735, suggesting that there should be one large circu-

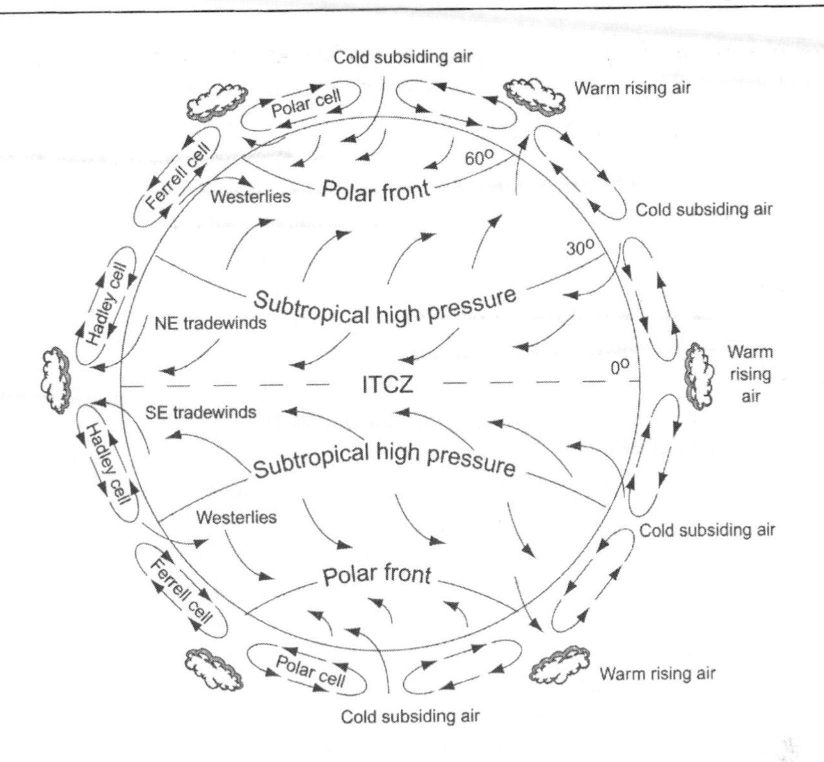

FIGURE 2.6. Earth's latitudinal atmospheric circulations are driven by rising air at the equator and subsiding air at the poles. These forces and the Coriolis forces produce three major cells of vertical atmospheric circulation (Hadley, Ferrell, and polar cells). Air warms and rises at the equator due to intense heating. After reaching the tropopause, the equatorial air moves poleward to about 30° N and S latitudes, where it descends and either returns to the equator, forming the Hadley cell, or moves poleward. Cold dense air at the poles subsides and moves toward the equator until it encounters poleward-moving air at about 60° latitude. There the air rises and moves either poleward to replace air that has subsided at the poles (the polar cell) or moves toward the equator to form the Ferrell cell. Also shown are the horizontal patterns of atmospheric circulation, consisting of the prevailing surface winds (the easterly trade winds in the tropics and the westerlies in the temperate zones). The boundaries between these zones are either low-pressure zones of rising air (the intertropical conversion zone, ITCZ, and the polar front) or high-pressure zones of subsiding air (the subtropical high pressure belt and the poles).

lation cell in the Northern Hemisphere and another in the Southern Hemisphere, driven by atmospheric heating and uplift at the equator and subsidence at the poles. Based on observations, Ferrell proposed in 1865 the conceptual model that we still use today, although the actual dynamics are much more complex. This model describes atmospheric circulation as a series of three circulation cells in each hemisphere. (1) The **Hadley cell** is driven by expansion and uplift of equatorial air. (2) The **polar cell** is driven by subsidence of cold converging air at the poles. (3) The intermediate **Ferrell cell** is driven indirectly by dynamical processes (Fig. 2.6). The Ferrell cell is actually the long-term average transport caused by weather systems in the mid-latitudes rather than a stable permanent atmospheric feature. The chaotic motion of these mid-latitude weather systems creates a net poleward transport of heat. These three cells subdivide the atmosphere into three distinct circulations: tropical air masses between the equator and 30° N and S, temperate air masses between 30 and 60° N and S, and polar air masses between 60° N and S and the poles (Fig. 2.6). The latitudinal location of these cells moves seasonally in response to latitudinal changes in surface heating by the sun.

Earth's rotation causes winds to deflect to the right in the Northern Hemisphere and to the left in the Southern Hemisphere. Earth and its atmosphere complete one rotation about Earth's axis every day. The direction of rotation is from west to east. Because the atmosphere in equatorial regions is farther from Earth's axis of rotation than that at higher latitudes, there is a corresponding poleward decrease in the linear velocity of the atmosphere as it travels around Earth. As parcels of air move north or south, they tend to maintain their **angular momentum** (M_a), just as your car tends to maintain its momentum when you try to stop or turn on an icy road.

$$M_a = mvr \quad (2.2)$$

where m is the mass, v is the velocity, and r is the radius of rotation. If the mass of a parcel of air remains constant, its velocity is inversely related to the radius of rotation. We know, for example, that skaters can increase their speed of rotation by pulling their arms close to their bodies, which reduces their effective radius. Air that moves from the equator toward the poles encounters a smaller radius of rotation around Earth's axis. Therefore, to conserve angular momentum, it moves more rapidly (i.e., moves from west to east), *relative to Earth's surface*, as it moves poleward (Fig. 2.6). Conversely, air moving toward the equator encounters an increasing radius of rotation around Earth's axis and, to conserve angular momentum, moves more slowly (i.e., moves from east to west), *relative to Earth's surface*. This causes the air to be deflected to the right, relative to Earth's surface, in the Northern Hemisphere and to the left in the Southern Hemisphere (Fig. 2.6), creating clockwise patterns of atmospheric circulation in the Northern Hemisphere and counterclockwise patterns in the Southern Hemisphere. This conservation of angular momentum that causes air to change its direction, relative to Earth's surface, is known as the **Coriolis force**. The Coriolis force is a pseudo-force that arises only because Earth is rotating, and we view the motion relative to Earth's surface. Similar Coriolis forces act on ocean currents, creating clockwise ocean circulation in the Northern Hemisphere and counterclockwise circulation in the Southern Hemisphere.

The interaction of vertical and horizontal motions of the atmosphere create Earth's **prevailing winds** (i.e., the most frequent wind directions). As air in the Hadley cell moves poleward along the tropopause, it is deflected by the Coriolis force to a westerly direction—that is, it blows from the west (Fig. 2.6). This prevents the poleward-moving air from reaching the poles, as it was supposed to do in Hadley's one-cell circulation model. This results in an accumulation of air and a belt of high pressure at about 30° N and S latitude, which causes the air to sink. Some of this subsiding air returns toward the equator at the surface, completing the Hadley circulation cell (Fig. 2.6). The interaction between motions induced by the pole-to-equator temperature gradient and the Coriolis force explains why there are three atmospheric circulation cells in each hemisphere rather than just one, as Hadley had proposed. At the boundaries between the major cells of atmospheric circulation, there are relatively sharp gradients of temperature and pressure that, together with the Coriolis force, generate strong winds over a broad height range in the upper troposphere. These are the subtropical and polar **jet streams**. The Coriolis force explains why these winds blow in a westerly direction (i.e., from west to east).

At the surface, the direction of prevailing winds depends on whether air is moving toward or away from the equator. In the tropics, surface air in the Hadley cell moves from 30° N and S toward the equator. In the Northern Hemisphere, this air is deflected to the right by the Coriolis force and forms the northeast **trade winds** (i.e., surface winds that blow from the northeast). Equatorward flow from 30° S is deflected to the left and forms the southeast trade winds. Thus equatorial winds blow predominantly from the east. The region where surface air from the Northern and Southern Hemispheres converges is called the **intertropical convergence zone** (ITCZ). Here the rising air creates a zone with light winds and high humidity (Fig. 2.6), an area known to

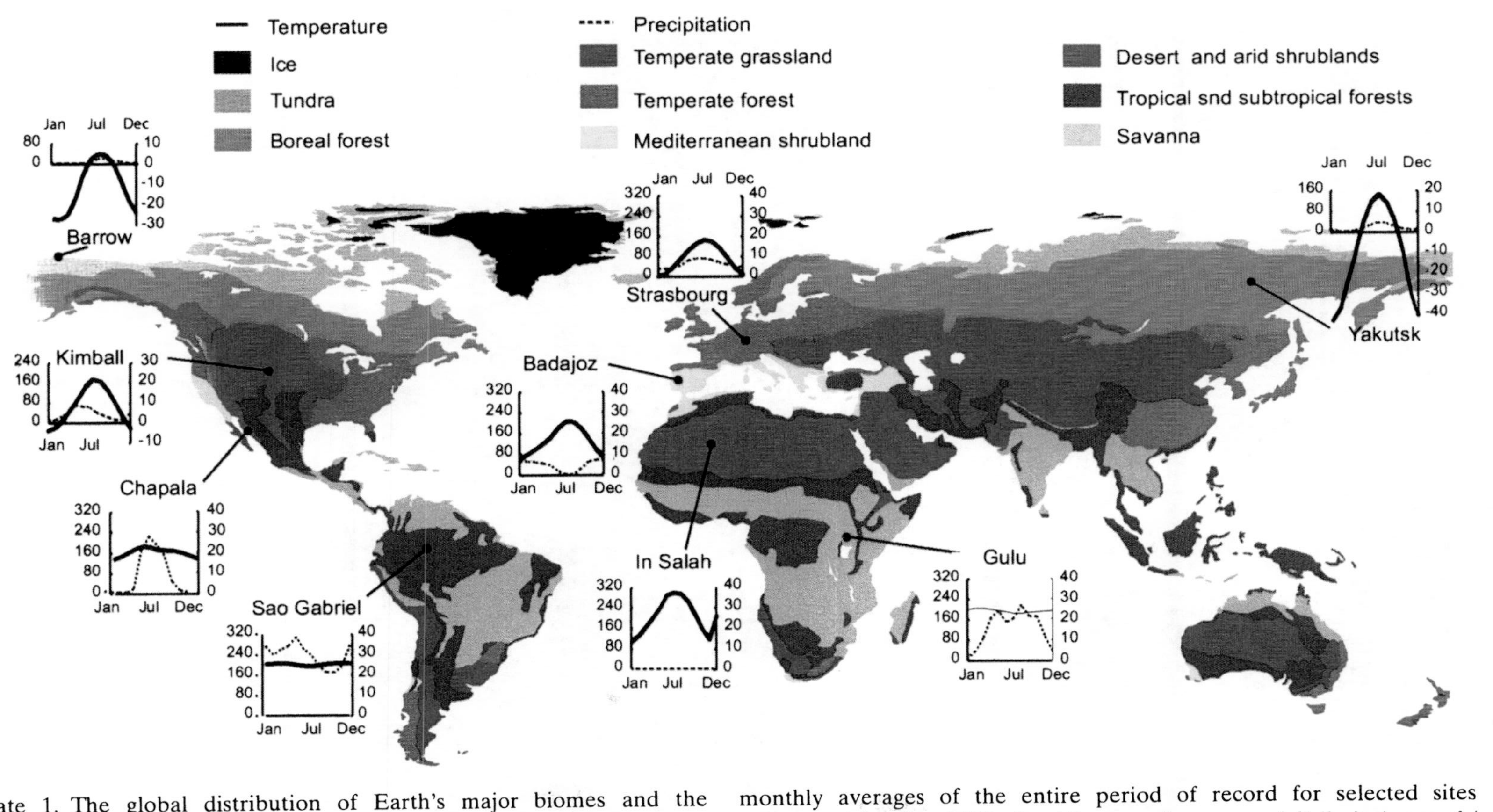

Plate 1. The global distribution of Earth's major biomes and the seasonal patterns of monthly average temperature and precipitation at one representative site for each biome (Bailey 1998). Climate data are monthly averages of the entire period of record for selected sites through the year 2000 (http://www.ncdc.noaa.gov/ol/climite/research/ghcn/ghcn.hlml).

Color Plate II

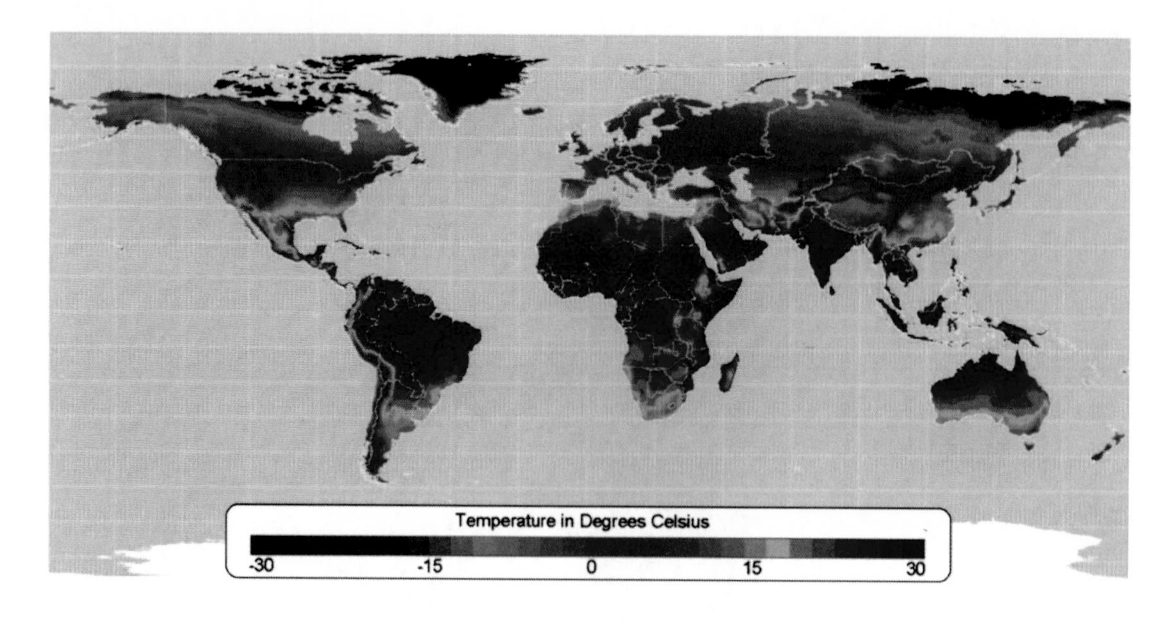

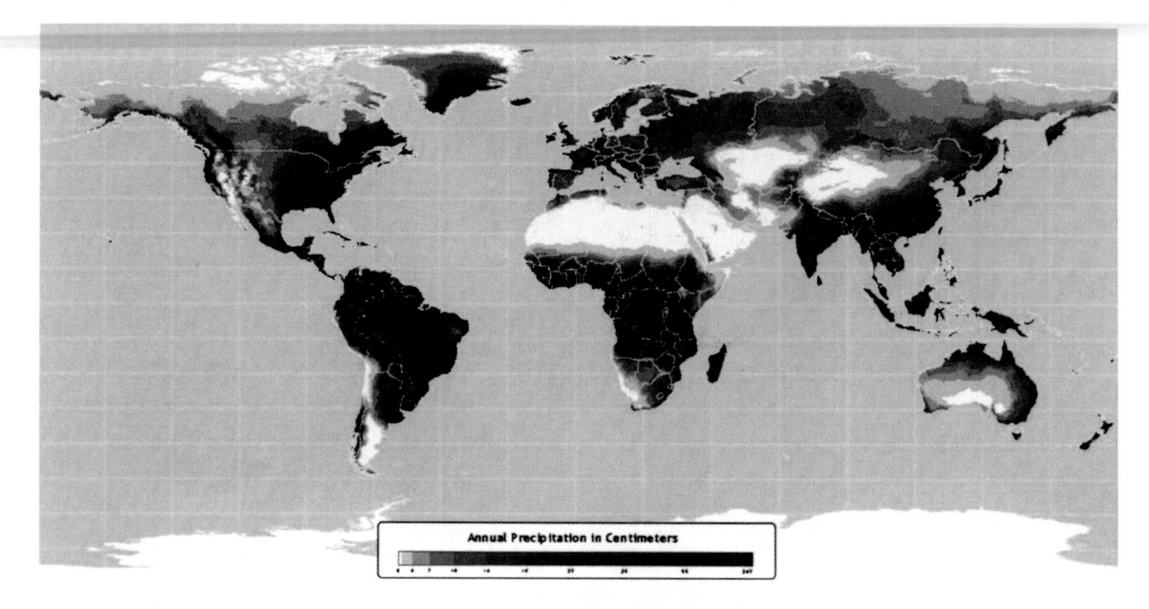

Plate 2. The global pattern of mean annual temperature and total annual precipitation (New et al. 1999). Temperature is highest at the equator and lowest at the poles and at high elevations. (Reproduced with permission from the Atlas of the Biosphere <http://atlas.sage.wisc.edu>.)

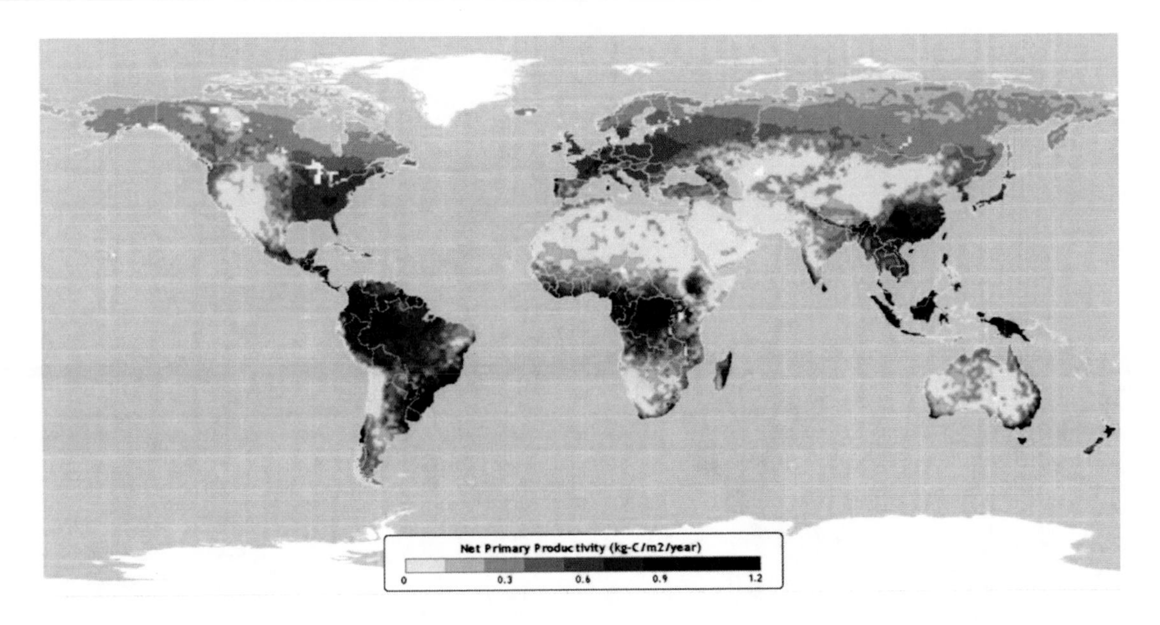

Plate 3. The global pattern of net primary productiviy (Foley et al. 1996, Kucharik et al. 2000). The patterns of productivity correlates more closely with precipitation than with temperature, including a strong role of moisture in regulating the productivity of the biosphere. (Reproduced with permission from the Atlas of the Biosphere <http://atlas.sage.wisc.edu>.)

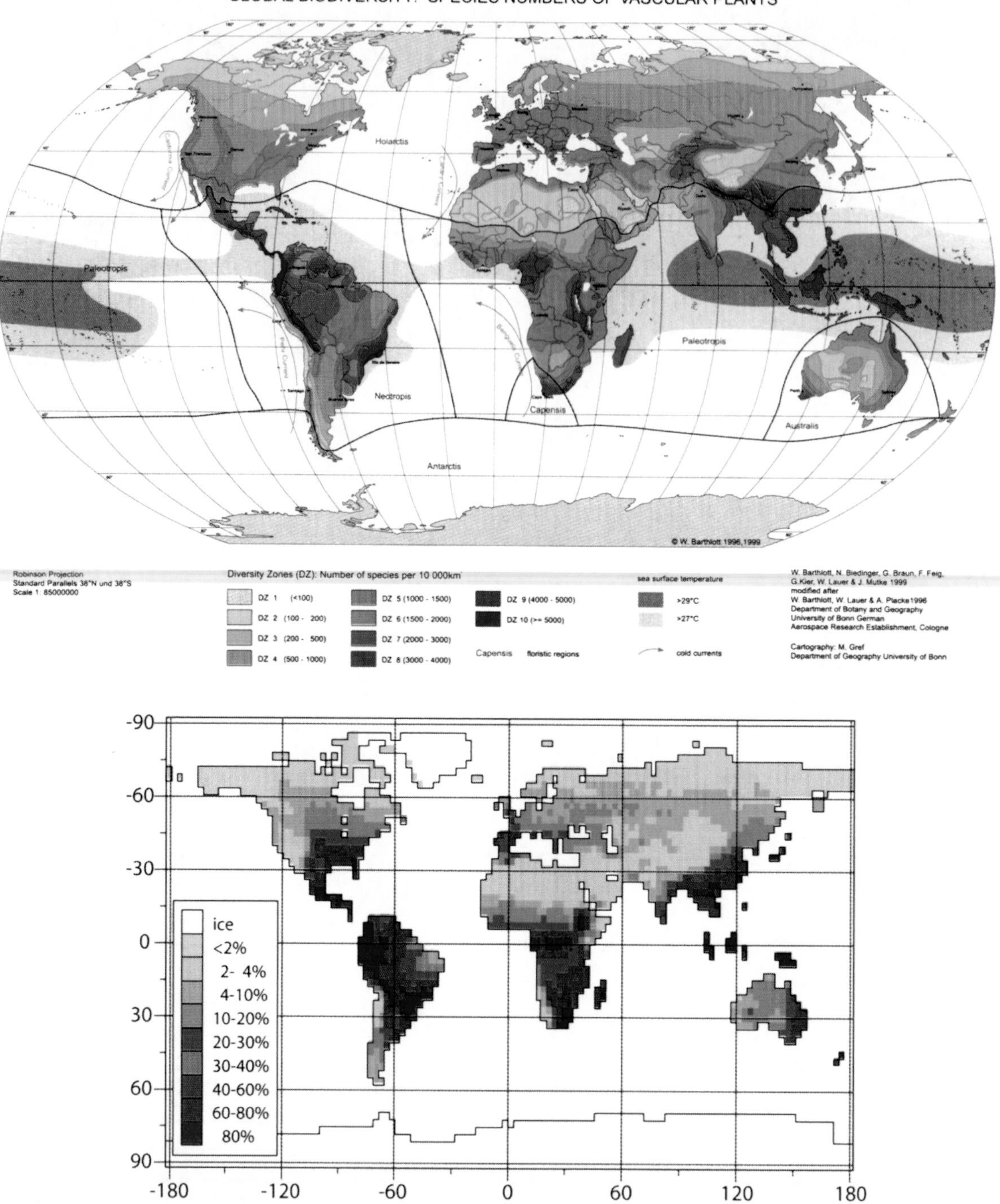

Plate 4. Global distribution of species richness based on observations and on model simulations, which use climate as a filter to reduce the number of allocation strategies (Kleiden and Mooney 2000). Reprinted with permission from *Global Change Biology*.

early sailors as the **doldrums**. Subsiding air at 30° N and S latitudes also produces relatively light winds, an area known as the **horse latitudes**. The surface air that moves poleward from 30° to 60° N and S is deflected toward the east by the Coriolis forces, forming the prevailing westerlies, or surface winds that blow from the west.

The locations of the ITCZ and of each circulation cell shifts seasonally because the zone of maximum solar radiation input varies from summer to winter due to Earth's 23.5° tilt with respect to the plane of its orbit around the sun. The seasonal changes in the location of these cells contribute to the seasonality of climate.

The uneven distribution of land and oceans on Earth's surface creates an uneven pattern of heating that modifies the general latitudinal trends in climate. At 30° N and S, air descends more strongly over cool oceans than over the relatively warm land because the air is cooler and more dense over the ocean than over the land. The greater subsidence over the oceans creates high-pressure zones over the Atlantic and Pacific Oceans (the Bermuda and Pacific highs, respectively) and over the southern oceans (Fig. 2.7). At 60° N, where air is rising, there are semipermanent low-pressure zones over Iceland and the Aleutian Islands (the Icelandic and Aleutian lows, respectively). These lows are actually time averages of mid-latitude storm tracks rather than stable features of the circulation. In the Southern Hemisphere, there is little land at 60° S; therefore, there is a trough of low pressure instead of distinct centers. Air that subsides in high-pressure centers spirals outward in a clockwise direction in the Northern Hemisphere and in a counterclockwise direction in the Southern Hemisphere (Fig. 2.7) due to an interaction between Coriolis forces and the pressure gradient force produced by the subsiding air. Winds spiral inward toward low-pressure centers in a counterclockwise direction in the Northern Hemisphere and in a clockwise direction in the Southern Hemisphere. Air in the low-pressure centers rises to balance the subsiding air in high-pressure centers. The long-term average of these vertical and horizontal motions produces the vertical circulation described by the Ferrell cell (Fig. 2.6) and a horizontal pattern of high- and low-pressure centers commonly observed on weather charts (Fig. 2.7).

These deviations from the expected easterly or westerly direction of prevailing winds are organized on a planetary scale and are known as **planetary waves**. These waves are influenced by both land–ocean heating contrasts and the locations of large mountain ranges, such as the Rockies and the Himalayas. These mountain barriers force the Northern Hemisphere westerlies vertically upward and to the north. Downwind of the mountains, air descends and moves to the south, forming a trough, much like the standing waves in the rapids of a fast-moving river that are governed by the location of rocks in the riverbed. Temperatures are comparatively low in the troughs, due to the southward movement of polar air, and are comparatively high in the ridges. The trough over eastern North America downwind of the Rocky Mountains (Fig. 2.7), for example, results in relatively cool temperatures and a more southerly location of the arctic treeline in eastern North America. Although planetary waves have preferred locations, they are not static. Changes in their location or in the number of waves alter regional patterns of climate. These step changes in the circulation pattern are referred to as **climate modes**.

Planetary waves and the distribution of major high- and low-pressure centers explain many details of horizontal motion in the atmosphere and therefore the patterns of ecosystem distribution. The locations of major high- and low-pressure centers, for example, explain the movement of mild moist air to the west coasts of continents at 60° N and S, where the temperate rainforests of the world occur (the northwestern United States and southwestern Chile, for example) (Fig. 2.7). The subtropical high-pressure centers at 30° N and S cause cool polar air to move toward the equator on the west coasts of continents, creating dry mediterranean climates at 30° N and S (Fig. 2.7). On the east coasts of continents subtropical highs cause warm moist equatorial air to move northward at 30° N and S, creating a moist subtropical climate (Fig. 2.7).

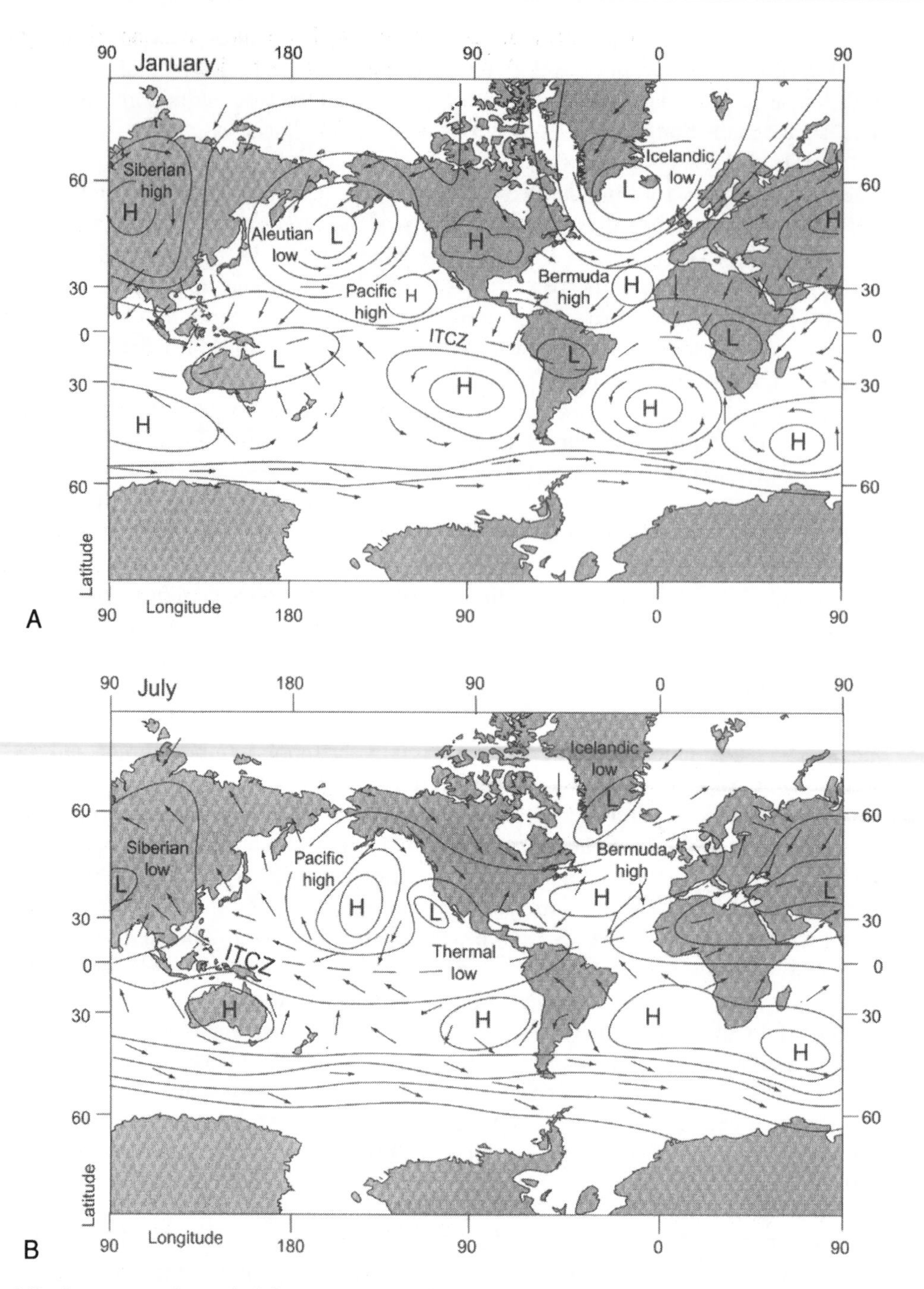

FIGURE 2.7. Average surface wind-flow patterns and the distribution of sea level pressure for January (**A**) and July (**B**). (Redrawn from *Essentials of Meteorology: An Invitation to the Atmosphere, 2nd edition*, by C. Ahrens © 1998, by permission of Brooks/Cole, an imprint of the Wadsworth Group, a division of Thomson Learning; Ahrens 1998.)

The Oceans

Ocean Structure

Oceans maintain rather stable layers with limited vertical mixing between them. The sun heats the ocean from the top, whereas the atmosphere is heated from the bottom. Because warm water is less dense than cold water, oceans maintain rather stable layers that do not readily mix. The uppermost warm layer of **surface water**, which interacts directly with the atmosphere, extends to depths of 75 to 200 m, depending on the depth of wind-driven mixing. Most primary production, detrital production, and decomposition take place in the surface waters (see Chapter 10). Another major difference between atmospheric and oceanic circulation is that density of ocean waters is determined by both temperature and salinity, so, unlike warm air, warm water can sink, if it is salty enough.

There are relatively sharp gradients in temperature (**thermocline**) and salinity (**halocline**) between warm surface waters of the ocean and cooler more saline waters at intermediate depths (200 to 1000 m) (Fig. 2.8). These two vertical gradients cause the surface waters to be less dense than deep water, creating a stable vertical stratification. The deep layer therefore mixes with the surface waters slowly over hundreds to thousands of years. These deeper layers nonetheless play critical roles in element cycling, productivity, and climate because they are long-term sinks for carbon and the sources of nutrients that drive ocean production (see Chapters 10 and 15). **Upwelling** areas, where deep waters move rapidly to the surface, support high levels of primary and secondary productivity (marine invertebrates and vertebrates) and are the locations of many of the world's major fisheries.

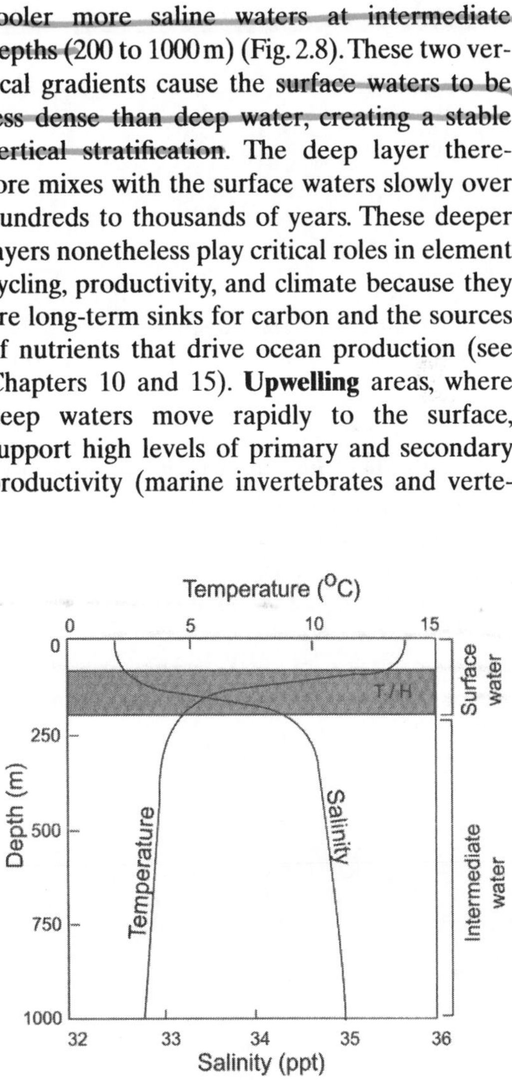

FIGURE 2.8. Typical vertical profiles of ocean temperature and salinity. The thermocline (T) and halocline (H) are the zones where temperature and salinity, respectively, decline most strongly with depth. These transition zones usually coincide approximately.

Ocean Circulation

Ocean circulation plays a critical role in Earth's climate system. On average, ocean circulation accounts for 40% of the latitudinal heat transfer from the equator to the poles, with the remaining 60% of heat transfer occurring through the atmosphere. The ocean is the dominant heat transporter in the tropics, and the atmosphere plays the stronger role at midlatitudes. The surface currents of the oceans are driven by surface winds and therefore show global patterns (Fig. 2.9) that are generally similar to those of the prevailing surface winds (Fig. 2.7). The ocean currents are, however, deflected 20 to 40° relative to the wind direction by Coriolis forces. This deflection and the edges of continents cause ocean currents to be more circular (termed **gyres**) than the winds that drive them. In equatorial regions, currents flow east to west, driven by the easterly trade winds, until they reach the continents, where they split and flow poleward along the western boundaries of the oceans, carrying warm tropical water to higher latitudes. On their way poleward, currents are deflected by Coriolis forces. Once the water reaches the high latitudes, some returns in surface currents toward the tropics along the eastern edges of ocean basins (Fig. 2.9), and some continues poleward.

Deep ocean waters show a circulation pattern quite different from the wind-driven surface circulation. In the polar regions, especially in the winter off southern Greenland and off Antarctica, cold air cools the surface waters, increasing their density. Formation of sea ice, which excludes salt from ice crystals (**brine rejection**), increases the salinity of surface waters, also increasing their density. The high density of these cold saline waters causes them to sink. This **downwelling** to form the North Atlantic deep water off of Greenland, and the Antarctic bottom water off of Antarctica drives the global **thermohaline circulation** in the middle and deep ocean that ultimately transfers water between the major ocean basins (Fig. 2.10). The descent of cold dense water at high

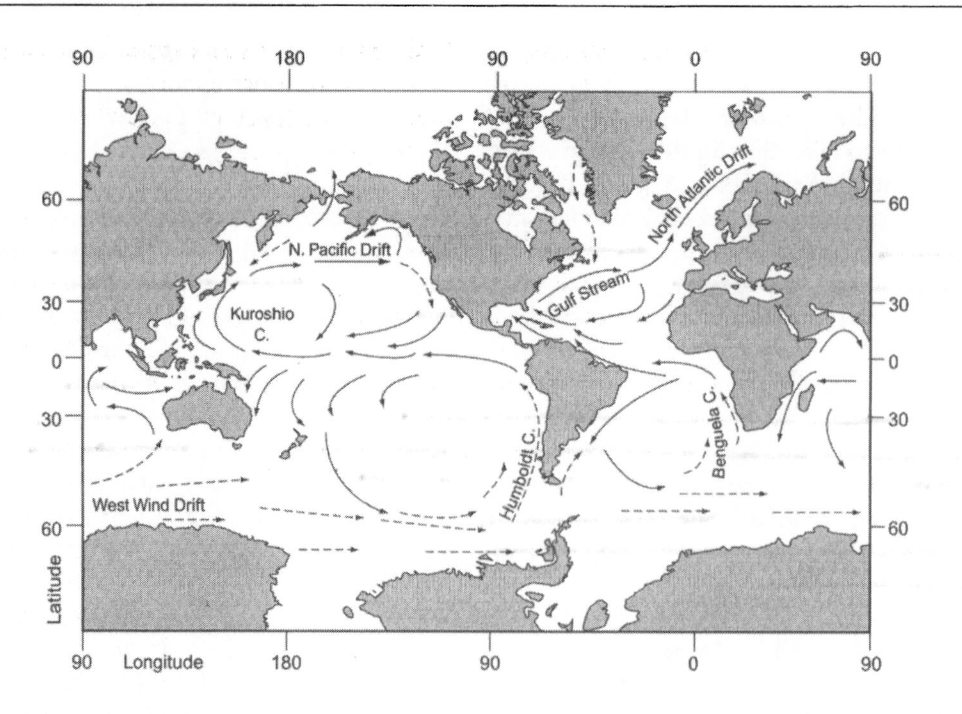

FIGURE 2.9. Major surface ocean currents. Warm currents are shown by solid arrows and cold currents by dashed arrows. (Redrawn from *Essentials of Meteorology: An Invitation to the Atmosphere, 2nd edition*, by C. Ahrens © 1998, by permission of Brooks/Cole, an imprint of the Wadsworth Group, a division of Thomson Learning; Ahrens 1998.)

latitudes is balanced by the upwelling of deep water on the eastern margins of ocean basins at lower latitudes, where along-shore surface currents are deflected offshore by Coriolis forces and easterly trade winds. Net poleward movement of warm surface waters balances the movement of cold deep water toward the equator. Because thermohaline circulation controls latitudinal heat transport, changes in its strength have significant effects on climate. In addition, it transfers carbon to depth, where it remains for centuries (see Chapter 15).

Oceans, with their high heat capacity, heat up and cool down much less rapidly than does land and thus have a moderating influence on the climate of adjacent land. Wintertime tempera-

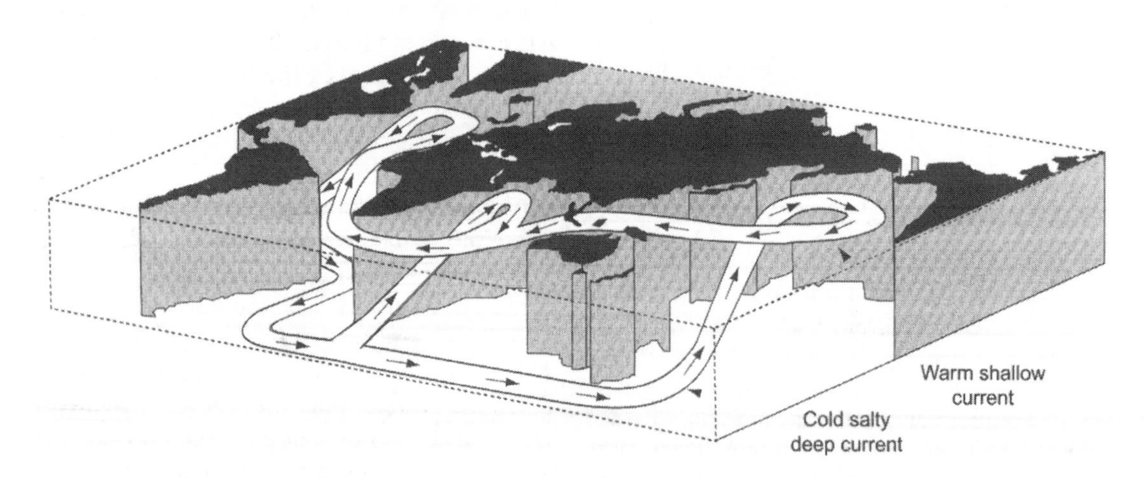

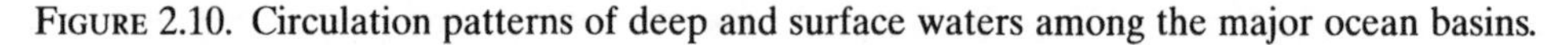
FIGURE 2.10. Circulation patterns of deep and surface waters among the major ocean basins.

tures in Great Britain and western Europe, for example, are much milder than at similar latitudes on the east coast of North America due to the warm North Atlantic drift (the poleward extension of the Gulf Stream) (Fig. 2.9). Conversely, cold upwelling currents, or currents moving toward the equator from the poles, cool adjacent land masses in summer. The cold California current, for example, which runs north to south along the west coast of the United States, keeps summer temperatures in northern California lower than the U.S. east coast at similar latitudes. Windward coastal situations are typically more strongly influenced by prevailing onshore winds and thus have more moderate temperatures than do coastal situations that are downwind of predominantly continental air. New York City, on the eastern edge of North America, therefore experiences relatively mild winters compared to inland cities like Minneapolis, but its winter temperatures are lower than those of cities on the western edge of the continent. These temperature differences play critical roles in determining the kinds of ecosystems that occur over different parts of the globe.

Landform Effects on Climate

The spatial distribution of land, water, and mountains modify the general latitudinal trends in climate. The greater heat capacity of water compared to land influences atmospheric circulation at local to continental scales. The seasonal reversal of winds (**monsoon**) in eastern Asia, for example, is driven largely by the differential temperature response of the land and the adjacent seas. During the Northern Hemisphere winter, the land is colder than the ocean, giving rise to cold dense air that flows southward across India to the ocean (Fig. 2.7). In summer, however, the land heats relative to the ocean. The heating over land forces the air to rise, in turn drawing in moist surface air from the ocean. Condensation of water vapor in the rising moist air produces large amounts of precipitation. Northward migration of the trade winds in summer enhances onshore flow of air, and the mountainous topography of northern India enhances vertical motion, increasing the proportion of water vapor that is converted to precipitation. Together, these seasonal changes in winds give rise to predictable seasonal patterns of temperature and precipitation that strongly influence the structure and functioning of ecosystems.

At scales of a few kilometers, the differential heating between land and ocean produces **land and sea breezes**. During the day, strong heating over land causes air to rise, drawing in cool air from the ocean. The rising of air over the land increases the height at which a given pressure occurs, causing this upper air to move from land toward the ocean. The resulting increase in the mass of atmosphere over the ocean augments the surface pressure, which causes surface air to flow from the ocean toward the land. The resulting circulation cell is identical in principle to that which occurs in the Hadley cell (Fig. 2.6) or Asian monsoon (Fig. 2.7). At night, when the ocean is warmer than the land, air rises over the ocean, and the surface breeze blows from the land to the ocean, causing the circulation cell to reverse. The net effect of sea breezes is to reduce temperature extremes and increase precipitation on land near oceans or large lakes.

Mountain ranges affect local atmospheric circulation and climate through several types of **orographic effects**—that is, effects due to presence of mountains. As winds carry air up the windward sides of mountains, the air cools, and water vapor condenses and precipitates. Therefore, the windward side tends to be cold and wet. When the air moves down the leeward side of the mountain, it contains little moisture, creating a **rain shadow**, or a zone of low precipitation downwind of the mountains. The rain shadow of the Rocky Mountains extends 1500 km to the east, resulting in a strong west-to-east gradient in annual precipitation from Colorado (300 mm) to Illinois (1000 mm) (see Fig. 14.1) (Burke et al. 1989). Deserts or desert grasslands (steppes) are often found immediately downwind of the major mountain ranges of the world. Mountain systems can also influence climate by channeling winds through valleys. The Santa Anna winds of southern California occur when high pressure over the interior deserts causes warm dry winds to be funneled through valleys toward the Pacific coast. These winds create extremely dry conditions that promote intense wildfires.

Sloping terrain creates unique patterns of microclimate at scales ranging from ant hills to mountain ranges. Slopes facing the equator (south-facing slopes in the Northern Hemisphere and north-facing slopes in the Southern Hemisphere) receive more radiation than opposing slopes and thus have warmer drier conditions. In cold or moist climates, the warmer microenvironment on slopes facing the equator provides conditions that enhance productivity, decomposition, and other ecosystem processes, whereas in dry climates, the greater drought on these slopes limits production. Microclimatic variation associated with slope and **aspect** (the compass direction in which a slope faces) allows stands of an ecosystem type to exist hundreds of kilometers beyond its major zone of distribution. These outlier populations are important sources of colonizing individuals during times of rapid climatic change and are therefore important in understanding species migration and the long-term dynamics of ecosystems.

Topography also influences climate through drainage of cold dense air. When air cools at night, it becomes more dense and tends to flow downhill (**katabatic winds**) into valleys, where it accumulates. This can produce strong temperature **inversions** (cool air beneath warm air, a vertical temperature profile reversed from the typical pattern in the troposphere of decreasing temperature with increasing elevation; Fig. 2.3). Inversions occur primarily at night and in winter, when there is insufficient heating from the sun to promote convective mixing. Clouds also tend to inhibit the formation of winter and nighttime inversions because they increase longwave emission to the surface. Increases in solar heating or windy conditions, such as might accompany the passage of frontal systems, break up inversions. Inversions are climatically important because they increase the seasonal and diurnal temperature extremes experienced by ecosystems in low-lying areas. In cool climates, inversions greatly reduce the length of the frost-free growing season.

Vegetation Influences on Climate

Vegetation influences climate through its effect on the surface energy budget. Climate is quite sensitive to regional variations in the vegetation and moisture content of Earth's surface. The albedo (the fraction of the incident shortwave radiation reflected from a surface) determines the quantity of solar energy absorbed by the surface, which is subsequently available for transfer to the atmosphere as longwave radiation and turbulent fluxes. Water generally has a low albedo, so lakes and oceans absorb considerable solar energy. At the opposite extreme, snow and ice have a high albedo and hence absorb little solar radiation, contributing to the cold conditions required for their persistence. Vegetation is intermediate in albedo, and values generally decrease from grasslands (with their highly reflective standing dead leaves) to deciduous forests to dark conifer forests (see Chapter 4). Recent land use changes have substantially altered regional albedo by increasing the area of exposed bare soil. The albedo of soil depends on soil type and wetness but is often higher than that of vegetation, particularly in dry climates. Consequently, overgrazing may increase albedo, reducing energy absorption and the transfer of energy to the atmosphere. This leads to cooling and subsidence, which can reduce precipitation and the capacity of vegetation to recover from overgrazing (Charney et al. 1977). The large magnitude of many land-surface feedbacks to climate suggests that land-surface change can be an important contributor to regional climatic change (Chase et al. 2000).

The energy absorbed by a soil or vegetation surface is transferred to the atmosphere via longwave radiation and turbulent fluxes of latent and sensible heat. The partitioning of energy among these pathways has important climatic consequences (see Chapter 4). Sensible heat fluxes and longwave radiation directly heat the atmosphere at the point at which the energy transfer occurs. Latent heat transfers water vapor to the atmosphere. Water vapor represents stored energy, which is released when the water vapor condenses to form clouds or precipitation. The energy released to the atmosphere by condensation typically occurs some distance downwind from the point at which the water evaporated. Ecosystem structure influences the efficiency with which sensible heat and latent heat are transferred to the atmosphere. Wind passing over tall uneven canopies creates mechanical turbulence that increases the efficiency of heat transfer from the surface to the atmosphere (see Chapter 4). Smooth surfaces, in contrast, tend to heat up because they transfer their heat less efficiently to the atmosphere, only by convection and not by mechanical turbulence.

The effects of vegetation structure on the efficiency of water and energy exchange influence regional climate. Between 25 and 40% of the precipitation in the Amazon basin comes from water that is recycled from land by evapotranspiration (Costa and Foley 1999). Simulations by climate models suggest that, if the Amazon basin were completely converted from forest to pasture, South America would have a permanently warmer drier climate (Shukla et al. 1990). The shallow roots of grasses would absorb less water than the deep tree roots, leading to lower transpiration rates (Fig. 2.11). Pastures would therefore release more of the absorbed solar radiation as sensible heat, which directly warms the atmosphere. The simulations also suggests that warming and drying of air caused by widespread conversion from forest to pasture would reduce the transport of moisture from the adjoining oceans, causing a permanent reduction in precipitation—conditions that favor persistence of pastures over forests. These simulations do an excellent job of exploring the consequences of such vegetation effects on processes that are well understood. There are still many uncertainties, however. Changes in cloudiness, for example, can have either a positive or a negative effect on radiative forcing, depending on the clouds' properties and height. Because these models do a poor job of simulating the processes that produce clouds, the simulations should be viewed as a way to synthesize the net effect of the processes that we understand, rather than as predictions of the future.

At high latitudes, tree-covered landscapes absorb more solar radiation before snow melt, due to their low albedo, than does snow-covered tundra. Model simulations suggest that the northward movement of the treeline 6000 years ago could have reduced the regional albedo and increased energy absorption sufficiently to explain half of the climate warming that occurred at that time (Foley et al. 1994). The warmer regional climate would, in turn, favor tree reproduction and establishment at the treeline (Payette and Filion 1985), providing a positive feedback to regional warming (see Chapter 12). Predictions about the impact of future climate on vegetation should therefore also consider ecosystem feedbacks to climate.

Albedo, energy partitioning between latent and sensible heat fluxes, and surface structure also influence the amount of longwave radia-

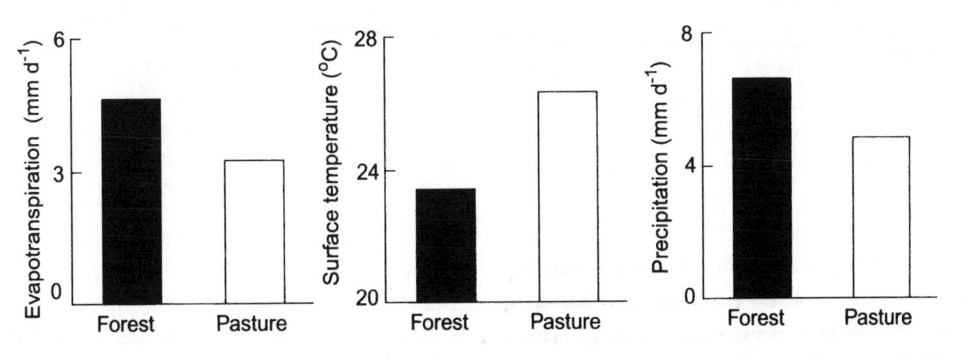

FIGURE 2.11. Simulations, using a general circulation model, of changes in evapotranspiration, surface air temperature, and precipitation that would occur if the rain forests of South America were replaced by pasture (Shukla et al. 1990).

tion transferred to the atmosphere (Fig. 2.2). This is due to the dependence of longwave radiation on surface temperature. Surface temperature tends to be high when the surface absorbs large amounts of incoming radiation (low albedo), has little water to evaporate, and/or has a smooth surface that is inefficient in transferring turbulent fluxes of sensible and latent heat to the atmosphere (see Chapter 4). Deserts, for example, experience large net longwave energy losses because their dry smooth surfaces lead to high surface temperatures, and there is little moisture to support evaporation that would otherwise cool the soil.

Temporal Variability in Climate

Long-Term Changes

Long-term climatic change is driven primarily by changes in solar input and changes in atmospheric composition. Earth's climate is a dynamic system that has changed repeatedly, producing frequent, and sometimes abrupt, changes in climate, manifested by a series of dramatic glacial epochs (Fig. 2.12) and sea level changes. Volcanic eruptions and asteroid impacts contributed to these changes by influencing the absorption or reflection of solar energy. Mountain building and erosion and continental drift have modified the patterns of atmospheric and ocean circulation. The primary force responsible for the evolution of Earth's climate, however, has been changes in the input of solar radiation, which has increased over much of the past 4 billion years, as the sun has matured (Fig. 2.13) (Schlesinger 1997). On shorter time scales, solar input has varied primarily due to predictable alterations in Earth's orbit (Fig. 2.14).

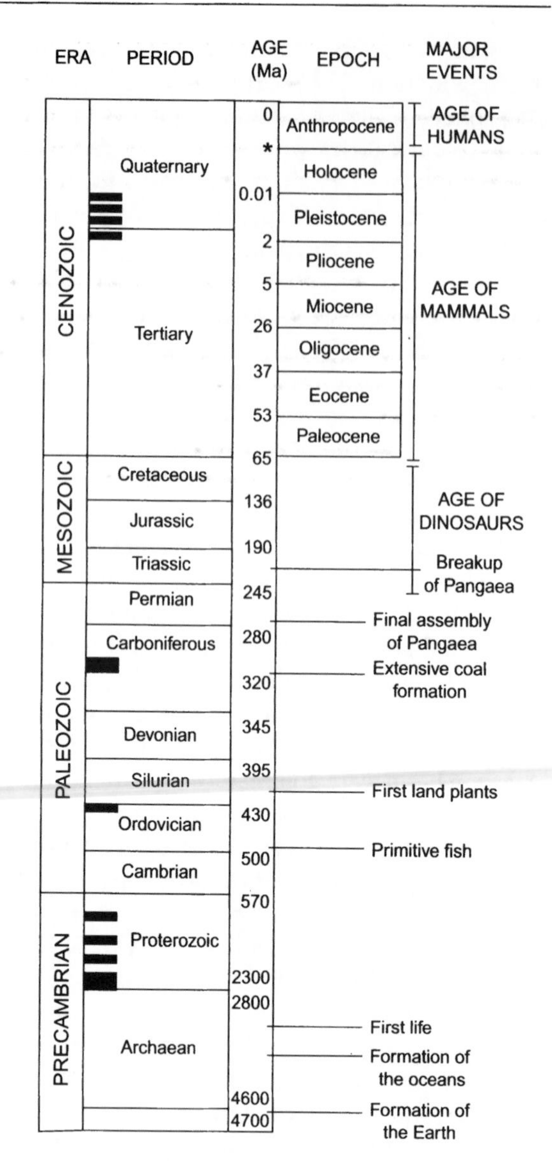

FIGURE 2.12. Geological time periods in Earth's history showing major glacial events (solid bars) and ecological events that strongly influenced ecosystem processes. Note the changes in time scale (Ma = millions of years). The most recent geologic epoch (the Anthropocene) began about 1750 with the beginning of the Industrial Revolution and is characterized by human domination of the biosphere. (Modified with permission from Sturman and Tapper, 1996.)

Three types of variations in Earth's orbit influence the amount of solar radiation received at the surface. These variations can be described by three parameters: **eccentricity** (the degree of ellipticity of Earth's orbit around the sun), **tilt** (the angle between Earth's axis of rotation and the plane of its orbit around the sun), and **precession** (a "wobbling" in Earth's axis of rotation with respect to the stars, determining the date during the year when solstices and equinoxes occur). At present, Earth's orbit is nearly circular (minimal eccentricity), leading to relatively small seasonal variation in solar input related to Earth–sun distance (Sturman

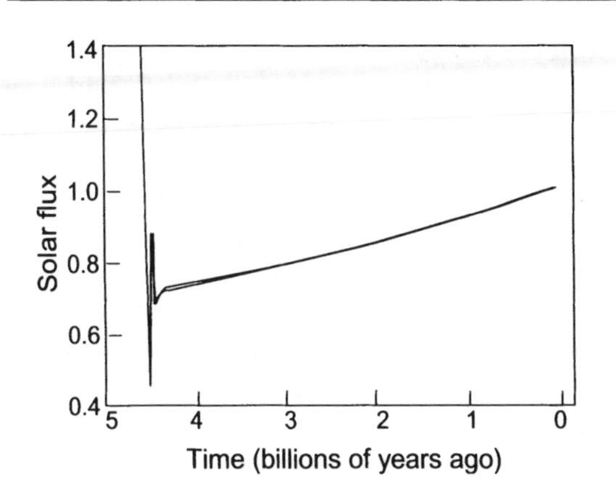

FIGURE 2.13. Solar flux (relative to present solar radiation) received by Earth since the beginning of the solar system (Graedel and Crutzen 1995).

and Tapper 1996). Tilt determines the strength of the seasons. Earth's tilt is presently intermediate (23.5°), providing an intermediate degree of seasonality. Earth's precession places the solstices in December and June, causing the Northern Hemisphere winters to be mild and Southern Hemisphere summers to be relatively warm. At times in the past, the solstices occurred at other times of year. The periodicities of these orbital parameters (eccentricity, tilt, and precession) are approximately 100,000, 41,000, and 23,000 years, respectively. Together they produce **Milankovitch cycles** of solar input that correlate with the glacial and interglacial cycles over the last 800,000 years, as determined by isotopic analyses of ocean sediments and ice cores.

The chemistry of ice and trapped air bubbles provide a paleorecord of the climate when the ice formed. The Vostok ice core, drilled at Vostok Station in Antarctica in the 1980s and 1990s, indicates considerable climate variability over the past 400,000 years, in large part related to the Milankovitch cycles (see Fig. 15.2). Analysis of bubbles in this and other ice cores indicates that past warming events have been associated with increases in CO_2 and CH_4 concentrations, providing circumstantial evidence

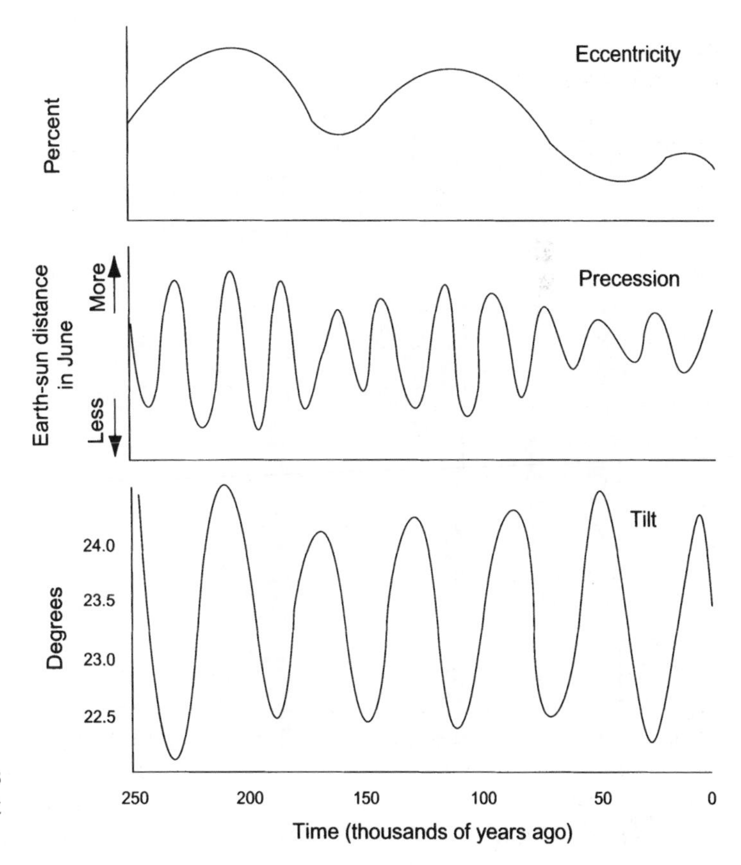

FIGURE 2.14. Long-term variations in three orbital parameters that result in glacial patterns.

for a past role of radiatively active gases in climate change. The unique feature of the recent anthropogenic increases in these gases is that they are occurring during an interglacial period, when Earth's climate is already relatively warm. The Vostok record suggests that the CO_2 concentration of the atmosphere is higher now than at any time in the last 400,000 years. Fine-scale analysis of ice cores from Greenland suggests that large changes from glacial to interglacial climate can occur in decades or less. Such rapid transitions in the climate system to a new state may be related to sudden changes in the strength of the thermohaline circulation that drives oceanic heat transport from the equator to the poles.

Tree ring records, obtained from living and dead trees, provide information about the climate during the past several thousand years. The width of tree rings gives a record of temperature and moisture, and the chemical composition of wood reflects the characteristics of the atmosphere at the time the wood was formed. Pollen preserved in low-oxygen sediments of lakes provides a history of plant taxa and climate over the past 10,000 years or more (Fig. 2.15). Pollen records from networks of sites can be used to construct maps of species distributions at various times in the past and can provide a history of species migrations across continents after climatic changes (COHMAP 1988). Other proxy records provide measures of temperature (species composition of Chironomids), precipitation (lake level), pH, and geochemistry.

The combination of paleoclimate proxies indicates that climate is inherently variable over all time scales. Atmospheric, oceanic, and other environmental changes that are occurring now due to human activities must be viewed as overlays on the natural climate variability that stems from long-term changes in Earth's surface characteristics and orbital geometry.

Earth's climate is now warmer than at any time in the last 1000 years (Fig. 2.16) and perhaps much longer. This warming is most pro-

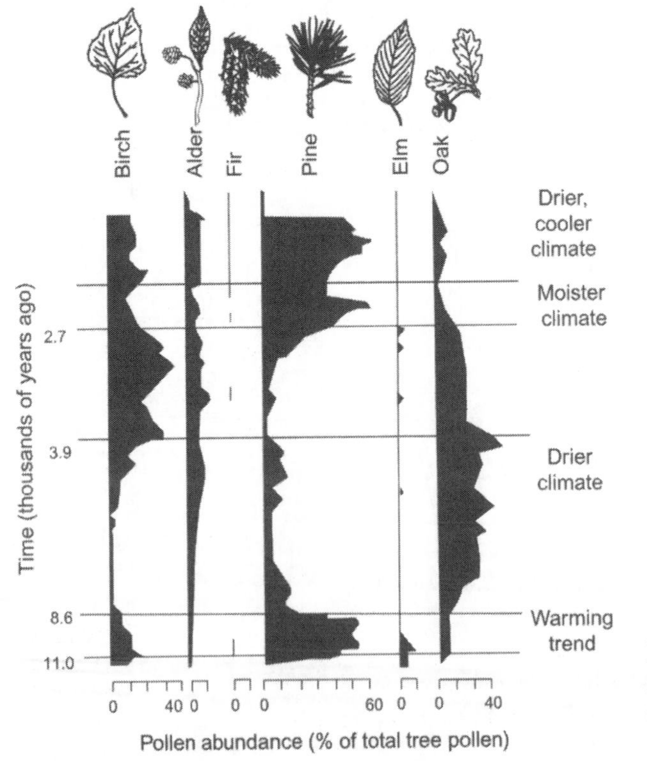

FIGURE 2.15. Pollen profile from a bog in northwestern Minnesota showing changes in the dominant tree species over the past 11,000 years. (McAndrews 1966).

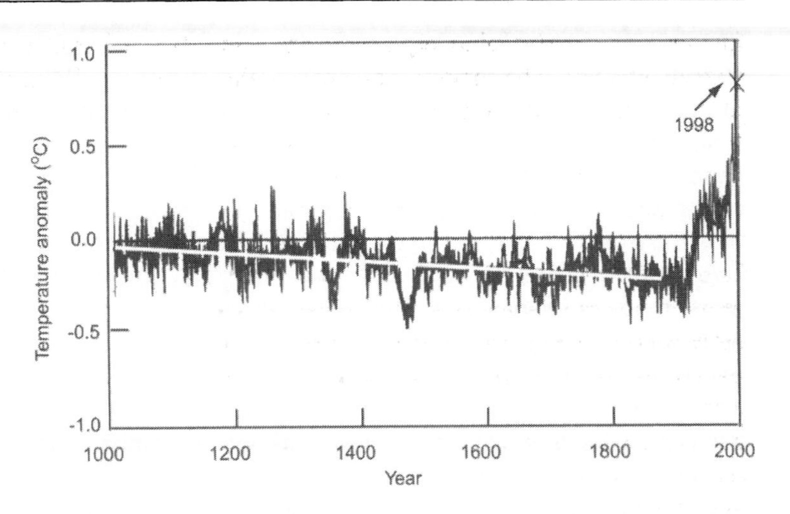

FIGURE 2.16. Time course of the average surface temperature of the Northern Hemisphere over the last 1000 years. The data are presented as a 40-year running average of the difference (anomaly) in temperature between each year and the average 1902–1980 temperature. Note that 1998 was the warmest year of the last millennium. (Redrawn with permission from *Geophysical Research Letters*; Mann et al. 1999.)

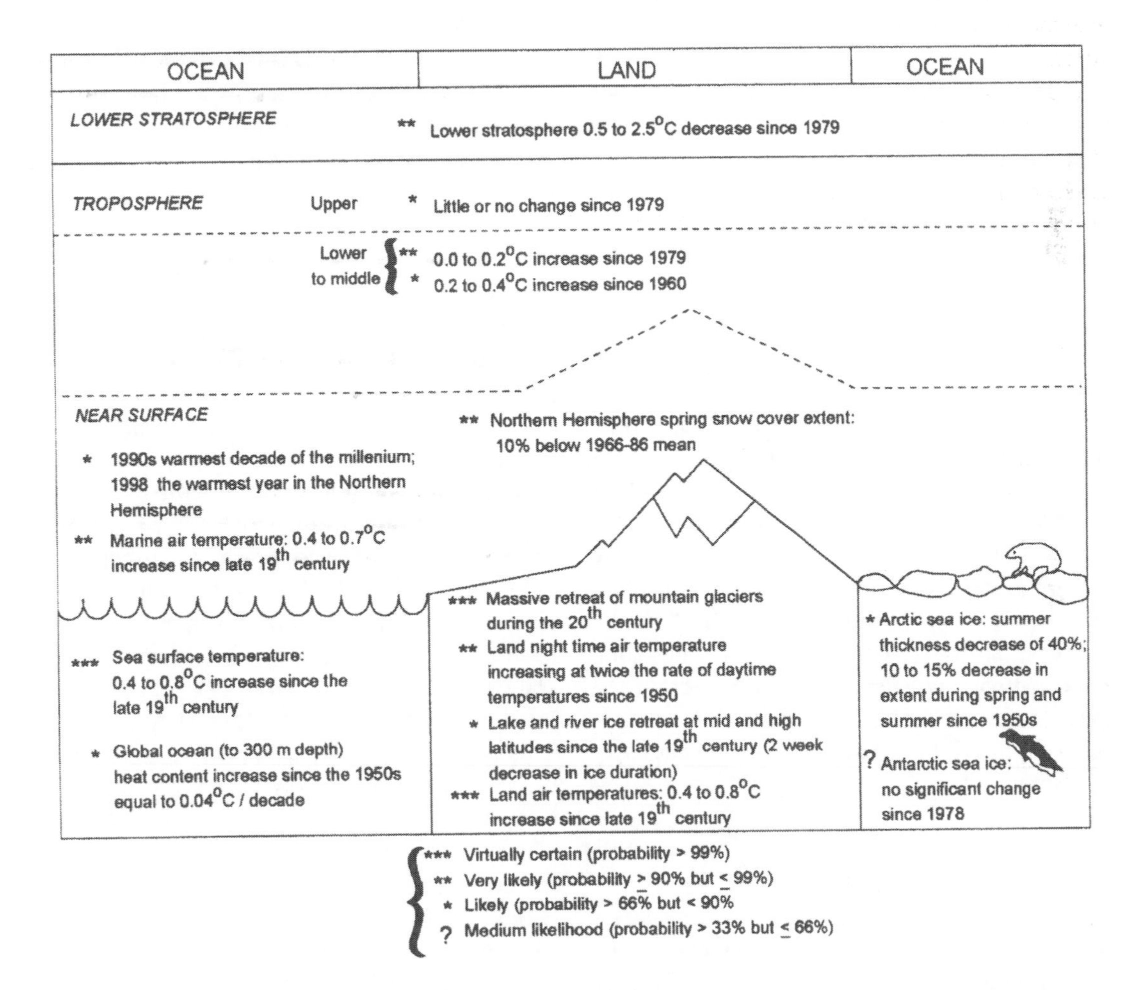

FIGURE 2.17. Sources of evidence that Earth's climate is warming. Reduction in stratospheric ozone causes less energy to be absorbed by the stratosphere; this causes the stratosphere to cool and allows more energy to penetrate to Earth's surface. Warming is most pronounced at Earth's surface and has caused the surface air and oceans to warm and glaciers and sea ice to melt. (Adapted from IPCC Assessment Report 2001; Folland et al. 2001.)

nounced near Earth's surface, where its ecological effects are greatest (Fig. 2.17). Some of the recent warming reflects an increase in solar input, but most of the warming results from human activities that increase the concentrations of radiatively active gases in the atmosphere (Fig. 2.18). Climate models and recent observations suggest that warming will be most pronounced in the interiors of continents, far from the moderating effects of oceans, and at high latitudes. The high-latitude warming reflects a positive feedback. As climate warms, the snow and sea ice melt earlier in the year, which replaces the reflective snow or ice cover with a low-albedo land or water surface. These darker surfaces absorb more radiation and transmit this energy to the atmosphere, which warms the climate. Those land-surface changes that produce a darker or drier surface also contribute to greater heat transfer to the atmosphere and to the warming of surface climate.

As climate warms, the air has a higher capacity to hold water vapor, so there is greater evaporation from oceans and other moist surfaces. In areas where rising air leads to condensation, there is greater precipitation. Continental interiors and rain shadows on the lee sides of mountains are less likely to experience large precipitation increases. Consequently, soil moisture and runoff to streams and rivers are likely to increase in coastal regions and mountains and to decrease in continental interiors. The complex controls and nonlinear feedbacks in the climate system make detailed climate projections problematic and are active areas of research.

Interannual Climate Variability

Much of the interannual variation in climate is associated with large-scale changes in the atmosphere–ocean system. Superimposed on the long-term climate variability are interannual variations that have been noted by farmers, fishermen, and naturalists for centuries. Some of this variability exhibits repeating geographic and temporal patterns. One of these phenomena that has received considerable attention is the **El Niño/southern oscillation** (ENSO) (Webster and Palmer 1997, Federov and Philander 2000). ENSO events are part of a large-scale, air–sea interaction that couples atmospheric pressure changes (the southern oscillation) with changes in ocean temperature (El Niño) over the equatorial Pacific Ocean. ENSO events have occurred, on average, every 3 to 7 years over the past century, with considerable irregularity (Trenberth and Haar 1996). No events occurred between 1943 and 1951, for example, and three major events occurred between 1988 and 1999.

In most years, the easterly trade winds push the warm surface waters of the Pacific westward so the layer of warm surface waters is deeper in the western Pacific than it is in the east (Figs. 2.9 and 2.19). The resulting warm waters in the western Pacific are associated with a low pressure center and promote convection and high rainfall in Indonesia. The offshore movement of surface waters in the eastern Pacific promotes upwelling of colder, deeper water off the coasts of Ecuador and Peru. These cold, nutrient-rich waters support a productive fishery (see Chapter 10) and promote subsidence of upper air, leading to the development of a high pressure center and low precipitation. At times, however, the Pacific high-pressure and Indonesian low-pressure centers weaken, and the easterly trades weaken. The warm surface waters then move eastward, forming a deep layer of warm water in the eastern Pacific. This weakens or shuts down the upwelling of cold water, promoting atmospheric convection and rainfall in coastal Ecuador and Peru. The colder waters in the western Pacific, in contrast, inhibit convection, leading to droughts in Indonesia, Australia, and India. This pattern is commonly termed El Niño. Periods in which the "normal" pattern is particularly strong are termed **La Niña**. The trigger for changes in this ocean–atmosphere system is unknown but may involve large-scale ocean waves, known as **Kelvin waves**,

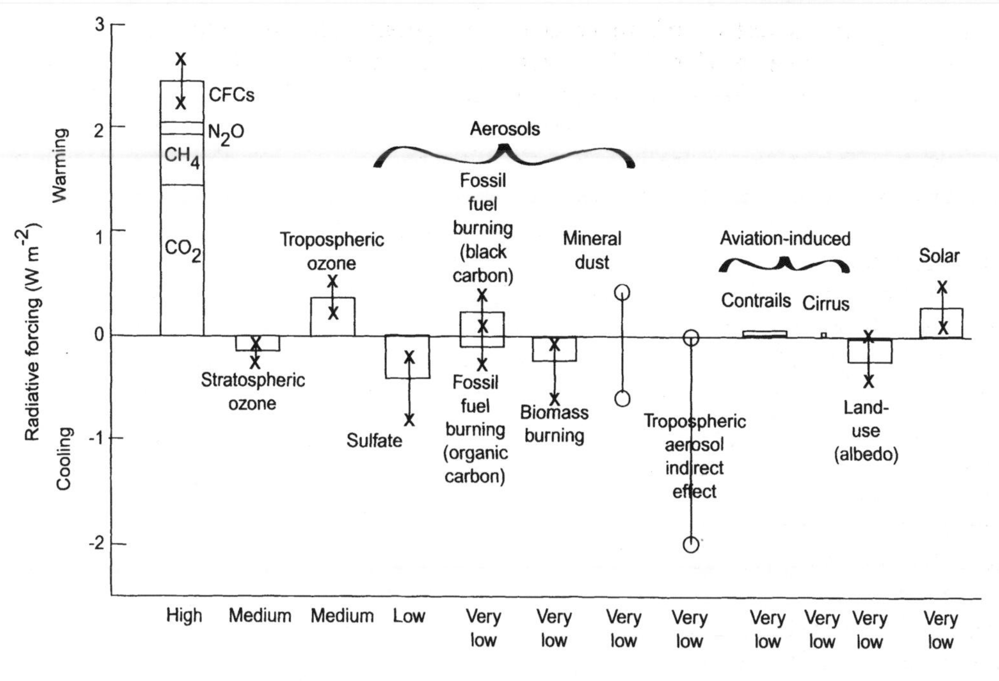

FIGURE 2.18. Major changes in the climate system that have caused Earth's climate to warm between 1750 and 2000. Some changes in the climate system lead to net warming; others lead to net cooling. The largest single cause of climate warming is probably the increased concentration of atmospheric CO_2, primarily as a result of burning fossil fuels. (Adapted from IPCC Assessment Report 2001; Ramaswamy et al. 2001.)

that travel back and forth across the tropical Pacific.

ENSO events have widespread climatic, ecosystem, and societal consequences. Strong El Niño phases cause dramatic reductions in anchovy fisheries in Peru and reproductive failure and mortality in sea birds and marine mammals. Extremes in precipitation linked to ENSO cycles are also evident in areas distant from the tropical Pacific. El Niño events bring hot, dry weather to the Amazon Basin, potentially affecting tree growth, soil carbon storage, and fire probability. Northward extension of warm tropical waters to the northern Pacific brings rains to coastal California and high winter temperatures to Alaska. An important lesson from ENSO studies is that strong climatic events in one portion of the globe have climatic consequences throughout the globe due to the dynamic interactions (termed **teleconnections**) associated with atmospheric circulation and ocean currents.

The Pacific North America (PNA) pattern is another large-scale pattern of climate variability. The positive mode of the PNA is characterized by above-average atmospheric pressure with warm dry weather in western North America and below-average pressure and low temperatures in the east. Variability in the PNA (or PNA-like) pattern is loosely linked to ENSO phases. Another large-scale climate pattern is the North Atlantic oscillation (NAO). Positive phases of the NAO are associated with a strengthening of the pressure gradient between the Icelandic low-pressure and the Bermuda high-pressure systems (Fig. 2.7). This increases heat transport to high latitudes by wind and ocean currents, leading to a warming of Scandinavia and western North America and a cooling of eastern Canada. Although the

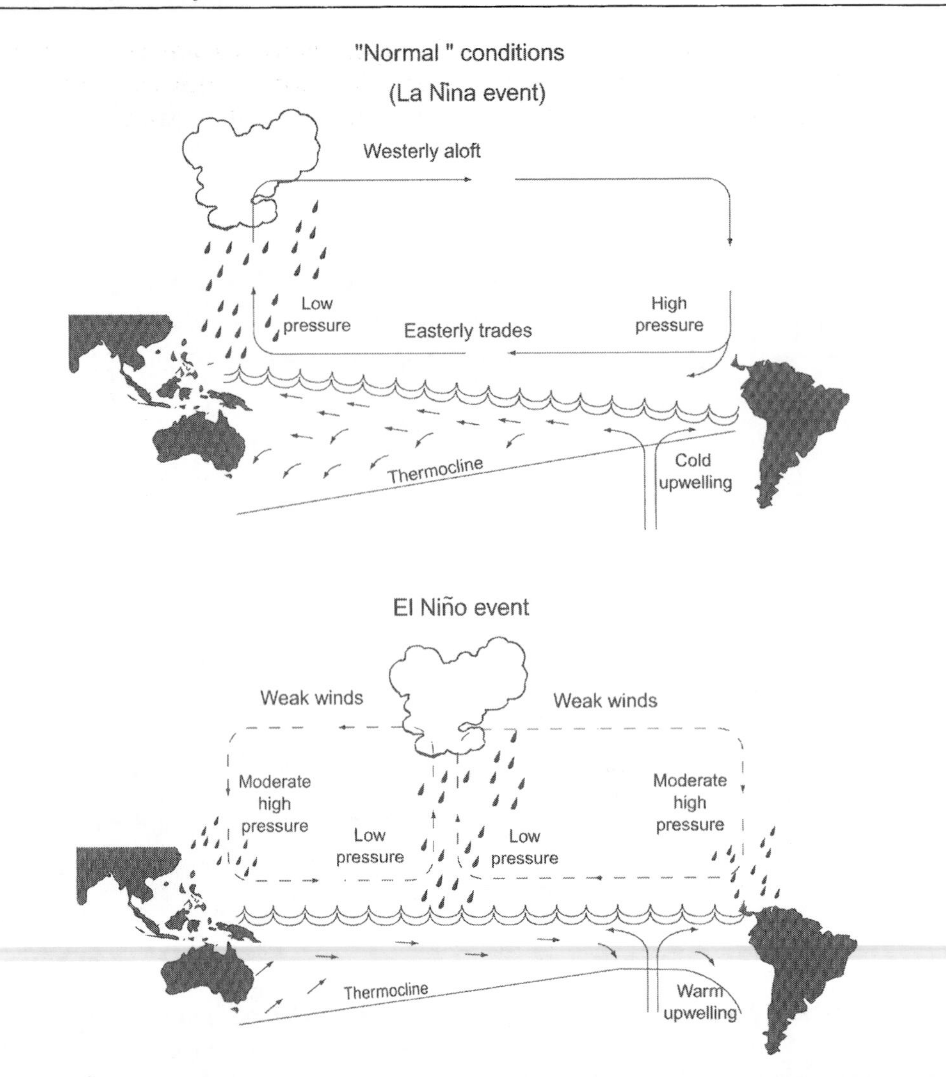

FIGURE 2.19. Walker circulation of the ocean and atmosphere in the tropical Pacific between South America and Indonesia during "normal" years and during El Niño years. In normal years, strong easterly trade winds push surface ocean waters to the west, producing deep, warm waters and high precipitation off the coast of Southeast Asia and cold, upwelling waters and low precipitation off the coast of South America. In El Niño years, however, weak easterly winds allow the surface waters to move from west to east across the Pacific Ocean, leading to cooler surface waters and less precipitation in Southeast Asia and warmer surface waters and more precipitation off South America.

factors that initiate these large-scale climate features are poorly understood, the patterns themselves and their ecosystem consequences are becoming more predictable. Future climatic changes will likely be associated with changes in the frequencies of certain phases of these large-scale climate patterns rather than any simple linear trend in climate. Climate warming, for example, might increase the frequency of ENSO events and positive phases of the NAO.

Seasonal and Daily Variations

Seasonal and daily variations in solar input have profound but predictable effects on climate and ecosystems. Perhaps the most obvious variations in the climate system are the

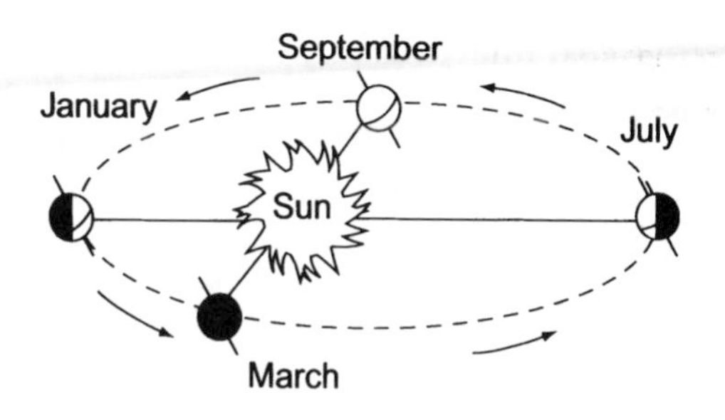

FIGURE 2.20. Earth's orbit around the sun, showing that the zone of greatest heating (the ITCZ) is south of the equator in January, north of the equator in July, and at the equator in March and September.

patterns of seasonal and diurnal change. The axis of Earth's rotation is fixed at 23.5° relative to its orbital plane about the sun. This tilt in Earth's axis results in strong seasonal variations in day length and the solar **irradiance**—that is, the quantity of solar energy received at Earth's surface per unit time. During the spring and autumn equinoxes, the entire Earth's surface receives approximately 12h of daylight (Fig. 2.20). At the Northern Hemisphere summer solstice, the sun's rays strike Earth most directly in that hemisphere, and day length is maximized. At the Northern Hemisphere winter solstice, the sun's rays strike Earth most obliquely in that hemisphere, and day length is minimized. The summer and winter solstices in the Southern Hemisphere are 6 months out of phase from those in the Northern Hemisphere. Variations in incident radiation become increasingly pronounced as latitude increases. Thus tropical environments experience relatively small seasonal differences in solar irradiance and day length, whereas they are maximized in the Arctic and Antarctic. Above the Arctic and Antarctic Circles, there are 24h of daylight at the summer solstice, and the sun never rises at the winter solstice. The relative homogeneity of temperature and light throughout the year in the tropics contributes to their high productivity and diversity. At higher latitudes, the length of the warm season strongly influences the life forms and productivity of ecosystems.

Variations in light and temperature also play an important role in determining the types of plants that grow in a given climate and the rates at which biological processes occur. Many biological processes are temperature dependent, and slower rates occur at lower temperatures. Diurnal variations in day length (**photoperiod**) provide important cues that allow organisms to prepare for seasonal variations in climate.

Relationship of Climate to Ecosystem Distribution and Structure

Climate is the major determinant of the global distribution of biomes. The major types of ecosystems (Plate 1) and their productivity show predictable relationships to climatic variables such as temperature and moisture (Holdridge 1947, Whittaker 1975) (Fig. 2.21; Plate 2). An understanding of the causes of geographic patterns of climate, as presented in this chapter, therefore allows us to predict the distribution of Earth's major biomes with their characteristic patterns of productivity and diversity (Plates 3 and 4).

Tropical wet forests occur from 12° N to 3° S and correspond to the ITCZ. Day length and

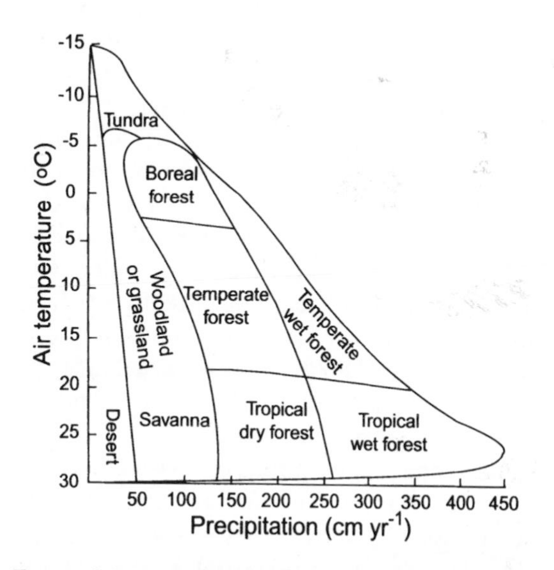

FIGURE 2.21. Distribution of major biomes in relation to mean annual temperature and precipitation. (Redrawn from *Communities and Ecosystems, 2nd edition*, by R.H. Whittaker © 1975, by permission of Pearson Education, Inc., Upper Saddle River, NJ; Whittaker 1975.)

solar angle show little seasonal change, leading to constant high temperatures. High solar radiation and convergence of the easterly trade winds at the ITCZ promote strong convective uplift, leading to high precipitation (175 to 400 cm annually). Periods of relatively low precipitation seldom last more than 1 to 2 months. **Tropical dry forests** occur north and south of tropical wet forests. Tropical dry forests have more pronounced wet and dry seasons because of seasonal movement of ITCZ over (wet season) and away from these forests (dry season). **Tropical savannas** occur between the tropical dry forests and the deserts. These savannas are warm and have low precipitation that is highly seasonal. **Subtropical deserts** at 25 to 30° N and S have a warm dry climate because of the subsidence of air in the descending limb of the Hadley cell.

Mid-latitude deserts, grasslands, and shrublands occur in the interiors of continents, particularly in the rain shadow of mountain ranges. They have low and unpredictable precipitation, low winter temperatures, and greater temperature extremes than tropical deserts. As precipitation increases, there is a gradual transition from desert to grassland to shrubland. **Temperate wet forests** occur on the west coasts of continents at 40 to 65° N and S, where westerlies blowing across a relatively warm ocean provide an abundant moisture source and migrating low pressure centers associated with the polar front promote high precipitation. Winters are mild, and summers are cool. **Temperate forests** occur in the mid-latitudes, where there is sufficient precipitation to support trees. The polar front migrates north and south of these forests from summer to winter, producing a strongly seasonal climate. **Mediterranean shrublands** are situated on the west coasts of continents. In summer, subtropical oceanic high pressure centers and cold upwelling coastal currents produce a warm dry climate. In winter, as wind and pressure systems move toward the equator, storms produced by polar fronts provide unpredictable precipitation.

The boreal forest (taiga) occurs in continental interiors at 50 to 70° N. The winter climate is dominated by polar air masses and the summer climate by temperate air masses, producing cold winters and mild summers. The distance from oceanic moisture sources results in low precipitation. The subzero mean annual temperatures lead to **permafrost** (permanently frozen ground) that restricts drainage and creates poorly drained soils and peatlands in low-lying areas. **Arctic tundra** is a zone north of the polar front in both summer and winter, resulting in a climate that is too cold to support growth of trees. Short cool summers restrict biological activity and limit the range of life forms that can survive.

Vegetation structure varies with climate both between and within biomes. Each biome type is dominated by predictable growth forms of plants. Tropical wet forests, for example, are dominated by broad-leaved evergreen trees, whereas areas that are periodically too cold or dry for growth are dominated by deciduous forests or, under more extreme conditions, by tundra or desert, respectively. Biomes are not discrete units with sharp boundaries but vary continuously in structure along climatic gradients. Along a moisture gradient in the tropics, for example, vegetation changes from evergreen tall trees in the wettest sites to a mix of evergreen and deciduous trees in areas that see the beginnings of a seasonal drought (Ellenberg 1979) (Fig. 2.22). As the climate becomes still drier, the stature of the trees and shrubs is reduced because there is less light competition and more competition for water. This leads to a shrubless desert with herbaceous perennial herbs in dry habitats. With extreme drought, the dominant life form becomes annuals and bulbs (herbaceous perennials in which aboveground parts die during the dry season). A similar gradient of growth forms, leaf types and life forms occurs along moisture gradients at other latitudes.

The diversity of growth forms within some ecosystems can be nearly as great as the diversity of dominant growth forms across biomes. In tropical wet forests, for example, continuous seasonal growth in a warm moist climate produces large trees with dense canopies that intercept and compete for a large fraction of the incoming radiation. Light then becomes the main driver of diversity within the ecosystem. Plants that can reach the canopy and have

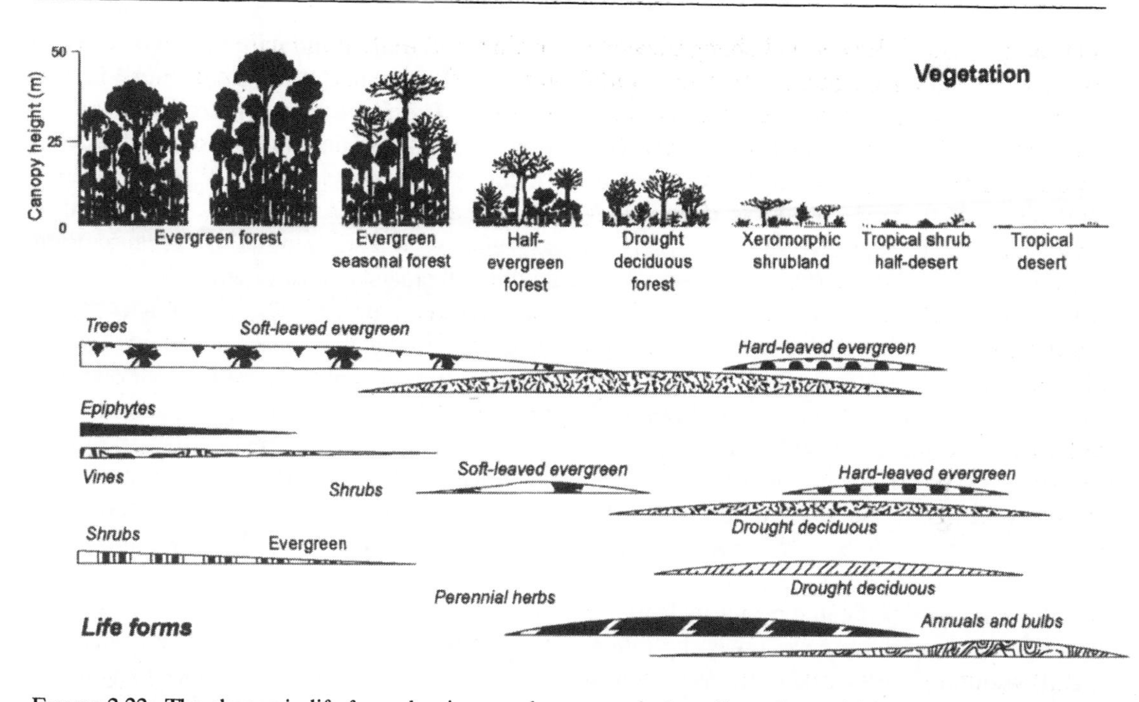

FIGURE 2.22. The change in life form dominance along a tropical gradient along which precipitation changes but temperature is relatively constant. (Redrawn with permission from *Journal of Ecology*; Ellenberg 1979.)

access to light compete effectively with tall trees. These growth forms include vines, which parasitize trees for support without investing carbon in strong stems. In this way vines can grow quickly into the canopy. Epiphytes are also common in the canopies of tropical wet forests, where they receive abundant light but, because their roots are restricted to the canopy, their growth is often water limited. Epiphytes have therefore evolved various specializations to trap water and nutrients. There is a wide range of subcanopy trees, shrubs, and herbs that are adapted to grow slowly under the low-light conditions beneath the canopy (Fig. 2.22). Light is therefore probably the general driver of structural diversity in the dense forests of wet tropical regions.

What determines structural diversity where moisture, rather than light, is limiting? Deserts, particularly warm deserts, have a great diversity of plant forms, including evergreen and deciduous small trees and shrubs, succulents, herbaceous perennials, and annuals. These growth forms do not show a well-defined vertical partitioning but show consistent horizontal patterns related to moisture availability. Competition for water results in diverse strategies for gaining, storing, and using the limited water supply. This leads to a wide range of rooting strategies and capacities to evade or endure drought.

Species diversity declines from the tropics to high latitudes and in many cases from low to high elevation. Species-rich tropical areas support more than 5000 species in a 10,000 km^2 area, whereas in the high arctic there are less than 200 species in the same area (Plate 4). Many animal groups show similar latitudinal patterns of diversity, in part because of their dependence on the underlying plant diversity. Climate, the evolutionary time available for species radiation, productivity, disturbance frequency, competitive interactions, land area available, and other factors have all been hypothesized to contribute to global patterns of diversity (Heywood and Watson 1995). Global patterns of diversity correlate most clearly with some dimension of climate and related ecosystem processes. Models that include only climate, acting as a filter on the plant functional types that can occur in a region, can reproduce the general global patterns of structural and species diversity (Kleiden and Mooney 2000). The actual causes for geographic patterns of species diversity are undoubtedly more complex, but these models and other analyses suggest that human-induced changes in climate, land use, and invasions of exotic species may alter future patterns of diversity.

Summary

The balance between incoming and outgoing radiation determines Earth's energy budget. The atmosphere transmits about half of the incoming shortwave solar radiation to Earth's surface but absorbs 90% of the outgoing longwave radiation emitted by Earth. This causes the atmosphere to be heated primarily from the bottom and generates convective motion in the atmosphere. Large-scale patterns of atmospheric circulation occur because the tropics receive more energy from the sun than they emit to space, whereas the poles lose more energy to space than they receive from the sun. The resulting circulation cells transport heat from the equator to the poles to balance these inequalities. In the process, they create three relatively distinct air masses in each hemisphere, a tropical air mass (0 to 30° N and S), a temperate air mass (30 to 60° N and S) and a polar air mass (60 to 90° N and S). There are four major areas of high pressure (the two poles and 30° N and S), where air descends and precipitation is low. The subtropical high pressure belts are the zones of the world's major deserts. There are three major zones of low pressure (the equator and 60° N and S), where air rises and precipitation is high. These areas support the tropical rain forests at the equator and the temperate rain forests of western North and southern South America. Ocean currents account for about 40% of the latitudinal heat transport from the equator to the poles. These currents are driven by surface winds and by the downwelling of cold saline waters at high latitudes, balanced by upwelling at lower latitudes.

Regional and local patterns of climate reflect heterogeneity in Earth's surface. Uneven heating between the land and the oceans modifies the general latitudinal patterns of climate by generating zones of prevailing high and low pressure. These pressure centers and the location of major mountain ranges guide storm tracks that strongly influence regional patterns of climate. Oceans and large lakes also moderate climate on adjacent lands because their high heat capacity causes them to heat or cool more slowly than land. These heating contrasts produce predictable seasonal winds (monsoon) and daily winds (land–sea breezes) that influence the adjacent land. Mountains also create heterogeneity in precipitation and in the quantity of solar radiation intercepted.

Vegetation influences climate through its effects on surface albedo, which determines the quantity of incoming radiation absorbed by the surface and energy released to the atmosphere via longwave radiation and turbulent fluxes of latent and sensible heat. Sensible heat fluxes and longwave radiation directly heat the atmosphere, and latent heat transfers water vapor to the atmosphere, influencing local temperature and moisture sources for precipitation.

Climate is variable over all time scales. Long-term variations in climate are driven largely by changes in solar input and atmospheric composition. Superimposed on these long-term trends are predictable daily and seasonal patterns of climate, as well as repeating patterns such as those associated with El Niño/southern oscillation. These oscillations cause widespread changes in the geographic patterns of climate on time scales of years to decades. Future changes in climate may reflect changes in the frequencies of these large-scale climate modes.

Review Questions

1. Describe the energy budget of Earth's surface and the atmosphere. What are the major pathways by which energy is absorbed by Earth's surface? By the atmosphere? What are the roles of clouds and radiatively active gases in determining the relative importance of these pathways?
2. Why is the troposphere warmest at the bottom but the stratosphere is warmest at

the top? How does each of these atmospheric layers influence the environment of ecosystems?

3. Explain how unequal heating of Earth by the sun and the resulting atmospheric circulation produce the major latitudinal climate zones, such as those characterized by tropical forests, subtropical deserts, temperate forests, and arctic tundra.
4. How does the rotation of Earth (and the resulting Coriolis forces) and the separation of Earth's surface into oceans and continents influence the global patterns of climate?
5. How does the chemical composition of Earth's atmosphere influence climate?
6. What causes the global pattern in surface ocean currents? Why are the deep water ocean currents different from those at the surface? What is the nature of the connection between deep and surface ocean currents?
7. How does ocean circulation influence climate at global, continental, and local scales?
8. How does topography affect climate at continental and local scales?
9. What are the major causes of long-term changes in climate? How would you expect future climate to differ from that of today in 100 years? 10,000 years? 2 billion years? Explain your answers.
10. Explain how the interannual variations in climate of Indonesia, Peru, and California are interconnected. Would these patterns influence eastern North America or Europe? How?
11. Explain the climatic basis for the global distribution of each major biome type. Use maps of global winds and ocean currents to explain these distributions.
12. Describe the climate of your birthplace. Using your understanding of the global climate system, explain why that location has its particular climate.

Additional Reading

Ahrens, C.D. 1998. *Essentials of Meteorology: An Invitation to the Atmosphere.* Wadsworth, Belmont, CA.

Bradshaw, M., and R. Weaver. 1993. *Physical Geography.* Mosby, St. Louis.

Graedel, T.E., and P.J. Crutzen. 1995. *Atmosphere, Climate, and Change.* Scientific American Library, New York.

Oke, T.R. 1987. *Boundary Layer Climates.* 2nd ed. Methuen, London.

Skinner, B.J., S.C. Porter, and D.B. Botkin. *The Blue Planet: An Introduction to Earth System Science.* 2nd ed. Wiley, New York.

Sturman, A.P., and N.J. Tapper. 1996. *The Weather and Climate of Australia and New Zealand.* Oxford University Press, Oxford, UK.

3
Geology and Soils

Within a given climatic regime, soil properties are the major control over ecosystem processes. This chapter provides background on the factors regulating the soil properties that most strongly influence ecosystem processes.

Introduction

Soils form a thin film over Earth's surface in which geological and biological processes intersect. A unique feature of terrestrial ecosystems is that vegetation acquires its resources from two quite different environments—the air and the soil. The soil is a multiphasic system consisting of solids, liquids, and gases, with solids typically occupying about half the soil volume, and liquids and gases each occupying 15 to 35% of the volume (Ugolini and Spaltenstein 1992). The physical soil matrix provides a source of water and nutrients to plants and microbes and is the physical support system in which terrestrial vegetation is rooted. It provides the medium in which most decomposer organisms and many animals live. For these reasons, the physical and chemical characteristics of soils strongly influence all aspects of ecosystem functioning, which, in turn, feed back to influence the physical, structural and chemical properties of soils (see Fig. 1.3). Soils play such an integral role in ecosystem processes that it is difficult to separate the study of soils from that of ecosystem processes. In open-water (**pelagic**) ecosystems, phytoplankton cannot directly tap resources from sediments, so soil processes provide nutrient resources to plants only indirectly through mixing of the water column.

Soils are also a critical component of the total Earth System. They play a key role in the giant global reduction–oxidation cycles of carbon, nitrogen, and sulfur. Soils mediate many of the key reactions in these cycles and provide essential resources to biological processes that drive these cycles.

Soils represent the intersection of the *bio*, *geo*, and *chemistry* in biogeochemistry. Many of the later chapters in this book address the short-term dynamics of soil processes, particularly those processes that occur on time scales of hours to centuries. This chapter emphasizes soil processes that occur over longer time scales or that are strongly influenced by physical and chemical interactions with the environment. This is essential background for understanding the dynamics of ecosystems.

Controls over Soil Formation

The soil properties of an ecosystem result from the dynamic balance of two opposing forces: soil formation and soil loss. State factors differ in their effects on these opposing processes and therefore on soil and ecosystem properties

(Dokuchaev 1879, Jenny 1941, Amundson and Jenny 1997).

Parent Material

The physical and chemical properties of rocks and the rates at which they are uplifted and weathered strongly influence soil properties. The dynamics of the rock cycle, operating over billions of years, govern the variation and distribution of geological materials on Earth's surface. The rock cycle describes the cyclic process by which rocks are formed and **weathered**—that is, the chemical and physical alteration of rocks and minerals near Earth's surface (Fig. 3.1). The rock cycle produces minerals that buffer the biological acidity that accounts for much of rock weathering but also provides many of the nutrients that allow biology to produce this acidity. The compounds produced by weathering move via rivers to the oceans where they are deposited to form sediments, which are then buried to form **sedimentary rocks. Igneous rocks** form when **magma** from the molten core of Earth moves upward toward the surface in cracks or volcanoes. Either sedimentary or igneous rocks can be modified under heat or pressure to form **metamorphic rocks.** With additional heat and pressure, metamorphic rocks melt and become magma. Any of these rock types can be raised to the surface via uplift during mountain-building episodes, after which the material is again subjected to weathering and erosion (Fig. 3.1). The timing

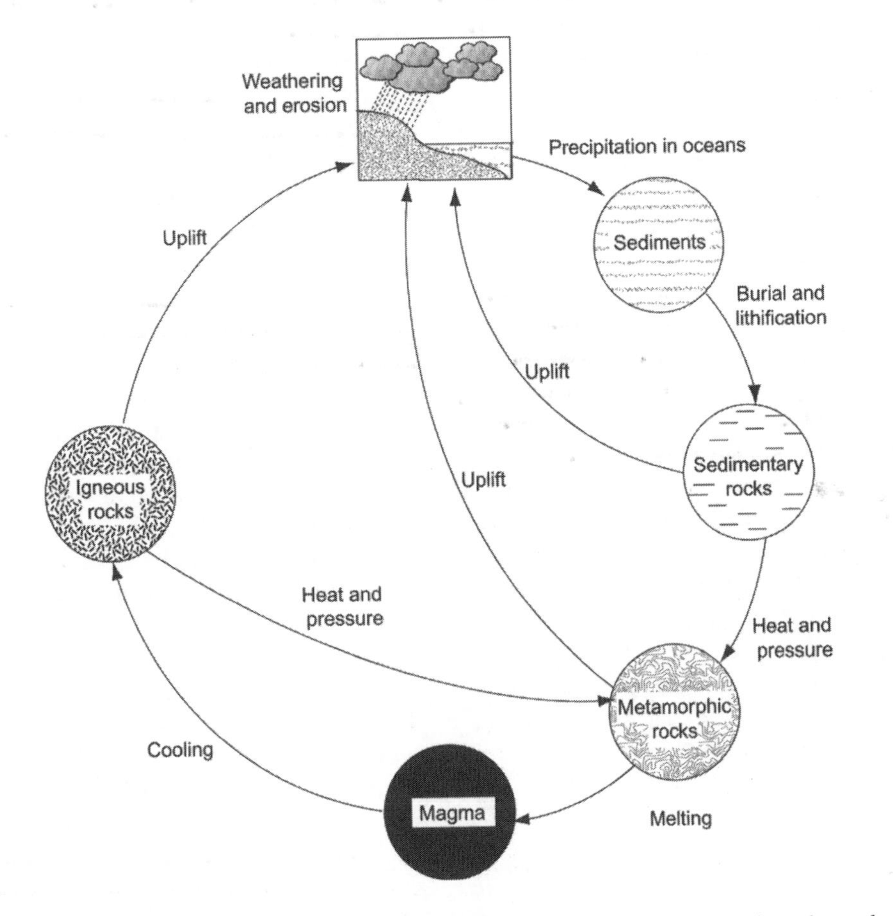

FIGURE 3.1. The rock cycles as proposed by Hutton in 1785. Rocks are weathered to form sediment, which is then buried. After deep burial, the rocks undergo metamorphosis, melting, or both. Later, they are deformed and uplifted into mountain chains, only to be weathered again and recycled. (Redrawn with permission from *Earth* by Frank Press and Raymond Siever © 1974, 1978, 1982, and 1986 by W.H. Freeman and Company; Press and Siever 1986.)

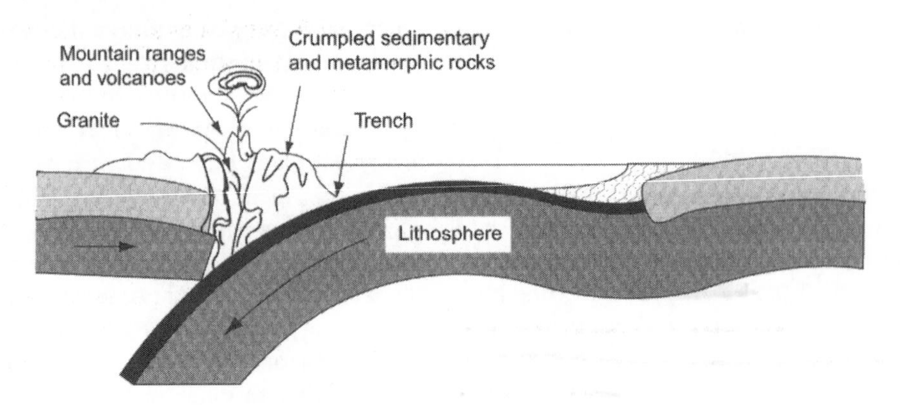

FIGURE 3.2. Cross-section of a zone of plate collision in which the oceanic plate is subducted beneath a continental plate, forming an ocean trench in the zone of subduction and mountains and volcanoes in the zone of uplift. (Redrawn with permission from *Earth* by Frank Press and Raymond Siever © 1974, 1978, 1982, and 1986 by W.H. Freeman and Company; Press and Siever 1986.)

and locations of uplift and the kinds of rock uplifted ultimately determine the distribution of different kinds of bedrock across Earth's surface.

Plate tectonics are the driving forces behind rock formation. The **lithosphere**, or crust—the strong outermost shell of Earth that rides on partially molten material beneath—is broken into large rigid plates, each of which moves independently. Where the plates converge and collide, portions of the lithosphere buckle downward and are **subducted**, leading to the formation of ocean trenches, and the overriding plate is **uplifted**, causing the formation of mountain ranges and volcanoes (Fig. 3.2). Regions of plate collision and active mountain building coincide with Earth's major earthquake belts. The Himalayan Mountains, for example, are still rising due to the collision of the Indian subcontinent with Asia. If plates converge in one place, they must diverge or separate elsewhere. Eurasia, Africa, and the Americas were once the single supercontinent of Pangaea, 200 million years ago. The mid-Atlantic and mid-Pacific ridges are zones of active divergence of today's ocean plates.

Climate

Temperature and moisture influence rates of chemical reactions that in turn govern the rate and products of weathering and therefore the development of soils from rocks. Temperature and moisture also influence biological processes, such as the production of organic matter by plants and its decomposition by microorganisms and therefore the amount and quality of organic matter in the soil (see Chapters 5 to 7). Soil carbon, for example, increases along elevational gradients as temperature decreases (Vitousek 1994b) and decreases in rain shadows downwind of mountain ranges (Burke et al. 1989). Precipitation is a major pathway by which many materials enter ecosystems. **Oligotrophic** (nutrient-poor) bogs are isolated from mineral soils and depend entirely on precipitation to supply new minerals. The movement of water is also crucial in determining whether the products of weathering accumulate or are lost from a soil. In summary, climate affects virtually all soil properties at scales ranging from local to global.

Topography

Topography influences soils through its effect on climate and differential transport of fine soil particles. The attributes of topography that are important for ecosystem processes include the site's topographic position on a **catena** or hillslope complex, the aspect of the slope, and the relationship between the site and hydrologic pathways (Amundson and Jenny 1997). Characteristics such as soil depth, texture, and mineral content vary with hillslope position. Erosional processes preferentially move fine-

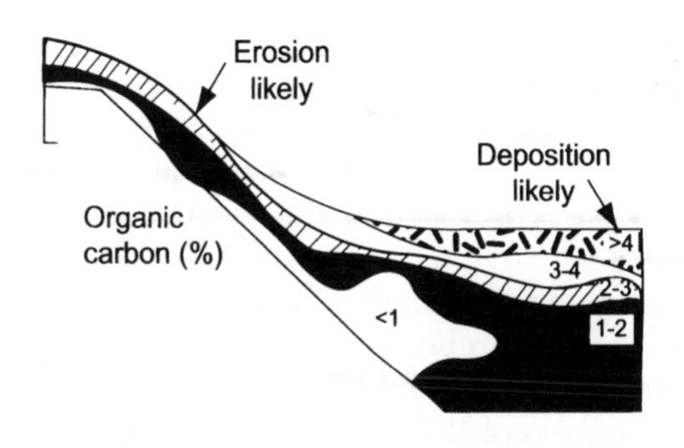

FIGURE 3.3. Relationship between hillslope position, likelihood of erosion or deposition, and soil organic carbon concentration. (Redrawn with permission from Oxford University Press; Birkeland 1999.)

grained materials downslope and deposit them at lower locations. Depositional areas at the base of slopes and in valley bottoms therefore tend to have deep fine-textured soils with a high soil organic content (Fig. 3.3) and high water-holding capacity. Depositional areas supply more soil resources to plant roots and microbes and provide greater physical stability than do higher slope positions. For these reasons valley bottoms typically exhibit higher rates of most ecosystem processes than do ridges or shoulders of slopes. Soils in lower slope positions in sagebrush ecosystems, for example, have greater soil moisture, higher soil organic matter content, and higher rates of nitrogen mineralization and gaseous losses than do upslope soils (Burke et al. 1990, Matson et al. 1991).

Slope position also determines patterns of snow redistribution in cold climates, with deepest accumulations beneath ridges and in the protected lower slopes. These differential accumulations alter effective precipitation and length of growing season sufficiently to influence plant and microbial processes well into the summer.

Finally, the aspect of a slope influences solar input (see Chapter 2) and therefore soil temperature, rates of evapotranspiration, and soil moisture. At high latitudes and in wet climates, these differences in soil environment reduce rates of decomposition and mineralization on poleward-facing slopes (Van Cleve et al. 1991). At low latitudes and in dry climates, however, the greater retention of soil moisture on poleward-facing slopes allows a longer growing season and supports forests, whereas slopes facing the equator are more likely to support desert or shrub vegetation (Whittaker and Niering 1965).

Time

Many soil-forming processes occur slowly, so the time over which soils develop influences their properties. Rocks and minerals are weathered over time, and important nutrient elements are transferred among soil layers or transported out of the ecosystem. Hillslopes erode, and valley bottoms accumulate materials, and biological processes add organic matter and critical nutrient elements like carbon and nitrogen. Phosphorus availability is high early in soil development and becomes progressively less available over time due to its losses from the system and to its fixation in mineral forms that are unavailable to plants (Fig. 3.4) (Walker and Syers 1976). This process required millions of years of soil development in Hawaii, despite a warm moist climate (Crews et al. 1995) and resulted in a change from nitrogen limitation of plant growth on young soils to phosphorus limitation on older soils (Vitousek et al. 1993).

Some changes in soil properties happen relatively quickly. Retreating glaciers and river floodplains often deposit phosphorus-rich till. If seed sources are available, these soils are colonized by plants with symbiotic nitrogen-fixing microbes, allowing such ecosystems to accumulate their maximum pool sizes of carbon and nitrogen within 50 to 100 years (Crocker and Major 1955, Van Cleve et al. 1991). Other soil-forming processes occur slowly. Young marine terraces in coastal California have relatively high phosphorus availability but low carbon and nitrogen content. Over hundreds of thousand of years, these terraces accumulate organic matter and nitrogen, causing a change from coastal grassland to productive redwood forest (Jenny et al. 1969). Over millions of years, silicates are leached out, leaving behind a hardpan of iron and aluminum oxides with low fertility and seasonally anaerobic soils. The pygmy cypress forests that develop on these old ter-

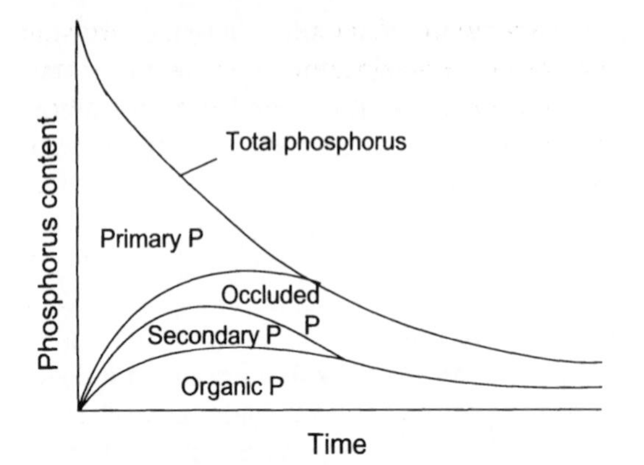

FIGURE 3.4. The generalized effects of long-term weathering and soil development on the distribution and availability of phosphorus (P). Newly exposed geologic substrate is relatively rich in weatherable minerals, which release phosphorus. This release leads to accumulation of both organic and readily soluble forms (secondary phosphorus, such as calcium phosphate). As primary minerals disappear and secondary minerals capable of sorbing phosphorus accumulate, an increasing proportion of the phosphorus remaining in the system is held in unavailable (occluded) forms. Availability of phosphorus to plants peaks relatively early in this sequence and declines thereafter. (Redrawn with permission from *Geoderma*, Vol. 15 © 1976 Elsevier Science; Walker and Syers 1976.)

races have very low productivity. The phenolic compounds produced by these trees as defenses against herbivores also retard decomposition, further reducing soil fertility (Northup et al. 1995) (see Chapter 13).

Potential Biota

The past and present organisms at a site strongly influence soil chemical and physical properties. Most soil development occurs in the presence of live organisms (Ugolini and Spaltenstein 1992). There are often clear associations between vegetation and soils. The organic acids in the litter of many coniferous species, for example, acidify the soil. This, in combination with the characteristically low quality of conifer litter, leads to slower decomposition in conifer than in deciduous forests (Van Cleve et al. 1991) (see Chapter 7). It is frequently difficult, however, to separate the chicken from the egg. Did the vegetation determine soil properties or vice versa?

One approach to determining vegetation effects on soils has been to plant monocultures or species mixes into initially homogeneous sites. Rapidly growing grasses in a nitrogen-poor perennial grassland enhanced the nitrogen mineralization of soils within 3 years (Wedin and Tilman 1990) (see Fig. 12.5), as did deep-rooted forbs in an annual grassland (Hooper and Vitousek 1998). Another approach has been to examine the consequences of species invasions or extinctions on soil processes. The invasion of a non-native nitrogen fixer into Hawaiian rain forests, for example, increased nitrogen inputs to the system more than fivefold, altering the characteristics of soils and the colonization and competitive balance among native plant species (Vitousek et al. 1987) (see Fig. 12.3).

Animals also influence soil properties. Earthworms, termites, and invertebrate shredders strongly influence decomposition rates (see Chapter 7) and therefore soil properties that are influenced by soil organic content.

Human Activities

Since the 1950s, the tripling of the human population and associated agricultural and industrial activities have strongly influenced soil development worldwide. Human activities influence soils directly through changes in nutrient inputs, irrigation, alteration of soil microenvironment through land use change (see Chapter 14), and increased erosional loss of soils. Human activities indirectly affect soils through changes in atmospheric composition and through the deletions and additions of species. Today and in the future, human activities will affect ecosystem properties both directly and through their effects on other interactive controls (see Chapters 14 and 16).

Controls over Soil Loss

Soil formation depends on the balance between deposition, erosion, and soil development. Soil thickness varies with hillslope position, with erosion dominating on steep hillslopes, deposition in valley bottoms, and soil development on gentle slopes and terraces where the lateral transport of materials is minimal (Fig. 3.3). Much of Earth's surface is in hilly or mountainous terrain where erosion and deposition are important processes. In these landscape positions, soils may have developed for only a few thousand years, even though soils on flat terraces of the same geomorphic age may have developed for millions of years. Erosion is an important process, because it removes the products of weathering and biological activity. In young soils, these erosional losses reduce soil fertility by removing clays and organic matter that store water and nutrients. On highly weathered landscapes, however, erosion renews soil fertility by removing the highly weathered remnants (sands and iron oxides) that contribute little to soil fertility and by exposing less weathered materials that provide a new source of essential nutrients.

Average regional erosion rates vary by two to three orders of magnitude among areas with different topography and climate (Table 3.1) (Milliman and Meade 1983). Erosion rates tend to approach rates of tectonic uplift, so regions with active tectonic uplift and steep slopes generally have higher erosion rates than flat weathered terrain. In steep terrain, soil creep and other processes unrelated to rainfall can be the dominant erosional processes. Climatic zones with high rainfall also tend to have high erosion rates. Some semiarid zones and polar montane regions with active uplift also have occasional episodes of active erosion during intense rains, because there is little vegetation to protect soils from erosion. Land use changes that reduce vegetation cover can increase erosion rates by several orders of magnitude, causing meters of soil to be lost in a few years. This compares to background erosion rates of 0.1 to 1.0 mm per 1000 years in many areas. High erosion rates caused by land use change are only temporary in a geologic sense. Once the soil mantle is gone, erosion rates decline to values determined by climate, bedrock type, and slope. Much of the erosion on natural landscapes probably occurs during rare periods of unusually high rainfall or low vegetation cover rather than during average conditions.

The dominant erosional processes depend on topography, surface material properties, and the pathways by which water leaves the landscape. **Mass wasting** is a major erosional process in most regions. This is the downslope movement of soil or rock material under the influence of gravity without the direct aid of other media such as water, air, or ice. The rate of mass wasting depends on the hillslope gradient, length, and curvature. Some mass wasting occurs rapidly, for example, in rockfalls, landslides, or debris flows. Landslides can transport sediment volumes ranging from a few cubic

TABLE 3.1. Climatic and topographic effects on long-term erosion rates.

Climate zone	Relief	Erosion rate[a] (mm per 1000 yr)
Glacial	Gentle (ice sheets)	50–200
	Steep (valleys)	1000–5000
Polar montane	Steep	10–1000
Temperate maritime	Mostly gentle	5–100
Temperate continental	Gentle	10–100
	Steep	100–200+
Mediterranean	—	10–?
Semiarid	Gentle	100–1000
Arid	—	10–?
Wet subtropics	—	10–1000?
Wet tropics	Gentle	10–100
	Steep	100–1000

[a] Erosion rates are estimated from average sediment yields of rivers in different climatic and topographic regimes.
Data From Selby (1993).

meters to cubic kilometers. The probability of a mass-wasting event depends on the balance between the driving forces for downslope movement and the forces that resist this movement. Gravity is the major driving force for mass wasting. The gravitational force (or **stress**) can be divided into two components: one parallel to the slope, which drives mass wasting, and one perpendicular to the slope, which increases the friction between the material and the bedrock (Fig. 3.5). The steeper the slope, the greater the downhill component of the force and therefore the greater the probability of mass wasting.

Many factors influence the **strength** of a soil mass (i.e., the amount of force required to initiate slope failure) (Selby 1993). These include the sliding friction between the material and some well-defined plane and the internal friction caused by the friction among individual grains within the soil matrix. In some cases, there is a well-defined plane along which materials can slide, such as the movement of soils over a frozen soil layer, but commonly it is the internal friction that largely determines the resistance to mass wasting. Cohesion among soil particles and water molecules enhances the internal friction that resists mass wasting. A small amount of water enhances cohesion among particles, explaining why sand castles are easier to make with moist than with dry sand. A high water content, however, exerts pressure on the grains, making them more buoyant and reducing the frictional strength. They become unstable, leading to liquefaction of the soil mass, which can flow downslope. Fine-particle soils have lower slope thresholds of instability and are more likely to lead to slope failure than are coarse-textured soils. Roots also increase the resistance of soils to downslope movement, so deforestation and other land use changes that reduce root biomass increase the probability of landslides.

Mass wasting on soil-mantled, well-vegetated gentle slopes occurs slowly through soil creep. Displacement of surface soil particles by freeze–thaw events or the movement of soils brought to the surface by burrowing animals, for example, is likely to cause a net downslope movement of soil. These small-scale processes contribute to erosion rates of 0.1 mm yr^{-1} or less.

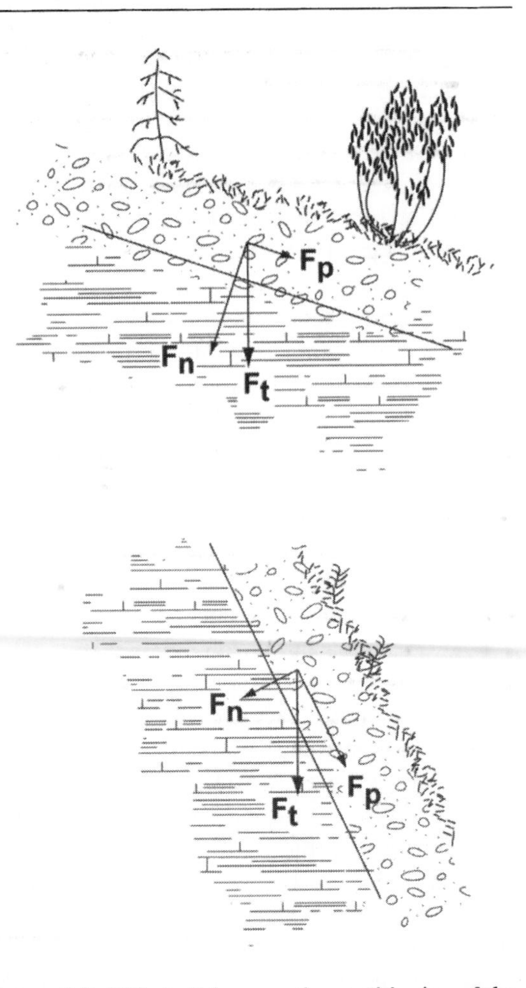

FIGURE 3.5. Effect of slope on the partitioning of the total gravitational force (F_t) into a component that is normal to the slope (F_n)—and therefore contributes to friction that resists erosion—and a component that is parallel to the slope (F_p)—and therefore promotes erosion. Steep slopes have a larger F_p value and therefore a greater tendency to erode.

The pathways by which water leaves the landscape strongly influence erosion. Water typically leaves a landscape by one of several pathways: groundwater flow, shallow subsurface flow, or overland flow (when precipitation rate exceeds infiltration rate) (see Fig. 14.6). The relative importance of these pathways is strongly influenced by topography, vegetation, and material properties such as the hydraulic conductivity of soils. Groundwater and shallow subsurface flow dissolve and remove ions and small particles. At the opposite extreme, overland flow causes erosion primarily by surface sheet wash, rills, and rain splash. This typically occurs in arid and semiarid soil-mantled landscapes or on disturbed ground. Overland flow rates of 0.15 to $3\,cm\,s^{-1}$ are sufficient to suspend clay and silt particles and move them downhill (Selby 1993). As water collects into gullies, its velocity, and therefore erosion potential, increases. Vegetation and litter layers greatly increase infiltration into the soil by reducing the velocity with which raindrops hit the soil, thereby preventing surface compaction by the raindrops. Vegetated soils are also less compact because roots and soil animals create channels in the soil. In these ways vegetation and a litter layer substantially increase infiltration and therefore groundwater and subsurface flow.

Wind is an important agent of erosion in areas where wind speeds are high at the soil surface, for example, where vegetation removal exposes the soil surface to strong winds. Some agricultural areas in China have lost meters of soil to wind erosion and have become an important source of iron to the phytoplankton in the Pacific Ocean (see Chapter 10).

Glaciers are an important erosional pathway in cold, moist climates. Glacial rivers often carry large sediment loads and produce a braided river valley with meandering stream channels that are important locations of primary succession.

Erosion in one location must be balanced by deposition elsewhere. Deposition can range from slow rates of dust or loess input to siltation events during floods to massive moraines or debris accumulations at the base of slopes.

Development of Soil Profiles

Soils develop through the addition of materials to the system, transformation of those materials in the system, transfer down and up in the soil profile, and loss of materials from the system (Fig. 3.6).

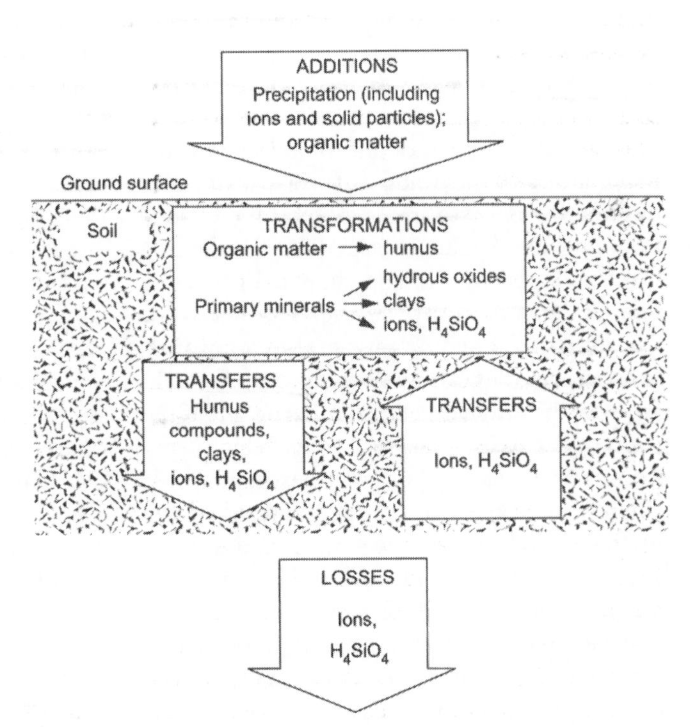

FIGURE 3.6. Processes leading to additions, transformations, transfers, and losses of materials from soils. Silica is H_4SiO_4. (Redrawn with permission from *Soils and Geomorphology* by Peter W. Birkeland, © 1999 Oxford University Press, Inc.; Birkeland 1999.)

Additions to Soils

Additions to the soil system can come from outside or inside the ecosystem. Inputs from *outside* the ecosystem include precipitation and wind, which deposit ions and dust particles, and floods and tidal exchange, which deposit sediments and solutes (see Chapter 9). The source of these materials determines their size distribution and chemistry, leading to the development of soils with specific textural and chemical characteristics. Organisms *within* the ecosystem add organic matter and nitrogen to the soil, including the aboveground and belowground portions of plants, animals, and soil microbes.

Soil Transformations

Within the soil, materials are transformed through an interaction of physical, chemical, and biological processes. Freshly deposited dead organic matter is transformed in the soil by **decomposition** to soil organic matter, releasing carbon dioxide and nutrients such as nitrogen and phosphorus (see Chapter 7). More recalcitrant organic compounds undergo physicochemical interactions with soil minerals that contribute to the long-term storage of soil organic matter.

Weathering is the change of parent rocks and minerals to produce more stable forms. This occurs when rocks and minerals become exposed to physical and chemical conditions different from those under which they formed (Ugolini and Spaltenstein 1992). Weathering involves both physical and chemical processes and is influenced by characteristics of the parent material and by temperature, moisture, and the activities of organisms. **Physical weathering** is the fragmentation of parent material without chemical change. This can occur when rocks are fractured by earthquakes or when stresses are relieved due to erosional loss of the weight of overlying rock and soil. In addition, soil particles and rock fragments are abraded by wind or are ground against one another by glaciers, landslides, or floods. Rocks also fragment when they expand and contract during freeze–thaw, heating–cooling, or wetting–drying cycles or when roots grow into rock fissures. Fire is a potent force of physical weathering because it rapidly heats the rock surface to a high temperature while leaving the deeper layers cool. Physical weathering is especially important in extreme and highly seasonal climates. Wherever it occurs, it opens channels in rocks for water and air to penetrate, increasing the surface area for chemical weathering reactions.

Chemical weathering occurs when parent rock materials react with acidic or oxidizing substances, usually in the presence of water. During chemical weathering, **primary minerals** (minerals present in the rock or unconsolidated parent material before chemical changes have occurred) are dissolved and altered chemically to produce more stable forms, ions are released, and **secondary minerals** (products that are formed through the reaction of materials released during weathering) are formed. Some primary minerals can be hydrolyzed by water, producing new minerals plus ions in solution. Hydrolysis reactions, however, typically include both water (H_2O) and an acid. Carbonic acid (H_2CO_3) is the most important acid involved in chemical weathering. The CO_2 concentration in most soils is 10- to 30-fold higher than in air, due to the low diffusivity of gases in soil and the respiration of plants, soil animals, and microorganisms. Weathering rates are particularly high adjacent to roots because of the high rates of biological activity and CO_2 production in **the rhizosphere**. Carbon dioxide dissolves and reacts with water to form carbonic acid, which then ionizes to produce a hydrogen ion (H^+) and a bicarbonate ion (HCO_3^-). Carbonic acid, for example, attacks potassium feldspar ($KAlSi_3O_8$), which is converted into a secondary mineral, kaolinite ($Al_2Si_2O_5(OH)_4$), by the removal of soluble silica (SiO_2) and potassium ion (K^+) (Eq. 3.1). Kaolinite can, under the right conditions, undergo another dissolution to form another secondary mineral gibbsite ($Al(OH)_3$).

$$2KAlSi_3O_8 + 2\left(H^+ + HCO_3^-\right) + H_2O \rightarrow Al_2Si_2O_5(OH)_4 + 4SiO_2 + 2K^+ + 2HCO_3^- \tag{3.1}$$

Plant roots and microbes secrete many organic acids into the soil, which influence chemical weathering through their contribution to soil acidity and their capacity to **chelate** ions. In the chelation process, organic acids combine with metallic ions, such as ferric iron (Fe^{3+}) and aluminum (Al^{3+}), making them soluble and mobile. Chelation lowers the concentration of inorganic ions at the mineral surface, so dissolved and primary mineral forms are no longer in equilibrium with one another. This accelerates the rate of weathering.

The physical and chemical properties of rock minerals determine their susceptibility to weathering and the chemical products that result. Sedimentary rocks like shale that form by chemical precipitation, for example, have more basic cations like calcium (Ca^{2+}), sodium (Na^+) and potassium (K^+) than does igneous rock and tends to produce soils with a relatively high pH and a high capacity to supply mineral cations to plants. Igneous rocks weather in the reverse order in which they crystallize during formation (Birkeland 1999). Olivine, for example, is one of the first minerals to crystallize as magma cools. It has a high energy of formation and weathers easily. Feldspar forms and weathers more slowly than olivine, and quartz is one of the last minerals to form (explaining why it forms crystals) and is highly resistant to weathering (Table 3.2). Secondary minerals such as the silicate clay minerals and iron and aluminum oxides are among the most resistant minerals to weathering. Textural differences in parent material also influence the rate of chemical breakdown, with fine-grained rocks weathering more slowly than coarse-grained rocks.

Warm climates promote chemical weathering because temperature speeds chemical reactions

TABLE 3.2. Stability of common minerals under weathering conditions at Earth's surface.

Most stable	Fe^{3+} oxides	Secondary mineral
	Al^{3+} oxides	Secondary mineral
	Quartz	Primary mineral
	Clay minerals	Secondary mineral
	K^+ feldspar	Primary mineral
	Na^+ feldspar	Primary mineral
	Ca^{2+} feldspar	Primary mineral
Least stable	Olivine	Primary mineral

Data from Press and Siever (1986).

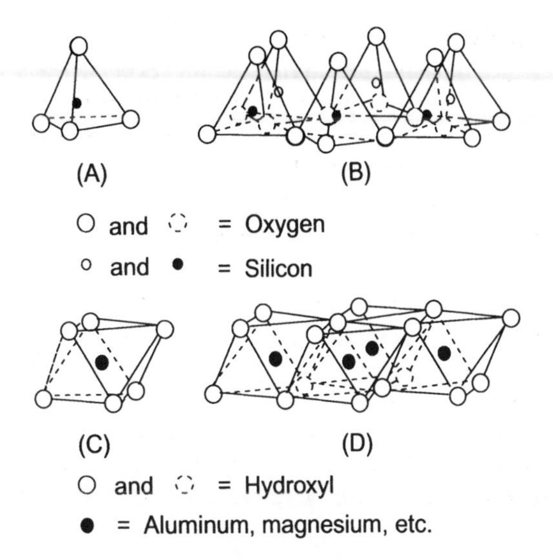

FIGURE 3.7. The molecular structure of a simple clay layer. **A**, A tetrahedral unit. **B**, A tetrahedral sheet. **C**, An octahedral unit. **D**, An octahedral sheet. (Redrawn with permission from *Clay Mineralology* by R.E. Grim, © 1968 McGraw-Hill Companies; Grim 1968.)

by increasing the kinetic energy of reactants. The activities of plants and microorganisms are also more rapid under warm conditions. Wet conditions promote weathering through their direct effects on weathering reactions and their effects on biological processes. Not surprisingly, the hot wet conditions of humid tropical climates yield the highest rates of chemical weathering.

The secondary minerals formed in weathering reactions play critical roles in soils and ecosystem processes. In temperate soils, weathering products include layered silicate clay minerals. These small particles (less than 0.002 mm) are hydrated silicates of aluminum, iron, and magnesium arranged in layers to form a crystalline structure. Two types of sheets make up these minerals: A tetrahedral sheet consists of units composed of one silicon atom surrounded by four atomic oxygen (O^-) groups (Fig. 3.7A,B). An octahedral sheet consists of units having six oxygen (O^-) or hydroxide (OH^-) ions surrounding an Al^{3+}, magnesium ion (Mg^{2+}), or Fe^{3+} ion (Fig. 3.7C,D). Various combinations of these sheets give rise to a wide variety of clay minerals. Montmorillonite and illite, for example, have 2:1 ratios of silica to aluminum-dominated layers. Kaolinite, a more strongly weathered clay mineral, has a 1:1 ratio of the two. The structure and concentration of these clay minerals influences the cation exchange capacity, water-holding capacity, and other characteristics of soils.

Secondary minerals that form in soils can be either **crystalline**, with highly regular arrangements of atoms, as in silicate clay minerals, or **amorphous**, with no regular arrangement of atoms. In many volcanic soils forming on ash deposits, allophane ($Al_2O_3 \cdot 2SiO_2 \cdot nH_2O$) is an amorphous mineral that is produced relatively early in weathering but is then transformed to crystalline aluminum oxide minerals like $Al(OH)_3$ with time. In many tropical soils, weathering has occurred in place for millions of years in a humid climate with high leaching rates. Here the relatively mobile ions of silicon (Si) and Mg^{2+} as well as Ca^{2+}, K^+, and Na^+ are preferentially leached, leaving behind the less mobile ions Al^{3+} and Fe^{3+}. Silicate clay minerals are therefore no longer present, and the clay-size particles are composed of iron and aluminum oxides. These oxides form when Fe^{2+} or Al^{2+} are released in the weathering of iron- or aluminum-bearing minerals and are then oxidized in solution and react with anions to form a precipitate. Hematite and gibbsite are examples of the oxides produced.

Soil Transfers

Vertical transfers of materials through soils generate distinctive soil profiles—that is, the vertical layering of soils. These transfers typically occur by **leaching** (the downward movement of dissolved materials) and particulate transport in water. Soluble ions that are added in precipitation or released by weathering in upper layers of the soil profile can move downward in solution until a change in chemical environment causes them to become reactants in chemical processes or until dehydration causes them to precipitate out of solution. The amounts of silica and base cations in secondary minerals therefore frequently increase with depth. These cations are leached from upper layers (termed **horizons**) and form new minerals under the new conditions of pH and ionic content encountered at depth. Chelated complexes of organic compounds and iron or aluminum ions are also water soluble and can move in water to deeper layers of the soil profile. Slight changes in ionic content and the microbial breakdown of the organic matter are among the processes that can cause the metal ions to precipitate as oxides. Clay-size particles like silicates and iron and aluminum oxides can also be transported downward in solution, sometimes forming deep horizons with high clay content in wet climates. Soil texture affects the rate and depth of leaching (Fig. 3.8) and thus the translocation and accumulation of materials in soil profiles. Constituents released during weathering of coarse-textured glacial till, for example, may be leached from the soil before they have a chance to react chemically to form secondary minerals.

Soils of arid and semiarid environments also accumulate materials in specific horizons. These systems often have hard calcium (or magnesium) carbonate-rich **calcic** horizon or caliche. Downward-moving soil water carries dissolved Ca^{2+} and bicarbonate (HCO_3^-). Precipitation as calcium carbonate ($CaCO_3$) occurs under conditions of increasing pH, which drives reaction 3.2 to the left. Precipitation can also occur under saturating concentrations of carbonate and with evaporation of soil water.

$$CaCO_3 + H_2CO_3 \leftrightarrow Ca^{2+} + 2HCO_3^- \quad (3.2)$$

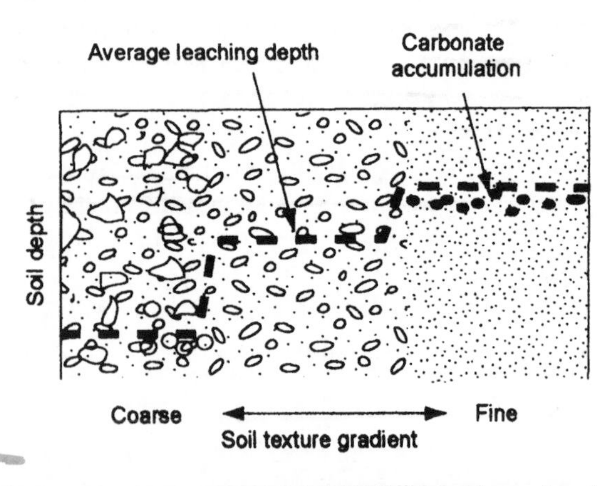

FIGURE 3.8. Hypothetical depth of leaching related to the texture of the original parent material. (Redrawn with permission from *Soils and Geomorphology* by Peter W. Birkeland, © 1999 Oxford University Press, Inc.; Birkeland 1999.)

Although most of the transfers in soils occur through the downward movement of water, materials can also move upward in water. The capillary rise of water from a shallow water table, for example, transfers water and ions from lower to upper soil layers (see Chapter 4). Because capillary water movement depends on adhesive properties of soil particles, the potential distance for capillary rise is greater in clay soils, which have small pore sizes, than in sandy soils (Birkeland 1999). Soluble ions or compounds may accumulate in layers at the top of the capillary fringe. Salt pans, for example, form at the soil surface in low-lying areas of deserts. Minerals that are added to soils in irrigation water in dry regions can also accumulate at the soil surface as the water evaporates. This process of **salinization** has led to widespread abandonment of farmland in dry regions of the world. In western Australia, for example, widespread removal of vegetation for croplands caused the soils to become so saline that many of these croplands have been abandoned. Salinization occurs naturally in deserts, when streams drain into **salt flats**, where the water evaporates rather than runs off.

Some minerals accumulate at the top of a water table that forms when downward percolation of water is impeded by an impermeable soil layer. Poor drainage often leads to low oxygen availability because oxygen diffuses 10,000 times more slowly in water than in air and is readily depleted in water-logged soils by root and microbial respiration. Low oxygen concentration creates reducing conditions that convert ions with multiple oxidation states to their reduced form. Iron and manganese, for example, are more soluble in their reduced (Fe^{2+} and Mn^{2+}, respectively) than in their oxidized state (Fe^{3+} and Mn^{4+}, respectively). Fe^{2+} and Mn^{2+} diffuse through waterlogged soils to the surface of the water table, where there is sufficient oxygen to convert them to their oxidized forms. There they precipitate out of solution to form a distinct layer that is rich in iron and manganese. This layering of iron and manganese is particularly pronounced in lake sediments, where there is a strong gradient in oxygen concentration from the sediment surface. The conversion from ferric (Fe^{3+}) to ferrous (Fe^{2+}) iron gives rise to the characteristic gray and bluish colors of waterlogged **gley soils**.

Soils that are subjected to repeated wetting and drying and saturation during some seasons can also develop characteristic accumulations. Plinthite layers (sometimes called **laterite**) in tropical soils, for example, are layers of iron-rich minerals that have hardened irreversibly on exposure to repeated saturation and drying cycles. Depending on their location within the profile, these layers can impede water drainage and root growth.

The actions of plant roots and soil animals transfer materials up and down the soil profile (Paton et al. 1995). Organic matter inputs to soil occur primarily at the surface and in upper soil horizons. When plants die, the minerals acquired by deep roots are also deposited on or near the soil surface. This contributes, for example, to the base-rich soils and unique ground flora beneath deep-rooted oak trees in southern Sweden (Andersson 1991). Tree windthrow, which occurs when large trees are toppled by strong winds, also redistributes roots and associated soil upward. Finally, animals such as gophers transfer materials up and down in the soil profile as they tunnel and feed on plant roots. Earthworms in temperate soils and termites in tropical soils are particularly important in transferring surface organic matter deep into the soil profile and, at the same time, bringing mineral soil from depth to the surface (see Chapter 12). These processes play critical roles in the redistribution of nutrients and in the control of net primary productivity.

Losses from Soils

Materials are lost from soil profiles primarily as solutions and gases. The quantity of minerals leached from an ecosystem depends on both the amount of water flowing through the soil profile and its solute concentration. Many factors influence these concentrations, including plant demand, microbial mineralization rate, cation- or anion-exchange capacity, and previous losses via leaching or gas fluxes. As water moves through the soil, exchange reactions with mineral and organic surfaces replace loosely bound ions on the exchange complex with ions that bind more tightly. In this way monovalent cations such as Na^+, ammonium ion (NH_4^+), and K^+ and anions such as chloride ion (Cl^-) and nitrate ion (NO_3^-) are readily released from the exchange complex into the soil solution and are particularly prone to leaching loss. The maintenance of charge balance of soil solutions requires that the leaching of negatively charged anions be accompanied by an equal charge of positive ions (cations). Inputs of sulfuric acid (H_2SO_4) in acid rain therefore increase leaching losses of readily exchangeable base cations like Na^+, NH_4^+, and K^+, which leach downward with sulfate ion (SO_4^{2-}).

Materials can also be lost from soils as gases. Gas emissions depend on the rate of production of the gas by microbes, the diffusional paths through soils, and the exchange at the soil–air interface (Livingston and Hutchinson 1995). The controls over these losses are discussed in Chapter 9.

Soil Horizons and Soil Classification

Ecosystem differences in additions, transformations, transfers, accumulations, and losses give rise to distinct soils and soil profiles. Soils include organic, mineral, gaseous, and aqueous constituents arranged in a relatively predictable vertical structure. The number and depth of **horizons** (layers) and the characteristics of each layer in a soil profile vary widely among soils. Nonetheless, a series of typical horizons can be described for many soils (Fig. 3.9). The organic horizon, or **O horizon**, consists of organic material that accumulates above the mineral soil. This layer of organic material is derived from the **litter** of dead plants and animals. The O horizon can be subdivided based on the degree of decomposition that the majority of material has undergone, with the lower portion of the O horizon being more decomposed. The **A horizon** is the uppermost mineral soil horizon. Being adjacent to the organic layers, it typically contains substantial quantities of organic matter and is therefore dark in color. The O and A horizons are the zones of most active plant and microbial processes and therefore have highest nutrient supply rates (see Chapter 9). Many soils in wet climates have an **E horizon** beneath the A horizon that is strongly leached. Most clay minerals and iron and aluminum oxides have been leached from the horizon, leaving behind resistant minerals like quartz, in addition to sand and silt-size particles. The **B horizon** beneath the A and E horizons is the zone of maximum accumulation of iron and aluminum oxides and clays. Salts and precipitates sometimes also accumulate here, especially in arid and semiarid environments. The **C horizon** lies beneath the A and B horizons. Although it may accumulate some of the

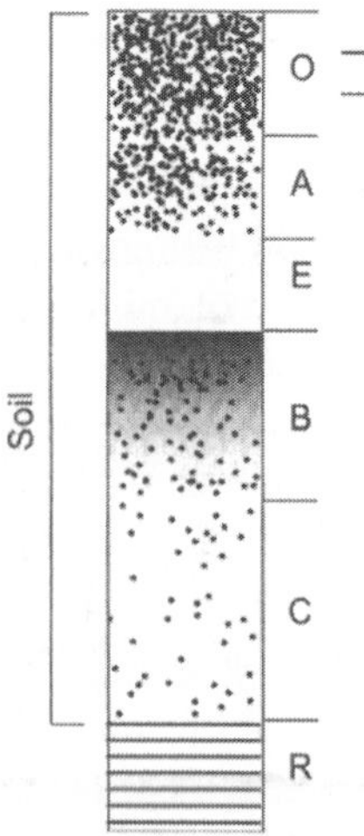

FIGURE 3.9. A generic soil profile showing the major horizons that are formed during soil development. Density of dots reflects concentration of soil organic matter.

leached material from above, it is relatively unaffected by soil-forming processes. C horizons typically include a significant portion of unweathered parent material. Finally, at some depth, there is an unweathered bedrock layer, the **regolith** (R).

Despite the large variation among the world's soils, they can be classed into major groups that share many of the same properties because they have formed in response to similar soil-forming factors and processes. Soil classification systems rely on the diagnostic characteristics of specific horizons and on organic matter content, base saturation, and properties that indicate wetness or dryness. The U.S. Soil Taxonomy recognizes 12 major soil groupings, called **soil orders** (Table 3.3) (Brady and Weil 1999). Most agronomic and ecosystem studies classify soils to the level of a **soil series**, a group of soil profiles with similar profile characteristics such as type, thickness, and properties of the soil horizons. Soil series can be further subdivided into **types**, based on the texture of the A horizon, and into **phases**, based on information such as landscape position, stoniness, and salinity. A comparison of soil profiles from the major soil orders illustrates the effect of different climatic regimes on soil development (Figs. 3.10 and 3.11).

Entisols are soils with minimal soil development. They occur either because the soils are recent or processes that disrupt soil development dominate over processes that form soils. This is the most widespread soil type in the world, making up 16% of the ice-free surface. **Inceptisols**, in which the soil profile has only begun to develop, occupy an additional 10% of the ice-free surface. Rock and shifting sand account for another 14% of the ice-free surface. Thus about 40% of the ice-free surface of Earth shows minimal soil development.

Histosols are highly organic soils that develop in any climate zone under conditions in which poor drainage restricts oxygen diffusion into the soil, leading to slow rates of decomposition and accumulation of organic matter. There is a well-developed O horizon of undecomposed organic material where most plants are rooted. The high water table prevents the vertical leaching required for soil devel-

TABLE 3.3. Names of the soil orders in the United States soil taxonomy and their characteristics and typical locations.

Soil Order	Area (% of ice-free land)	Major Characteristics	Typical Occurrence
Entisols	16.3	no well-developed horizons	sand deposits, plowed fields
Inceptisols	9.9	weakly developed soils	young or eroded soils
Histosols	1.2	highly organic; low oxygen	peatland, bog
Gelisols	8.6	presence of permafrost	tundra, boreal forest
Andisols	0.7	from volcanic ejecta; moderately developed horizons	recent volcanic areas
Aridisols	12.1	dry soils with little leaching	arid areas
Mollisols	6.9	deep dark-colored A horizon with >50% base saturation	grasslands, some deciduous forests
Vertisols	2.4	high content (>30%) of swelling clays; crack deeply when dry	grassland with distinct wet and dry seasons
Alfisols	9.7	sufficient precipitation to leach clays into a B horizon; >50% base saturation	humid forests; shrublands
Spodosols	2.6	sandy leached (E) horizon; acidic B horizon; surface organic accumulation	cold wet climates, usually beneath conifer forests
Ultisols	8.5	clay-rich B horizon, low base saturation	wet tropical or subtropical climate; forest or savanna
Oxisols	7.6	highly leached horizon with low clay; highly weathered on old landforms	hot humid tropics beneath forests
Rock and sand	14.1		

Data from Miller and Donahue (1990) and Brady and Weil (2001).

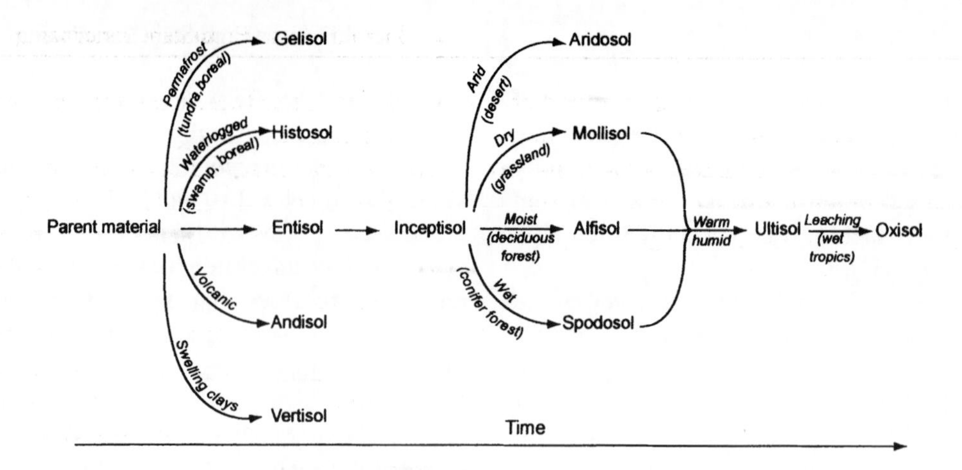

FIGURE 3.10. Relationships among the major soil orders, showing the conditions under which they form, the relative time required for formation, and the types of ecosystems with which they are commonly associated (Birkeland 1999).

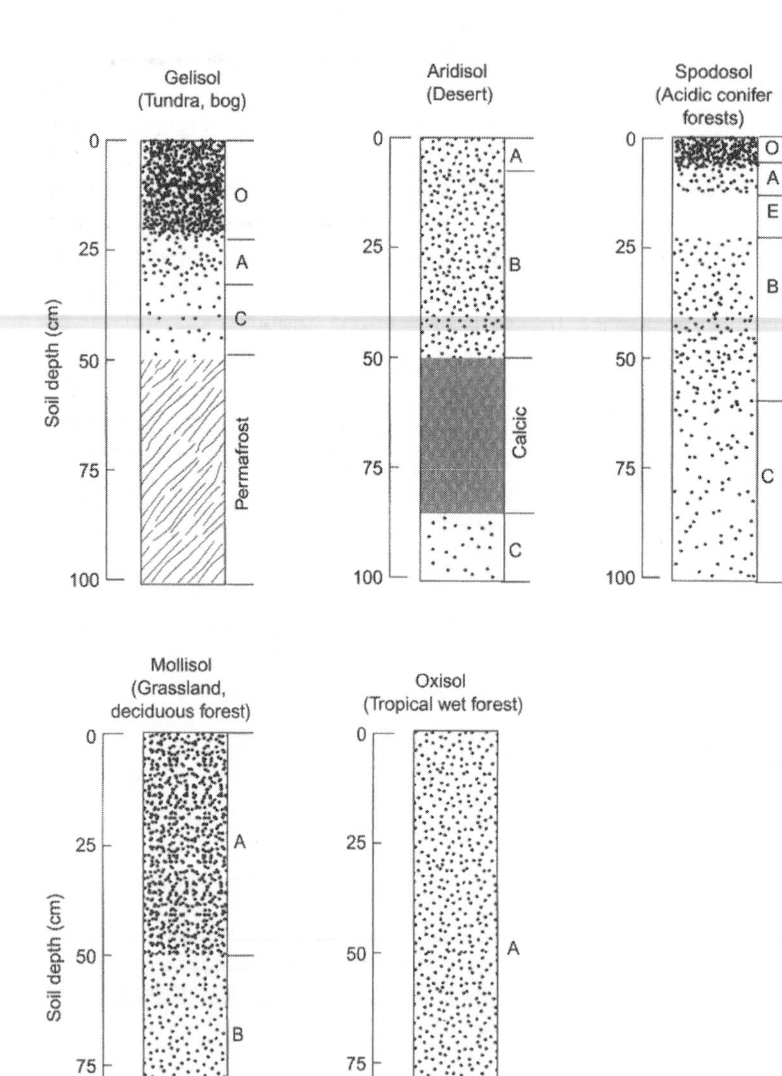

FIGURE 3.11. Typical profiles of five contrasting soil orders showing differences in the types and depths of horizons. Symbols same as in Figure 3.9.

opment, so these soils have weak development of mineral soil horizons. **Gelisols** are organic soils that develop in climates with a mean annual temperature below 0°C and are underlain by a layer of permanently frozen soil (**permafrost**).

Andisols are young soils that occur on volcanic substrates and tend to produce amorphous clays.

Aridisols, as the name implies, develop in arid climates. The low rainfall minimizes leaching and weathering and may allow accumulation of soluble salts. There is no surface O horizon. The shallow A horizon has little organic matter due to low productivity and rapid decomposition. Low precipitation results in a poorly differentiated B horizon. Many of these soils develop a calcic layer of calcium and magnesium carbonates that precipitate at depth because there is insufficient water to leach them out of the system. Desert calcic layers can greatly reduce root penetration, restricting the roots of many desert plants to surface soils. Aridisols are a common world soil type, accounting for 12% of the terrestrial surface (Miller and Donahue 1990).

Vertisols are characterized by swelling and shrinking clays. They tend to occur in warm regions with a moist to dry climate.

Mollisols are fertile soils that develop beneath grasslands and some deciduous forests. They have a deep organic-rich A horizon with a high nutrient content that grades into a B horizon. Due to their high fertility, mollisols have been extensively cultivated and support the major grain-growing regions of the world. They account for 25% of soils in the United States and 7% of soils worldwide (Miller and Donahue 1990).

Spodosols (or podzols by the older terminology) are highly leached soils that develop in cold wet climates, usually beneath conifer stands. Beneath the A horizon is a highly leached, almost white, E horizon and a dark brown or black B horizon, where leaching products accumulate. These soils are acidic with a high sand and low clay content. **Alfisols** develop beneath temperate and subtropical forests, especially in deciduous forests that receive less precipitation. They are less strongly leached than spodosols and therefore have a higher clay content and lower acidity.

Ultisols develop in warm, humid areas where there is substantial leaching. These soils often have a high clay content, a low base saturation, and low fertility. **Oxisols** are the most highly weathered leached soils. They occur on old landforms in the wet tropics. The A horizon is so highly weathered that it contains iron oxides but very little clay and has extremely low fertility. This horizon often extends several meters in depth.

Soil Properties and Ecosystem Functioning

Spatial and temporal variations in soil development result in large variations in soil properties. These soil properties, in turn, modulate the availability of water and nutrient resources for plant growth and therefore the cycling of water and nutrients through ecosystems. In this section, we discuss a few of the important properties.

Soil texture is defined by the relative proportions of soil particles of different sizes, ranging from **clay**-size particles (less than 0.002 mm) to **silt** (0.002 to 0.05 mm), and **sand** (0.05 to 2.0 mm) (Fig. 3.12). **Loam** soils contain substantial proportions of at least two size classes of particles. Rocks and gravel are large (greater than 2 mm), unweathered primary minerals that can occupy a substantial proportion of the volume of many soils. Sand- and silt-size particles consist of unweathered primary minerals and some secondary minerals released by weathering. Clay-size particles consist of a larger proportion of secondary minerals, including layered silicate clays and other small crystalline and amorphous minerals. Parent material has a large effect on soil texture because rocks differ in their rates of physical and chemical weathering.

Soil texture also depends on the balance between soil development that occurs in place, deposition from wind or water, and erosional loss of materials. When soils weather in place, the conversion of primary minerals to clay-size secondary minerals increases the proportion of

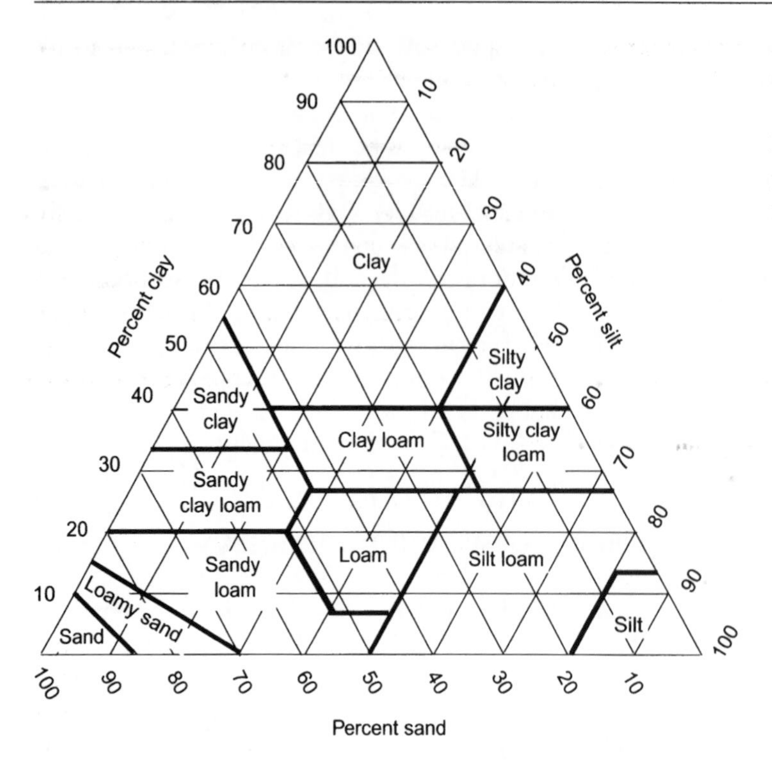

FIGURE 3.12. Percentages of sand, silt, and clay in the major soil textural classes. (Redrawn with permission from *Soils and Geomorphology* by Peter W. Birkeland, © 1999 Oxford University Press, Inc.; Birkeland 1999.)

fine soil particles. For this reason, high-latitude soils, with their slow rates of chemical weathering, frequently have a lower clay content than do temperate or tropical soils of similar bedrock and age. Fine particles are, however, more susceptible to erosion by wind or water than are large particles. Water erosion transports clays from hilltops to valley bottoms, producing fine-textured soils in river valleys. If the river valleys are poorly vegetated, as in braided rivers that drain glaciated landscapes, wind can then move fine particles back to hillslopes to form **loess** soils with a high silt content. Over millions of years, minerals dissolve and are lost from the soil.

Texture is important primarily because it determines the total surface area in a volume of soil. Fine-grained particles in the soil matrix have greater surface-to-volume ratio. Soils with fine textures and large surface areas hold more water by adsorption of water films to soil particles.

The packing of small particles in the soil matrix results in greater water-holding pore space between particles. Under intermediate levels of rainfall, ecosystems growing on sandy soils tend to be more **xeric** (characterized by plants that are tolerant of dry conditions) than those occurring on finer-textured soils. In the case of soils with silicate clay particles, the increased surface area also represents greater surface charge and thus greater cation exchange capacity, as described later. Many other soil properties such as bulk density, nutrient content, water-holding capacity, and redox potential are related to soil texture, so texture can be a good predictor of many ecosystem properties (Parton et al. 1987).

Soil structure reflects the aggregation of soil particles into larger units. **Aggregates** form when soil particles become cemented together and then crack into larger units as soils dry or freeze. There are many types of glue that bind soil particles together to form aggregates. These include organic matter, iron oxides, polyvalent cations, clays, and silica. Soil aggregates range in size from less than 1 mm to greater than 10 cm in diameter.

Soil texture strongly affects the formation of soil aggregates. Sandy soils form few aggregates, whereas loam and clay soils form a range of aggregate sizes. Polysaccharides secreted by roots and bacteria are important sources of organic matter that binds soil particles together. Fungal hyphae contribute strongly to aggregation in many soils. For these reasons, the loss of soil organic matter and its associated microbes can lead to a loss of soil structure, which contributes to further soil degradation. Earthworms and other soil invertebrates contribute to aggregate formation by ingesting soil and producing feces that retain a coherent structure. Plant species and their microbial associates differ in the capacity of their exudates to form aggregates, so soil texture, organic matter content, and species composition all influence soil structure.

The cracks and channels between aggregates are important pathways for water infiltration, gaseous diffusion, and root growth, thus affecting water availability, soil aeration, oxidation–reduction processes, and plant growth. The fine-scale heterogeneity created by soil aggregates is critical to the functioning of soils. Slow gas diffusion through the partially cemented pores within aggregates creates anaerobic conditions immediately adjacent to aerobic surfaces of soil pores. This allows the occurrence in well-drained soils of anaerobic processes such as denitrification, which requires the products of aerobic processes (nitrification, in this case) (see Chapter 9).

Compaction by animals and machinery fills the cracks and pores between aggregates. Plowing reduces aggregation through mechanical disruption of aggregates and through loss of soil organic matter and associated cementing activity of microbial exudates and fungal hyphae (Fisher and Binkley 2000). The loss of soil structure through compaction prevents rapid infiltration of rainwater and leads to increased overland flow and erosion.

Bulk density is the mass of dry soil per unit volume, usually expressed in grams per cubic centimeter ($g\,cm^{-3}$; equivalent to megagrams per cubic meter, or $Mg\,m^{-3}$). Bulk density varies with soil texture and soil organic matter content. Bulk densities of mineral soils (1.0 to $2.0\,g\,cm^{-3}$) are typically higher than those of organic soil horizons (0.05 to $0.4\,g\,cm^{-3}$). Fine-textured soils have higher internal surface area and more pore space than coarser-textured soils, and thus their bulk densities tend to be lower. If they are compacted, however, bulk densities of clay soils can be higher than those of coarse-textured soils. Bulk density strongly influences the nutrient and water characteristics of a site. Organic soils, for example, frequently have highest *concentrations* (percent of dry mass) of carbon in surface horizons with low bulk density but the greatest *quantities* (grams per cubic centimeter) of carbon at depth, where bulk density is greater. The quantity of nutrient per unit volume is calculated by multiplying the percentage concentration of the nutrient times soil bulk density. Volumetric nutrient content is generally more relevant than nutrient concentration in describing the quantity of nutrients directly available to plants and microbes.

Water is a critical resource for most ecosystem processes. In soils, water is held in pore spaces as films of water adsorbed to soil particles. The soil is **water-saturated** when all pore spaces are filled with water. Under these conditions water drains under the influence of gravity (**saturated flow**) until, often after several days, the adhesive forces that hold water in films on soil particles equals the gravitational pressure. At this point, called **field capacity**, water no longer freely drains.

At water contents below field capacity, water moves through the soil by **unsaturated flow** in response to gradients of **water potential**—that is, the potential energy of water relative to pure water (see Chapter 4). When plant roots absorb water from the soil to replace water that is lost in transpiration, there is a reduction in the thickness of water films adjacent to roots, which causes the remaining water to adhere more tightly to soil particles. The net effect is to reduce the soil water potential at the root surface. Water moves along water films through the soil pores toward the root in response to this gradient in water potential. Plants continue to transpire, and water continues to move toward the root until some minimal water potential is reached, when roots can no longer remove water from the particle surfaces. This point is called the **permanent wilting point**. **Water-holding capacity** is the difference in water content between field capacity and permanent wilting point (see Fig. 4.5). Water-holding capacity is substantially enhanced by presence of clay and soil organic matter because of the large surface area of these materials. The water-holding capacity of an organic soil might, for example, be 300% (300 g H_2O per 100 g dry soil), while that of a clay soil may be 30% and that of a sandy soil could be less than 20%. On a volumetric basis, water-holding capacity is normally highest in loam soils. One consequence of this difference is that, for a given amount of rainfall, sandy soils are wetted more deeply than clay soils but retain less water in soil horizons that are accessible to plants. The water-holding characteristics of soils help determine the amount of water available for plant uptake and growth and for microbial processes, including decomposition and nutrient cycling and loss.

Oxidation–reduction reactions involve the transfer of electrons from one reactant to another, yielding chemical energy that can be used by organisms (Lindsay 1979). In these reactions, the energy source gives up one or more electrons (**oxidation**). These electrons are transferred to electron acceptors (**reduction**). A handy mnemonic is: "LEO the lion says GER," where *LEO* stands for loss of electrons—oxidation, and *GER* stands for gain of electrons—reduction. **Redox potential** is the electrical potential of a system due to the tendency of substances in it to lose or accept electrons (Schlesinger 1997, Fisher and Binkley 2000). There is a wide range of redox potentials among soils due to their ionic and chemical compositions. One important set of redox reactions, which occurs inside the mitochondria of live eukaryotic cells, is the transfer of electrons from carbohydrates through a series of reactions to oxygen. This series of reactions releases the energy needed to support cellular growth and maintenance. Many other redox reactions occur in the cells of soil organisms, when electrons are transferred from electron donors to acceptors (Table 3.4). The greatest amount of energy can be harvested by organisms by transferring electrons to oxygen. However, under anaerobic conditions, which commonly occur in flooded soils with high organic matter contents or in aquatic sediments, electrons must be transferred to other electron acceptors; thus progressively less energy is released with the transfer to each of the following electron acceptors:

$$O_2 > NO_3^- > Mn^{4+} > Fe^{3+} > SO_4^{2-} > CO_2 > H^+ \quad (3.3)$$

As soil redox potential declines, the preferred electron carriers are gradually consumed (Table 3.4). As oxygen becomes depleted, for example, the redox reaction that generates the most energy is denitrification (transfer of electrons to nitrate), followed by reduction of Mn^{4+} to Mn^{2+}, then reduction of Fe^{3+} to Fe^{2+}, then reduction of SO_4^{2-} to hydrogen sulfide (H_2S), then reduction of CO_2 to methane (CH_4). Thus poorly aerated soils with high sulfate concentrations (e.g., salt marshes) are less likely to reduce CO_2 to CH_4 than are similar soils with lower SO_4^{2-} concentrations.

Many soil organisms carry out only one or a few redox reactions, although certain bacteria can couple the reduction of Mn^{4+} and Fe^{3+} directly to the oxidation of simple organic substrates (Schlesinger 1997). Temporal and spatial variations in soil redox potential alter the types of redox reactions that occur primarily by altering the competitive balance among these organisms. Organisms that derive more energy from their redox reactions (e.g., denitrifiers compared to methane producers) will be competitively superior, when they have an adequate supply of electron acceptors.

Soil organic matter content is a critical component of soils, affecting rates of weathering and soil development, soil water-holding capacity, soil structure, and nutrient retention. In addition, soil organic matter provides the energy and carbon base for heterotrophic soil organisms (see Chapter 7) and is an important reservoir of essential nutrients required for plant growth (see Chapter 8). Soil organic matter originates from dead plant, animal, and microbial tissues, but includes a range of materials from new, undecomposed plant tissues to resynthesized humic substances that are thousands of years old, whose origins are chemically and physically unrecognizable (see Chapter 7). Because soil organic matter is important to so many soil properties, loss of soil organic matter

TABLE 3.4. Sequence of H^+-consuming redox reactions that occur with progressive declines in redox potential.

Reaction[a]	Redox potential (mV)	Energy release[b] (Kcal mol^{-1} per e^-)
Reduction of O_2 $O_2 + 4H^+ + 4e^- \rightarrow 2H_2O$	812	29.9
Reduction of NO_3^- $NO_3^- + 2H^+ + 2e^- \rightarrow NO_2^- + H_2O$	747	28.4
Reduction of Mn^{4+} to Mn^{2+} $MnO_2 + 4H^+ + 2e^- \rightarrow Mn^{2+} + 2H_2O$	526	23.3
Reduction of Fe^{3+} to Fe^{2+} $Fe(OH)_3 + 3H^+ + e^- \rightarrow Fe^{2+} + 3H_2O$	−47	10.1
Reduction of SO_4^{2-} to H_2S $SO_4^{2-} + 10H^+ + 8e^- \rightarrow H_2S + 4H_2O$	−221	5.9
Reduction of CO_2 to CH_4 $CO_2 + 8H^+ + 8e^- \rightarrow CH_4 + 2H_2O$	−244	5.6

[a] The reactions at the top of the table occur in soils with high redox potential and release more energy (and are therefore favored) when the electron acceptors are available. The reactions at the bottom of the table release less energy and therefore occur only if other electron acceptors are absent or have already been consumed by redox reactions. Abbreviations include electrons (e^-), nitrite ion (NO_2^-), manganese dioxide (MnO_2), Ferric hydroxide ($Fe(OH)_3$), organic matter (CH_2O), universal gas constant (R), temperature (T), and equilibrium constant (K).
[b] Assumes coupling to the oxidation reaction: $CH_2O + H_2O \rightarrow CO_2 + 4H^+ + 4e^-$ and that the energy released – RT ln(K).
Data from Schlesinger 1997.

as a consequence of inappropriate land management is a major cause of land degradation and loss of biological productivity.

The negative log of the hydrogen ion (H^+) activity (effective concentration) in solution is referred to as **pH**, which is a measure of the active acidity of the system. The pH strongly affects nutrient availability through its effects on cation exchange (see next paragraph) and the solubility of phosphate compounds and micronutrients such as iron, zinc, copper, and manganese.

Cation exchange capacity (CEC) reflects the capacity of a soil to hold exchangeable cations on negatively charged sites on the surfaces of soil minerals and organic matter. Cation exchange occurs when a loosely held cation on a negatively charged site exchanges with a cation in solution. Values for CEC vary more than 100-fold among clay minerals. Crystalline clay minerals typically have a negative or neutral charge under ambient soil pH. The negative charge originates from unsatisfied negative charges along the interlayer surfaces of the silicate clay lattices, especially in the 2:1 clays, and from hydroxide (—OH) groups that are exposed on the edges of 1:1 clay particles. Soil organic matter has very high CEC that originates from the presence of —OH and carboxyl (—COOH) groups at the surfaces of organic compounds and within particles of humic materials. Soil organic matter contributes substantially to the total CEC of some soils. For example, organic matter accounts for most CEC in tropical soils that consist primarily of iron and aluminum oxides and 1:1 silicate clay minerals, because there is little CEC in the mineral matrix. High-latitude soils also derive a large proportion of their CEC from organic matter due to their high organic content and low clay content. The pool of exchangeable cations in the soil is much larger than the soil solution pool and represents the major short-term store of cations for plant and microbial uptake.

Exchangeable cations are attracted to the negatively charged surfaces. **Base saturation** is the percentage of the total exchangeable cation pool that is accounted for by **base cations** (the nonhydrogen, nonaluminum cations). The identity of the cations on the exchange sites depends on the concentrations of cations in the soil solution and on the strength with which different cations are held to the exchange complex. In general, cations occupy exchange sites and displace other ions in the sequence

$$H(Al^{3+}) > H^+ > Ca^{2+} > Mg^{2+} > K^+ \approx NH_4^+ > Na^+ \quad (3.4)$$

so leached soils tend to lose Na^+ and NH_4^+ but retain Al^{3+} and H^+. This displacement series is a consequence of differences among ions in charge and hydrated radius. Ions with more positive charges bind more tightly to the exchange complex than do ions with a single charge. Ions with a smaller hydrated radius have their charge concentrated in a smaller volume and tend to bind tightly to the exchange complex.

Minerals like the iron and aluminum oxides found in many tropical soils have surface charges that vary between positive and negative, depending on pH. At the low pH conditions typical of these soils, the net charge is sometimes positive (Uehara and Gillman 1981), so they attract anions, creating an anion exchange capacity. As with cations, **anion absorption** depends on the concentration of anions and their relative capacities to be held or to displace other anions. Anions generally occupy exchange sites and displace other ions in the sequence

$$PO_4^{3-} > SO_4^{3-} > Cl^- > NO_3^- \quad (3.5)$$

so leached soils tend to lose NO_3^- and Cl^- but retain phosphate (PO_4^{3-}) and sulfate (SO_4^{3-}). This retention reflects both anion exchange and the formation of covalent bonds that are not readily broken.

The high CEC and base saturation found in many soils, especially in many temperate soils, provide **buffering capacity** that keeps the soils from becoming acid. When additional H^+ is added to the system in solution (e.g., in acid rain), it exchanges with cations that were held on cation exchange sites on clay minerals and soil organic matter. Buffering capacity allows the pH in forest soils to remain relatively constant for long periods despite chronic exposure to acid rain. When the buffering capacity is exceeded, the soil pH begins to drop, which can solubilize aluminum hydroxides ($Al(OH)_x$), Al^{3+}, and other cations, with potentially toxic effects on both terrestrial and downstream aquatic ecosystems (Schulze 1989, Aber et al. 1998). In many tropical soils, the relatively low CEC does not function as efficiently to buffer soil solution chemistry. Additions of acids to these already acidic unbuffered systems releases aluminum in solution more readily, making these soils potentially toxic to many plants and microbes.

Summary

Five state factors control the formation and characteristics of soils. Parent material is generated by the rock cycle, in which rocks are formed, uplifted, and weathered to produce the materials from which soil is derived. Climate is the factor that most strongly determines the rates of soil-forming processes and therefore rates of soil development. Topography modifies these rates at a local scale through its effects on microclimate and the balance between soil development and erosion. Organisms also strongly influence soil development through their effects on the physical and chemical environment. Time integrates the impact of all state factors in determining the long-term trajectory of soil development. In recent decades, human activities have modified the relative importance of these state factors and substantially altered Earth's soils.

The development of soil profiles represents the balance between profile development, soil mixing, erosion, and deposition. Profile development occurs through the input, transformation, vertical transfer, and loss of materials from soils. Inputs to soils come from both outside the ecosystem (e.g., dust or precipitation inputs) and inside the ecosystem (e.g., litter inputs). The organic matter inputs are decomposed to produce CO_2 and nutrients or are transformed into recalcitrant organic compounds. The carbonic acid derived from CO_2 and the organic acids produced during decomposition convert primary minerals into clay-size secondary minerals, which have greater surface area and cation exchange capacity. Water moves these secondary minerals and the soluble weathering products down through the soil profile until new chemical conditions cause them to become reactants or to precipitate out of solution. Leaching of materials into groundwater or erosion and gaseous losses to the atmosphere are the major avenues of loss of materials from soils. The net effect of these processes is to form soil horizons that vary with climate, parent material, biota, and soil age. These horizons have distinctive physical, chemical, and biological properties.

Review Questions

1. What processes are responsible for the cycling of rock material in Earth's crust?
2. Over a broad geographic range, what are the state factors that control soil formation? How might interactive controls modify the effects of these state factors?
3. What processes determine erosion rate? Which of these processes are most strongly influenced by human activities?
4. What processes cause soil profiles to develop? Explain how differences in climate, drainage, and biota might affect profile development.
5. What are the processes involved in physical and chemical weathering? Give examples of each. How do plants and plant products contribute to each?
6. How is soil texture defined? How does it affect other soil properties? Why does it influence ecosystem processes so strongly?
7. What is cation exchange capacity (CEC), and what determines its magnitude in temperate soils? How would you expect the determinants of CEC to differ among histosol, alfisol, and oxisol soils?
8. In a warm climate, how will soil processes and properties differ between sites with extremely high and extremely low precipitation? In a moist climate, how will soil processes and properties differ between sites with extremely high and extremely low soil temperature?
9. If global warming caused only an increase in temperature, how would you expect this to affect soil properties after 100 years? After 1 million years?

Additional Reading

Amundson, R., and H. Jenny. 1997. On a state factor model of ecosystems. *BioScience* 47:536–543.

Birkeland, P.W. 1999. *Soils and Geomorphology*. 3rd ed. Oxford University Press, New York.

Brady, N.C., and R.R. Weil. 1999. *The Nature and Properties of Soils*. 12th ed. Prentice-Hall, Upper Saddle River, NJ.

Jenny, H. 1941. *Factors of Soil Formation*. McGraw-Hill, New York.

Selby, M.J. 1993. *Hillslope Materials and Processes*. Oxford University Press, Oxford, UK.

Ugolini, F.C., and H. Spaltenstein. 1992. Pedosphere. Pages 123–153 *in* S.S. Butcher, R.J. Charlson, G.H. Orians, and G.V. Wolfe, editors. *Global Biogeochemical Cycles*. Academic Press, London.

Part II
Mechanisms

4
Terrestrial Water and Energy Balance

The hydrologic cycle is the matrix in which all other biogeochemical cycles function. This chapter describes the controls over the hydrologic cycle and ecosystem energy balance, which drives the hydrologic cycle.

Introduction

Water and solar energy are essential for the functioning of the Earth System. Since neither is distributed evenly around the globe, the mechanisms by which they are redistributed (the global hydrologic cycle and energy budget) are important (see Chapter 2). These processes are so tightly intertwined that they cannot be treated separately (Box 4.1). Solar energy drives the hydrologic cycle through the vertical transfer of water from Earth to the atmosphere via **evapotranspiration**, the sum of evaporation from surfaces and **transpiration**, which is the water loss from plants. Conversely, evapotranspiration accounts for 75% of the turbulent energy transfer from Earth to the atmosphere and is therefore a key process in Earth's energy budget (see Fig. 2.2). The hydrologic cycle also controls Earth's biogeochemical cycles by influencing all biotic processes, dissolving nutrients, and transferring them within and among ecosystems. These nutrients provide the resources that support growth of organisms. The movement of materials that are dissolved and suspended in water links ecosystems within a landscape.

The importance of the hydrologic cycle raises concerns about the extent to which it has been modified by human activities. Humans currently use half of Earth's readily available fresh water, which is about half of the annual mean runoff in regions accessible to people. Water use is projected to increase to 70% by 2050 (Postel et al. 1996). This human use of fresh water affects land and water management; the movement of pollutants among ecosystems; and, indirectly, ecosystem processes in unmanaged ecosystems. Land use changes have altered terrestrial water and energy budgets sufficiently to change regional and global climate (Chase et al. 2000). Finally, human activities alter the capacity of the atmosphere to hold water vapor. Water vapor is the major greenhouse gas. It is transparent to solar radiation but absorbs longwave radiation from Earth (see Fig. 2.1) and thus provides an insulative thermal blanket. Climate warming caused by emissions of other greenhouse gases increases the quantity of water vapor in the atmosphere and therefore the efficiency with which the atmosphere traps longwave radiation. This **water vapor feedback** explains why climate responds so sensitively to emissions of other greenhouse gases (see Chapter 2). Warming accelerates the hydrologic cycle, increasing evaporation and rainfall at the global scale (see Chapter 15). Warming also causes the sea level to rise, due primarily to the thermal expansion of the oceans and secondarily to melting of

Box 4.1. Properties of Water that Link Water and Energy Budgets

The unique properties of water are critical to understanding the linkages between energy and water budgets. Evapotranspiration is one of the largest terms in both the water and the energy budgets of ecosystems, so factors governing the magnitude of evapotranspiration determine the strength of the linkage between the water and energy cycles. How much energy is required to melt ice, to warm water, and to evaporate water? What determines the quantity of water vapor that the atmosphere can hold before precipitation occurs?

Due to its high **specific heat**—the energy required to warm 1 g of a substance by 1°C—water changes temperature relatively slowly for a given energy input (Table 4.1). Consequently, the summer temperature near large water bodies fluctuates less and is generally cooler than in inland areas. A wet surface also heats more slowly but evaporates more water than a dry surface.

Considerable amounts of energy are absorbed or released when water changes state. The energy required to change 1 g of ice to liquid water, the **heat of fusion**, is 0.33 MJ kg^{-1}. More than seven times that energy (2.45 MJ kg^{-1}) is required to change 1 g of liquid water to water vapor at 20°C, the **heat of vaporization**. Changes between liquid and vapor therefore generally have greater effects on ecosystem energy budgets than do changes between liquid and solid.

A consequence of these properties of water is that energy is released from a transpiring leaf as the water changes from a liquid to vapor, causing the leaf to cool. Conversely, the atmosphere generally becomes warmer when water condenses to form cloud droplets because energy is released when water changes phase from a vapor to a liquid state. This energy provides the additional buoyancy that forms tall thunderhead clouds (see Chapter 2).

The **water vapor density** (or absolute humidity) is a measure of the mass of water per volume of dry air. The amount of water vapor that can be held in air without it becoming saturated increases greatly as temperature rises (Fig. 4.1). Consequently, as climate warms, the water-holding capacity of the atmosphere increases in a greater-than-linear fashion. The **relative humidity** (RH) is the ratio of the actual amount of water held in the atmosphere compared to the maximum amount that could be held at that temperature. Since the maximum potential vapor density is quite sensitive to temperature, the same relative humidity can occur at very different vapor densities (Fig. 4.1). Relative humidity alone is therefore not a good indicator of the absolute water vapor content of the air.

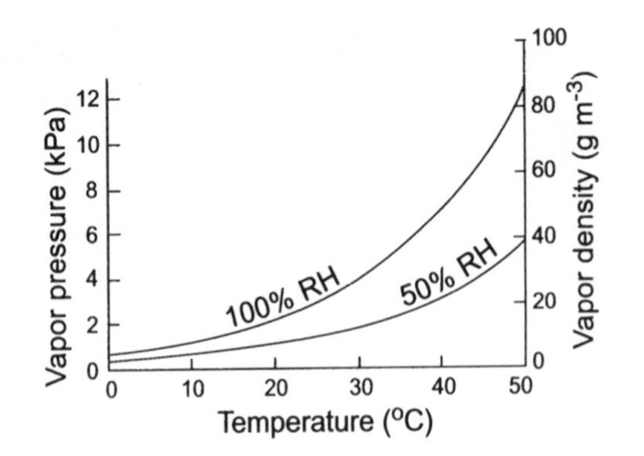

FIGURE 4.1. Relationship between temperature and the water-holding capacity of the atmosphere at 50% and 100% relative humidity (RH). Note that relative humidity is not a good predictor of evaporation rate. The same relative humidity can occur at a series of different vapor pressures.

TABLE 4.1. Specific heat of various materials.[a]

Substance	(kJ kg^{-1} °C^{-1})
Ice	2.1
Water	4.2
Steam	2.0
Dry sand	0.8
Peat soils	1.8
Air	1.0

[a] It takes four times the energy to raise the temperature of an equal mass of water to that of air.

Vapor pressure is the partial pressure exerted by water molecules in the air. The driving force for evaporation is the difference in vapor pressure between the air immediately adjacent to an evaporating surface and that of the air with which it mixes. The air immediately adjacent to an evaporating surface is approximately saturated at the temperature of the surface. The **vapor pressure deficit** (VPD) is the difference between actual vapor pressure and the vapor pressure of air at the same temperature and pressure that is saturated with water vapor. This term is loosely used to describe the difference in vapor pressure between air immediately adjacent to an evaporating surface and the bulk atmosphere, although strictly speaking, the air masses are at different temperatures.

glaciers and ice caps. A rising sea level endangers the coastal zone, where most of the major cities of the world are located. Given the key role of water and energy in ecosystem and global processes, it is critical that we understand the controls over water and energy exchange and the extent to which they have been modified by human actions.

Surface Energy Balance

Solar Radiation Budget

The energy absorbed by a surface is the balance between incoming and outgoing radiation. Here we focus on ecosystem-scale radiation budgets, although the general principles apply at any scale, ranging from the surface of a leaf to the surface of the globe (see Fig. 2.2). The two major components of the radiation budget are shortwave radiation (K), the high-energy radiation emitted by the sun, and longwave radiation (L), the thermal energy emitted by all bodies (see Chapter 2). **Net radiation** (R_{net}) is the balance between the inputs and outputs of shortwave and longwave radiation, measured as watts per meter squared ($W\,m^{-2}$).

$$R_{net} = (K_{in} - K_{out}) + (L_{in} - L_{out}) \qquad (4.1)$$

On a clear day, **direct radiation** from the sun accounts for most of the shortwave input to an ecosystem (see Fig. 2.2). Additional input of shortwave radiation comes as **diffuse radiation**, which is scattered by particles and gases in the atmosphere, or as **reflected radiation** from clouds and surrounding landscape units such as lakes, dunes, or snowfields. At noon on a clear day in a nonpolluted atmosphere, direct radiation accounts for 90% of incoming shortwave radiation to an ecosystem, and diffuse radiation becomes proportionately greater on cloudy days or near dawn or dusk when sun angles are lower. For Earth as a whole, direct and diffuse radiation each account for about half of incoming shortwave radiation (see Fig. 2.2).

The proportion of the incoming shortwave radiation that is absorbed depends on the **albedo** (α) or shortwave reflectance of the ecosystem surface. Ecosystem albedos vary at least 10-fold, ranging from highly reflective surfaces such as fresh snow to dark surfaces such as wet soils (Table 4.2). Conifer canopies, for

TABLE 4.2. Typical values of albedo of major surface types on earth.

Surface type	Albedo
Oceans and lakes	0.03–0.10[a]
Sea ice	0.30–0.45
Snow	
Fresh	0.75–0.95
Old	0.40–0.70
Arctic tundra	0.15–0.20
Conifer forest	0.09–0.15
Broadleaf forest	0.15–0.20
Agricultural crops	0.18–0.25
Grassland	0.16–0.26
Savanna	0.18–0.23
Desert	0.20–0.45
Bare soil	
Wet, dark	0.05
Dry, dark	0.13
Dry, light	0.40

[a] Albedo of water increases greatly (from 0.1 to 1.0) at solar angles <30°.

Data from Oke (1987), Sturman and Tapper (1996), and Eugster et al. (2000).

example, have a lower albedo than deciduous forests, and grasslands with large amounts of standing dead leaves have relatively high albedo. The albedo of a complex canopy is less than that of individual leaves, because much of the light reflected or transmitted by one leaf is absorbed by other leaves and stems. For this reason, deep, uneven canopies of conifer forests have a low albedo. Changes in ecosystem albedo explain in part why high-latitude regions are projected to warm more rapidly than low latitudes. As climate warms, snow and sea ice will melt earlier in the spring, replacing a reflective snow-covered surface with a dark absorptive surface. This process, together with the resulting change in temperature, is referred to as the **snow (or ice) albedo feedback**. Over longer time scales, the northward movement of trees into tundra causes an additional reduction in regional albedo in winter because the dark forest canopy masks the underlying snow-covered surface. With the low sun angles typical of high latitudes, this effect is significant even with sparse canopies. As the treeline moves north, the land surface absorbs more energy, which is then transferred to the atmosphere, causing a positive feedback to regional warming (Foley et al. 1994). Albedo also varies diurnally; it is about twice as high in early morning and evening as at midday. The diurnal changes in absorbed radiation are therefore greater than one would expect from diurnal variations in incoming radiation.

Ecosystem Radiation Budget

The amount of longwave radiation emitted by an object depends on its temperature and its **emissivity**, a coefficient that describes the capacity of a body to emit radiation. Most absorbed radiation is emitted (emissivity about 0.98 in vegetated ecosystems), so differences among ecosystems in longwave radiation balance depend primarily on the temperature of the sky, which determines L_{in}, and the surface temperature of the ecosystem, which determines L_{out}.

$$R_{net} = (K_{in} - K_{out}) + (L_{in} - L_{out})$$
$$= (1-\alpha)K_{in} + \sigma\left(\varepsilon_{sky}T_{sky}^{4} - \varepsilon_{surf}T_{surf}^{4}\right) \quad (4.2)$$

where α is the surface albedo, σ is the Stefan-Boltzman constant ($5.67 \times 10^{-8}\,\mathrm{W\,m^{-2}\,K^{-4}}$), T is absolute temperature (K), and ε is emissivity. Clouds are warmer than space and effectively trap longwave emissions from the surface, so ecosystems receive more longwave radiation under cloudy than under clear conditions. This explains why cloudy nights are warmer than clear ones and why deserts are generally cold at night, despite the high inputs of solar energy during the day.

Longwave radiation emitted by the ecosystem depends on surface temperature, which, in turn, depends on the quantity of radiation received by the surface and the efficiency with which this energy is transmitted into the air and soil. Surfaces that absorb a large amount of radiation, due to high solar inputs and/or low albedo tend to be warmer and therefore emit more longwave radiation. Dry surfaces and leaves with low transpiration rates also tend to be warm because they are not cooled by the evaporation of water. Desert sands, recent burn scars, and city pavements, for example, are generally hot. Similarly, a well-watered lawn is much cooler than an ecosystem that is dry or is dominated by plants with low transpiration rates. Because L_{out} is a function of temperature raised to the fourth power (Eq. 4.2), surface temperature has a powerful multiplying effect on L_{out}.

Canopy structure also influences surface temperature and surface energy exchange through its effect on the efficiency of energy dissipation. The irregular surface of vegetation slows down airflow unevenly, creating **mechanical turbulence**. Tall uneven canopies such as conifer forests are aerodynamically more **rough** than are short smooth canopies. The mechanical turbulence generated by airflow across the vegetation surface creates eddies, which sweep down into the canopy, transporting bulk air inward and canopy air out. These eddies transfer energy away from the surface and mix it with the atmosphere (Jarvis and McNaughton 1986). Air flowing across short, smooth canopies such as grasslands or crops tends to be less turbulent, so these canopies are less efficient in shedding the energy that they absorb—that is, they are less tightly **coupled** to the bulk

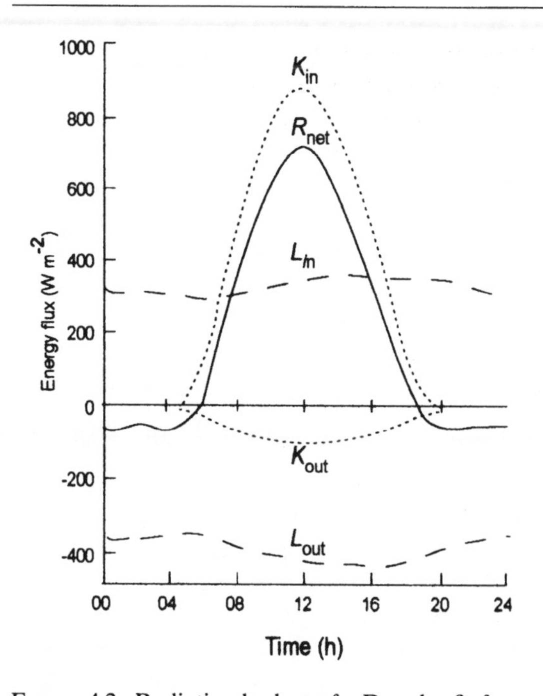

FIGURE 4.2. Radiation budget of a Douglas fir forest during the summer. (Redrawn with permission from Methuen; Oke 1987.)

atmosphere. Smooth canopies therefore tend to have higher surface temperatures during the day and greater longwave emissions than do forest canopies. In general, albedo and surface temperature have the greatest impact on radiation balance and therefore net radiation (Eq. 4.2). Shortwave radiation varies much more strongly from day to night than does longwave radiation (Fig. 4.2).

Energy Partitioning

The net energy absorbed by an ecosystem (net radiation) is approximately balanced by energy that is released to the atmosphere or conducted into the soil. There are two major ways in which energy can be stored (S) in an ecosystem, an increase in the temperature of biomass and the conversion of light to chemical energy through photosynthesis. These two forms of energy storage are generally less than 10% of net radiation on a daily basis. Thus, although the energy trapped by photosynthesis is the major energetic engine that drives the carbon cycle of ecosystems, it is only a small part of the total energy budget of the ecosystem. Because ecosystem energy storage is usually small, energy input at the surface approximately equals energy loss over a day.

Net radiation is partitioned primarily among three major avenues of energy exchange between the ecosystem and the atmosphere–soil: ground heat flux (G), latent heat flux (or evapotranspiration, LE), and sensible heat flux (convective heating, H). The **latent heat of vaporization** (L; 2.45 MJ kg^{-1} at 20°C) is a constant describing the quantity of energy required to evaporate a given mass of water, and evapotranspiration rate (E) is the rate of water transfer from a land or water surface to the atmosphere. Net radiation is positive when directed toward the surface; H, LE, G, and ΔS are positive when directed away from the surface.

$$R_{net} = H + LE + G + \Delta S \quad (4.3)$$

Available energy is that portion of R_{net} that is neither stored nor conducted into the ground; it is the energy available for turbulent exchange with the atmosphere as H and LE.

$$\begin{aligned} \text{Available energy} &= R_{net} - (G + \Delta S) \\ &= H + LE \quad (4.4) \end{aligned}$$

Ground heat flux, the heat transferred from the surface into and out of the soil, is negligible over a day in most temperate and tropical ecosystems, because the heat conducted into the soil during the day is balanced by heat conducted back to the surface at night. The magnitude of ground heat flux depends on the thermal gradient between the soil surface and deep soils and the thermal conductivity of soils, which is greatest in soils that are wet and have a high bulk density. In contrast to temperate soils, permafrost regions of the arctic and boreal forest have substantial ground heat flux (10 to 20% of net radiation), due to the strong thermal gradient between the soil surface and the permafrost. Clear lakes and the ocean, which transmit substantial radiation to depth, also exchange substantial energy as "ground" heat flux.

Latent heat flux is the energy transferred to the atmosphere when water is transpired by plants or evaporates from leaf or soil surfaces. Latent heat flux, which is measured in energy units, is identical to evapotranspiration, which is measured in water units. This heat is transported from the surface into the atmosphere by convection and is released to the atmosphere when water vapor condenses to form cloud droplets, often at considerable distances from the point at which evaporation occurred. Dewfall represents a small latent heat flux from the atmosphere to the ecosystem at night under conditions of high relative humidity and cold leaf or soil surfaces. Note that latent heat flux is the process that transfers water from the ecosystem to the atmosphere. Equation 4.3 therefore also relates ecosystem water loss to the energy budget that drives evapotranspiration. It is because of the conservation of energy and mass that the energy and water cycles intersect.

Sensible heat flux is the heat that is initially transferred to the near-surface atmosphere by conduction and to the bulk atmosphere by convection; it is controlled in part by the temperature differential between the surface and the overlying air. Air close to the surface becomes warmer and more buoyant than the air immediately above it, causing this parcel of air to rise, the process of **convective turbulence**. Mechanical turbulence is caused by winds blowing across a rough surface; it generates eddies that transport warm moist air away from the surface and bring cooler drier air from the bulk atmosphere back toward the surface. Surface turbulence is the major process that transfers latent and sensible heat between the surface and the atmosphere (see Chapter 2).

There are important interactions between latent and sensible heat fluxes from ecosystems. The consumption of heat by the evaporation of water cools the surface, thereby reducing the temperature differential between the surface and the air that drives sensible heat flux. Conversely, the warming of surface air by sensible heat flux increases the quantity of water vapor that the air can hold and causes convective movement of moist air away from the evaporating surfaces. Both of these processes increase the vapor pressure gradient that drives evaporation. Because of these interdependencies, surface moisture has a strong effect on the **Bowen ratio**—that is, the ratio of sensible to latent heat flux.

Bowen ratios range from less than 0.1 for tropical oceans to great than 10 for deserts, indicating that either latent heat flux or sensible heat flux can dominate the turbulent energy transfer from ecosystems to the atmosphere, depending on the nature of the ecosystem and the climate. In general, ecosystems with abundant moisture have higher rates of evapotranspiration and therefore lower Bowen ratios than do dry ecosystems. Similarly, ecosystems dominated by rapidly growing plants, which have high transpiration rates (see Chapter 5), have proportionately lower sensible heat fluxes and low Bowen ratios (Table 4.3). Strong winds and/or rough canopies, which generate surface turbulence, tend to prevent a temperature buildup at the surface and therefore reduce sensible heat flux and Bowen ratio. For these reasons, energy partitioning varies substantially both seasonally and among ecosystems. The Bowen ratio determines the strength of the linkage between the energy budget and the hydrologic cycle, because it is inversely related to the proportion of net radiation that drives water loss from ecosystems: the lower the Bowen ratio, the tighter the linkage between the energy budget and the hydrologic cycle.

The spatial patterning of ecosystems influences energy partitioning (see Chapter 14). Heating contrasts between adjacent ecosystems create convective turbulence; this turbulence is therefore much greater at boundaries between ecosystems with contrasting energy budgets than in the centers of ecosystems. Most evaporation from large lakes, for example,

TABLE 4.3. Representative Bowen ratios for different vegetation types.

Surface type	Bowen ratio
Desert	>10
Semiarid landscape	2–6
Arctic tundra	0.3–2.0
Temperate forest and grassland	0.4–0.8
Boreal forest	0.5–1.5
Forest, wet canopy	–0.7–0.4
Water-stressed crops	1.0–1.6
Irrigated crops	–0.5–0.5
Tropical rain forest	0.1–0.3
Tropical ocean	<0.1

Data from Jarvis (1976), Oke (1987), and Eugster et al. (2000).

occurs near their edges, rather than in the center, where the overlying air is so stable that it saturates rapidly and supports a relatively low evaporation rate. A swamp with interspersed patches of vegetation would have greater convective turbulence and overall evaporation than homogeneous water or moist vegetation. When ecosystem patches that differ strongly in energy partitioning are larger in diameter than the depth of the planetary boundary layer (greater than about 10 km), they can modify mesoscale atmospheric circulations and cloud and precipitation patterns (Pielke and Avisar 1990, Weaver and Avissar 2001).

Seasonal Energy Exchange

Over sufficiently long time scales, energy outputs are tightly coupled to inputs because there is little energy storage at Earth's surface. Most ecosystems have a limited capacity for long-term energy storage in vegetation and surface soils. Consequently, energy losses to the atmosphere closely track solar inputs on both a daily and a seasonal basis, although the form in which this energy is lost (sensible vs. latent heat flux) varies, depending on moisture availability. Ground heat fluxes are usually negligible when averaged over 24 h (Oke 1987). Important exceptions to this generalization are water bodies such as lakes and oceans, in which the solar inputs often penetrate to depth, resulting in substantial warming of the water, and some high-latitude regions. Water bodies often absorb substantial energy in spring and early summer, when the solar angle is greatest, causing the water to warm. There is a net release of energy in the autumn, moderating the temperatures of adjacent terrestrial surfaces (see Chapter 2). This seasonal pattern of energy exchange drives the annual or semiannual turnover of lakes (see Chapter 10). Permafrost contributes to a seasonal imbalance in energy absorption and release in cold climates. In the arctic, for example, 10 to 20% of the energy absorbed during summer is consumed by thawing of frozen soil. This energy is released back to the atmosphere the next winter, when the soil refreezes (Chapin et al. 2000a).

Snow-covered surfaces experience threshold changes in energy exchange at the time of snowmelt (Liston 1999). The high albedo of snow-covered surfaces minimizes energy absorption until snowmelt occurs, at which time there is a dramatic increase in the energy absorbed by the surface and transferred to the atmosphere. This may result in abrupt increases in regional air temperature after snowmelt. Leaf out also alters energy exchange by both changing albedo and increasing evapotranspiration at the expense of sensible heat flux.

Water Inputs to Ecosystems

Precipitation is the major water input to most terrestrial ecosystems. Global and regional controls over precipitation therefore determine the quantity and seasonality of water inputs to most ecosystems (see Chapter 2). In ecosystems that receive some precipitation as snow, however, the water contained in the snowpack does not enter the soil until snow melt, often months after the precipitation occurs. This causes the seasonality of water input to soils to differ from that of precipitation.

Vegetation in some ecosystems, particularly in riparian zones, accesses additional groundwater that flows laterally through the ecosystem. Desert communities of **phreatophytes** (deep-rooted plants that tap groundwater), for example, may absorb sufficient groundwater that the ecosystem loses more water in transpiration than it receives in precipitation. Lakes and streams also receive most of their water inputs from groundwater or runoff that drains from adjacent terrestrial ecosystems. Water inputs to freshwater ecosystems are therefore only indirectly linked to precipitation.

In ecosystems with frequent fog, **canopy interception** of fog increases the water inputs to ecosystems, when cloud droplets that might not otherwise precipitate are deposited on leaf surfaces and are absorbed by leaves or drip from the canopy to the soil. The coastal redwood trees of California, for example, depend on fog-derived water inputs during summer, when precipitation is low, but fog occurs frequently.

Similarly, in areas that are climatically marginal for Australian rain forests, the capture of fog and mist by trees can augment rainfall by 40% (Hutley et al. 1997). In most ecosystems, however, most of the precipitation that is intercepted by the canopy evaporates before it reaches the soil, so canopy interception generally *reduces* the effectiveness of precipitation that enters the ecosystem rather than increasing water inputs.

Water Movements Within Ecosystems

Basic Principles of Water Movement

The soil behaves like a bucket that is filled by precipitation and emptied by evapotranspiration and runoff. The soil is the major water storage reservoir of ecosystems. When water inputs from precipitation exceed the capacity of the soil to hold water, the excess drains to groundwater or runs off over the ground surface (Fig. 4.3), just as water added to a full bucket spills over the edges instead of being retained in the bucket. The water losses from the ecosystem move laterally to other ecosystems such as streams and lakes. Evaporation from the soil surface and transpiration by plants are the other major avenues of water loss from the soil reservoir. These processes continue only as long as the soil contains water that plants can tap, just as evaporation from a bucket continues only as long as the bucket contains water.

Water is stored in soil primarily in pores between soil particles, so the water-holding capacity of a soil depends on its total pore volume. Pore volume, in turn, depends on soil depth and the proportion of the soil volume occupied by pores, the spaces between soil particles. Shallow soils on ridgetops hold less water than deep valley-bottom soils; rocky or sandy soils, in which soil solids occupy much of the soil volume, hold less water than fine-textured soils; and soils compacted by intensive grazing or farm machinery hold less water than noncompacted soils (see Chapter 3). As described later, water is held most effectively by small soil pores that have a large surface-to-volume ratio.

Water moves along a gradient from high to low potential energy. The energy status of water depends on its concentration and various pressures. The pressures in natural systems can be

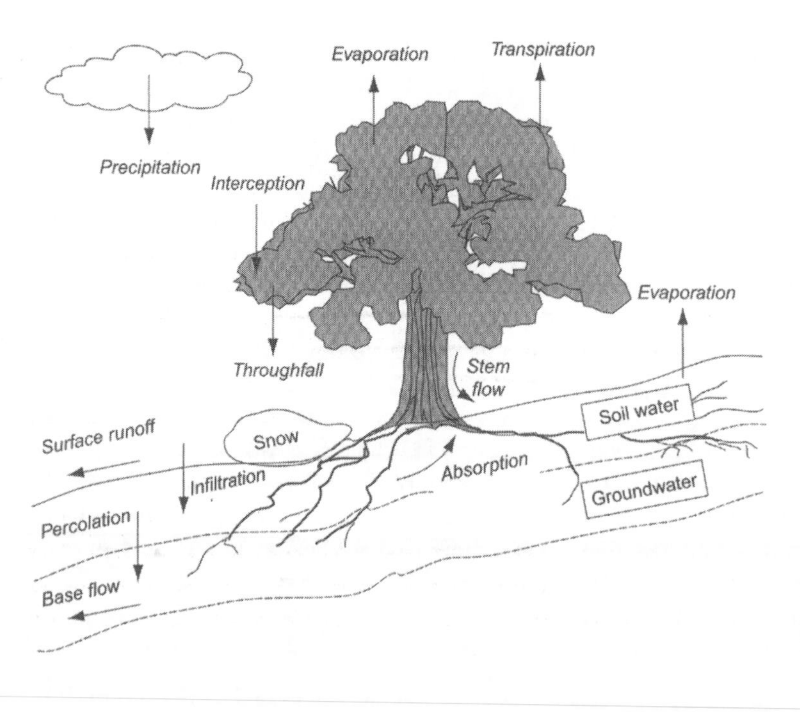

FIGURE 4.3. Water balance of an ecosystem (Waring and Running 1998).

described in terms of either hydrostatic pressures or matric forces (Passioura 1988). The major hydrostatic pressures in natural systems are (1) gravitational pressure, which depends on height and (2) pressures that are generated by physiological processes in organisms. The gravitational pressure is higher at the top of a tree than in the roots, so plants must move water to leaves against this gravitational force. Plants can generate either positive pressures, (e.g., the turgor pressure that maintains the rigidity of plant cells) or negative pressures (which move water from the roots to leaves). Matric forces result from the adsorption of water to the surfaces of cells or soil particles. The thinner the water film, the more tightly the water molecules are held to surfaces by matric forces.

In most cases water moves in response to some combination of these forces. We can consider all these forces simultaneously by expressing them in units of **water potential**—that is, the potential energy of water relative to pure water at the soil surface. The total water potential (ψ_t) is the sum of the individual potentials.

$$\Psi_t = \Psi_p + \Psi_o + \Psi_m \quad (4.5)$$

The **pressure potential** (ψ_p) is generated by gravitational forces and physiological processes of organisms; the **osmotic potential** (ψ_o) reflects presence of substances dissolved in water; and the **matric potential** (ψ_m) is caused by adsorption of water to surfaces. In some treatments, matric potential is considered a component of pressure potential (Passioura 1988). By convention, the water potential of pure water under no pressure at the soil surface is given a value of zero. Water potentials are positive if they have a higher potential energy than this reference and negative if they have a lower potential energy.

Water Movement from the Canopy to the Soil

In closed-canopy forests, the canopy intercepts a substantial proportion of incoming precipitation (Fig. 4.3). Intercepted precipitation can be evaporated directly back to the atmosphere, absorbed by the leaves, drip to the ground (**throughfall**), or run down stems to the ground (**stem flow**). **Interception** is the fraction of precipitation that does not reach the ground. It commonly ranges from 10 to 50% for closed-canopy ecosystems (Waring and Running 1998). After light rain or snowfalls, a substantial proportion of the precipitation may evaporate and return directly to the atmosphere without entering the soil. For this reason, canopy heterogeneity generates heterogeneity in water input to soil and therefore soil moisture availability. Throughfall is the process that delivers most of the water from the canopy to the soil.

The capacity of the canopy to intercept and store water differs among ecosystems. It depends primarily on canopy surface area, particularly the surface area of leaves (Fig. 4.4). Forests, for example, frequently store 0.8, 0.3, and 0.25 mm of precipitation on leaves, branches, and stems, respectively. Conifer forests typically store about 15% of precipitation, whereas deciduous forests store 5 to 10% (Waring and Running 1998). Epiphytes, which are rooted in the canopy, depend completely on canopy interception for their water supply and increase canopy interception. Factors such as stand age and epiphyte load influence canopy interception through their effects on canopy surface area.

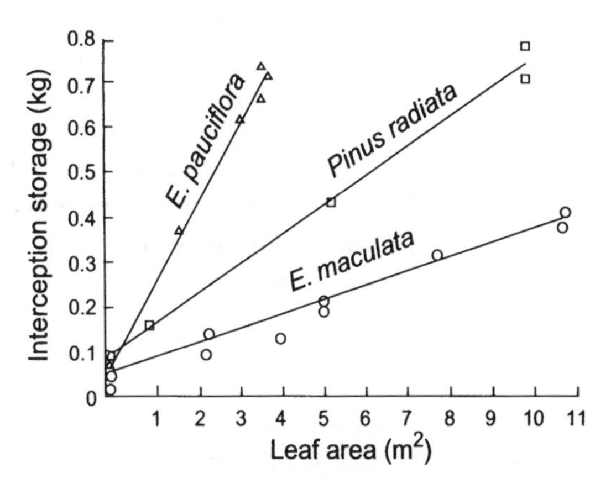

FIGURE 4.4. Interception storage capacity of *Eucalyptus* spp. and *Pinus radiata* with different leaf areas. (Redrawn with permission from *Journal of Hydrology*, Vol. 42 © 1979 Elsevier Science; Aston 1979.)

The bark texture and architecture of stems and trunks influence the amount and direction of stem flow. Trees and shrubs with smooth bark have greater stem flow (about 12% of precipitation) than do rough-barked plants such as conifers (about 2% of precipitation) (Waring and Running 1998). In the *Eucalyptus* mallee in southwestern Australia as much as 25% of the incoming precipitation runs down stems, due to the parachute architecture of these shrubs. The stem flow then penetrates to depth in the soil profile through channels at the soil–root interface (Nulsen et al. 1986).

Water Movement Within the Soil

Pressure gradients associated with gravity and matric forces control most water movement through soils. The rate of water flow through the soil (J_s) depends on the driving force (the gradient in water potential) and the resistance to water movement. This resistance, in turn, depends on the **hydraulic conductivity** (L_s) of the soil, and the path length (l) of the column through which the water travels.

$$J_s = L_s \frac{\Delta \Psi_t}{l} \tag{4.6}$$

This simple relationship describes most of the patterns of water movement through soils, including the **infiltration** of rainwater or snow melt into the soil and the movement of water from the soil to plant roots. Soils differ strikingly in hydraulic conductivity due to differences in soil texture and aggregate structure (see Chapter 3). For this reason water moves much more readily through sandy soils than through clay soils or compacted soils. The rate of water flow in saturated soils, for example, differs by three orders of magnitude between fine and coarse soils (less than 0.25 to greater than 250 mm h^{-1}).

Infiltration of rainwater into the soil depends not only on hydraulic conductivity but also on preferential flow through **macropores** created by cracks in the soil or channels produced by plant roots and soil animals (Dingman 2001). Variation in flow paths in the surface few millimeters of soil can have large effects on infiltration. Impaction by raindrops on an unprotected mineral soil, for example, can reduce hydraulic conductivity dramatically. For this reason the presence of a surface moss or litter layer, which prevents impaction by raindrops, is one of the most important factors determining whether water enters the soil or flows over the surface. Any time that precipitation rate exceeds the infiltration rate, water accumulates on the surface and **overland flow** may occur, leading to erosional loss of soil.

Some soils have horizons of low hydraulic conductivity that prevent water **percolation** to depth. For example, **calcic** (caliche) layers in deserts, **permafrost** in arctic and boreal ecosystems, and **hardpans** in highly weathered soils are horizons with such low hydraulic conductivity that the water table remains close to the surface, rather than moving into a deep groundwater pool (see Chapter 3).

Once water enters the soil, it moves downward under the force of gravity until the matric forces, which account for the adsorption of water to soil particles, exceed the gravitational potential. Water that is not retained by matric forces drains through the soil to groundwater. The **field capacity** of a soil is the quantity of water retained by a saturated soil after gravitational water has drained. The large surface area per unit soil volume in fine-textured soils explains their high field capacity. A clay soil, for example, with its high proportion of small particles (Table 4.4), holds four times more water than a sandy soil. Organic matter also enhances the field capacity of soils, because of its hydrophilic nature and its effects on soil structure. For this reason the soils beneath shrubs in deserts, which have higher organic

TABLE 4.4. Typical pore size distribution of different soil types.

	Pore space (% of soil volume)		
Particle size (μm)	Sand	Loam	Clay
>30	75	18	6
0.2–30	22	48	40
<0.2	3	34	53

content, retain more water than do soils that receive less litter input.

At field capacity, the water potential of a soil is about –0.03 MPa—that is, close to the water potential of pure water (0.00 MPa). As a soil dries, the films of soil water become thinner, and the remaining water is held more tightly to particle surfaces. The **permanent wilting point** is the soil water potential (about –1.5 MPa) at which most mesic plants wilt because they cannot obtain water from soils. Many drought-adapted plants, however, can obtain water from soils at water potentials as low as –3.0 to –8.0 MPa (Larcher 1995). A second consequence of thin water films in dry soils is that water cannot move directly across water-filled soil pores but must move around the edges of the air-filled pores along a much longer, more tortuous path. For this reason, the hydraulic conductivity of soil declines dramatically as the soil dries. The difference in the water content between field capacity and permanent wilting point (water-holding capacity) provides an estimate of the plant-available water, although some of this water is held in such small pores that it moves slowly to roots (Fig. 4.5). Vegetation often extracts 65 to 75% of the plant-available water before there are signs of water stress (Waring and Running 1998). The total quantity of water available to vegetation is the available water content per unit soil volume times the volume exploited by roots.

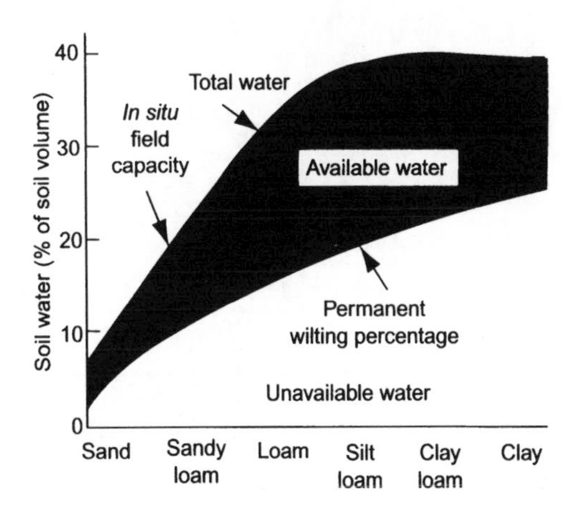

FIGURE 4.5. Plant-available water as a function of soil texture. (Redrawn with permission from Academic Press; Kramer and Boyer 1995.)

Water Movement from Soil to Roots

Water moves from the soil to the roots of transpiring plants by flowing from high to low water potential. Water moves from the soil into the root whenever the root has a lower water potential than the surrounding soil. Movement of water into the root along a water potential gradient causes the water film on adjacent soil particles to become thinner. The localized reduction in water potential near the root causes water to move along soil films toward the root. In this way a root can access most available water within a radius of about 6 mm. As the soil dries, hydraulic conductivity declines, and the root accesses water less rapidly. In saline soils, the osmotic potential of the soil solution reduces total soil water potential, so roots with a given water potential can absorb less water than from nonsaline soils.

A continuous pathway for water movement from the soil to the root is provided by root hairs and mycorrhizal hyphae that extend into the soil and by carbohydrates secreted by the root that maximize contact between the root and the soil. When this root–soil contact is interrupted by the shrinking of drying soil or the consumption of root cortical cells by soil animals, the root can no longer absorb water.

Rooting depth reflects a compromise between water and nutrient availability. Most plant roots are in the upper soil horizons where nutrient inputs are greatest and where nutrients are generally most available (see Chapter 8). In a given ecosystem, short-lived herbs are generally more shallow rooted than long-lived shrubs and trees and depend more on surface moisture (Fig. 4.6). In arid ecosystems surface evaporation and transpiration dry out the surface soils. For this reason, deserts, arid shrublands, and tropical savannas have many species with deep roots (Fig. 4.7). Phreatophytes are an extreme example of deep-rooted plants. These desert plants produce roots that extend to the water table, often a depth of tens of meters. These plants have no physiological adaptations to drought and have high transpiration rates. Even wet ecosystems such as tropical rain forests

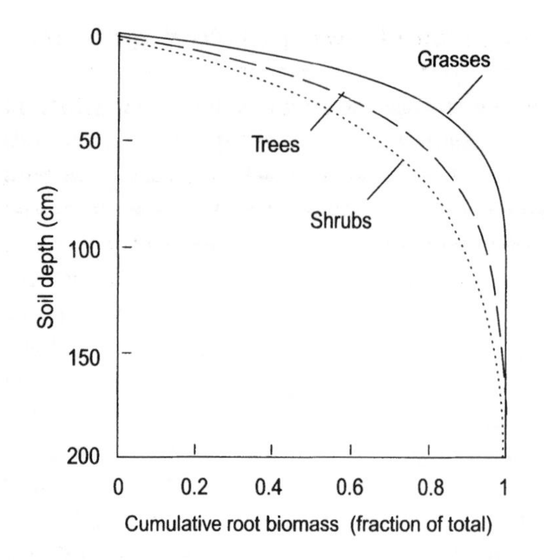

FIGURE 4.6. The cumulative fraction of roots found at different soil depths for three plant growth forms averaged over all biomes. (Redrawn with permission from *Oecologia*; Jackson et al. 1996.)

have dry seasons. This may explain the occurrence in such forests of deep-rooted trees that tap water from depths of more than 8m (Nepstad et al. 1994). Deep-rooted plants may be more common than generally appreciated. Deep roots often extend into cracks in bedrock, where they tap water as it drains through rock channels to groundwater.

Rooting depth has important ecosystem consequences because it determines the soil volume that can be exploited by vegetation (see Chapter 12). California grassland soils below a 1-m depth, for example, remain moist even at the end of the summer drought, whereas an adjacent chaparral shrub community uses water to a depth of 2m (Fig. 4.8). This greater rooting depth contributes to the longer growing season and greater productivity of the chaparral. Even in the chaparral, species differences in rooting depth (Box 4.2) lead to differences in moisture supply and drought stress.

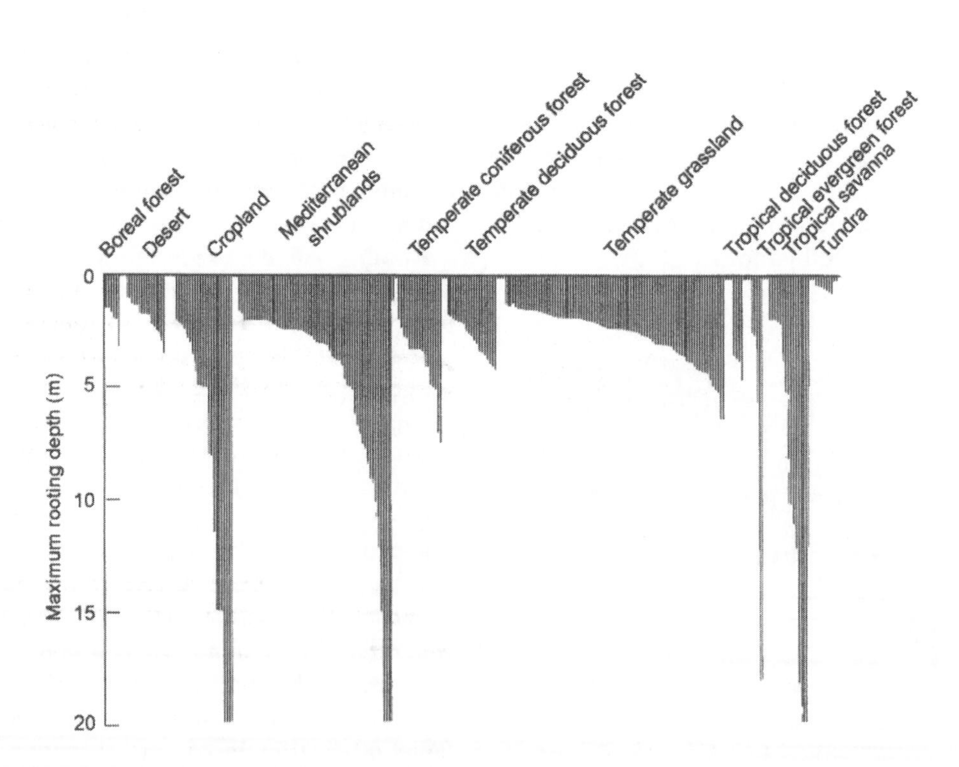

FIGURE 4.7. Maximum rooting depths of selected species in the major biome types of the world showing that species in each biome differ widely. Woody species in dry environments are often deeply rooted. (Redrawn with permission from *Oecologia*; Canadell et al. 1996.)

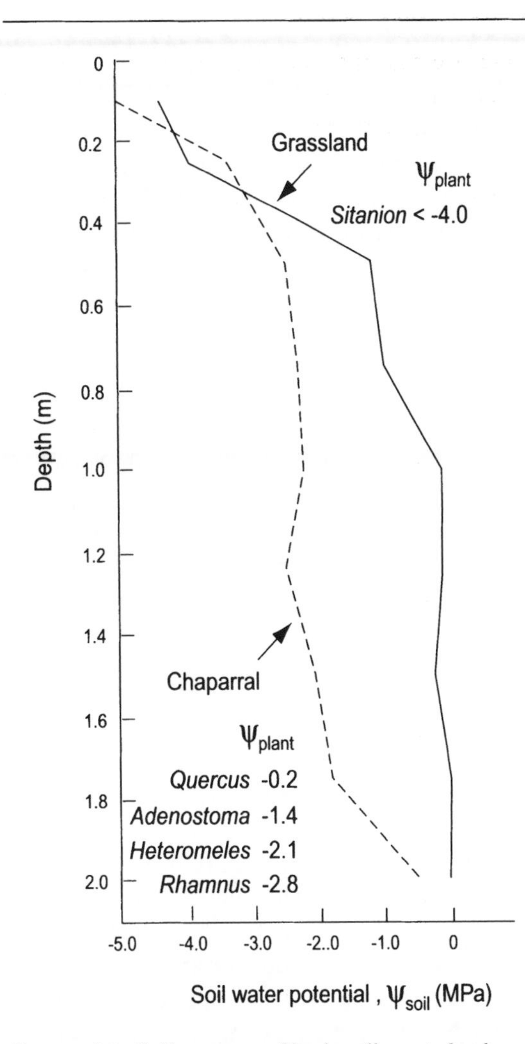

FIGURE 4.8. Soil water profiles in adjacent shrub and grassland communities at the end of the summer drought period. Predawn water potentials are a good index of the soil moisture and the degree of drought stress experienced by the plant. (Redrawn with permission from *Oecologia*; Davis and Mooney 1986.)

Water Movement Through Plants

The vapor-pressure gradient from the leaf surface to the atmosphere is the driving force for water movement through plants. Water transport from the soil through the plant to the atmosphere takes place in a soil-plant-atmosphere continuum that is interconnected by a continuous film of liquid water. Water moves from the soil through the plant to the atmosphere along a gradient in water potential. The low partial pressure of water vapor in air relative to that inside the leaves is the major driving force for water loss from leaves, which in turn drives water transport along a pressure gradient from the roots to the leaves, which in turn drives water movement from the soil into the plant. The rate of water movement through the plant (J_p) (Eq. 4.7) is determined by the water-potential gradient (the driving force; $\Delta\Psi_t$) and the resistance to water movement, just as described for water movement through soils (Eq. 4.6). As in soils, the resistance to water movement through the plant depends on hydraulic conductivity (L_p) and path length (l).

$$J_p = L_p \frac{\Delta\Psi_t}{l} \tag{4.7}$$

The movement of water into and through the plant is driven entirely by the physical process of evaporation from the leaf surface and involves no expenditure of metabolic energy by the plant. This contrasts with the acquisition of carbon and nutrients for which the plant must expend considerable metabolic energy (see Chapters 5 and 8).

Roots

Water moves through roots along a water-potential gradient from moist soils to the atmosphere during the day and sometimes to dry surface soils at night. In moist soils, the cell membranes, which are composed of hydrophobic lipids, provide the greatest resistance to water movement through roots (see Fig. 8.3). This membrane resistance to water flow is greatest under conditions of low root temperature or flooding, so plants that are not adapted to these conditions experience substantial water stress in cold or saturated soils. In dry soils the contact between the root and the soil accounts for the greatest resistance to water flux through the root. Plants overcome this resistance primarily by increasing allocation to the production of new roots (see Chapter 6).

In dry environments, there is a strong vertical gradient in soil water potential due to the low soil water potential in dry surface soils. However, water moves slowly through the soil because of the low hydraulic conductivity of dry soils. During the day, when plants lose water by transpiration, plant water potential is lower than soil water potential, so water moves from the soil into the plant, particularly from deep soils where water is most available (highest soil water potential) (Fig. 4.11). At night, when transpiration ceases, plant water potential equilibrates with the water potential of deep soils. When surface soils are drier than those at depth, the water potential gradient is from deep soils through roots to shallow soils. Because roots have much higher hydraulic conductivity than soils, this gradient in water potential drives **hydraulic lift**, the vertical movement of water from deep to shallow soils through roots along a water potential gradient (Caldwell et al. 1998). Hydraulic lift occurs in most arid ecosystems and in many moist forests. Sugar maple trees, for example, acquire all their moisture from deep roots, but 3 to 60% of the water used by shallow-rooted herbs in these forests comes from water that has been hydraulically lifted by the maple trees (Dawson 1993). In the Great Basin deserts of western North America, 20 to 50% of the water used by shallow-rooted grasses comes from water that is hydraulically lifted by deep-rooted sagebrush shrubs. The water provided by hydraulic lift stimulates decomposition and mineralization in dry shallow soils, augmenting the supplies of both water and nutrients to shallow-rooted species. Because deep-rooted plants both provide water to and remove water and nutrients from shallow soils, hydraulic lift complicates the interpretation of species interactions in many

Box 4.2. Isotopic Signatures of Water Sources

The source of water used by plants can be determined from the isotopic composition of plant water. The ratio of the concentration of deuterium (D) to hydrogen (H) provides a useful signature of different water sources. Evaporation discriminates against the heavier isotope (deuterium), causing the isotopic ratio of D:H to decline (become more negative) relative to the water source that gave rise to evaporation (Fig. 4.9). Condensation, on the other hand, raises the D:H ratio, causing rainfall to have a less negative hydrogen isotopic ratio than its parent air mass. The D:H ratio of water vapor remaining in the atmosphere therefore declines with sequential rainfall events. There is also a linear relationship between air temperature at the time of precipitation and the hydrogen isotope ratio, so summer precipitation has a higher D:H ratio than winter precipitation. These phenomena generate a characteristic signature that identifies different sources of water that plants use (Fig. 4.10). The isotopic ratios of xylem water, for

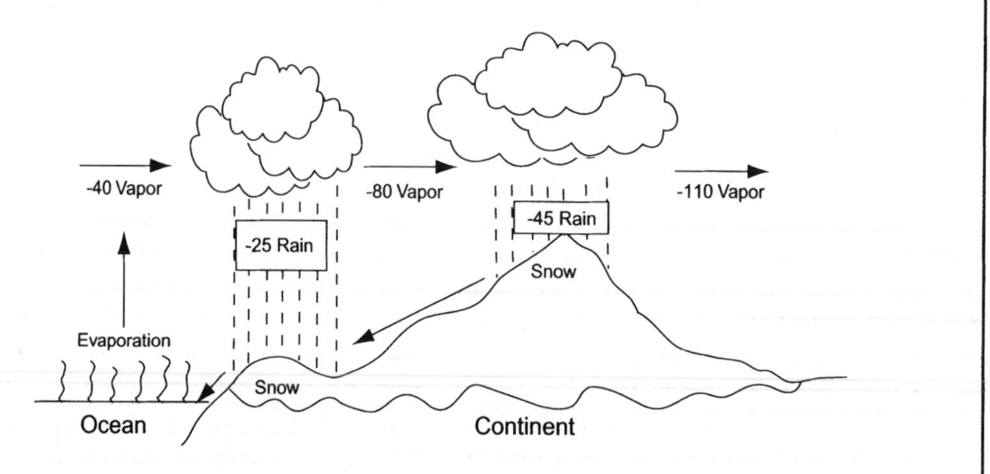

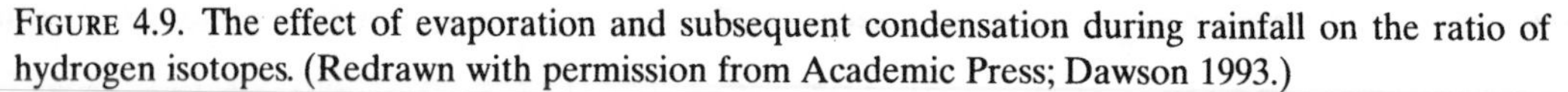
FIGURE 4.9. The effect of evaporation and subsequent condensation during rainfall on the ratio of hydrogen isotopes. (Redrawn with permission from Academic Press; Dawson 1993.)

example, show that some plants derive most of their water from fog, whereas others use soil water or ground water. These measurements also differentiate water derived from winter vs. summer precipitation (Dawson 1993). Similarly, D:H ratios of stream water identify the relative contributions of soil water from recent precipitation events vs. ground water. Oxygen isotope ratios in water show patterns of variation similar to those of hydrogen.

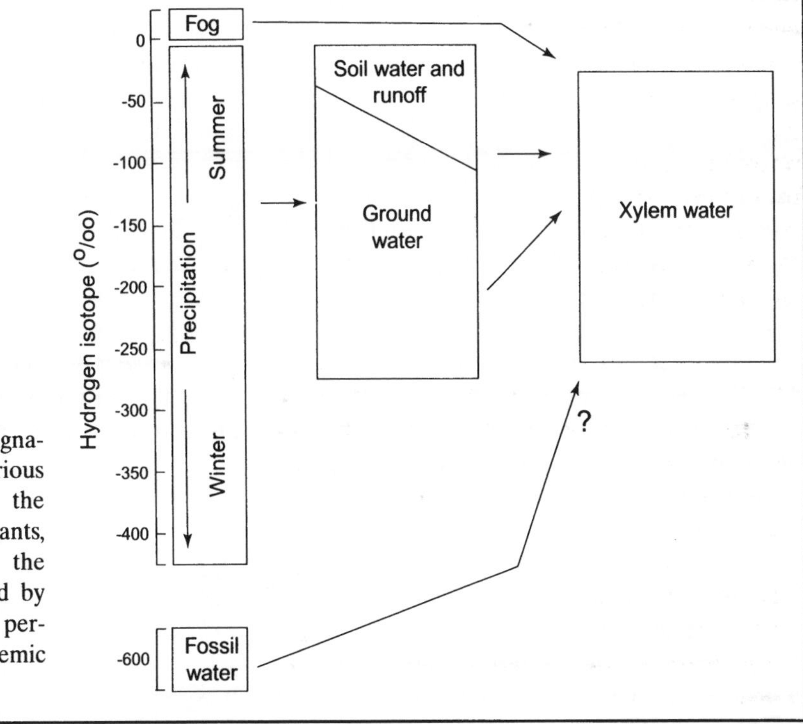

FIGURE 4.10. Isotopic signature of water from various sources. By sampling the water in the xylem of plants, one can determine the main water supply used by a plant. (Redrawn with permission from Academic Press; Dawson 1993.)

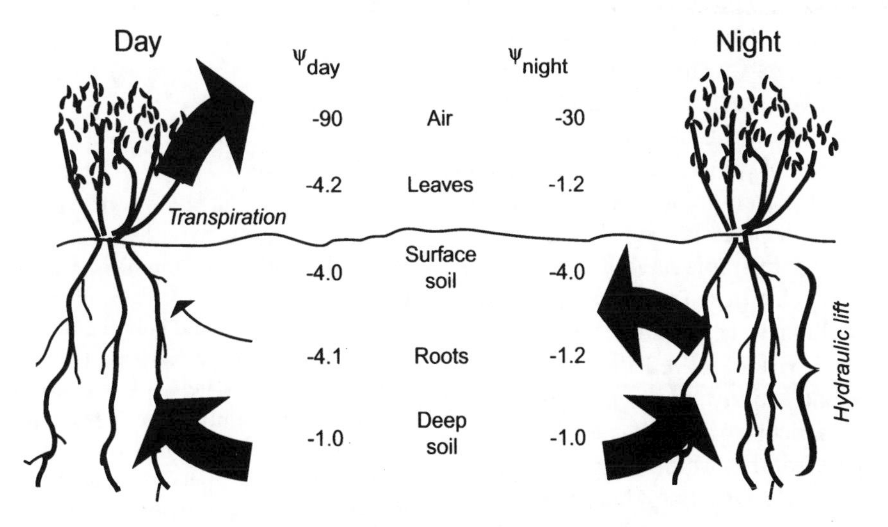

FIGURE 4.11. Patterns of soil water potential and water movement in arid environments during the day and at night. During the day, water moves from soils (especially deep soils) to the atmosphere in response to the strong water potential gradient from the plant to the atmosphere. At night, water moves from wet soils at depth to dry surface soils through the root system, the process of hydraulic lift.

ecosystems. When surface soils are wetter than deep soils after rain, roots provide an avenue to recharge deep soils, perhaps explaining how deep-rooted desert plants grow through dry soils to the water table (Burgess et al. 1998). Thus roots provide an avenue for rapid water transport from soil of high to low water potential, regardless of the vertical direction of the water-potential gradient.

Stems

Water moves through stems to replace water lost by transpiring leaves. The water-conducting tissues in the xylem are narrow capillaries of dead cells that extend from the roots to the leaves. Water is "sucked up" through these capillary tubes in response to the water-potential gradient created by transpirational water loss. The cohesion of water molecules to one another and their adhesion to the walls of the narrow capillary tubes allow these water columns to be raised under tension (a negative water potential) as much as 100m in tall trees.

There is a tradeoff between hydraulic conductivity of stems and their risk of **cavitation**—that is, the breakage of water columns under tension. Hydraulic conductivity of stems varies with the fourth power of capillary diameter, so a small increase in vessel diameter greatly increases hydraulic conductivity. For example, vines, which have relatively small stems and rely on other plants for their physical support, have large-diameter xylem vessels. This allows rapid water transport through narrow stems but increases the risk of cavitation and may explain why vines are most common in moist environments such as tropical forests. The stems of tropical vines, for example have hydraulic conductivities and velocities of sap flow that are 50- to 100-fold higher than those of conifers (Larcher 1995). Broad-leaved deciduous trees are intermediate. Many plants in moist environments, particularly herbaceous plants, function close to the water potential where cavitation occurs, suggesting that they invest just enough in water transport tissues to allow water transport for the growing season (Sperry 1995). Plants from dry environments produce stems with a larger safety factor—that is, stems that resist cavitation at much lower water potential than the plants commonly experience (Fig. 4.12). Small roots are generally more vulnerable to cavitation than are stems and may function as a "hydraulic fuse" that localizes failure in relatively cheap and replaceable parts of the plant (Jackson et al. 2000).

Plants in cold environments suffer cavitation from freezing. Trees adapted to these cold environments typically produce abundant small-diameter vessels that can, in some species, refill after cavitation (diffuse-porous species). In contrast, many trees in warm environments produce both small- and large-diameter vessels that cannot be refilled after cavitation and therefore function for only a single growing season (ring-porous species).

The water transported by a stem depends on both the hydraulic conductivity of individual conducting elements and the total quantity of

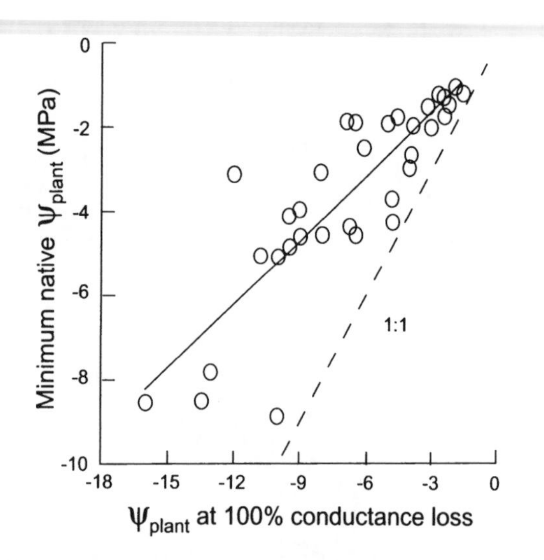

FIGURE 4.12. The relationship between the water potential at which a plant loses all xylem conductivity due to embolism and the minimum water potential observed in nature. Each datum point represents a different species. The 1:1 line is the line expected if there were no safety factor—that is, if each species lost all conductivity at the lowest water potential observed in nature. Species that naturally experience low water potentials exhibit a greater margin of safety (i.e., a greater departure from the 1:1 line). (Redrawn with permission from Academic Press; Sperry 1995.)

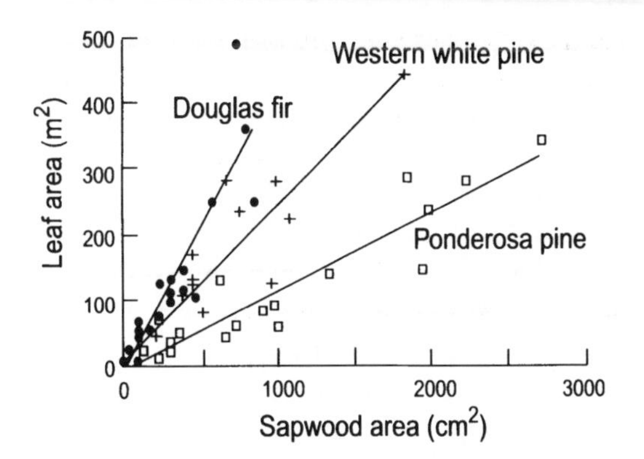

FIGURE 4.13. Leaf area versus sapwood cross-sectional area for three forest trees. Ponderosa pine, which typically occupies dry sites, has smaller vessels and therefore supports less leaf area per unit sapwood than does Douglas fir from moist sites. (Redrawn with permission from *Canadian Journal of Forest Research*; Monserud and Marshall 1999.)

conducting tissue (the **sapwood**). There is a strong linear relationship between the cross-sectional area of sapwood and the leaf area supported by a tree (Fig. 4.13). However, the slope of this relationship varies strikingly among species and environments. Drought-resistant species generally have less leaf area per unit of sapwood than do drought-sensitive species because of the small vessel diameter of drought-resistant species. The ratio of leaf area to sapwood area, for example, is generally more than twice as great in trees from mesic environments as in trees from dry environments, due to the smaller-diameter conducting elements in dry environments (Margolis et al. 1995). Any factor that enhances the productivity of a tree increases its ratio of leaf area to sapwood area. This ratio increases, for example, with improvements in nutrient or moisture status and is greater in dominant than subdominant individuals of a stand.

Water storage in stems buffers the plant from imbalances in water supply and demand. The water content of trunks of trees generally decreases during the day, causing water uptake by roots to lag behind transpirational water loss by about 2h (Fig. 4.14). The quantity of water stored in sapwood is substantial, equivalent to as much as 5 to 10 days of transpiration. This sapwood water, however, exchanges relatively slowly, so stores of water in sapwood seldom account for more than 10% of transpiration. In dry tropical forests, where trees lose their leaves during the dry season, this stored water is critical to support flowering during the dry season. Trees in these forests that have low-density wood and large stem water storage can flower during the dry season, whereas trees with high-density wood and low stem water storage can flower only during the wet season (Borchert 1994). Water storage in desert succulents may allow transpiration to continue for

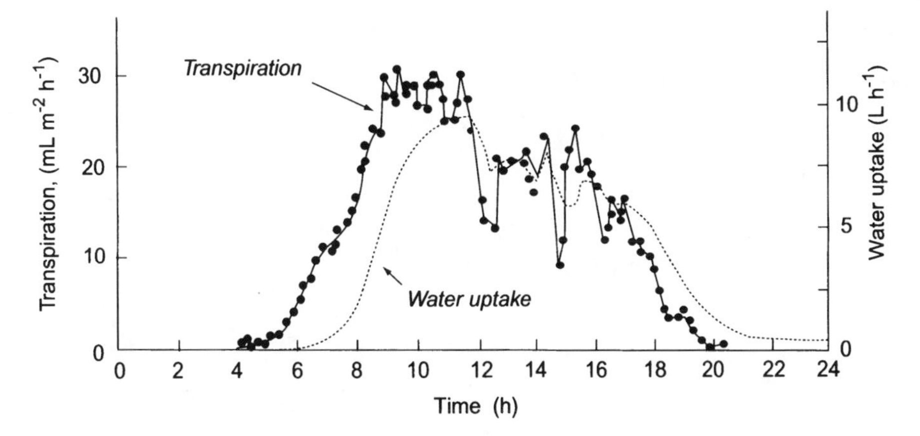

FIGURE 4.14. Diurnal time course of water uptake and water loss by Siberian larch. During morning, transpiration is supported by water loss from stems, creating a lower water potential in stems and roots, which generates the water potential gradient to absorb water from the soil. The water stored in stems is replenished at night. (Redrawn with permission from *BioScience*; Schulze et al. 1987.)

several weeks after water uptake from the soil has ceased.

Leaves

Water loss from leaves is controlled by the evaporative potential of the air, the water supply from the soil, and the regulation of water loss by leaves. Soil water supply and the evaporative potential of the air are the major environmental controls over water loss from leaves (Fig. 4.15). **Stomata** are pores in the leaf surface that can be opened or closed to regulate the entry of CO_2 into leaves and the loss of water from leaves. Stomata are the valves that determine the resistance to water movement between the soil and the air. The low hydraulic conductivity of dry soils minimizes the amount of water that can move directly from dry soil to the air by surface evaporation. The extensive root systems of plants and the high hydraulic conductivity of plant xylem make plants an effective conduit for moving water from the soil to the atmosphere. Plants adjust the size of stomatal openings to regulate the loss of water from leaves. Because stomatal conductance also determines the rate of CO_2 entry into leaves, there is an inevitable tradeoff between carbon gain and water loss by leaves (see Chapter 5).

Diurnal and climatic differences in air temperature and humidity determine the driving force for transpiration. Air inside the leaf is always saturated with water vapor because it is adjacent to moist cell surfaces. On a sunny day, air temperature rises to a maximum shortly after midday, allowing the air to hold more water. This rise in air temperature and the radiation absorbed by the leaf increases the temperature of the leaf and therefore the water vapor concentration inside the leaf. The water vapor concentration of the external air generally increases less than that inside the leaf. The resulting increase in the gradient in water vapor concentration between the inside and the outside of the leaf increases the transpirational water loss from the leaf. In evening the temperature decreases, causing a decline in the water vapor concentration inside the leaf and a decline in transpiration. Variations in weather or climate that cause an increase in air temperature and/or a decrease

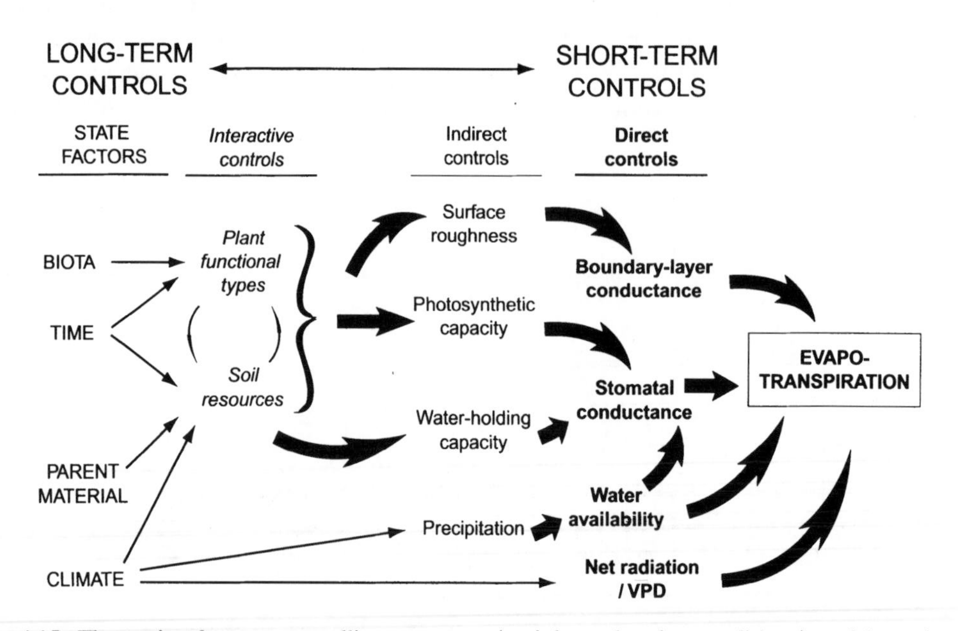

FIGURE 4.15. The major factors controlling evapotranspiration from a plant canopy. The major short-term controls over evapotranspiration include water availability, net radiation, and stomatal conductance (mainly under dry conditions) and boundary layer conductance (mainly under moist conditions), which in turn depend on climate, parent material, soil resources and disturbance.

in atmospheric moisture content also enhance the driving force for transpirational water loss. The evaporative potential of desert air is therefore extremely high because it is both hot and dry. Cloud forests generally have low evaporative potential because the air is saturated and because clouds reduce radiation input. Cold climates have low evaporative potential because cold air holds relatively little water vapor.

Stomatal conductance is the major control that plants exert over water loss from a leaf. Some plants reduce stomatal conductance when leaves are exposed to warm dry air that would otherwise cause high transpirational water loss. This response of stomatal conductance to dry air makes water loss from leaves less sensitive to dry atmospheric conditions than we might expect. Species differ considerably in their sensitivity of stomatal conductance to the evaporative potential of the air. Both the mechanism and the ecological patterns in the sensitivity of stomatal conductance to atmospheric humidity are poorly understood.

Stomatal conductance declines in response to drought because plants sense the soil moisture content of their root systems. Roots exposed to low soil moisture produce **abscisic acid** (ABA), a hormone that is transported to leaves and causes a reduction in stomatal conductance. Plants from mesic habitats are particularly sensitive to low soil water potential, closing stomates in response to soil drying before they experience large changes in plant water potential. Plants from dry environments show less response of stomatal conductance to soil drying and therefore continue to absorb and lose water, as the soil dries. Drought-adapted plants in arid ecosystems therefore maintain greater physiological activity in dry soils than do plants adapted to moist habitats; and, in the process, they transfer more water to the atmosphere under dry conditions.

There are important differences among species in stomatal conductance. Stomatal conductance under favorable conditions is highest in rapidly growing plants adapted to moist fertile soils (Körner et al. 1979, Schulze et al. 1994) (see Chapter 5).

Water Losses from Ecosystems

Water inputs are the major determinant of water outputs from ecosystems. The water loss from ecosystems equals the inputs in precipitation (P) adjusted for any changes in water storage (ΔS). The major avenues of loss are evapotranspiration (E) and runoff (R).

$$P \pm \Delta S = E + R \qquad (4.8)$$

Just as in the case of carbon and energy, the changes in water storage are generally small relative to inputs and outputs, when averaged over long time periods.

$$P \approx E + R \qquad (4.9)$$

where ΔS is small (i.e., input $\approx$ output).

Consequently, the quantity of water entering the ecosystem largely determines water output, just as gross primary production (GPP; carbon input) is the major determinant of ecosystem respiration (carbon output) (see Chapters 6 and 7). The route by which water leaves an ecosystem depends on the partitioning between evapotranspiration and runoff. This partitioning has a critical effect on regional hydrologic cycles because water that returns to the atmosphere is available to support precipitation in the same or other ecosystems. Runoff supplies the water input to aquatic ecosystems and provides most of the water used by humans. In a sense, runoff is the "leftovers" of water that entered in precipitation and was not transferred to the atmosphere by evapotranspiration. In summary, controls over evapotranspiration largely determine the partitioning between evapotranspiration and runoff.

Evaporation from Wet Canopies

Evaporation of water intercepted by the canopy is most important in ecosystems with a high surface roughness. Forests have high rates of evaporation from wet canopies, primarily because the effective mixing that occurs in rough forest canopies promotes rapid evaporation from each leaf (Kelliher and Jackson 2001). The large water-storage capacity of forest canopies is less important in explaining the quantity of water evaporated from wet canopies. The evaporation rate from a wet canopy depends primarily on the climatic conditions that drive evaporation (primarily vapor pressure deficit) and the degree to which environmental conditions in the canopy are coupled by turbulence to conditions in the atmosphere. Turbulence, in turn, is greatest in ecosystems with a tall aerodynamically rough canopy. In forests, which are tightly coupled to atmospheric conditions, wet canopy evaporation is largely independent of net radiation and is similar during the day and night. In grasslands, which are less tightly coupled to the atmosphere, wet canopy evaporation depends on net radiation as well as vapor pressure deficit and is greater during the day than at night. Due to differences in canopy roughness, forests have greater wet canopy evaporation than do shrublands or grasslands, and conifer forests evaporate more water from wet canopies than do deciduous forests.

Climate is the other factor that governs evaporation from wet canopies. Climate determines the frequency with which the canopy intercepts precipitation or dew and the conditions that drive evaporation. Ecosystems in wet climates generally have greater canopy evaporation because of the more frequent capture of rainfall by the canopy, even though the low vapor pressure deficit of wet climates causes this evaporation to occur slowly. The frequency of rainfall and dew formation is probably at least as important as total precipitation in governing the annual flux of wet canopy evaporation. Canopy evaporation increases exponentially with temperature because of the temperature effects on vapor pressure deficit (McNaughton 1976), so ecosystems generally lose more intercepted water through canopy evaporation in warm than in cold climates. Despite these generalizations, the interactions among multiple controls over wet canopy evaporation are so complex that they are best addressed through physically based models that consider all these factors simultaneously (Monteith and Unsworth 1990, Waring and Running 1998).

Canopies that intercept precipitation as snow or ice frequently store twice as much water equivalent as when precipitation is received in liquid form. Snow interception and subsequent **sublimation** (vaporization of a solid) is greatest in ecosystems with a high leaf area index (LAI), the quantity of leaf area per unit ground area. Most snow usually falls to the ground, where low net radiation and low wind speeds minimize sublimation. In tundra, however, where there is no canopy in winter to shade the snow and in continental boreal forests with low precipitation and low wind speeds, sublimation can account for 30% and 50%, respectively, of winter precipitation (Liston and Sturm 1998, Pomeroy et al. 1999, Sturm et al. 2001).

Evapotranspiration from Dry Canopies

Soil moisture directly limits evapotranspiration rate in dry soils. Plant water potential and transpiration rate are surprisingly insensitive to water availability until plants have depleted about 75% of the plant-available soil water (Fig. 4.16). Evapotranspiration is therefore relatively insensitive to precipitation in moist environments (Fig. 4.17). As soils dry, however, their hydraulic conductivity declines. This creates a relatively abrupt threshold of soil moisture, below which the rate of water supply to roots declines and plants experience water stress (lower water potential) (Fig. 4.16). Under these circumstances, stomatal conductance declines below its physiological maximum, causing a decline in evapotranspiration, just as described earlier for individual leaves. Evapotranspiration rates are therefore generally low in deserts, even though climatic conditions could support a high evapotranspiration rate, if moisture were available.

When soil moisture is adequate, vegetation structure and climate govern evapotranspiration rate. There are two conductances, boundary layer conductance and surface conductance, that govern ecosystem effects on evapotranspiration. The **boundary layer conductance** (also termed aerodynamic conductance) is a measure of the *physical* controls over water vapor transfer from the ecosystem to the atmosphere. It depends on wind speed and the size and number of roughness elements, such as trees. Boundary layer conductance is greatest when surface turbulence mixes large quantities of air from the bulk atmosphere with air inside the

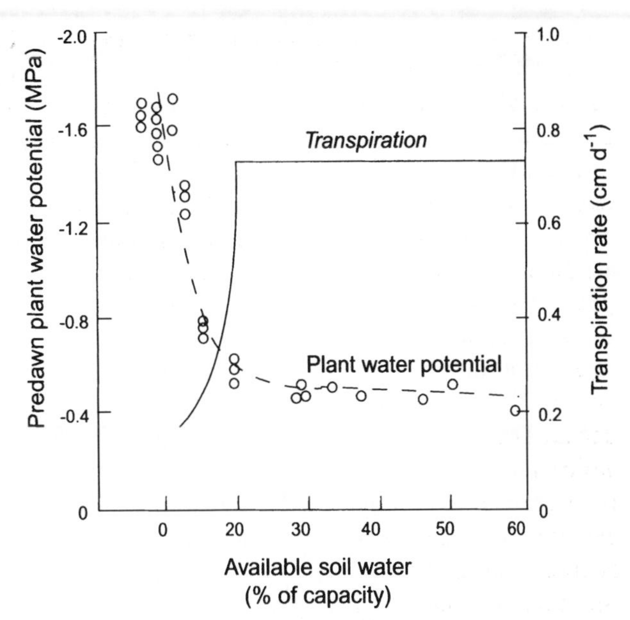

FIGURE 4.16. Response of plant water potential and transpiration to soil moisture (Sucoff 1972, Gardner 1983, Waring and Running 1998). Soil moisture has little effect on plant water potential or transpiration until about 75% of the available water has been removed from the rooting zone.

canopy. Turbulent mixing couples the evaporation at the leaf or soil surfaces with the moisture content of the bulk air of the atmosphere. Boundary layer conductance is therefore higher in ecosystems such as forests with tall, aerodynamically rough canopies than in grasslands or crops. Forests therefore transport water more effectively to the atmosphere and reduce soil moisture more rapidly than do grasslands.

Surface conductance is a measure of the potential of leaf and soil surfaces in the ecosystem to lose water. *Under moist conditions,* surface conductance is surprisingly insensitive to vegetation properties (Kelliher et al. 1995). In sparse vegetation, evaporation from the soil surface is the major avenue of water loss. As LAI increases, transpiration increases (more leaf area to transpire); this is counteracted by a decrease in soil evaporation (more shading and less turbulent exchange at the soil surface). Consequently, surface conductance is relatively insensitive to LAI.

Vegetation affects surface conductance primarily through its effects on stomatal conductance (Kelliher et al. 1995). Even this influence,

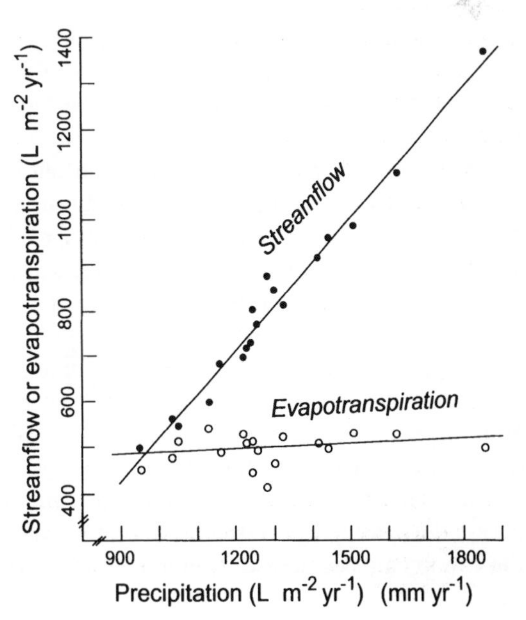

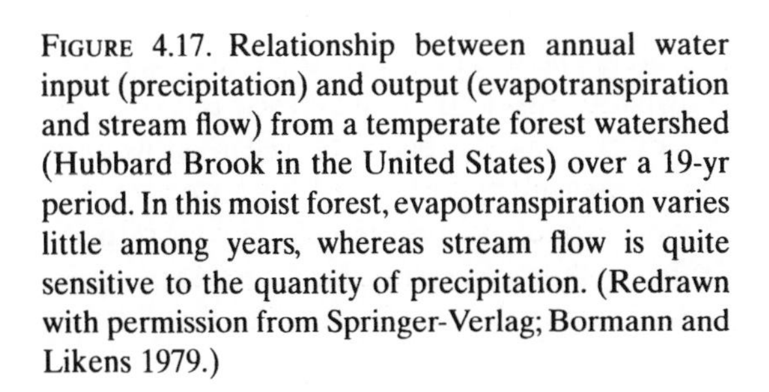
FIGURE 4.17. Relationship between annual water input (precipitation) and output (evapotranspiration and stream flow) from a temperate forest watershed (Hubbard Brook in the United States) over a 19-yr period. In this moist forest, evapotranspiration varies little among years, whereas stream flow is quite sensitive to the quantity of precipitation. (Redrawn with permission from Springer-Verlag; Bormann and Likens 1979.)

however, is often relatively small. Maximum stomatal conductance of individual leaves is relatively similar among natural ecosystems (Körner 1994, Kelliher et al. 1995). Woody and herbaceous ecosystems, for example, have similar stomatal conductance of individual leaves (Körner 1994) and similar surface conductance of entire ecosystems (Kelliher et al. 1995). Crops, however, which have about 50% higher stomatal conductance than does natural vegetation, also have about 50% higher surface conductance (Schulze et al. 1994, Kelliher et al. 1995). There are currently insufficient data to know whether ecological variation in stomatal conductance associated with gradients in soil fertility causes similar variation in surface conductance and therefore evapotranspiration from ecosystems.

In summary, aerodynamic roughness, which depends on plant height and the number of roughness elements, is the main way in which vegetation influences evapotranspiration from dry canopies under conditions of adequate water supply. Stomatal conductance exerts additional control in some ecosystems. It also has an increasingly important control over evapotranspiration as soil moisture declines. In other words, stomatal conductance accounts for temporal variation in evapotranspiration in response to variation in soil moisture, but surface roughness is the major factor explaining differences in evapotranspiration among moist ecosystems.

Vegetation structure also influences the relative importance of different climatic variables in regulating evapotranspiration. In aerodynamically rough, well-mixed canopies such as open-canopied forests, the moisture content of the air within the canopy is similar to that of the bulk air above the canopy. Under these well-coupled conditions, evapotranspiration is determined more by the moisture content of the air (and the accompanying stomatal response) than by net radiation (Waring and Running 1998). In short canopies, by contrast, the air adjacent to the leaves is mixed less readily with the bulk air above the canopy, allowing evapotranspiration to increase the moisture content within the canopy environment. In other words, the canopy air becomes decoupled from conditions in the bulk atmosphere. In these smooth canopies, evapotranspiration is determined more by net radiation than by the moisture content of the bulk air, just as when canopies are wet. The **decoupling coefficient** is a measure of the degree to which a canopy is decoupled from the atmosphere (Table 4.5) (Jarvis and McNaughton 1986). It is determined primarily by canopy height. In summary, net radiation is the dominant environmental control over evapotranspiration in short canopies, whereas the vapor pressure deficit is the dominant control in tall canopies, when water is freely available (Waring and Running 1998).

TABLE 4.5. Decoupling coefficient of vegetation canopies in the field under conditions of adequate moisture supply.

Vegetation	Decoupling coefficient[a]
Alfalfa	0.9
Strawberry patch	0.85
Permanent pasture	0.8
Grassland	0.8
Tomato field	0.7
Wheat field	0.6
Prairie	0.5
Cotton	0.4
Heathland	0.3
Citrus orchard	0.3
Forest	0.2
Pine woods	0.1

[a] A completely smooth surface has a decoupling coefficient of 1.0, and a canopy in which the air is identical to that in the atmosphere has a decoupling coefficient of zero.
Data from Jarvis and McNaughton (1986) and Jones (1992).

Changes in Storage

Water inputs that exceed outputs replenish water that is stored in soil and groundwater. Water that enters the soil is retained until the soil reaches field capacity. Additional water moves downward to groundwater. In cold climates in winter, most of the precipitation input is stored above ground in the snowpack. The snowpack substantially increases the quantity of water that an ecosystem can store and the residence time of water in the ecosystem. Stored water supports evapotranspiration at times when evapotranspiration exceeds precipitation; the declines in soil moisture during periods of dry weather draw down water storage. The seasonal recharge and depletion of stored water are important controls over evapotranspiration and net primary production (NPP) in many ecosystems.

Groundwater—the water beneath the rooting zone—is a large pool that is inaccessible to plants in many ecosystems. The size of this pool depends on the depth to impermeable layers and the porosity of materials in this layer. Porosity governs the pore volume available to hold water and the resistance to lateral drainage of water. The groundwater pool has a relatively constant size, so when new water enters groundwater from the top, it displaces older water that drains laterally to streams, rivers and oceans. The time lag between inputs to groundwater and outputs can be substantial (months to millennia) because of the large size of this pool. Groundwater therefore generally has an isotopic composition quite different from that of precipitation or soil water (Fig. 4.10).

People modify groundwater pools by changing the vegetation and associated rooting depth and by tapping groundwater to support human activities. Introduction of deep-rooted exotic species in arid regions increases the pool of water available to support evapotranspiration by vegetation. This can cause the water table to drop. The introduction of deep-rooted *Tamarix* in North American deserts, for example, caused the water table to drop enough that desert ponds have dried, endangering endemic fish species (Berry 1970).

Removal of vegetation causes the water table to rise because surface water is no longer tapped to support evapotranspiration. The clearing of heathlands for agriculture in western Australia, for example, reduced the depth of the rooting zone, causing saline groundwater to rise close to the surface. This reduced the productive potential of the crops, further reducing evapotranspiration and the depth to groundwater. Finally, evaporation from the soil surface increased soil salinity to the point that soils no longer supported crop growth in many areas, nor could they be recolonized by native heath vegetation (Nulsen et al. 1986). In this way human modification of vegetation permanently altered the hydrologic cycle and all aspects of ecosystem structure and functioning.

Expansion of human populations into arid regions is frequently subsidized by tapping groundwater that would otherwise be unavailable to surface organisms. Wells that provide water to animals in African savannas, for example, attract high densities of animals, which overgraze and alter the composition of nearby vegetation. Often 80 to 90% of the water in populated arid areas is used to support irrigated agriculture. Irrigated agriculture is highly productive because warm temperatures and high solar radiation support a high productivity, when the natural constraints of water limitation are removed. The substantial cost of irrigated agriculture often requires that it be intensively managed with fertilizers and pesticides (see Chapter 16). These irrigated lands are important sources of fruits, vegetables, cotton, rice, and other high-value crops. Conversion of arid regions to irrigated agriculture, however, reduces the amount of water available for runoff. Human use of water in the arid southwestern United States, for example, converted the Rio Grande River from a major river to a small stream with intermittent flow during some times of year. Irrigation also increases soil evaporation, which increases soil salinity in a fashion similar to that described for western Australia.

In cases where evapotranspiration of irrigated agriculture exceeds precipitation, there is not only a decrease in runoff but also a depletion of the groundwater pool. The Ogallala aquifer in the north-central United States, for example, accumulated water when the climate was much wetter than today. Tapping of this "fossil" water has increased the depth to the water table substantially. Continued drawdown of this aquifer cannot be sustained indefinitely, because current water sources cannot replenish it as rapidly as it is being depleted to support irrigation.

Runoff

Runoff is the difference between precipitation inputs, changes in storage, and losses to evapo-

transpiration (Eq. 4.8). Average runoff from an ecosystem depends primarily on precipitation and evapotranspiration because long-term changes in storage are usually negligible. Runoff responds to variation in precipitation much more strongly than does evapotranspiration (Fig. 4.17) because it constitutes the leftovers after the water demands for evapotranspiration and groundwater recharge have been met. Runoff is therefore greater in wet than in dry climates or seasons. Over hours to weeks, runoff generally increases after rainfall events and decreases during dry periods. Changes in water storage buffer this linkage between precipitation and runoff. The recharge of soil moisture in grasslands, shrublands, and dry forests, for example, may prevent large increases in flow from occurring after rainfall when soils are dry. In ecosystems with a small capacity to store water such as deserts with coarse-textured soils and a calcic layer or ecosystems underlain by permafrost, runoff responds almost immediately to precipitation, and rainstorms can cause flash floods. Conversely, slowly draining groundwater provides a continued source of water to streams (**base flow**) even at times when there is no precipitation.

In ecosystems that develop a snowpack in winter, precipitation inputs are stored in the ecosystem during winter, causing winter stream flows to decline, regardless of the seasonality of precipitation. Much of the water stored in the snowpack can move directly to streams, when the snow melts, causing large spring runoff events. Glacial rivers, for example, have greatest runoff in midsummer, when temperatures are highest, whereas nonglacial rivers in the same climate zone have peak flow in early spring following spring snow melt (Fig. 4.18).

Flow in streams and rivers integrates the precipitation, evapotranspiration, and changes in storage throughout the watershed. In large rivers, the seasonal variations in flow often reflect patterns of precipitation and evapotranspiration that occur upstream, hours to weeks previously. These integrative effects of runoff from large watersheds make this an effective indicator of temporal changes in the hydrologic cycle.

Seasonal variations in stream flow are a major determinant of the structure and seasonality of ecosystem processes in streams and rivers. Periods of high flow in small streams, for example, scour stream channels, removing

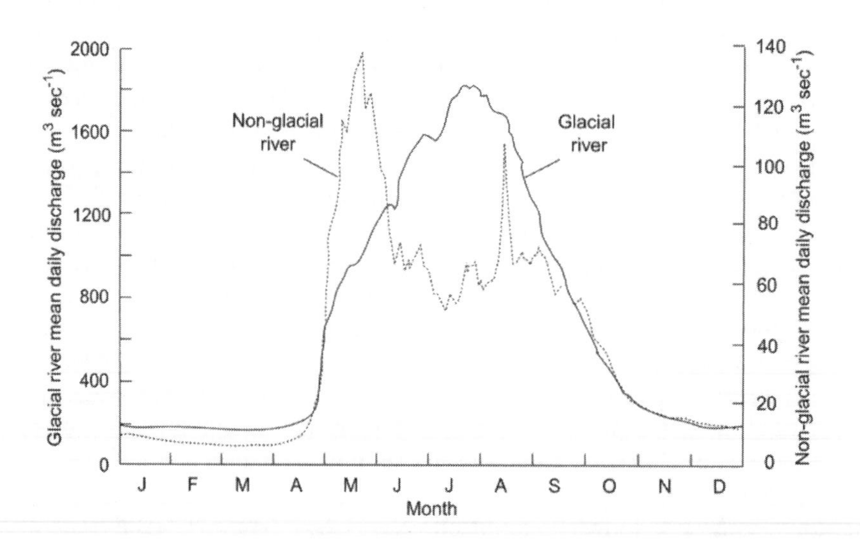

FIGURE 4.18. Average daily discharge from a glacial and a clearwater river. Runoff from the clearwater river peaks at snow melt, whereas the glacial river has peak discharge when temperatures are warmest in midsummer. (Redrawn with permission from Phyllis Adams; Adams 1999.)

or redistributing sediments, algae, and detritus (Power 1992a). In larger rivers, high flow events may lead to predictable patterns of bank erosion and deposition. Dams that reduce the intensity of high-flow events dramatically alter the natural disturbance regime and functioning of freshwater ecosystems (see Chapter 14).

Vegetation strongly influences the quantity of runoff. Because evapotranspiration is such a large component of the hydrologic budget of an ecosystem, any vegetation change that alters evapotranspiration inevitably affects runoff. Deforestation of a watershed, for example, can double runoff (see Fig. 13.13). As vegetation regrows during succession, runoff returns to preharvest levels. Regional changes in land cover can have long-term effects on regional hydrology. Watersheds that lose forest cover exhibit increased runoff, whereas those that gain forest cover through reforestation show less runoff (Trimble et al. 1987) (Fig. 4.19). More subtle vegetation changes also alter runoff. Conifer forests produce less runoff than deciduous forests because of their greater leaf area their higher rates and longer season for evapotranspiration (Swank and Douglass 1974).

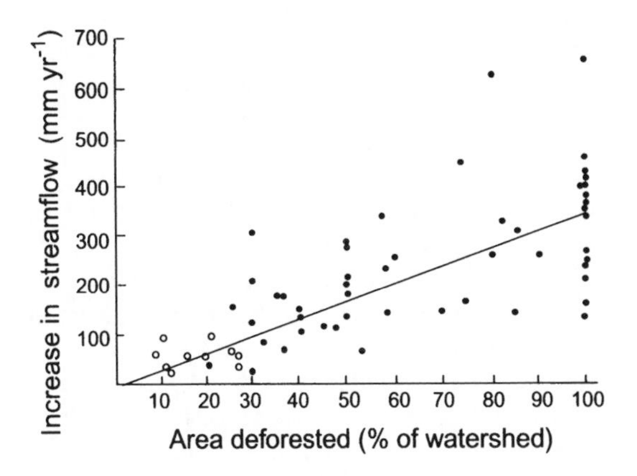

FIGURE 4.19. Influence of deforestation on changes in stream flow in the southeastern United States. Stream flow increases linearly with the proportion of the watershed that is deforested. Also included in this dataset are watersheds that show reduced stream flow in response to increases in forest cover. (Redrawn with permission from *Water Resource Research*; Trimble et al. 1987.)

Summary

The energy and water budgets of ecosystems are inextricably linked because net radiation is an important driving force for evapotranspiration, and evapotranspiration is a large component of both water and energy flux from ecosystems. Net radiation is the balance between incoming and outgoing shortwave and longwave radiation. Ecosystems affect net radiation primarily through albedo (shortwave reflectance), which depends on the reflectance of individual leaves and other surfaces and on canopy roughness, which depends primarily on canopy height and complexity. Most absorbed energy is released to the atmosphere as latent heat flux (evapotranspiration) and sensible heat flux. Latent heat flux cools the surface and transfers water vapor to the atmosphere, whereas sensible heat flux warms the surface air. The Bowen ratio, the ratio of sensible to latent heat flux, determines the strength of the coupling of the water cycle to the energy budget.

Water enters terrestrial ecosystems primarily as precipitation and leaves as evapotranspiration and runoff. Water moves through ecosystems in response to gradients in water potential, which is determined by pressure potential, osmotic potential, and matric potential. Water enters the ecosystem and moves down through the soil in response to gravity. Available water in the soil moves along a film of liquid water through the soil-plant-atmosphere continuum in response to a gradient in water potential that is driven by transpiration (evaporation from the cell surfaces inside leaves). Evapotranspiration from canopies depends on the driving forces for evaporation (net radiation and vapor pressure deficit of the air) and two conductance terms, boundary layer and surface conductances. Boundary layer conductance depends on the degree to which the canopy is coupled to the atmosphere, which varies with canopy height and aerodynamic roughness. Surface conductance is mainly influenced by the average stomatal conductance of leaves in the canopy. Stomatal and surface conductances are relatively similar among natural ecosystems but are somewhat higher in crop systems. Climate influences evapotranspiration by determining the driving forces for evapotranspiration and the water availability in the soil, which determines stomatal conductance. Vegetation influences evapotranspiration through plant height and aerodynamic roughness, which govern boundary layer conductance, and through stomatal conductance, which influences surface conductance and the plant response to soil moisture.

The partitioning of water loss between evapotranspiration and runoff depends primarily on water storage in the rooting zone and the rate of evapotranspiration. Runoff is the leftover water that drains from the ecosystem at times when precipitation exceeds evapotranspiration plus increases in water storage. Human activities alter the hydrologic cycle primarily through changes in land cover and use, which affect evapotranspiration and soil water storage.

Review Questions

1. What climatic and ecosystem properties govern energy input to an ecosystem?
2. What are the major avenues by which energy absorbed by an ecosystem is exchanged with the atmosphere? What determines the total energy exchange? What determines the relative importance of the pathways by which energy is exchanged?
3. What are the consequences of transpiration for ecosystem energy exchange and for the linkage between energy and water budgets of an ecosystem?
4. How might global changes in climate and land use alter the components of energy exchange in an ecosystem?
5. What are the major pathways of water movement in an ecosystem? What determines the balance among these pathways, for example, between evaporation, transpiration, and runoff? How do climate, soils, and vegetation influence the pools and fluxes of water in an ecosystem?
6. What are the mechanisms driving water uptake and loss from plants? How do plant properties influence water uptake and loss?
7. How do the controls over water loss from plant canopies differ from the controls at the level of individual leaves?
8. Describe how grassland and forests differ in properties that influence wet canopy evaporation, transpiration, soil evaporation, infiltration, and runoff. What will be the consequences for runoff and for regional climate of a policy that encourages the replacement of grasslands with forests so as to increase terrestrial carbon storage?

Additional Reading

Campbell, G.S., and J.M. Norman. 1998. *An Introduction to Environmental Biophysics*. Springer-Verlag, New York.

Dawson, T.E. 1993. Water sources of plants as determined from xylem-water isotopic composition: Perspectives on plant competition, distribution, and water relations. Pages 465–496 *in* J.R. Ehleringer, A.E. Hall, and G.D. Farquhar, editors. *Stable Isotopes and Plant Carbon-Water Relations*. Academic Press, San Diego, CA.

Jarvis, P.G., and K.G. McNaughton. 1986. Stomatal control of transpiration: Scaling up from leaf to region. *Advances in Ecological Research* 15:1–49.

Jones, H.G. 1992. *Plants and Microclimate: A Quantitative Approach to Environmental Plant Physiology*. Cambridge University Press, Cambridge, UK.

Kelliher, F.M., R. Leuning, M.R. Raupach, and E.-D. Schulze. 1995. Maximum conductances for evaporation from global vegetation types. *Agricultural and Forest Meteorology* 73:1–16.

Monteith, J.L., and M. Unsworth. 1990. *Principles of Environmental Physics*. 2nd ed. Arnold, London.

Oke, T.R. 1987. *Boundary Layer Climates*. 2nd ed. Methuen, London.

Schulze, E.-D., F.M. Kelliher, C. Körner, J. Lloyd, and R. Leuning. 1994. Relationship among maximum stomatal conductance, ecosystem surface conductance, carbon assimilation rate, and plant nitrogen nutrition: A global ecology scaling exercise. *Annual Review of Ecology and Systematics* 25:629–660.

Sperry, J.S. 1995. Limitations on stem water transport and their consequences. Pages 105–124 *in* B.L. Gartner, editor. *Plant Stems: Physiology and Functional Morphology*. Academic Press, San Diego, CA.

Waring, R.H., and S.W. Running. 1998. *Forest Ecosystems: Analysis at Multiple Scales*. Academic Press, New York.

5
Carbon Input to Terrestrial Ecosystems

Photosynthesis by plants provides the carbon and energy that drive most biological processes in ecosystems. This chapter describes the controls over this carbon input.

Introduction

The energy fixed by photosynthesis directly supports plant growth and produces organic matter that is consumed by animals and soil microbes. The carbon derived from photosynthesis makes up almost half of the organic matter on Earth; hydrogen and oxygen account for most of the remainder. Human activities have radically modified the rate at which carbon enters the terrestrial biosphere by changing most of the controls over this process. We have increased by 30% the quantity of atmospheric CO_2 to which all terrestrial plants are exposed. On a regional scale we have altered the availability of water and nutrients, the major resources that determine the capacity of plants to use atmospheric CO_2. Finally, through changes in land cover and the introduction and extinction of species, we have changed the regional distribution of the carbon-fixing potential of the terrestrial biosphere. Because of the central role that carbon plays in the climate system (see Chapter 2) and the biosphere, it is critical that we understand the factors that regulate its cycling through vegetation and ecosystems. We address carbon inputs to terrestrial ecosystems through photosynthesis in this chapter and inputs to aquatic ecosystems in Chapter 10. In Chapters 6 and 7, we explore the carbon losses from plants and ecosystems, which, together with photosynthesis, govern the patterns of accumulation and loss of carbon in ecosystems.

Overview

The availability of water and nutrients is the major factor governing carbon input to ecosystems. Photosynthesis is the process by which most carbon and chemical energy enter ecosystems. The proximate controls over photosynthesis by a single leaf are the availability of reactants such as light energy and CO_2; temperature, which governs reaction rates; and the availability of nitrogen, which is required to produce photosynthetic enzymes (Fig. 5.1). Photosynthesis at the scale of ecosystems is termed **gross primary production** (GPP). Like photosynthesis by individual leaves, GPP varies diurnally and seasonally in response to changes in light, temperature, and nitrogen supply. Differences among ecosystems in annual GPP, however, are determined primarily by the quantity of leaf area and the length of time that this leaf area is photosynthetically active. Leaf area and photosynthetic season, in turn, depend on the availability of soil resources (water and nutrients), climate, and time since disturbance.

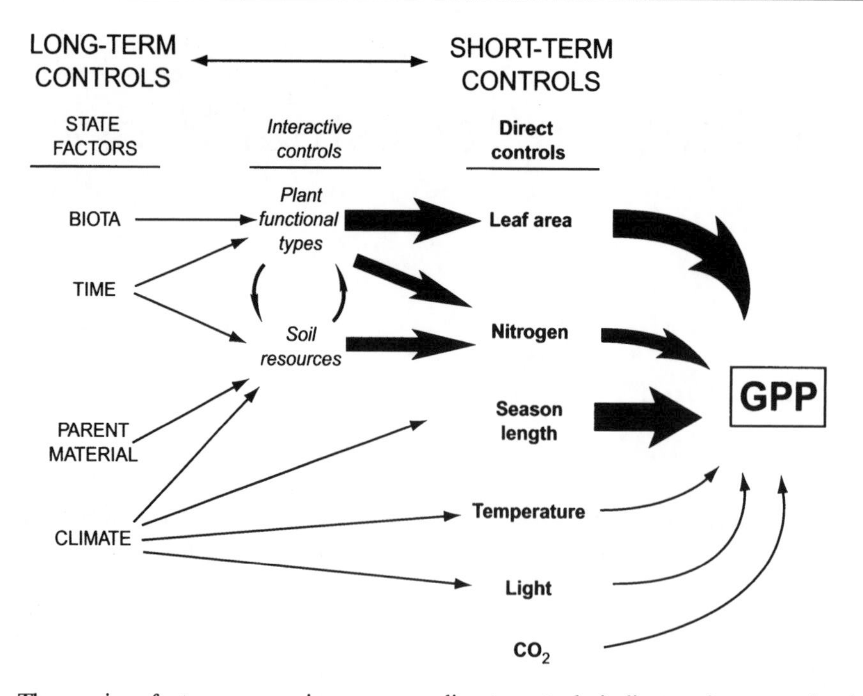

FIGURE 5.1. The major factors governing gross primary production (GPP) in ecosystems. These controls range from the direct controls, which determine the diurnal and seasonal variations in GPP, to the interactive controls and state factors, which are the ultimate causes of ecosystem differences in GPP. Thickness of the arrows associated with direct controls indicates the strength of the effect. The factors that account for most of the variation in GPP among ecosystems are leaf area and length of the photosynthetic season, which are ultimately determined by the interacting effects of soil resources, climate, vegetation, and disturbance regime.

In this chapter we explore the mechanisms behind these causal relationships.

Carbon and energy are linked as they move through ecosystems because the same processes govern their entry into ecosystems in photosynthesis, transfer within ecosystems, and loss from ecosystems. Photosynthesis uses light energy (i.e., radiation in the visible portion of the spectrum) to reduce CO_2 and produce carbon-containing organic compounds. This organic carbon and its associated energy are then transferred among components within the ecosystem and are eventually released to the atmosphere by respiration or combustion.

The energy content of organic matter differs among carbon compounds, but for whole tissues, it is relatively constant at about $20\,kJ\,g^{-1}$ of ash-free dry mass (Golley 1961, Larcher 1995). The carbon concentration of organic matter is also variable but averages about 45% in herbaceous tissues and 50% in wood (Gower et al. 1999). Both the carbon and energy contents of organic matter are greatest in materials such as seeds and animal fat that have high lipid content and are lowest in tissues with high concentrations of minerals or organic acids (Fig. 5.2). Because of the relative constancy of the carbon and energy contents of organic matter, carbon, energy, and biomass have been used interchangeably as currencies of the carbon and energy dynamics of ecosystems. The preferred units differ among fields of ecology, depending on the processes that are of greatest interest or are measured most directly. Production studies, for example, typically focus on biomass; trophic studies, on energy; and gas exchange studies, on carbon.

Photosynthetic Pathways

C_3 Photosynthesis

The rate of carbon input to ecosystems depends on the response of photosynthetic biochemistry

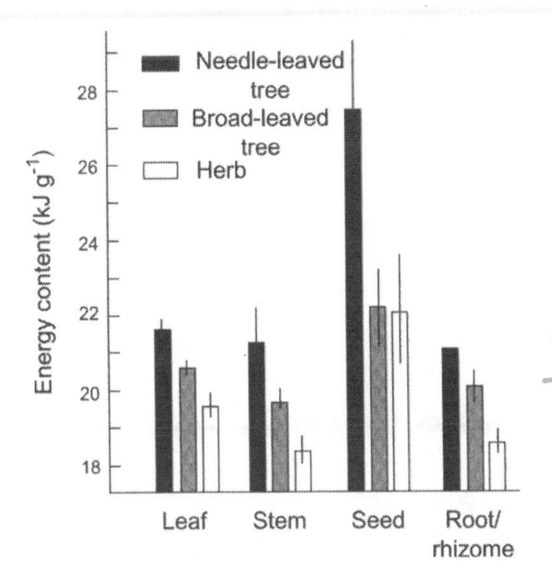

FIGURE 5.2. Energy content of major tissues in conifer trees, dicotyledonous trees, and dicotyledonous herbs. Compounds that contribute to a high energy content include lipids (seeds), terpenes and resins (conifers), proteins (leaves), and lignin (woody tissues). Values are expressed per gram of ash-free dry mass. (Redrawn with permission from Springer-Verlag; Larcher 1995.)

to environment. A brief overview of the biochemistry of photosynthesis provides a mechanistic basis for understanding the environmental controls over carbon input to ecosystems.

There are two major groups of reactions in photosynthesis. The **light-harvesting reactions** transform light energy into a temporary form of chemical energy. The **carbon-fixation reactions** use the products of the light-harvesting reactions to convert CO_2 into sugars, a more permanent form of chemical energy that can be stored, transported, or metabolized. In light, both groups of reactions occur simultaneously in the **chloroplasts**, which are organelles inside the **mesophyll cells** (photosynthetic cells) of green leaves (Fig. 5.3). In the light-harvesting reactions, **chlorophyll** (a light-absorbing pigment) captures energy from visible light. Absorbed radiation is converted to chemical energy (NADPH and ATP), and oxygen is produced as a waste product (Fig. 5.3). Visible radiation accounts for only 40% of incoming solar radiation (see Chapter 2), which places an upper limit on the potential efficiency of photosynthesis in converting solar radiation into chemical energy.

The carbon-fixation reactions of photosynthesis use the chemical energy (ATP and NADPH) from the light-harvesting reactions to reduce CO_2 to sugars. The rate-limiting step in the carbon-fixation reactions is the reaction of a five-carbon sugar (ribulose-bisphosphate; RuBP) with CO_2 to form two three-carbon sugars. Because the initial products of photosynthesis are three-carbon sugars, this photosynthetic pathway is known as **C_3 photosynthesis**. The initial attachment of CO_2 to a carbon skeleton is catalyzed by the enzyme ribulose-bisphosphate carboxylase-oxygenase (**Rubisco**). The rate of this reaction is generally limited by the products of the light reaction and by the concentration of CO_2 in the chloroplast. A surprisingly high concentration of Rubisco is required for carbon fixation. Rubisco accounts for about 25% of leaf nitrogen, and other photosynthetic enzymes make up an additional 25%. The remaining enzymatic steps in the carbon-fixation reactions use ATP and NADPH from the light-harvesting reactions to regenerate RuBP as a carbon acceptor to sustain further photosynthesis (Fig. 5.3). The most notable features of the carbon-fixation reactions are (1) their large nitrogen requirement for Rubisco and other photosynthetic enzymes; (2) their dependence on the products of the light-harvesting reactions (ATP and NADPH), which in turn depend on **irradiance** (the light received by the leaf); and (3) their frequent limitation by CO_2 supply to the chloroplast. The basic biochemistry of photosynthesis therefore dictates that this process must be sensitive to light and CO_2 supplies over time scales of milliseconds to minutes and sensitive to nitrogen supply over time scales of days to weeks (Fig. 5.1).

Rubisco is both a **carboxylase**, which initiates the carbon-fixation reactions of photosynthesis, and an **oxygenase**, which catalyzes the reaction between RuBP and oxygen (Fig. 5.3). The oxygenase activity initiates a series of steps that breaks down sugars to CO_2. This process of **photorespiration** immediately respires away 20

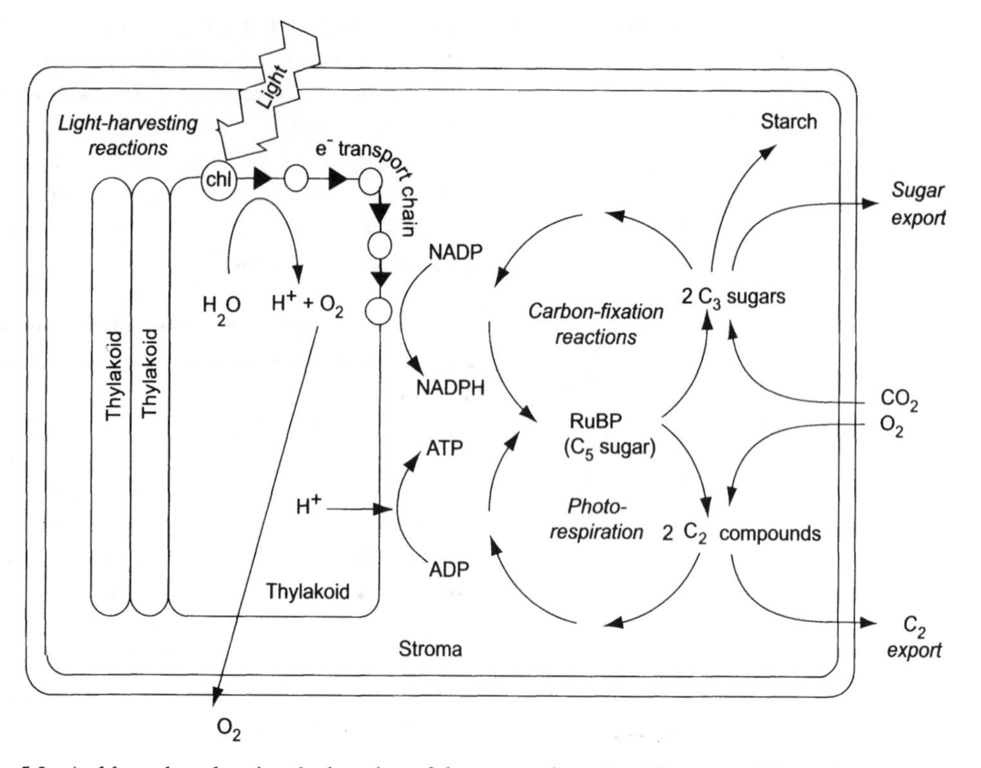

FIGURE 5.3. A chloroplast showing the location of the major photosynthetic reactions. The light-harvesting reactions occur in the **thylakoid** membranes; chlorophyll (chl) absorbs visible light and funnels it to reaction centers (Photosystems I and II). In Photosystem II, water is split to H^+ and O_2, and the resulting electrons are then passed down an electron-transport chain inside the thylakoid, ultimately to NADP, producing NADPH. During this process, protons move across the thylakoid membrane to the stroma, and the proton gradient drives the synthesis of ATP. ATP and NADPH provide the energy to regenerate ribulose-bisphosphate (RuBP) within the carbon fixation reactions. RuBP reacts either with CO_2 to produce sugars and starch (carbon-fixation reactions of photosynthesis) or with O_2 to produce two-carbon intermediates (photorespiration). These two-carbon intermediates are exported from the chloroplast to mitochondria or peroxisomes, where they are again converted to sugars, with loss of CO_2 and ATP. Through either carbon fixation or photorespiration, ADP and NADP again become reactants available to produce additional ATP and NADPH. The net effect of photosynthesis is to convert light energy into chemical energy (sugars and starches) that is available to support plant growth and maintenance.

to 40% of the carbon fixed by C_3 photosynthesis and regenerates ADP and NADP in the process. Why do C_3 plants have such an inefficient system of carbon acquisition, by which they immediately lose a third of the carbon that they acquire from photosynthesis? Although we have no definite answer to this question, the most likely explanation is that photorespiration acts as a safety valve. It provides a supply of reactants (ADP and NADP) to the light reaction under circumstances in which an inadequate supply of CO_2 limits the rate at which these reactants can be regenerated by carbon-fixation reactions. In the absence of photorespiration, continued light harvesting produces oxygen radicals that destroy photosynthetic pigments.

Plants have additional lines of defense against excessive energy capture, which are at least as important as photorespiration. One such **photoprotection** mechanism involves pigments that change from one form to another in the **xanthophyll cycle**. When excess excitation energy is present and cannot be processed to generate ATP and NADPH, xanthophyll pigment is converted to a form that can receive excess absorbed energy from the excited chlorophyll (Demming-Adams and Adams

1996). The energy is then harmlessly dissipated as heat. The xanthophyll cycle processes much of the energy that is not used for carbon fixation under high light and serves as further protection against photodestruction of photosynthetic pigments under conditions of high light.

Net photosynthesis is the net rate of carbon gain measured at the level of individual leaves. It is the balance between simultaneous CO_2 fixation and leaf respiration in the light (both photorespiration and mitochondrial respiration). The overall efficiency of converting light energy into sugars is about 6% under optimal conditions at low light, but is closer to 1% under most field conditions.

The CO_2 used in photosynthesis diffuses along a concentration gradient from the atmosphere outside the leaf into the chloroplast. Carbon dioxide first diffuses across a layer of relatively still air close to the leaf surface (the plant **boundary layer**) and then through the **stomata** (small pores in the leaf surface), the diameter of which is regulated by the plant (Fig. 5.4). Once inside the leaf, CO_2 diffuses through air spaces between cells, dissolves in water on the cell surfaces, and diffuses the short distance from the cell surface to the chloroplast. Stomata are the largest (and most variable) component of the total resistance to CO_2 diffusion. The thin flat shape of most leaves and the abundance of air spaces inside leaves maximize the rate of CO_2 diffusion from the bulk air to the chloroplast.

Cell walls inside the leaf are covered with a thin film of water, which facilitates the efficient transfer of CO_2 from the air to the interior of cells. This water readily evaporates, and water vapor diffuses out through the stomata across the boundary layer to the atmosphere. This process is called **transpiration**. The open stomata that are necessary for plants to gain carbon are an avenue for water loss. In other words, plants face an inevitable trade-off between CO_2 uptake (which is necessary to drive photosynthesis) and water loss (which must be replaced by absorption of water from the soil). This trade-off can be as high as 400 moles of water lost for each mole of CO_2 absorbed. Plants regulate CO_2 uptake and water loss by changing the size of stomatal openings, which regulates **stomatal conductance**, the flux of water vapor or CO_2 per unit driving force (i.e., for a given concentration gradient). When plants reduce stomatal conductance to conserve water, photosynthesis

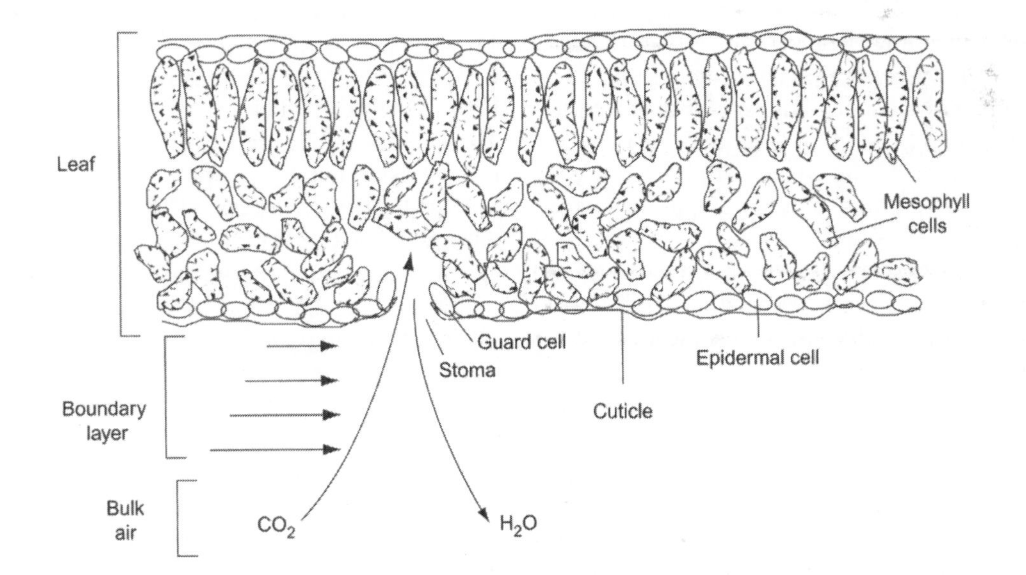

FIGURE 5.4. Cross-section of a leaf showing the diffusion pathways of CO_2 and H_2O into and out of the leaf, respectively. Length of the horizontal arrows outside the leaf is proportional to wind speeds in the boundary layer.

declines, reducing the efficiency with which plants convert light energy to carbohydrates. Plant regulation of CO_2 delivery to the chloroplast is therefore a compromise between maximizing photosynthesis and minimizing water loss and depends on the relative supplies of CO_2, light, and mineral nutrients.

C_4 Photosynthesis

C_4 photosynthesis adds a set of carbon-fixation reactions that enable some plants to increase photosynthetic water use efficiency in dry environments. About 85% of vascular plant species fix carbon by the **C_3 photosynthetic pathway**, in which Rubisco is the primary carboxylating enzyme. The first products of C_3 photosynthesis are three-carbon sugars. About 5% of the global flora photosynthesize by the **C_4 photosynthetic pathway**. C_4 species dominate many warm high-light environments, particularly tropical grasslands and savannas. C_4-dominated ecosystems account for nearly a third of the ice-free terrestrial surface (see Table 6.5) and are therefore quantitatively important in the global carbon cycle. In C_4 photosynthesis, phosphoenolpyruvate (PEP) is first carboxylated by **PEP carboxylase** in mesophyll cells to produce four-carbon organic acids (Fig. 5.5). These organic acids are transported to specialized **bundle sheath cells**, where they are decarboxylated. The CO_2 released from the organic acids then enters the normal C_3 pathway of photosynthesis to produce sugars that are exported from the leaf. There are three ecologically important features of the C_4 photosynthetic pathway.

First, C_4 acids move to the bundle sheath cells, where they are decarboxylated, concentrating CO_2 at the site where Rubisco fixes carbon. This increases the efficiency of carboxylation by Rubisco because it increases the concentration of CO_2 relative to O_2, which would otherwise compete for the active site of the enzyme. Apparent photorespiration measured at the leaf level is low in C_4 plants because most of the Rubisco reacts with CO_2 rather than with O_2 and because the PEP carboxylase in the mesophyll cells scavenges any photorespired CO_2 that escapes from the bundle sheath cells. The high efficiency of Rubisco in C_4 plants reduces the quantity of Rubisco (and therefore nitrogen) required for C_4 photosynthesis.

Second, PEP carboxylase is more effective than Rubisco in drawing down the concentration of CO_2 inside the leaf. This increases the CO_2 concentration gradient between the external air and the internal air spaces of the leaf. A C_4 plant can therefore absorb CO_2 with more tightly closed stomata than can a C_3 plant, thus reducing water loss.

Third, the net cost of regenerating the carbon acceptor molecule (PEP) of the C_4 pathway is two ATPs for each CO_2 fixed, a 30% increase in the energy requirement of photosynthesis compared to C_3 plants.

The major advantages of the C_4 photosynthetic pathway are that less water is lost and less nitrogen is required to maintain a given rate of photosynthesis compared to C_3 plants. C_4 plants therefore often have a high CO_2 fixation rate under high-light, low-nitrogen conditions. Moreover, due to their lack of photorespiration, which is quite temperature sensitive, C_4 plants can maintain higher rates of net photosynthesis at high temperatures than can C_3 plants; this explains their success in warm environments. The main disadvantage of the C_4 pathway is the additional energy cost for each carbon fixed by photosynthesis. The C_4 pathway is therefore most advantageous in warm, high-light conditions, such as tropical grasslands. The C_4 pathway occurs in 18 plant families and has evolved independently at least 30 times (Kellogg 1999). C_4 species first became abundant in the late Miocene 6 to 8 million years ago, probably triggered by the global decline in atmospheric CO_2 concentration (Cerling 1999). C_4 grasslands expanded during glacial periods, when CO_2 concentrations declined, and retracted at the end of glacial periods, when atmospheric CO_2 concentration increased, suggesting that the evolution of C_4 photosynthesis was tightly tied to variations in atmospheric CO_2 concentration. However, there is little geographic variation in atmospheric CO_2 concentration, so the current global geographic distribution of C_4 plants appears to be controlled primarily by temperature and by the

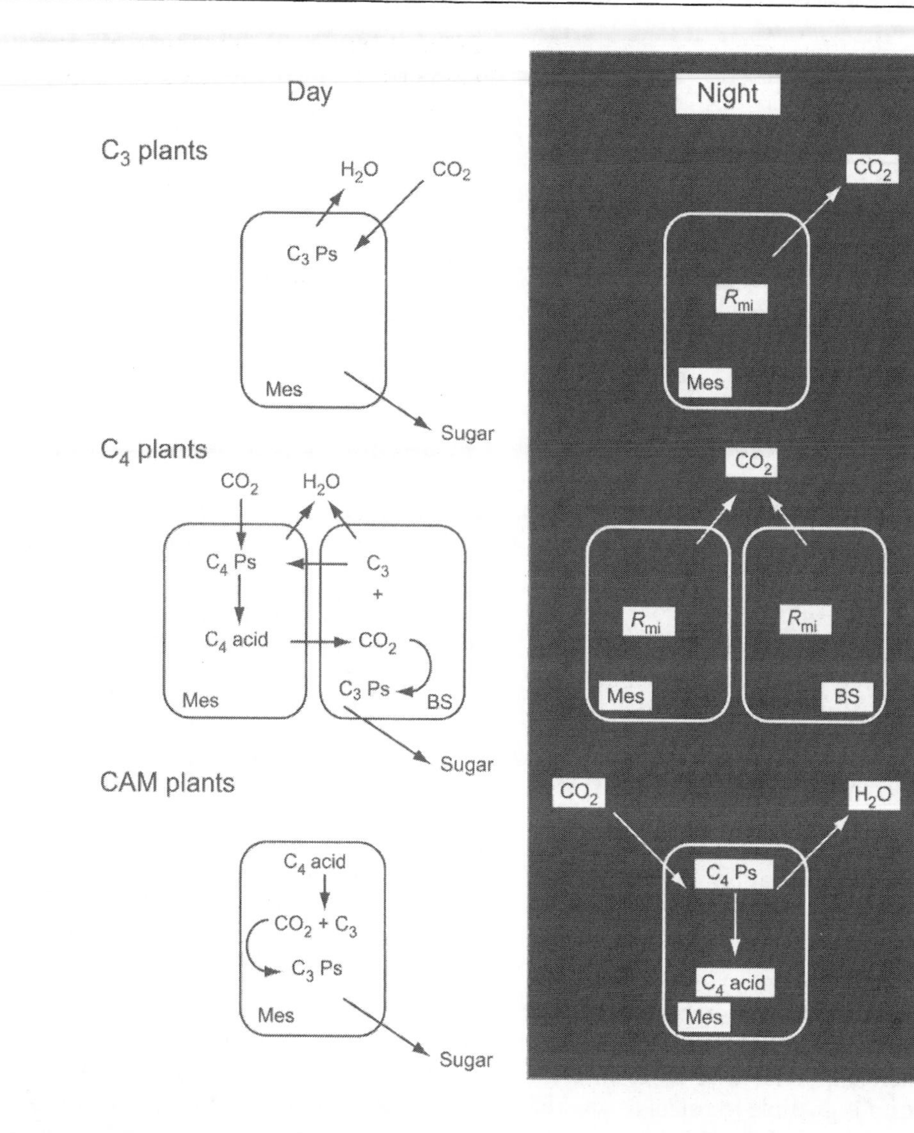

FIGURE 5.5. Cellular location and temporal cycle of CO_2 fixation and water exchange in leaves with C_3, C_4, and CAM photosynthetic pathways. In C_3 and CAM plants, all photosynthesis occurs in mesophyll (Mes) cells. In C_4 plants, C_4 carbon fixation (C_4 Ps) occurs in mesophyll cells and C_3 fixation (C_3 Ps) occurs in bundle sheath (BS) cells. Mitochondrial respiration (R_{mi}) occurs at night. Exchanges of CO_2 and water vapor with the atmosphere occur during the day in C_3 and C_4 plants and at night in CAM plants.

availability of light, water, and nitrogen. C_4 photosynthesis is absent from most woody plants.

C_4 plants have an isotopic signature that enables us to track their past and present role in ecosystems. C_4 plants incorporate much more ^{13}C than do C_3 plants during photosynthesis (Box 5.1) and therefore have a distinct isotopic signature that remains with any organic matter produced by this photosynthetic pathway. Isotopic measurements are important tools in studying ecological processes in ecosystems in which the relative abundance of C_3 and C_4 plants has changed over time (Ehleringer et al. 1993).

Crassulacean Acid Metabolism Photosynthesis

Crassulacean acid metabolism (CAM) enables ecosystems to gain carbon under extremely dry

Box 5.1. Carbon Isotopes

The three isotopic forms of carbon (^{12}C, ^{13}C, and ^{14}C) differ in the number of neutrons but have the same number of protons. The additional atomic mass causes the heavier isotopes to react more slowly in some reactions, particularly in the carboxylation of CO_2 by Rubisco. The lightest of these isotopes (^{12}C) is preferentially fixed by carboxylating enzymes. C_3 plants generally have a relatively high CO_2 concentration inside the leaf, due to their high stomatal conductance. Under these circumstances, Rubisco **discriminates** against the heavier isotope ^{13}C, causing $^{13}CO_2$ to accumulate within the leaf. $^{13}CO_2$ therefore diffuses out of the leaf through the stomata along a concentration gradient of $^{13}CO_2$ at the same time that $^{12}CO_2$ is diffusing into the leaf. In C_4 and crassulacean acid metabolism (CAM) plants, in contrast, PEP carboxylase has such a high affinity for CO_2 that it reacts with most of the CO_2 that enters the leaf, resulting in relatively little discrimination against $^{13}CO_2$. Consequently the ^{13}C concentrations of CAM and C_4 plants are much higher (less negative) than those of C_3 plants (Table 5.1).

This difference in isotopic composition among C_3, C_4, and CAM plants remains in any organic compounds derived from these plants. Thus it is possible to calculate the relative proportions of C_3 and C_4 plants in the diet of animals by measuring the ^{13}C content of their tissues; this can be done even in fossil bones such as those of early humans. Changes in the isotopic composition of fossil bones clearly indicate changes in diet. In situations in which vegetation has changed from C_3 to C_4 dominance (or vice versa), the organic matter in plants differs in its isotopic composition from that of the soil. Changes in the carbon isotope composition of soil organic matter over time therefore provides a tool for estimating the rates of turnover of soil organic matter that formed beneath the previous vegetation.

TABLE 5.1. Representative ^{13}C concentrations (‰) of atmospheric CO_2 and selected plant and soil materials.

Material	$\delta^{13}C$ (‰)[a]
Pee Dee limestone standard	0.0
Atmospheric CO_2	–8
Plant material	
Unstressed C_3 plant	–27
Water-stressed C_3 plant	–25
Unstressed C_4 plant	–13
Water-stressed C_4 plant	–13
CAM plant[b]	–27 to –11
Soil organic matter	
Derived from unstressed C_3 plants	–27
Derived from C_4 or CAM plants	–13

[a] The concentrations are expressed relative to an internationally agreed-on standard (Pee Dee belemnite): $\delta^{13}C_{std} = 1000\left(\frac{R_{sam}}{R_{std}} - 1\right)$, where $\delta^{13}C$ is the isotope ratio in delta units relative to a standard, and R_{sam} and R_{std} are the isotope abundance ratios of the sample and standard, respectively (Ehleringer and Osmond 1989).

[b] Values of –11 under conditions of CAM photosynthesis. Many CAM plants switch to C_3 photosynthesis under favorable moisture regimes, giving an isotopic ratio similar to that of unstressed C_3 plants.

Data from O'Leary (1988) and Ehleringer and Osmond (1989).

conditions. Succulent plant species in dry environments, including many epiphytes in tropical forests, gain carbon through CAM photosynthesis. CAM accounts for just a small proportion of terrestrial carbon gain, because it is active only under extremely dry conditions. Even in these environments, some CAM plants switch to C_3 photosynthesis when sufficient water is available.

In CAM photosynthesis, plants close their stomata during the day, when high tissue temperatures and low relative humidity of the air would otherwise cause large transpirational water loss (Fig. 5.5). At night, they open their stomata, and CO_2 enters the leaf and is fixed by PEP carboxylase. The resulting C_4 acids are stored in vacuoles until the next day when they are decarboxylated, releasing CO_2 to be fixed by normal C_3 photosynthesis. Thus, in CAM plants, there is a *temporal* (day–night) separation of C_3 and C_4 carbon dioxide fixation, whereas in C_4 plants there is a *spatial* separation of C_3 and C_4 carbon dioxide fixation between bundle sheath and mesophyll cells. CAM photosynthesis is energetically expensive, like C_4 photosynthesis; it therefore occurs primarily in dry high-light environments, such as deserts, shallow rocky soils, and canopies of tropical forests. CAM photosynthesis allows some plants to gain carbon under extremely dry conditions that would otherwise preclude carbon fixation in ecosystems.

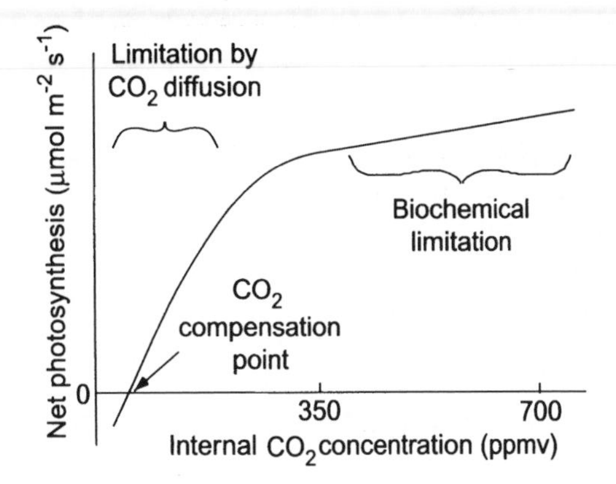

FIGURE 5.6. Relationship of the net photosynthetic rate to the CO_2 concentration inside the leaf. Photosynthetic rate is limited by the rate of CO_2 diffusion into the chloroplast in the linear portion of the CO_2-response curve and by biochemical processes at higher CO_2 concentrations. The CO_2 compensation point is the minimal CO_2 concentration at which the leaf shows a net gain of carbon.

Net Photosynthesis by Individual Leaves

Basic Principle of Environmental Control

Plants adjust the components of photosynthesis so physical and biochemical processes co-limit carbon fixation. Photosynthesis operates most efficiently when the rate of CO_2 diffusion into the leaf matches the biochemical capacity of the leaf to fix CO_2. Plants regulate the components of photosynthesis to achieve this balance, as seen from the response of photosynthesis to the CO_2 concentration inside the leaf (Fig. 5.6). When the internal CO_2 concentration is low, photosynthesis increases linearly with increasing CO_2 concentration. Under these circumstances, the leaf has more carbon fixation capacity than it can effectively use, and photosynthesis is limited by the rate of diffusion of CO_2 into the leaf. The plant can increase photosynthesis only by opening stomatal pores. Alternatively, if CO_2 concentration inside the leaf is high, photosynthesis shows little response to variation in CO_2 concentration. In this case, photosynthesis is limited by the rate of carboxylation of RuBP (the asymptote in Fig. 5.6), and changes in stomatal opening have little influence on photosynthesis. At high internal CO_2 concentrations, carboxylation may be limited by (1) insufficient light (or light-harvesting pigments) to provide energy; (2) insufficient nitrogen invested in photosynthetic enzymes to process the ATP, NADPH, and CO_2 present in the chloroplast; or (3) insufficient phosphate or sugar phosphates to synthesize RuBP.

Under a wide variety of circumstances, plants adjust the components of photosynthesis so CO_2 diffusion and biochemistry are about equally limiting to photosynthesis (Farquhar and Sharkey 1982). Plants make this adjustment by altering stomatal conductance, which occurs within minutes, or by changing the concentrations of light-harvesting pigments or photosynthetic enzymes, which occurs over days to weeks. The general principle of co-limitation of photosynthesis by biochemistry and diffusion provides the basis for understanding most of the adjustments by individual leaves to minimize the environmental limitations of photosynthesis. Stomatal conductance is regulated so photosynthesis occurs near the break point of the CO_2-response curve (Körner et al. 1979) (Fig. 5.6), at which CO_2 supply and carbon-fixation capacity are about equally limiting to photosynthesis.

Light Limitation

Leaves adjust stomatal conductance and photosynthetic capacity to maximize carbon gain in

different light environments. Leaves experience large fluctuations (10- to 1000-fold) in incident light due to changes in sun angle, cloudiness, and the location of **sunflecks** (patches of direct sunlight that penetrate a plant canopy) (Fig. 5.7). Leaf chloroplasts

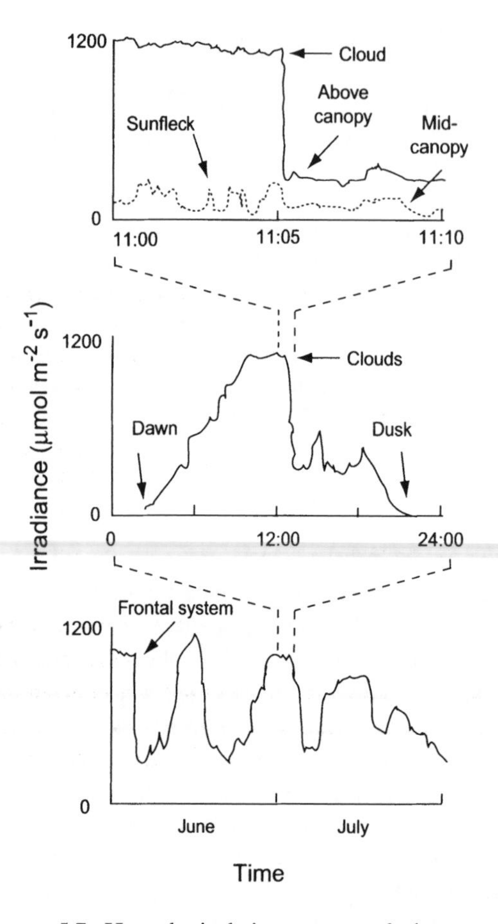

FIGURE 5.7. Hypothetical time course of photosynthetically active radiation above and below the canopy of a temperate forest over minutes, hours, and months. Over the course of a few minutes, light at the top of the canopy varies with cloudiness. Below the canopy, light also varies due to the presence of sunflecks of direct irradiance, which can last tenths of seconds to minutes. During a day, there are large changes in light due to changes in solar angle, with smaller fluctuations caused by passing clouds. Convective activity often increases cloudiness in the afternoon. During the growing season, the major causes of variation in light are seasonal changes in the solar angle and the passage of frontal systems. Some times of year have greater frequency of cloudiness than others due to changes in directions of the prevailing winds and the passage of frontal systems.

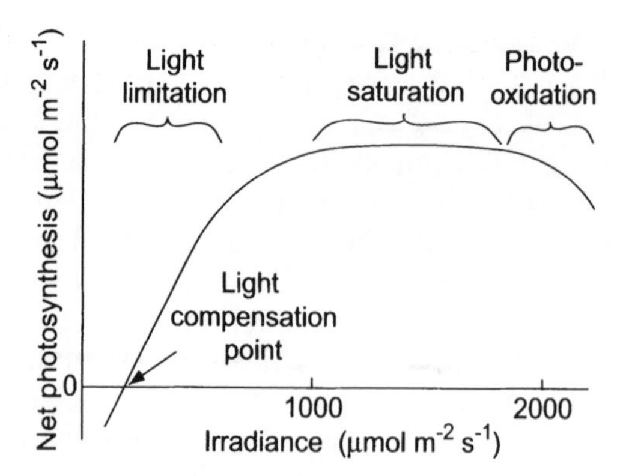

FIGURE 5.8. Relationship of net photosynthetic rate to photosynthetically active radiation and the processes that limit photosynthesis at different irradiance. The linear increase in photosynthesis in response to increased light (in the range of light limitation) indicates relatively constant light use efficiency.

respond to changes in light availability over minutes by changing both the levels of metabolites, which influence the activity of photosynthetic enzymes, and the stomatal conductance, which influences CO_2 supply and water loss (Pearcy 1990, Chazdon and Pearcy 1991). Stomatal conductance increases in high light, when CO_2 demand is high, and decreases in low light, when photosynthetic demand for CO_2 is low. These stomatal adjustments result in a relatively constant CO_2 concentration inside the leaf, as expected from our hypothesis of co-limitation of photosynthesis by biochemistry and diffusion. It allows plants to conserve water under low light and to maximize carbon uptake at high light.

At low light, when the supply of ATP and NADPH from the light-harvesting reactions limits the rate of carbon fixation, net photosynthesis increases linearly with increasing light (Fig. 5.8). The slope of this line (the **quantum yield** of photosynthesis) is a measure of the efficiency with which plants use absorbed light to produce sugars. The quantum yield is similar among all C_3 plants at low light in the absence of environmental stress. In other words, all C_3 plants have a relatively constant **photosynthetic light use efficiency** (LUE) (about 6%) of converting absorbed visible light (**photosynthetically active radiation**, PAR) into chemical energy under these low-light conditions. At high irradiance, photosynthesis becomes **light saturated**—that is, it no longer responds to changes in light supply, due to the finite capacity of the light-harvesting reactions to capture light. As a result, light energy is converted less efficiently into sugars at high light. Photosynthesis may decline at extremely high light as a result of **photo-oxidation** of photosynthetic enzymes and pigments.

Over longer time scales (days to months) plants respond to variations in light availability by producing leaves with different photosynthetic properties. This physiological adjustment by an organism in response to a change in some environmental parameter is known as **acclimation**. Leaves at the top of the canopy (**sun leaves**) have more cell layers, are thicker, and therefore have greater photosynthetic capacity per unit leaf area than do **shade leaves** produced under low light (Terashima and Hikosaka 1995, Walters and Reich 1999). The respiration rate of a tissue depends on its protein content (see Chapter 6), so the low photosynthetic capacity and protein content of shade leaves are associated with a lower respiration rate per unit area than in sun leaves. For this reason, shade leaves maintain a more positive carbon balance (photosynthesis minus respiration) under lower light than do sun leaves (Fig. 5.9). The net effect of acclimation to variation in light availability is to extend the range of light availability over which vegetation maintains a relatively constant LUE—that is, a relatively constant relationship between absorbed PAR and net photosynthesis.

Plants can also produce shade leaves as a result of **adaptation**, the genetic adjustment by a population to maximize performance in a particular environment. Species that are *adapted* to high light and are intolerant of shade typically have a higher photosynthetic capacity per unit mass or area than do shade-tolerant species, even in the shade (Walters and Reich 1999). The main disadvantage of the high protein and photosynthetic rate of shade-intolerant species is that these species also have a higher respiration rate than do shade-tolerant species, due to their higher protein content. In addition, shade-intolerant species produce short-lived leaves, so they must continuously produce new leaves to maintain their leaf area. The net carbon balance of individual leaves of shade-tolerant and shade-intolerant species in the shade is similar, but shade-intolerant species often have a less favorable *whole-plant* carbon balance in the shade owing to their more rapid leaf turnover. Shade-intolerant species have a higher carbon-gaining capacity at high light.

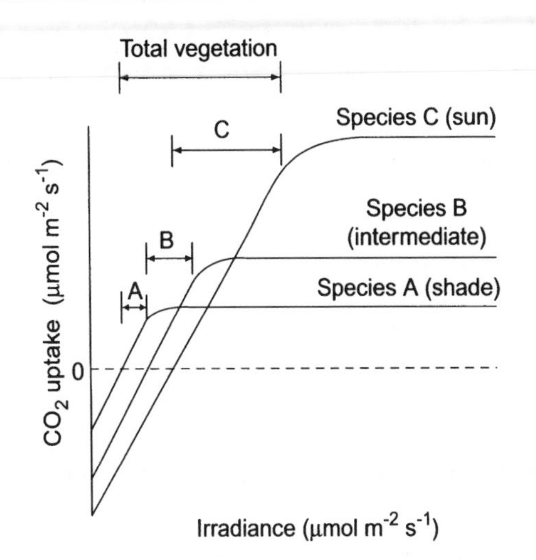

FIGURE 5.9. Light-response curves of net photosynthesis in plants acclimated to low, intermediate, and high light. Horizontal arrows show the range of irradiance over which net photosynthesis is positive and responds linearly to irradiance for each species and for the vegetation as a whole. Acclimation increases the range of irradiance over which net photosynthesis responds linearly to light (i.e., has a constant LUE).

Variations in leaf angle also increase the efficiency with which a plant canopy uses light. At high light, plants produce leaves that are steeply angled, so they absorb less light (see Chapter 4). This is advantageous because it reduces the probability of overheating or photo-oxidation of photosynthetic pigments at the top of the canopy. At the same time, it allows more light to penetrate to lower leaves. Leaves at the bottom of the canopy, on the other hand, are more horizontal in orientation to maximize light capture. The variations in leaf angle and photosynthetic properties optimize the performance of individual leaves, which explains why these patterns have evolved. This optimization at the level of individual plants translates into a maximization of the LUE and carbon fixation by the canopy as a whole. This is one of many examples in ecosystem ecology in which selection for optimal performance at the level of individuals gives rise to predictable patterns at the level of ecosystems.

Leaf area is the major factor governing the light environment experienced by individual leaves within the canopy. There is a maximum leaf area that plants in an ecosystem can attain, because light is a directional resource that is greatest at the top of the canopy and decreases exponentially within the canopy, according to the following equation:

$$I = I_0 e^{-kL} \quad (5.1)$$

where I is the **irradiance** (the quantity of radiant energy received at a surface per unit time) at any point in the canopy, I_o is the irradiance at the top of the canopy; k is the extinction coefficient, and L is the projected **leaf area index** (LAI; the leaf area per unit of ground area) above the point of measurement.

LAI is a key parameter governing ecosystem processes because it determines the light attenuation through a canopy and strongly influences the capacity of vegetation to gain carbon and transfer water and energy to the atmosphere (see Chapter 4). LAI has been defined in two ways: (1) **Projected LAI** is the leaf area projected onto a horizontal plane. (2) **Total LAI** is the total surface area of leaves, including the upper and lower surface of flat leaves and the cylindrical surface of conifer needles. Total LAI is approximately twice the value of projected LAI, except in the case of conifer needles, for which the projected leaf area is multiplied by π (3.14) to get total leaf area. Total LAI is particularly useful in describing the effective leaf area of conifer forests, in which leaves are more cylindrical than flat. A flat leaf cannot absorb more photons than move through a horizontal plane, because light is a directional resource. Conifer needles can, however, absorb considerably more light per unit of projected leaf area than flat leaves. Conifer needles are particularly effective in absorbing diffuse light, which provides a more uniform illumination of the overall canopy. Diffuse light makes up a larger proportion of total irradiance at low sun angles and under cloudy conditions. This may explain the predominance of conifers in high-latitude forests, where sun angles are low and in temperate rain forests, where conditions are usually cloudy. Unfortunately, there is no consistent agreement on whether data should be expressed as projected LAI or total LAI, and many review papers do not specify clearly which definition of LAI is being used. Micrometeorologists measuring radiation transfer and ecologists working with broad-leaved forests tend to use projected LAI, whereas ecophysiologists interested in carbon exchange and ecologists working in conifer forests tend to use total LAI. Each definition of LAI has advantages for addressing particular questions. The important thing is to specify which definition is being used.

Projected LAI varies widely among ecosystems but typically has values of 1 to 8 m^2 leaf m^{-2} ground for ecosystems with a closed canopy. The **extinction coefficient** is a constant that describes the exponential decrease in irradiance through a canopy. It is low for vertically inclined or small leaves (e.g., 0.3 to 0.5 for grasses), allowing substantial light penetration into the canopy, but high for near-horizontal leaves (0.7 to 0.8). Clumping of leaves around stems, as in conifers, and variable leaf angles result in intermediate values for k. Equation 5.1 indicates that light is distributed unevenly in an ecosystem and that the leaves near the top of the canopy capture most of the available light. Irradiance at the ground surface of a forest, for example, is often only 1 to 2% of that at the top of the canopy. At very low irradiance, leaf respiration completely offsets photosynthetic carbon gain, resulting in zero net photosynthesis, the **light compensation point** of the leaf (Fig. 5.8). A mature shaded leaf typically does not import carbon from the rest of the plant, so the leaf senesces and dies if it falls below the light compensation point for extended periods of time. This puts an upper limit on the leaf area that an ecosystem can support, regardless of how favorable the climate and the supply of soil resources may be.

Do differences in light availability explain the differences among ecosystems in carbon gain? In midsummer, when plants of most ecosystems are photosynthetically active, the daily input of visible light is nearly as great in the Arctic as in the tropics but is spread over more hours and is more diffuse at high latitudes (Billings and Mooney 1968). The greater daily carbon gain in the tropics than at high latitudes is therefore unlikely to be a simple function of the light available to drive photosynthesis. Neither can variation in light availability due to cloudiness explain differences among ecosystems in energy capture. The most productive ecosystems on Earth, the tropical and temperate rain forests, have a high frequency of cloudiness, whereas arid grasslands and deserts, which are less cloudy and receive nearly 10-fold more light annually, are less productive. Seasonal and interannual variations in irradiance can, however, contribute to temporal variation in carbon gain by ecosystems. Aerosols emitted by volcanic eruptions and fires, for example, can reduce solar irradiance and photosynthesis over large areas in particular years. In summary, light availability strongly influences daily and seasonal patterns of carbon input and the distribution of photosynthesis within the canopy, but it is only a minor factor explaining regional variations in carbon inputs to ecosystems (Fig. 5.1).

CO_2 Limitation

Changes in stomatal conductance by leaves minimize the effects of CO_2 supply on photosynthesis. The free atmosphere is sufficiently well mixed that its CO_2 concentration varies globally by only 4%—not enough to cause significant regional variation in photosynthesis. In dense canopies, photosynthesis reduces CO_2 concentration somewhat. The shade leaves that experience this low CO_2, however, tend to be light limited and therefore are relatively unresponsive to CO_2 concentration. Consequently, even this decline in CO_2 concentration within the canopy causes little variation in whole-ecosystem photosynthesis (Field 1991). In other words, CO_2 differs from other resources required by plants in that plant uptake of CO_2 does not greatly deplete the availability of this resource to other plants (Rastetter and Shaver 1992).

Although spatial variation in CO_2 concentration does not explain much of the global variation in photosynthetic rate, the continued worldwide increases in atmospheric CO_2 concentration (see Fig. 6.11) could cause a general increase in carbon gain by ecosystems. A doubling of the CO_2 concentration to which leaves are exposed, for example, leads to a 30 to 50% increase in photosynthetic rate (Curtis and Wang 1998). Enhancement of photosynthesis by addition of CO_2 is most likely to occur in ecosystems dominated by C_3 plants, whose photosynthetic rate is not CO_2 saturated at current atmospheric concentrations (Fig. 5.6). The magnitude of this stimulation of photosynthesis by rising atmospheric CO_2 concentration is, however, uncertain. Herbaceous plants and deciduous trees (but not conifers) sometimes acclimate to increased CO_2 concentration by reducing photosynthetic capacity and stomatal conductance (Ellsworth 1999, Mooney et al. 1999), as expected from our hypothesis of co-limitation of photosynthesis by biochemistry and diffusion. In other cases, acclimation has no effect on photosynthetic rate and stomatal conductance (Curtis and Wang 1998). The **down regulation** of CO_2 uptake in response to elevated CO_2 enables plants to gain similar amounts of carbon while minimizing water loss. It also causes photosynthesis to respond less strongly to elevated CO_2 than we might expect from a simple extrapolation of a CO_2-response curve of photosynthesis (Fig. 5.6).

Over the long term, indirect effects of elevated CO_2 may become important. In dry environments, for example, the reduced stomatal conductance caused by elevated CO_2 leads to a decline in transpiration, which reduces the rate at which water is lost from the soil and increases soil moisture. Elevated CO_2 often has a greater effect on plant growth through the reduction of moisture limitation than through a direct stimulation of photosynthesis by elevated CO_2. These indirect effects complicate the prediction of long-term responses of ecosystem carbon gain to elevated CO_2. Given that the atmospheric CO_2 concentration has increased 30% (by 90 parts per million by volume; ppmv) since the beginning of the Industrial Revolution, it is important to understand and predict these indirect effects of elevated CO_2 on carbon gain by ecosystems.

Photosynthesis in C_4 plants is relatively unresponsive to CO_2 concentration in the short term because PEP carboxylase is highly effective in drawing down CO_2 concentration inside the leaf. This leads to the prediction that C_4 plants might be displaced by C_3 plants if atmospheric CO_2 concentration continues to increase, just as occurred in tropical grasslands when CO_2 concentration rose at the end of the last glaciation (Cerling 1999). C_4 plants are, however, often just as sensitive to the indirect effects of CO_2 (e.g., increased soil moisture) as are C_3 plants, so the long-term effects of elevated CO_2 on the competitive balance of C_3 and C_4 plants are difficult to predict (Mooney et al. 1999).

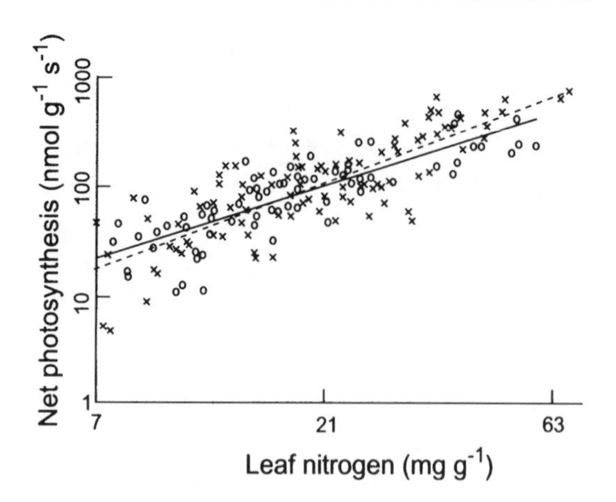

FIGURE 5.10. Relationship between leaf nitrogen concentration and maximum photosynthetic capacity for plants from Earth's major biomes. Open circles and the solid regression line are for 11 species from six biomes using a common methodology. Exes and the dashed regression line are data from the literature. (Redrawn with permission from *Proceedings of the National Academy of Sciences U. S. A.*, Vol. 94 © 1997 National Academy of Sciences, USA; Reich et al. 1997.)

Nitrogen Limitation and Photosynthetic Capacity

Plant species differ 10- to 50-fold in their photosynthetic capacity. Photosynthetic capacity is the photosynthetic rate per unit leaf mass measured under favorable conditions of light, moisture, and temperature. It is a measure of the carbon-gaining potential *per unit of biomass invested in leaves*. Photosynthetic capacity correlates strongly with leaf nitrogen concentration (Fig. 5.10) (Field and Mooney 1986, Reich et al. 1997, Reich et al. 1999) because photosynthetic enzymes account for a large proportion of the nitrogen in leaves (Fig. 5.1). Many ecological factors can lead to a high leaf nitrogen concentration and therefore a high photosynthetic capacity. Plants growing in high-nitrogen soils, for example, have higher tissue nitrogen concentrations and photosynthetic rates than do the same species growing on less fertile soils.

This acclimation of plants to a high nitrogen supply contributes to the high photosynthetic rates in agricultural fields and other ecosystems with a rapid nitrogen turnover. Many species differ in their nitrogen concentration, even when growing in the same soils. Species adapted to productive habitats usually produce leaves that are short lived and have high tissue nitrogen concentrations and high photosynthetic rates. Nitrogen-fixing plants also typically have high leaf nitrogen concentrations and correspondingly high photosynthetic rates. Environmental stresses that cause plants to produce leaves with a low leaf nitrogen concentration result in low photosynthetic capacity. In summary, regardless of the cause of variation in leaf nitrogen concentration, there is always a strong positive correlation between leaf nitrogen concentration and photosynthetic capacity (Field and Mooney 1986) (Fig. 5.10).

Plants with a high photosynthetic capacity have a high stomatal conductance, in the absence of environmental stress (Fig. 5.11), as expected from our hypothesis of co-limitation of photosynthesis by biochemistry and diffusion. This enables plants with a high photosynthetic capacity to gain carbon rapidly, at the cost of high rates of water loss. Conversely, species with a low photosynthetic capacity conserve water as a result of their lower stomatal conductance.

There appears to be an unavoidable trade-off between traits that maximize photosynthetic rate and traits that maximize leaf longevity (Fig. 5.12) (Reich et al. 1997, Reich et al. 1999). Many species of plants that grow in low-nutrient environments produce long-lived

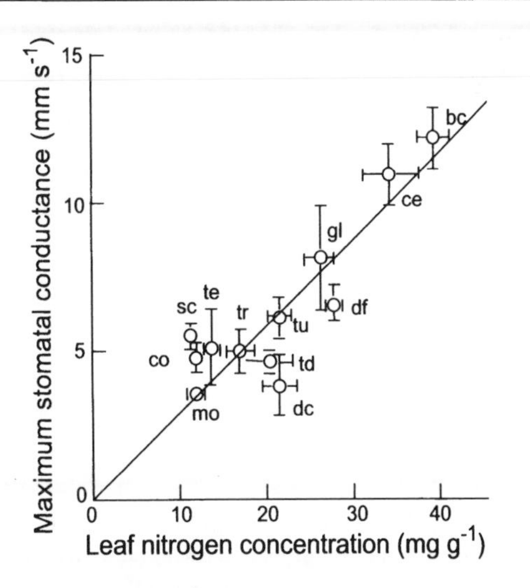

FIGURE 5.11. Relationship between leaf nitrogen concentration and maximal stomatal conductance of plants from Earth's major biomes. Each point and its standard error represent a different biome. bc, Broad-leaved crops; ce, cereal crops; co, evergreen conifer forest; dc, deciduous conifer forest; df, tropical deciduous forest; gl, grassland; mo, monsoonal forest; sc, sclerophyllous shrub; td, temperate deciduous broad-leaved forest; te, temperate evergreen broad-leaved forest; tr, tropical rain forest; tu, herbaceous tundra. (Redrawn with permission from the *Annual Review of Ecology and Systematics*, Vol. 25 © 1994 by Annual Reviews, www.AnnualReviews; Schulze et al. 1994.)

leaves because there are insufficient nutrients to support rapid leaf turnover (Chapin 1980). Shade-tolerant species also produce longer-lived leaves than do shade-intolerant species (Walters and Reich 1999). Long-lived leaves typically have a low leaf nitrogen concentration and a low photosynthetic capacity; they must therefore photosynthesize for a relatively long time to break even in their lifetime carbon budget (Chabot and Hicks 1982, Gulmon and Mooney 1986, Reich et al. 1997). To survive, long-lived leaves must have sufficient structural rigidity to withstand drought and/or winter desiccation. These structural requirements cause leaves to be dense—that is, to have a small surface area per unit of biomass, termed **specific leaf area** (SLA). Long-lived leaves must also be well defended against herbivores and pathogens, if they are to persist. This requires substantial allocation to lignin, tannins, and other compounds that deter herbivores, but also contribute to tissue mass and a low SLA. Many woody plants in dry environments also produce long-lived leaves. For the same reasons, these leaves typically have a low SLA and a low photosynthetic capacity (Reich et al. 1999).

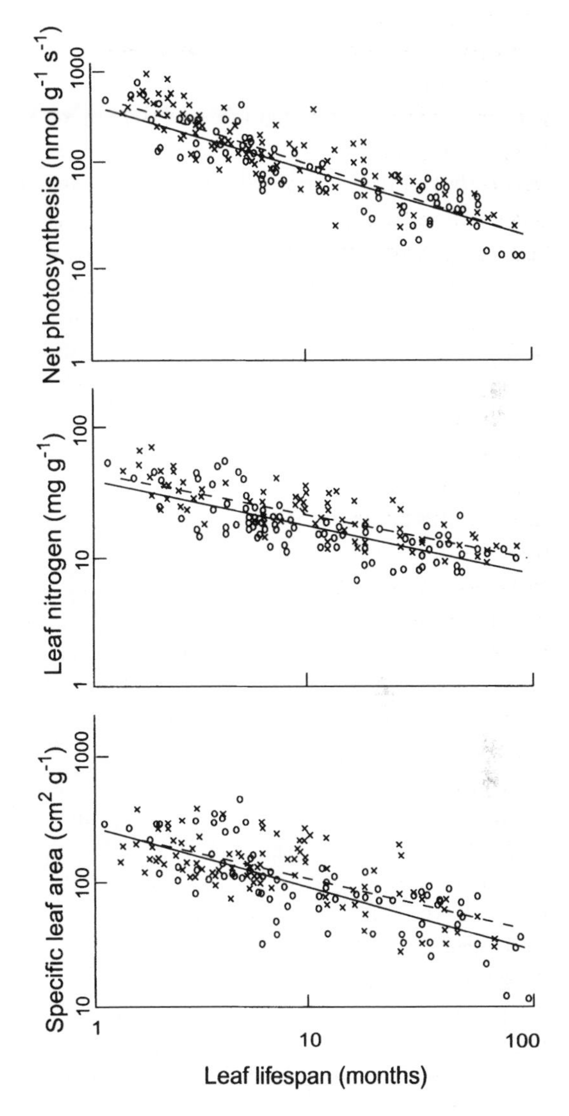

FIGURE 5.12. The effect of leaf life span on photosynthetic capacity (photosynthetic rate measured under favorable conditions), leaf nitrogen concentration, and specific leaf area. Symbols same as in Figure 5.10. (Redrawn with permission from *Proceedings of the National Academy of Sciences U. S. A.*, Vol. 94 © 1997 National Academy of Sciences, USA; Reich et al. 1997.)

The broad relationship among species with respect to photosynthetic rate and leaf life span is similar in all biomes; a 10-fold decrease in leaf life span gives rise to about a 5-fold increase in photosynthetic capacity (Reich et al. 1999). Species with long-lived leaves, low photosynthetic capacity, and low stomatal conductance are common in all low-resource environments, including those that are dry, infertile, or shaded.

Plants in productive environments, in contrast, produce short-lived leaves with a high tissue nitrogen concentration and a high photosynthetic capacity; this allows a large carbon return per unit of biomass invested in leaves, if sufficient light is available. These leaves have a high SLA, which maximizes the quantity of leaf area displayed and the light captured per unit of leaf mass. The resulting high rates of carbon gain support a high maximum relative growth rate in the absence of environmental stress or competition from other plants (Fig. 5.13) (Schulze and Chapin 1987). Many early successional habitats, such as recently abandoned agricultural fields or postfire sites, have sufficient light, water, and nutrients to support high growth rates and are characterized by species with short-lived leaves, high tissue nitrogen concentration, high SLA, and high photosynthetic rates (see Chapter 13). Even in late succession, environments with high water and nutrient availability are characterized by species with relatively high nitrogen concentrations and photosynthetic rates. Plants in these habitats can grow quickly to replace leaves removed by herbivores or to fill canopy gaps produced by death of branches or individuals.

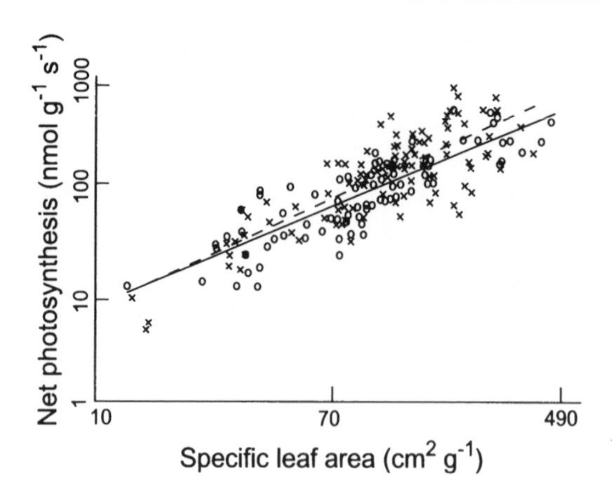

FIGURE 5.14. The relationship between SLA and photosynthetic capacity. The consistency of this relationship makes it possible to use SLA as an easily measured index of photosynthetic capacity. Symbols same as in Figure 5.10. (Redrawn with permission from *Proceedings of the National Academy of Sciences U. S. A.*, Vol. 94 © 1997 National Academy of Sciences USA; Reich et al. 1997.)

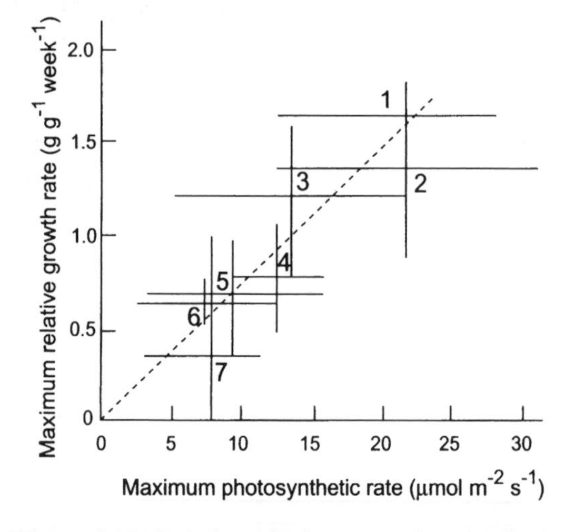

FIGURE 5.13. Relationship between photosynthetic rate and relative growth rate for major plant growth forms. 1, Agricultural crop species; 2, herbaceous sun species; 3, grasses and sedges; 4, summer deciduous trees; 5, evergreen and deciduous dwarf shrubs; 6, herbaceous shade species and bulbs; 7, evergreen conifers. (Redrawn with permission from Springer-Verlag; Schulze and Chapin 1987.)

In summary, plants produce leaves with a continuum of photosynthetic characteristics, ranging from short-lived thin leaves with a high nitrogen concentration and high photosynthetic rate to long-lived dense leaves with a low nitrogen concentration and low photosynthetic rate. These correlations among traits are so consistent that SLA is often used in ecosystem comparisons as an easily measured index of photosynthetic capacity (Fig. 5.14).

There is only modest variation in photosynthetic capacity per unit leaf area because leaves with a high photosynthetic capacity per unit leaf biomass also have a high SLA. Photosynthetic capacity calculated *per unit leaf area* (P_{area}) is a measure of the capacity of leaves to capture a unit of incoming radiation. It is calculated by dividing photosynthetic capacity per unit leaf mass (P_{mass}) by SLA.

$$P_{area} = \frac{P_{mass}}{SLA} \qquad (5.2)$$

where $(g\,cm^{-2}\,s^{-1}) = (g\,g^{-1}\,s^{-1})/(cm^2\,g^{-1})$.

There is relatively little variation in P_{area} among plants from different ecosystems (Lambers and Poorter 1992). In productive habitats, both mass-based photosynthesis and SLA are high (Fig. 5.12). In unproductive habitats both of these parameters are low, resulting in modest variation in area-based photosynthetic rate. To the extent that P_{area} varies among plants, it is highest in species with short-lived leaves (Reich et al. 1997). There is, however, no strong pattern of photosynthetic rate per unit area among ecosystems (Lambers and Poorter 1992). Mass-based photosynthetic capacity is a good measure of the physiological potential for photosynthesis (the photosynthetic rate per unit of biomass invested in leaves). Area-based photosynthetic capacity is a good measure of the effectiveness of these leaves at the ecosystem scale (photosynthetic rate per unit of available light). Variation in soil resources has a much greater effect on the quantity of leaf area produced than on the photosynthetic capacity per unit leaf area.

Water Limitation

Water limitation reduces the capacity of individual leaves to match CO_2 supply with light availability. Water stress is often associated with high light because sunny conditions correlate with low precipitation (low water supply) and with low humidity (high rate of water loss). High light also leads to an increase in leaf temperature and water vapor concentration inside the leaf and therefore greater water loss by transpiration (see Chapter 4). The high-light conditions in which a plant would be expected to increase stomatal conductance to minimize CO_2 limitations to photosynthesis are therefore often the same conditions in which the resulting transpirational water loss is greatest and most detrimental to the plant. This trade-off between a response that maximizes carbon gain (stomata open) and one that minimizes water loss (stomata closed) is typical of the physiological compromises faced by plants whose physiology and growth may be limited by more than one environmental resource (Mooney 1972). When water supply is abundant, leaves typically open their stomata in response to high light, despite the associated high rate of water loss. As leaf water stress develops, stomatal conductance declines to reduce water loss. This decline in stomatal conductance reduces photosynthetic rate and the efficiency of using light to fix carbon (i.e., LUE) below levels found in unstressed plants.

Plants that are acclimated and adapted to dry conditions reduce their photosynthetic capacity and leaf nitrogen content toward a level that matches the low stomatal conductance that is necessary to conserve water in these environments (Wright et al. 2001). A high photosynthetic capacity provides little benefit if the plant must maintain a low stomatal conductance to minimize water loss. Conversely, low nitrogen availability or other factors that constrain leaf nitrogen concentration result in leaves with low stomatal conductance (Fig. 5.11). This strong correlation between photosynthetic capacity and stomatal conductance maintains the balance between photosynthetic capacity and CO_2 supply—that is, the co-limitation of photosynthesis by diffusional and biochemical processes. In addition to their low photosynthetic capacity and low stomatal conductance, plants in dry areas minimize water stress by reducing leaf area (by shedding leaves or producing fewer new leaves). Some drought-adapted plants produce leaves that minimize radiation absorption; their leaves reflect most incoming radiation or are steeply inclined toward the sun (see Chapter 4) (Ehleringer and Mooney 1978). High radiation absorption is a *disadvantage* in dry environments because it increases leaf temperature, which increases respiratory carbon loss (see Chapter 6) and transpirational water loss (see Chapter 4).

Thus there are several mechanisms by which plants in dry environments reduce radiation absorption and photosynthetic capacity to conserve water and carbon. The low leaf area, the reflective nature of leaves, and the steep angle of leaves are the main factors accounting for the low absorption of radiation and low carbon inputs in dry environments. In other words, plants adjust to dry environments primarily by altering leaf area and radiation absorption rather than by altering photosynthetic capacity per unit leaf area.

Water use efficiency (WUE) of photosynthesis is defined as the carbon gain per unit of water lost. Water use is quite sensitive to the size of stomatal openings, because stomatal conductance has slightly different effects on the rates of CO_2 entry and water loss. Water leaving the leaf encounters two resistances to flow: the stomata and the boundary layer of still air on the leaf surface (Fig. 5.4). Resistance to CO_2 diffusion from the bulk air to the site of photosynthesis includes the same stomatal and boundary layer resistances *plus* an additional internal resistance associated with diffusion of CO_2 from the cell surface into the chloroplast and any biochemical resistances associated with carboxylation. Because of this additional resistance to CO_2 movement into the leaf, any change in stomatal conductance has a *proportionately* greater effect on water loss than on carbon gain. In addition, water diffuses more rapidly than does CO_2 because of its smaller molecular mass and because of the steeper concentration gradient that drives diffusion across the stomata. For all these reasons, as stomata close, water loss declines to a greater extent than does CO_2 absorption. The low stomatal conductance of plants in dry environments results in less photosynthesis per unit of time but greater carbon gain per unit of water loss—that is, greater WUE. Plants in dry environments also enhance WUE by maintaining a somewhat higher photosynthetic capacity than would be expected for their stomatal conductance, thereby drawing down the internal CO_2 concentration and maximizing the diffusion gradient for CO_2 entering the leaf (Wright et al. 2001). Carbon isotope ratios in plants provide an integrated measure of WUE during plant growth because the ^{13}C concentration of newly fixed carbon increases under conditions of low internal CO_2 concentration (Ehleringer 1993) (Box 5.1). C_4 and CAM photosynthesis are additional adaptations that augment WUE of ecosystems.

Temperature Effects

Extreme temperatures limit carbon uptake. Photosynthetic rate is typically highest near leaf temperatures commonly experienced on sunny days (Fig. 5.15). Leaf temperature may differ substantially from air temperature due to the cooling effects of transpiration, the effects of leaf surface properties on light absorption, and the influence of adjacent surfaces on the thermal and radiation environment of the leaf (see Chapter 4). At low temperatures, photosynthesis is limited directly by temperature, as are all chemical reactions. At high temperatures, photosynthesis also declines, due to increased photorespiration and, under extreme conditions, enzyme inactivation and destruction of photosynthetic pigments. Temperature extremes often have a greater effect on photosynthesis than does average temperature because of damage to photosynthetic machinery (Waring and Running 1998).

There are several factors that minimize the sensitivity of photosynthesis to temperature. The enzymatically controlled carbon-fixation reactions are typically more sensitive to low

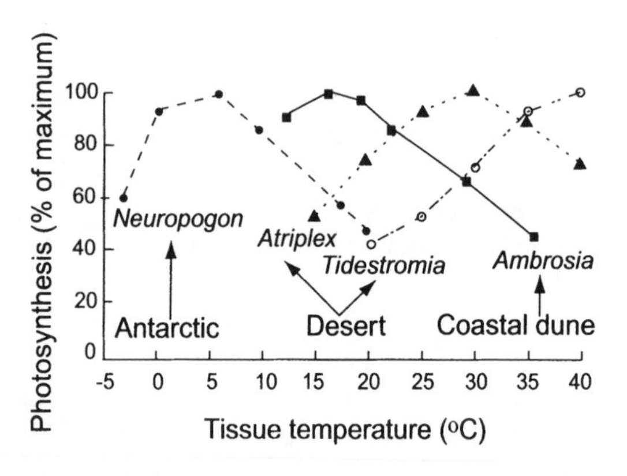

FIGURE 5.15. Temperature response of photosynthesis in plants from contrasting temperature regimes. Species include antarctic lichen (*Neuropogon acromelanus*), a cool coastal dune plant (*Ambrosia chamissonis*), an evergreen desert shrub (*Atriplex hymenelytra*), and a summer-active desert perennial (*Tidestromia oblongifolia*). (Redrawn with permission from Blackwell Science, Ltd; Mooney 1986.)

temperature than are the biophysically controlled light-harvesting reactions. Carbon fixation reactions therefore tend to limit photosynthesis at low temperature. Plants adapted to cold climates compensate for this by producing leaves with high concentrations of leaf nitrogen and photosynthetic enzymes, which enable carboxylation to keep pace with the energy supply from the light-harvesting reactions (Berry and Björkman 1980). This explains why arctic and alpine plants typically have high leaf nitrogen concentrations despite low soil nitrogen availability (Körner and Larcher 1988). Plants in cold environments also have hairs and other morphological traits that raise leaf temperature above air temperature (Körner 1999). In hot environments with an adequate water supply, plants produce leaves with high photosynthetic rates. The associated high transpiration rate can cool the leaf, so leaf temperatures are much lower than air temperatures. In hot, dry environments, plants close stomata to conserve water, and the cooling effect of transpiration is reduced. Plants in these environments often produce small leaves that shed heat effectively and maintain temperatures close to air temperature (see Chapter 4). In summary, despite the sensitivity of photosynthesis to short-term variation in temperature, leaf properties minimize the differences in leaf temperature among ecosystems, and plants acclimate and adapt so there is no clear relationship between temperature and average photosynthetic rate in the field, when ecosystems are compared.

Pollutants

Pollutants reduce carbon gain primarily by reducing leaf area or photosynthetic capacity. Many pollutants, such as sulfur dioxide (SO_2) and ozone, reduce photosynthesis through their effects on growth and the production of leaf area. Pollutants also directly reduce photosynthesis by entering the stomata and damaging the photosynthetic machinery, thereby reducing photosynthetic capacity (Winner et al. 1985). Plants then reduce stomatal conductance to balance CO_2 uptake with the reduced capacity for carbon fixation. This reduces the entry of pollutants into the leaf, reducing the vulnerability of the leaf to further injury. Plants growing in low-fertility or dry conditions are preadapted to pollutant stress because their low stomatal conductance minimizes the quantity of pollutants entering leaves. These plants are therefore less affected by pollutants than are rapidly growing crops and other plants with high stomatal conductance.

Gross Primary Production

Gross primary production is the sum of the photosynthesis by all leaves measured at the ecosystem scale. It is typically integrated over time periods of days to a year ($g\,C\,m^{-2}$ of ground yr^{-1}) and is the process by which carbon and energy enter ecosystems. GPP is generally estimated from simulation models rather than measured directly, because it is impossible to measure the net carbon exchange of all the leaves in the canopy in isolation from other ecosystem components (e.g., respiration by stems and soil) and without modifying the vertical gradient in environment. The results of these modeling studies suggest that most conclusions derived from leaf-level measurements of net photosynthesis can be extended to the ecosystem scale, when the vertical profiles of photosynthetic capacity and environment are considered. Measurement of whole-ecosystem carbon exchange provides another way to estimate GPP (see Chapter 6).

Canopy Processes

The vertical profile of leaf photosynthetic properties in a canopy maximizes GPP. In most closed-canopy ecosystems, photosynthetic capacity decreases exponentially through the canopy in parallel with the exponential decline in irradiance (Eq. 5.1) (Hirose and Werger 1987). This matching of photosynthetic capacity to light availability is the response we would expect from individual leaves within the canopy, because it maintains the co-limitation of photosynthesis by diffusion and biochemical processes in each leaf. It also serves to maximize GPP in closed-canopy ecosystems. The matching of photosynthetic capacity to light availability occurs through the preferential transfer of nitrogen to leaves at the top of the canopy. At least three processes cause this to happen. (1) Sun leaves at the top of the canopy develop more cell layers than shade leaves and therefore contain more nitrogen per unit leaf area. (2) New leaves are produced primarily at the top of the canopy, causing nitrogen to be transported to the top of the canopy (Field 1983, Hirose and Werger 1987). (3) Leaves at the bottom of the canopy senesce when they become shaded below their light compensation point. Much of the nitrogen resorbed from these senescing leaves (see Chapter 8) is transported to the top of the canopy to support the production of young leaves with high photosynthetic capacity. The accumulation of nitrogen at the top of the canopy is most pronounced in dense canopies, which develop under circumstances of high water and nitrogen availability (Field 1991). In environments in which leaf area is limited by water, nitrogen, or time since disturbance, there is less advantage to concentrating nitrogen at the top of the canopy, because light is abundant throughout the canopy. In these canopies, light availability, nitrogen concentrations, and photosynthetic rates are more uniformly distributed through the canopy.

Canopy-scale relationships between light and nitrogen appear to occur even in multispecies communities. In a single individual, there is an obvious selective advantage to optimizing nitrogen distribution within the canopy because this provides the greatest carbon return per unit of nitrogen invested in leaves. We know less about the factors governing carbon gain in multispecies stands. In such stands, the individuals at the top of the canopy account for most of the photosynthesis and may be able to support greater root biomass to acquire more nitrogen, compared to smaller subcanopy or understory individuals (Hikosaka and Hirose 2001). This specialization among individuals probably contributes to the vertical scaling of nitrogen and photosynthesis that is observed in multispecies stands.

Vertical gradients in other environmental variables often reinforce the maximization of carbon gain near the top of the canopy. The canopy modifies not only light availability but also other variables that influence photosynthetic rate, including wind speed, temperature, relative humidity, and CO_2 concentration (Fig. 5.16). The most important of these effects is the decrease in wind speed from the free atmosphere to the ground surface. The friction of air moving across Earth's surface causes wind speed to decrease exponentially from the free atmosphere to the top of the canopy. In other words, Earth's surface creates a boundary layer similar to that which develops around individual leaves (Fig. 5.4). Wind speed continues to decrease from the top of the canopy to the ground surface in ways that depend on canopy structure. Smooth canopies, characteristic of crops or grasslands, show a gradual decrease in wind speed from the top of the canopy to the ground surface, whereas rough canopies, characteristic of many forests, create more friction and turbulence, which increase the vertical mixing of air within the canopy (see Chapter 4) (McNaughton and Jarvis 1991). For this reason, gas exchange in rough canopies is more tightly **coupled** to conditions in the free atmosphere than in smooth canopies.

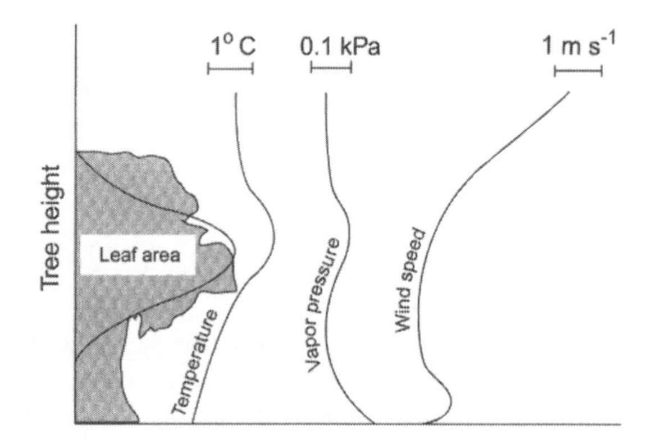

FIGURE 5.16. Typical vertical gradients in temperature, vapor pressure, and wind speed in a forest. Temperature is highest in the midcanopy where most energy is absorbed. The increase in wind speed at the bottom of the canopy occurs in open forests, where there is little understory. (Redrawn with permission from Academic Press; Landsberg and Gower 1997.)

Wind speed is important because it reduces the thickness of the boundary layer of still air around each leaf, producing steeper gradients in temperature and in concentrations of CO_2 and water vapor from the leaf surface to the atmosphere. This speeds the diffusion of CO_2 into the leaf and the loss of water from the leaf, enhancing both photosynthesis and transpiration. A reduction in thickness of the leaf boundary layer also brings leaf temperature closer to air temperature. The net effect of wind on photosynthesis is generally positive at moderate wind speeds and adequate moisture supply, enhancing photosynthesis at the top of the canopy. When low soil moisture or a long pathway for water transport from the soil to the top of the canopy reduces water supply to the uppermost leaves, as in tall forests, the uppermost leaves reduce their stomatal conductance, causing the zone of maximum photosynthesis to shift farther down in the canopy. Although multiple environmental gradients within the canopy have complex effects on photosynthesis, they probably enhance photosynthesis near the top of canopies in ecosystems with sufficient water and nutrients to develop dense canopies.

Canopy properties extend the range of light availability over which the LUE of the canopy remains constant. The light-response curve of canopy photosynthesis, measured in closed canopies (total LAI greater than about 3), saturates at higher irradiance than does photosynthesis by a single leaf (Fig. 5.17) (Jarvis and Leverenz 1983). This increase in the efficiency of converting light energy into fixed carbon occurs for several reasons. The less horizontal angle of leaves at the top of the canopy reduces the probability of light saturation of the upper leaves and increases the light penetration into the canopy. The clumped distribution of leaves in shoots, branches, and crowns also increases light penetration into the canopy. Conifer canopies are particularly effective in distributing light through the canopy due to the clumping of needles around stems. This could explain why conifer forests frequently support a higher LAI than deciduous forests. The light compensation point also decreases from the top to the bottom of the canopy (Fig. 5.9), so lower leaves

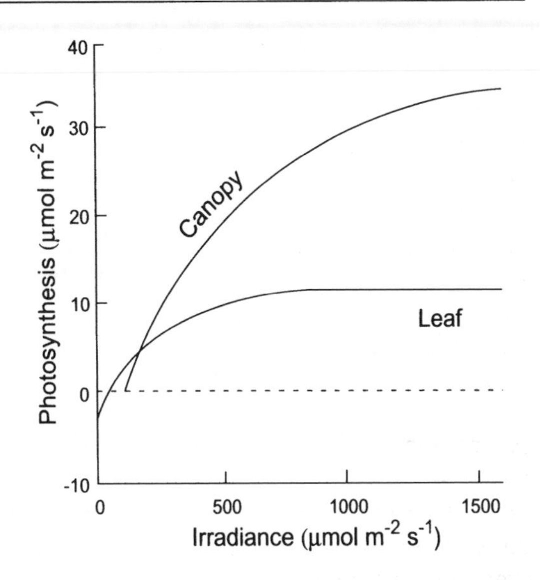

FIGURE 5.17. Light-response curve of a single leaf and a forest canopy. Canopies maintain a constant LUE (linear response of photosynthesis to light) over a broader range of light availability than do individual leaves. (Reprinted with permission from *Advances in Ecological Research*; Ruimy et al. 1996.)

maintain a positive carbon balance, despite the relatively low light availability. In crop canopies, where water and nutrients are highly available, the linear relationship between canopy carbon exchange and irradiance (i.e., constant LUE) extends up to irradiance typical of full sunlight. In other words, there is no evidence of light saturation, and LUE remains constant over the full range of natural light intensities (Fig. 5.18) (Ruimy et al. 1996).

Satellite Estimates of GPP

The similarity among ecosystems of canopy LUE provides a basis for estimating carbon input to ecosystems globally, using remote sensing. An important conclusion of leaf- and canopy-level studies of photosynthesis is that there are many factors that cause ecosystems to converge toward a relatively similar efficiency of converting light energy into carbohydrates. (1) All C_3 plants have a similar quantum yield (LUE) at low to moderate irradiance. (2) Penetration of light and vertical variations in photo-

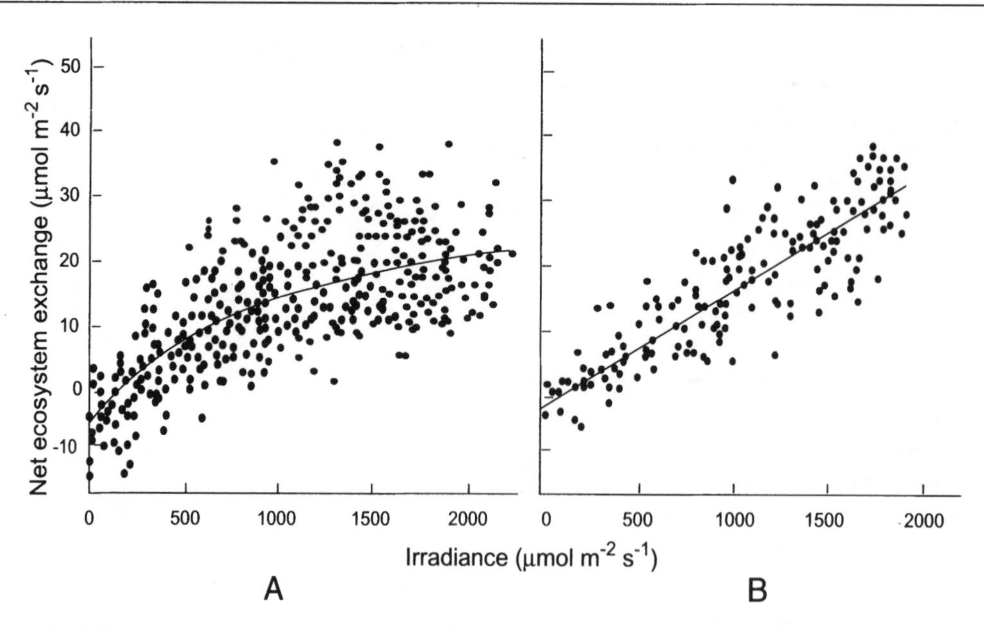

FIGURE 5.18. Effect of vegetation type and irradiance on net ecosystem exchange in forests (**A**) and crops (**B**). Forests maintain a relatively constant LUE up to 30 to 50% of full sun, although there is considerable variability. Crops maintain a constant light LUE over the entire range of naturally occurring irradiance. (Redrawn with permission from *Advances in Ecological Research*; Ruimy et al. 1996.)

synthetic properties through a canopy extend the range of irradiance over which the LUE remains constant. (3) Light use efficiency is reduced primarily by short-term environmental stresses that cause plants to reduce stomatal conductance. Over the long term, however, plants respond to such stresses by reducing the concentrations of photosynthetic pigments and enzymes so photosynthetic capacity matches stomatal conductance and by reducing leaf area. In other words, plants in low-resource environments reduce the amount of light absorbed more strongly than they reduce the efficiency with which absorbed light is converted to carbohydrates. Modeling studies suggest that light use efficiency varies about twofold among ecosystems (Field 1991), although this is difficult to demonstrate conclusively because GPP cannot be directly measured.

If LUE is indeed similar among ecosystems, GPP could be estimated by determining the quantity of light absorbed by ecosystems, which can be measured from satellites. Leaves at the top of the canopy have a disproportionately large effect on the light that is both absorbed and reflected by the ecosystem. The reflected radiation can also be measured by satellites. This similarity in bias between the vertical distribution of absorbed and reflected radiation makes satellites an ideal tool for estimating canopy photosynthesis. The challenge, however, is to estimate the fraction of absorbed radiation that has been absorbed by leaves rather than by soil or other nonphotosynthetic surfaces. Vegetation has a different spectrum of absorbed and reflected radiation than does the atmosphere, water, clouds, and bare soil. This occurs because chlorophyll and associated light-harvesting pigments or accessory pigments, which are concentrated at the canopy surface, absorb visible light (VIS) effectively. The optical properties that result from the cellular structure of leaves, however, makes them highly reflective in the near infrared (NIR) range. Ecologists have used these unique properties of vegetation to generate an index of vegetation "greenness": the **normalized difference vegetation index (NDVI)**.

$$\text{NDVI} = \frac{(\text{NIR} - \text{VIS})}{(\text{NIR} + \text{VIS})} \tag{5.3}$$

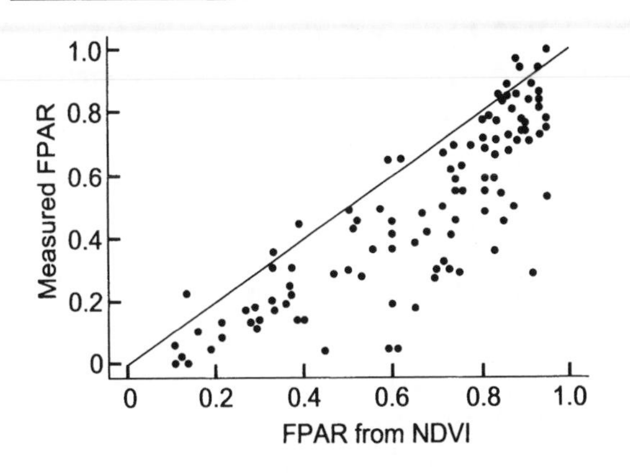

FIGURE 5.19. Relationship between the fraction of photosynthetically active radiation (FPAR) absorbed by vegetation estimated from satellite measurements of NDVI (*x*-axis) and FPAR measured in the field (*y*-axis). Data were collected from a wide range of ecosystems, including temperate and tropical grasslands and temperate and boreal conifer forests. Satellites provide an approximate measure of the photosynthetically active radiation absorbed by vegetation and therefore the carbon inputs to ecosystems. (Redrawn with permission from *Journal of Hydrometeorology*; Los et al. 2000.)

Sites with a high rate of carbon gain generally have a high NDVI because of their high chlorophyll content (low reflectance of VIS) and high leaf area (high reflectance of NIR). Species differences in leaf structure also influence infrared reflectance (and therefore NDVI). Conifer forests, for example, generally have a lower NDVI than deciduous forests despite their greater leaf area. Consequently, NDVI must be used cautiously when comparing ecosystems dominated by different types of plants (Verbyla 1995).

NDVI is ecologically useful because it is proportional to the amount of light energy absorbed by photosynthetic tissues, when structurally similar stands of vegetation are compared (Fig. 5.19). This index of absorbed radiation is most useful in ecosystems with low to moderate NDVI, because the relationship saturates at high NDVI. To the extent that light use efficiency is similar among ecosystem types, the summation of absorbed radiation through the season of active plant growth should be a reasonable index of carbon input to ecosystems.

Controls over GPP

Ecosystem differences in GPP are determined primarily by leaf area index and the duration of the photosynthetic season and secondarily by the environmental controls over photosynthesis.

Leaf Area

Variation in soil resource supply accounts for much of the spatial variation in leaf area and GPP among ecosystem types. Analysis of satellite imagery shows that about 70% of the ice-free terrestrial surface has relatively open canopies (Graetz 1991) (Fig. 5.20). GPP correlates closely with leaf area below a total LAI of about 8 (projected LAI of 4) (Schulze et al. 1994), suggesting that leaf area is a critical determinant of GPP on most of Earth's terrestrial surface (Fig. 5.1). GPP is less sensitive to

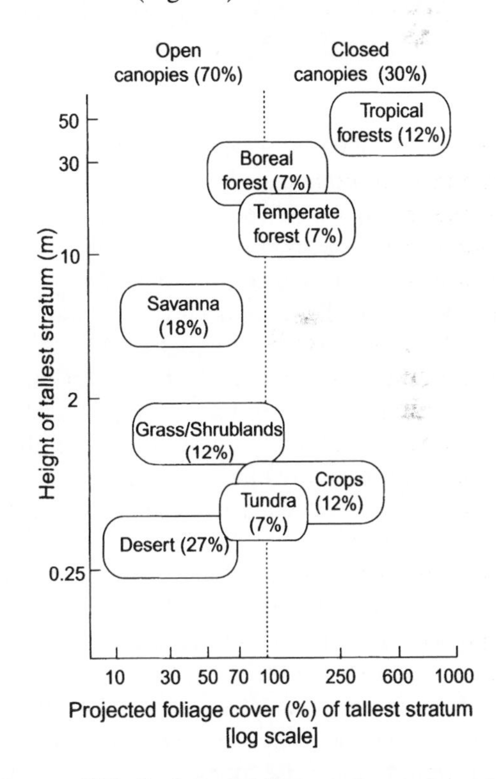

FIGURE 5.20. Projected foliage cover and canopy height of the major biomes. Typical values for that biome and the percentage of the terrestrial surface that it occupies are shown. The vertical line shows 100% canopy cover. (Reprinted with permission from *Climatic Change*, Vol. 18 © 1991 Kluwer Academic Publishers; Graetz 1991.)

LAI in dense canopies, because the leaves in the middle and bottom of the canopy contribute relatively little to GPP. The availability of soil resources, especially water and nutrient supply, is a critical determinant of LAI for two reasons: Plants in high-resource environments produce a large amount of leaf biomass, and leaves produced in these environments have a high SLA—that is, a large leaf area per unit of leaf biomass. As discussed earlier, a high SLA maximizes light capture and therefore carbon gain per unit of leaf biomass (Fig. 5.12) (Lambers and Poorter 1992, Reich et al. 1997). In low-resource environments, plants produce fewer leaves, and these leaves have a lower SLA. Ecosystems in these environments have a low LAI and therefore a low GPP.

Disturbances, herbivory, and pathogens reduce leaf area below levels that resources can support. Soil resources and light extinction through the canopy determine the upper limit to the leaf area that an ecosystem can support. However, many factors regularly reduce leaf area below this potential LAI. Drought and freezing are climatic factors that cause plants to shed leaves. Other causes of leaf loss include physical disturbances (e.g., fire and wind) and biotic agents (e.g., herbivores and pathogens). After major disturbances the remaining plants may be too small, have too few meristems, or lack the productive potential to produce the leaf area that could potentially be supported by the climate and soil resources of a site. For this reason, LAI tends to increase with time after disturbance to an asymptote and then (at least in forests) often declines in late succession (see Chapter 13).

Human activities increasingly affect the leaf area of ecosystems in ways that cannot be predicted from climate. Overgrazing by cattle, sheep, and goats, for example, directly removes leaf area and causes shifts to vegetation types that are less productive and have less leaf area than would otherwise occur in that climate zone. Acid rain and other pollutants also cause leaf loss. Nitrogen deposition can stimulate leaf production above levels that would be predicted from climate and soil type. Because of human activities, LAI cannot be estimated simply from correlations with climate. Fortunately, satellites provide the opportunity to estimate LAI directly. This information is an important input to global models that calculate regional patterns of carbon input to terrestrial ecosystems.

Length of the Photosynthetic Season

The length of the photosynthetic season accounts for much of the ecosystem differences in GPP. Most ecosystems experience times that are too cold or too dry for significant photosynthesis to occur. During winter in cold climates and times with negligible soil water in dry climates, plants either die (annuals), lose their leaves (deciduous plants), or become physiologically dormant (some evergreen plants). During these times, there is negligible carbon uptake by the ecosystem, regardless of light availability and CO_2 concentration. In a sense, the nonphotosynthetic season is simply a case of extreme environmental stress. Conditions are so severe that plants gain negligible carbon. At high latitudes and altitudes and in dry ecosystems, this is probably *the* major constraint on carbon inputs to ecosystems (Fig. 5.1; see Chapter 6) (Körner 1999). For annuals and deciduous plants, the lack of leaf area is sufficient to explain the absence of photosynthetic carbon gain in the nongrowing season. Lack of water or extremely low temperatures can, however, prevent even evergreen plants from gaining carbon. Some evergreen species partially disassemble their photosynthetic machinery during the nongrowing season. These plants require some time following the return of favorable environmental conditions to reassemble their photosynthetic machinery (Bergh and Linder 1999), so not all early-season irradiance is used efficiently to gain carbon. In tropical ecosystems, however, where conditions are more continuously favorable for photosynthesis, leaves maintain their photosynthetic machinery from the time they are fully expanded until they are shed. Models that simulate GPP often define the length of the photosynthetic season in terms of thresholds of minimum temperature or moisture below which plants do not produce leaves or do not photosynthesize.

Environmental Controls over Growing-Season GPP

Environmental controls over GPP during the growing season are identical to those described for net photosynthesis of individual leaves. As described earlier, variations in light, temperature, and moisture account for much of the diurnal and seasonal variations in net photosynthesis. These factors have a particularly strong effect on leaves at the top of the canopy, which account for most GPP. The thinner boundary layer and greater distance for water transport from roots, for example, makes the uppermost leaves particularly sensitive to variation in temperature, soil moisture, and relative humidity. Consequently, the response of GPP to diurnal and seasonal variations in environment is similar to that described for net photosynthesis by individual leaves. Variation in these environmental factors explains much of the diurnal and seasonal variations in GPP. Soil resources (nutrients and moisture) influence GPP through their effects on both photosynthetic potential and leaf area. Photosynthetic potential is closely matched with the light-harvesting capacity of leaves. Soil resources therefore influence GPP primarily by determining the capacity of canopies to capture light (through variations in leaf area and photosynthetic potential) rather than through variations in the efficiency of converting light to carbohydrates. The *seasonal changes* in GPP are sensitive to variations in the photosynthetic rate of individual leaves, due to variations in light and temperature; this causes variation in light use efficiency. In contrast, differences in GPP *among ecosystem types and among successional stages* may be determined more strongly by differences in the quantity of light absorbed as a result of differences in leaf area and photosynthetic potential.

Summary

Most carbon enters terrestrial ecosystems through photosynthesis mediated by plants. The light-harvesting reactions of photosynthesis transform light energy into chemical energy, which is used by the carbon-fixation reactions to convert CO_2 to sugars. The enzymes that carry out these reactions account for about half of the nitrogen in the leaf. Plants adjust the components of photosynthesis so physical and biochemical processes co-limit carbon fixation. At low light, for example, plants reduce the quantity of photosynthetic machinery per unit leaf area by producing thinner leaves. As atmospheric CO_2 concentration increases, plants reduce stomatal conductance. The net effect of these adjustments is that ecosystem carbon gain is relatively insensitive to differences among ecosystems in light or CO_2 availability. The major environmental factors that explain differences among ecosystems in carbon gain are the length of time during which conditions are suitable for photosynthesis and the soil resources (water and nutrients) available to support the production and maintenance of leaf area. Environmental stresses such as inadequate water supply, extreme temperatures, and pollutants reduce the efficiency with which plants use light to gain carbon. Plants respond to these stresses by reducing leaf area and nitrogen content so as to maintain a relatively constant efficiency in the use of light to fix carbon.

Review Questions

1. How do plants with different photosynthetic pathways differ in their photosynthetic responses to water and nitrogen? What is the biochemical basis for these differences in response to environment?
2. How does each major environmental variable (CO_2, light, nitrogen, water, temperature, pollutants) affect photosynthetic rate in the short term? How do the photosynthetic properties of individual leaves change in response to changes in these factors to optimize photosynthetic performance?
3. How does the response of photosynthesis to one environmental variable (e.g., water or nitrogen) affect the response to other environmental variables (e.g., light, CO_2, or pollutants)? Considering these interactions among environmental variables, how might

anthropogenic increases in nitrogen inputs affect the response of Earth's ecosystems to rising atmospheric CO_2?
4. How do environmental stresses affect light use efficiency in the short term? How does vegetation adjust to maximize LUE in stressful environments over the long term?
5. What factors are most important in explaining differences among ecosystems in GPP? Over what time scale does each of these factors have its greatest impact on GPP? Explain your answers.
6. What factors most strongly affect leaf area and photosynthetic capacity of vegetation?
7. How do the factors regulating photosynthesis in a forest canopy differ from those in individual leaves? How does availability of soil resources (water and nutrients) and the structure of the canopy influence the importance of these canopy effects?

Additional Reading

Ehleringer, J.R., and C.B. Field, editors. 1993. *Scaling Physiological Processes: Leaf to Globe*. Academic Press, San Diego, CA.

Field, C., and H.A. Mooney. 1986. The photosynthesis-nitrogen relationship in wild plants. Pages 25–55 *in* T.J. Givnish, editors. *On the Economy of Plant Form and Function*. Cambridge University Press, Cambridge, UK.

Lambers, H., F.S. Chapin III, and T. Pons. 1998. *Plant Physiological Ecology*. Springer-Verlag, Berlin.

Larcher, W. 1995. *Physiological Plant Ecology*. Springer-Verlag, Berlin.

Mooney, H.A. 1972. The carbon balance of plants. *Annual Review of Ecology and Systematics* 3:315–346.

Reich, P.B., M.B. Walters, and D.S. Ellsworth. 1997. From tropics to tundra: Global convergence in plant functioning. *Proceedings of the National Academy of Sciences U. S. A.* 94:13730–13734.

Ruimy, A., P.G. Jarvis, D.D. Baldocchi, and B. Saugier. 1996. CO_2 fluxes over plant canopies and solar radiation: A review. *Advances in Ecological Research* 26:1–68.

Sage, R.F., and R.K. Monson, editors. 1999. *C_4 Plant Biology*. Academic Press, San Diego, CA.

Schulze, E.-D., F.M. Kelliher, C. Körner, J. Lloyd, and R. Leuning. 1994. Relationship among maximum stomatal conductance, ecosystem surface conductance, carbon assimilation rate, and plant nitrogen nutrition: A global ecology scaling exercise. *Annual Review of Ecology and Systematics* 25:629–660.

6
Terrestrial Production Processes

The balance between gross primary production and ecosystem respiration determines the net carbon accumulation by the biosphere. This chapter describes the factors that regulate the carbon balance of terrestrial vegetation and ecosystems.

Introduction

The carbon balance of vegetation and ecosystems governs the productivity of the biosphere and the impact of ecosystems on the Earth System. Plant production determines the amount of energy available to sustain all organisms, including humans. We depend on plant production directly for food and fiber and indirectly because of the critical role of vegetation in all ecosystem processes (see Chapter 16). Much of the carbon produced by plants eventually moves to the soil, where it influences the capacity of soils to retain water and nutrients and therefore to support plant production (see Chapter 3). Carbon cycling through ecosystems also directly affects Earth's climate by modifying the concentration of atmospheric CO_2 (see Chapter 2). Because of the many critical roles of carbon balance in the biosphere and the Earth System, the recent rapid change in carbon cycling of plants and ecosystems is an issue of fundamental societal importance.

Overview

Carbon that enters ecosystems as gross primary production (GPP) accumulates within the ecosystem, returns to the atmosphere via respiration or disturbance, or is transported laterally to other ecosystems. About half of GPP is respired by plants to provide the energy that supports their growth and maintenance (Schlesinger 1997, Waring and Running 1998). **Net primary production** (NPP) is the net carbon gain by vegetation and equals the difference between GPP and plant respiration. Plants lose carbon through several pathways besides respiration (Fig. 6.1). The largest of these releases is typically the transfer of carbon from plants to the soil. This occurs through **litterfall** (the shedding of plant parts and death of plants), **root exudation** (the secretion of soluble organic compounds by roots into the soil), and carbon transfers to microbes that are symbiotically associated with roots (e.g., mycorrhizae and nitrogen-fixing bacteria). These carbon transfers from plants to soil eventually give rise to **soil organic matter** (SOM), which is typically the largest pool of ecosystem carbon. Herbivores also remove carbon from plants. Herbivory often accounts for 5 to 10% of NPP in terrestrial ecosystems but can be less than 1% in some forests or greater than 50% in some grasslands (see Chapter 11). Herbivores account for most of the carbon loss from plants in aquatic ecosystems (see Chapter 10). Plants also release carbon to the atmosphere through emission of volatile organic compounds or by

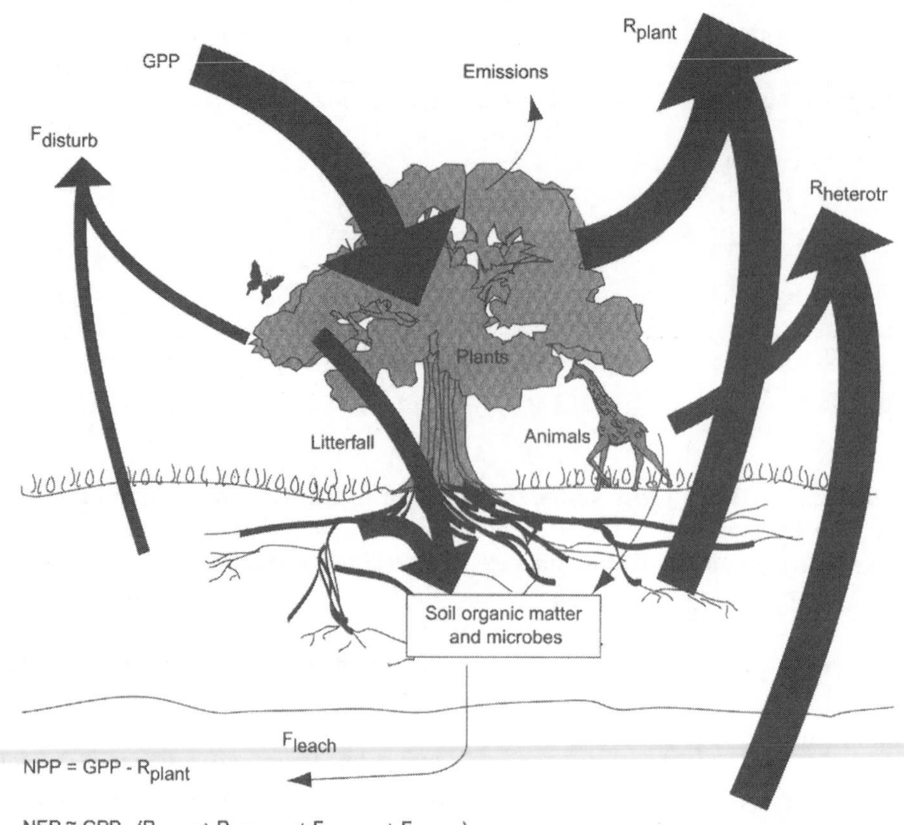

FIGURE 6.1. Overview of the major carbon fluxes of an ecosystem. Carbon enters the ecosystem as gross primary production (GPP), through photosynthesis by plants. Roots and aboveground portions of plants return about half of this carbon to the atmosphere as plant respiration (R_{plant}). Net primary production (NPP) is the difference between carbon gain by GPP and carbon loss through R_{plant}. Most NPP is transferred to soil organic matter as litterfall, root death, root exudation, and root transfers to symbionts; some NPP is eaten by animals and sometimes is lost from the ecosystem through disturbance. Animals also transfer some carbon to soils through excretion and mortality. Most carbon entering the soil is lost through microbial respiration (which, together with animal respiration, is termed heterotrophic respiration; $R_{heterotr}$). Additional carbon is lost from soils through leaching and disturbance. Net ecosystem production (NEP) is the net carbon accumulation by an ecosystem; it equals the carbon inputs from GPP minus the various avenues of carbon loss: respiration, leaching, and disturbance. If an ecosystem were at steady state, in the absence of disturbance, carbon inputs in GPP would approximately balance the carbon outputs in plant respiration (about 50% of GPP), heterotrophic respiration (40 to 50% of GPP), and leaching (0 to 10% of GPP). Most ecosystems, however, generally show either a net gain or net loss of carbon (i.e., positive or negative NEP, respectively), due to an imbalance between GPP and the various avenues of carbon loss.

combustion in fires. Volatile emissions typically account for less than 1% of NPP but give plants their distinctive smells, which govern the behavior of many herbivores and are an important component of atmospheric chemistry. Fire is rare in some ecosystems but can be the major avenue of carbon release from plants in many ecosystems in some years. Finally, carbon can be removed from vegetation by human harvest or other disturbances.

The carbon balance of ecosystems depends not only on the carbon balance of vegetation but also on the respiration of **heterotrophs**, organisms that eat live or dead organic matter.

Heterotrophic respiration by microbes and animals converts organic matter to CO_2, which is lost from the ecosystem to the atmosphere. In some ecosystems fire transforms additional organic matter to CO_2, which moves to the atmosphere. Finally carbon leaches from ecosystems in dissolved and particulate forms and moves laterally through erosion and deposition of soil, movement of animals, etc. These lateral fluxes of carbon from terrestrial ecosystems are critical energy subsidies to aquatic ecosystems and constitute a significant component of the carbon budgets of many ecosystems.

Plant Respiration

Physiological Basis of Respiration

Respiration provides the energy for a plant to acquire nutrients and to produce and maintain biomass. At the ecosystem scale, plant respiration includes mitochondrial respiration of nonphotosynthetic organs at all times and the respiration of leaves at night (Schlesinger 1997). Leaf respiration in the light is included within GPP (see Chapter 5). Plant respiration is not "wasted" carbon. It serves the essential function in plants of providing energy for growth and maintenance, just as it does in animals and microbes. We can separate total plant respiration (R_{plant}) into three functional components: growth respiration (R_{growth}), maintenance respiration (R_{maint}), and the respiratory cost of ion uptake (R_{ion}).

$$R_{plant} = R_{growth} + R_{maint} + R_{ion} \quad (6.1)$$

Each of these components of respiration involves mitochondrial oxidation of carbohydrates to produce ATP. They differ only in the functions for which ATP is used by the plant. Separation of respiration into these components allows us to understand the ecological controls over plant respiration.

All plants are similar in their efficiency of converting sugars into new biomass. Growth of new tissue requires biosynthesis of many classes of chemical compounds, including cellulose, proteins, nucleic acids, and lipids (Table 6.1). The carbon cost of synthesizing each compound includes the carbon that is incorporated into that compound plus the carbon oxidized to CO_2 to provide the ATPs that drive biosynthesis. These carbon costs can be calculated for each class of compound from a knowledge of their biosynthetic pathways (Penning de Vries et al. 1974). The cost of producing 1 g tissue can then be calculated from the concentration of each class of chemical compound in a tissue and its carbon cost of synthesis.

There is a threefold range in the carbon cost of synthesis of the major classes of chemical

TABLE 6.1. Concentration and carbon cost of major chemical constituents in a sedge leaf.

Component	Concentration (%)	Cost ($mg\,C\,g^{-1}$ product)	Total cost ($mg\,C\,g^{-1}$ tissue)[a]
Sugar	11.9	438	52
Nucleic acid	1.2	409	5
Polysaccharide	9.0	467	42
Cellulose	21.6	467	101
Hemicellulose	31.0	467	145
Amino acid	0.9	468	4
Protein	9.7	649	63
Tannin	4.8	767	37
Lignin	4.2	928	39
Lipid	5.7	1212	69
Total cost			557

[a] The four most expensive constituents account for 37% of the cost of synthesis but only 24% of the mass of the tissue. The total cost of production (557 $mg\,C\,g^{-1}$ tissue) is equivalent to 1.23 g carbohydrate per gram of tissue, indicating growth respiration consumes about 25% more carbon than accumulates in biomass.

Reprinted with permission from *American Naturalist*; Chapin (1989).

compounds found in plants (Table 6.1). The most expensive compounds in plants are proteins, tannins, lignin, and lipids. In general, metabolically active tissues, such as leaves, have high concentrations of proteins, tannins, and lipids. The tannins and lipophilic substances such as terpenes serve primarily to defend protein-rich tissues from herbivores and pathogens (see Chapter 11). Structural tissues have high lignin and low protein, tannin, and lipid concentrations. Leaves of rapidly growing species with high protein concentration have higher tannin and lower lignin concentrations than leaves with low protein concentrations. Consequently, most plant tissues contain some expensive constituents, although the nature of these constituents differs among plant parts and species. In fact, the carbon cost of plant growth is surprisingly similar across species, tissue types, and ecosystems (Chapin 1989, Poorter 1994) (Fig. 6.2). On average, growth respiration is about 25% of the carbon incorporated into new tissues (Table 6.1) (Waring and Running 1998). The rates of growth and therefore of growth respiration measured at the ecosystem scale (grams of carbon per square meter per day) increase when temperature and moisture favor growth, but growth respiration is always a nearly constant fraction of NPP, regardless of environmental conditions.

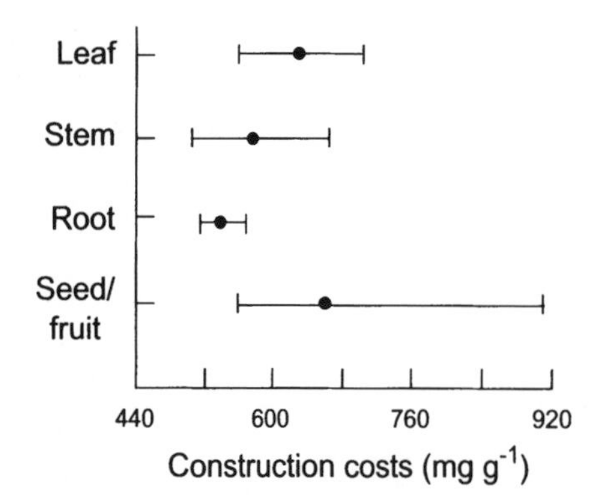

FIGURE 6.2. Range of construction costs for a survey of leaves ($n = 123$), stems ($n = 38$), roots ($n = 35$), and seeds or fruits ($n = 31$). Values are means with 10th and 90th percentiles. The carbon cost of producing new biomass differs little among plant parts, except for seeds and fruits that store lipid. These tissues have a higher cost of synthesis than do other plant parts. (Redrawn with permission from SPB Academic; Poorter 1994.)

The total respiratory cost of ion uptake probably correlates with NPP. Ion transport across membranes is energetically expensive and may account for 25 to 50% of root respiration (Lambers et al. 1998). Several factors cause this cost of ion uptake to differ among ecosystems. The quantity of nutrients absorbed is greatest in productive environments, although the respiratory cost per unit of absorbed nutrients may be greater in slow-growing plants from unproductive environments (Lambers et al. 1998). The respiratory cost of nitrogen uptake and use depends on the form of nitrogen absorbed, because nitrate must be reduced to ammonium (an exceptionally expensive process) before it can be incorporated into proteins or other organic compounds. The cost of nitrate reduction is also variable among plant species and ecosystems, depending on whether the nitrate is reduced in the roots or the leaves (see Chapter 8). In general, we expect R_{ion} to correlate with the total quantity of ions absorbed and therefore to show a positive relationship with NPP.

Maintenance respiration: How variable is the cost of maintaining plant biomass? All live cells, even those that are not actively growing, require energy to maintain ion gradients across cell membranes and to replace proteins, membranes, and other constituents. Maintenance respiration provides the ATP for these maintenance and repair functions. Laboratory experiments suggest that about 85% of maintenance respiration is associated with the turnover of proteins (about 6% turnover per day), explaining why there is a strong correlation between protein concentration and whole-tissue respiration rate in nongrowing tissues (Penning de Vries 1975). We therefore expect maintenance respiration to be greatest in ecosystems with high tissue nitrogen concentrations and/or a large plant biomass and thus to be greatest in productive ecosystems. Simulation models suggest that maintenance respiration may account for about half of total plant respiration; the other half is associated with growth and ion uptake (Lambers et al. 1998). These proportions may vary with environment and plant growth rate and are difficult to estimate precisely.

Maintenance respiration depends on environment as well as tissue chemistry. It increases with temperature because proteins and membrane lipids turn over more rapidly at high temperatures. Drought also imposes short-term metabolic costs associated with synthesis of osmotically active organic solutes (see Chapter 4). These effects of environmental stress on maintenance respiration are the major factors that alter the partitioning between growth and respiration and therefore are the major sources of variability in the efficiency of converting GPP into NPP. Maintenance respiration increases during times of environmental change but, following acclimation, maintenance respiration returns to values close to those predicted from biochemical composition (Semikhatova 2000). Over the long term, therefore, maintenance respiration is not strongly affected by environmental stress.

Plant respiration is a relatively constant proportion of GPP, when ecosystems are compared. Although the respiration rate of any given plant increases exponentially with ambient temperature, acclimation and adaptation counterbalance this direct temperature effect on respiration. Plants from hot environments have lower respiration rates at a given temperature than do plants from cold places (Billings and Mooney 1968). The net result of these counteracting temperature effects is that plants from different thermal environments have similar respiration rates when measured at their mean habitat temperature (Semikhatova 2000).

In summary, studies of the basic components of respiration associated with growth, ion uptake, and maintenance suggest that total plant respiration should be a relatively constant fraction of GPP. These predictions are consistent with the results of model simulations of plant carbon balance. These modeling studies indicate that total plant respiration is about half (48 to 60%) of GPP when a wide range of ecosystems is compared (Ryan et al. 1994, Landsberg and Gower 1997) (see Fig. 6.6). Variation in maintenance respiration is the most likely cause for variability in the efficiency of converting GPP into NPP. There are too few detailed studies of ecosystem carbon balance to know how variable this efficiency is among seasons, years, and ecosystems. The current view that the efficiency of converting GPP to NPP is relatively constant may reflect insufficient data or could emerge as an important ecosystem generalization.

Net Primary Production

What Is NPP?

Net primary production is the net carbon gain by vegetation. It is the balance between the carbon gained by gross primary production and carbon released by plant mitochondrial respiration.

$$NPP = GPP - R_{plant} \quad (6.2)$$

Like GPP, NPP is generally measured at the ecosystem scale, usually over relatively long time intervals, such as a year (grams biomass or grams carbon per square meter per year). NPP includes the new biomass produced by plants, the soluble organic compounds that diffuse or are secreted by roots into the soil (**root exudation**), the carbon transfers to microbes that are symbiotically associated with roots (e.g., mycorrhizae and nitrogen-fixing bacteria), and the volatile emissions that are lost from leaves to the atmosphere (Clark et al. 2001a). Most field measurements of NPP document only the new plant biomass produced and therefore probably underestimate the true NPP by at least 30% (Table 6.2). Root exudates are rapidly taken up and respired by microbes adjacent to roots and are generally measured in field studies as a portion of root respiration. Volatile emissions are also rarely measured but are generally a small fraction (less than 1 to 5%) of NPP and thus are probably not a major source of error (Guenther et al. 1995). Some biomass aboveground and belowground dies or is removed by herbivores before it can be measured, so even the new biomass measured in field studies is an underestimate of biomass production. For some purposes, these errors

TABLE 6.2. Major components of NPP and representative values of their relative magnitudes.

Components of NPP[a]	% NPP
New plant biomass	40–70
Leaves and reproductive parts (fine litterfall)	10–30
Apical stem growth	0–10
Secondary stem growth	0–30
New roots	30–40
Root secretions	20–40
Root exudates	10–30
Root transfers to mycorrhizae	10–30
Losses to herbivores and mortality	1–40
Volatile emissions	0–5

[a] Seldom, if ever, have all of these components been measured in a single study.

may not be too important. A frequent objective of measuring NPP, for example, is to estimate the rate of biomass increment. Root exudates, transfers to symbionts, losses to herbivores, and volatile emissions are lost from plants and therefore do not directly contribute to biomass increment. Consequently, failure to measure these components of NPP does not bias estimates of biomass accumulation. However, these losses of NPP from plants fuel other ecosystem processes such as herbivory, decomposition, and nutrient turnover and so are important components of the overall carbon dynamics of ecosystems and a critical carbon source for microbes.

Some components of NPP, such as root production, are particularly difficult to measure and have sometimes been assumed to be some constant ratio (e.g., 1:1) of aboveground production (Fahey et al. 1998). Fewer than 10% of the studies that report ecosystem NPP actually measure components of belowground production (Clark et al. 2001a). Estimates of aboveground NPP sometimes include only large plants (e.g., trees in forests) and exclude understory shrubs or mosses, which can account for a substantial proportion of NPP in some ecosystems. Most published summaries of NPP do not explicitly state which components of NPP have been included (or sometimes even whether the units are grams of carbon or grams of biomass). For these reasons, considerable care must be used when comparing data on NPP or biomass among studies.

Physiological Controls over NPP

Photosynthesis, NPP, and respiration: Who is in charge? NPP is the balance of carbon gained by GPP and the carbon lost by respiration of all plant parts (Fig. 6.1). However, this simple equation (Eq. 6.2) does not tell us whether the conditions governing photosynthesis dictate the amount of carbon that is available to support growth or whether conditions influencing growth rate determine the magnitude of photosynthesis. On short time scales (seconds to days), environmental controls over photosynthesis (e.g., light and water availability) strongly influence photosynthetic carbon gain. However, on weekly to annual time scales, plants appear to adjust leaf area and photosynthetic capacity so carbon gain matches the soil resources that are available to support growth (see Fig. 5.1). Plant carbohydrate concentrations are usually lowest when environmental conditions favor rapid growth (i.e., carbohydrates are drawn down by growth) and tend to accumulate during periods of drought or nutrient stress or when low temperature constrains NPP (Chapin 1991b). If the products of photosynthesis directly controlled NPP, we would expect high carbohydrate concentrations to coincide with rapid growth or to show no consistent relationship with growth rate.

Results of growth experiments also indicate that growth is not simply a consequence of the controls over photosynthetic carbon gain. Plants respond to low availability of water, nutrients, or oxygen in their rooting zone by producing hormones that reduce growth rate. The decline in growth subsequently leads to a decline in photosynthesis (Gollan et al. 1985, Chapin 1991b, Davies and Zhang 1991). The general conclusion from these experiments is that plants actively sense the resource supply in their environment and adjust their growth rate accordingly. These changes in growth rate then change the **sink strength** (demand) for carbohydrates and nutrients, leading to changes in photosynthesis and nutrient uptake (Chapin 1991b, Lambers et al. 1998). The resulting changes in growth and nutrition determine the leaf area index (LAI) and photosynthetic capacity, which, as we have seen, largely account for ecosystem differences in carbon input (Gower et al. 1999) (see Fig. 5.1).

The feedbacks to photosynthesis are not 100% effective: Leaf carbohydrate concentrations increase during the day and decline at night, allowing plants to maintain a relatively constant supply of carbohydrates to nonphotosynthetic organs. Similarly, carbohydrate concentrations increase during periods (hours to weeks) of sunny weather and decline under cloudy conditions. Over these short time scales, the conditions affecting photosynthesis are the primary determinants of the carbohydrates available to support growth. The short-term controls over photosynthesis by environment probably determine the hourly to weekly patterns of NPP, whereas soil resources govern annual carbon gain and NPP.

Environmental Controls over NPP

The climatic controls over NPP are mediated primarily through the availability of belowground resources. At a global scale, the largest ecosystem differences in NPP are associated with variation in climate. NPP is greatest in warm moist environments, where tropical rain forests occur, and is least in climates that are dry (e.g., deserts) or cold (e.g., tundra) (see plates 1 to 3). NPP correlates most strongly with precipitation; NPP is highest at 2 to $3\,m\,yr^{-1}$ of precipitation (typical of rain forests) and declines at extremely high precipitation (Gower 2002, Schuur et al. 2001). When dry ecosystems (i.e., deserts) are excluded, NPP also increases exponentially with increasing temperature (Fig. 6.3). The largest differences in NPP reflect biome differences in both climate and vegetation structure. When ecosystems are grouped into biomes, there is a 14-fold range in average NPP (Table 6.3; Fig. 6.4). Do these correlations of NPP with climate reflect a simple direct effect of temperature and moisture on plant growth, or are other factors involved?

Extensive research in temperate grasslands illustrates some of the complexities of environmental controls over NPP. When grassland sites are compared across precipitation gradients, the average NPP of a site increases with

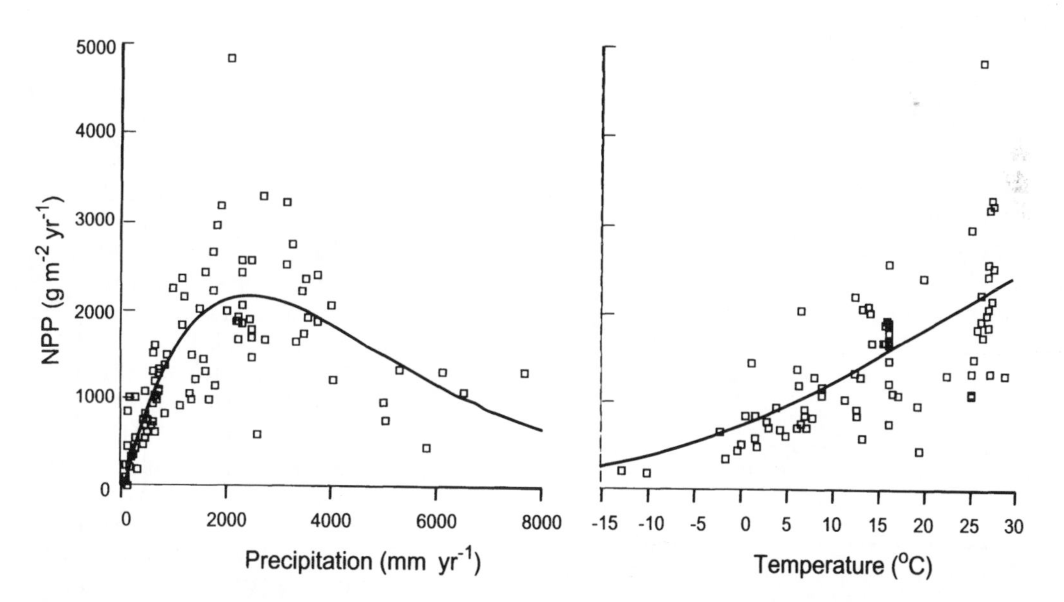

FIGURE 6.3. Correlation of NPP (in units of biomass) with temperature and precipitation. NPP is greatest in warm moist environments such as tropical forests and lowest cold or dry ecosystems such as tundra and deserts. In tropical forests, NPP declines at extremely high precipitation ($>3\,m\,yr^{-1}$), due to indirect effects of excess moisture, such as low soil oxygen and loss of nutrients through leaching. (Figure modified from Schuur Unpublished; data from Lieth 1975, Clark et al. 2001b, and Schuur et al. 2001.)

TABLE 6.3. Net primary production of the major biome types based on biomass harvests[a].

Biome	Aboveground NPP ($g m^{-2} yr^{-1}$)	Belowground NPP ($g m^{-2} yr^{-1}$)	Belowground NPP (% of total)	Total NPP ($g m^{-2} yr^{-1}$)
Tropical forests	1400	1100	0.44	2500
Temperate forests	950	600	0.39	1550
Boreal forests	230	150	0.39	380
Mediterranean shrublands	500	500	0.50	1000
Tropical savannas and grasslands	540	540	0.50	1080
Temperate grasslands	250	500	0.67	750
Deserts	150	100	0.40	250
Arctic tundra	80	100	0.57	180
Crops	530	80	0.13	610

[a] NPP is expressed in units of dry mass. NPP estimated from harvests excludes NPP that is not available to harvest as a result of consumption by herbivores, root exudation, transfer to mycorrhizae, and volatile emissions.
Data from Saugier et al. (2001).

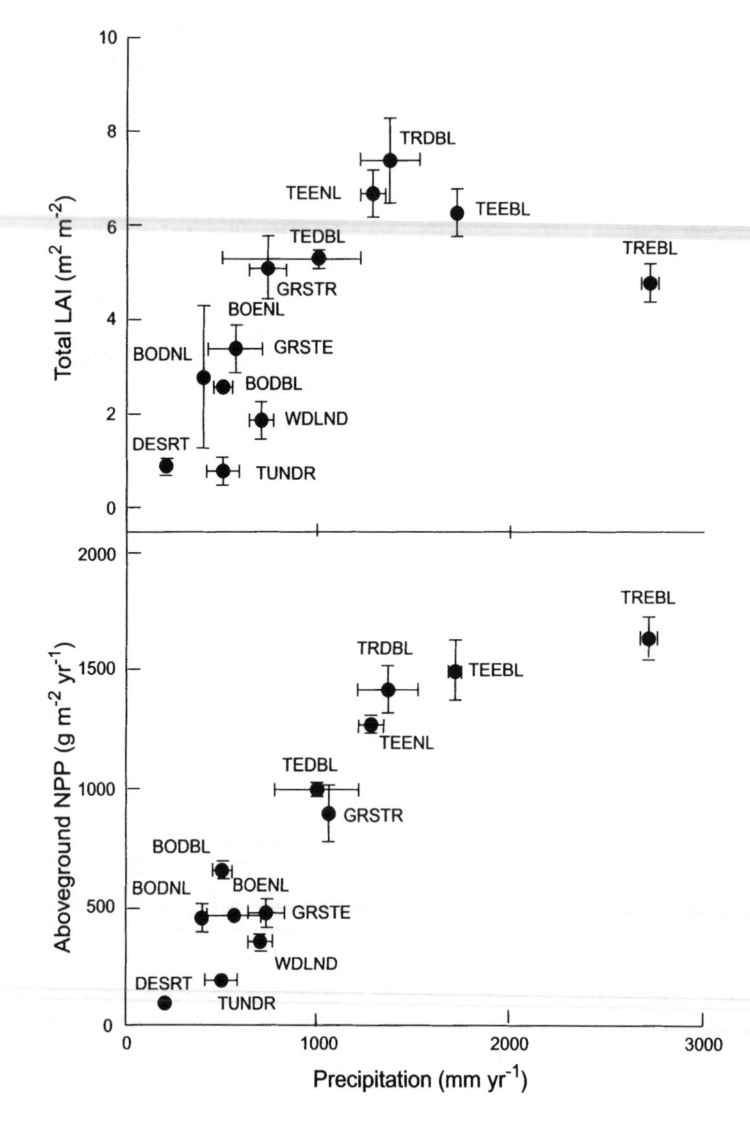

FIGURE 6.4. LAI and aboveground net primary production of major biome types as a function of precipitation. DESRT, desert; TUNDR, tundra; BOENL, boreal evergreen needle-leafed; WDLND, woodland; BODBL, boreal deciduous broadleaved; BODNL, boreal deciduous needleleaved; GRSTE, temperate grassland; TEDBL, temperate deciduous broadleaved; TEENL, temperate evergreen needle-leafed; TRDBL, tropical deciduous broadleafed; TREBL, tropical evergreen broadleaved; GRSTR, tropical grassland; and TEEBL, temperate evergreen broad-leafed. (Redrawn with permission from Blackwell Scientific; Gower 2002.)

increasing precipitation (Fig. 6.5) (Sala et al. 1988, Lauenroth and Sala 1992), just as observed across biomes (Fig. 6.3 and 6.4). In any single grassland site, NPP also increases in years with high precipitation and responds to experimental addition of water, demonstrating that grassland NPP is water limited (Lauenroth et al. 1978). However, part of the water limitation reflects the effects of water on moisture-limited decomposition and therefore nutrient supply (see Chapters 7 and 8). Thus at least two resources (water and nutrients) limit the NPP of temperate grasslands, and the relative importance of these resources depends on climate and soil type. What about other resources? No one has tested whether addition of light would stimulate the productivity of any natural ecosystem. A doubling of atmospheric CO_2 stimulates grassland NPP by 10 to 30%, but most of this stimulation reflects the effects of CO_2 on water and nutrient availability rather than the direct effects of CO_2 on photosynthesis. Finally, species composition and biomass influence the response of grassland NPP to climate. Arid grasslands are never as productive in wet years as grasslands that regularly receive high moisture inputs, presumably because arid grasslands lack the plant species, biomass, or soil fertility to exploit effectively the years of high moisture (Fig. 6.5) (Lauenroth and Sala 1992). In grasslands, therefore, water appears to be the factor that most strongly controls NPP, but soil moisture determines NPP in at least three ways: through its direct stimulation of NPP, through its effects on nutrient supply, and through its effect on the species composition and productive capacity of the ecosystem.

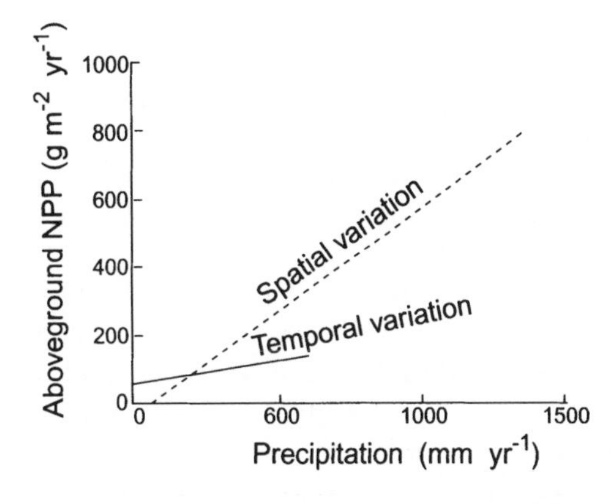

FIGURE 6.5. Correlation of grassland NPP with precipitation across grassland sites (spatial variation) and through time for a single site (temporal variation). A single site responds less to interannual variation in precipitation than would be expected from the relationship between average precipitation and average NPP across all sites, because a single site lacks the species and productive potential capable of exploiting high moisture availability. (Redrawn with permission from *Ecological Applications*; Lauenroth and Sala 1992.)

The controls over NPP in deserts are similar to those in grasslands: Desert NPP correlates closely with precipitation among sites, among years, and in response to water addition (Gutierrez and Whitford 1987). Even in deserts, however, NPP is greatest in patches with high nutrient availability (Schlesinger et al. 1990) and responds to added nitrogen, especially in experiments that also add water (Gutierrez and Whitford 1987), indicating a secondary limitation of desert NPP by nutrient supply.

In the tundra, where the climate correlations suggest that NPP should be temperature limited, NPP increases more in response to added nitrogen than to experimental increases in temperature (Chapin et al. 1995, McKane et al. 1997). Thus, in tundra, the climate–NPP correlation probably reflects the effects of temperature on nitrogen supply (see Chapter 9) or length of growing season more than a direct temperature effect on NPP (Chapin 1983). Similarly, NPP in the boreal forest correlates closely with soil temperature, but soil warming experiments demonstrate that this effect is mediated primarily by enhanced decomposition and nitrogen supply (Van Cleve et al. 1990).

Thus in ecosystems in which climate–NPP correlations suggest a strong climatic limitation of NPP, experiments and observations indicate that this is mediated primarily by climatic effects on belowground resources. What constrains NPP in warm, moist climates where temperature and moisture appear optimal for growth?

Tropical forests typically have higher NPP than other biomes (Fig. 6.4). Among tropical forests, litter production tends to correlate with the supply of nutrients, especially phosphorus (Vitousek 1984), suggesting that NPP in tropical forests may also be limited by the supply of belowground resources. NPP in tropical forests is maximal at intermediate levels of precipitation (Schuur et al. 2001). NPP in dry tropical forests is moisture limited, but in extremely wet climates (greater than 2 to $3\,m\,yr^{-1}$ of precipitation) NPP declines in response to increasing precipitation, probably due to oxygen limitation to roots and/or soil microbes and to leaching loss of essential nutrients. NPP in tropical forests is therefore probably also limited by the supply of belowground resources, including nutrients and sometimes water or oxygen.

In a temperate salt marsh, where water and apparently nutrients are abundant, NPP responds directly to increases in CO_2 (Drake et al. 1996), as do crops that receive a high nutrient supply. However, NPP is enhanced by nutrient additions even in the most fertile agricultural systems (Evans 1980), indicating the widespread occurrence of nutrient limitation to NPP (see Fig. 8.1).

In summary, experiments and observations in a wide range of ecosystems provide a relatively consistent picture. NPP is generally constrained by the supply of belowground resources. The factors determining the supply and acquisition of belowground resources are the major direct controls over NPP and therefore the carbon input to ecosystems. Only in the most productive ecosystems do other factors, such as CO_2 concentration, directly enhance the NPP of ecosystems.

The importance of belowground resources in controlling NPP is consistent with our earlier conclusion that GPP is governed more by leaf area and length of the photosynthetic season than by the direct effects of temperature and CO_2 on photosynthesis (see Chapter 5). In fact, modeling studies suggest that NPP is a surprisingly constant fraction (40 to 52%) of GPP across broad environmental gradients (Fig. 6.6) (Landsberg and Gower 1997, Waring and Running 1998). This is consistent with our conclusion that GPP and NPP are controlled by the same factors.

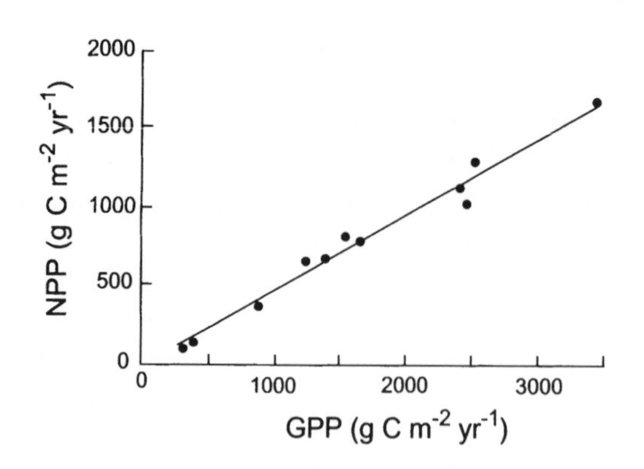

FIGURE 6.6. Relationship between GPP and NPP in 11 forests of the United States, Australia, and New Zealand (Williams et al. 1997). These forests were selected from a wide range of moisture and temperature conditions. GPP and NPP were estimated using a model of ecosystem carbon balance. The simulations suggest that all these forests show a similar partitioning of GPP between plant respiration (53%) and NPP (47%), despite large variations in climate. (Redrawn with permission from Academic Press; Waring and Running 1998.)

Allocation

Allocation of NPP

Patterns of biomass allocation minimize resource limitation and maximize resource capture and NPP. Our discussion of the controls over NPP suggests an interesting paradox: A high leaf area is necessary to maximize NPP, yet the major factors that constrain NPP are belowground resources. The plant is faced with a dilemma of how to distribute biomass between leaves (to maximize carbon gain) and roots (to maximize acquisition of belowground resources). Plants exhibit a consistent pattern of **allocation**—the distribution of growth among plant parts—that maximizes growth in response to the balance between aboveground and belowground resource supply rates (Enquist and Niklas 2001).

In general, plants allocate production to minimize limitation by any single resource. Plants allocate new biomass preferentially to roots when water or nutrients limit growth. They allocate new biomass preferentially to shoots when light is limiting (Reynolds and Thornley 1982).

Plants can increase acquisition of a resource by producing more biomass of the appropriate tissue, by increasing the activity of each unit of biomass, or by retaining the biomass for a longer time. A plant can, for example, increase carbon gain by increasing leaf area or photosynthetic rate per unit leaf area or by retaining the leaves for a longer time before they are shed. Similarly, a plant can increase nitrogen uptake by altering root morphology or by increasing root biomass, root longevity, nitrogen uptake rate per unit root, or extent of mycorrhizal colonization. Changes in allocation and root morphology have a particularly strong impact on nutrient uptake. It is the integrated activity (mass multiplied by acquisition rate per unit biomass multiplied by time) that must be balanced between shoots and roots to maximize growth and NPP (Garnier 1991). These allocation rules are key features of all simulation models of NPP.

Observations in ecosystems are generally consistent with allocation theory. Tundra, grasslands, and shrublands, for example, allocate a larger proportion of NPP belowground than do forests (Table 6.3) (Gower et al. 1999, Saugier et al. 2001).

Allocation Response to Multiple Resources

NPP in most ecosystems is limited most strongly by a single resource, but it also responds to other resources. If plants were perfectly successful in allocating biomass to acquire the most limiting resource, they would be equally limited by all resources (Bloom et al. 1985, Rastetter and Shaver 1992). As we have seen, this is seldom the case. NPP in most ecosystems responds most strongly to a particular resource, for example to water in deserts, grasslands, and arid shrublands; to nitrogen in tundra, boreal forests, and temperate forests; and to phosphorus in tropical wet and dry forests. Thus, as a first approximation, deserts are water-limited ecosystems, and temperate forests are nitrogen-limited ecosystems. In many ecosystems, however, NPP does respond to increased availability of more than one resource. Why does this occur?

The simplest view of environmental limitation is that growth is limited by a single resource at any moment in time. Another resource becomes limiting only when the supply of the first resource increases above the point of limitation (Liebig's **law of the minimum**). At least four processes contribute to the multiple resource limitation observed in many ecosystems: (1) Plants adjust resource acquisition to maximize capture of (and minimize limitation by) the most limiting resource. (2) Changes in the environment (e.g., rain storms, pulses of nutrient supply) change the relative abundance of resources so different factors limit NPP at different times. (3) Plants exhibit mechanisms that increase the supply of the most limiting resource. (4) Different resources limit different species in an ecosystem, so ecosystem-scale NPP responds to the addition of more than one resource. Each of these processes contributes to the response of ecosystems to multiple resources.

Plants adjust resource acquisition to maximize capture of (and minimize limitation by) the most limiting resource. As discussed earlier, plants adjust allocation of new production to roots vs. shoots to minimize limitation by belowground vs. aboveground resources, respectively. Plants also alter allocation within the root system to maximize capture of the most limiting belowground resource (Rastetter and Shaver 1992). In deserts, for example, nutrient availability is greatest close to the soil surface, whereas water supplies are generally more consistently available at depth. The amount of nutrient or water that a new root acquires therefore depends on the depth at which it is produced. To acquire water, some desert plants produce coarse, deep water roots, which effectively conduct water but are relatively ineffective in absorbing nutrients. Other plants produce only shallow roots and remain active only when surface water is available. The biochemical investment by roots is specific for each nutrient. Nitrogen uptake, for example, requires synthesis of specific enzymes to absorb nitrogen, reduce nitrate, and assimilate reduced nitrogen into amino acids, whereas different enzymes are required to absorb phosphorus (see Chapter 8).

Changes in the environment (e.g., rain storms, pulses of nutrient supply) change the relative abundance of resources so different factors limit NPP at different times. Most ecosystems probably experience temporal changes in the factor that most limits NPP because essential resources do not become equally available at the same time. Light, for example, decreases but water increases during rainy periods. Many ecosystems experience a pulse of nutrient availability at the beginning of the growing season, when temperatures may be suboptimal for growth. Because all the major factors that determine NPP change dramatically over several time scales, it would be surprising if there were not corresponding changes in the relative importance of these factors in limiting NPP.

Temporal changes in the limitation of NPP are buffered by storage. Plants accumulate carbohydrates or nutrients during times when their availability is high and use their stores to support growth when the supply declines (Chapin et al. 1990). Over seasonal time scales, plants use stored carbohydrates and nutrients to support their burst of spring growth and replenish these stores at other times when photosynthesis and nutrient uptake exceed the demands for growth (see Chapter 8). Plants vary substantially in their capacity to store water (see Chapter 4). In summary, storage enables plants to acquire resources when they are readily available and use them at times of low supply, thus reducing temporal variation in the nature of resource limitation.

Plants exhibit several mechanisms that increase the supply of the most limiting resource. Plants that have symbiotic associations with nitrogen-fixing microbes directly promote nitrogen inputs to ecosystems (see Chapter 8). Some ericoid and ectomycorrhizal associates of other plant species break down proteins and transport the resulting amino acids to plants (Read 1991). Some plants enhance the supply of phosphorus through the production of organic chelates that solubilize mineral phosphorus or through the production of phosphatases that cleave organic phosphates. Plants also exude carbohydrates that enhance mineralization near the root (see Chapter 9). These mechanisms will be described in more detail in later chapters.

Different resources limit different species, so ecosystem-scale NPP responds to the addition of more than one resource. Every species in an ecosystem has slightly different environmental requirements and therefore will be limited by different resource combinations (Tilman 1988). Tundra species, for example, differ in their response to temperature, light, and nutrients (Chapin and Shaver 1985) and in some cases to the addition of nitrogen vs. phosphorus. These differences among plant species in the factors limiting growth contribute to the co-existence of species in a variable environment (Tilman 1988). This may be particularly important in explaining why species differ in their productivity among years and why the productivity of ecosystems varies less among years than does the productivity of any of the component species (Chapin and Shaver 1985). Spatial heterogeneity in the supply of potentially limiting resources is another important reason why different plants may be limited by different resources.

Diurnal and Seasonal Cycles of Allocation

Photosynthesis and growth are highly resilient to daily and seasonal variations in the environment. Daily and seasonal variations in the environment are two of the most predictable perturbations experienced by ecosystems. Many organisms adjust their physiology and behavior based on innate **circadian** (about 24-h) **rhythms** that lead to 24-h cycles. For example, stomatal conductance and carbon gain show a circadian rhythm even under constant conditions because stomata have an innate 24-h cycle of stomatal opening and closing. Plants store starch in the leaves during the day and break it down at night, so the rate of carbohydrate transport to roots is nearly constant (Lambers et al. 1998). Thus belowground processes, such as root exudation and carbon transport to mycorrhizae, are buffered from diurnal variations in photosynthetic carbon gain.

Organisms adjust seasonally in response to changing **photoperiod** (day length). Many temperate plants, for example, exhibit a relatively predictable pattern of **phenology**, the seasonal timing of production and loss of leaves, flowers, fruits, etc. Plant leaves begin to senesce and reduce their rates of photosynthesis when day length or other environmental cues signal the characteristic onset of winter. Before senescence, plants transport carbohydrates and nutrients from leaves to storage organs to prevent their loss during senescence (Chapin et al. 1990). These stores provide resources to support plant growth the next spring, so NPP does not depend entirely on acquisition of new resources at times when no leaves are present. Other ecosystem processes change as either direct consequences of changes in environment (e.g., the decline in decomposition during winter due to lower temperatures) or indirect consequences of changes in other processes (e.g., the pulse of litter input to soil after leaf senescence). Ecosystem processes largely recover after each period of the cycle due to the predictable nature of diurnal and seasonal perturbations and the resilience of most processes to these changes. It is therefore unnecessary to consider explicitly the physiological basis of circadian and photoperiodic controls to predict ecosystem processes over longer time scales (see Chapter 13). In contrast to temperate ecosystems, tropical wet forests exhibit a less well-defined seasonality. Individual species frequently shed their leaves synchronously, but species differ in their timing of senescence, so the ecosystem as a whole shows no strong seasonal pulse of production and senescence.

The seasonality of plant growth depends on the seasonality of leaf area and factors regulating photosynthesis. Spring growth of plants is initially supported by stored reserves of carbon and nutrients that were acquired in previous years. Leaves quickly become a net source of carbon for the rest of the plant, and growth during the remainder of the growing season is largely supported by the current year's photosynthate. There is often competition among plant parts for allocation of a limited carbohydrate supply early in the growing season, resulting in a seasonal progression of production of different plant parts, for example, with leaves produced first, followed by roots, and then by wood (Kozlowski et al. 1991). Plants species differ, however, in their seasonal patterns of allocation. Plants with evergreen leaves may allocate NPP to root growth earlier than would deciduous plants, because they already have a leaf canopy that can provide carbon (Kummerow et al. 1983). Ring-porous temperate trees must first allocate carbon to xylem production in spring to develop a functional water transport system. The water columns in their large-diameter vessels **cavitate** (break) during winter freezing, so xylem vessels remain functional for only a single growing season. This large carbon requirement to rebuild xylem vessels each spring may explain the northern boundary of ring-porous species such as oaks (Zimmermann 1983). Seedlings in dry environments often depend entirely on their cotyledons for photosynthesis during the first weeks of growth and allocate all NPP to root growth to provide a dependable water supply. The allocation calendar of a plant provides a general seasonal framework for allocation. Variations in environment cause plants to modify this allocation calendar to achieve the appropriate balance of carbon, water, and nutrients.

In ecosystems with short growing seasons, such as arctic tundra, a substantial proportion of the "current" year's production is actually supported by resources that were acquired in previous years. In late summer, carbon and nutrient stores are replenished to support the next year's production. This seasonal pattern of storage buffers plant production from seasonal and interannual variations in the environment (Chapin et al. 1990). The seasonality of plant growth is constrained by the availability of leaf area early and late in the growing season but otherwise follows seasonal patterns of factors that govern photosynthesis (temperature, light, and moisture); the relative importance and timing of these seasonal controls differ among species and ecosystems.

Tissue Turnover

The balance between NPP and biomass loss determines the annual increment in plant biomass. Plants retain only part of the biomass that they produce. Some of this biomass loss is physiologically regulated by the plant, for example the senescence of leaves and roots. Senescence occurs throughout the growing season in grasslands and during autumn or at the beginning of the dry season in many trees. Other losses occur with varying frequency and predictability and are less directly controlled by the plant, such as the losses to herbivores and pathogens, wind-throw, and fire. The plant also influences these tissue loss rates through the physiological and chemical properties of the tissues it produces. Still other biomass transfers to dead organic matter result from mortality of individual plants. Given the substantial, although incomplete, physiological control over tissue loss, why do plants dispose of the biomass in which they have invested so much carbon, water, and nutrients to produce?

Tissue loss is an important mechanism by which plants balance resource requirements with resource supply from the environment. Plants depend on regular large inputs of carbon; water; and, to a lesser extent, nutrients to maintain vital processes. For example, once biomass is produced, it must have continued carbon inputs to support maintenance respiration. If the plant (or organ) cannot meet these carbon demands, the plant (or organ) dies. Similarly, if the plant cannot absorb sufficient water to replace the water that is inevitably lost during photosynthesis, it must shed transpiring organs (leaves) or die. The plant must therefore shed biomass whenever resources decline below some threshold needed for maintenance. Senescence is just as important as production in adjusting to changes in resource supply and is the only mechanism by which plants can reduce biomass when resources decline in abundance.

Senescence is the programmed breakdown of tissues. The location of senescence is physiologically controlled to eliminate tissues that are least useful to the plant. Grazing of aboveground tissues, for example, can cause a pulse of root mortality (Ruess et al. 1998), whereas grazing of belowground tissues reduces the longevity of leaves (Detling et al. 1980). Although the controls over senescence and mortality of belowground tissues are poorly understood, these patterns of senescence appear to maintain the functional balance between leaves and roots in response to a changing environment (Garnier 1991).

Growth and senescence together enable individual plants to explore new territory. Leaf and shoot growth generally occur at the top of the canopy or in canopy gaps, where light availability is highest. This is balanced by senescence of leaves and stems in less favorable light environments (Bazzaz 1996). This balance between biomass production and loss allows trees and shrubs to grow toward the light. Similarly, roots often proliferate in areas of nutrient enrichment or where there is minimal competition from other roots, and root death is greatest in zones of local water or nutrient depletion (see Chapter 8). This exploration of unoccupied habitat by plants would occur much less effectively if there were not senescence and loss of tissues in less favorable habitats to reduce maintenance requirements and to provide nutrient capital to produce new tissues. The exploration of new territory through synchronized growth and senescence reduces spatial variability in ecosystems by filling canopy gaps and exploiting nutrient-rich patches of soil.

Senescence causes tissue loss at times when maintenance costs greatly exceed resource gain. In seasonally variable environments, there are extended periods of time when temperature or moisture is predictably unfavorable. In these ecosystems, the cost of producing tissues that can withstand the rigors of this unfavorable period and of maintaining tissues when they provide negligible benefit to the plant may exceed the cost of producing new tissues when conditions again become favorable (Chabot and Hicks 1982). Arctic, boreal, and temperate ecosystems, for example, predictably experience seasons that are too cold for effective growth or resource acquisition. There is a pulse of autumn senescence of leaves and roots, often triggered by some combination of photoperiod and low temperature (Ruess et al. 1996). Dry ecosystems experience similar pulses of leaf and root senescence with the onset of drought. Senescence and tissue loss are therefore highly pulsed in most ecosystems and occur just before the period when conditions are least favorable for resource acquisition and growth. These seasonal pulses of senescence cause the greatest tissue loss in highly seasonal environments.

Leaf longevity varies among plant species from a few weeks to several years or decades. In general, plants in high-resource environments produce short-lived leaves with a high specific leaf area (SLA) and a high photosynthetic rate per leaf mass, but they have little resistance to environmental stresses and are poorly defended against herbivores. These disposable leaves are typically shed when conditions become unfavorable (winter or dry season) and are replaced the next spring. Both root and leaf longevity are greater in low-resource environments (Berendse and Aerts 1987) and lower at high latitudes than in the tropics (Fig. 6.7). The greater longevity of leaves from low-resource environments reduces the nutrient requirement by plants to maintain leaf area (see Chapter 8).

Senescence enables plants to shed parasites, pathogens, and herbivores. Because leaves and fine roots represent relatively large packets of nutrients and organic matter, they are constantly under attack from pathogens, parasites, and herbivores. **Phyllosphere fungi**, for example, begin colonizing and growing on leaves shortly after budbreak, initially as parasites and later as part of the decomposer community, when the leaf is shed (see Chapter 7). These fungi account for some of the mottled appearance of older leaves. Pathogenic root fungi are a major cause of reduced yields in agroecosystems and are common in natural ecosystems. Plants have a variety of mechanisms for detecting natural enemies and respond initially through the production of induced chemical defenses (see Chapter 11) and, in the case of severe attack, by shedding tissues.

Large unpredictable biomass losses occur in most ecosystems. Wind storms, fires, herbivore outbreaks, and epidemics of pathogens frequently cause large tissue losses that are unpredictable and occur before any programmed senescence of tissues and associated nutrient resorption can occur. These unpredictable biomass losses incur approximately twice the nutrient loss to the plant as that occurring after senescence (see Chapter 8). They often increase spatial heterogeneity of light and nutrient resources in the ecosystem through creation of gaps, which range in scale from the loss of individual leaves to the destruction of biomass over large regions. Most ecosystems are at some stage in the regrowth after such biomass losses.

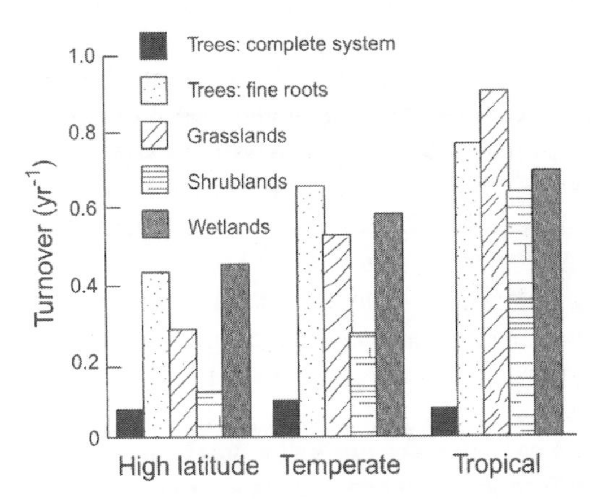

FIGURE 6.7. Synthesis of information on root turnover in major ecosystem types along a latitudinal gradient. (Redrawn with permission from *New Phytologist*; Gill and Jackson 2000.)

Global Distribution of Biomass and NPP

Biome Differences in Biomass

The plant biomass of an ecosystem is the balance between NPP and tissue turnover. NPP and tissue loss are seldom in perfect balance. NPP tends to exceed tissue loss shortly after disturbance; at other times tissue loss exceeds NPP. As ecosystems or landscapes approach steady state (see Chapter 14), however, there is often a consistent relationship between plant biomass and the climate or biome type that characterizes that ecosystem. Average plant biomass varies 50-fold among Earth's major

TABLE 6.4. Biomass distribution of the major terrestrial biomes[a].

Biome	Shoot ($g m^{-2}$)	Root ($g m^{-2}$)	Root (% of total)	Total ($g m^{-2}$)
Tropical forests	30,400	8,400	0.22	38,800
Temperate forests	21,000	5,700	0.21	26,700
Boreal forests	6,100	2,200	0.27	8,300
Mediterranean shrublands	6,000	6,000	0.5	12,000
Tropical savannas and grasslands	4,000	1,700	0.3	5,700
Temperate grasslands	250	500	0.67	750
Deserts	350	350	0.5	700
Arctic tundra	250	400	0.62	650
Crops	530	80	0.13	610

[a] Biomass is expressed in units of dry mass.
Data from Saugier et al. (2001).

terrestrial biomes (Table 6.4). Forests have the most biomass. Among forests, average biomass declines 5-fold from the tropics to the low-stature boreal forest, where NPP is low and stand-replacing fires frequently remove biomass. Deserts and tundra have only 1% as much aboveground biomass as do tropical forests. In any biome, disturbance frequently reduces plant biomass below levels that the climate and soil resources could support. Crops, for example, have a biomass similar to that of tundra or desert, despite more favorable growing conditions; regular removal of crop biomass by harvest prevents it from accumulating to levels that climate and soil resources could potentially support. When disturbance frequency declines, for example through fire prevention in grasslands and savannas, biomass often increases through invasion of shrubs and trees.

Patterns of biomass allocation reflect the factors that most strongly limit plant growth in ecosystems (Table 6.4). Between 70 and 80% of the biomass in forests is aboveground because forests characterize sites with relatively abundant supplies of water and nutrients, so light often limits the growth of individual plants. In shrublands, grasslands, and tundra, however, water or nutrients more severely limit production, and most biomass occurs belowground. Because of favorable water and nutrient regimes, crops maintain a smaller proportion of biomass as roots than do most unmanaged ecosystems.

Tropical forests account for about half of Earth's total plant biomass, although they occur on only 13% of the ice-free land area; other forests contribute an additional 30% of global biomass (Table 6.5). Nonforest biomes therefore account for less than 20% of total plant biomass, although they occupy 70% of the ice-free land surface. Crops for example, account for only 1% of terrestrial biomass, although they occupy more than 10% of the ice-free land area. Thus most of the terrestrial surface has relatively low biomass (see Fig. 5.20). This observation alone raises concerns about tropical deforestation, independent of the associated species losses.

Biome Differences in NPP

The length of the growing season is the major factor explaining biome differences in NPP. Most ecosystems experience times that are too cold or too dry for significant photosynthesis or for plant growth to occur. When NPP of each biome is adjusted for the length of the growing season, all forested ecosystems have similar NPP (about $5 g m^{-2} d^{-1}$), and there is only about a threefold difference in NPP between deserts and tropical forests (Table 6.6). These calculations suggest that the length of the growing season accounts for much of the biome differences in NPP (Gower et al. 1999, Körner 1999).

Leaf area accounts for much of the biome differences in carbon gain during the growing season. Average total LAI varies about sixfold among biomes; the most productive ecosystems generally have the highest LAI (Table 6.6; Fig. 6.4). When NPP is adjusted for differences

TABLE 6.5. Global distribution of terrestrial biomes and their total carbon in plant biomass[a].

Biome	Area ($10^6 km^2$)	Total C pool (Pg C)	Total NPP (Pg C yr^{-1})
Tropical forests	17.5	340	21.9
Temperate forests	10.4	139	8.1
Boreal forests	13.7	57	2.6
Mediterranean shrublands	2.8	17	1.4
Tropical savannas and grasslands	27.6	79	14.9
Temperate grasslands	15.0	6	5.6
Deserts	27.7	10	3.5
Arctic tundra	5.6	2	0.5
Crops	13.5	4	4.1
Ice	15.5		
Total	149.3	652	62.6

[a] Biomass is expressed in units of carbon, assuming that plant biomass is 50% carbon.
Data from Saugier et al. (2001).

in both length of growing season and leaf area, unproductive ecosystems, such as tundra or desert, do not differ consistently in NPP from more productive ecosystems. If anything, the less-productive ecosystems may have higher NPP per unit of leaf area and growing season length than do crops and forests. On average, plants in most biomes produce 1 to 3 g biomass m^{-2} leaf d^{-1} during the growing season. This is equivalent to a GPP of 1 to 3 g carbon m^{-2} leaf d^{-1}, because NPP is about half of GPP and biomass is about 50% carbon. Apparent differences among biomes in these values reflect a substantial uncertainty in the underlying data. At this point, there is little evidence for strong ecological patterns in NPP per unit leaf area and length of growing season.

LAI is both a cause and a consequence of differences in NPP. LAI is determined largely by the availability of soil resources (mainly water and nutrients), climate, and time since disturbance (see Chapter 5; Fig. 6.4). Tropical rain forests, for example, occur in a warm, moist climate that provides adequate water and nutrient release to support a large leaf area. These leaves remain photosynthetically active throughout the year, because there are no long periods of unfavorable weather causing massive leaf loss, and plants can tap stores of deep groundwater during dry months

TABLE 6.6. Productivity per day and per unit leaf area[a].

Biome	Season length[b] (days)	Daily NPP per ground area (g $m^{-2} d^{-1}$)	Total LAI[c] ($m^2 m^{-2}$)	Daily NPP per leaf area (g $m^{-2} d^{-1}$)
Tropical forests	365	6.8	6.0	1.14
Temperate forests	250	6.2	6.0	1.03
Boreal forests	150	2.5	3.5	0.72
Mediterranean shrublands	200	5.0	2.0	2.50
Tropical savannas and grasslands	200	5.4	5.0	1.08
Temperate grasslands	150	5.0	3.5	1.43
Deserts	100	2.5	1.0	2.50
Arctic tundra	100	1.8	1.0	1.80
Crops	200	3.1	4.0	0.76

[a] Calculated from Table 6.3. NPP is expressed in units of dry mass.
[b] Estimated.
[c] Data from Gower (2002).

(Woodward 1987). Deserts, in contrast, produce little leaf area because of inadequate precipitation and water storage, and arctic tundra is cold and supplies nitrogen too slowly to produce a large leaf area. In both deserts and tundra, the short growing season gives little time for leaf production, and unfavorable conditions between growing seasons limit leaf survival. The resulting low leaf area that generally characterizes these ecosystems is a major factor accounting for their low productivity (Table 6.6).

Disturbances modify the relationship between climate and NPP. There is substantial variability in NPP among sites within a biome. Some of this variability reflects variation in state factors such as climate and parent material. However, disturbance also affects NPP substantially, in part through changes in resource supply and LAI. Forest NPP, for example, frequently increases after disturbance until the canopy closes, and the available light is fully used (see Fig. 13.8) (Ryan et al. 1997). In later successional forests, NPP declines for a variety of reasons (see Chapter 13).

About 60% of the NPP of the biosphere occurs on land; the rest occurs in aquatic ecosystems (see Chapter 15). When summed at the global level, tropical forests account for about a third of Earth's terrestrial NPP; all forests account for about half of terrestrial NPP (Table 6.5). Grasslands and savannas account for an additional third of terrestrial NPP; these ecosystems are much more important in their contribution to terrestrial production than to biomass. Crops are 10-fold more productive than the global average; they account for about 10% of terrestrial production and occupy 1% of the ice-free land surface.

Net Ecosystem Production

Ecosystem Carbon Storage

Net ecosystem production (NEP) is the net accumulation of carbon by an ecosystem. It is the balance between carbon entering and leaving the ecosystem. Most carbon enters the ecosystem as gross primary production and leaves through several processes, including plant and heterotrophic respiration, leaching, plant volatile emissions, methane flux, and disturbance. Lateral transfers, such as erosion/deposition or animal movements, can be either carbon inputs to or outputs from the ecosystem (Fig. 6.8; Box 6.1). Alternatively, we can express the carbon input to ecosystems as NPP (the balance between GPP and plant respiration) and the respiratory carbon loss as heterotrophic respiration from microbes and animals ($R_{heterotr}$) (Box 6.1). This enables us to consider separately the fluxes associated with plants (carbon inputs) and with heterotrophs (carbon outputs).

NEP is ecologically important because it represents the increment in carbon stored by an ecosystem. When integrated globally, NEP determines the impact of the terrestrial biosphere on the quantity of CO_2 in the atmosphere, which strongly affects climate and the amount of carbon transferred to oceans (see Chapter 15). The components of NEP show large temporal variation. Disturbances such as fire or forest harvest are episodic events that dominate the carbon exchange of an ecosystem when they occur, but are less important at other times. GPP is an important process during the growing season, when conditions are suitable for photosynthesis but is negligible at other times of year. Heterotrophic respiration and leaching loss of carbon also vary seasonally and through succession (see Chapter 13). In ecosystems that have not recently experienced disturbance, NEP is a small net difference between two large fluxes: (1) photosynthetic carbon gain and (2) carbon loss through respiration (primarily plants and microbes) and leaching (Fig. 6.9). During the season of peak plant growth, NEP is positive because photosynthesis exceeds respiration. In winter, when photosynthesis is low, NEP is negative and is mainly due to heterotrophic respiration. Thus carbon uptake seldom balances carbon loss from ecosystems at any point in the seasonal cycle.

There is a necessary functional linkage between NPP and heterotrophic respiration. NPP provides the organic material that fuels heterotrophic respiration, and heterotrophic

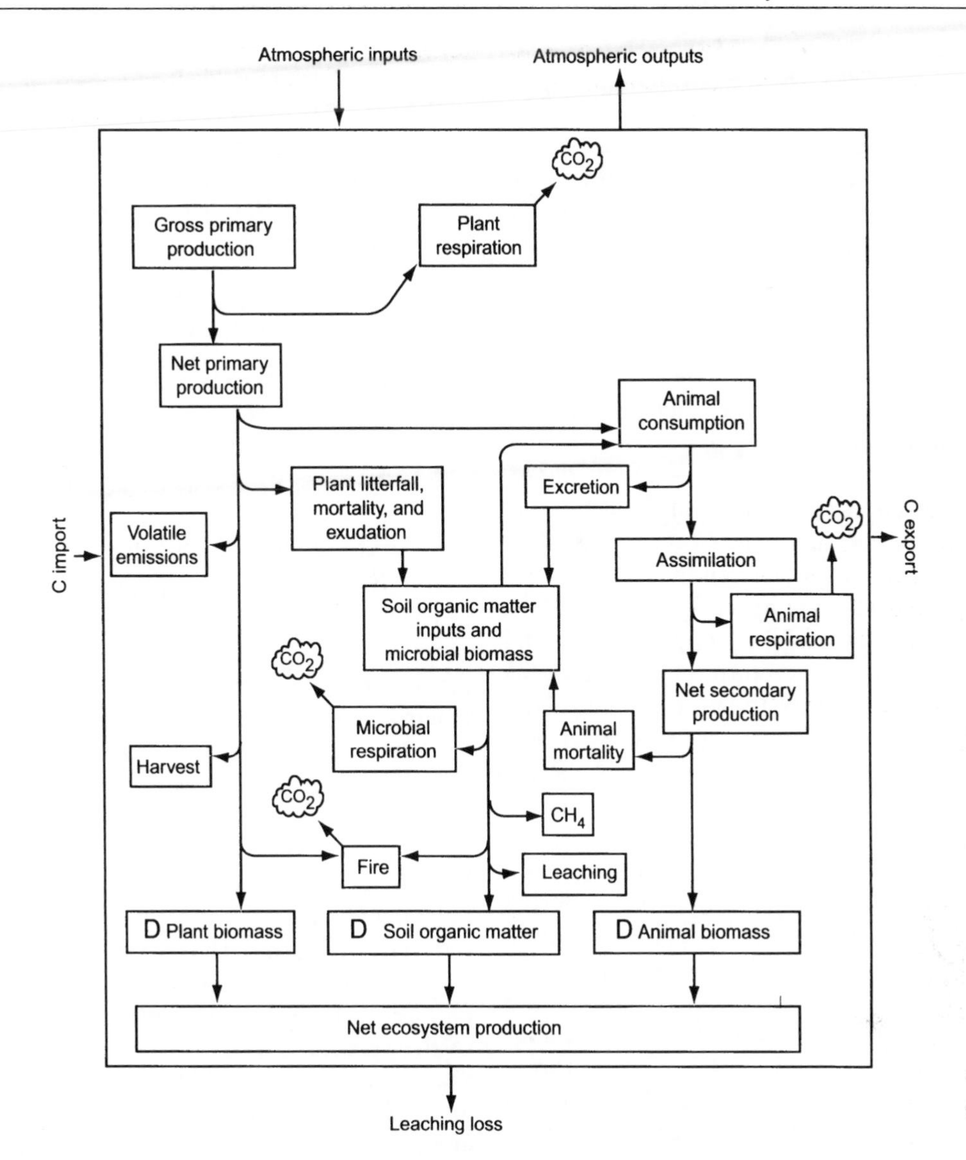

FIGURE 6.8. Overview of the carbon fluxes of an ecosystem. The large box represents the ecosystem, which exchanges carbon with the atmosphere, other ecosystems, and groundwater.

respiration releases the minerals that support NPP (Harte and Kinzig 1993). For these reasons, NPP and heterotrophic respiration tend to be closely matched in ecosystems at steady state. At steady state, by definition, NEP equals zero, regardless of carbon input or climate. In fact, peat bogs, which are quite unproductive, are the ecosystems with the greatest long-term carbon storage, because the anaerobic soil conditions characteristic of these bogs restrict decomposition more strongly than they restrict NPP.

Leaching

Leaching of dissolved organic carbon (DOC) and dissolved inorganic carbon (DIC) to groundwater and streams is a quantitatively important avenue of carbon loss from some ecosystems (Fig. 6.8). Groundwater is generally supersaturated with respect to CO_2 because of the high CO_2 concentration in the soil atmosphere. Some dissolved CO_2 leaches out of the ecosystem to groundwater and then moves to streams and lakes, where the excess CO_2 is

Box 6.1. Components of an Ecosystem Carbon Budget

Plant biomass accumulation is the balance between GPP and the losses of carbon from plants. An important motivation for studying primary production and decomposition (see Chapter 7) is to understand how these processes influence the net carbon gain or loss by an ecosystem. Here we summarize, in an ecosystem context, the major processes that were described previously (Fig. 6.8).

GPP is the net carbon gain by photosynthesis in the light and is the best measure of the carbon that enters ecosystems. Nonphotosynthetic organs (and leaves at night) respire (plant respiration; R_{plant}) some of the carbon fixed by GPP to support growth and maintenance. R_{plant} returns about half of the carbon fixed by GPP to the atmosphere. NPP is the net annual carbon gain by vegetation and equals the difference between GPP and plant respiration (Eq. 6.2).

The second largest avenue of carbon loss from plants, after respiration, is the carbon flux to the soil ($F_{pl\text{-}soil}$) through litterfall (the shedding of plant parts and plant mortality), secretion of soluble organic compounds by roots into the soil (root exudation), and carbon fluxes to microbes that are symbiotically associated with roots (e.g., mycorrhizae and nitrogen fixers). These carbon fluxes from plants to soil are the largest inputs to soil organic matter (SOM). Another important pathway of carbon loss from plants is herbivory (F_{herbiv}), the consumption of plants by animals (see Chapter 11). Plants also lose small amounts of carbon to the atmosphere by emission of volatile organic compounds (F_{emiss}). Finally, carbon can be lost from plants by fire ($F_{pl\text{-}fire}$) or human harvest (F_{harv}). Carbon losses due to fire or harvest differ from the other components of the plant carbon budget because they are highly episodic. In a typical year, these fluxes have values of zero, but, when they occur, they can be the dominant pathway of carbon loss from vegetation. The annual accumulation of plant biomass (ΔB_{plant}) depends on the balance between carbon gain through NPP (or GPP) and the various pathways of carbon loss over a given time (t) interval.

$$\frac{\Delta B_{plant}}{\Delta t} = NPP - (F_{pl\text{-}soil} + F_{herbiv} + F_{emiss} + F_{pl\text{-}fire} + F_{harv}) \quad (B6.1)$$

or

$$\frac{\Delta B_{plant}}{\Delta t} = GPP - R_{plant} - (F_{pl\text{-}soil} + F_{herbiv} + F_{emiss} + F_{pl\text{-}fire} + F_{harv}) \quad (B6.2)$$

As with plants, the accumulation of animal biomass (ΔB_{animal}) equals inputs minus outputs—that is, the plant biomass eaten (F_{herbiv}) minus losses to animal respiration (R_{animal}) and fluxes from animals to soils ($F_{anim\text{-}soil}$) as a result of excretion and mortality. Soil animals feed primarily on microbial biomass ($F_{micro\text{-}anim}$). When animals are eaten by other animals (including humans), some of the carbon remains in the animal biomass pool (but transferred to a different organism). The rest of the carbon is transferred to the soil as unconsumed biomass or feces or is transferred into or out of the ecosystem ($F_{lateral}$) (see Chapter 11). Transfers of carbon within the animal box are subsumed in the overall carbon budget equations.

$$\frac{\Delta B_{animal}}{\Delta t} = F_{herbiv} + F_{microb\text{-}anim} - (R_{animal} + F_{anim\text{-}soil}) \quad (B6.3)$$

SOM accumulates when the annual carbon inputs to SOM from plants ($F_{pl\text{-}soil}$) and animals ($F_{anim\text{-}soil}$) exceed the respiration of soil microorganisms, including soil animals (R_{microb}), and carbon losses due to consumption of microbes by soil animals ($F_{micro\text{-}anim}$), methane emissions to the atmosphere (F_{CH_4}), leaching of organic and inorganic carbon to groundwater (F_{leach}), and fire ($F_{soil\text{-}fire}$). Leaching of dissolved organic and inorganic carbon in groundwater can be a significant loss from ecosystems, although it is often poorly quantified. There can be lateral carbon fluxes ($F_{lateral}$) into or out of the ecosystem due to deposition or erosion.

$$\frac{\Delta SOM}{\Delta t} = F_{pl\text{-}soil} + F_{anim\text{-}soil} - (R_{microb} + F_{microb\text{-}anim} + F_{CH_4} + F_{leach} + F_{soil\text{-}fire}) \quad (B6.4)$$

The pool sizes of plants, animals, and SOM get larger when inputs exceed outputs and get smaller when outputs exceed inputs. Fire, for example, causes an instantaneous decrease in the pool size of plant and soil carbon, whereas these pools generally increase in size during succession after fire (see Chapter 13).

Ecosystem carbon accumulation depends primarily on the balance between carbon inputs through photosynthesis and carbon losses through respiration and disturbance. Net ecosystem production is the net annual carbon accumulation by the ecosystem and is the sum of the net carbon accumulation in plants, animals, and the soil plus lateral transfers of carbon among ecosystems ($F_{lateral}$) (Olsen 1963, Bormann et al. 1974, Aber and Melillo 1991).

$$NEP = \frac{(\Delta B_{plant} + \Delta B_{anim} + \Delta SOM)}{\Delta t} \pm F_{lateral} \quad (B6.5)$$

NEP is positive when carbon inputs to the ecosystem exceed carbon losses and is negative when losses exceed inputs. It is difficult, however, to measure NEP accurately from changes in the carbon pools in plants, animals and SOM, because the changes in these pools over short time intervals are small relative to measurement errors. NEP is therefore often estimated from changes in those fluxes by which carbon enters or leaves the ecosystem, ignoring the fluxes of carbon that occur within the ecosystem.

When carbon fluxes are aggregated at the ecosystem scale, some fluxes cancel out, because a loss from one component represents a gain by another component and therefore does not affect the total quantity of carbon in the ecosystem. Important fluxes that cancel out are consumption of plants and microbes by animals and the carbon fluxes from plants or animals to soil. Thus litterfall and herbivory alter ecosystem carbon budgets primarily by altering the location of the carbon within the ecosystem, not by altering *directly* the carbon inputs to or losses from the ecosystem. Even large changes in these fluxes such as occur during insect outbreaks or hurricanes are simply internal transfers of carbon within the ecosystem and need not be represented as separate terms in an overall budget of changes in carbon pools.

Some ecosystem fluxes can be aggregated at the ecosystem scale. **Ecosystem respiration** ($R_{ecosyst}$) is the combined respiration of plants, animals and microbes. It can be partitioned into plant respiration (R_{plant}), also termed **autotrophic respiration**, and **heterotrophic respiration** ($R_{heterotr}$)—that is, the respiration by organisms that gain their carbon by consuming organic matter rather than producing it themselves. Heterotrophs include animals and microbes.

$$R_{heterotr} = R_{animal} + R_{microb} \quad (B6.6)$$

$$R_{ecosyst} = R_{plant} + R_{heterotr} \quad (B6.7)$$

It is useful to treat separately those disturbances, such as fire and harvest, that directly remove carbon from ecosystems, because these disturbances are episodic in nature and frequently involve large fluxes that occur over a few hours to days.

$$F_{disturb} = F_{pl\text{-}fire} + F_{soil\text{-}fire} + F_{harv} \quad (B6.8)$$

Based on Equations B6.1 to B6.5 we can describe NEP in terms of carbon fluxes rather than changes in pool sizes.

$$\begin{aligned}
\text{NEP} &= \frac{(\Delta B_{\text{plant}} + \Delta B_{\text{animal}} + \Delta \text{SOM})}{\Delta t} \pm F_{\text{lateral}} \\
&= [\text{GPP} - R_{\text{plant}} - (F_{\text{pl-soil}} + F_{\text{herbiv}} \\
&\quad + F_{\text{emiss}} + F_{\text{pl-fire}} + F_{\text{harv}})] \\
&\quad + [\text{F}_{\text{herbiv}} + F_{\text{microb-anim}} \\
&\quad - (R_{\text{animal}} + F_{\text{anim-siol}})] \\
&\quad + [F_{\text{pl-soil}} + F_{\text{anim-soil}} - (R_{\text{microb}} \\
&\quad + F_{\text{microb-anim}} + F_{\text{CH4}} \\
&\quad + F_{\text{leach}} + F_{\text{soil-fire}})] \pm F_{\text{lateral}} \\
&= \text{GPP} \pm F_{\text{lateral}} - (R_{\text{plant}} + R_{\text{animal}} \\
&\quad + R_{\text{microb}}) - (F_{\text{pl-fire}} + F_{\text{soil-fire}} + F_{\text{harv}}) \\
&\quad - (F_{\text{leach}} + F_{\text{emiss}} + F_{\text{CH}_4}) \\
&= \text{GPP} \pm F_{\text{lateral}} - (R_{\text{ecosyst}} + F_{\text{disturb}} \\
&\quad + F_{\text{leach}} + F_{\text{emiss}} + F_{\text{CH}_4}) \qquad \text{(B6.9)}
\end{aligned}$$

Alternatively, we can use Equations 6.2 and B6.1 to B6.6 to express the carbon input as NPP and the respiratory carbon loss as heterotrophic respiration from microbes and animals (R_{heterotr}). This enables us to consider separately the fluxes associated with plants (carbon inputs) and with heterotrophs (carbon outputs).

$$\begin{aligned}
\text{NEP} &= \text{NPP} \pm F_{\text{lateral}} - (R_{\text{heterotr}} + F_{\text{disturb}} \\
&\quad + F_{\text{leach}} + F_{\text{emiss}} + F_{\text{CH}_4}) \qquad \text{(B6.10)}
\end{aligned}$$

Some of the terms in Equation B6.11 are usually small and can be ignored in some ecosystems some of the time. For example, emission of volatile organic compounds from plants are 0 to 5% of GPP (see Chapter 5), and net CH_4 flux from soils is substantial only in sites with anaerobic soils (0 to 8% of GPP). In ecosystems where all these fluxes are small, the NEP equation can be simplified.

$$\begin{aligned}
\text{NEP} &= \text{NPP} \pm F_{\text{lateral}} - (R_{\text{ecosyst}} + F_{\text{disturb}} \\
&\quad + F_{\text{leach}}) \quad \text{[in some ecosystems]} \qquad \text{(B6.11)}
\end{aligned}$$

or

$$\begin{aligned}
\text{NEP} &= \text{NPP} \pm F_{\text{lateral}} - (R_{\text{heterotr}} + F_{\text{disturb}} \\
&\quad + F_{\text{leach}}) \quad \text{[in some ecosystems]} \qquad \text{(B6.12)}
\end{aligned}$$

The **net ecosystem exchange** (NEE) of CO_2 can be estimated from measurements of CO_2 exchange above the canopy, further simplifying the terms that must be measured to estimate NEP.

$$\text{NEP} = \text{NEE} \pm F_{\text{lateral}} - (F_{\text{disturb}} + F_{\text{leach}}) \quad \text{[in some ecosystems]} \qquad \text{(B6.13)}$$

Over short time scales, in the absence of erosion and deposition, between episodes of fire or harvest, NEP simply reflects the net CO_2 exchange with the atmosphere and the leaching loss of carbon to groundwater.

$$\text{NEP} = \text{NEE} - F_{\text{leach}} \quad \text{[in the absence of fire, deposition, and erosion, harvest]} \qquad \text{(B6.14)}$$

Over long time scales and over large regions, the fluxes associated with disturbance, erosion, and deposition cannot be ignored, even though we seldom measure these fluxes directly (Harden et al. 2000). **Net biome production** (NBP) is NEP integrated at large spatial scales. It explicitly includes carbon removals by fire and harvest. NBP is the best measure of ecosystem carbon balance at the regional scale (Schulze et al. 2000, Randerson et al., in press), because it explicitly includes fluxes associated with disturbance, erosion, and deposition.

$$\begin{aligned}
\text{NBP} &= \text{NEP} = \text{NEE} \pm F_{\text{lateral}} \\
&\quad - (F_{\text{disturb}} + F_{\text{leach}}) \quad \text{[at regional scales]} \qquad \text{(B6.15)}
\end{aligned}$$

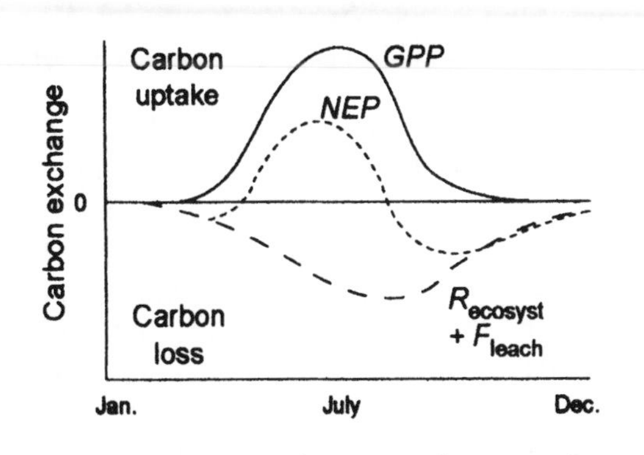

FIGURE 6.9. Representative seasonal pattern of gross primary production, ecosystem respiration, and net ecosystem production of an ecosystem. NEP is the difference between two large fluxes (carbon inputs in GPP and carbon losses, of which $R_{ecosyst}$ and F_{leach} are generally greatest). Annual CO_2 flux in this graph is at steady state because the NEP summed over the annual cycle is close to zero. Here, carbon losses due to disturbance are assumed to be zero.

released to the atmosphere (see Chapter 10). Approximately 20% of the CO_2 produced in arctic soils, for example, leaches to groundwater and is released from lakes and streams (Kling et al. 1991). Dissolved organic carbon is also lost from ecosystems by leaching to groundwater. Despite their importance, leaching losses of carbon to groundwater are seldom measured and therefore frequently ignored in ecosystem carbon budgets.

Lateral Transfers

Lateral transfer of carbon into or out of ecosystems can be important to the long-term carbon budgets of ecosystems. Carbon can move laterally in ecosystems through erosion and deposition by wind or water or by movement by animals (Fig. 6.8). These lateral transfers are usually so small that they are undetectable in any single year. Over long time periods or during extreme events, such as floods or landslides, these lateral transfers can, however, be quantitatively important. These transfers are typically more important for elements that are tightly cycled within ecosystems and have small annual inputs and losses (e.g., phosphorus) than they are for carbon.

Disturbance

Disturbance is an episodic cause of carbon loss from many ecosystems. Disturbances such as fire and harvest of plants or peat can be the dominant avenues of carbon losses from ecosystems in the years when they occur. In many cases the carbon losses with disturbance are so large that they become significant components of long-term carbon budgets (Fig. 6.8; Box 6.1). Carbon losses during fires in the Canadian boreal forest, for example, are equivalent to 10 to 30% of average NPP (Harden et al. 2000).

Controls over Net Ecosystem Production

NEP is determined by factors that cause an imbalance between carbon gain and loss. An ecosystem is never at equilibrium at any moment in time. NEP varies with season, time since disturbance, interannual variation in weather, and long-term trends in environment. High-latitude ecosystems, for example, are a net carbon source in warm years and a carbon sink in cool years (Goulden et al. 1998) because heterotrophic respiration responds to temperature more strongly than does photosynthesis in cold climates. We expect NEP to change in response to long-term changes in any factor that differs in its effects on GPP and the various avenues of carbon loss from ecosystems (e.g., plant respiration, heterotrophic respiration, disturbance, or leaching loss). Increased concentrations of atmospheric CO_2 or nitrogen inputs to ecosystems, for example, have greater direct effects on GPP than on decomposition, whereas a reduction in the soil moisture of poorly drained wetlands increases decomposition and fire probability more strongly than it affects NPP. Human activities are currently having greatest impact on precisely those environmental factors that we expect to affect plants and decomposers differentially and therefore to affect global terrestrial carbon storage (see Chapter 15).

Net carbon accumulation by an ecosystem depends more strongly on time since disturbance than on climate. The greatest causes of variation in NEP among ecosystems are cycles of disturbance and succession (see Chapter 13). Most disturbances initially cause a large negative NEP. Fire, for example, releases carbon directly by combustion. Fire also removes vegetation that transpires water and shades the ground surface. The warm, moist soils in recently burned sites promote decomposition and leaching loss and reduce plant biomass and therefore NPP. The net effect is a negative NEP during and in the first years after fire (Kasischke et al. 1995, Harden et al. 2000) (see Fig. 13.11). Agricultural tillage breaks up soil aggregates and increases access of soil microbes to soil organic matter. The resulting increase in decomposition can lead to a negative NEP after conversion of natural ecosystems to agriculture. During succession after disturbance there is typically a rapid increase in plant biomass and soil organic matter, because NPP increases more rapidly during succession than does decomposition (see Chapter 13). The NEP of a given stand therefore fluctuates dramatically through cycles of disturbance and succession. The spatial scale of these imbalances between NPP and decomposition ranges from individual gopher mounds or treefall gaps to large disturbances such as those caused by fire, forest harvest, or regional programs of land conversion to agriculture.

Because of the sensitivity of NEP to successional status, NEP estimated at the regional scale depends on the relative abundance of stands of different ages. NEP at the regional scale is termed **net biome production** (NBP) (Schulze et al. 2000). At times of increasing disturbance frequency, NBP is likely to be negative. Conversely, areas that have experienced widespread abandonment of agricultural lands in the last century, as in Europe or the northeastern United States, may experience a positive NBP. Inadequate information on the regional variation in disturbance frequency and NBP is one of the greatest causes of uncertainty in explaining recent changes in the global carbon cycle (see Chapter 15).

Net Ecosystem Exchange

Net ecosystem exchange (NEE) provides a direct measure of the net CO_2 exchange between ecosystems and the atmosphere. One of the greatest impediments to understanding the carbon dynamics of ecosystems is that we cannot directly measure most of the component processes at the ecosystem scale. GPP (net photosynthesis) cannot be readily separated from respiration by nonphotosynthetic plant parts. We can directly measure only some of the components of NPP, such as the accumulation of plant biomass. Decomposition is not easily separated from root respiration at the ecosystem scale. Many of the most pressing societal issues surrounding ecosystem carbon dynamics, however, revolve around NEP, the balance of carbon inputs and outputs. An important tool in improving our estimates of NEP has been an enhanced ability to measure NEE, which is the net exchange of CO_2 between the ecosystem and the atmosphere. NEE is the balance between GPP and **ecosystem respiration** ($R_{ecosyst}$), the sum of plant and heterotrophic respiration—that is, the total respiration by an ecosystem.

$$\begin{aligned} NEE &= GPP - (R_{plant} + R_{heterotr}) \\ &= GPP - R_{ecosyst} \end{aligned} \quad (6.3)$$

NEE, which excludes fluxes associated with disturbance and leaching, is the largest component of NEP in most ecosystems most of the time (Box 6.1).

GPP (net photosynthesis) is zero in the dark, so NEE is a direct measure of ecosystem respiration ($R_{ecosyst}$) under these conditions.

$$NEE_{dark} = -R_{ecosyst} \quad \text{[in the dark]} \quad (6.4)$$

The total diurnal $R_{ecosyst}$ can be estimated from simple models of $R_{ecosyst}$ as an exponential function of temperature (see Fig. 7.4). During the day, NEE is approximately equal to the sum of GPP and ecosystem respiration.

$$NEE_{light} \approx GPP - R_{ecosyst} \quad \text{[in the light]} \quad (6.5)$$

or

$$GPP \approx NEE_{light} + R_{ecosyst} \quad (6.6)$$

The results of this calculation are only approximate, because mitochondrial respiration in leaves declines in the light, when much of the energy for metabolism comes directly from carbon fixation (a component of GPP). Nonetheless, it is the closest thing to a direct measurement of GPP that is currently available.

A global network of sites measures NEE continuously in many of the world's ecosystems. These measurements show that, in the absence of disturbance, most temperate ecosystems that have been measured are net sinks for CO_2 (Fig. 6.10) (Valentini et al. 2000). There are at least four possible explanations for this important finding: (1) Ecosystems may typically be carbon sinks between episodes of disturbance, and disturbance may be the factor that brings NEP into balance at the regional scale. (2) Recent environmental changes, such as increased atmospheric CO_2 and nitrogen deposition, may have stimulated photosynthesis more than respiration. (3) Midsuccessional ecosystems with high NEP may have been overrepresented in the sampling network relative to the rest of the world. Many western European forests, where these studies were done, are productive midsuccessional sites that are developing after agricultural abandonment. (4) Carbon loss through leaching and other transfers may be an important component of the regional carbon balance. These nongaseous losses would not be detected in measurements of NEE.

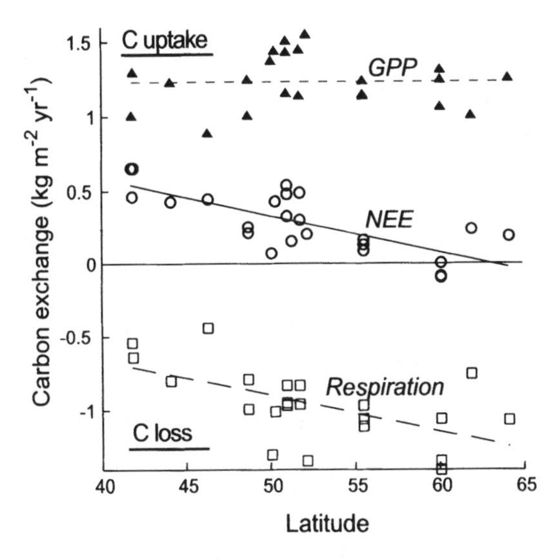

FIGURE 6.10. Latitudinal variation in annual gross primary production, net ecosystem exchange, and ecosystem respiration ($R_{ecosyst}$) among 12 naturally occurring European forests (Valentini et al. 2000). The greater NEE of low-latitude forests reflected their lower rates of ecosystem respiration. There was no latitudinal trend in GPP.

A second striking result of this study is that latitudinal variation in NEE reflects variations in ecosystem respiration rather than in GPP. Recent high-latitude warming could contribute to the greater respiration observed at high latitudes. At the few sites where there are long-term measurements of NEE, both respiration and photosynthesis contribute to interannual variations in NEE (Goulden et al. 1996, 1998). Only recently has NEE been measured in enough ecosystems to begin to identify regional patterns in NEE and their likely causes.

Global Patterns of NEE

Seasonal and latitudinal variations in the CO_2 concentration of the atmosphere provide a clear indication of global-scale patterns of NEE (Fung et al. 1987, Keeling et al. 1996b). At high northern latitudes, conditions are warm during summer, and photosynthesis exceeds total respiration (positive NEE), causing a decline in the concentration of atmospheric CO_2 (Fig. 6.11). Conversely, in winter, when photosynthesis is reduced by low temperature and shedding of leaves, respiration becomes the dominant carbon exchange (negative NEE), causing an increase in atmospheric CO_2. These seasonal changes in the balance between photosynthesis and respiration occur synchronously over broad latitudinal bands, giving rise to regular annual fluctuations in atmospheric CO_2, literally the breathing of the **biosphere** (i.e., all live organisms on Earth) (Fung et al. 1987).

Latitudinal variations in climate modify these patterns of annual carbon exchange. In contrast to the striking seasonality of NEE at north temperate and high latitudes, the concentration of atmospheric CO_2 remains nearly constant in the tropics, because carbon uptake by photosynthesis is balanced by approximately equal carbon loss by respiration throughout the year. In other words, NEE is close to zero throughout the year. There is also relatively weak seasonality of atmospheric CO_2 at high southern latitudes where oceans occupy most

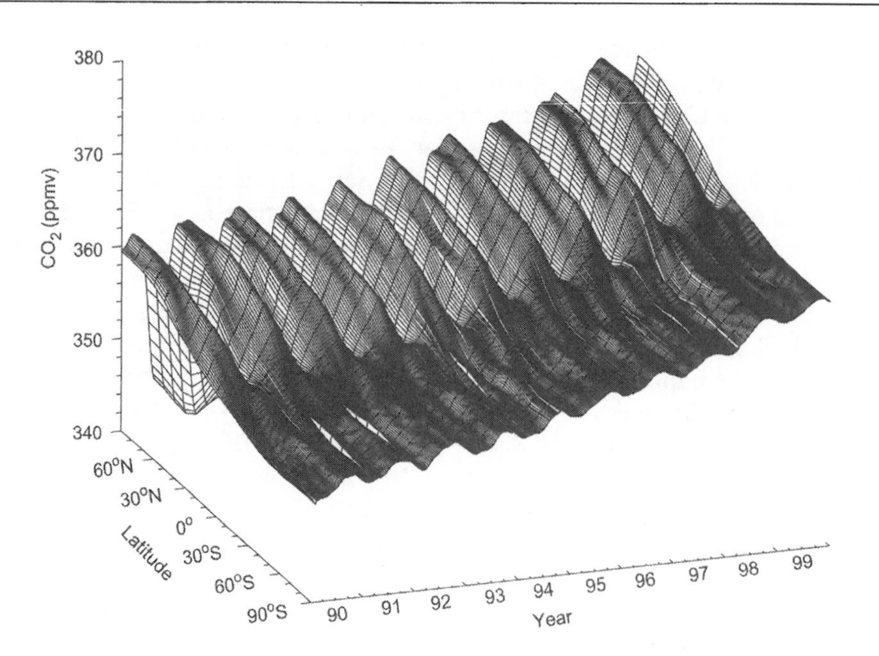

FIGURE 6.11. Seasonal and latitudinal variations in the concentration of atmospheric CO_2. Seasonal and latitudinal variations in CO_2 concentration reflect primarily the balance of terrestrial photosynthesis and respiration. The upward trend in concentration across years results from anthropogenic CO_2 inputs to the atmosphere. (Redrawn with permission from the National Oceanic and Atmospheric Administration, Climate Monitoring and Diagnostics Laboratory, Carbon Cycle-Greenhouse Gases.)

of Earth's surface. Carbon exchange in the oceans is largely determined by physical factors, such as wind, temperature, and CO_2 concentration in the surface waters (see Chapter 15), which show less seasonal variation. In summary, the global patterns of variation in atmospheric CO_2 concentration provide convincing evidence that carbon exchange by terrestrial ecosystems is large in scale and sensitive to climate.

The final general pattern evident in the atmospheric CO_2 record is a gradual increase in CO_2 concentration from one year to the next (Fig. 6.11), primarily a result of fossil fuel inputs to the atmosphere that began with the Industrial Revolution in the nineteenth century (see Chapter 15). The rising concentration of atmospheric CO_2 is an issue of international concern because CO_2 is a greenhouse gas that contributes to climate warming. Note that the interannual variation in CO_2 concentration caused by biospheric exchange is much larger than the annual CO_2 increase. If there were some way to increase net carbon uptake by ecosystems over the long term, this might reduce the rate of climate warming. There are therefore important societal reasons for understanding the controls over NEP in terrestrial ecosystems.

Summary

Plant respiration provides the energy to acquire nutrients and to produce and maintain biomass. All plants are similar in their efficiency of converting sugars into biomass. Therefore, ecosystem differences in plant respiration largely reflect differences in the amount and nitrogen content of biomass produced and, secondarily, in the effects of environmental stress, particularly temperature and moisture, on maintenance respiration. Most ecosystems appear to exhibit a similar efficiency of converting photosynthate (GPP) into NPP; about half of the carbon gain becomes NPP, and the other half is returned to the atmosphere as plant respiration.

Net primary production is the net carbon gained by vegetation. It includes new plant biomass produced, root exudation, carbon transfers to root symbionts, and the emission of volatile organic compounds by plants. Biome differences in NPP correlate with climate at the global scale largely because temperature and precipitation determine the availability of soil resources required to support plant growth. Plants actively sense the availability of these resources and adjust photosynthesis and NPP to match this resource supply. For this reason, NPP is greatest in environments with high availability of belowground resources. After disturbance, NPP is often reduced below levels that the environment can support. Plants maximize production by allocating new growth to tissues that acquire the most limiting resources. Constantly shifting patterns of allocation reduce the degree of limitation of NPP by any single resource and make NPP in most ecosystems responsive to more than one resource.

Tissue loss is just as important as NPP in explaining changes in plant biomass. Programmed loss of tissues provides a supply of plant resources that supports new production. Biomass and NPP are greatest in warm, moist environments and least in environments that are cold or dry. The length of the photosynthetic season and leaf area are the two strongest determinants of the global patterns in NPP. Most ecosystems have a similar (1 to 3 g biomass m^{-2} of leaf d^{-1}) daily NPP per unit leaf area.

Net ecosystem production is a measure of the rate of carbon accumulation in ecosystems. It correlates more strongly with time since disturbance than with environment. NEP is generally greatest in midsuccession, when ecosystems accumulate plant biomass and SOM. NEP is greater under conditions that promote NPP (e.g., elevated CO_2, N deposition) than under conditions that promote decomposition. Net biome production integrates NEP at the regional scale, taking account of regional patterns of disturbance and stand age. Human activities are altering most of the major controls over NEP at a global scale in ways that are likely to affect global climate.

Review Questions

1. What controls the partitioning of carbon between growth and respiration? Explain why the efficiency of converting sugars into new biomass is relatively constant.
2. What factors influence the variability in maintenance respiration?
3. Describe the multiple ways in which climate affects the NPP of grasslands or tundra.
4. There is generally a close correlation between GPP and NPP. Describe the mechanisms that account for short-term variations in GPP and NPP (e.g., diurnal and seasonal variations).
5. Describe the mechanisms that account for the relationship between GPP and NPP when ecosystems from different climatic regimes are compared.
6. How does allocation to roots vs. shoots respond to shade, nutrients, CO_2, grazing, or water?
7. How does variation in allocation influence resource limitation, resource capture, and NPP?
8. Why do plants senesce tissues in which they have invested carbon and nutrients rather than retaining tissues until they are removed by disturbance or herbivory? How does this physiologically programmed senescence influence NPP?
9. Describe the carbon budget of a plant and of an ecosystem in terms of GPP, respiration, and production. How would you expect each of these parameters to respond to changes in temperature, water, light, and nitrogen?
10. How do the controls over NEP differ from the controls over GPP and decomposition. Why are these controls different?

Additional Reading

Chapin, F.S. III. 1991. Integrated responses of plants to stress. *BioScience* 41:29–36.

Chapin, F.S. III, E.-D. Schulze, and H.A. Mooney. 1990. The ecology and economics of storage in plants. *Annual Review of Ecology and Systematics* 21:423–448.

Clark, D.A., S. Brown, D.W. Kicklighter, J.Q. Chambers, J.R. Thomlinson, and J. Ni. 2001. Measuring net primary production in forests: Concepts and field methods. *Ecological Applications* 11:356–370.

Lieth, H. 1975. Modeling the primary productivity of the world. Pages 237–263 *in* H. Lieth and R.H. Whittaker, editors. *Primary Productivity of the Biosphere*. Springer-Verlag, Berlin.

Poorter, H. 1994. Construction costs and payback time of biomass: A whole plant perspective. Pages 111–127 *in* J. Roy and E. Garnier, editors. *A Whole-Plant Perspective on Carbon-Nitrogen Interactions*. SPB Academic, The Hague.

Rastetter, E.B., and G.R. Shaver. 1992. A model of multiple element limitation for acclimating vegetation. *Ecology* 73:1157–1174.

Schlesinger, W.H. 1977. Carbon balance in terrestrial detritus. *Annual Review of Ecology and Systematics* 8:51–81.

Waring, R.H., and S.W. Running. 1998. *Forest Ecosystems: Analysis at Multiple Scales*. Academic Press, New York.

7
Terrestrial Decomposition

Decomposition breaks down dead organic matter, releasing carbon to the atmosphere and nutrients in forms that can be used for plant and microbial production. This chapter describes the key controls over decomposition and soil organic matter accumulation by ecosystems.

Introduction

Decomposition is the physical and chemical breakdown of detritus (i.e., dead plant, animal, and microbial material). Decomposition causes a decrease in detrital mass, as materials are converted from dead organic matter into inorganic nutrients and CO_2. If there were no decomposition, ecosystems would quickly accumulate large quantities of detritus, leading to a sequestration of nutrients in forms that are unavailable to plants and a depletion of atmospheric CO_2. Depletion of these resources in nondecomposing detritus would eventually cause many biological processes to grind to a halt. Although this has never occurred, there have been times such as the Carboniferous period (see Fig. 2.12) when decomposition did not keep pace with primary production, leading to vast accumulations of carbon-containing coal and oil. The balance between NPP and decomposition therefore strongly influences carbon cycling at ecosystem and global scales.

If the climate warming associated with anthropogenic CO_2 emissions were to cause even small changes in the balance between net primary prouction (NPP) and decomposition, the CO_2 concentration in the atmosphere would be greatly altered and therefore so would the rate of climate warming. Understanding the impacts of decomposition on carbon cycling is thus critical for making projections about the future state of Earth's climate.

Overview

The leaching, fragmentation, and chemical alteration of dead organic matter by decomposition produces CO_2 and mineral nutrients and a remnant pool of complex organic compounds that are resistant to further microbial breakdown. Decomposition is a consequence of interacting physical and chemical processes occurring inside and outside of living organisms. Decomposition results from three types of processes with different controls and consequences. (1) **Leaching** by water transfers soluble materials away from decomposing organic matter into the soil matrix. These soluble materials either are absorbed by organisms, react with the mineral phase of the soil, or are lost from the system in solution. (2) **Fragmentation** by soil animals breaks large pieces of organic matter into smaller ones, which provide a food source for soil animals and create fresh surfaces for microbial colonization.

Soil animals also mix the decomposing organic matter into the soil. (3) **Chemical alteration** of dead organic matter is primarily a consequence of the activity of bacteria and fungi, although some chemical reactions also occur spontaneously in the soil without microbial mediation.

Dead plant material (**litter**) and animal residues are gradually decomposed until their original identity is no longer recognizable, at which point they are considered **soil organic matter** (SOM). Litter consists primarily of compounds that are too large and insoluble to pass through microbial membranes. Microbes therefore secrete **exoenzymes** (extracellular enzymes) into their environment to initiate breakdown of litter. These exoenzymes convert macromolecules into soluble products that can be absorbed and metabolized by microbes. Microbes also secrete products of metabolism, such as CO_2 and inorganic nitrogen, and produce polysaccharides that enable them to attach to soil particles. When microbes die, their bodies become part of the organic substrate available for decomposition.

The controls over organic matter breakdown change radically once soil organic matter becomes incorporated into mineral soil. The soil moisture and thermal regimes of mineral soil are quite different from those in the litter layer. In the mineral soil, SOM can complex with clay minerals or undergo nonenzymatic chemical reactions to form more complex compounds. **Humus**, for example, is a complex mixture of chemical compounds with highly irregular structure containing abundant aromatic rings. Humus tends to accumulate in soils because exoenzymes cannot easily degrade its irregular structure (Oades 1989).

Decomposition is largely a consequence of the feeding activity of soil animals (fragmentation) and heterotrophic microbes (chemical alteration). The evolutionary forces that shape decomposition are those that maximize the growth, survival, and reproduction of soil organisms. Controls over decomposition are therefore best understood in terms of the controls over the activities of these organisms. The ecosystem consequences of decomposition are the **mineralization** of organic matter to inorganic components (CO_2, mineral nutrients, and water) and the **transformation** of organic matter into complex organic compounds that are **recalcitrant** (i.e., resistant to further microbial breakdown). In other words, decomposition occurs to meet the energetic and nutritional demands of decomposer organisms, not as a community service for the carbon cycle.

Leaching of Litter

Leaching is the rate-determining step for mass loss of litter when it first falls to the ground. Leaching is the physical process by which mineral ions and small water-soluble organic compounds dissolve in water and move through the soil. During leaf senescence, many of the compounds in a leaf are broken down and transported to other plant parts (see Chapter 8). This **resorption** process is still actively occurring when the leaf is shed, so the senesced leaf contains relatively high concentrations of water-soluble breakdown products that are readily leached. Leaching begins when tissues are still alive and is most important during tissue senescence and when litter first falls to the ground. Leaching losses from litter are proportionally more important for nutrients than for carbon. Leaching losses from fresh litter are greatest in environments with high rainfall and are negligible in dry environments. Compounds leached from leaves include sugars, amino acids, and other compounds that are **labile** (readily broken down) or are absorbed intact by soil microbes. Leachates frequently support a pulse of microbial growth and respiration during periods of high litterfall.

Litter Fragmentation

Fragmentation creates fresh surfaces for microbial colonization and increases the proportion of the litter mass that is accessible to microbial attack. Fresh detritus is initially covered by a protective layer of cuticle or bark on plants or of skin or exoskeleton on animals. These outer coatings are designed, in part, to protect tissues from microbial attack. Within plant tissues, the labile cell contents are further protected from microbial attack by lignin-impregnated cell walls. Fragmentation of litter greatly enhances microbial decomposition by piercing these protective barriers and by increasing the ratio of litter surface area to mass.

Animals are the main agents of litter fragmentation, although freeze–thaw and wetting–drying cycles can also disrupt the cellular structure of litter. Animals fragment litter as a by-product of their feeding activities. Bears, voles, and other mammals tear apart wood or mix the soil as they search for insects, plant roots, and other food. Soil invertebrates fragment the litter to produce particles that are small enough to ingest. Enzymes in animal guts digest the microbial "jam" that coats the surface of litter particles, providing energy and nutrients to support animal growth and reproduction.

Chemical Alteration

Fungi

Fungi are the main initial decomposers of terrestrial dead plant material and, together with bacteria, account for 80 to 90% of the total decomposer biomass and respiration. Fungi have networks of **hyphae** (i.e., filaments that enable them to grow into new substrates and transport materials through the soil over distances of centimeters to meters). Hyphal networks enable fungi to acquire their carbon in one place and their nitrogen in another, much as plants gain CO_2 from the air and water and nutrients from the soil. Fungi that decompose litter on the forest floor, for example, may acquire carbon from the litter and nitrogen from the mineral soil. Fungi are the principal decomposers of fresh plant litter, because they secrete enzymes that enable them to penetrate the cuticle of dead leaves or the suberized exterior of roots to gain access to the interior of a dead plant organ. Here they proliferate within and between dead plant cells. At a smaller scale, some fungi gain access to the nitrogen and other labile constituents of dead cells by breaking down the lignin in cell walls. This energy investment in lignin-degrading enzymes serves primarily to gain access to the relatively labile contents of the interior of cells.

Fungi produce hyphae with a dense concentration of cytoplasm when there is adequate substrate to support growth. The hyphae contain more vacuoles (and proportionally less cytoplasm) when resources are scarce. This flexible growth strategy enables fungi to grow into new areas to explore for substrate, even when current substrates are exhausted. A substantial proportion (perhaps 25%) of the carbon and nitrogen used to support fungal growth are transported from elsewhere in the hyphal network, rather than being absorbed from the immediate environment where the fungal growth occurs (Mary et al. 1996).

Fungi have enzyme systems capable of breaking down virtually all classes of plant compounds. They have a competitive advantage over bacteria in decomposing tissues with low nutrient concentrations because of their ability to import nitrogen and phosphorus. White-rot fungi specialize on lignin degradation in logs, whereas brown-rot fungi cleave some of the side-chains of lignin but leave the phenol units behind (giving the wood a brown color). White-rot fungi are generally outcompeted by more rapidly growing microbes when nitrogen is abundant, so nitrogen additions have little effect (or sometimes a negative effect) on white-rot fungal decomposition of wood.

Fungi account for 60 to 90% of the microbial biomass in forest soils, where litter frequently has a high lignin and low nitrogen concentration. They have a competitive advantage at low pH, which is also common in forest soils. Fungi make up about half the microbial biomass in grassland soils, where pH is higher and wood is absent. Most fungi lack a capacity for anaerobic metabolism and are therefore absent from or dormant in anaerobic soils and aquatic sediments.

Mycorrhizae are a symbiotic association between plant roots and fungi in which the plant gains nutrients from the fungus in return for carbohydrates (see Chapter 8). Although mycorrhizal fungi get most of their carbon from plant roots, they can also play a role in decomposition by breaking down proteins into amino acids, which are absorbed; amino acids both support fungal growth and are transferred to their host plants (Read 1991). Mycorrhizal fungi also produce cellulases to gain entry into plant roots, but it is uncertain whether these cellulases participate in decomposition of dead organic matter.

Bacteria

The small size and large surface to volume ratio of bacteria enable them to absorb soluble substrates rapidly and to grow and divide quickly in substrate-rich zones. This opportunist strategy explains the bacterial dominance in the **rhizosphere** (the zone of soil directly influenced by plant roots) and in dead animal carcasses, where labile substrates are abundant. Bacteria are also important in lysing and breaking down live and dead bacterial and fungal cells. The major functional limitation resulting from their small size is that each bacterium completely depends on the substrates that move to it. Some of these substrates are products of bacterial exoenzymes. These products diffuse to the bacterium along a concentration gradient created by the activity of the exoenzymes (which produce soluble substrates), and by the uptake of substrates by the bacterium (which reduces substrate concentrations at the bacterial surface). Other soluble substrates flow past the bacterium in water films moving through the soil. This water movement is driven by gradients in water potential associated with plant transpiration, evaporation at the soil surface, and gravitational water movement after rain (see Chapter 4). Water movement (and therefore the supply rate of substrates) is most rapid in **macropores** (relatively large air or water spaces between aggregates). Bacteria therefore often line the macropore surfaces and absorb substrates from the flowing water, just as fishermen net salmon migrating up a stream or an intertidal filter-feeder extracts organic particles from the water column. Macropores are also preferentially exploited by roots because of the reduced physical resistance to root elongation, providing an additional source of labile substrates to bacteria. Bacteria attached to the exposed surfaces of macropores are vulnerable to predation by protozoa and nematodes, which use the water films in macropores as highways to move through the soil. This leads to rapid bacterial turnover on exposed particle surfaces.

There is a wide range of bacterial types in soils. Rapidly growing gram-negative bacteria specialize on labile substrates secreted by roots. Actinomycetes are slow-growing, gram-positive bacteria that have a filamentous structure similar to that of fungal hyphae. Like fungi, actinomycetes produce lignin-degrading enzymes and can break down relatively recalcitrant substrates. They often produce fungicides to reduce competition from fungi.

The bacterial communities that coat soil aggregates exhibit a surprisingly complex structure. They are often present as **biofilms**, a microbial community embedded in a matrix of polysaccharides secreted by bacteria. This microbial "slime" protects bacteria from grazing by protozoa and reduces bacterial water stress by retaining water. The matrix also increases the efficiency of bacterial exoenzymes by preventing them from being swept away in moving water films. The bacteria in biofilms often act as a **consortium**—that is, a group of genetically unrelated bacteria, each of which produces only some of the enzymes required to break down complex macromolecules. The breakdown of these molecules to the point that soluble products are released requires the coordinated production of exoenzymes by several types of bacteria. This is analogous to an assembly line, in which the final product, such as a car or a television set, depends on the coordinated action of several consecutive steps; as with an assembly line, no bacterium benefits unless all the steps are in place to produce the final product. The evolutionary forces and population interactions that shape the composition of microbial consortia are virtually unknown. Consortia are particularly important in the breakdown of pesticides and other organic residues that humans have added to the environment.

Most bacteria are immobile and move passively through the soil, carried by soil water or animals. An important consequence of immobility is that a bacterial colony eventually exhausts the substrates in its immediate environment, especially in microenvironments within soil particles that have restricted water movement. When bacteria exhaust their substrate, they become inactive and reduce their respiration to negligible rates. Bacteria may remain inactive for years. Live bacteria have been recovered from permafrost that is tens of thousands of years old. Between 50 and 80% of the bacteria in soils are metabolically inactive (Norton and Firestone 1991). Inactive bacteria reactivate in the presence of labile substrates, for example, when a root grows through the soil and exudes carbohydrates. The inactive bacteria in soils represent a reservoir of decomposition potential analogous to the buried seed pool, which is an important source of plant colonizers after a disturbance. Like the buried seed pool, the enzymatic potential of these inactive bacteria may be different from the enzymes produced by the active bacterial community. Consequently, DNA probes or microbiological culturing techniques are better indices of what the soil *could* do (its metabolic potential) than of its actual metabolic activity at any given time.

Soil Animals

Soil animals influence decomposition by fragmenting and transforming litter, grazing populations of bacteria and fungi, and altering soil structure. The **microfauna** is made up of the smallest animals (less than 0.1 mm). They include nematodes; protozoans, such as ciliates and amoebae; and some mites (Fig. 7.1) (Wallwork 1976, Lousier and Bamforth 1990). Protozoans consist of a single cell and ingest their prey primarily by **phagocytosis**—that is, by enclosing them in a membrane-bound structure that enters the cell. Protozoans are usually mobile and are voracious predators of bacteria and other microfauna species (Lavelle et al. 1997). Nematodes are an abundant and trophically diverse group in which each species specializes on bacteria, fungi, roots, or other soil animals. Bacterial-feeding nematodes in forest litter, for example, can consume about 80 $g\,m^{-2}\,yr^{-1}$ of bacteria, resulting in the mineralization of 2 to 13 $g\,m^{-2}\,yr^{-1}$ of nitrogen—

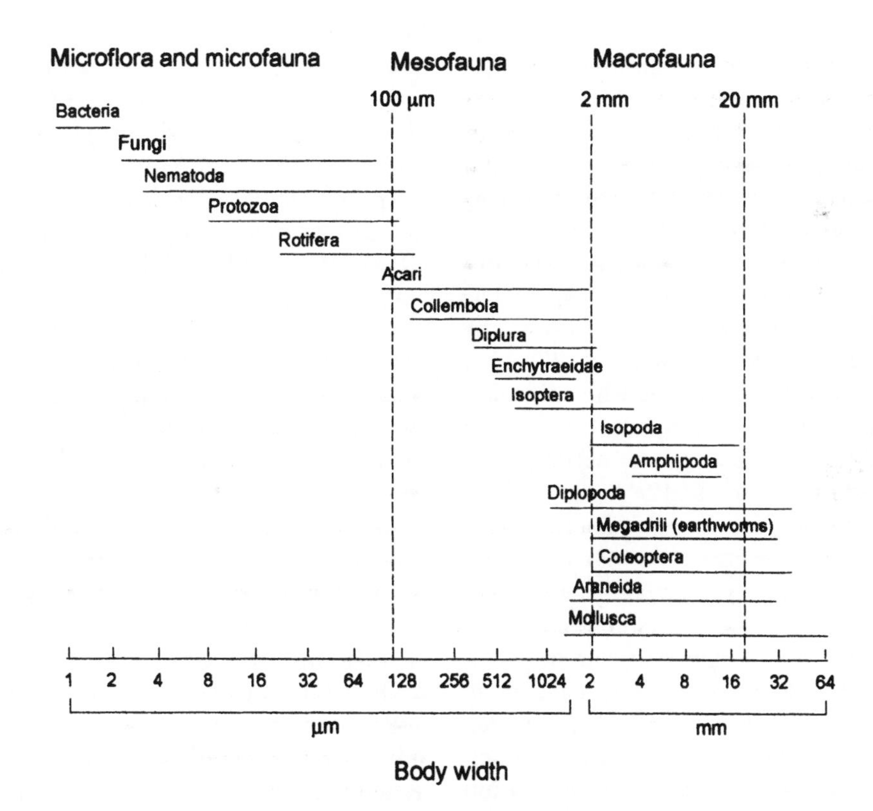

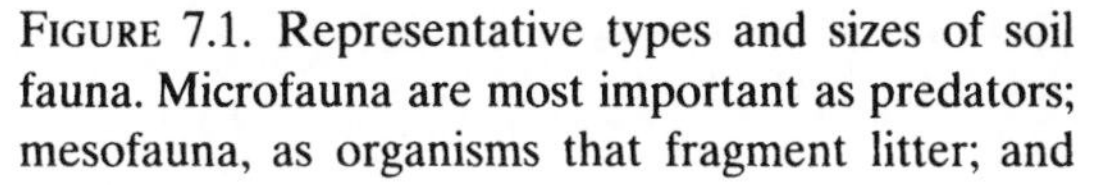

FIGURE 7.1. Representative types and sizes of soil fauna. Microfauna are most important as predators; mesofauna, as organisms that fragment litter; and macrofauna, as ecosystem engineers. (Redrawn with permission from Blackwell Scientific; Swift et al. 1979.)

a substantial proportion of the nitrogen that annually cycles through the soil (Anderson et al. 1981). Protozoans are particularly important predators in the rhizosphere and other soil microsites that have rapid bacterial growth rates (Coleman 1994). The preferential grazing by protozoa on bacteria (even on particular species of bacteria), alters the microbial community composition, tending to reduce bacterial to fungal ratios in these soils compared to soils from which protozoa are excluded. Protozoans and nematodes are aquatic animals that move through water films on the surface of soil particles and are therefore more sensitive to water stress than are fungi and the mesofauna and macrofauna species that fragment soil particles. Their populations fluctuate dramatically, both spatially and temporally, due to drying–wetting events and to predation (Beare et al. 1992). When protozoans die, their bodies are rapidly broken down by soil microbes, especially by bacteria.

The **mesofauna** includes a taxonomically diverse group of soil animals 0.1 to 2 mm in length (Fig. 7.1). They are the animals that have the greatest effect on decomposition. Mesofauna species fragment and ingest litter coated with microbial biomass, producing large amounts of fecal material that has greater surface area and moisture-holding capacity than the original litter (Lavelle et al. 1997). This altered litter environment is more favorable for decomposition. These organisms selectively feed on litter that has been conditioned by microbial activity. Collembola are small insects that feed primarily on fungi, whereas mites (Acari) are a more trophically diverse group of spiderlike animals that consume decomposing litter or feed on bacteria and/or fungi.

Large soil animals (the **macrofauna**), such as earthworms and termites, are **ecosystem engineers** that alter resource availability by modifying the physical properties of soils and litter (Jones et al. 1994). Some of them fragment litter, like the mesofauna species (Lavelle et al. 1997). Others burrow or ingest soil, reducing soil bulk density, breaking up soil aggregates, and increasing soil aeration and the infiltration of water (Beare et al. 1992). The passages created by earthworms create channels in the soil through which water and roots readily penetrate. They create patterns of soil structure that promote or constrain the activities of soil microbes and other soil animals. In temperate pastures, earthworms may process $4\,kg\,m^{-2}\,yr^{-1}$ of soil, moving 3 to 4 mm of new soil to the ground surface each year (Paul and Clark 1996). This is a geomorphic force that is, on average, orders of magnitude larger than landslides or surface soil erosion (see Table 3.1). Soil mixing by earthworms tends to disrupt the formation of distinct soil horizons. Once the soil enters the digestive tract of an earthworm, mixing and secretions by the earthworm stimulate microbial activity, so soil microbes act as gut mutualists. Many of the soil organisms are lysed and digested during passage through the gut; the resulting products are absorbed by the earthworm. Earthworms are most abundant in the temperate zone, whereas termites are the dominant ecosystem engineers in tropical soils. Termites eat plant litter directly, digest the cellulose with the aid of mutualistic protozoans in their guts, and mix the organic matter into the soil. Dung beetles in tropical grasslands perform a similar function with mammalian dung. This burial of surface organic matter places it in a humid environment where decomposition occurs more rapidly.

The soil fauna is critical to the carbon and nutrient dynamics of soils. Microbes contain 70 to 80% of the labile carbon and nitrogen in soils, so variations in predation rates of microbes by animals dramatically alter carbon and nitrogen turnover in soils. Soil animals have high respiration rates and metabolize much of the microbial carbon from their food to CO_2 to support their high energetic costs of movement. As a result, the microbial nitrogen and phosphorus acquired by soil animals generally exceeds their requirements for growth and reproduction. These nutrients are therefore excreted and become available for plant uptake (see Chapter 8). Soil animals account for only about 5% of soil respiration, so their major effect on decomposition is their enhancement of microbial activity through fragmentation (Wall et al. 2001), rather than their own processing of energy derived from detritus. Soil food webs are complex (see

Chapter 11), so many of the effects of soil animals on decomposition are indirect. Loss or exclusion of soil invertebrates can reduce decomposition rate (and therefore nutrient cycling) substantially, indicating the important role of animals in the decomposition process (Swift et al. 1979, Verhoef and Brussaard 1990).

Temporal and Spatial Heterogeneity of Decomposition

Temporal Pattern

The predominant controls over decomposition change with time. Decomposition is the consequence of the interactions of fragmentation, chemical alteration, and leaching. As soon as a leaf unfolds, it is colonized by aerially borne bacteria and fungal spores that begin breaking down the cuticle and leaf surfaces that are exposed by herbivores, pathogens, or physical breakage (Haynes 1986). This **phyllosphere decomposition** of live leaves is generally ignored because it is not readily separated from plant-controlled changes in leaf mass and chemistry. It does, however, provide a microbial innoculum that rapidly initiates decomposition of labile substrates when the leaf falls to the ground. Similarly, breakdown of the root cortex begins while the conducting tissues of roots still function in water and nutrient transport.

As litter decomposes, its mass decreases approximately exponentially with time. Leaf litter frequently loses 30 to 70% of its mass in the first year and another 20 to 30% of its mass in the next 5 to 10 years (Haynes 1986). An exponential decline in litter mass implies that a constant *proportion* of the litter is decomposed each year.

$$L_t = L_0 e^{-kt} \tag{7.1}$$

$$\ln\frac{L_t}{L_0} = -kt \tag{7.2}$$

where L_0 is the litter mass at time zero, and L_t is the mass at time t. The **decomposition constant**, k, is an exponent that characterizes the decomposition rate of a particular material. The mean **residence time**, or the time required for the litter to decompose under steady-state conditions, equals $\frac{1}{k}$. Residence time of litter can also be estimated as the average pool size of litter divided by the average annual input.

$$\frac{l}{k} = \frac{\text{litter pool}}{\text{litterfall}} \quad \text{or} \quad k = \frac{\text{litterfall}}{\text{litter pool}} \tag{7.3}$$

The estimation of residence times from pools and fluxes assumes that the ecosystem is in steady state, which is often not the case (see Chapter 1). Midsuccessional ecosystems, for example, generally receive more litterfall input than would occur at steady state, leading to an overestimate of k. Year-to-year variation in weather or directional changes in climate cause more rapid changes in litterfall than in the litter pool, also creating biases in estimates of residence time. The decomposition constant varies widely with substrate composition. Sugars, for example, have a residence time of hours to days, whereas lignin has a residence time of months to decades, depending on the ecosystem. Plant and animal tissues differ substantially in their chemical composition and therefore in their decay constants. Taken as a whole, litter generally has a residence time of months to years, and organic matter mixed with mineral soil has a residence time of years to centuries.

The exponential model of decomposition (Eq. 7.1), which implies a constant decomposition rate, is only a rough approximation of the pattern of decline in litter mass with time. The process is more accurately described by a curve with at least three phases (Fig. 7.2). During the first phase, leaching of cell solubles is the predominant process. Fresh litter can lose 5% of its mass in 24h due to leaching alone. The second phase of decomposition occurs more slowly and involves a combination of fragmentation by soil animals, chemical alteration by soil microbes, and leaching of decay products from the litter. Decomposition during this second phase is often measured as mass loss from dead leaves (Aerts 1997), roots (Berg et al. 1998), or twigs that are tethered on threads or placed in mesh **litter bags** and weighed periodically (Vogt et al. 1986, Robertson and Paul 2000). The exponential model of decomposition has been applied

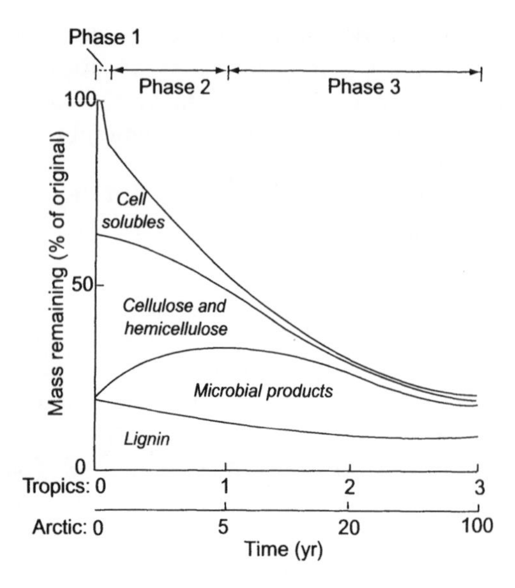

FIGURE 7.2. Representative time course of leaf-litter decomposition showing the major chemical constituents (cell solubles, cellulose and hemicellulose, microbial products, and lignin), the three major phases of litter decomposition, and the time scales commonly found in warm (tropical) and cold (arctic) environments. Leaching dominates the first phase of decomposition. Substrate composition of litter changes during decomposition because labile substrates, such as cell solubles, are broken down more rapidly than are recalcitrant compounds, such as lignin and microbial cell walls.

primarily to this second phase. The final phase of decomposition occurs quite slowly and involves the chemical alteration of organic matter that is mixed with mineral soil and the leaching of breakdown products to other soil layers. Decomposition during this final phase is often estimated from measurements of soil respiration or isotopic tracers (Schlesinger 1977, Trumbore and Harden 1997). The decomposition rate and decomposition constant (k in Eq. 7.1) gradually decline through these three phases of decomposition.

In seasonal environments, microbial respiration often occurs over a longer time period and peaks later in the season than does plant growth. Like plant growth, microbial respiration is favored by warm, moist conditions and is therefore greatest during the season of maximum plant growth. Heterotrophic respiration, however, typically begins earlier and ends later than does plant growth for at least three reasons: (1) Microbial respiration typically occurs over a broader range of temperatures (e.g., –10° to 40°C) and soil moistures than plant growth. (2) The soil is buffered from temperature extremes that aboveground parts of plants must cope with. (3) Because soils warm more slowly than the air, heterotrophic respiration generally lags behind gross primary production (GPP), with relatively low rates in early spring, when leaf growth is most active. Heterotrophic respiration continues in autumn and winter, long after leaf senescence. Microbial activity is also influenced by the seasonality of plant activity. Root turnover and exudation are often greatest in midseason when photosynthesis is high, contributing to the midseason peak in soil respiration. Autumn senescence provides an additional input of substrates that contributes to late-season soil respiration.

Spatial Pattern

Most decomposition occurs near the soil surface, where litter inputs are concentrated. Gravity carries most aboveground litter to the ground surface, where the initial decomposition and nutrient release occur. Roots therefore tend to grow in surface soils to access these nutrients. Thus root litter is also produced primarily in surface soils, reinforcing the surface localization of most decomposition. Deep roots are, however, not negligible, especially in dry environments. These "water roots" can be found to depths of 10 to 100 m, depending on the depth of the water table. Soil mixing by animals, especially termites and earthworms, and leaching of dissolved organic matter also transfer surface carbon to depth. About half of the soil organic carbon therefore is typically below 20 cm depth, even though only a third of the roots are below that depth (Fig. 7.3) (Jobbágy and Jackson 2000). Deep-soil decomposition therefore cannot be ignored. The deep-soil carbon is often older, more recalcitrant, and more strongly protected by complexes with soil minerals than is surface carbon (Trumbore and Harden 1997).

Decomposition rate is spatially heterogeneous at several scales. The litter layer above the mineral soil exhibits large daily changes in temperature and moisture. Decomposition in

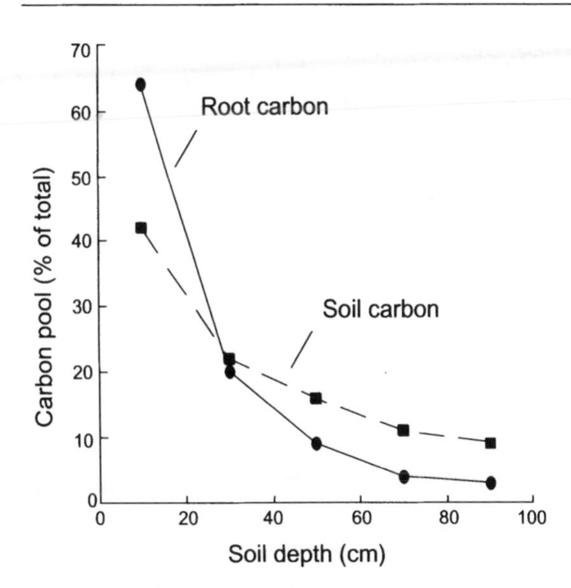

FIGURE 7.3. Globally averaged depth profiles of soil organic matter and roots in the first meter of soil. (Redrawn with permission from *Ecological Applications*; Jobbágy and Jackson 2000.)

this layer is dominated by fungi that import nitrogen from below. This is a radically different environment from the mineral soil, where temperature and moisture are more stable, some of the organic matter is humified and recalcitrant, and mineral soil surfaces bind dead organic matter and microbial enzymes. At a finer scale, the rhizosphere around roots is a carbon-rich microenvironment that supports much higher microbial activity than the bulk soil, which is virtually a nutritional desert. Finally, the interior of soil aggregates is more likely to be anaerobic than are the surfaces of soil pores. Movement within the soil of roots, water, and soil animals is constantly changing the spatial arrangement of these different environments for decomposition.

In some ecosystems, such as tropical forests, significant quantities of aboveground litter are caught on epiphytes and branches of the canopy. In these wet ecosystems, substantial decomposition, nutrient release, and nutrient uptake by rooted epiphytes occur in the canopy and short-circuit the soil phase (Nadkarni 1981). Some terrestrial litter and dissolved organic carbon (DOC) also enter streams and lakes, where they become energy sources for aquatic food webs (see Chapter 10). In unproductive ecosystems, the DOC that enters streams is so recalcitrant that it remains largely unprocessed, leading to the "black-water" rivers that characterize many tropical and boreal forests and temperate swamps.

Factors Controlling Decomposition

Decomposition is controlled by three types of factors: the physical environment, the quantity and quality of substrate available to decomposers, and the characteristics of the microbial community (Swift et al. 1979).

The Physical Environment

Temperature

Temperature affects decomposition directly by promoting microbial activity and indirectly by altering soil moisture and the quantity and quality of organic matter inputs to the soil. Rising temperature causes an exponential increase in microbial respiration over a broad temperature range (Fig. 7.4), speeding up the mineralization of organic carbon to CO_2. This temperature response is similar to that observed in respiration of most organisms. At moderate temperatures, most of the respiratory energy supports microbial growth. As temperature increases, however, an increasing proportion of the energy is used for maintenance and may not lead to a corresponding increase in microbial production. Microbial community composition also changes in response to temperature toward a community dominated by individuals that are adapted and acclimated to higher temperatures. There are therefore several physiological and community changes that account for the deceptively simple response of microbial respiration to temperature. Continuously high temperature, and therefore rapid decomposition, explain why many tropical forests have a small litter pool despite their high productivity (Fig. 7.5).

Temperature also affects decomposition through freeze–thaw events. Freezing kills many of the microbes present in decomposing litter and SOM, releasing soluble organic mate-

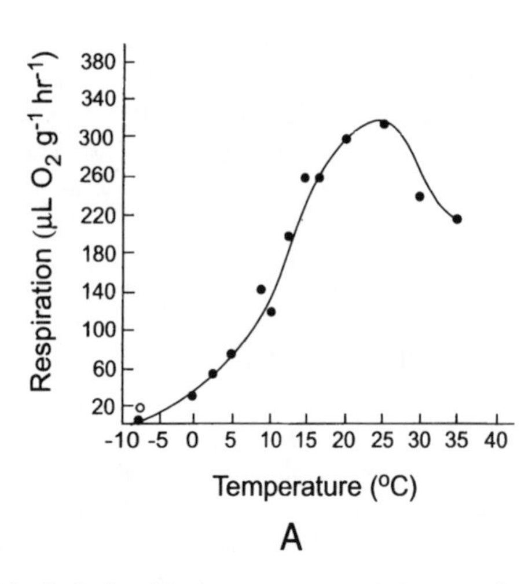

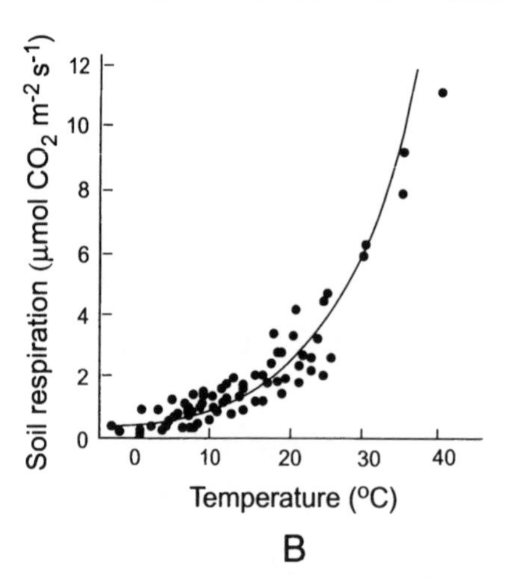

FIGURE 7.4. Relationship between temperature and soil respiration in (**A**) laboratory incubations of tundra soils and (**B**) field measurements of soil respiration in 15 studies, where data have been fitted to have the same respiration rate at 10°C. (**A**, Flanagan and Veum 1974. **B**, Redrawn with permission from *Functional Ecology*; Lloyd and Taylor 1994.)

rials into the soil. This pulse of available substrate can support rapid decomposition and nitrogen mineralization the following spring (Lipson et al. 1999). Freezing and thawing also stimulates decomposition by physically disrupting soil aggregates and the cellular structure of litter, thereby exposing fresh surfaces to decomposition. In some arctic ecosystems the decomposition that occurs during autumn, winter, and spring accounts for most of the annual litter mass loss (Hobbie and Chapin 1996).

Temperature has many indirect effects on decomposition (Fig. 7.6). High temperature reduces soil moisture by increasing evaporation and transpiration. Soil drying reduces decomposition in dry climates but accelerates it where soils are wet enough to restrict oxygen supply.

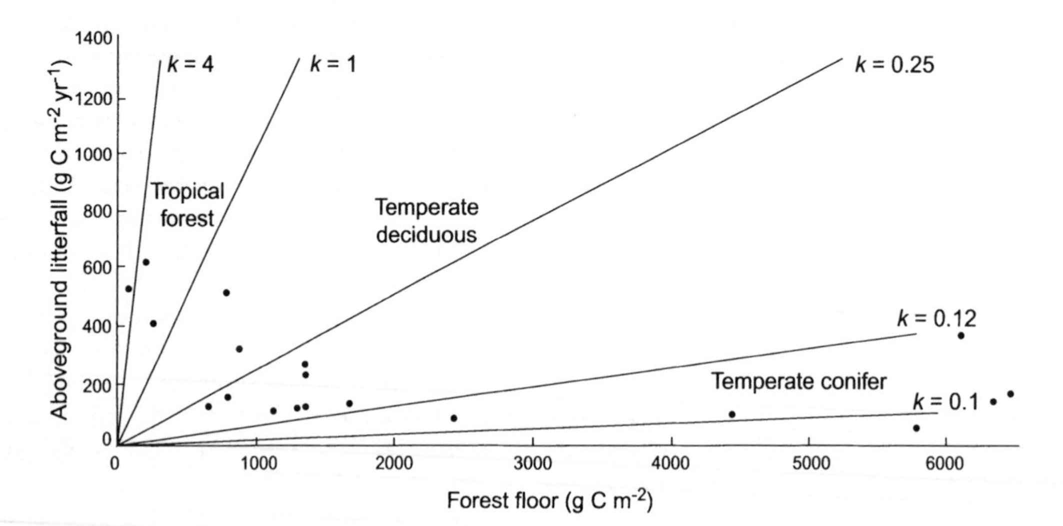

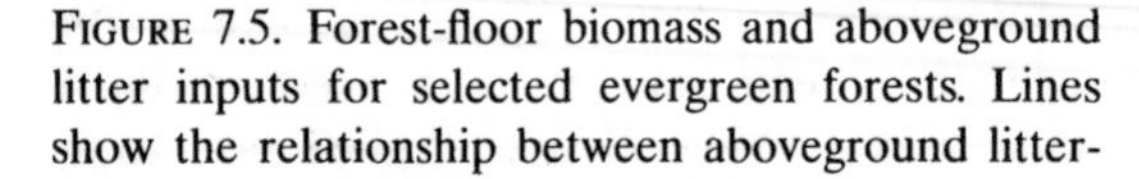

FIGURE 7.5. Forest-floor biomass and aboveground litter inputs for selected evergreen forests. Lines show the relationship between aboveground litterfall and forest floor mass for selected decomposition constants. (Redrawn with permission from *Ecology*; Olsen 1963.)

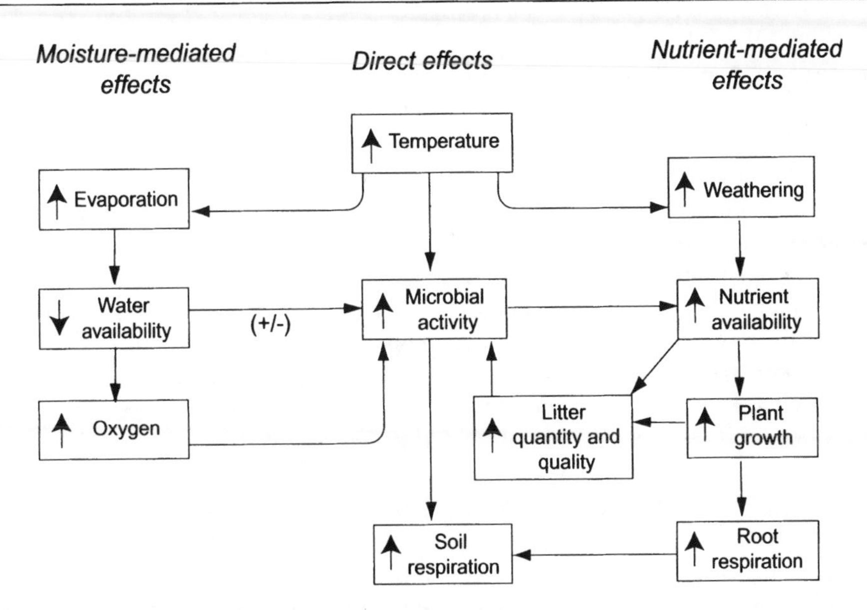

FIGURE 7.6. Direct and indirect effects of temperature on soil respiration.

The stimulation of microbial activity by warm temperatures also initiates a series of feedback loops that influence decomposition. The consumption of oxygen by microbial and root respiration constrains decomposition in wet soils or wet microsites (e.g., the interior of soil aggregates). On the other hand, the nutrients released by decomposition at high temperatures increase the quantity and quality of litter produced by plants, altering the substrate available for decomposition. High temperatures also increase the rate of chemical weathering, which in the short term enhances nutrient supply. In cold climates, low temperature leads to a layer of permanently frozen soils (permafrost) that restricts drainage and therefore decomposition. Most of the indirect temperature effects enhance soil respiration at warm temperatures and contribute to the more rapid decomposition observed in warm climates.

Moisture

Carbon accumulation is greatest in wet soils because decomposition is more restricted by high soil moisture but is less restricted by low soil moisture than is NPP. Decomposers, like plants, are most productive under warm moist conditions, provided sufficient oxygen is available. This accounts for the high decomposition rates in tropical forests (Gholz et al. 2000). The decomposition rate of mineral soil generally declines at soil moistures less than 30 to 50% of dry mass (Haynes 1986), due to the reduction in thickness of moisture films on soil surfaces and therefore the rate of diffusion of substrates to microbes (Stark and Firestone 1995). Osmotic effects further restrict the activity of soil microbes under conditions of extremely low soil moisture or salt accumulation. Bacteria function at lower water availability than do plant roots, so decomposition continues in soils that are too dry to support plant activity. The high concentrations of osmotic metabolites synthesized by microbes in dry or saline conditions create severe osmotic gradients after soil wet-up, causing many microbial cells to burst. This results in pulses of nutrient availability after the first rains. Even short-term drying–wetting cycles, such as rain storms or the daily formation and evaporation of dew, can strongly influence decomposition in the litter layer and surface soils. The net effect of drying–wetting cycles is the stimulation of decomposition, if the cycles are infrequent (as generally occurs in soils). Frequent moisture fluctuations, as in the litter layer, can, however, reduce microbial population numbers to an extent that decomposition rates are reduced (Clein and Schimel 1994). Drying–wetting cycles tend to stimulate the decomposition of labile substrates (e.g., hemicellulose), which are broken down largely by rapidly growing bacteria, and to retard the decomposition of recalcitrant ones (e.g., lignin) (Haynes 1986), which are broken down by slow-growing fungi.

Decomposition is also reduced at high soil moisture contents (e.g., greater than 100 to 150% of soil dry mass in mineral soils) (Haynes 1986). Oxygen diffuses 10,000 times more slowly through water than through air, so water acts as an effective barrier to oxygen supply to decomposers in wet soils or in wet microsites within well-drained soils. Oxygen limitation to decomposition can occur for many reasons, including topographic controls over drainage, presence of hardpans or permafrost, high clay content, or compaction by animals and agricultural equipment. Irrigation or rain events can lead to short-term oxygen depletion. In warm environments,.the solubility of oxygen in water is low, and oxygen is rapidly depleted by root and microbial respiration, making decomposition particularly sensitive to high soil moisture. Decomposition is also frequently oxygen limited in bogs and wetlands and in arctic tundra, where permafrost prevents drainage over a wide range of topographic situations. NPP is frequently less limited by high soil moisture than is decomposition, because many plants that are adapted to these conditions transport oxygen from leaves to the roots. The large accumulations of SOM in histosol soils of swamps and bogs at all latitudes clearly indicate the importance of oxygen to decomposition.

Decaying logs create their own unique microenvironment and generally have a higher moisture content than does the adjacent surface litter. Log decomposition rate may therefore be limited by oxygen supply at times when microbes in neighboring surface litter are moisture limited. The decomposition rate of logs generally decreases with increasing log diameter, because large logs generally have more moisture and less oxygen than small ones.

Soil Properties

All else being equal, decomposition occurs more rapidly in neutral than in acidic soils due to a variety of interacting factors. Fungi tend to predominate in acidic soils (Haynes 1986). The increase in bacterial abundance and the overall increase in decomposition rate at higher pH probably reflects a complex of interacting factors, including changes in plant species composition and associated changes in the quantity and quality of litter. Many factors can acidify soils, including cation leaching, acid deposition, and the accumulation of organic acids in soil organic matter during succession. Alternatively, pH can increase in response to dust input (Walker et al. 1998), particularly in deserts, braided river valleys of glacial landscapes, and degraded agricultural lands. Regardless of the cause of the change in acidity and associated plant species composition, low pH tends to be associated with low decomposition rates.

Clay minerals reduce the decomposition rate of soil organic matter, thereby increasing soil organic content. Clays alter the physical environment of soils by increasing water-holding capacity (see Chapter 3). The resulting restriction in oxygen supply can reduce decomposition in wet clay soils. Even at moderate soil moisture, clays enhance organic accumulation by binding soil organic matter (making it less accessible to microbial enzymes); binding microbial enzymes (reducing their effectiveness in breaking down substrates); and binding the soluble products of exoenzyme activity (making these products less available for absorption by soil microbes). This binding of organic matter to clays occurs because the high density of negatively charged sites on clay minerals attract the positive charges on the organic matter (amine groups) or form bridges with polyvalent cations (Ca^{2+}, Fe^{3+}, Al^{3+}, Mn^{4+}) that bind to negative groups (e.g., carboxyl groups) on organic matter (Fig. 7.7). The net effect of this binding by clay minerals is to protect soil organic matter and reduce its decomposition rate. SOM protection by clay minerals is most important in ecosystems such as grasslands and tropical forests, in which decomposition is relatively rapid and where soil animals rapidly mix fresh litter with mineral soil. Mineral protection of SOM is less important in conifer forests and tundra in which much of the decomposi-

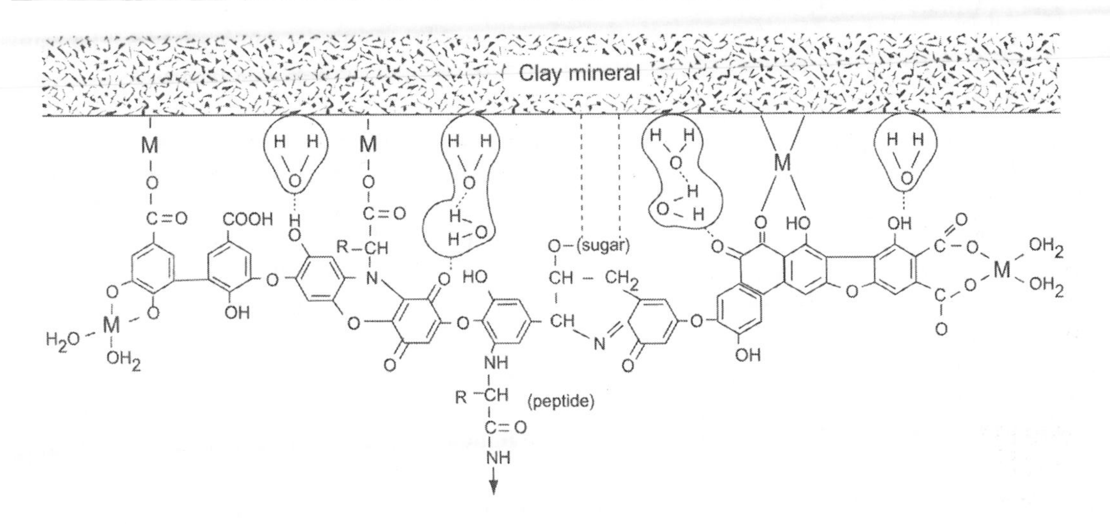

FIGURE 7.7. The interactions between soil organic matter and clay particles, as mediated by water (H—O—H) and metal ions (M). (Redrawn with permission from *Human Chemistry, 2nd edition* by F.J. Stevenson, © John Wiley & Sons, Inc.)

tion occurs above the mineral soil in a well-developed organic mat (O horizon).

Both the type and the quantity of clay influence decomposition. Many tropical clay minerals have a high aluminum concentration that binds tightly to organic matter through covalent bonds. Clays with a multilayered lattice structure bind organic compounds between the silicate layers, making them particularly effective in SOM protection (see Chapter 3).

Soil Disturbance

Soil disturbance increases decomposition by promoting aeration and exposing new surfaces to microbial attack. The mechanism by which disturbance stimulates decomposition is basically the same at all scales, ranging from the movement of earthworms through soils to tillage of agricultural fields. Disturbance disrupts soil aggregates so the organic matter contained within them becomes more exposed to oxygen and microbial colonization. This disturbance effect explains why the introduction of European earthworms to the northeastern United States considerably speeded forest decomposition rates and why plowing causes rapid organic matter loss from grassland or forest soils after conversion to agriculture (see Fig. 14.12). This disturbance effect is most pronounced in warm wet soils, where the increased aeration has greatest effect on decomposition. A soil converted to irrigated cotton, for example, lost half its organic content in 3 to 5 years (Haynes 1986), reversing a period of centuries to millennia that were required to accumulate soil organic matter. The loss of organic matter and disruption of aggregates by plowing eventually impedes the drainage of water, the growth of roots, and the mineralization of soil nutrients.

Substrate Quality and Quantity

Litter

Carbon quality of substrates may be the predominant chemical control over decomposition. There is a 5-fold to 10-fold range in decomposition rate of litter in a given climate, due to differences in substrate **quality**—that is, susceptibility of a substrate to decomposition measured under standardized conditions. Animal carcasses decompose more rapidly than plants; leaves decompose more rapidly than wood; deciduous leaves decompose more rapidly than evergreen leaves; and leaves from high-nutrient environments decompose more rapidly than leaves from infertile sites (Figs. 7.8 and 7.9). These differences in decomposition rate are a logical consequence of the types of chemical compounds

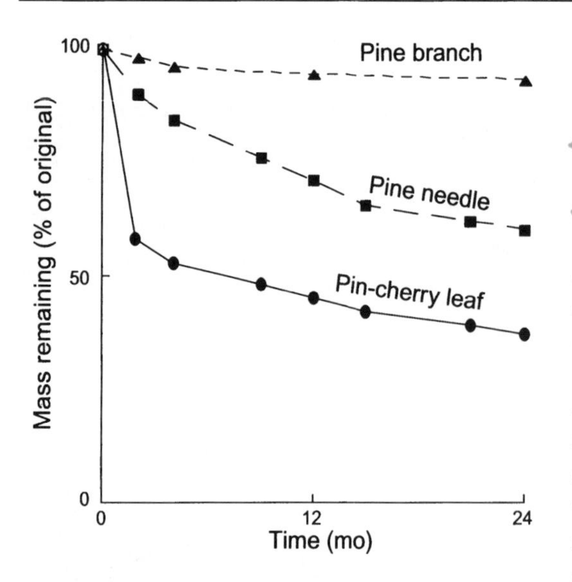

FIGURE 7.8. Time course of decomposition of a deciduous leaf, a conifer needle, and wood in a Canadian temperate forest (MacLean and Wein 1978).

present in litter. These compounds can be categorized roughly as labile metabolic compounds, such as sugars and amino acids; moderately labile structural compounds, such as cellulose and hemicellulose; and recalcitrant structural material, such as lignin and cutin. Rapidly decomposing litter generally has higher concentrations of labile substrates and lower concentrations of recalcitrant compounds than does slowly decomposing litter.

Five interrelated chemical properties of organic matter determine substrate quality (J. Schimel, personal communication, 2001): the size of molecules, the types of chemical bonds, the regularity of structures, the toxicity, and the nutrient concentrations. (1) Large molecules cannot pass through microbial membranes so they must be processed extracellularly by exoenzymes. This limits the degree of control that a given microbe can exert over the detection of substrate availability, the delivery of enzymes in response to substrate supply, and the efficient use of breakdown products. Due to differences in molecular size, sugars and amino

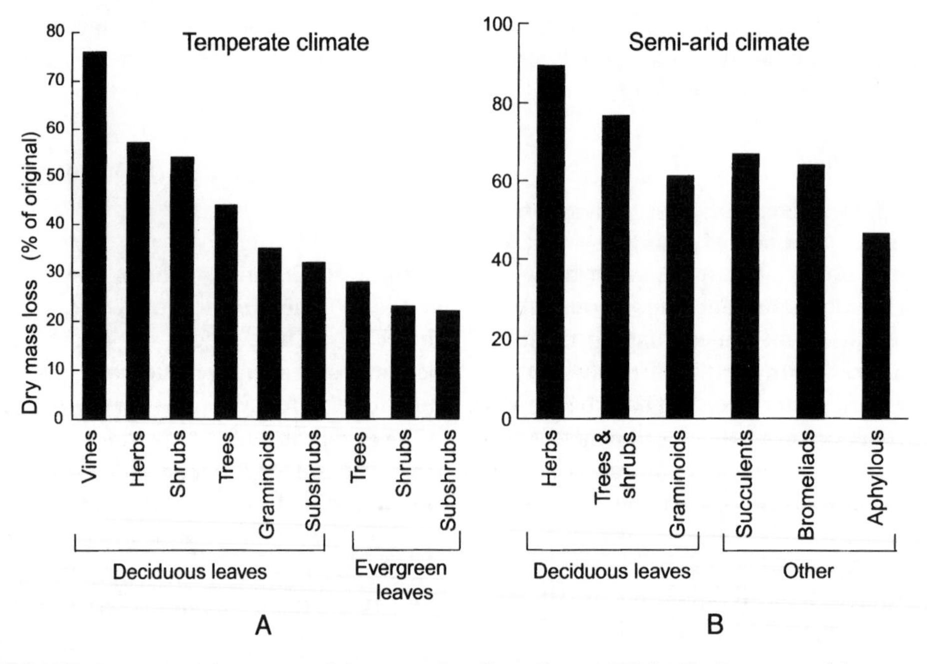

FIGURE 7.9. **A**, Decomposition rate of leaves of British deciduous and evergreen plant species. (Redrawn with permission from *Journal of Ecology*; Cornelissen 1996). **B**, Decomposition rate of deciduous plants and aridzone plants in Argentina. (N. Perez; Perez-Harguindeguy et al. 2000.)

acids are metabolized more readily than cellulose and proteins, respectively. (2) Some chemical bonds are easier to break than others. Ester linkages that bind phosphate to organic skeletons and peptide bonds that link amino acids to form proteins, for example, are easier to break than the double bonds of aromatic rings. For these reasons, the nitrogen in proteins is much more available to microbes than the nitrogen contained in aromatic rings. (3) Compounds like lignin that have a highly irregular structure do not fit the active sites of most enzymes, so they are broken down much more slowly than are compounds like cellulose, which consist of chains of regularly repeating glucose units. (4) Some soluble compounds such as phenolics and alkaloids are toxic and kill or reduce the activity of microbes that absorb them. (5) Organic nitrogen and phosphorus are the major sources of nutrients for supporting microbial growth, so organic matter, such as straw, that contains low concentrations of these elements may not provide sufficient nutrients to allow microbes to use fully the carbon present in the litter.

All of these chemical properties influence decomposition, but their relative importance is not well understood. Nonetheless, any of these properties can serve as a *predictor* of decomposition rate because the properties tend to be strongly correlated with one another. The ratio of carbon concentration to nitrogen concentration (**C:N ratio**), for example, has frequently been used as an index of litter quality, because litter with a low C:N ratio (high nitrogen concentration) generally decomposes quickly (Enríquez et al. 1993, Gholz et al. 2000). However, neither the nitrogen concentration of the litter nor the nitrogen availability in the soil *directly* influences the decomposition rate in most natural ecosystems (Haynes 1986, Prescott 1995, Prescott et al. 1999, Hobbie and Vitousek 2000); this suggests that C:N ratio is not the chemical property that directly controls decomposition in these ecosystems. This contrasts with agricultural residues such as straw, which have a low nitrogen concentration and a high concentration of moderately labile carbon sources like cellulose and hemicellulose. Nitrogen concentration appears to limit directly the decomposition rate of organic matter primarily when labile carbon substrates are available to support microbial growth (Haynes 1986). This is more likely to occur in the rhizosphere than in fresh litter. Under other circumstances, carbon lability rather than nitrogen may be the primary control over decomposition rate (Hobbie 2000). Despite our uncertainty of the mechanistic role of C:N ratio in decomposition, many biogeochemical models use this ratio as a predictor of decomposition rate when different ecosystem types are compared (see Chapter 9).

In recalcitrant litter, the concentration of lignin or the lignin:N ratio is often a good predictor of decomposition rate (Berg and Staaf 1980, Melillo et al. 1982, Taylor et al. 1989) (Fig. 7.10), again suggesting an important role of carbon quality in determining decomposition rates of litter. The carbon quality of litter is probably best defined in terms of the classes of organic compounds present and the enzymatic potential of the decomposer community, as described later. This information is available for so few ecosystems, however, that more readily measured properties, such as C:N ratio or

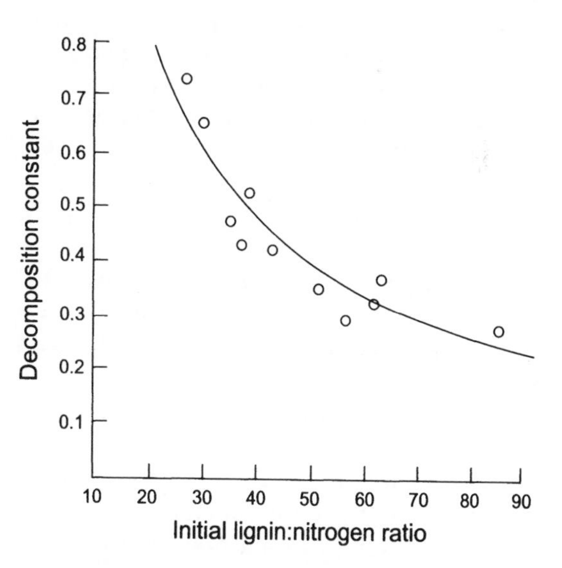

FIGURE 7.10. Relationship between the decomposition constant and the lignin:N ratio of litter. (Redrawn with permission from *Ecology*; Mellillo et al. 1982.)

lignin : N ratio, are frequently used as predictors of decomposition rate.

The effects of litter quality on decomposition rate often depend on the age of the litter. High-quality litter, for example, loses its labile carbon so quickly that the remaining old litter may have a lower decomposition potential than litter that initially had a low litter quality and slow decomposition rate (Berg and Ekbohm 1991).

The availability of belowground resources is the major ecological control over litter quality. Rapidly growing plants from high-resource sites typically produce litter that decomposes quickly because the same morphological and chemical traits that promote NPP also regulate decomposition (Hobbie 1992). Both NPP and decomposition are enhanced by a high allocation to leaves and by the production of leaves with a short life span. These tissues decompose rapidly because they have high concentrations of labile compounds such as proteins and low concentrations of recalcitrant cell-wall components such as lignin (Reich et al. 1997). Consequently, species from productive sites produce litter that decomposes rapidly (Cornelissen 1996) (Fig. 7.9). Species differences in litter quality make up an important mechanism by which plant species affect ecosystem processes (see Chapter 12) (Hobbie 1992) and are excellent predictors of landscape patterns of litter decomposition (Flanagan and Van Cleve 1983).

Soil Organic Matter

Both the age and the initial quality of SOM influence its rate of decomposition. As litter decomposes, its decomposition rate declines, because microbes first consume the more labile substrates, leaving progressively more recalcitrant compounds in the remaining litter (Fig. 7.2). Through fragmentation by soil invertebrates and these chemical alterations, the litter becomes converted to soil organic matter. As microbes die, chitin and other recalcitrant components in their cell walls comprise an increasing proportion of the litter mass (actually litter plus microbial mass), and nonenzymatic reactions produce recalcitrant humic compounds. All these processes contribute to a gradual reduction in organic matter quality as SOM ages. The C : N ratio also declines as decomposition proceeds, because carbon is respired away, and some of the mineralized nitrogen is incorporated into humus. The decline in C : N ratio is not, however, an indicator of increased nitrogen availability, because the nitrogen becomes incorporated into aromatic rings and other chemical structures that are recalcitrant. In summary, in SOM, as in litter, the carbon quality is a better predictor of decomposition rate than is the C : N ratio or the nitrogen concentration of SOM (Berg and Staaf 1980, Melillo et al. 1982).

Site differences in nutrient availability influence SOM decomposition primarily through their effects on the carbon quality of litter and SOM, rather than through direct nutrient effects on SOM decomposition. Sites with high productivity and litter quality typically produce a low-lignin SOM that decomposes readily (Van Cleve et al. 1983). As in the case of fresh litter, SOM decomposition rate does not show a consistent response to nutrient addition (Haynes 1986, Fog 1988), suggesting that nutrients seldom directly regulate SOM decomposition. Decomposition of SOM increases in response to nitrogen addition primarily when the organic matter consists of labile carbon substrates, for example when straw is plowed into agricultural soils (Mary et al. 1996) or when root exudation is enhanced by elevated CO_2 (Hu et al. 2001) (see Chapter 9).

The heterogeneous nature of SOM makes it difficult to identify the chemical controls over its decomposition. It is a mixture of organic compounds of different ages and chemical compositions. Components of SOM include fragments of recently shed root and leaf litter, together with soil organic matter that is thousands of years old (Oades 1989). These different aged components of SOM can be separated by density centrifugation, because recently produced particles are less dense than older ones and are less likely to be bound to mineral particles. Soils in which a large proportion of the SOM is in the light fraction generally have higher decomposition rates (Robertson and Paul 2000). Alternatively, soil can be chemically separated into distinct fractions, such as water-soluble compounds, humic acids, and fulvic acids, that differ in average age and ease of breakdown. SOM as a whole typically has a residence time of 20 to 50 years, although this can range from 1 to 2 years in cultivated fields to thousands of years in environments with slow decomposition rates. Even in a single soil, different chemical fractions of SOM have residence times ranging from days to thousands of years. Computer simulations of decomposition rate capture ecosystem carbon dynamics more effectively when they distinguish among these different soil carbon pools (Parton et al. 1993, Clein et al. 2000).

Decomposition in the rhizosphere is more rapid than in bulk soil for reasons that are poorly understood. The rhizosphere makes up virtually all the soil in fine-rooted grasslands, where the average distance between roots is about 1 mm, whereas forests are less densely rooted (often 10 mm between roots) (Newman 1985). Roots alter the chemistry of the rhizosphere by secreting carbohydrates and absorbing nutrients. These processes are most active in the region behind the tips of actively growing roots (Fig. 7.11) (Jaeger et al. 1999b). The growth of bacteria in the zone of exudation (Norton and Firestone 1991) is supported by abundant carbon availability (20 to 40% of NPP; see Table 6.2) and is therefore limited most strongly by nutrients (Cheng et al. 1996). Bacteria must acquire their nutrients for growth by breaking down SOM. In other words, plant roots use carbon-rich exudates to "prime" the decomposition process in the rhizosphere, just as you might use water to prime a pump. Microbial immobilization of nutrients in the rhizosphere benefits the plant only if these nutrients are subsequently released and become available to the root. Two processes may contribute to the release of nutrients from rhizosphere microbes: First, protozoa and nematodes may graze the populations of rhizosphere bacteria, using bacterial carbon to support their high energetic demands and excreting the excess nutrients (Clarholm 1985). Second, as the root matures and exudation rate declines, those bacteria that survive predation may become energy limited and break down

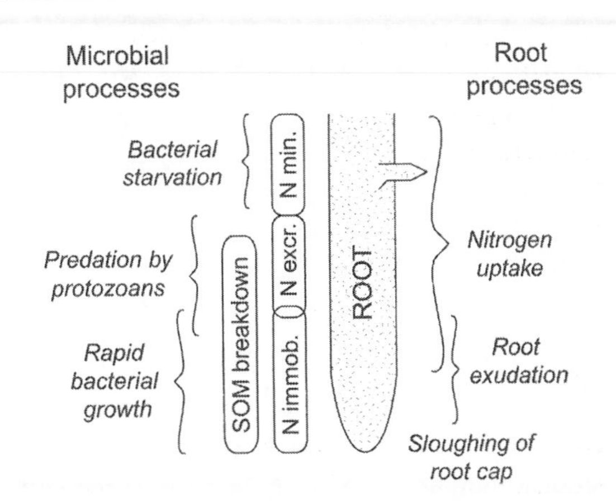

FIGURE 7.11. Root and microbial processes in the rhizosphere and the resulting effects on soil organic matter breakdown and nitrogen dynamics in the rhizosphere. N immob., N immobilization by rhizosphere microbes; N excr., N excretion by protozoa; N min., N miniralization by carbon-starved microbes.

nitrogen-containing compounds to meet their energy demands, releasing the nitrogen into the rhizosphere as ammonium. The relative contribution of grazing and starvation in these processes is unknown, but net nitrogen mineralization in the rhizosphere has been estimated to be 30% higher than in bulk soil. Rhizosphere decomposition occurs most readily in soils with relatively labile soil carbon and low soil lignin (Bradley and Fyles 1996) and therefore may occur to a greater extent in grasslands or early successional communities than in mature forests. Rhizosphere decomposition may be more sensitive to factors influencing plant carbohydrate status (e.g., light and grazing) than to soil environment (Craine et al. 1999), so the nature of controls over decomposition (soil environment vs. plant carbohydrate status) could differ substantially among ecosystems. However, the extent of rhizosphere decomposition and the nature of its ecological controls are not well characterized under field conditions, so it is difficult to evaluate its ecological importance.

Mycorrhizal fungi are functionally an extension of the root system, allowing the root–fungal symbiosis to absorb nutrients at a distance from the root. The **mycorrhizosphere** around mycorrhizal fungal hyphae rapidly moves plant carbon into the bulk soil through a combination of hyphal turnover and exudation (Norton et al. 1990). This may prime the decomposition process, just as occurs in the rhizosphere of roots, although nothing is known about this process under field conditions.

Microbial Community Composition and Enzymatic Capacity

Soil enzyme activity depends on microbial community composition and the nature of the soil matrix. We have seen that soil animals strongly affect decomposition through their effects on soil structure, litter fragmentation, and microbial community composition. The composition of the microbial community is important in turn because it influences the types and rates of enzyme production and thus the rates at which substrates are broken down. Enzymes that break down common substrates like proteins and cellulose are produced by so many types of microbes that these enzymes occur universally in soils (Schimel 2001). Enzymes involved in processes that occur only in specific environments, such as denitrification or methane production and oxidation, appear more sensitive to microbial community composition (Gulledge et al. 1997, Schimel 2001).

Soil enzyme activity is also influenced by the rates at which enzymes are inactivated in soils, either by degradation by soil proteases or by binding to soil minerals. Binding of an enzyme to the external surface of roots or microbes frequently prolongs enzyme activity in the soil, whereas binding to mineral particles can alter the enzyme configuration or block the active site of the enzyme, thereby reducing its activity. A brief description of a few soil enzyme systems illustrates some of the microbial and soil controls over exoenzyme activity.

Most soil microbes, including ericoid and ectomycorrhizal fungi, produce enzymes (proteases and peptidases) that break down proteins into amino acids. These breakdown products are readily absorbed by microbes and used either to produce microbial protein or to provide respiratory energy. Because proteases are subject to attack by other proteases, their lifetime in the soil is short, and soil protease activity tends to mirror microbial activity. Phosphatases, which cleave phosphate from organic phosphate compounds, are, however, more long lived, so their activity in soil is correlated more strongly with the availability of organic phosphate in soil than with microbial activity (Kroehler and Linkins 1991).

Cellulose is the most abundant chemical constituent of plant litter. It consists of chains of glucose units, often thousands of units in length; but none of this glucose is available until acted on by exoenzymes. Cellulose breakdown requires three separate enzyme systems (Paul and Clark 1996): **Endocellulases** break down the internal bonds to disrupt the crystalline structure of cellulose. **Exocellulases** then cleave off disaccharide units from the ends of chains, forming cellobiose, which is then absorbed by microbes and broken down intracellularly to glucose by **cellobiase**. Some soil microbes, including most fungi, can produce the entire suite of cellulase enzymes. Other organisms, such as some bacteria, produce only some cellulase enzymes and must function as part of microbial consortia to gain energy from cellulose breakdown.

Lignin is degraded slowly because only some organisms (primarily fungi) produce the necessary enzymes, and these microbes produce enzymes only when other more labile substrates are unavailable. Lignin forms nonenzymatically by condensation reactions with phenols and free radicals, creating an irregular structure that does not fit the specificity required by the active site of most enzymes. For this reason, lignin-degrading enzymes use free radicals, which have a low specificity for substrates. Oxygen is required to generate these free radicals, so lignin breakdown does not occur in anaerobic soils. Decomposers generally invest more energy in producing lignin-degrading enzymes than they gain by metabolizing the breakdown products of lignin (Coûteaux et al. 1995). Lignin is apparently degraded primarily to provide access to labile compounds such as cellulose, hemicellulose, and protein, which actually meet the energetic and nitrogen requirements of the decomposers. Some of the enzymes involved in lignin breakdown may also function in the formation and breakdown of soil humus.

Long-Term Storage of Soil Organic Matter

In climates that are favorable for decomposition, humus is the major long-term reservoir of soil carbon. Up to this point, we have focused primarily on the factors controlling the breakdown and loss of soil organic matter. Equally important are the processes that transform soil organic matter into relatively recalcitrant humus, allowing its accumulation in soils. Soil humus decomposes slowly for several reasons. As with lignin, its highly irregular structure is not efficiently attacked by a single enzyme system. Its large size and highly cross-linked form make most of the structure inaccessible to soil enzymes.

Its tendency to bind with soil minerals protects it from enzymatic attack. Much of the SOM in soil is therefore not good "food" for microbes, despite its high nitrogen content. Humus also constitutes a large reservoir of nitrogen in many ecosystems. This nitrogen turns over extremely slowly, except when disturbance increases the rate of humus decomposition. As the carbon in humus is respired away, the nitrogen is released, providing an important nutrient source to support ecosystem recovery after disturbance (see Chapter 13). The sensitivity of humus to breakdown following disturbance makes ecosystems with large humus accumulations, such as tropical forests or grasslands, particularly vulnerable to carbon loss after such changes.

The formation of humus by **humification** occurs through a combination of biotic and abiotic processes (Zech and Kogel-Knabner 1994). The following five steps have been implicated in humus formation (Fig. 7.12), although the relative importance of factors governing these steps is poorly understood.

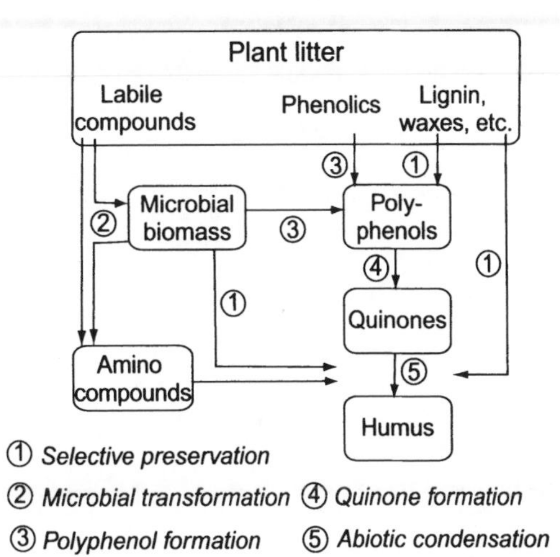

FIGURE 7.12. Principle pathways of humus formation. See text for details.

1. **Selective preservation.** Decomposition selectively degrades labile compounds in detritus, leaving behind recalcitrant materials like waxes, cutins, suberin, lignin, chitin, and microbial cell walls. Partial microbial breakdown of these recalcitrant leftovers often produces compounds with reactive groups and side chains that are common reactants in the nonspecific soil reactions that occur during humification.
2. **Microbial transformation.** Enzymatic breakdown of SOM produces low molecular weight water-soluble products, some of which participate in humus formation. Amino compounds, such as amino acids from protein breakdown and sugar amines from degradation of microbial cell walls, are particularly important in humification (see Step 5).
3. **Polyphenol formation.** Soluble phenolic compounds are important reactants in humus formation. They come from at least three sources (Haynes 1986): microbial degradation of plant lignin, the synthesis of phenolic polymers by soil microbes from simple nonlignin plant precursors, and polyphenols produced by plants as defenses against herbivores and pathogens.
4. **Quinone formation.** The polyphenol oxidase and peroxidase enzymes produced by fungi to break down lignin and other phenolic compounds also convert polyphenols into highly reactive compounds called quinones (Fig. 7.13).
5. **Abiotic condensation.** The quinones spontaneously undergo condensation reactions with many soil compounds, especially compounds

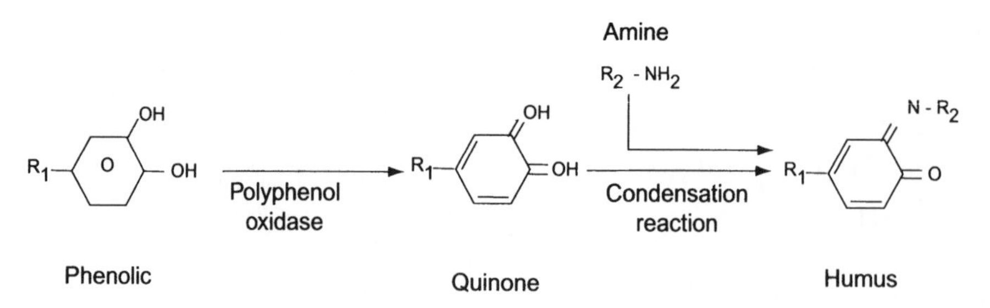

FIGURE 7.13. Reactions that occur during humus formation.

that have amino groups (with which they react most readily) or that are abundant (such as recalcitrant compounds that accumulate in soils).

The chemical nature of humus differs among ecosystems due to differences in the raw materials available for humification (Haynes 1986, Paul and Clark 1996). Forests produce a predominance of **humic acids**, which are large, relatively insoluble compounds with extensive networks of aromatic rings and few side chains. Phenolic-rich plants, which are common in many forests, provide many of the phenolic precursors for humic acids. **Humin** contains more long-chain nonpolar groups derived from cutin and waxes than do humic acids and are also relatively insoluble. **Fulvic acids** are more water soluble because of their extensive side chains and many charged groups. Their more open structure binds readily to other organic and inorganic materials. Grasslands have a more balanced mixture of fulvic and humic acids than do forests, perhaps because grassland plants produce fewer polyphenolic precursors for humus formation. The nitrogen content of humic acids (4%) is fivefold greater than in fulvic acids, but most of this nitrogen is in ring structures that are not readily broken down.

Environmentally protected organic matter accumulates in cold and wet environments. In environments in which low oxygen availability or low temperature inhibits decomposition, organic matter accumulates in a relatively nondecomposed state. This organic matter accumulates, not because it is recalcitrant, but because conditions constrain the activity of decomposers more strongly than they constrain carbon inputs by plants. Ecosystems with some of the largest carbon stores, such as wetlands and tundra, have soil organic matter that is highly labile and decomposes quickly, once the environmental limitations to decomposition are released. This makes carbon balance of these ecosystems quite vulnerable to global environmental change.

Decomposition at the Ecosystem Scale

Aerobic Heterotrophic Respiration

Aerobic heterotrophic respiration is the major avenue of carbon loss from ecosystems. It is the sum of aerobic respiration by soil microbes, which is equivalent to stand-level decomposition (discussed above), and the respiration by animals. Microbes and animals are grouped together as **heterotrophs** because they derive their energy and carbon from organic matter produced by plants (see Chapter 11). Decomposition accounts for most heterotrophic respiration, but animal respiration is also a significant avenue of carbon loss from some ecosystems (see Fig. 6.1). We discuss the factors regulating the consumption of plants and microbes by animals in Chapter 11.

The controls over stand-level decomposition are similar to the controls over GPP and NPP. As in the case of GPP and NPP, we cannot directly measure stand-level decomposition under undisturbed field conditions. Soil respiration includes the CO_2 respired by soil microbes, soil animals, and roots, and these cannot be separated by direct measurements. Isotopic tracers and ecosystem models are two tools that have proven particularly valuable in estimating decomposition at the ecosystem scale (Box 7.1). Both of these approaches indicate that stand-level decomposition rate depends not only on environment, as discussed earlier, but also on the amount and quality of recent carbon inputs to soils (Fig. 7.14). The quantity of carbon input to soils, in turn, generally depends on NPP. Carbon quality is also highest in productive stands.

Since GPP and NPP are important determinants of stand-level decomposition rate, it is not surprising that the controls over stand-level decomposition are similar to those for GPP and NPP. In other words, decomposition is ultimately controlled by the availability of soil resources, disturbance regime, and climate (Fig. 7.14). Measurements of soil respiration, which includes both heterotrophic and root respiration, are consistent with this generalization. Soil respiration correlates closely with NPP (Raich and Schlesinger 1992) (Fig. 7.15). Carbon loss through soil respiration is about 25% higher than carbon inputs through NPP, suggesting that about 25% of soil respiration derives from roots, and the rest comes from decomposition (Raich and Schlesinger 1992). Both NPP and decomposition are higher in the tropics than in the arctic and higher in rain forests than in deserts, due to similar environmental sensitivities of plants and decomposers. Likewise, plant species that are highly productive produce litter of higher quality than do species of low potential productivity. Habitats dominated by productive species are therefore characterized by high rates of litter decomposition (Hobbie 1992), high concentrations of labile carbon, and high microbial biomass (Zak et al. 1994), all contributing to the high stand-level decomposi-

Box 7.1. Isotopes and Soil Carbon Turnover

The quantity of soil carbon differs dramatically among ecosystems (Post et al. 1982). The total quantity of carbon in an ecosystem, however, gives relatively little insight into its dynamics. Tropical forests and tundra, for example, have similar quantities of soil carbon, despite their radically different climates and productivities. The simplest measure of soil carbon turnover is its residence time estimated from the pool size and carbon inputs (Eq. 7.3). These measurements show that, even though tropical forests and arctic tundra have similar size soil carbon pools, the turnover may be 500 times more rapid in the tropical forest. More sophisticated approaches to estimating soil carbon turnover using carbon isotopes (Ehleringer et al. 2000) lead to a similar conclusion. In the tropics, 85% of the ^{14}C that entered ecosystems during the era of nuclear testing in the 1960s has been converted to humus, whereas this proportion is only 50% in temperate soils and approximately 0% in boreal soils (Trumbore 1993, Trumbore and Harden 1997). This comparison clearly indicates more rapid turnover of soil organic matter in the tropics than at high latitudes.

Carbon isotopes can also be used to estimate the impacts of land use change on carbon turnover in situations in which the vegetation change is associated with a change in carbon isotopes. In Hawaii, for example, replacement of C_3 forests by pastures dominated by C_4 grasses causes a gradual change in the carbon isotope ratio of soil organic matter from values similar to C_3 plants toward values similar to C_4 plants (Townsand et al. 1995). This information can be used to estimate the quantity of the original forest carbon that remains in the ecosystem:

$$\%C_{S1} = \frac{C_{S2} - C_{V2}}{C_{V1} - C_{V2}} \times 100 \qquad \text{(B7.1)}$$

where $\%C_{S1}$ is the percentage of soil derived from the initial ecosystem type, C_{S2} is the ^{13}C content of soil from the second soil type, C_{V2} is the ^{13}C content of soil from the second vegetation type, and C_{V1} is the ^{13}C content of vegetation from the initial ecosystem type.

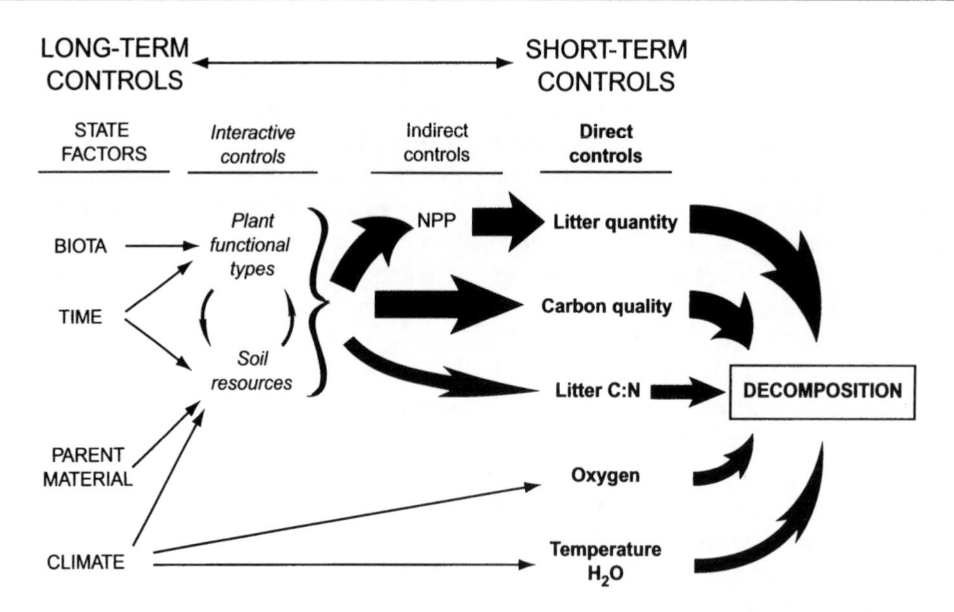

FIGURE 7.14. The major factors governing decomposition at the ecosystem scale. These controls range from proximate controls that determine the seasonal variations in decomposition to the state factors and interactive controls that are the ultimate causes of ecosystem differences in decomposition. Thickness of the arrows indicates the strength of the direct and indirect effects. The factors that account for most of the variation in decomposition among ecosystems are the quantity and carbon quality of litter inputs, which are ultimately determined by the interacting effects of soil resources, climate, vegetation, and disturbance regime.

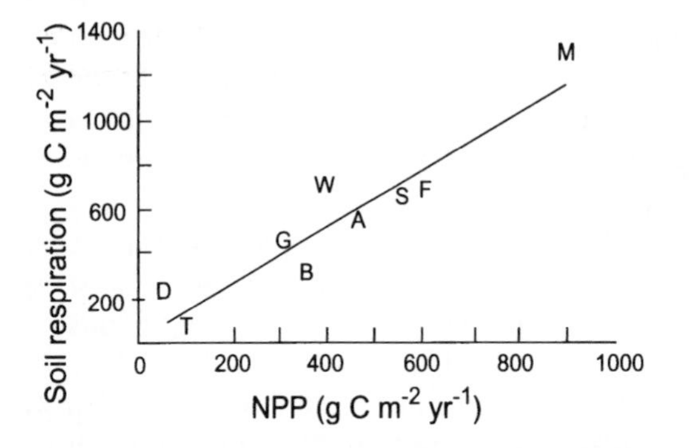

FIGURE 7.15. Relationship between mean annual soil respiration rate and mean annual NPP for Earth's major biomes. Root respiration probably accounts for the 25% greater soil respiration than NPP at any point along this regression line. A, agricultural lands; B, boreal forest and woodland; D, desert scrub; F, temperate forest; G, temperate grassland; M, moist tropical forest; S, tropical savanna and dry forest; T, tundra; W, mediterranean woodland and heath. (Redrawn with permission from *Tellus*; Raich and Schlesinger 1992.)

tion rates of productive sites. The relative importance of the direct effects of climate on decomposition vs. its indirect effects mediated by availability of soil resources and the quantity and quality of litter inputs remains to be determined.

Decomposition and carbon inputs to soils are seldom precisely in balance. Disturbances, herbivore outbreaks, and other events periodically cause organic matter inputs to soils to differ substantially from NPP (see Chapter 13). After hurricanes, for example, large inputs of plant material to the soil cause a pulse of decomposition to coincide with a sharp dip in NPP. In addition, decomposition is generally less sensitive to drought and more sensitive to low oxygen and to warm temperatures than is NPP, so seasonal or interannual variations in weather cause stand-level decomposition to be greater or less than NPP at any moment in time.

Stand-level decomposition shows little relationship with the total quantity of organic matter in soils because most soil carbon is either relatively recalcitrant or decomposes slowly because of an unfavorable soil environment (low temperature or low oxygen availability). Most decomposition derives from relatively recent litter inputs rather than from older soil organic matter. Consequently, total soil organic content is not a good measure of the food available for microbes or a good predictor of stand-level decomposition (Clein et al. 2000). In fact, the largest soil carbon accumulations frequently occur in ecosystems with slow decomposition, such as peat bogs.

The activity of soil microbes is more important than microbial biomass in determining decomposition rate. In boreal forests, for example, greatest decomposition occurs in soils with large inputs of high-quality litter. Microbial biomass is a relatively constant proportion (about 2%) of total soil carbon and therefore has the largest pool size (in grams per square meter) in those stands with the largest quantities of soil carbon; these are the stands with lowest productivity and slowest decomposition (Vance and Chapin 2001). In agricultural soils, microbial biomass also tends to be higher in extremely wet or dry soils, where decomposition is slow, than in moderately moist soils with higher decomposition rate (Insam 1990). Since most microbial biomass is inactive, it is probably more important as a reservoir of nutrients (see Chapter 9) than as a predictor of decomposition rate. This differs from the controls over carbon inputs to ecosystems, where the quantities of plant biomass and leaf area are extremely important determinants of GPP. Microbial processes like nitrification, which are conducted by a restricted number of microbial groups, on the other hand, appear to be sensitive to the population sizes of these groups (see Chapter 9).

Anaerobic Heterotrophic Respiration

Decomposition in anaerobic environments occurs slowly and produces energy inefficiently. Most of this chapter has focused on aerobic decomposition, which consumes oxygen and returns carbon to the atmosphere as CO_2. Wetlands, estuaries, and sediments beneath lakes and oceans, however, occupy vast areas of Earth's surface. These are environments where oxygen supply frequently limits decomposition rate, so organisms must use other electron acceptors to derive energy from organic matter. Oxygen is the preferred electron acceptor, when it is available, because it provides the most energy return per unit of organic matter oxidized. Progressively less energy is released with transfer to each of the following electron acceptors (see Chapter 3):

$$O_2 > NO_3^- > Mn^{4+} > Fe^{3+} > SO_4^{2-} > CO_2 > H^+ \qquad (7.4)$$

where NO_3^- is nitrate ion and SO_4^{2-} is sulfate ion. As oxygen becomes depleted by aerobic decomposition, denitrifiers gain a competitive advantage. They use most of the metabolic machinery associated with aerobic respiration to transfer electrons from organic matter to nitrate, producing the gases nitrous oxide (N_2O) and di-nitrogen (N_2), as waste products (see Chapter 9). The availability of nitrate is generally limited in anaerobic environments because nitrification, which produces nitrate, is an aerobic process. As the supply of nitrate becomes depleted, other bacteria, using other electron acceptors, gain a competitive advantage. Decomposition shifts to fermenters that break down labile organic compounds to acetate, other simple organic compounds, and hydrogen. These fermentation products are then used by sulfate reducers or methanogens, depending on the availability of sulfate, which transfer electrons to sulfate or CO_2 to produce hydrogen sulfide or methane, respectively. Estuaries, salt marshes, and ocean sediments frequently have enough marine-derived sulfate to make sulfate reduction the dominant pathway of anaerobic metabolism. The supply of sulfate is limited in many terrestrial environments, however, so the production of methane by a specialized group of Archaea known as **methanogens** becomes quantitatively important. Wetlands, such as swamps and rice paddies, are therefore important methane sources.

Methane emission from soils to the atmosphere is of global concern. Methane is 20-fold more effective in absorbing infrared radiation than is CO_2. Moreover, its concentration in the atmosphere has risen dramatically in recent decades, in part as a result of increased area of rice paddies and reservoirs (see Fig. 15.3). Even in wetlands, methane accounts for only 5 to 15% of the carbon released to the atmosphere by decomposers. Methane is thus quantitatively more important in its role as a greenhouse gas than as a path of carbon loss from ecosystems (see Fig. 6.8).

Methane is even more highly reduced than are carbohydrates, so it is an effective energy source for organisms that have access to oxygen. Another group of bacteria (**methanotrophs**) that occur in the surface soils of wetlands use this methane as an energy source and consume much of the methane before it diffuses to the atmosphere. The enzyme system that converts ammonium to nitrate also reacts with methane, causing well-aerated soils to be a net sink for methane. There are therefore important transfers between methane producers and consumers that occur both vertically within poorly drained ecosystems and horizontally from lowland to upland ecosystems.

Summary

Decomposition is the conversion of dead organic matter into CO_2 and inorganic nutrients through the action of leaching, fragmentation, and chemical alteration. Leaching removes soluble materials from decomposing organic matter. Fragmentation by soil animals breaks large pieces of organic matter into smaller ones that provide a food source for soil animals and create fresh surfaces for microbial colonization. Fragmentation also mixes the decomposing organic matter into the soil. Chemical alteration of dead organic matter is primarily a consequence of the activity of bacteria and fungi, although some chemical reactions occur spontaneously in the soil without microbial mediation.

Decomposition rate is regulated by physical environment, substrate quality, and the composition of the microbial community (including soil animals). Carbon chemistry is a strong determinant of litter quality; labile substrates, such as sugars and proteins, decompose more rapidly than recalcitrant ones, such as lignin and microbial cell walls. Nitrogen and phosphorus supply can also constrain the decomposition of labile carbon substrates, such as agricultural residues and root exudates. Plants in high-resource environments produce litter with high litter quality and therefore rapid decomposition rates. Decomposition rate declines with time, as recalcitrant substrates are depleted. Soil animals strongly influence decomposition by fragmenting litter, consuming soil microbes, and mixing the litter into mineral soil. The environmental factors that favor NPP (warm, moist, fertile soils) also promote decomposition so there is no clear relationship between the amount of carbon that accumulates in soils with either NPP or decomposition rate.

Review Questions

1. What is decomposition, and why is it important to the functioning of ecosystems?
2. What are the three major processes that give rise to decomposition? What are the major controls over each of these processes? Which of these processes is directly responsible for most of the mass loss from decomposing litter?
3. What are the major similarities and differences between bacteria and fungi in the ways in which they decompose dead organic matter? How do these two groups of decomposers differ in their response to moisture and nutrients? Why?
4. What roles do soil animals play in decomposition? How does this role differ between protozoans and earthworms?
5. Why do decomposer organisms secrete enzymes into the soil rather than breaking down dead organic matter inside their bodies?
6. What chemical traits determine the quality of soil organic matter? How do carbon quality and the C:N ratio differ between

the litter of plants growing on fertile vs. infertile soils?

7. Describe the mechanisms by which temperature and moisture affect decomposition rate.
8. How do roots influence decomposition rate? How does decomposition in the rhizosphere differ from that in the bulk soil? Why?
9. How do soil properties and disturbance affect decomposition rate?
10. How is humus formed? What are its precursors? How long does humus remain in the soil of undisturbed ecosystems? Why is humus formation important to the functioning of ecosystems?

Additional Reading

Anderson, J.M. 1991. The effects of climate change on decomposition processes in grassland and coniferous forests. *Ecological Applications* 1:326–347.

Beare, M.H., R.W. Parmelee, P.F. Hendrix, W. Cheng, D.C. Coleman, and D.A. Crossley Jr. 1992. Microbial and faunal interactions and effects on litter nitrogen and decomposition in agroecosystems. *Ecological Monographs* 62:569–591.

Coûteaux, M.-M., P. Bottner, and B. Berg. 1995. Litter decomposition, climate and litter quality. *Trends in Ecology and Evolution* 10:63–66.

Fog, K. 1988. The effect of added nitrogen on the rate of decomposition of organic matter. *Biological Review* 63:433–462.

Haynes, R.J. 1986. The decomposition process: Mineralization, immobilization, humus formation, and degradation. Pages 52–126 *in* R.J. Haynes, editors. *Mineral Nitrogen in the Plant-Soil System*. Academic Press, Orlando, FL.

Mary, B., S. Recous, D. Darwis, and D. Robin. 1996. Interactions between decomposition of plant residues and nitrogen cycling in soil. *Plant and Soil* 181:71–82.

Oades, J.M. 1989. An introduction to organic matter in mineral soils. Pages 89–159 *in* J.B. Dixon, and S.B. Weed, editors. *Minerals in Soil Environments*. Soil Science Society of America, Madison, WI.

Paul, E.A., and F.E. Clark. 1996. *Soil Microbiology and Biochemistry*. 2nd ed. Academic Press, San Diego, CA.

Swift, M.J., O.W. Heal, and J.M. Anderson. 1979. *Decomposition in Terrestrial Ecosystems*. Blackwell Scientific, Oxford, UK.

Zech, W., and I. Kogel-Knabner. 1994. Patterns and regulation of organic matter transformation in soils: Litter decomposition and humification. Pages 303–335 *in* E.-D. Schulze, editor. *Flux Control in Biological Systems: From Enzymes to Populations and Ecosystems*. Academic Press, San Diego, CA.

8 Terrestrial Plant Nutrient Use

Nutrient uptake, use, and loss by plants are key steps in the mineral cycling of ecosystems. This chapter describes the factors that regulate nutrient cycling through vegetation.

Introduction

Nutrient cycling in ecosystems involves highly localized exchanges between plants, soil, and soil microbes. In contrast to carbon, which is exchanged with a well-mixed atmospheric pool, nutrients are absorbed by plants and returned to the soil largely within the extent of the root system of an individual plant. More than 90% of the nitrogen and phosphorus absorbed by plants of most ecosystems comes from the recycling of nutrients that were returned from vegetation to soils in previous years (Table 8.1). The controls over nutrient uptake and use must therefore be examined at a more local scale than for carbon. Individual ecosystems, and indeed individual plants, have strong local effects on nutrient supply (Hobbie 1992, Van Breemen and Finzi 1998). The patterns of nutrient cycling beneath a deep-rooted oak that absorbs calcium from depth and produces a cation-rich litter, for example, may be quite different from those beneath a shallow-rooted pine that absorbs less cations and produces more organic acids (Andersson 1991).

Nutrient supply constrains the productivity of the terrestrial biosphere. Experimental addition of nutrients increases productivity of virtually every ecosystem, indicating the widespread importance of nutrients in constraining terrestrial production. Within any climatic zone, there is usually a strong positive correlation between soil fertility and plant production. Even in agricultural ecosystems, where nutrients are added regularly, production usually responds to nutrient addition. The major factors responsible for increased productivity of agriculture during the Green Revolution were increased rates of fertilizer, water, and pesticide application and breeding of crops capable of using water and nutrients, mostly through greater allocation to harvestable tissues (Fig. 8.1) (Evans 1980). There have been no major changes in photosynthetic rate during crop evolution. Given the widespread occurrence of nutrient limitation (Vitousek and Howarth 1991), an understanding of the controls over acquisition, use, and loss of nutrients by vegetation is essential to characterizing the controls over plant production and other ecosystem processes.

Overview

The quantity of nutrients that cycle through vegetation depends on the dynamic balance between nutrient supply from the soil and

TABLE 8.1. Major sources of nutrients that are absorbed by plants.

Nutrient	Source of plant nutrient (% of total)		
	Deposition/fixation	Weathering	Recycling
Temperate forest (Hubbard Brook)			
Nitrogen	7	0	93
Phosphorus	1	<10?	>89
Potassium	2	10	88
Calcium	4	31	65
Tundra (Barrow)			
Nitrogen	4	0	96
Phosphorus	4	<1	96

Data from Whittaker et al. (1979) and Chapin et al. (1980b).

nutrient demand by vegetation. The balance of nutrients required to support maximal growth is similar for most plants (Ingestad and Ågren 1988). Any nutrient present in less than the optimal balance is likely to limit growth, so plants invest preferentially in absorption of the nutrients that most strongly limit growth. Nutrients that accumulate in excess of plant requirements are absorbed more slowly. Nutrient ratios in plants therefore converge toward a common ratio. This pattern was first observed in the ocean (Redfield 1958) but also tends to occur on land. The consequence of this convergence toward a common nutrient ratio is that the nutrient that most strongly limits growth determines cycling rates of all nutrients. This element stoichiometry (Elser and Urabe 1999) defines patterns of cycling of most nutrients in ecosystems. The key to understanding nutrient cycling is thus to determine the factors controlling the cycling of the most strongly limiting element. These cycling rates may be constrained by either the supply of that nutrient from soil or its demand by vegetation. Supply rate of the growth-limiting nutrient could be constrained, for example, by climatic factors or by the chemical nature of parent material. Vegetation demand for the most limiting nutrient could be constrained, for example, by water limitation of growth or by the amount of biomass present after disturbance. In this chapter we first explore the controls over nutrient uptake by vegetation, then the relationship between nutrient content and production, and finally the controls over nutrient loss from vegetation.

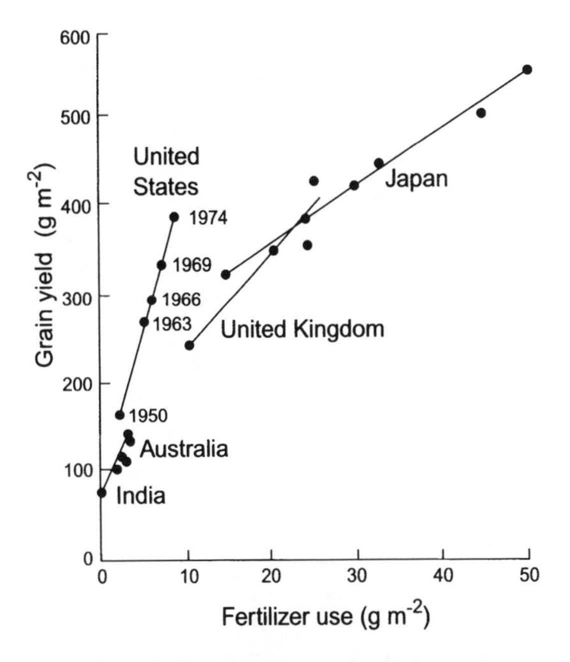

FIGURE 8.1. Response of grain yield of cereal crops to fertilizer addition. These studies were conducted during the Green Revolution. Yield is most responsive to low nutrient addition rates, often saturating with further nitrogen additions. (Redrawn with permission from *American Scientist*; Evans 1980.)

Nutrient Movement to the Root

Nutrients contact the root surface by three mechanisms: diffusion, mass flow, and root interception. Roots absorb only those nutrients that are in direct contact with live cells. Because roots constitute only a small proportion (much less than 1%) of the belowground volume, nutrients must first move from the **bulk soil** (i.e., the soil that is not in direct contact with roots) to the root surface before plants can absorb them.

Diffusion

Diffusion is the process that delivers most nutrients to plant roots. It is the movement of molecules or ions along a concentration gradient. **Nutrient uptake** and mineralization provide the driving forces for **diffusion** to the root surface by reducing nutrient concentration at the root surface (uptake) and increasing the concentration elsewhere in the soil (mineralization). Mineralization and other inputs to the pool of soluble nutrients are the main controls over the quantity of nutrients available to diffuse to the root surface (see Chapter 9).

Cation exchange capacity (CEC) of soils also influences the pool of nutrients available to diffuse to the root and the volume of soil that the root exploits. Soils with a high CEC store more available cations per unit soil volume—that is, they have a high **buffering capacity**—but retard the rate of nutrient movement to the root surface through exchange reactions (see Chapter 3). The root therefore draws more nutrients from a given volume of soil in soils with a high CEC but acquires these nutrients more slowly. The net effect of a high cation exchange capacity is to increase the supply of nutrients available to a root under conditions of high base saturation—that is, where the exchange complex has abundant cations. Anion exchange capacity is generally much lower than cation exchange capacity (see Chapter 3), so most anions, like nitrate, diffuse more rapidly in soils than do cations. Some anions, like phosphate, however, tend to precipitate readily from the soil solution and therefore diffuse slowly to the root surface.

Rates of diffusion differ strikingly among ions, due to differences in **charge density** (i.e., the charge per unit hydrated volume of the ion). Charge density, in turn, depends on the number of charges per ion and the hydrated radius of the ion. Divalent cations, like calcium and magnesium, are bound more tightly to the exchange complex and diffuse more slowly than do monovalent cations, such as ammonium and potassium. Ions of a given charge also differ slightly in diffusion rates because of differences in radius and number of water molecules that are loosely bound to the ion. Soil particle size and moisture determine the path length of diffusion from the bulk soil to the root surface. Ions diffuse through water films that coat the surface of soil particles. The higher the water content and the smaller the particle size, the more direct the diffusion path from the bulk soil to the root surface. Moist soils therefore permit more rapid diffusion than dry soils, and soils with a high clay content allow more rapid diffusion than do coarse-textured sandy soils.

Each absorbing root creates a **diffusion shell**, or a cylinder of soil that is depleted in the nutrients absorbed by the root. This diffusion shell constitutes the zone of soil directly influenced by plant uptake. The root accesses a relatively large volume of soil for those ions that diffuse rapidly. Nitrate, for example, which diffuses rapidly, is typically depleted in a shell 6 to 10 mm in radius around each absorbing root, whereas ammonium is depleted over a radius of less than 1 to 2 mm, and phosphate is depleted over a radius of less than 1 mm. It therefore takes a higher root density to fully exploit the soil for phosphate or ammonium than for nitrate. The root densities in many ecosystems are high enough to exploit most of the soil volume for nitrate but only a small proportion of the soil volume for ammonium or phosphate. The major way in which a plant can enhance uptake of ions that diffuse slowly is to increase root length and therefore the proportion of the soil that it exploits.

Mass Flow

Mass flow of nutrients to the root surface augments the supply of ions. Mass flow is the movement of dissolved nutrients to the root surface in flowing soil water. Transpirational water loss by plants is the major mechanism that causes mass flow of soil solution to the root surface. Mass flow can be an important mechanism for supplying nutrients that are abundant in the soil solution or that the plant needs in small quantities. Calcium, for example, is present in such a high concentration in many soils that the plant demands for calcium are completely met by mass flow of calcium from the bulk soil to the root surface (Table 8.2). Corn, for example, receives fourfold more calcium by mass flow to the root than is acquired by the plant. Plants that receive too much calcium by mass flow actively secrete calcium from roots into the soil solution, creating a diffusion gradient *away* from the root surface toward the bulk soil. Other nutrients are required in such small quantities by plants (**micronutrients**) that the needs of the plant can be completely met by mass flow (Table 8.2). However, mass flow is not sufficient for supplying the nutrients that are required by plants in large quantities but are present in low concentrations in the soil solution, such as nitrogen, phosphorus, and potassium. These **macronutrients** (i.e., nutrients required in large quantities) are supplied primarily by diffusion. The quantity of nutrients carried to the root surface in the transpiration stream in natural ecosystems, for example, is a small proportion of the nutrients required to support primary production (Table 8.2), so most of the nutrients must be supplied by some other mechanism—primarily diffusion. Even in agricultural soils, in which soil solution concentrations are much higher, mass flow supplies less than 10% of those nutrients that typically limit plant production. Diffusion therefore, rather than mass flow, is the major mechanism that supplies potentially limiting nutrients (nitrogen, phosphorus, and usually potassium) to plants. Diffusion becomes proportionately more important in supplying nutrients as soil fertility declines (Table 8.2).

Saturated flow of water through soils supplies additional nutrients and replenishes diffusion shells. Saturated flow is the movement of water through soil in response to gravity (see Chapter 3). After a heavy rain, water drains ver-

TABLE 8.2. Mechanisms by which nutrients move to the root surface.

Nutrient	Quantity absorbed by the plant ($g\,m^{-2}$)	Mechanism of nutrient supply (% of total absorbed)		
		Root interception	Mass flow	Diffusion
Sedge tundra (natural ecosystem)				
Nitrogen	2.2		0.5	99.5
Phosphorus	0.14		0.7	99.3
Potassium	1.0		6	94
Calcium[a]	2.1		250	0
Magnesium	4.7		83	17
Corn crop (agricultural ecosystem)				
Nitrogen	19	1	79	20
Phosphorus	4	2	4	94
Potassium	20	2	18	80
Calcium[a]	4	150	413	0
Magnesium[a]	4.5	33	244	0
Sulfur	2.2	5	95	0
Iron	0.2		53	
Manganese[a]	0.03		133	0
Zinc	0.03		33	
Boron[a]	0.02		350	0
Copper[a]	0.01		400	0
Molybdenum[a]	0.001		200	0

[a] Mass flow of these elements is sufficient to meet the total plant requirement, so no additional nutrients must be supplied by diffusion. The amount supplied by mass flow is calculated from the concentration of the nutrients in the bulk soil solution multiplied by the rate of transpiration. The amount supplied by diffusion is calculated by difference; other forms of transport to the root (e.g., mycorrhizae) may also be important but are not included in these estimates.

Data from Barber (1984), Chapin (1991a), and Lambers et al. (1998).

tically through the soil by saturated flow whenever the water content exceeds the soil's water-holding capacity. Because nutrient availability and mineralization rates are generally highest in the uppermost soils, this vertical flow of water redistributes nutrients and replenishes diffusion shells surrounding roots. Both root growth and vertical soil water movement occur preferentially in soil cracks, quickly eliminating diffusion shells around these roots. Saturated flow is also important in ecosystems where there is regular horizontal flow of ground water across an impermeable soil layer. Deep-rooted species in tundra underlain by permafrost, for example, have 10-fold greater nutrient uptake and productivity in areas of rapid subsurface flow than in areas without lateral groundwater flow (Chapin et al. 1988). The high productivity of trees and shrubs in riparian ecosystems results in part because their roots often extend to the water table and to groundwater beneath the stream (the **hyporrheic zone**), where roots tap the saturated flow of nutrients through the rooting zone.

Root Interception

Root interception is not an important mechanism for directly supplying nutrients to roots. As roots elongate into new soil, they intercept available nutrients in this unoccupied soil. The quantity of available nitrogen, phosphorus, and potassium per unit soil volume is, however, always less than the quantity of nutrients required to construct the root, so root interception can never be an important mechanism of nutrient supply to the shoot. Root growth is critical, not because it intercepts nutrients, but because it explores new soil volume and creates new root surface to which nutrients can move by diffusion and mass flow.

Nutrient Uptake

Nutrient uptake. Who is in charge? Three factors govern nutrient uptake by vegetation: nutrient supply rate from the soil, root length, and root activity. Just as with photosynthesis, several factors influence nutrient uptake at the ecosystem scale. Our main conclusions in this section are as follows: (1) Nutrient supply rate is the major factor accounting for differences among ecosystems in nutrient uptake at steady state. In other words, nutrient supply by the soil rather than plant traits determines biome differences in nutrient uptake by vegetation. (2) Plant traits such as root length and root activity strongly influence total nutrient uptake by vegetation in ecosystems in which biomass is increasing rapidly after disturbance. (3) Root length is the major factor governing which plants in an ecosystem are most successful in competing for a limited supply of nutrients.

Nutrient Supply

Nutrient uptake by vegetation at steady state is driven primarily by nutrient supply. There are unresolved debates about the relative importance of soil and plant characteristics in determining stand-level rates of nutrient absorption. There are several lines of evidence, however, suggesting that nutrient supply exerts primary control over nutrient uptake by vegetation at steady state. The most direct evidence for the controlling role of nutrient supply in driving uptake by vegetation is that most ecosystems respond to nutrient addition with increased uptake and net primary production (NPP) (Fig. 8.1). This differs strikingly from the controls over photosynthesis, where ecosystem differences in carbon uptake are determined primarily by capacity of vegetation to acquire carbon (leaf area and photosynthetic capacity), rather than by the supplies of CO_2 or light (see Chapter 5).

Simulation models support the conclusion that plant uptake is more sensitive to nutrient supply and to the volume of soil exploited by roots than to the kinetics of nutrient uptake, particularly for immobile ions like phosphate. At low nutrient supply rates, for example, variation in factors affecting diffusion (diffusion coefficient and buffering capacity) and root length (elongation rate) have a much greater effect on nutrient uptake than do kinetics (maximum and minimum capacity for

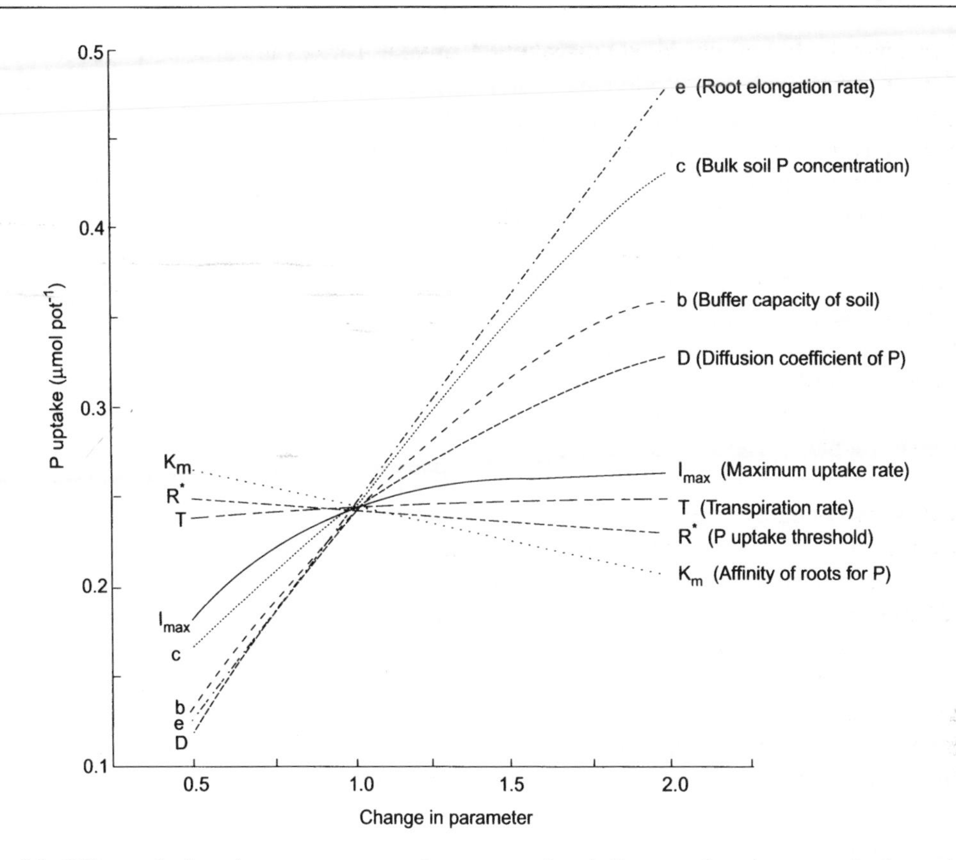

FIGURE 8.2. Effect of changing parameter values (from 0.5 to 2.0 times the standard value) in a model that simulates phosphate uptake by roots of soybean. The factors that have greatest influence on phosphate uptake are plant parameters that determine the quantity of roots (*e*) and soil parameters that influence phosphate supply from the soil (*c*, *b*, and *D*). (Redrawn with permission from *Annual Review of Plant Physiology*, Volume 36 © 1985 by Annual Reviews www.AnnualReviews.org; Clarkson 1985.)

uptake or affinity of roots for nutrients) or factors influencing mass flow (transpiration rate) (Fig. 8.2).

Development of Root Length

Root biomass differs less among ecosystems than does aboveground biomass at steady state, due to counteracting effects of production and allocation. In favorable environments, total plant biomass is large, but most of this biomass is allocated aboveground to compete for light. In dry or infertile environments, on the other hand, there is less biomass but a larger proportion of it is allocated belowground (see Chapter 6). The net result of these counteracting effects is that fine root biomass is probably more similar among ecosystems at steady state than is aboveground biomass. Thus the large differences among ecosystems in nutrient uptake and NPP reflect differences in nutrient supply more than differences in root biomass. Production of new root biomass enhances nutrient uptake by vegetation only if there are zones in the soil that are not yet exploited by roots—that is, where total root length is low enough that diffusion shells among adjacent roots do not overlap. This is most likely to occur after disturbance, just as new leaf production aboveground enhances carbon gain primarily when the leaf canopy is sparse (Craine et al., in press). Even with a fully developed "root canopy," increased root growth by an individual plant may be advantageous because it increases the proportion of the total nutrient supply captured by that plant.

Roots grow preferentially in resource hot spots. Root growth in the soil is not random. Roots that encounter microsites of high nutrient availability branch profusely (Hodge et al. 1999), allowing plants to exploit preferentially zones of high nutrient availability. This explains why root length is greatest in surface soils (Fig. 7.3), where nutrient inputs and mineralization are greatest, even though roots tend to be **geotropic** (i.e., grow vertically downward). This exploitation of nutrient hot spots ensures that plants maximize the nutrient return for a given investment in roots and reduces the fine-scale heterogeneity in soil nutrient concentrations. At a finer scale, **root hairs**, the elongate epidermal cells of the root that extend out into the soil, increase in length (e.g., from 0.1 to 0.8 mm) in response to a reduction in the supply of nitrate or phosphate (Bates and Lynch 1996). Both of these responses increase the length and surface area of roots available for nutrient uptake. Exploitation of hot spots does not always occur, however (Robinson 1994), and may be more pronounced in fast-growing than in slow-growing species (Huante et al. 1998).

Root length is a better predictor of nutrient uptake than is root biomass. Root length correlates closely with nutrient acquisition in short-term studies of nutrient uptake by plants from soils. Roots with a high **specific root length** (SRL; i.e., root length per unit mass) maximize their surface area per unit root mass and therefore the volume of soil that can be explored by a given investment in root biomass. We know much less about the morphology and physiology of roots in soil than of leaves. The limited available data suggest, however, that herbaceous plants (especially grasses) often have a greater SRL than woody plants and that there is a wide range in SRL among roots in any ecosystem. Much of the variation in SRL reflects the multiple functions of belowground organs. Roots can have a high SRL either because they have a small diameter or because they have a low tissue density (mass per unit volume). Some belowground stems and coarse roots have large diameters to store carbohydrates and nutrients or to transport water and nutrients and play a minor role in nutrient uptake. There may also be a tradeoff between SRL and longevity among fine roots, with high-density roots being less prone to desiccation and herbivory than low-density roots. Both the leaves and roots of slowly growing species often have high tissue density, low rates of resource acquisition (carbon and nutrients, respectively) but greater longevity than do leaves and roots of more rapidly growing species (Craine et al. 2001).

Mycorrhizae

Mycorrhizae increase the volume of soil exploited by plants. Mycorrhizae are symbiotic relationships between plant roots and fungal hyphae, in which the plant acquires nutrients from the fungus in return for carbohydrates that constitute the major carbon source for the fungus. About 80% of angiosperm plants, all gymnosperms, and some ferns are mycorrhizal (Wilcox 1991). These mycorrhizal relationships are important across a broad range of environmental and nutritional conditions, including fertilized crops (Allen 1991, Smith and Read 1997). With respect to nutrient uptake, mycorrhizal hyphae basically serve as an extension of the root system into the bulk soil, often providing 1 to 15 m of hyphal length per centimeter of root—that is, an increase in absorbing length of two to three orders of magnitude. Because the nutrient transport through hyphae occurs more rapidly than by diffusion along a tortuous path through soil water films, mycorrhizae reduce the diffusion limitation of uptake by plants. The small diameter of mycorrhizal hyphae (less than 0.01 mm) compared to roots (generally 0.1 to 1 mm) enables plants to exploit more soil with a given biomass investment in mycorrhizal hyphae than for the same biomass invested in roots. Plants typically invest 4 to 20% of their gross primary production (GPP) in supporting mycorrhizal hyphae (Lambers et al. 1996). Most of this carbon supports mycorrhizal respiration rather than fungal biomass, so a given carbon investment in mycorrhizal biomass can represent a large carbon cost to the plant. Mycorrhizae are most important in supplementing the nutrients that diffuse slowly through soils, particularly phosphate and potentially ammonium in ecosystems with low rates of nitrification. Although laboratory experiments show that plants consistently exclude mycorrhizae from roots under high-nutrient conditions, the extensive distribution of mycorrhizae across a wide range of soil fertilities, including most crop ecosystems, suggests that mycorrhizae continue to provide a net benefit to plants even in relatively fertile soils.

There are a range of mycorrhizal types, but the most common are **arbuscular mycorrhizae** (AM; also termed **vesicular arbuscular mycorrhizae**, VAM) and **ectomycorrhizae**. AM fungi grow through the cell walls of the **root cortex** (i.e., the layers of root cells involved in nutrient uptake), much as does a root pathogenic fungus. In contrast to root pathogens, AM produce **arbuscules**, which are highly branched treelike structures produced by the fungus and surrounded by the plasma membrane of the root cortical cells. Arbuscules are the structures through which nutrients and carbohydrates are exchanged between the fungus and the plant. AM are most common in herbaceous communities, such as grasslands, and in phosphorus-limited tropical forests and early successional temperate forests. Many AM associations are relatively nonspecific and can occur even with ecotmycorrhizal plant species shortly after disturbance. AM are generally eliminated after ectomycorrhizae colonize the roots of these species.

In a given ecosystem type, AM associations are best developed under conditions of phosphorus limitation, where they short-circuit the diffusion limitation of uptake (Allen 1991, Read 1991). Their effectiveness in overcoming phosphorus limitation may contribute to the nitrogen limitation in many temperate ecosystems (Grogan and Chapin 2000). The AM symbiosis is a dynamic interaction between plant and fungus, in which both roots and hyphae turn over rapidly. Under conditions in which plant growth is carbon limited, as in young seedlings or in shaded or highly fertile conditions, mycorrhizae may act as parasites and reduce plant growth (Koide 1991). Under these conditions, the plant reduces the number of infection points in new roots. As older roots die, this reduces the proportion of colonized roots, thus decreasing the carbon drain from the plant. AM associations might be viewed as a balanced parasitism between root and fungus that is carefully regulated by both partners.

Ectomycorrhizae are relatively stable associations between roots and fungi that occur primarily in woody plants. The exchange organ is a **mantle** or **sheath** of fungal hyphae that surround the root plus additional hyphae that grow through the cell walls of the cortex (the **Hartig net**). Roots respond to ectomycorrhizal colonization by reducing root elongation and increasing branching, forming short, highly branched rootlets. Fungal tissue accounts for about 40% of the volume of these root tips. As with AM, ectomycorrhizae involve an exchange of nutrients and carbohydrates between the fungus and the plant. In contrast to AM, ectomycorrhizae generally prolong root longevity. Ectomycorrhizae also differ from AM in that they have proteases and other enzymes that attack organic nitrogen compounds. The fungus then absorbs the resulting amino acids and transfers them to the plant (Read 1991). Ectomycorrhizae therefore enhance both nitrogen and phosphorus uptake by plants.

There are other mycorrhizal associations that differ functionally from AM and ectomycorrhizae. Fine-rooted heath plants in the families Ericaceae and Epacridaceae, for example, form mycorrhizae in which the fungal tissue accounts for 80% of the root volume. These mycorrhizae, like ectomycorrhizae, hydrolyze organic nitrogen and transfer the resulting amino acids to their host plants. Many nonphotosynthetic orchids totally depend on their mycorrhizae for carbon as well as nutrients. Their mycorrhizal fungi generally form links between the orchid and some photosynthetic plant species, especially conifers. In this case, the plant is clearly parasitic on the fungus.

As with the orchid–fungal association, ectomycorrhizae and AM often attach to several host plants, often of different species. Carbon and nutrients can be transferred among plants through this fungal network, although relatively few studies have shown a *net* transfer of carbon among plants (Simard et al. 1997). If these fungal connections among plants cause large net transfers of carbon and nutrients, they could alter competitive interactions, with potentially important ecosystem consequences. The net carbon transfer from canopy plants to shaded seedlings (Simard et al. 1997), for example, could promote the establishment of shade-tolerant tree seedlings, which would increase total ecosystem carbon demand and perhaps net photosynthesis (GPP). The quantitative significance of these transfers in natural ecosystems is totally unknown.

Root Uptake Properties

Active transport is the major mechanism by which plants absorb potentially limiting nutrients from the soil solution at the root surface. Plant roots acquire nutrients from the soil solution primarily by **active transport**, an energy-dependent transport of ions across cell membranes against a concentration gradient. Due to the high concentrations of ions and metabolites inside plant cells, there is a constant leakage out of the root along a concentration gradient. Phosphate, for example, leaks from roots at about a third of the rate at which it is absorbed from the soil. This passive leakage of ions, sugars, and other metabolites probably accounts for much of the exudation from fine roots. Ions that enter the root move passively by diffusion and mass flow through the cell walls of the **cortex** toward to the interior of the root (Fig. 8.3). As nutrients move through the cortical cell walls toward the center of the root, adjacent cortical cells absorb these nutrients by active transport. Nutrients can move through the cell walls only as far as the **endodermis**, a **suberin**-coated (wax-coated) layer of cells between the cortex and the xylem. Once nutrients have been absorbed by cortical cells, they move through a chain of interconnected cells to the endodermis, where they are secreted into the dead xylem cells that transport the nutrients to the shoot in the transpiration stream. As much as 30 to 50% of the carbon budget of the root goes to supporting nutrient

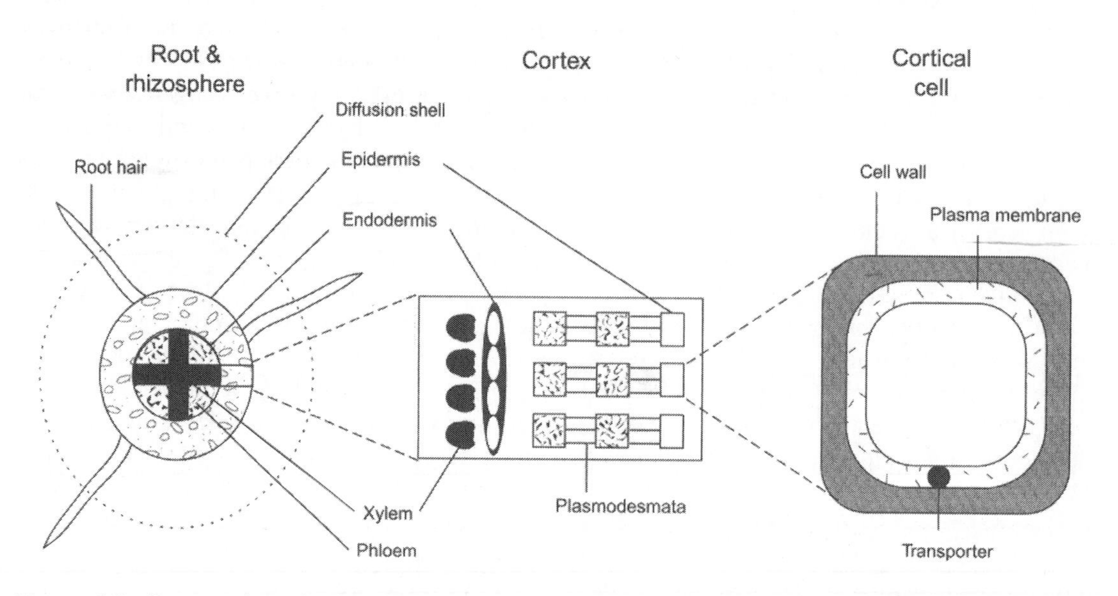

FIGURE 8.3. Cross-section of a root at three scales. The rhizosphere (or diffusion shell) is the zone of soil influenced by the root. The cortex has an outer layer of cells (the epidermis), some of which are elongated to form root hairs. The cortex is separated from the transport tissues (xylem and phloem) by a layer of wax-impregnated cells (the endodermis). Each cortical cell absorbs ions that diffuse through the pore spaces in the cell wall to the cell membrane. Membrane-bound proteins (transporters) transport ions across the cell membrane by active transport. Ions move from the outermost cortical cells toward the endodermis either through the cell walls or through the cytoplasmic connections between adjacent cortical cells (plasmodesmata).

TABLE 8.3. Preference ratios for plant absorption of different forms of nitrogen[a] when all forms are equally available.

Species	$NH_4^+:NO_3^-$ preference[b]	Glycine : NH_4^+ preference[b]	References
Arctic vascular plants	1.1	2.1 ± 0.6 (12)	Chapin et al. (1993), Kielland (1994)
Arctic nonvascular plants		5.0 ± 1.5 (2)	Kielland (1997)
Boreal trees	19.3 ± 5.8 (4)	1.3	Chapin et al. (1986b), Kronzucker et al. (1997), Näsholm et al. (1998)
Alpine sedges	3.9 ± 1.3 (12)	1.5 ± 0.4 (11)	Raab et al. (1999)
Temperate heath		1.0	Read and Bajwa (1985)
Salt marsh	1.3		Morris (1980)
Mediterranean shrub	1.2		Stock and Lewis (1984)
Barley	1.0		Chapin et al. (1993)
Tomato	0.6		Smart and Bloom (1988)

[a] Assumes all forms of nitrogen are equally available.
[b] A preference ratio >1 indicates that the first form of nitrogen is absorbed preferentially over the second. Number of species studied in parentheses. Research shows that many plants preferentially absorb glycine (a highly mobile amino acid) over ammonium and preferentially absorb ammonium over nitrate, when all forms are equally available.

absorption, indicating the large energetic cost of nutrient uptake. Elements required in small quantities are often absorbed simply by mass flow or diffusion into the root cortical cells (Table 8.2).

Some plant species tap pools of nutrients that are unavailable to other plants. Although all plants require the same suite of nutrients in similar proportions, nitrogen is available in several forms (nitrate, ammonium, amino acids, etc.) that differ in availability among ecosystems. Many species preferentially absorb ammonium (and perhaps amino acids) over nitrate, when all nitrogen forms are equally available (Table 8.3). Species differ, however, in their relative preference for these nitrogen forms, frequently showing a high capacity to absorb the forms that are most abundant in the ecosystems to which they are adapted. Many species that occupy highly organic soils of tundra and boreal forest ecosystems, for example, preferentially absorb amino acids (Chapin et al. 1993, Näsholm et al. 1998), although even agricultural species use amino acid nitrogen (Näsholm et al. 2000). An important community consequence of species differences in nitrogen preference is that nitrogen represents several distinct resources for which species can compete. Species in the same community often have quite different isotopic signatures of tissue nitrogen, because they acquire nitrogen from different sources—either different chemical fractions (nitrate, ammonium, organic nitrogen) or different soil depths (Nadelhoffer et al. 1996) (Fig. 8.4). Changes in species composition could therefore alter the nitrogen pools that are used to support primary production. In most cases, the species present are capable of using the prevailing forms of available nitrogen. If human activities alter the prevailing form of available nitrogen, for example through nitrate deposition in coniferous forests, this novel form of nitrogen may be used less effectively by the extant vegetation.

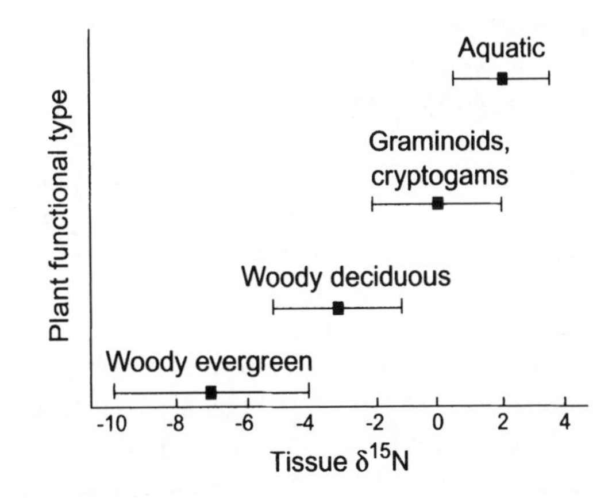

FIGURE 8.4. Concentration of ^{15}N in tissues from different growth forms of boreal plants. (Figure provided by K. Kielland; Kielland 1999.)

Spruces, for example, preferentially use ammonium over nitrate (Kronzucker et al. 1997), so nitrate entering in acid rain might leach from the ecosystem and pollute groundwater, lakes, and streams rather than being effectively absorbed by plants.

The three major forms of nitrogen differ in their carbon cost of incorporation into biomass. The carbon cost of incorporating amino acids is minimal, whereas ammonium must be attached to a carbon skeleton (the process of **assimilation**) before it is useful to the nitrogen economy of the plant. Finally, nitrate must be reduced to ammonium, which is energetically expensive, before it can be assimilated. Most plants reduce some of the nitrate in leaves, using excess reducing power from the light reaction. In this case, the high energy cost of nitrate reduction does not detract from energy available for other plant processes. High availability of light or nitrate usually increases the proportion of nitrate reduced in leaves. Species also differ in their capacity to reduce nitrate in leaves, with species adapted to high-nitrate environments usually having a higher capacity to reduce nitrate in the leaves. Tropical and subtropical perennials and many annual plants typical of disturbed habitats, for example, reduce a substantial proportion of their nitrate in leaves, whereas temperate perennials reduce most nitrate in the roots (Lambers et al. 1998). Nitrogen availability is usually so limited in temperate and high-latitude terrestrial environments that the relative availability of nitrogen forms is more important than cost of assimilation in determining which forms of nitrogen are used by plants. Plants usually absorb whatever they can get.

Plant species also differ in the pools of phosphorus they can tap. Roots of some plant species produce phosphatase enzymes that release inorganic phosphate for absorption by plant roots. The dominant sedge in arctic tussock tundra, for example, meets about 75% of its phosphorus requirement by absorbing the products of its phosphatase enzymes (Kroehler and Linkins 1991). Other plants, particularly those in dry environments, secrete chelates such as citrate that diffuse from the root into the bulk soil. These chelates bind iron from insoluble iron phosphate complexes, which solubilizes phosphate; soluble phosphate then diffuses to the root (Lambers et al. 1998). Some plants, particularly Australian and South African heath plants in the Proteaceae, produce dense clusters of roots (**proteoid roots**) that are particularly effective in secreting chelates and solubilizing iron phosphates. There are many classes of chelates (**siderophores**) produced by plant roots, although the benefit to the plant from these secretions is poorly known. Plants therefore differ in the soil phosphorus pools they can exploit, but we have only a rudimentary understanding of the ecosystem consequences of these species differences.

Species differences in rooting depth and density influence the pool of nutrients that can be absorbed by vegetation. Grasslands and forests growing adjacent to one another on the same soil often differ greatly in annual nutrient uptake and productivity, because the more deeply rooted forest trees exploit a larger soil volume and therefore a larger pool of water and available nutrients than do shallow-rooted species (see Chapter 12). In summary, there are several mechanisms by which species composition influences the quantity and form of nutrients acquired by vegetation.

Root uptake capacity increases in response to plant demand for nutrients. When the aboveground environment favors rapid growth and associated high demand for nutrients, plant roots respond by synthesizing more transport proteins in root cortical cells, thus increasing the capacity of the root to absorb nutrients. Species that have an inherently high relative growth rate or experience conditions that support rapid growth therefore have a high capacity to absorb nutrients (Chapin 1980). High light and warm air temperatures, for example, increase root uptake capacity, whereas shade, drought, and phenologically programmed periods of reduced growth lead to a low uptake capacity. The rates of nutrient uptake by vegetation are therefore influenced both by soil factors that determine nutrient supply and by plant factors that determine nutrient demand. There is a close correlation in the field between nutrient uptake and NPP (Fig. 8.5). It is difficult, however, to separate cause from effect in explaining this correlation.

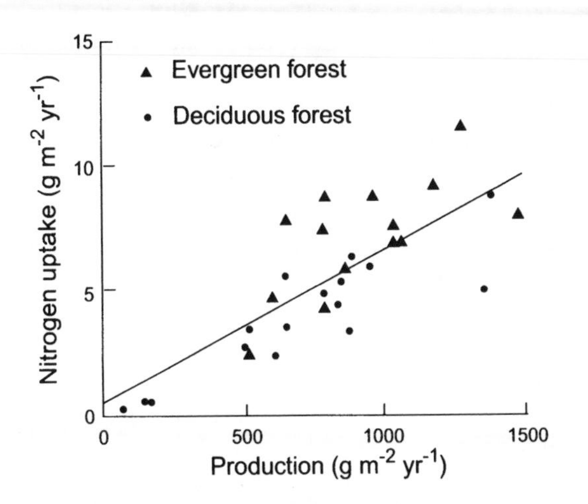

FIGURE 8.5. Relationship between nitrogen uptake of temperate and boreal coniferous and deciduous forests and NPP. (Redrawn with permission from Academic Press; Chapin 1993b.)

Changes in root uptake kinetics fine-tune the capacity of plants to acquire specific nutrients. Ion-transport proteins are specific for particular ions. In other words, ammonium, nitrate, phosphate, potassium, and sulfate are each transported by a different membrane-bound protein that is individually regulated (Clarkson 1985). Plants induce the synthesis of additional transport proteins for those ions that specifically limit plant growth. Roots of a phosphorus-limited plant therefore have a high capacity to absorb phosphate, whereas roots of a nitrogen-limited plant have a high capacity to absorb nitrate and ammonium (Table 8.4). Nitrate reductase, the enzyme that reduces nitrate to ammonium (the first step before nitrate-nitrogen can be incorporated into amino acids for biosynthesis) is also specifically induced by presence of nitrate. There are therefore several adjustments that plants make to improve resource balance. Plants first alter the root to shoot ratio to improve the balance between acquisition of belowground and aboveground resources. Plants then regulate the location of root growth to exploit hot spots of nutrient availability. Finally, plants adjust their capacity to absorb specific nutrients, which brings the plant nutrient ratios closer to values that are optimal for growth.

Nutrient ratios define a stoichiometry of nutrient cycles in ecosystems. The oceanographer Redfield (1958) noted that algae with a ratio of nitrogen to phosphorus (N:P ratio) greater than 14:1 tend to respond to phosphorus addition, whereas algae with a lower N:P ratio are nitrogen limited. This ratio is now referred to as the **Redfield ratio**. The important implication of nutrient ratios is that, *if they are constant*, the element that most strongly constrains production by vegetation defines the quantities of all elements that are cycled through vegetation. Marine algae with high N:P ratios, for example, preferentially absorb phosphorus. They absorb nitrogen and other nutrients in proportion (the Redfield ratio) to the phosphorus that they are able to acquire. Algae with low N:P ratios preferentially absorb nitrogen and absorb phosphorus and other nutrients in proportion to the nitrogen that they are able to acquire. The most strongly limiting element therefore determines the cycling rates of *all* elements. Experiments in terrestrial ecosystems suggest that terrestrial plants also adjust their mineral nutrition to converge on the same Redfield ratio. Heath plants with N:P ratios in leaves of less than 14:1 generally respond to experimental additions of nitrogen, whereas plants with N:P ratios greater than 16:1 generally respond to added phosphorus but not to nitrogen (Fig. 8.6). This

TABLE 8.4. Effect of environmental stresses on rate of nutrient absorption by barley.

Stress	Ion absorbed	Uptake rate by stressed plants (% of control)
Nitrogen	ammonium	209
	nitrate	206
	phosphate	56
	sulfate	56
Phosphorus	phosphate	400
	nitrate	35
	sulfate	70
Sulfur	sulfate	895
	nitrate	69
	phosphate	32
Water	phosphate	32
Light	nitrate	73

Data from Lee (1982), Lee and Rudge (1987), and Chapin (1991a).

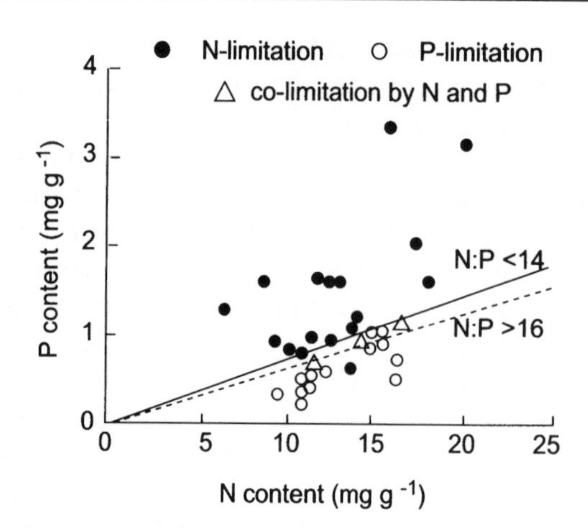

FIGURE 8.6. Relationship between nitrogen and phosphorus concentration of leaves in heath plants. Each datum point represents a site where nutrient addition experiments show that plant growth is limited by nitrogen, phosphorus, or both. Plants with an N:P ratio less than 14 respond primarily to nitrogen, whereas plants with an N:P ratio greater than 16 respond primarily to phosphorus. (Redrawn with permission from *Journal of Applied Ecology*; Koerselman and Mueleman 1996.)

principle of optimal element ratios is used in agriculture to determine which nutrients limit crop growth so nutrient additions can be matched to plant requirements.

Element ratios are more variable among terrestrial plants they are among phytoplankton because of the greater capacity for nutrient storage. Terrestrial plants have storage organs (e.g., stems) and organelles (e.g., vacuoles) in which they store nutrients that are nonlimiting to growth. In this way, terrestrial plants can take advantage of short-term pulses of nutrient supply. Nitrogen and phosphorus, for example, often show an autumn pulse of availability, when leaves are shed and leached by rain, and a spring pulse, after a winter season when decomposers are more active than plants (see Chapter 6). Plants absorb these nutrients at times of abundant supply, altering the ratios of elements in their tissues. These nutrients are then drawn out of storage at times when the demands for growth exceed uptake from the soil. In arctic tundra, for example, each year's production is supported primarily by nutrients absorbed in previous years, and uptake serves primarily to replenish these stores (Chapin et al. 1980a). In one field experiment, tundra plants that were provided with only distilled water grew just as rapidly as did plants with free access to soil nutrients, indicating that nutrient stores were sufficient to support an entire season's production (Jonasson and Chapin 1985). Even in the ocean and freshwater ecosystems, element ratios of algae can be variable due to storage of nonlimiting nutrients in vacuoles (see Chapter 10).

Nutrient uptake alters the chemical properties of the rhizosphere. Nutrient absorption by plant roots reduces the concentration of nutrients adjacent to the root. This depletion of soluble nutrients by root uptake can be substantial for nutrients that diffuse readily and create large diffusion shells. The *pool sizes* of dissolved nutrients in the soil solution are therefore a poor indicator of nutrient availability; dissolved nutrient pools can be small because of low mineralization rates or rapid uptake. Plant nutrient uptake is a critical control over ecosystem retention of mobile nutrients such as nitrate. Forest clearing or crop removal, for example, makes soils more prone to nitrate leaching into groundwater and streams (Bormann and Likens 1979).

A second major consequence of plant nutrient absorption is a change in rhizosphere pH. Whenever a root absorbs an excess of cations, it secretes hydrogen ions (H^+) into the rhizosphere to maintain electrical neutrality. This H^+ secretion acidifies the rhizosphere. Except for nitrogen, which can be absorbed either as a cation (NH_4^+) or an anion (NO_3^-), the ions absorbed in greatest quantities by plants are cations (e.g., Ca^{2+}, K^+, Mg^{2+}), with phosphate and sulfate being the major anions (Table 8.2). When plants absorb most nitrogen as NH_4^+, their cation uptake greatly exceeds anion uptake, and they secrete H^+ into the rhizosphere to maintain charge balance, causing acidification of the rhizosphere. When plants absorb most nitrogen as NO_3^-, their cation–anion absorption is more nearly balanced, and roots have less effect on rhizosphere pH. Ammonium tends to be the dominant form of inorganic nitrogen in acidic soils, whereas nitrate makes up a larger proportion of inorganic nitrogen in basic soils (see Chapter 9). The uptake process therefore tends to make acidic soils more acidic.

Roots also alter the nutrient dynamics of the rhizosphere through large carbon inputs from root death, the sloughing of mucilaginous carbohydrates from **root caps**, and the exudation of organic compounds by roots. These carbon inputs to soil may account for 10 to 30% of the GPP (see Chapter 6). Root exudation provides a labile carbon source that stimulates the growth of bacteria, which acquire their nitrogen by mineralizing organic matter in the rhizosphere (see Chapter 7). This nitrogen becomes available to plant roots when bacteria are grazed by protozoans or bacteria become energy starved due a reduction in root exudation (see Fig. 7.12). Plants are sometimes effective competitors with microbes for soil nutrients, for example, when plant carbon status is enhanced by added CO_2 (Hu et al. 2001). We know relatively little, however, about factors that govern competition for nutrients between plants and microbes.

Nutrient Use

Nutrients absorbed by plants are used primarily to support the production of new tissues (NPP). Plants store nutrients only when nutrient uptake exceeds the requirements for production (Chapin et al. 1990). Nutrients absorbed by roots move upward in the xylem and are then recirculated in the phloem to sites where production or storage occurs. The hormonal balance of the plant governs the patterns of carbon and nutrient transport in the phloem (and therefore the allocation of nutrients within the plant).

Nitrogen is incorporated primarily into proteins with lesser amounts in nucleic acids and lipoproteins (Chapin and Kedrowski 1983, Chapin et al. 1986a). Phosphorus is incorporated preferentially into sugar phosphates involved in energy transformations (photosynthesis and respiration), nucleic acids, and phospholipids (Chapin et al. 1982). Calcium is an important component of cell walls (calcium pectate). Potassium is important in osmotic regulation. Other cations (e.g., magnesium and manganese) serve as cofactors for enzymes (Marschner 1995).

The highest concentrations of nitrogen, phosphorus, and potassium typically occur in leaves (and to a lesser extent in fine roots) because of the importance of these elements in metabolism. Wood, in contrast, has low concentrations of all elements, (especially nitrogen and phosphorus) because of the large proportion of xylem, which consists of dead cells. Calcium, which is associated with cell walls, forms a higher proportion of total plant nutrients in wood than other tissues. Roots are intermediate in their tissue concentrations. The quantity of elements allocated to each tissue depends on tissue concentrations and on biomass allocation.

Nutrient supply affects growth more than it affects nutrient concentration. Plants respond to increased supply of a limiting nutrient in laboratory experiments primarily by increasing plant growth, giving a linear relationship between rate of nutrient accumulation and plant growth rate (Fig. 8.7) (Ingestad and Ågren 1988). Plants also respond to increased nutrient supply in the field primarily through increased NPP (Fig. 8.5), with proportionately

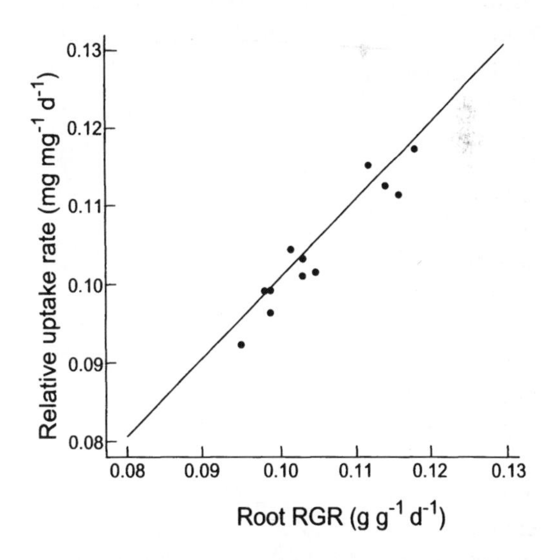

FIGURE 8.7. The rate of nitrogen uptake in tobacco as a function of the relative growth rate of roots (RGR). (Redrawn with permission from *Botanical Gazette*, Vol. 139 © 1978 University of Chicago Press; Raper et al. 1978.)

less increase in tissue nutrient concentration. Tissue nutrient concentrations increase substantially only when other factors begin to limit plant growth. The sorting of species by habitat also contributes to the responsiveness of nutrient uptake and NPP to variations in nutrient supply observed across habitats. Species such as trees that use large quantities of nutrients dominate sites with high nutrient supply rates, whereas infertile habitats are dominated by species with lower capacities for nutrient absorption and growth. Despite these physiological and species adjustments, tissue nutrient concentrations in the field generally increase with an increase in nutrient supply.

Nutrient use efficiency is greatest where production is nutrient limited. Differences among plants in tissue nutrient concentration provide insight into the quantity of biomass that an ecosystem can produce per unit of nutrient. **Nutrient use efficiency** (NUE) is the ratio of nutrients to biomass lost in litterfall (i.e., the inverse of nutrient concentration in plant litter) (Vitousek 1982). This ratio is highest in unproductive sites (Fig. 8.8), suggesting that plants are more efficient in producing biomass per unit of nutrient acquired and lost when nutrients are in short supply. Several factors contribute to this pattern (Chapin 1980). First, tissue nutrient concentration tends to decline as soil fertility declines, as described earlier. Individual plants that are nutrient limited also produce tissues more slowly and retain these tissues for a longer period of time, resulting in an increase in average tissue age. Older tissues have low nutrient concentrations, causing a further decline in concentration (i.e., increased NUE). Finally, the dominance of infertile soils by species with long-lived leaves that have low nutrient concentrations further contributes to the high NUE of ecosystems on infertile soils.

Plants maximize NUE in infertile soils by reducing nutrient loss more than by increasing nutrient productivity. There are at least two ways in which a plant might maximize biomass gained per unit of nutrient (Berendse and Aerts 1987): through a high **nutrient productivity** (a_n)—that is, a high instantaneous rate of carbon uptake per unit nutrient, and through a long **residence time** (t_r)—that is, the average time that the nutrient remains in the plant.

$$\mathrm{NUE} = a_n \times t_r \qquad \left[\mathrm{g\ biomass\ (g\,N)^{-1}} = \mathrm{g\ biomass\ (g\,N)^{-1}\ yr^{-1} * yr}\right] \qquad (8.1)$$

Species characteristic of infertile soils have a long residence time of nutrients but a low nutrient productivity (Chapin 1980, Lambers and Poorter 1992), suggesting that the high NUE in unproductive sites results primarily

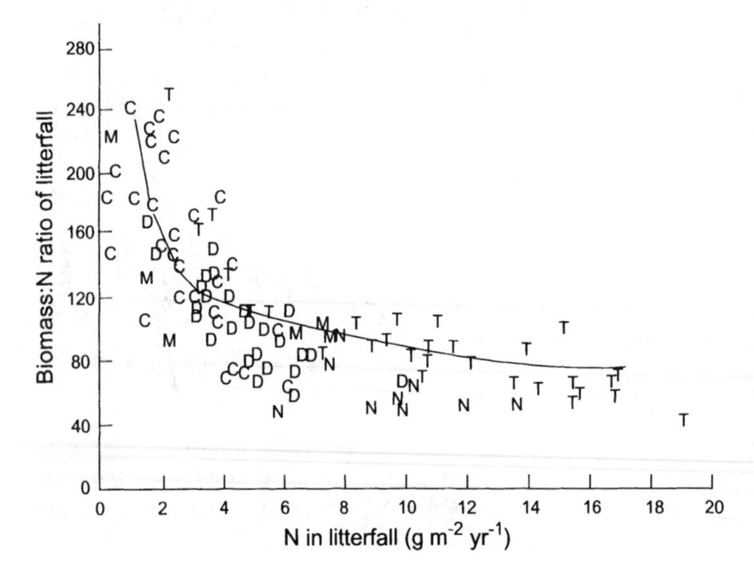

FIGURE 8.8. Relationship between the amount of nitrogen in litterfall and nitrogen use efficiency (ratio of dry mass to nitrogen in that litterfall). C, conifer forests; D, temperate deciduous forests; M, mediterranean-type ecosystems; N, temperate sites dominated by symbiotic, nitrogen fixers; T, evergreen tropical forests. (Redrawn with permission from *American Naturalist*; Vol. 119 © 1982 University of Chicago Press; Vitousek 1982.)

from traits that reduce nutrient loss rather than traits that promote a high instantaneous rate of biomass gain per unit of nutrient (Table 8.5). Shading also reduces tissue loss more strongly than it reduces the rate of carbon gain (Walters and Reich 1999).

There is an innate physiological trade-off between nutrient residence time and nutrient productivity. This occurs because the traits that allow plants to retain nutrients reduce their capacity to grow rapidly (Chapin 1980, Lambers and Poorter 1992). Plants with a high nutrient productivity grow rapidly and have high photosynthetic rates, which is associated with thin leaves, a high specific leaf area, and a high tissue nitrogen concentration (see Chapter 5). Conversely, a long nutrient residence time is achieved primarily through slow rates of replacement of leaves and roots. Leaves that survive a long time have more structural cells to withstand unfavorable conditions and higher concentrations of lignin and other secondary metabolites that deter pathogens and herbivores. Together these traits result in dense leaves with low tissue nutrient concentrations and therefore low photosynthetic rates per gram of biomass (see Chapter 5). The high NUE of plants on infertile soils therefore reflects their capacity to retain tissues for a long time rather than a capacity to use nutrients more effectively in photosynthesis.

Little is known about the trade-offs between root longevity and nutrient uptake rate. Nutrient uptake declines as roots age, lose root hairs, and become suberized; so trade-offs between physiological activity and longevity that have been well documented for leaves probably also exist for roots.

The trade-off between NUE and rate of resource capture explains the diversity of plant types along resource gradients. Low-resource environments are dominated by species that conserve nutrients through low rates of tissue turnover, high NUE, and the physical and chemical properties necessary for tissues to persist for a long time. These stress-tolerant plants outcompete plants that are less effective at nutrient retention in environments that are dry, infertile, or shaded (Chapin 1980, Walters and Field 1987). A high NUE and associated traits constrain the capacity of plants to capture carbon and nutrients. In high-resource environments, therefore, species with high rates of resource capture, rapid growth rates, rapid tissue turnover, and consequently low NUE, outcompete plants with high NUE. In other words, neither a rapid growth rate nor a high NUE is universally advantageous, because there are inherent physiological trade-offs among these traits. The relative benefits to the plant of efficiency vs. rapid growth depends on environment.

TABLE 8.5. Nitrogen use efficiency and its physiological components in a heathland evergreen shrub and a grass[a].

Process	Evergreen shrub	Grass
Nitrogen productivity ($g\,biomass\,g^{-1}\,N\,yr^{-1}$)	77	110
Mean residence time (yr)	1.2	0.8
NUE ($g\,biomass\,g^{-1}\,N$)	90	89

[a] Species are an evergreen shrub (*Erica tetralix*) and a co-occurring deciduous grass (*Molinia caerulea*) that is adapted to higher soil fertility. These two species have similar NUE, which is achieved by a high nitrogen productivity in the high-nutrient-adapted species and by a high mean residence time in the low-nutrient-adapted species. Data from Berendse and Aerts (1987).

Nutrient Loss from Plants

The nutrient budget of plants, particularly long-lived plants, is determined just as much by nutrient loss as by nutrient uptake. The potential avenues of nutrient loss from plants include tissue senescence and death, leaching of dissolved nutrients from plants, consumption of tissues by herbivores, loss of nutrients to parasites, exudation of nutrients into soils, and catastrophic loss of nutrients from vegetation by fire, wind-throw, and other disturbances. Nutrient loss from plants is an internal transfer within ecosystems (the transfer from plants to soil). After this transfer to soil, nutrients are potentially available for uptake by microbes or plants or may be lost from the ecosystem. Nutrient loss from *plants* to soil therefore has very different consequences from nutrient loss from the *ecosystem* to the atmosphere or groundwater.

Senescence

Tissue senescence is the major avenue of nutrient loss from plants. Plants reduce senescence loss of nutrients primarily by reducing tissue turnover, particularly in low-resource environments. The leaves of grasses and evergreen woody plants, for example, show greater leaf longevity in low-nutrient or low-water environments than in high-resource environments (Chapin 1980). Species differences in tissue turnover strengthen this pattern of high tissue longevity in low-resource environments. The proportion of evergreen woody species increases with decreasing soil fertility, reducing the rate of leaf turnover at the ecosystem level. All else being equal, a reduction in tissue turnover causes a corresponding reduction in the turnover of the associated tissue nutrients. This reduction in tissue turnover is probably the single most important adaptation for nutrient retention in low-nutrient habitats (Chapin 1980, Lambers and Poorter 1992).

Nutrient **resorption** is the transfer of soluble nutrients out of a senescing tissue through the phloem. It plays a crucial, but poorly understood, role in nutrient retention by plants. Plants resorb, on average, about half of their nitrogen, phosphorus, and potassium from leaves before leaves are shed at senescence (Table 8.6), so nutrient resorption is quantitatively important to plant nutrient budgets. Although there is a large variation (0 to 90%) in the proportion of leaf nutrients resorbed both among species and across sites, there is no consistent relationship between the proportion of nutrients resorbed and plant nutrient status (Chapin and Kedrowski 1983, Aerts 1995). Efficient nutrient resorption is promoted by presence of an active sink, for example when new leaf production coincides with leaf senescence in evergreens. The relatively high resorption efficiency of graminoids may also reflect their pattern of growth, in which production of new leaves is linked to senescence of older leaves (Table 8.6). Drought reduces the efficiency of nutrient resorption (Pugnaire and Chapin 1992, Aerts and Chapin 2000). One possible explanation for the lack of clear environmental controls over nutrient resorption is that nutrient resorption may be influenced by so many factors that no single environmental control is readily identified. Nutrient resorption may also be such an important trait that it is equally expressed in all plants (i.e., there may be no strong trade-offs that make effective resorption disadvantageous in high-resource sites). There have been too few studies of nutrient resorption from roots or wood to draw firm conclusions about factors controlling nutrient resorption from these tissues, although resorption from roots and wood appears to be much less than from leaves and may not occur at all. Resorbed nutrients are transferred to other plant parts (e.g., seeds, storage organs, or leaves at the top of the canopy) to support growth at other times or parts of the plant. Some nutrients, such as

TABLE 8.6. Nitrogen and phosphorus resorption efficiency of different growth forms.

Growth form	Resorption efficiency (% of maximum pool)[a]	
	Nitrogen	Phosphorus
All data	50.3 ± 1.0 (287)	52.2 ± 1.5 (226)
Evergreen trees and shrubs	46.7 ± 1.6 (108) (a)	51.4 ± 2.3 (88) (a)
Deciduous trees and shrubs	54.0 ± 1.5 (115) (b)	50.4 ± 2.0 (98) (a)
Forbs	41.4 ± 3.7 (33) (a)	42.4 ± 7.1 (18) (a)
Graminoids	58.5 ± 2.6 (31) (b)	71.5 ± 3.4 (22) (b)

[a] Means ± SE. Number of species in parentheses. Letters indicate statistical difference between growth forms ($P < .05$).
Data from Aerts (1995).

calcium and iron are immobile in the phloem so plants cannot resorb these nutrients from senescing tissues. Because these nutrients seldom limit plant growth, their lack of resorption has little direct nutritional impact on plants, except where acid rain greatly reduces their availability (Aber et al. 1998, Driscoll et al. 2001).

Leaching Loss from Plants

Leaching of nutrients from leaves is an important secondary avenue of nutrient loss from plants. Leachates account for about 15% of the annual nutrient return from aboveground plant parts to the soil. Rain dissolves nutrients on leaf and stem surfaces and carries these to the soil as **throughfall** (water that drops from the canopy) or **stem flow** (water that flows down stems). Stem flow typically has high concentrations of nutrients due to leaching of the stem surface; however, only a small amount of water moves by this pathway. Throughfall typically accounts for 90% of the nutrients leached from plants. Although plants with high nutrient status lose more nutrients per leaf, the *proportion* of nutrients recycled by leaching is surprisingly similar across a wide range of ecosystems (Table 8.7). Leaching loss is most pronounced for those nutrients that are highly soluble or are not resorbed. As much as 50% of the calcium and 80% of the potassium in an apple leaf, for example, can be leached within 24 h. Leaching rate is highest when rain first contacts a leaf and then declines exponentially with time. Ecosystems with very different rainfall regimes may therefore return similar proportions of nutrients to the soil through leaching vs. senescence. Although leaching loss is quantitatively important to plant nutrient budgets, there are no clear adaptations to minimize leaching loss. The thick cuticle of evergreen leaves was once thought to reduce leaching loss and explain the presence of evergreen leaves in wet, nutrient-poor forests. There is no evidence, however, that leaching loss is related to cuticle thickness. Like nutrient resorption, leaching loss from plants is a quantitatively important term in plant nutrient budgets that is not well understood. Biologists understand the acquisition of carbon and nutrients by plants much better than the loss of these resources.

Plant canopies can also absorb soluble nutrients from precipitation. Canopy uptake from precipitation is greatest in ecosystems where there is strong growth limitation by a given nutrient. In Germany, for example, nitrogen inputs in precipitation are so high that forest growth has switched from nitrogen to phosphorus limitation. These forest canopies absorb phosphorus directly from precipitation.

TABLE 8.7. Nutrients leached from the canopy (throughfall) as a percentage of the total aboveground nutrient return from plants to the soil.

	Throughfall (% of annual return)[a]	
Nutrient	Evergreen forests	Deciduous forests
Nitrogen	14 ± 3	15 ± 3
Phosphorus	15 ± 3	15 ± 3
Potassium	59 ± 6	48 ± 4
Calcium	27 ± 6	24 ± 5
Magnesium	33 ± 6	38 ± 5

[a] Means ± SE, for 12 deciduous and 12 evergreen forests. Data from Chapin (1991b).

Herbivory

Herbivores are sometimes a major avenue of nutrient loss from plants. Herbivores consume a relatively small proportion (1 to 10%) of plant production in many ecosystems. In ecosystems such as productive grasslands, however, herbivores regularly eat a large proportion of plant production; and, during outbreaks of herbivore population, herbivores may consume most aboveground production (see Chapter 11). Herbivory has a much larger impact on plant nutrient budgets than the biomass losses suggest, because herbivory precedes resorption, so vegetation loses approximately twice as much nitrogen and phosphorus per unit biomass to herbivores than it does through senescence. Animals also generally feed on tissues that are rich in nitrogen and phosphorus, thus maximizing the nutritional impact of herbivory on plants. There has therefore been strong selection for effective chemical and morphological defenses that deter herbivores and pathogens. These defenses are best developed in tissues that are long lived and in environments where there is an inadequate supply of nutrients to readily replace nutrients lost to herbivores (Coley et al. 1985, Gulmon and Mooney 1986, Herms and Mattson 1992). Most nutrients transferred from plants to herbivores are rapidly returned to the soil in feces and urine, where they quickly become available to plants. In this way herbivory speeds up nutrient cycling (see Chapter 11), particularly in ecosystems that are managed for grazing. Nutrients are susceptible to loss from the ecosystem in situations in which overgrazing reduces plant biomass to the point that plants cannot absorb the nutrients returned to the soil by herbivores.

Other Avenues of Nutrient Loss from Plants

Other avenues of nutrient loss are poorly known. Although laboratory studies suggest that root exudates containing amino acids may be a significant component of the plant carbon budget (Rovira 1969), the magnitude of nitrogen loss from plants by this avenue is unknown. Other avenues of nutrient loss from plants include plant parasites such as mistletoe and nutrient transfers by mycorrhizae from one plant to another. Although these nutrient transfers may be critical to the nutrient distribution among species in the community, they do not greatly alter nutrient retention or loss by vegetation as a whole.

Disturbances cause occasional large pulses of nutrient release. Fire, wind, disease epidemics, and other catastrophic disturbances cause massive nutrient losses from vegetation when they occur (see Chapter 13), but these losses are small (less than 1% of nutrients cycled through vegetation) when averaged over the entire disturbance cycle. Even in fire-prone savannas and grasslands, fires generally burn during the dry season after senescence has occurred and burn more litter than live plant biomass. Most plant nutrients in these ecosystems are stored belowground during times when fires are likely to occur.

Summary

Nutrient availability is a major constraint on the productivity of the terrestrial biosphere. Whereas carbon acquisition by plants is determined primarily by plant traits (leaf area and photosynthetic capacity), nutrient uptake is usually governed more strongly by environment (the rate of supply by the soil) rather than by plant traits. In early succession, however, plant traits can have a significant impact on nutrient uptake by vegetation at the ecosystem level. Diffusion is the major process that delivers nutrients from the bulk soil to the root surface. Mass flow of nutrients in flowing soil water augments this nutrient supply for abundant nutrients or nutrients that are required in small amounts by plants. Root biomass differs less among ecosystems than does aboveground biomass because those ecosystems that are highly productive and produce a large aboveground biomass have a relatively low allocation to roots.

Plants adjust their capacity to acquire nutrients in several ways. Preferential allocation to roots under conditions of nutrient limitation maximizes the root length available to absorb nutrients. Root growth is concentrated in hot spots of relatively high nutrient availability, maximizing the nutrient return for roots that are produced. Plants further increase their capacity to acquire nutrients through symbiotic associations with mycorrhizal fungi. Plants alter the kinetics of nutrient uptake in response to their demand for nutrients. Plants that grow rapidly, due either to a favorable environment or a high relative growth rate, have a high capacity to absorb nutrients. Plants adjust the absorption of specific nutrients by maximizing the capacity to absorb those elements that most strongly limit growth. In the case of nitrogen, which is frequently the most strongly limiting nutrient, plants typically absorb whatever forms are available in the soil. When all forms are equally available, most plants preferentially absorb ammonium or amino acids rather than nitrate. Nitrate absorption is often important, however, because of its high mobility.

There is an inevitable trade-off between the maximum rate of nutrient investment in new growth and the efficiency with which nutrients are used to produce biomass. Plants produce biomass most efficiently per unit of nutrient under nutrient-limiting conditions. Nutrient use efficiency is maximized by prolonging tissue longevity—that is, by reducing the rate at which nutrients are lost. Senescence is the major avenue by which nutrients are lost from plants. Plants minimize loss of growth-limiting nutrients by resorbing about half of the nitrogen, phosphorus, and potassium from a leaf before it is shed. About 15% of the annual nutrient return from aboveground plant parts to the soil comes as leachates, primarily as through-fall that drips from the canopy. Herbivores can also be important avenues of nutrient loss because they preferentially feed on nutrient-rich tissues and consume these tissues before resorption can occur. For these reasons plants lose more than twice as much nutrients per unit of biomass to herbivores than through senescence. Other factors that cause occasional large nutrient losses from vegetation include disturbances (e.g., fire and wind) and diseases that kill plants.

Review Questions

1. Mass flow, diffusion, and root interception are three processes that deliver nutrients to the root surface. Describe the mechanism by which each process works. What is the relative importance of these three processes in providing nutrients to plants? How does soil fertility influence the relative importance of these processes?
2. How do soil and plant properties influence rates of diffusion and mass flow? How can the plant maximize these transport processes?
3. What is the major mechanism by which plants get nutrients into the plant, once they have arrived at the root surface?
4. How do plants compensate for (a) low availability of all nutrients, (b) spatial variability in nutrients within the soil (localized hot spots), and (c) imbalance between nutrients required by plants (e.g., nitrogen vs. phosphorus availability)?
5. What is the rhizosphere? How do plants influence the rhizosphere?
6. How does plant growth rate affect nutrient uptake?
7. What are the major mechanisms by which mycorrhizae increase nutrient uptake by plants? Under what circumstances are mycorrhizae most strongly developed?
8. What are the major processes involved in converting nitrogen from nitrate to a form that is biochemically useful to the plant? Why is nitrogen acquisition by plants so energetically expensive?
9. Why are nitrogen and carbon flows in plants so tightly linked? What happens to nutrient uptake when carbon gain is restricted? What happens to carbon gain when nutrient uptake is restricted? What are the mechanisms by which these adjustments occur?
10. What is nitrogen use efficiency? What are the physiological causes of differences in NUE, and what are the ecosystem consequences?
11. Why do all aquatic and terrestrial plants tend to show a similar balance of nutrients (the Redfield ratio)? How can you use this information to estimate which nutrient is most strongly limiting to plants in a particular site?
12. What are the major differences in types of species that occur in fertile vs. infertile soils? What are the advantages and disadvantages of each plant strategy in each soil type?
13. What are the major avenues of nutrient loss from plants? How do all plants minimize this nutrient loss? What are the major adaptations that minimize nutrient loss from plants that are adapted to infertile soils?

Additional Reading

Aerts, R. 1995. Nutrient resorption from senescing leaves of perennials: Are there general patterns? *Journal of Ecology* 84:597–608.

Chapin, F.S. III. 1980. The mineral nutrition of wild plants. *Annual Review of Ecology and Systematics* 11:233–260.

Hobbie, S.E. 1992. Effects of plant species on nutrient cycling. *Trends in Ecology and Evolution* 7:336–339.

Ingestad, T., and G.I. Ågren. 1988. Nutrient uptake and allocation at steady-state nutrition. *Physiologia Plantarum* 72:450–459.

Nye, P.H., and P.B. Tinker. 1977. *Solute Movement in the Soil-Root System*. University of California Press, Berkeley.

Read, D.J. 1991. Mycorrhizas in ecosystems. *Experientia* 47:376–391.

Vitousek, P.M. 1982. Nutrient cycling and nutrient use efficiency. *American Naturalist* 119:553–572.

9
Terrestrial Nutrient Cycling

Nutrient cycling involves nutrient inputs to, and outputs from, ecosystems and the internal transfers of nutrients within ecosystems. This chapter describes these nutrient dynamics.

Introduction

Human impact on nutrient cycles has fundamentally changed the regulation of ecosystem processes. Rates of cycling of carbon (see Chapters 5 to 7) and water (see Chapter 4) are ultimately regulated by the availability of belowground resources, so changes in the availability of these resources fundamentally alter all ecosystem processes. The combustion of fossil fuels has released large quantities of nitrogen and sulfur oxides to the atmosphere and increased their inputs to ecosystems (see Chapter 15). Fertilizer use and the cultivation of nitrogen-fixing crops have further increased the fluxes of nitrogen in agricultural ecosystems (Galloway et al. 1995, Vitousek et al. 1997a). Together these human impacts have doubled the natural background rate of nitrogen inputs to the terrestrial biosphere. The resulting increases in plant production may be large enough to affect the global carbon cycle. Anthropogenic disturbances such as forest conversion, harvest, and fire increase the proportion of the nutrient pool that is available and therefore vulnerable to loss from the ecosystem. Some of these losses occur by leaching of dissolved elements to groundwater, causing a depletion of soil cations, an increase in soil acidity, and increases in nutrient inputs to aquatic ecosystems. Gaseous losses of nitrogen influence the chemical and radiative properties of the atmosphere, causing air pollution and enhancing the greenhouse effect (see Chapter 2). Changes in the cycling of nutrients therefore dramatically affect the interactions among ecosystems as well as the carbon cycle and the climate of Earth.

Overview

Nutrient cycling involves the entry of nutrients to ecosystems, their internal transfers between plants and soils, and their loss from ecosystems. Nutrients enter ecosystems through the chemical weathering of rocks, the biological fixation of atmospheric nitrogen, and the deposition of nutrients from the atmosphere in rain, wind-blown particles, or gases. Fertilization is an additional nutrient input in managed ecosystems. Internal cycling processes include the conversion of nutrients from organic to inorganic forms, chemical reactions that change elements from one ionic form to another, biological uptake by plants and microorganisms, and exchange of nutrients on surfaces within the soil matrix. Nutrients are lost from ecosystems by leaching, trace-gas emission, wind and water erosion, fire, and the removal of materials in harvest.

Most of the nitrogen and phosphorus required for plant growth in unmanaged ecosystems is supplied by the decomposition of plant litter and soil organic matter. Inputs and outputs in these ecosystems are a small fraction of the quantity of nutrients that cycle internally, producing relatively **closed systems** with conservative nutrient cycles. Human activities tend to increase inputs and outputs relative to the internal transfers and make the element cycles more open.

There are important differences among elements in their patterns of biogeochemical cycling. We approach this by first describing the cycling of nitrogen, the element that most frequently limits plant production; we then compare its cycling to that of other elements.

Nitrogen Inputs to Ecosystems

Under natural conditions, nitrogen fixation is the main pathway by which new nitrogen enters terrestrial ecosystems. Earth's atmosphere contains an abundant well-distributed pool of nitrogen; 78% of the atmosphere's volume is dinitrogen (N_2). Thus, although all organisms are literally bathed in nitrogen, it is the element that most frequently limits the growth of plants and animals. This paradox occurs because N_2 is unavailable to most organisms. Only certain types of bacteria, known as **nitrogen fixers**, have the capacity to break the triple bonds of N_2 and **fix** it into ammonium (NH_4^+), which they use for their own growth. These bacteria occur either as free-living forms in soils, sediments, and waters; or within nodules formed on the roots of certain vascular plants; or in lichens as symbiotic associations with fungi. Nitrogen fixed by nitrogen-fixing plants becomes available to other plants in the community primarily through the production and decomposition of litter.

Biological Nitrogen Fixation

The characteristics of nitrogenase, the enzyme that catalyzes the reduction of N_2 to NH_4^+, dictate much of the biology of nitrogen fixation. The reduction of N_2 catalyzed by nitrogenase has a high-energy requirement and therefore occurs only where the bacterium has an abundant carbohydrate supply. The enzyme is denatured in the presence of oxygen, so organisms must protect the enzyme from contact with oxygen.

Groups of Nitrogen Fixers

Nitrogen-fixing bacteria in symbiotic association with plants have the highest rates of nitrogen fixation. This occurs because plants can provide the abundant carbohydrates needed to meet the high energy demand of nitrogen fixation. The most common symbiotic nitrogen fixers are *Rhizobium* species associated with legumes (soybeans, peas, etc.) and *Frankia* species (actinomycete bacteria) associated with alder, *Ceanothus*, and other nonlegume woody species (Table 9.1). These plant-associated symbiotic nitrogen-fixing bacteria usually reside in root nodules, where the nitrogenase enzyme is protected from oxygen. In legumes, this is done by leghemoglobin. This oxygen-binding pigment is similar to hemoglobin, which transports oxygen in the bloodstream of vertebrate animals. Nitrogen-fixing bacteria in nodules are heterotrophic and depend on carbohydrates from plants to meet the energy requirements of nitrogen fixation. The energetic requirement for nitrogen fixation can be about 25% of the gross primary production (GPP) under laboratory conditions, two to four times higher than the cost of acquiring nitrogen from soils (Lambers et al. 1998). The relative costs of nitrogen fixation and nitrogen uptake under field conditions are more difficult to estimate because of the uncertain costs of mycorrhizal association and root exudation. When inorganic nitrogen is naturally high or is added to soils, nitrogen-fixing plants generally reduce their capacity for nitrogen fixation and acquire their nitrogen from available forms in the soil solution.

Some heterotrophic nitrogen-fixing bacteria are free living and get their organic carbon from the environment (Table 9.1). These bacteria have highest nitrogen-fixation rates in soils

TABLE 9.1. Organisms and associations involved in dinitrogen fixation.

Type of association[a]	Key characteristics	Representative genera
Heterotrophic nitrogen fixers		bacteria
Associative		
Nodulated (symbiotic)	legume	*Rhizobium*
	nonlegume woody plants	*Frankia*
Non-nodulated	rhizosphere	*Azotobacter, Bacillus*
	phyllosphere	*Klebsiella*
Free living	aerobic	*Azotobacter, Rhizobium*
	facultative aerobic	*Bacillus*
	anaerobic	*Clostridium*
Phototrophic nitrogen fixers		cyanobacteria
Associative	lichens	*Nostoc, Calothrix*
	liverworts (*Marchantia*)	*Nostoc*
	mosses	*Holosiphon*
	gymnosperms (*Cycas*)	*Nostoc*
	water fern (*Azolla*)	*Nostoc*
Free living	cyanobacteria	*Nostoc, Anabaena*
	purple nonsulfur bacteria	*Rhodospirillium*
	sulfur bacteria	*Chromarium*

[a] Nitrogen-fixing microbes are heterotrophic bacteria if they get their organic carbon from the environment. They are phototrophic bluegreen algae if they produce it themselves through photosynthesis. Among both groups of microbes, some forms are typically associated with plants and others are free living. Note that the same microbial genus can have both associative and free-living forms.
Data from Paul and Clark (1996).

or sediments with high concentrations of organic matter, which provide the carbon substrate that fuels nitrogen reduction. Other heterotrophic nitrogen fixers occur in the rhizosphere and depend on root exudation and root turnover for their carbon supply. Heterotrophic nitrogen fixers typically have highest fixation rates in aerobic environments, because aerobic respiration yields much more energy per gram of substrate than does anaerobic respiration. Aerobic heterotrophs have various mechanisms that reduce oxygen concentration in the vicinity of nitrogenase. Some have high respiration rates that scavenge oxygen around the bacterial cells. Others clump together or produce slime to reduce oxygen diffusion to the enzyme. Some heterotrophs fix nitrogen in low-oxygen situations and therefore require no specialized adaptations to prevent denaturation of nitrogenase by oxygen. These organisms, however, have low rates of nitrogen fixation due to low energy availability.

There are many free-living nitrogen-fixing **phototrophs** that produce their own organic carbon by photosynthesis. These include cyanobacteria (blue-green algae) that occur in aquatic systems and on the surface of many soils. Many phototrophs have specialized non-photosynthetic cells, called **heterocysts**, that protect nitrogenase from denaturation by the oxygen produced during photosynthesis in adjacent photosynthetic cells.

There are also associative (symbiotic) nitrogen-fixing phototrophs. For example, nitrogen-fixing lichens are composed of green algae or cyanobacteria as the photosynthetic symbiont, cyanobacteria that fix nitrogen, and fungi that provide physical protection. These lichens provide an important nitrogen input in many early successional ecosystems. The small freshwater fern *Azolla* and cyanobacteria such as *Nostoc* form a phototrophic association that is common in rice paddies and tropical aquatic systems.

Legumes and other symbiotic nitrogen fixers have the highest rates of nitrogen fixation, typically 5 to 20 $g\,m^{-2}\,yr^{-1}$. Phototrophic symbionts such as *Nostoc* in association with *Azolla* in rice paddies often fix $10 g m^{-2} yr^{-1}$. When *Nostoc* is a free-living phototroph, it fixes about $2.5 g m^{-2} yr^{-1}$. In contrast, free-living heterotrophs fix only 0.1 to $0.5 g m^{-2} yr^{-1}$, a quantity similar to the input from nitrogen deposition in unpolluted environments.

Causes of Variation in Nitrogen Fixation

Biotic and abiotic constraints on nitrogen fixation lead to nitrogen limitation in most temperate terrestrial and many marine ecosystems. The rate of nitrogen fixation varies widely among ecosystems, in part reflecting the types of nitrogen fixers that are present. Even within a single type of nitrogen-fixing system, however, there are large ranges in fixation rates. What causes this variation? If nitrogen limits growth in many ecosystems, why does nitrogen fixation not occur almost everywhere? One would expect nitrogen fixers to have a competitive advantage over other plants and microbes that cannot fix their own nitrogen. Why don't nitrogen fixers respond to nitrogen limitation by fixing nitrogen until nitrogen is no longer limiting in the ecosystem? Several factors constrain nitrogen fixation, thereby maintaining nitrogen limitation in many ecosystems (Vitousek and Howarth 1991).

Energy availability constrains nitrogen fixation rates in closed-canopy ecosystems. The cost of nitrogen fixation (3 to 6 g carbon g^{-1} N, not including the cost of nodule production) by symbiotic and autotrophic nitrogen fixers is high relative to that of absorbing ammonium or nitrate. Nitrogen fixation is therefore largely restricted to the high-light environments that occur in early succession or where nutrients or drought limit canopy development. As canopies close during succession, energy becomes limiting to the establishment of nitrogen-fixing plants. These plants could fix nitrogen if they were in the canopy, but the cost of nitrogen fixation makes it difficult for them to grow through shade to the canopy. Leguminous trees are common in tropical forests and savannas. In savannas, where nitrogen cycles are relatively open due to large nitrogen losses in fires, leguminous trees are heavily nodulated and fix substantial quantities of nitrogen (Högberg and Alexander 1995). Leguminous trees in tropical forests, however, are seldom nodulated and may contribute little to the nitrogen inputs to tropical forests. Nitrogen fixation in aquatic systems is most common in shallow waters or waters with low turbidity where light reaches benthic cyanobacterial mats. When phosphorus availability is adequate, these mats have high fixation rates.

Nonsymbiotic heterotrophic nitrogen-fixing bacteria are also limited by the availability of labile organic carbon. When available carbon is scarce, there is no benefit to heterotrophic nitrogen fixation. Decaying wood, which has low nitrogen and high levels of organic carbon, often has substantial rates of heterotrophic nitrogen fixation. Heterotrophic nitrogen fixation also occurs in anaerobic sediments, but the gaseous loss of nitrogen by **denitrification**—that is the conversion of nitrate to gaseous forms, usually exceeds the gains from nitrogen fixation.

Nitrogen fixation in many ecosystems, especially in the tropics, is limited by the availability of some other nutrient, such as phosphorus. Nitrogen fixers have a high requirement for ATP and other phosphorus-containing compounds to support the energy transformations associated with nitrogen fixation. The growth of nitrogen fixers therefore often becomes phosphorus limited before that of other plants. Other elements that can limit nitrogen fixation include molybdenum, iron, and sulfur, which are essential cofactors of nitrogenase. Pasture legumes are limited by molybdenum in parts of Australia, for example, due to low molybdenum availability in the highly weathered soils, particularly at low pH. Nitrogen fixers may be limited by iron in marine ecosystems, at times when other phytoplankton are nitrogen limited (see Chapter 14). Phosphorus, iron, sulfur, or molybdenum may, in these cases, be the ultimate "master element" that limits production, even though nitrogen is the factor to which primary production responds most strongly in short-term experiments.

Grazing of nitrogen-fixing organisms often constrains their capacity to support continuously high nitrogen fixation rates. The high protein content typical of nitrogen fixers enhances their palatability to many herbivores, although nitrogen-based defenses such as alkaloids, which occur in many nitrogen-fixing plants, deter other herbivores (see Chapter 11). The resulting intense grazing on many nitrogen-fixing plants reduces their capacity to compete with other plants, causing their abundance and nitrogen inputs to the ecosystem to decline. Areas from which grazers are excluded frequently have more nitrogen-fixing plants and greater nitrogen inputs to the ecosystem and ultimately more productivity and biomass (Ritchie et al. 1998).

Nitrogen Deposition

Nitrogen is deposited in ecosystems in particulate, dissolved, and gaseous forms. All ecosystems receive nitrogen inputs from atmospheric deposition. These inputs are smallest (often 0.2 to $0.5\,g\,m^{-2}yr^{-1}$) in ecosystems downwind from pollution-free open oceans (Hedin et al. 1995). Nitrogen inputs to these coastal ecosystems derive primarily from organic particulates and nitrate (NO_3^-) in sea-spray evaporates and from ammonia (NH_3) volatilized from seawater. In inland areas, nitrogen derives from the volatilization of NH_3 from soils and vegetation and from dust produced by wind erosion of deserts, unplanted agricultural fields, and other sparsely vegetated ecosystems. Lightning also produces nitrate, contributing to atmospheric deposition.

Human activities are now the major source of nitrogen in deposition in many areas of the world. The application of urea or ammonia fertilizer leads to volatilization of NH_3, which is then converted to ammonium (NH_4^+) in the atmosphere and deposited in rainfall. Domestic animal husbandry has also substantially increased emissions of NH_3 to the atmosphere. The emission of nitric oxides (NO and NO_2, together known as NO_x) from fossil fuel combustion, biomass burning, and volatilization from fertilized agricultural systems have dwarfed natural sources at the global scale: 80% of all NO_x flux is anthropogenic (Delmas et al. 1997). Nitrogen derived from these sources can be transported long distances downwind from industrial or agricultural areas before being deposited. "Arctic haze" over the Arctic Ocean and Canadian High Arctic islands, for example, derives primarily from pollutants produced in eastern Europe. Inputs of anthropogenic sources of nitrogen to ecosystems can be quite large, for example, 1 to $2\,g\,m^{-2}yr^{-1}$ in the northeastern United States and 5 to $10\,g\,m^{-2}yr^{-1}$ in central Europe—10- to 100-fold greater than background levels of nitrogen deposition. The highest rates are similar to the amounts annually absorbed by vegetation and cycled through litterfall (see Chapter 8). Most ecosystems have a substantial capacity to store added nitrogen in soils and vegetation. Once these reservoirs become saturated, however, nitrogen losses to the atmosphere and groundwater can be substantial. The nitrogen cycle in some polluted ecosystems has changed from being more than 90% closed (see Table 8.1) to being just as open as the carbon cycle, in which the amount of nitrogen or carbon annually cycled by vegetation is similar to the amount that is annually gained and lost from the ecosystem.

Climate and ecosystem structure determine the processes by which nitrogen is deposited in ecosystems. Deposition occurs by three processes. (1) Wet deposition delivers nutrients dissolved in precipitation. (2) Dry deposition delivers compounds as dust or aerosols by sedimentation (vertical deposition) or impaction (horizontal deposition or direct absorption of gases such as nitric acid, HNO_3, vapor). (3) Cloud-water deposition delivers nutrients in water droplets onto plant surfaces immersed in fog. Although data are most available for wet deposition because it is most easily measured, wet and dry deposition are often equally important sources of nitrogen inputs (Fig. 9.1). Wet deposition of nitrogen is typically greater in wet than in dry ecosystems. Dry deposition of nitrogen, however, shows no clear correlation with climate, although arid ecosystems receive a larger *proportion* of their nitrogen inputs by dry deposition. Cloud-water deposition is greatest on cloud-covered mountaintops or regions with coastal fog. The relative importance of wet, dry, and cloud-water deposition also depends on ecosystem structure. Conifer canopies, for example, tend to collect more dry

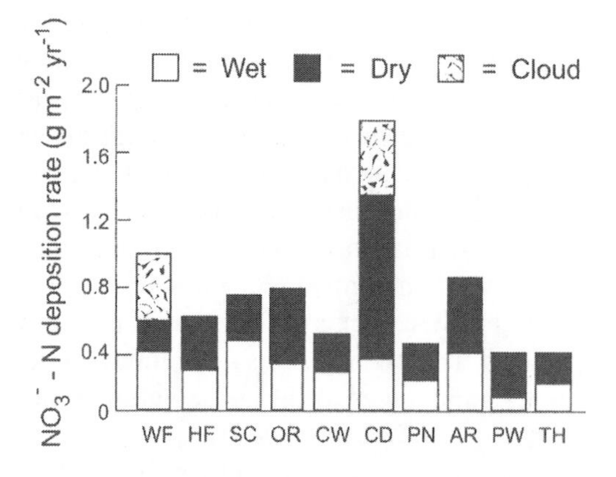

FIGURE 9.1. Wet, dry, and cloud-water deposition of nitrogen in a variety of ecosystems. WF, Whiteface Mountain, New York; HF, Huntington Forest, New York; SC, State College, Pennsylvania; OR, Oak Ridge, Tennessee; CW, Coweta, North Carolina; CD, Clingman's Dome, North Carolina; PN, Panola, Georgia; AR, Argonne, Illinois; PW, Pawnee, Colorado; and TH, Thompson, Washington. (Redrawn with permission from *Ecological Applications*; Lovett 1994).

deposition and cloud-water deposition than do deciduous canopies because of their greater leaf surface area. Their rough canopies also cause moisture-laden air to penetrate more deeply within the forest canopy and therefore to contact more leaf surfaces (see Chapter 4).

The form of nitrogen deposition determines its ecosystem consequences. NO_3^- and NH_4^+ are immediately available for biological uptake by plants and microbes, whereas some organic nitrogen must first be mineralized. Nitrate inputs such as nitric acid (and ammonium inputs, if followed by **nitrification**, the conversion of ammonium to nitrate) acidify the soil when nitrate accompanied by base cations leaches from the ecosystem. Sulfuric acid, which often accompanies fossil-fuel sources of NO_x, also acidifies soils. Organic nitrogen compounds make up about a third of the total nitrogen deposition, but their chemical nature varies among ecosystems (Neff et al. 2002). In coastal areas, for example, organic nitrogen is deposited primarily as marine-derived reduced compounds such as amines. In inland areas affected by air pollution, most organic nitrogen enters as oxidized organic nitrogen compounds that result from the reaction of organic compounds and NO_x in the atmosphere.

Weathering of sedimentary rocks may contribute to the nitrogen budgets of some ecosystems. Sedimentary rocks, which make up 75% of the rocks exposed on Earth's surface, sometimes contain substantial nitrogen. In one watershed underlain by high-nitrogen sedimentary rocks, for example, rock weathering contributed about 2 g N m^{-2} yr^{-1}(Holloway et al. 1998), similar to the quantities that entered from the atmosphere. In most ecosystems, however, rock weathering is thought to provide only a small nitrogen input to ecosystems.

Internal Cycling of Nitrogen

Overview of Mineralization

In natural ecosystems, most nitrogen absorbed by plants becomes available through the decomposition of organic matter. Most (more than 99%) soil nitrogen is contained in dead organic matter derived from plants, animals, and microbes. As microbes break down this dead organic matter during decomposition (see Chapter 7), the nitrogen is released as **dissolved organic nitrogen** (DON) through the action of exoenzymes (Fig. 9.2). Plants and mycorrhizal fungi can absorb some DON, using it to support plant growth. Decomposer microbes also absorb DON, using it to support their nitrogen and/or their carbon requirements for growth. When DON is insufficient to meet this nitrogen requirement, microbes absorb additional inorganic nitrogen (NH_4^+ or NO_3^-) from the soil solution. **Immobilization** is the removal of inorganic nitrogen from the available pool by microbial uptake and chemical fixation. Studies using ^{15}N-labeled NH_4^+ or NO_3^- indicate that microbes take up and immobilize both forms of nitrogen, although NH_4^+ uptake is typically greater (Vitousek and Matson 1988, Fenn et al. 1998). Microbial growth is often carbon limited. Under these circumstances, microbes break down the DON, use the carbon skeleton to support their energy requirements for growth and maintenance, and secrete NH_4^+ into the soil. This process is termed **nitrogen mineralization** or **ammonifi-**

cation because ammonium is the immediate product of this process. In some ecosystems, some or all NH_4^+ is converted to nitrite (NO_2^-) and then to nitrate (NO_3^-). The conversion from ammonium to nitrate is termed **nitrification**.

Production and Fate of Dissolved Organic Nitrogen

The conversion from insoluble organic nitrogen to dissolved organic nitrogen is the initial and typically the rate-limiting step in nitrogen mineralization (Fig. 9.2). The relatively large pool of nonsoluble organic nitrogen in soils suggests that this initial step in nitrogen mineralization is the rate-limiting step. All of the organic nitrogen that is eventually mineralized to NH_4^+ or NO_3^- must first become dissolved before it can be absorbed by microbes and mineralized. The flux through the DON pool is therefore usually large, relative to other nitrogen fluxes, even in ecosystems in which its concentration is low

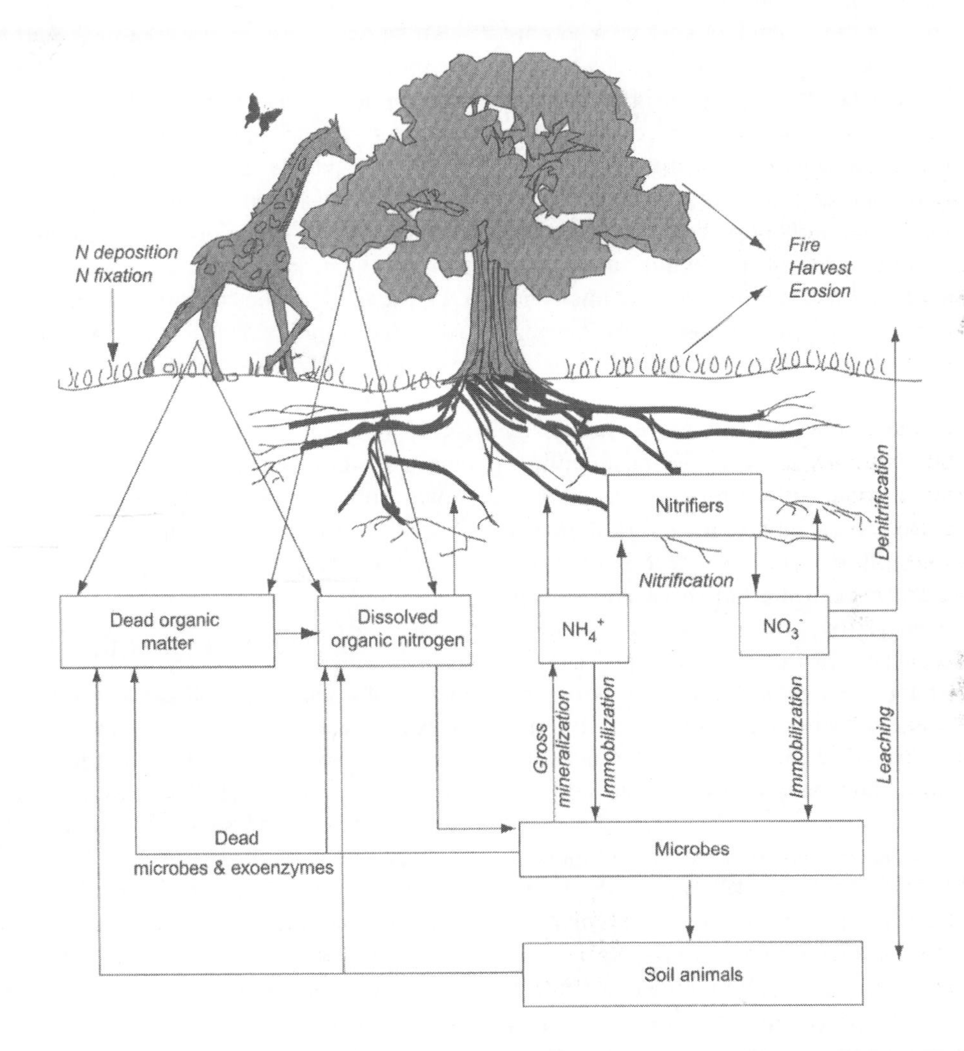

FIGURE 9.2. Simplified terrestrial nitrogen cycle. Both plants and microbes take up dissolved organic nitrogen, NH_4^+, and NO_3^- and release dead organic matter and DON. Microbes also release ammonium when they absorb more nitrogen than they require for growth. Nitrifiers are a specialized microbial group that either converts ammonium to nitrite or nitrite to nitrate. Nitrogen is consumed by animals when they eat plants or soil microbes and is returned to the soil as dead organic matter and DON. Nitrogen is lost from the ecosystem by denitrification, erosion, harvest, or fire. Nitrogen enters the ecosystem through nitrogen deposition or nitrogen fixation.

(Eviner and Chapin 1997). The breakdown of particulate organic nitrogen is carried out in parallel with the breakdown and use of particulate organic carbon and is therefore controlled by the same organisms and factors that control decomposition (see Fig. 7.14). These controls include the quantity and chemical nature of the substrate, the composition of the microbial community, and the environmental factors regulating the activity of soil microbes and animals (see Chapter 7).

Most nitrogen in dead organic matter is contained in complex polymers such as proteins, nucleic acids, and chitin (from fungal cell walls and insect exoskeletons) that are too large to pass through microbial membranes. Microbes must therefore secrete exoenzymes such as proteases, ribonucleases, and chitinases to break down the large polymers into small water-soluble subunits, such as amino acids and nucleotides that can be absorbed by microbial cells. Urease is another important exoenzyme that breaks down urea from animal urine or fertilizer into CO_2 and NH_3. The microbial enzymes are themselves subject to attack by microbial proteases, so microbes must continually invest nitrogen in exoenzymes to acquire nitrogen from their environment, a potentially costly trade-off. Exoenzymes often bind to soil minerals and organic matter. This can inactivate the enzyme, if the shape of the active site is altered, or can protect the enzyme against attack from other exoenzymes, lengthening the time that the enzyme remains active in the soil (see Chapter 7). Proteases are produced by mycorrhizal and saprophytic fungi and by bacteria.

Plants, mycorrhizal fungi, or decomposer microbes can absorb DON. When plants absorb DON directly or through their mycorrhizal fungi, no mineralization is required to provide this nitrogen to plants. Although we know that direct uptake of organic nitrogen by plants occurs in many ecosystems (Read 1991, Kielland 1994, Näsholm et al. 1998, Lipson et al. 1999, Raab et al. 1999), we know little about the relative importance of this pathway vs. the flux of nitrogen through mineralization. In many cases, microbes probably have a competitive advantage over plant roots because the microbes producing exoenzymes are closest to the site of enzymatic activity. In this case, we expect nitrogen mineralization to be the major fate of DON in soils. In organic soils, where DON concentrations are high, plants compete well for amino acids (Schimel and Chapin 1996) and meet a significant proportion (about 65% in one study) of their nitrogen requirement through direct uptake of DON (Lipson et al. 2001) (see Chapter 8). Even in agricultural ecosystems, 20% of plant nitrogen uptake can be met by DON.

DON is a chemically complex mixture of compounds, only a few percent of which consists of amino acids and other labile forms of nitrogen. The labile DON that is absorbed by microbial cells can be incorporated directly into microbial proteins and amino acids or transformed to other organic compounds that support microbial growth and respiration. DON can also be adsorbed onto the soil exchange complex or leached from the ecosystem in groundwater. Amino acids contain both positively and negatively charged groups (NH_2^+ and COO^-, respectively). Small neutrally charged amino acids, such as glycine, are most mobile in soils and are most readily absorbed by both plants and microbes (Kielland 1994).

Production and Fate of Ammonium

The net absorption or release of ammonium by microbes depends on their carbon status. When microbial growth is carbon limited, microbes use the carbon from DON to support growth and respiration and secrete NH_4^+ as a waste product into the soil solution. This process of ammonification is the mechanism by which nitrogen is mineralized. Other nitrogen-limited microbes may absorb, or **immobilize**, some of this ammonium and use it for growth. Mineralization and microbial uptake occur simultaneously in soils so a given unit of nitrogen can cycle between microbial release and uptake many times before it is used by plants or undergoes some other fate. **Gross mineralization** is the *total* amount of nitrogen released via mineralization (regardless of whether it is subsequently immobilized or not). **Net miner-**

alization is the *net* accumulation of inorganic nitrogen in the soil solution over a given time interval. Net mineralization occurs when microbial growth is limited more strongly by carbon than by nitrogen, whereas net immobilization occurs when microbial communities are nitrogen limited.

Net nitrogen mineralization is an excellent measure of the nitrogen supply to plants in ecosystems with rapid nitrogen turnover, where there is little competition for nitrogen between plants and microbes. The annual net mineralization in the deciduous forests of eastern North America, for example, approximately equals nitrogen uptake by vegetation (Nadelhoffer et al. 1992). In less fertile ecosystems, such as arctic tundra, net nitrogen mineralization rate substantially underestimates the amount of nitrogen that is annually acquired by plants (Nadelhoffer et al. 1992). There are at least two explanations for this apparent discrepancy. (1) Plant roots and their mycorrhizal fungi, which are excluded in net mineralization assays, may be good competitors with saprophytic microbes for mineralized nitrogen in the real world, so this assay may underestimate the quantity of nitrogen that would be mineralized in the presence of roots (Stark 2000). (2) Plants and their mycorrhizal fungi that absorb amino acids may also not depend heavily on nitrogen mineralization to meet their nitrogen requirements in infertile ecosystems due to absorption of DON, so the low rates of net nitrogen mineralization in these ecosystems may be an accurate reflection of small fluxes that naturally occur through this pathway.

Nitrogen mineralization rate is controlled by the availability of DON and inorganic nitrogen, the activity of soil microbes, and their relative demands for carbon and nitrogen. The quantity and quality of organic matter, including its relative amounts and forms of carbon and nitrogen, that enter the soil are the major determinants of the substrate available for decomposition (see Chapter 7) and therefore the substrate available for nitrogen mineralization (Fig. 9.3). Thus the state factors and interactive controls that promote productivity and

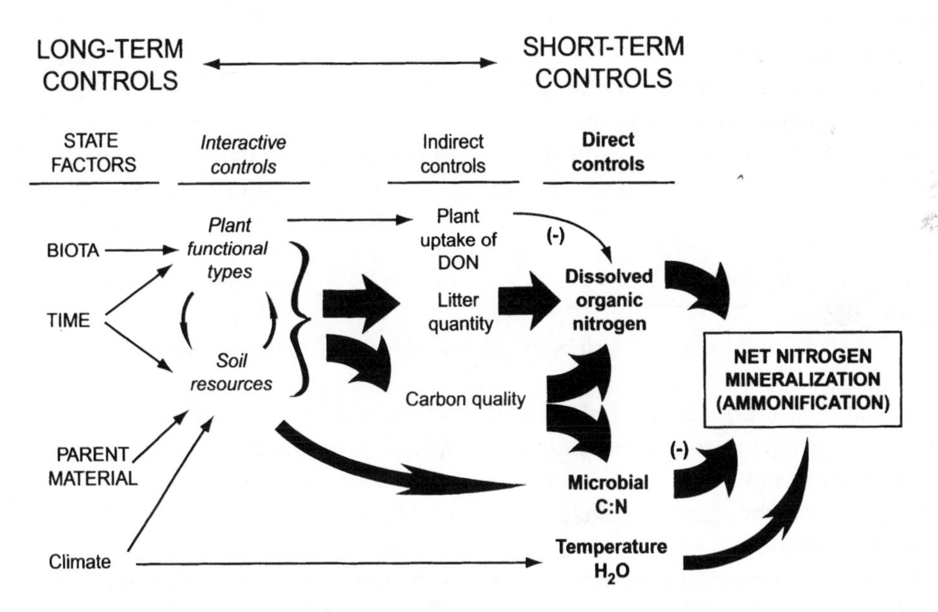

FIGURE 9.3. The major factors controlling ammonification (net nitrogen mineralization) in soils. These controls range from the proximate control over nitrogen mineralization (the concentration of DON, physical environment, and microbial carbon to nitrogen ratio) to the state factors and interactive controls that ultimately determine the differences among ecosystems in mineralization rates. Thickness of the arrows indicates the strength of effects. The influence of one factor on another is positive unless otherwise indicated (–).

high litter quality (see Fig. 7.14) also promote nitrogen mineralization.

Environmental conditions that promote microbial activity enhance both gross and net nitrogen mineralization. Net nitrogen mineralization rates are therefore generally higher in tropical than in temperate forest soils, whereas arctic soils show net nitrogen immobilization (a net decrease in the concentrations of inorganic nitrogen) during the growing season (Nadelhoffer et al. 1992). Even within a biome, factors that improve the soil temperature and moisture environment for microbial activity enhance net nitrogen mineralization. Across the Great Plains of the United States, for example, rates of mineralization are positively related to precipitation and temperature. Recently deforested areas also typically have higher rates of net nitrogen mineralization than do undisturbed forests, at least in part due to warmer, moister soils (Matson and Vitousek 1981). Soil moisture that is high enough to restrict microbial activity also restricts net nitrogen mineralization.

Why do favorable litter quality, moisture, and temperature lead to net nitrogen mineralization rather than immobilization of nitrogen in a growing microbial biomass? Several factors contribute to this pattern. First, the limitation of microbial activity by carbon availability, particularly in environments where nitrogen is readily available and litter nitrogen is relatively high, causes microbes to use some DON to meet their carbon requirements for growth and maintenance, secreting the ammonium as a waste product. Second, warm temperatures increase maintenance respiration and therefore the carbon demands for microbial activity. Finally, increases in microbial productivity promote predation by soil animals, causing greater microbial turnover and release of nitrogen to the soil.

Substrate quality influences nitrogen mineralization rate, not only through its effects on carbon quality, which governs decomposition rate (see Chapter 7) but also through its effects on the balance between carbon and nitrogen limitation of microbial growth. The carbon to nitrogen (C:N) ratio in microbial biomass is about 10:1. As microbes break down organic matter, they incorporate about 40% of the carbon from their substrates into microbial biomass and return the remaining 60% of the carbon to the atmosphere as CO_2 through respiration. With this 40% growth efficiency, microbes require substrates with a C:N ratio of about 25:1 to meet their nitrogen requirement (Box 9.1). At higher C:N ratios, microbes import nitrogen to meet their growth requirements, and at lower C:N ratios nitrogen exceeds microbial growth requirements and is secreted into the litter and soil. In practice, microbes vary in their C:N ratio (5 to 10 in bacteria and 8 to 15 in fungi) and in their growth

Box 9.1. Estimation of Critical C:N Ratio for Net Nitrogen Mineralization

The critical C:N ratio that marks the dividing line between net nitrogen mineralization and net nitrogen uptake by microbes can be calculated from the growth efficiency of microbial populations and the C:N ratios of the microbial biomass and their substrate. Assume, for example, that the microbial biomass has a growth efficiency of 40% and a C:N ratio of 10:1. If the microbes break down 100 units of carbon, they will incorporate 40 units of carbon into microbial biomass and respire 60 units of carbon as CO_2. The 40 units of microbial carbon require 4 units of nitrogen to produce a microbial C:N ratio of 10:1 (= 40:4). If the 100 units of original substrate is to supply all of this nitrogen, its initial C:N ratio must have been 25:1 (= 100:4). At higher C:N ratios, microbes must absorb additional inorganic nitrogen from the soil to meet their growth demands. At lower C:N ratios, microbes excrete excess nitrogen into the soil.

efficiency. Bacteria typically have a lower growth efficiency than do fungi. All microbes convert substrates into biomass less efficiently with less labile substrates or with greater environmental stress. Nonetheless, 25:1 is often considered the critical C:N ratio above which there is no net nitrogen release from decomposing organic matter. Note that there is a clear mechanistic effect of C:N ratio on the net immobilization or mineralization of nitrogen, whereas its effect on decomposition (see Chapter 7) is much less certain. The growth efficiency of microbes (about 40%) is similar to that of plants (net primary production averages about 47% of GPP; see Chapter 6), despite quite different mechanisms of acquiring carbon and nitrogen from the environment. This similarity may reflect a common underlying biochemistry of costs of synthesis and costs of maintenance.

The ammonium produced by nitrogen mineralization has several potential fates. In addition to being absorbed by plants or microbes, ammonium readily adsorbs to the negatively charged surfaces of soil minerals and organic matter (see Chapter 3), resulting in relatively low concentrations of NH_4^+ (often less than 1 ppm) in the soil solution. When plant and microbial uptake depletes NH_4^+ from the soil solution, this shifts the equilibrium between dissolved and exchangeable pools, and adsorbed ions go back into solution from the exchange complex. The cation exchange complex thus serves as a storage reservoir of readily available NH_4^+ and other cations. NH_4^+ can also be fixed in the interlayer portions of certain aluminosilicate clays or complexed with stabilized soil organic matter, making it less available to plants or microbes. As long as the organic complex remains protected, the NH_4^+ in the complex is unavailable to plants and microbes. Finally, NH_4^+ can be volatilized to ammonia gas (NH_3) or oxidized, mainly by bacteria, to NO_2^- and NO_3^-.

Production and Fate of Nitrate

Nitrification is the process by which NH_4^+ is oxidized to NO_2^- and subsequently to NO_3^-. Unlike ammonification, which is carried out by a broad suite of decomposers, most nitrification is carried out by a restricted group of **nitrifying bacteria**. There are two general classes of nitrifiers. **Autotrophic nitrifiers** use the energy yield from NH_4^+ oxidation to fix carbon used in growth and maintenance, analogous to the way plants use solar energy to fix carbon via photosynthesis. **Heterotrophic nitrifiers** gain their energy from breakdown of organic matter.

Autotrophic nitrifiers include two groups, one that converts ammonium to nitrite (e.g., *Nitrosolobus* and other "nitroso-" genera) and another that converts nitrite to nitrate (e.g., *Nitrobacter* and other "nitro-" genera). These autotrophic nitrifiers are obligate aerobes that synthesize structural and metabolic carbon compounds by reducing CO_2 and using energy from NH_4^+ or NO_2^- oxidation to drive CO_2 fixation. In most systems, these two groups occur together, so NO_2^- typically does not accumulate in soils. NO_2^- is most likely to accumulate in dry forest and savanna ecosystems during the dry season, when the activity of *Nitrobacter* is restricted, and in some fertilized ecosystems, where nitrogen inputs are high relative to plant and microbial demands.

Although autotrophic nitrification predominates in many ecosystems, heterotrophic nitrification can be important in ecosystems with low nitrogen availability or acidic soils. Many heterotrophic fungi and bacteria, including actinomycetes, produce NO_2^- or NO_3^- from NH_4^+. Some also use organic nitrogen in the process. Because heterotrophs obtain their energy from organic materials, it is not clear what advantage they gain from the NO_3^- oxidation process.

Nitrification has multiple effects on ecosystem processes. The oxidation of NH_4^+ to NO_2^-, which occurs in the first step of nitrification, produces 2 moles of H^+ for each mole of NH_4^+ consumed and therefore tends to acidify soils. The mono-oxygenase that catalyzes this step has a broad substrate specificity and also oxidizes many chlorinated hydrocarbons, suggesting a role of nitrifiers in the breakdown of pesticide residues. Finally, nitric oxide (NO) and nitrous oxide (N_2O), which are produced during nitrification (Fig. 9.4), are gases that have important effects on atmospheric chemistry.

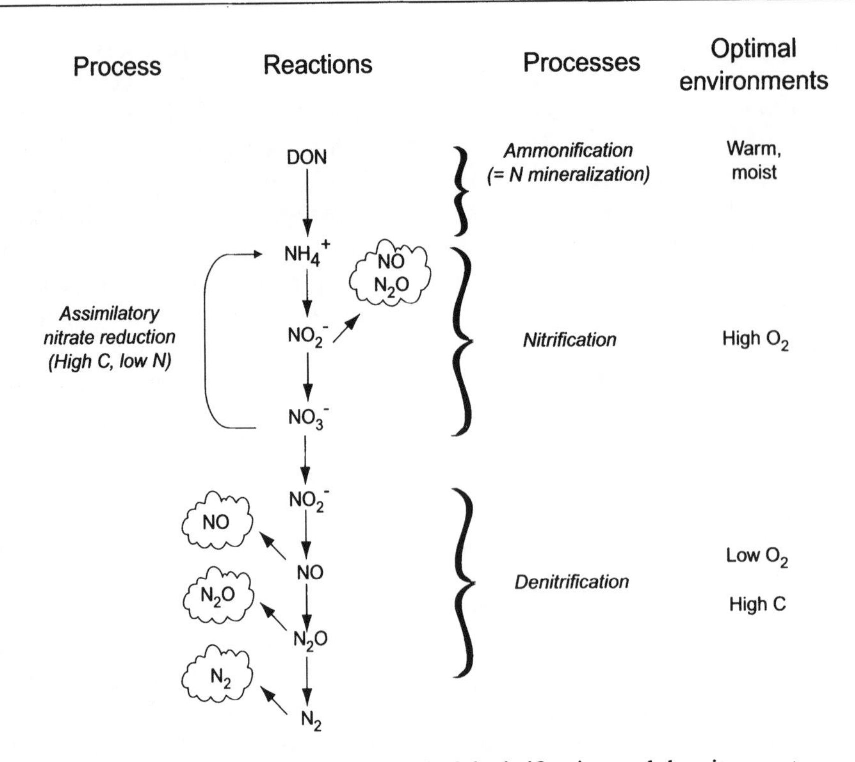

FIGURE 9.4. Pathways of autotrophic nitrification and of denitrification and the nitrogen trace gases emitted by these pathways (Firestone and Davidson 1989).

The availability of NH_4^+ is the most important direct determinant of nitrification rate (Fig. 9.5) (Robertson 1989). The NH_4^+ concentration must be high enough, at least in certain soil microsites, to allow nitrifiers to compete with other soil microbes. This is particularly important for autotrophic nitrifiers, which rely on NH_4^+ as their sole energy source. NH_4^+ supply, in turn, is regulated by the effects of substrate quality and environment on ammonification rate (Fig. 9.3). Fertilizer inputs and ammonium deposition are additional sources of ammonium to many ecosystems. Conversely, plant roots reduce NH_4^+ concentration in the soil solution, thereby competing with nitrifiers for NH_4^+. Many productive ecosystems have high nitrification rates despite low average NH_4^+ concentrations in the bulk soil, perhaps because of spatial heterogeneity in NH_4^+ concentration. Nitrification rates can also be substantial, even when NO_3^- concentrations in soils are low, because nitrate is relatively mobile and can be absorbed by plants or microbes, leached from soils, and denitrified as rapidly as it is produced.

Nitrifier populations are often too small in infertile soils to support significant nitrification. When ammonium substrate becomes available (e.g., through additions of nitrogen, or increases in mineralization rates), nitrifier populations can increase in size, and nitrification rates can increase. The response can be rapid in some soils but show a long delay in others (Vitousek et al. 1982). Secondary metabolites have been hypothesized to inhibit nitrification in some ecosystems, including those in late succession (Rice 1979), but the decline in nitrification in late succession is generally best explained by a decline in ammonium supply rather than as toxicity to nitrifiers (Pastor et al. 1984, Schimel et al. 1996). Limitation of nitrifiers by other resources is another possible cause of slow or delayed nitrification. In most cases, however, the availability of ammonium ultimately governs nitrification rate through its effects on both the population density and activity of nitrifying bacteria.

Oxygen is an important additional factor controlling nitrification because most nitrifiers

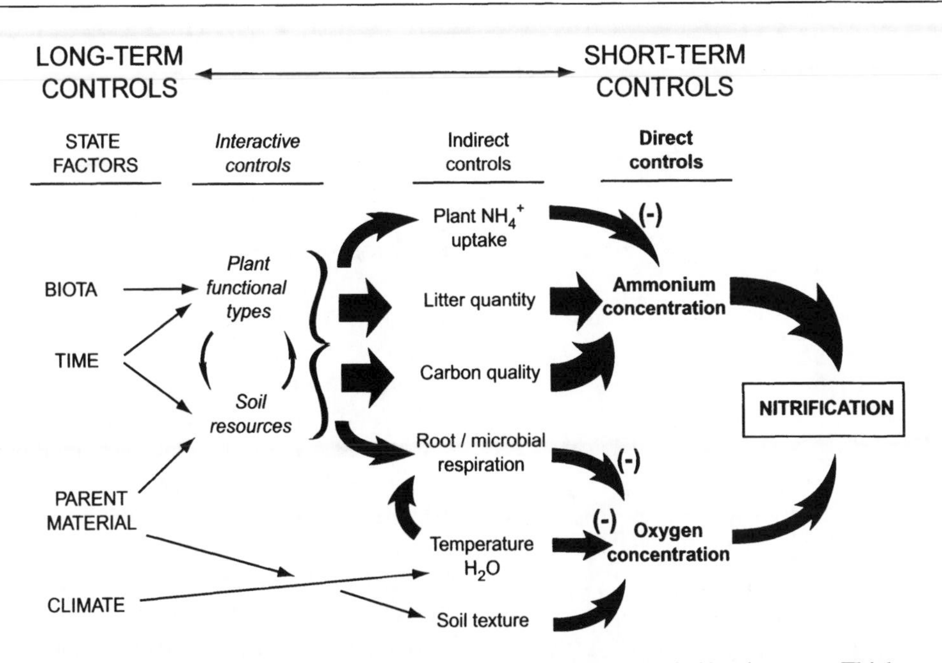

FIGURE 9.5. The major factors controlling nitrification in soils (Robertson 1989). These controls range from concentrations of reactants that directly control nitrification to the interactive controls, such as climate and disturbance regime, that are the ultimate determinants of nitrification rate. Thickness of the arrows indicates the strength of effects. The influence of one factor on another is positive unless otherwise indicated (–).

require oxygen for oxidation of NH_4. Oxygen availability, in turn, is influenced by many factors, including soil moisture, soil texture, soil structure, and respiration by microbes and roots (Fig. 9.5).

Nitrifier activity is sensitive to temperature. It does, however, continue at low rates at low temperatures, so over a long winter season, substantial nitrification can occur, particularly in nitrogen-rich agricultural soils. Nitrification rates are slow in dry soils primarily because thin water films restrict NH_4^+ diffusion to nitrifiers (Stark and Firestone 1995). Under extremely dry conditions, low water potential further restricts the activity of nitrifiers. The importance of acidity in regulating nitrification rates is uncertain. In laboratory cultures of agricultural soils, maximum nitrification rates occur between pH 6.6 and 8.0 and are negligible below pH 4.5 (Paul and Clark 1996). Many natural ecosystems with acidic soils, however, have substantial nitrification rates, even at pH 4 (Stark and Hart 1997).

The fraction of mineralized nitrogen that is oxidized to nitrate varies widely among ecosystems. In many unpolluted temperate coniferous and deciduous forests, nitrification is only a small proportion of net mineralization (e.g., 0 to 4%) but, as ecosystems receive increasing nitrogen deposition, the fraction of nitrification can increase to 25% (McNulty et al. 1990). In tropical forests, in contrast, net nitrification is typically nearly 100% of net mineralization, even in sites with low rates of net mineralization and without inputs of additional nitrogen (Vitousek and Matson 1988) (Fig. 9.6). In tropical ecosystems, plant and microbial growth are frequently limited by nutrients other than nitrogen, and their demand for nitrogen is low, so nitrifiers have ready access to NH_4^+.

The potential fates of nitrate are absorption by plants and microbes, exchange on anion exchange sites, and loss from ecosystems via denitrification or leaching. Because nitrate is relatively mobile in soil solutions, it readily moves to plant roots by mass flow or diffusion (see Chapter 8) or can be leached from the soil. Microbes also absorb nitrate and use it for synthesis of amino acids through **assimilatory**

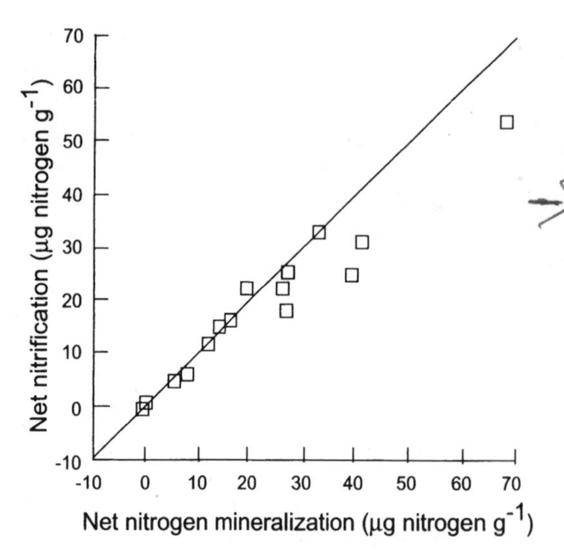

FIGURE 9.6. The relationship between net nitrogen mineralization and net nitrification (per gram of dry soil for a 10-day incubation) across a range of tropical forest ecosystems (Vitousek and Matson 1984). Nearly all nitrogen that is mineralized in these systems is immediately nitrified. In contrast, nitrification is frequently less than 25% of net mineralization in temperate ecosystems.

nitrate reduction (Fig. 9.4). This process is energetically expensive and occurs primarily when microbes are nitrogen limited but have an adequate energy supply. The low nitrate concentrations observed in many acidic conifer forest soils reflect a combination of low nitrification rates and nitrate absorption by soil microbes and plants (Stark and Hart 1997).

Although NO_3^- is more mobile than most cations, it can be held on exchange sites of soils with a high anion exchange capacity (see Chapter 3). All mineral soils have a variable charge that depends on pH. In the layered-clay silicate minerals typical of the temperate zone, the zero point of charge (below which pH the charge is positive and above which it is negative) is typically near pH 2, well below the pH of most soils. In some soils, especially those in the tropics, iron and aluminum oxide minerals have a positive surface charge at their typical pH. In these soils, there is sufficient anion exchange capacity to prevent leaching losses of nitrate after disturbance (Matson et al. 1987). In most soils, the strength of the anion adsorption is $PO_4^{3-} > SO_4^{3-} > Cl^- > NO_3^-$, so NO_3^- is desorbed relatively easily.

Temporal and Spatial Variability

Fine-scale ecological controls cause large temporal and spatial variability in nitrogen cycling. Nitrogen transformation rates in soils are notoriously variable, with rates often differing by an order of magnitude between adjacent soil samples or sampling dates (Robertson et al. 1997). This variability reflects the fine temporal and spatial scales over which controlling factors vary. Anaerobic conditions that support denitrification (discussed later) in the interiors of soil aggregates, for example, can occur within millimeters of aerobic soil pores. Fine roots create zones of rhizosphere soils with high carbon and low soluble nitrogen concentrations adjacent to bulk soil, where carbon-limited soil microbes mineralize organic nitrogen to meet their energy demands. In densely rooted microsites, plants deplete concentrations of NH_4^+ below levels that can sustain nitrification, whereas nitrification can be substantial in adjacent non-root microsites. The effects of this fine-scale spatial heterogeneity on nitrogen cycling are difficult to study, so we know only qualitatively of their importance.

Temporal variability in environment and extreme events have a strong influence on nitrogen mineralization. Drying–wetting and freeze–thaw events, for example, burst many microbial cells and release pulses of nutrients. For this reason, the first rains after a long dry-season often cause a pulse of nitrification and nitrate leaching (Davidson et al. 1993). The spring runoff after snowmelt in northern or mountain ecosystems also frequently carries with it a pulse of nutrient loss to streams because of both freeze–thaw events and the absence of plant uptake of nitrogen during winter. For example, 90% of the annual nitrogen input to Toolik Lake in arctic Alaska, occurs in the first 10 days of snowmelt (Whalen and Cornwell 1985).

The seasonality of nitrogen mineralization often differs from the seasonality of plant nitrogen uptake. In ecosystems in which plants are dormant for part of the year, soil microbes continue to mineralize nitrogen during the dormant season; this leads to an accumulation of available nitrogen that plants use when they become active. In temperate forests, for example, mineralization during winter (even beneath a snowpack) creates a substantial pool of available nitrogen that is not absorbed by plants until the following spring. This asynchrony between microbial activity and plant uptake is particularly important in low-nutrient environments, where microbes may immobilize nitrogen during the season of most active plant growth, effectively competing with plants for nitrogen (Jaeger et al. 1999b). In soils that freeze or dry, the death of microbial cells provides additional labile substrates that support net mineralization by the remaining microbes when conditions again become suitable for microbial activity. Plant storage of nutrients to support spring growth is particularly important in low-fertility ecosystems, because nutrients are not dependably available from the soil at times when the environment favors plant growth (Chapin et al. 1990).

Pathways of Nitrogen Loss

Gaseous Losses of Nitrogen

Ammonia volatilization, nitrification, and denitrification are the major avenues of gaseous nitrogen loss from ecosystems. These processes release nitrogen as ammonia gas, nitrous oxide, nitric oxide, and dinitrogen. Gas fluxes are controlled by the rates of soil processes and by soil and environmental characteristics that regulate diffusion rates through soils. Once in the atmosphere, these gases can be chemically modified and deposited downwind.

Ecological Controls

Ammonia gas (NH_3) can be emitted from soils and scenescing leaves. In soils, it is emitted as a consequence of the pH-dependent equilibrium between NH_4^+ and NH_3. At pH values greater than 7, a significant fraction of NH_4^+ is converted to NH_3 gas.

$$NH_4^+ + OH^- \leftrightarrow NH_3 + H_2O \qquad (9.1)$$

Ammonia then diffuses from the soil to the atmosphere. This diffusion is most rapid in coarse dry soils with large air spaces. In dense canopies, some of the NH_3 emitted from soils is absorbed by plant leaves and incorporated into amino acids.

NH_3 flux is low from most ecosystems because NH_4^+ is maintained at low concentrations by plant and microbial uptake and by binding to the soil exchange complex. NH_3 fluxes are substantial, however, in ecosystems in which NH_4^+ accumulates due to large nitrogen inputs. In grazed ecosystems, for example, urine patches dominate the aerial flux of NH_3. Agricultural fields that are fertilized with ammonium-based fertilizers or urea often lose 20 to 30% of the added nitrogen as NH_3, especially if fertilizers are placed on the surface. Nitrogen-rich basic soils are particularly prone to NH_3 volatilization because of the pH effect on the equilibrium between NH_4^+ and NH_3. Leaves also emit NH_3 during senescence, when nitrogen-containing compounds are broken down for transport to storage organs. Fertilization and domestic animal husbandry have substantially increased the flux of NH_3 to the atmosphere (see Chapter 15).

The production of NO and N_2O during nitrification depends primarily on the rate of nitrification. The conversion of NH_4^+ to NO_3^- by nitrification produces some NO and N_2O as byproducts (Fig. 9.4), typically at a NO to N_2O ratio of 10 to 20. The quantities of NO and N_2O released during nitrification are correlated with the total flux through the nitrification pathway, suggesting that nitrification acts like a leaky pipe (Firestone and Davidson 1989), in which a small proportion (perhaps 0.1 to 10%) of the nitrogen "leaks out" as trace gases during nitrification.

The reduction of nitrate or nitrite to gaseous nitrogen by denitrification occurs under conditions of high nitrate and low oxygen. Many types of bacteria contribute to biological denitrification. They use NO_3^- or NO_2^- as an electron acceptor to oxidize organic carbon for energy when oxygen concentration is low. Most denitrifiers are facultative anaerobes and use oxygen rather than NO_3^-, when oxygen is available. In addition to biological denitrification,

chemodenitrification converts NO_2^- (nitrite) abiotically to nitric oxide gas (NO) where NO_2^- accumulates in the soil at low pH. Chemodenitrification is typically much less important than biological denitrification.

The sequence of NO_3^- reduction is $NO_3^- \rightarrow NO_2^- \rightarrow NO \rightarrow N_2O \rightarrow N_2$. The last three products, particularly N_2O and N_2, are released as gases to the atmosphere (Fig. 9.4). Most denitrifiers have the enzymatic potential to carry out the entire reductive sequence but produce variable proportions of N_2O and N_2, depending in part on the relative availability of oxidant (NO_3^-) versus reductant (organic carbon). When NO_3^- is relatively more abundant than labile organic carbon, more N_2O than N_2 is produced. Other factors that favor N_2O over N_2 production include low pH, low temperature, and high oxygen. Although NO is often released during denitrification in laboratory incubations, this seldom occurs in nature because its diffusion to the air is impeded by water-filled pore spaces. Some of the NO that is produced serves as a substrate for further reduction to N_2O or N_2 by denitrifying bacteria.

The three conditions required for significant denitrification are low oxygen, high nitrate concentration, and a supply of organic carbon (Fig. 9.7) (Del Grosso et al. 2000). In most nonflooded soils, oxygen availability exerts the strongest control over denitrification. Oxygen supply is reduced by high soil water content, which impedes the diffusion of oxygen through soil pores. Soil moisture, in turn, is controlled by other environmental factors such as slope position, soil texture, and the balance between precipitation and evapotranspiration. Soil oxygen concentration is also sensitive to its rate of consumption by soil microbes and roots. It is consumed most quickly in warm, moist environments.

The second major control over denitrification is an adequate supply of the substrate NO_3^-. Because nitrification is a primarily an aerobic process, the low-oxygen conditions that are optimal for denitrification frequently limit NO_3^- supply. Some wetlands, for example, have

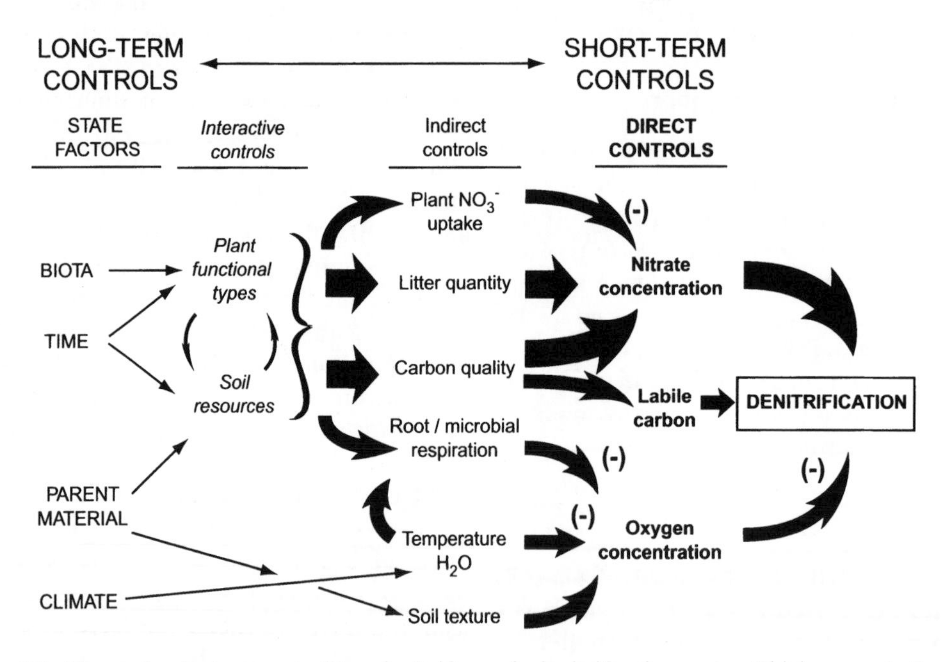

FIGURE 9.7. The major factors controlling denitrification in soils. These controls range from concentrations of substrates that directly control nitrification to the interactive controls such as climate and disturbance regime that are the ultimate determinants of denitrification rate. Thickness of the arrows indicates the strength of effects. The influence of one factor on another is positive unless otherwise indicated (–).

low denitrification rates despite their saturated soils and large quantities of organic matter due to low availability of nitrate. Wetlands support high denitrification rates only if (1) they receive NO_3^- from outside the system (lateral transfer); (2) they have an aerobic zone above an anaerobic zone (vertical transfer), as in partially drained wetlands; or (3) they go through cycles of flooding and drainage (temporal separation), as in many rice paddies. At a finer scale, denitrification can occur within soil aggregates or other anaerobic microsites (e.g., pieces of soil organic matter) in moderately well drained soils due to fine-scale heterogeneity in soil oxygen concentration and nitrification rate. Finally, the availability of organic carbon substrates can limit denitrification because the process is carried out primarily by heterotrophic bacteria. Long-term cultivation of agricultural soils, for example, can reduce soil organic matter concentrations sufficiently to limit denitrification.

Fires also account for large gaseous losses of nitrogen. The amount and forms of nitrogen volatilized during fire depend on the temperature of the fire. Fires with active flames produce considerable turbulence, are well supplied with oxygen, and release nitrogen primarily as NO_x. Smoldering fires release nitrogen in more reduced forms, such as ammonia (Goode et al. 2000). About a third of the nitrogen is emitted as N_2. Fire is an important part of the nitrogen cycle of many ecosystems. Fire suppression in some areas and biomass burning in others have altered the natural patterns of nitrogen cycling in many ecosystems.

Atmospheric Roles of Nitrogen Gases

The four nitrogen gases have different roles and consequences for the atmosphere. NH_3 that enters the atmosphere reacts with acids and thus neutralizes atmospheric acidity.

$$NH_3 + H_2SO_4 \rightarrow (NH_4)_2SO_4 \qquad (9.2)$$

In this reaction, NH_3 is converted back to NH_4^+, which can be deposited downwind on the surface of dry particles or as NH_4^+ dissolved in precipitation. Ammonia volatilization and deposition transfer nitrogen from one ecosystem to another.

In the atmosphere, the nitrogen oxides are in equilibrium with one another due to their rapid interconversion. NO_x is very reactive, and its concentration regulates several important atmospheric chemical reactions. At high NO_x concentrations, for example, carbon monoxide (or methane and nonmethane hydrocarbons) are oxidized, producing tropospheric ozone (O_3), an important component of photochemical smog in urban, industrial, and agricultural areas.

$$CO + 2O_2 \rightarrow CO_2 + O_3 \qquad (9.3)$$

When NO_x concentrations are low, the oxidation of CO consumes O_3.

$$CO + O_3 \rightarrow CO_2 + O_2 \qquad (9.4)$$

In addition to its role as a catalyst that alters atmospheric chemistry and generates pollution, NO_x can be transported long distances and alter the functioning of ecosystems downwind. In the form of nitric acid, it is a principle component of acid deposition and adds both available nitrogen and acidity to the soil. In its gaseous NO_2 form, it can be absorbed through the stomata of leaves and be used in metabolism (see Chapter 5). It can also be deposited in particulate form, another type of inadvertent fertilization.

In contrast to the highly reactive NO_x, nitrous oxide (N_2O) has an atmospheric lifetime of 150 years and is not chemically reactive in the troposphere. The low reactivity of N_2O contributes to a different environmental problem. N_2O is a greenhouse gas that is 200 times more effective per molecule than is CO_2 in absorbing infrared radiation (see Chapter 2). In addition, N_2O in the stratosphere reacts with excited oxygen in presence of ultraviolet radiation to produce NO, which catalyzes the destruction of stratospheric ozone (O_3).

Given that the atmosphere is already 78% N_2, dinitrogen emissions to the atmosphere via denitrification have no significant atmospheric effects, although these losses may influence ecosystem nitrogen pools. Atmospheric N_2 has a turnover time of thousands of years.

Solution Losses

Nitrate and dissolved organic nitrogen account for most of the solution loss of nitrogen from ecosystems. Undisturbed ecosystems that receive low atmospheric inputs generally lose relatively little nitrogen, and these small losses occur primarily as dissolved organic nitrogen (Hedin et al. 1995). Although nitrate is also highly mobile, plants and microbes absorb most nitrate before it leaches below the rooting zone of intact ecosystems. Disturbance, however, often improves the environment for mineralization by increasing soil moisture and temperature and reduces the biomass of vegetation available to absorb nutrients (see Chapter 12). At the Hubbard Brook Forest in the northeastern United States, for example, all vegetation was removed from an experimental watershed to examine the consequences of devegetation. There were large losses of nitrate, calcium, and potassium to the groundwater and streams when vegetation regrowth was prevented (Bormann and Likens 1979) (Fig. 9.8). Once vegetation began to regrow, however, the accumulating plant biomass absorbed most of the mineralized nutrients, and stream nutrient concentrations returned to their preharvest levels. Additions of fertilizer nitrogen or nitrogen deposition that exceed plant and microbial nitrogen demands also increase nitrate leaching. Increased nitrate leaching is one of the characteristics of **nitrogen saturation**, the changes that occur in ecosystem functioning when anthropogenic nitrogen additions relieve nitrogen limitation to plants and microbes (Aber et al. 1998). Anthropogenic nitrogen inputs are generally correlated with nitrogen outputs via leaching (Tietema and Beier 1995, Fenn et al. 1998) (Fig. 9.9).

Nitrate loss to groundwater can have important consequences for human health and for the ecological integrity of aquatic ecosystems. Nitrite, which forms from nitrate under reducing conditions, can reduce the capacity of hemoglobin in animals to transport oxygen, producing anemia, especially in infants. Groundwater in areas of intensive agriculture

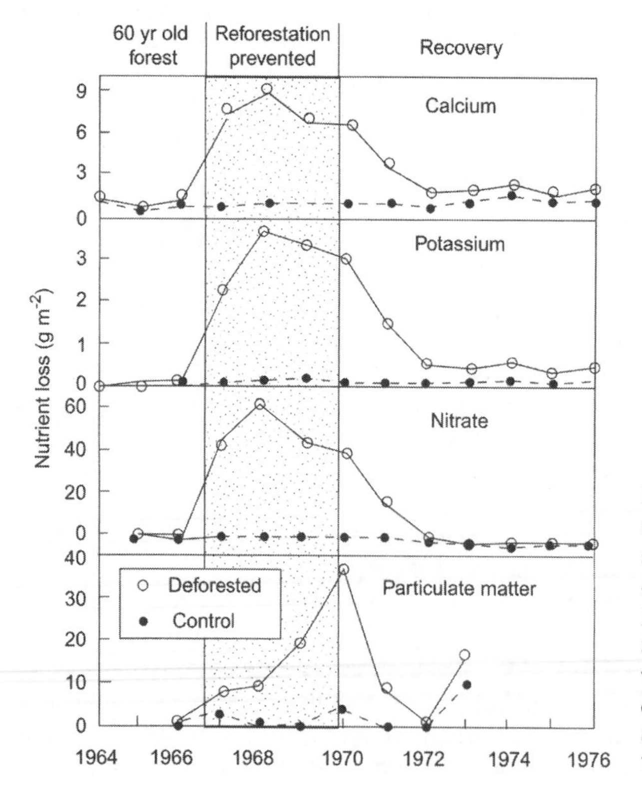

FIGURE 9.8. Losses of calcium, potassium, nitrate, and particulate organic matter in stream water before and after deforestation of an experimental watershed at Hubbard Brook Forest in the northeastern United States. The shaded area shows the time interval during which vegetation was absent due to cutting of trees and herbicide application. (Redrawn with permission from Springer-Verlag; Bormann and Likens 1979.)

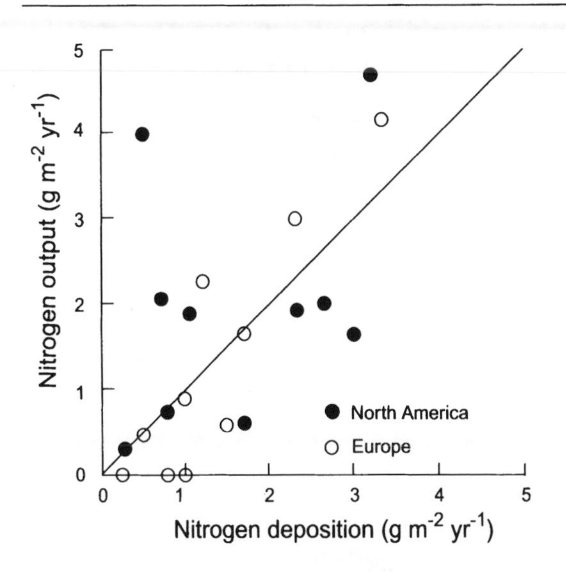

FIGURE 9.9. Comparisons of inputs from nitrogen deposition and nitrogen outputs in solution from forests of North America and Europe. There is a strong relationship between inputs and outputs in nitrogen-saturated ecosystems. (Data from Tietema and Beier 1995, Fenn et al. 1998.)

often has nitrate concentrations that exceed public health standards.

Nitrogen leached from terrestrial ecosystems moves in groundwater to lakes and rivers. The movement of nitrate to the North Atlantic Ocean from major rivers has increased 6- to 20-fold in the past century (Howarth et al. 1996), primarily due to increased inputs of fertilizer, atmospheric deposition, nitrogen fixation by crops, and food imports (see Chapter 15). Nitrate in coastal marine systems frequently increases productivity and detritus accumulation. The resulting stimulation of decomposition can reduce oxygen concentrations sufficiently to kill fish, particularly in winter, when primary production is temperature limited (see Chapter 14). The nitrogen loading from agricultural and urban systems in the Mississippi drainage, for example, has produced a "dead zone" where this river enters the Gulf of Mexico (Mitsch et al. 2001).

Solutions that move through the soil must maintain a balanced charge, with negatively charged ions like nitrate balanced by cations. Therefore, every nitrate ion that leaches from soil carries with it a cation such as calcium, potassium, and ammonium to maintain charge balance. When cation loss by leaching exceeds the rate of cation supply by weathering plus deposition, the net loss of cations can lead to cation deficiency (Driscoll et al. 2001). After these nutrient cations are depleted, nitrate takes with it H^+ and/or Al^{3+}, which are deleterious to downstream ecosystems. Nitrification also generates acidity:

$$2NH_4^+ + 3O_2 \rightarrow 2NO_2^- + H_2O + H^+ \quad (9.5)$$

The hydrogen ion released in this reaction exchanges with other ions on cation exchange sites in the soil, making these cations more vulnerable to leaching loss.

Erosional Losses

Erosion is a natural pathway of nitrogen loss that often increases dramatically after land use changes. As with leaching, erosional losses of nitrogen include both organic and inorganic forms, although organic forms associated with soil aggregates and particles are most important. In some ecosystems, especially those on unstable slopes or in areas exposed to high winds, erosion is a dominant natural pathway of nitrogen loss.

Other Element Cycles

Phosphorus

Weathering of primary minerals is the major source of new phosphorus to ecosystems. In contrast to nitrogen, whose major source is the atmosphere, phosphorus enters ecosystems primarily by weathering of rocks (Fig. 9.10). The weathering of phosphorus-containing apatite by the carbonic acid generated from soil respiration, for example, releases phosphorus in available forms (Eq. 9.6) that can be taken up directly by plants or microorganisms or be adsorbed or precipitated (see Chapter 3).

$$Ca_5(PO_4)_3 + 4H_2CO_3 \rightarrow 5Ca^{2+} + 3HPO_4^{2-} + 4HCO_3 + H_2O \quad (9.6)$$

Phosphorus inputs from weathering depend on the mineralogy of the parent material, the

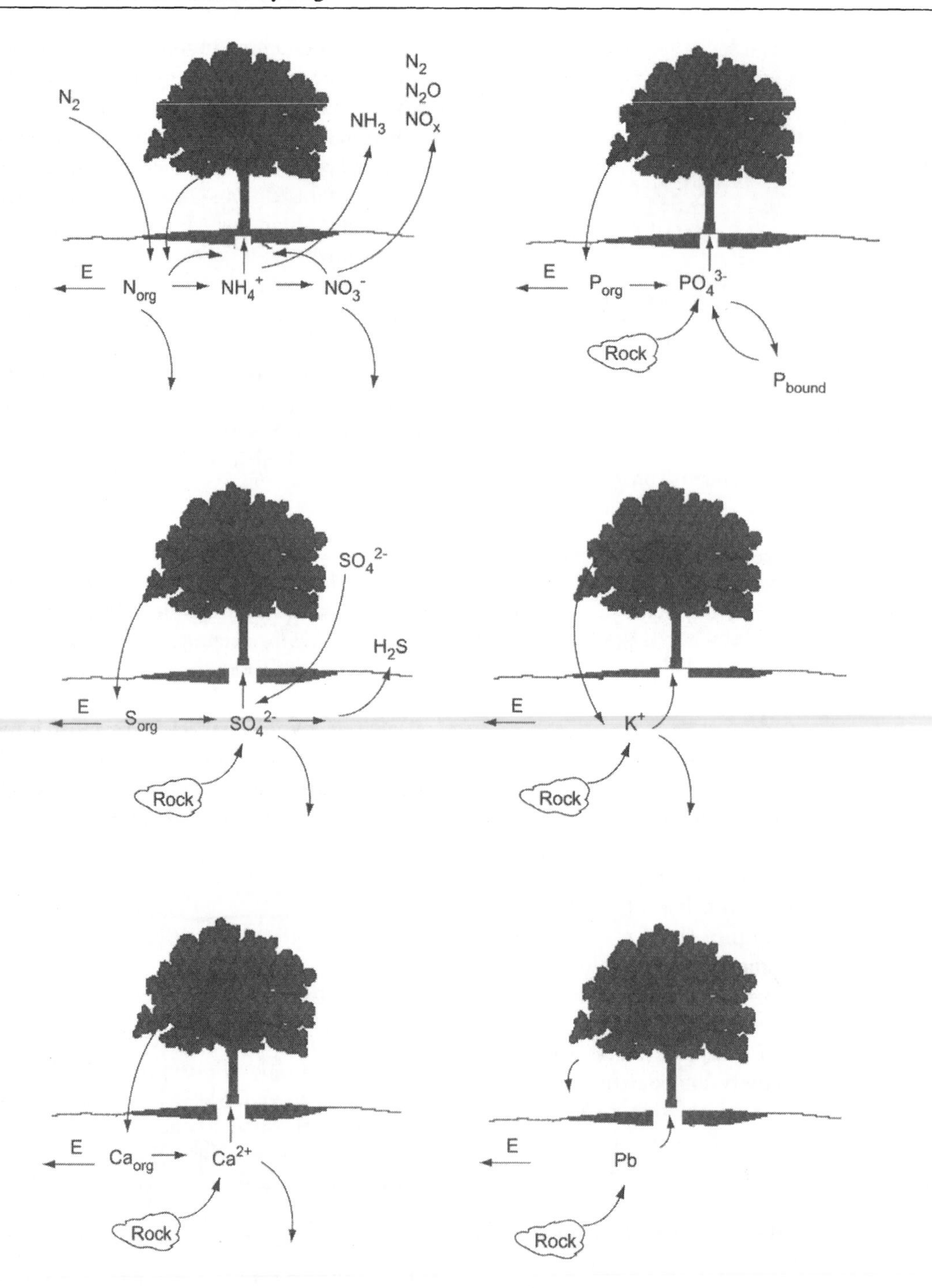

FIGURE 9.10. Comparison of natural element cycles with respect to the relative importance of internal recycling, inputs, and outputs. Inputs of nitrogen come primarily from the atmosphere, whereas inputs of phosphorus, potassium, calcium, and most unessential elements such as lead (Pb) come primarily from rocks. Sulfur comes from both the atmosphere and rocks. Over long time scales, atmospheric inputs of all elements can be important. Element losses occur through leaching, erosion (E), and, in the case of nitrogen and sulfur, gaseous emission. Subscripts indicate organic (org) or bound forms of the element.

climate, and the landscape age. Marine-derived calcareous rocks have relatively high phosphorus content and weather readily (Eq. 9.6). Over millions of years, phosphorus-containing minerals become depleted, leading to increasing phosphorus limitation, as landscapes age. The effects of landscape age are most pronounced in the tropics where weathering occurs most rapidly. In areas downwind of deserts or agricultural areas, the deposition of phosphorus in rainfall or dry deposition often represents a significant input to ecosystems, particularly in old landscapes where inputs from weathering are low.

As with nitrogen, the internal cycling of phosphorus in ecosystems requires the cleaving of bonds with organic matter to produce a form that is water soluble and can be absorbed by microbes and plants. Phosphorus turnover is somewhat less tightly linked to decomposition than is nitrogen, because the ester linkages that bind phosphorus to carbon (C—O—P) can be cleaved without breaking down the carbon skeleton. Nitrogen, in contrast, is directly bonded to the carbon skeletons of organic matter (C—N) and is generally released by breakdown of the carbon skeleton into amino acids and other forms of dissolved organic nitrogen.

The decomposition process that breaks down the organic matter exposes the C—O—P bonds to enzymatic attack. Low soil phosphorus availability induces plants and microbes to invest nitrogen in enzymes to acquire phosphorus. Plant roots and their mycorrhizal associates—particularly arbuscular mycorrhizae, which are abundant in grasslands, tropical forests, and many other systems—produce phosphatases that cleave ester bonds in organic matter to release phosphate (PO_4^{3-}). There is therefore tight cycling of phosphorus between organic matter and plant roots in many ecosystems. In tropical forests, for example, mats of mycorrhizal roots form in the litter layer and produce phosphatases that cleave phosphate from organic matter. Mycorrhizal roots directly absorb much of this phosphate before it interacts with the mineral phase of the soil. Plant and microbial phosphatases are induced by low soil phosphate. This contrasts with protease, whose activity correlates more strongly with microbial activity than with concentrations of soil organic nitrogen.

Microbial biomass frequently accounts for 20 to 30% of the organic phosphorus in soils (Smith and Paul 1990, Jonasson et al. 1999), much larger than the proportion of microbial carbon (about 2%) or nitrogen (about 4%). Microbial biomass is therefore an important reservoir of potentially available phosphorus, particularly in ecosystems with highly basic or acidic soils. Microbial phosphorus is potentially more available than inorganic phosphate because it is protected from reactions with the mineral phase of soils, as described later. Although the biogeochemical literature emphasizes the importance of C:N ratios, C:P ratios of dead organic matter can also be critical. In ecosystems with low phosphorus availability, the C:P ratio of dead organic matter controls the balance between phosphorus mineralization and immobilization and therefore the supply of phosphorus to plants.

Chemical reactions with soil minerals play a key role in controlling phosphorus availability in soils. Unlike nitrogen, phosphorus undergoes no oxidation–reduction reactions in soils and has no important gas phases or atmospheric components. In addition, many of the reactions that control phosphorus availability are geochemical rather than biological in nature. Phosphate is the main form of available phosphorus in soils. Theoretically, soil pH determines the most common form of phosphate in the soil solution:

$$\underset{4}{H_2PO_4^-} \leftrightarrow H^+ + \underset{10}{HPO_4^{2-}} \leftrightarrow 2H^+ + \underset{14}{PO_4^{3-}} \quad [\text{pH range}] \qquad (9.7)$$

This is important because the less highly charged forms of phosphate ($H_2PO_4^-$) are more mobile in soil and are therefore more available to plants and microbes. The actual effects of pH depend, however, on the concentrations of other ions and minerals present in the soil matrix (Fig. 9.11). At low pH, for example, where $H_2PO_4^-$ should dominate, aluminum, iron, and manganese are also quite soluble and react with $H_2PO_4^-$ to form insoluble compounds:

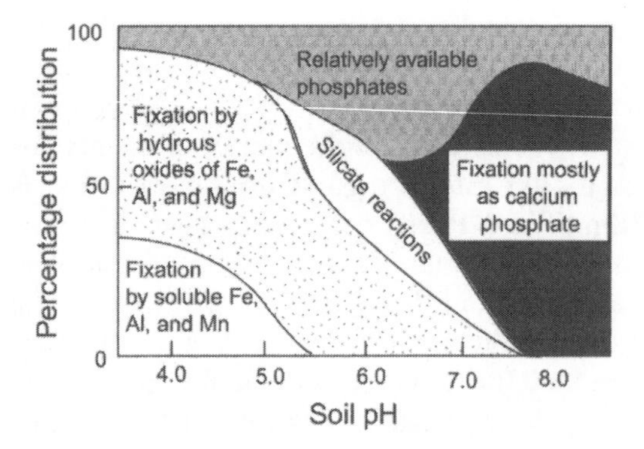

FIGURE 9.11. Effect of pH on the major forms of phosphorus present in soils. The low solubility of phosphorus compounds at low and high pH result in a relatively narrow window of phosphate availability near pH 6.5. (Redrawn from *Nature and Properties of Soils, 13th edition* by N.C. Brady and R.R. Weil © 2001, by permission of Pearson Education, Inc., Upper Saddle River, NJ; Brady and Weil 2001.)

$$Al^{3+} + H_2PO_4^- + 2H_2O \leftrightarrow 2H^+ + Al(OH)_2H_2PO_4 \quad (9.8)$$

soluble insoluble

Very little of the phosphorus in soils is soluble at any time because many inorganic and organic mechanisms retain phosphorus in insoluble forms. Inorganic mechanisms are strongly pH dependent. At low pH, phosphorus can be sorbed onto the surfaces of clays and oxides of iron and aluminum. Phosphate is initially electrostatically attracted to positively charged sites on minerals through anion exchange. Once there, phosphate can become increasingly tightly bound (and correspondingly unavailable to plants) as it forms one or two covalent bonds with the metals on the mineral surface. Phosphorus can also bind with soluble minerals (especially iron oxides) to form insoluble precipitates. Chemical precipitates of phosphorus with these oxides and phosphate sorption on oxide surfaces, explain why highly weathered tropical soils (Oxisols and Ultisols) have extremely low phosphorus availability and why the growth of forests on those soils is typically phosphorus limited (see Chapter 3). The silicate clay minerals that dominate temperate soils fix phosphate to a lesser extent than do the oxides of tropical oxisols.

In soils with high concentrations of exchangeable calcium and calcium carbonate ($CaCO_3$), which typically occur at high pH, calcium phosphate precipitates, reducing phosphate availability in solution:

$$Ca(H_2PO_4)_2 + 2Ca^{2+} \rightarrow Ca_3(PO_4)_2 + 4H^+ \quad (9.9)$$

soluble insoluble

At high pH, phosphate combines with Ca to form (in order of decreasing availability) monocalcium, dicalcium, and tricalcium phosphates. Precipitation of calcium phosphate is one of the main reasons that phosphate fertilizer immediately becomes unavailable in calcium-rich temperate agricultural ecosystems. Due to the precipitation reactions that occur at high and low pH, phosphorus is most available in a narrow range around pH 6.5 (Fig. 9.11). Organic compounds in the soil also regulate, both directly and indirectly, phosphorus binding and availability. Charged organic compounds, for example, can compete with phosphate ions for binding sites on the surfaces of oxides, thereby decreasing phosphorus fixation. Organic compounds can also chelate metals and prevent their reaction with phosphate. On the other hand, organic compounds form complexes with iron, aluminum, and phosphate that protect these compounds from enzymatic attack. In tropical allophane soils, these complexes form a major sink for phosphorus.

Much of the phosphorus that precipitates as iron, aluminum, and calcium compounds is essentially unavailable to plants and is referred to as **occluded** phosphorus. During soil development, primary minerals gradually disappear as a result of weathering and erosional loss. The mass of phosphate in soils tends to shift from mineral, organic, and nonoccluded forms to occluded and organically bound forms, causing a shift from nitrogen to phosphorus limitation in ecosystems over long time scales (see Fig. 3.4) (Crews et al. 1995).

The tight binding of phosphate to organic matter or to soil minerals in most soils causes 90% of the phosphorus loss to occur through surface runoff and erosion of particulate phosphorus rather than through leaching of soluble phosphate to groundwater (Tiessen 1995). Two thirds of the dissolved phosphorus that enters groundwater is organic and therefore less reactive with soil minerals. The productivity of aquatic ecosystems is so sensitive to phosphorus additions that even small additions in groundwater can cause large changes in their functioning (see Chapter 13).

Sulfur

Sulfur cycling is particularly complex because it undergoes oxidation–reduction reactions, like nitrogen, and has both gaseous and mineral sources (Fig. 9.10). Rock weathering, which is the primary natural source of sulfur in most ecosystems, is being increasingly supplemented by atmospheric inputs in the form of acid rain. Combustion of fossil fuels produces gaseous sulfur dioxide (SO_2), which dissolves in cloud droplets to produce sulfuric acid (H_2SO_4), a strong acid that is responsible for much of the lake acidification downwind of industrial areas (see Chapter 15). Sulfur in plant tissues is both carbon and ester bonded, so microbial mineralization includes immobilization and release processes similar to those of nitrogen and phosphorus. Like nitrogen, inorganic sulfur undergoes oxidation–reduction reactions and is therefore sensitive to oxygen availability in the environment. In anaerobic soils, sulfate acts as an electron acceptor that allows microbes to metabolize organic carbon for energy, with hydrogen sulfide being produced as a byproduct. In aerobic environments, however, reduced sulfur can be an important energy source for bacteria. The productivity associated with deep-sea vents, for example, is based entirely on the oxidation of hydrogen sulfide (H_2S) from the vents. Sulfur is a component of most enzymes, including the nitrogenase of nitrogen fixers, so low availability of sulfur in highly weathered soils in unpolluted areas can limit nitrogen inputs to ecosystems and therefore plant production and nutrient turnover. Superphosphate fertilizer has a high sulfur concentration, so vegetation responses to application of phosphate fertilizers in some ecosystems may include a response to sulfur (Eviner et al. 2000). Sulfur compounds in the atmosphere play critical roles as aerosols, which increase the albedo of the atmosphere and therefore cause climatic cooling (see Chapter 2).

Essential Cations

Rock weathering is the primary avenue for element inputs of potassium, calcium, magnesium, and manganese, the cations required in largest amounts by plants. As with nitrogen, phosphorus, and sulfur, the quantities of these cations cycling in ecosystems from soils to plants and back to soils are much larger than are annual inputs to and losses from ecosystems. The availability of cations in the soil solution is largely governed by exchange reactions and depends on the cation exchange capacity of the soil and its base saturation (see Chapter 3), which, in turn, is influenced by parent material and weathering characteristics. Calcium is an important structural component of plant and fungal cell walls. Its release and cycling therefore depends on decomposition in a way somewhat similar to that of nitrogen and phosphorus (Fig. 9.11). Potassium, on the other hand, occurs primarily in cell cytoplasm and is released through the leaching action of water moving through live and dead organic material. Magnesium and manganese are intermediate between calcium and potassium in their cycling characteristics.

Potassium limits plant production in some ecosystems, but calcium concentration in the soil solution of most ecosystems is so high that it is actively excluded by plant cells during the uptake process (see Chapter 8). Availability of calcium and other cations may be low enough to limit plant production on the old, highly weathered tropical soils or where acid rain has leached most available calcium from soils.

These cations have no gaseous phase, but atmospheric transfers of these elements (and of essential micronutrients) in dust can be an important pathway of loss from deserts and agricultural areas that experience wind erosion and an important input to the open ocean and to ecosystems on highly weathered parent materials. Cations can also be lost via leaching. Nitrate or other anions that are leached from ecosystems must be accompanied by cations to maintain electrical neutrality. Thus high nitrate leaching rates that occur in nitrogen-saturated sites or as a result of excessive nitrogen fertilization are accompanied by high losses of cations. The declines in forest production observed in Europe and the eastern United States in response to acid rain is at least partly a consequence of calcium and magnesium deficiencies induced by cation leaching (Schulze 1989, Aber et al. 1998, Driscoll et al. 2001).

Nonessential Elements

The cycling of nonessential elements is dominated by the balance between inputs from weathering, precipitation, and dust and outputs in leaching. Vegetation plays a relatively small role in the balance between inputs and outputs of elements such as chloride, mercury, and lead that are not required by organisms (Fig. 9.11). Consequently, external cycling of elements (ecosystem inputs and outputs) dominates the cycling of nonessential elements, whereas internal cycling through vegetation dominates the cycling of essential elements. The cycling of nonessential elements is therefore not strongly affected by successional changes in vegetation activity, whereas the losses of essential elements decline dramatically during early succession when organic matter and associated nutrients are accumulating in plant and microbial biomass (see Fig. 13.12) (Vitousek and Reiners 1975).

Interactions Among Element Cycles

Across broad gradients in nutrient availability, the supply rate of the most strongly limiting nutrient determines the rate of cycling through vegetation of all essential nutrients. Nutrient absorption by vegetation depends on a dynamic balance between the rate of supply of nutrients in soil and nutrient demands by vegetation (see Chapter 8). Most plants exhibit a limited range of element ratios. The element that most strongly limits plant growth has greatest impact on net primary production (NPP). Absorption of other elements is then adjusted to maintain relatively constant nutrient ratios in vegetation.

Despite this general synchrony of cycling rates across broad gradients in nutrient supply rates, there are important differences in factors governing supply of different elements. Some phases of the cycles are controlled by the same factors, such as uptake and release by vegetation, but other components of the cycles, such as input and losses, occur by separate pathways with different controlling factors. Consequently, essential elements do not cycle perfectly in tandem in ecosystems. Differential cycling leads to the opportunity for one element cycle to directly influence another. The cycling of nitrogen, phosphorus, and sulfur through vegetation and dead organic matter, for example, is not perfectly coupled. Each nutrient may be absorbed when it is most abundant and stored until it is required to support growth (see Chapter 8). Availabilities of these elements also differ in their amounts and timing, due to their differential dependence on inputs, on decomposition, and on reactions with soil minerals. Over long time scales, differential availabilities and plant requirements for these nutrient elements set up the potential for interactions among them.

The importance of interactions among cycles is illustrated by the changes in element cycling during long-term soil and ecosystem development. Newly exposed substrates such as fresh glacial till, lava, or sand dune contain substantial concentrations of phosphorus, major cations (Ca, Mg, K), and metals (Fe, Cu, Zn, Mo) in primary minerals but low concentrations of nitrogen. When exposed to water and acidity, these minerals weather and release elements into biologically available forms. The biological availability of these rock-derived elements increases rapidly during early primary succession. Over time, a fraction of these elements is lost via leaching and/or bound into insoluble or physically protected forms, and their availability declines.

Over the very long term (thousands to millions of years, depending on the element, the rock, and the climate), the primary minerals in soil are depleted, and most of the elements they supplied are lost or irreversibly bound. Phosphorus availability in particular decreases to the point that ecosystems on extremely old soils in humid regions are strongly phosphorus limited (see Fig. 3.5) (Walker and Syers 1976), until some geological disturbance such as a landslide provides a new source of unweathered minerals. This pattern of phosphorus decrease in old soils has been observed in developmental sequences in radically different climates, such as those found in New Zealand, Australia, and Hawaii (Vitousek and Farrington 1997).

In contrast to phosphorus, fixed nitrogen is nearly absent from most nonsedimentary rocks, so young soils must accumulate it from the atmosphere. Although sedimentary rocks contribute some nitrogen by weathering (0.01 to $2\,g\,m^{-2}\,yr^{-1}$) (Holloway et al. 1998), much of this may be lost in groundwater rather than entering ecosystems. The combination of substantial inputs of phosphorus and other elements with no nitrogen give nitrogen fixers a strong advantage early in soil and ecosystem development. Indeed, many early successional systems are dominated by symbiotic nitrogen fixers (Chapin et al. 1994). Where nitrogen fixers occur, nitrogen accumulates relatively quickly. Where nitrogen fixers are sparse or absent, nitrogen enters from deposition and accumulates slowly. Nitrogen continues to accumulate in ecosystems until nitrogen availability comes into approximate equilibrium with other resources, including phosphorus (Walker and Syers 1976). Nitrogen limits forest growth on young substrates in Hawaii, for example; phosphorus limits growth on old substrates; and nitrogen and phosphorus are both relatively available in intermediate-aged sites (Vitousek and Farrington 1997).

Why does phosphorus rather than other rock-derived elements limit biological processes in the long run? The major cations, especially calcium, are absorbed by organisms in much larger quantities than is phosphorus and are more readily leached from soils. On the Hawaiian sequence, rock-derived calcium, magnesium, and potassium virtually disappear within 100,000 years but do not limit forest production anywhere on the sequence (Vitousek and Farrington 1997). Atmospheric inputs of cations prevent these elements from becoming limiting. Marine-derived aerosols containing calcium, magnesium, and potassium are deposited on forests in Hawaii through rain and cloud droplets. Phosphorus concentrations in marine aerosols are low, however, because high phosphorus demands by marine organisms maintain a low concentration in surface waters. The atmospheric inputs of calcium are 10-fold less than weathering inputs in young sites but are more than 1000-fold greater than weathering inputs in older sites (Chadwick et al. 1999). In continental interiors, dust from semiarid and other sparsely vegetated areas is a major source of cations. Even in Hawaii, dust from Asia, over 6000 km away, is an important input of phosphorus, especially during glacial times, when vegetation cover was sparse and wind speeds were high (Chadwick et al. 1999) (Box 9.2). In situ weathering of parent material is

Box 9.2. Geochemical Tracers to Identify Source of Inputs to Ecosystems

Geochemical tracers have been used to identify dust and to determine its rate of input to the Hawaiian Islands. Hawaiian rocks are derived from Earth's mantle, whereas Asian dust comes from the crust. These two sources differ in the ratio of two isotopes of neodynium, in the ratio of europium to other lanthanide elements, and in the ratio of thorium to halfnium. All of these elements are relatively immobile in soils, so changes over time in the isotopic or elemental ratios can be used to calculate time-integrated inputs of Asian dust. Knowing the phosphorus content of the dust, it is then possible to calculate phosphorus inputs by this pathway. Atmospheric inputs of phosphorus are much lower than weathering for the first million years or more of soil development. However, by 4 million years, rock-derived phosphorus has nearly disappeared, and Asian dust provides most of the phosphorus input to the soil. The biological availability of phosphorus is low in old sites, but it would be much lower were it not for inputs of Asian dust, most of it transported more than 10,000 years ago (Chadwick et al. 1999).

therefore not always the dominant input of minerals to ecosystems.

Summary

Nutrients enter ecosystems through the chemical weathering of rocks, the biological fixation of atmospheric nitrogen, and the deposition of nutrients from the atmosphere in rain, windblown particles, or gases. Human activities have greatly increased these inputs, particularly of nitrogen and sulfur, through combustion of fossil fuels, addition of fertilizers, and planting of nitrogen-fixing crops. Unlike carbon, the internal recycling of essential plant nutrients is much larger than the annual inputs and losses from the ecosystem, producing relatively closed nutrient cycles.

Most nutrients that are essential to plant production become available to plants due to the microbial release of elements from dead organic matter during decomposition. Microbial exoenzymes break down the large polymers in particulate dead organic matter into soluble compounds and ions that can be absorbed by microbes or plant roots. The net mineralization of nutrients depends on the balance between the microbial immobilization of nutrients to support microbial growth and the secretion of nutrients that exceed microbial requirements for growth. The first product of nitrogen mineralization is ammonium. Ammonium can be converted to nitrate by autotrophic nitrifiers that use ammonium as a source of reducing power or by heterotrophic nitrifiers. Both plants and microbes use DON, ammonium, and nitrate in varying proportions as nitrogen sources, when their growth is nitrogen limited. Soil minerals and organic matter also influence nutrient availability to plants and microbes through exchange reactions (primarily with soil cations, except in some tropical soils that have a substantial anion exchange capacity), the precipitation of phosphorus with soil minerals, and the incorporation of nitrogen into humus.

Nutrients are lost from ecosystems through the leaching of elements out of the ecosystem in solution, emissions of gases, loss of nutrients adsorbed on soil particles in wind or water erosion, and the removal of materials in harvest. Human activities, as with nutrient inputs, often increase nutrient losses from terrestrial ecosystems.

Review Questions

1. What are the relative magnitudes of atmospheric inputs and mineralization from dead organic matter in supplying the annual nitrogen uptake by vegetation?
2. If Earth is bathed in dinitrogen gas, why is the productivity of so many ecosystems limited by availability of nitrogen? What is biological nitrogen fixation? What factors influence when and where it occurs?
3. What are the mechanisms by which nitrogen is deposited from the atmosphere into terrestrial ecosystems?
4. What are the major steps in the mineralization of litter nitrogen to inorganic forms? What microbial processes mediate each step and what are the products of each step? Which of these processes are extracellular and which are intracellular?
5. What ecological factors account for differences among ecosystems in annual net nitrogen mineralization? How does each of these factors influence microbial activity?
6. What determines the balance between nitrogen mineralization and nitrogen immobilization in soils?
7. What factors determine the balance between plant uptake and microbial uptake of dissolved organic and inorganic nitrogen in soils?
8. How do ammonium and nitrate differ in mobility in the soil? Why? How does this influence plant uptake and susceptibility to leaching loss?
9. What is denitrification and what regulates it? What are the gases that can be produced, and what are their roles in the atmosphere?
10. What is the main mechanism by which phosphorus enters ecosystems?
11. What factors control availability of phosphorus for plant uptake? Why is phosphorus availability low in many tropical soils?

12. Why are mycorrhizae so important for plant acquisition of phosphorus?
13. What is the main pathway of phosphorus loss from terrestrial ecosystems?

Additional Reading

Andreae, M.O., and D.S. Schimel. 1989. *Exchange of Trace Gases between Terrestrial Ecosystems and the Atmosphere*. John Wiley, New York.

Howarth, R.W., editor. 1996. *Nitrogen Cycling in the North Atlantic Ocean and Its Watershed*. Kluwer, Dordrecht.

Paul, E.A., and F.E. Clark. 1996. *Soil Microbiology and Biochemistry*. 2nd Edition. Academic Press, San Diego, CA.

Sala, O.E., R.B. Jackson, H.A. Mooney, and R.W. Howarth, editors. 2000. *Methods in Ecosystem Science*. Springer-Verlag, New York.

Schlesinger, W.H. 1997. *Biogeochemistry. An Analysis of Global Change*. Academic Press, San Diego, CA.

Tiessen, H., editor. 1997. *Phosphorus in the Global Environment: Transfers, Cycles and Management* [Scope Vol. 54]. John Wiley, New York.

Vitousek, P.M., J.D. Aber, R.W. Howarth, G.E. Likens, P.A. Matson, D.W. Schindler, W.H. Schlesinger, and G.D. Tilman. 1997. Human alteration of the global nitrogen cycle: Sources and consequences. *Ecological Applications* 7:737–750.

Vitousek, P.M., and R.W. Howarth. 1991. Nitrogen limitation on land and in the sea: How can it occur? *Biogeochemistry* 13:87–115.

10
Aquatic Carbon and Nutrient Cycling

Aquatic ecosystems differ radically from their terrestrial counterparts in physical environment and therefore in controls over ecosystem processes. This chapter describes the major differences in carbon and nutrient cycling between terrestrial and aquatic ecosystems.

Introduction

The same general principles govern carbon and nutrient cycling in aquatic and terrestrial ecosystems, but the physical differences between water and air result in radically different ecological controls. Many of the basic principles of ecosystem ecology were developed in aquatic ecosystems, including the concepts of trophic dynamics and energy flow (Lindeman 1942), the interactions among biogeochemical cycles (Redfield 1958), and species effects on ecosystem processes (Redfield 1958, Carpenter and Kitchell 1993). These concepts are broadly applicable to terrestrial ecosystems. Nonetheless, many of the ecological patterns and dynamics of aquatic and terrestrial ecosystems are quite different, largely due to the differences between air and water as the basic support medium. In this chapter we explore the consequences of this physical difference and describe the broad similarities and differences in carbon and nutrient cycling between aquatic and terrestrial ecosystems.

Aquatic ecosystems are just as structurally and functionally diverse as are terrestrial ecosystems. We therefore initially focus our aquatic–terrestrial comparison on **pelagic** (open water) marine ecosystems, which differ most dramatically from terrestrial ecosystems and then discuss in less detail the differences between marine ecosystems and lakes and streams.

Ecosystem Properties

The differences in physical properties between water and air result in fundamental differences in structure and functioning between aquatic and terrestrial ecosystems. Due to the greater density of water than air, the physical support for photosynthetic organisms is greater in water than on land (Table 10.1). The primary producers in pelagic ecosystems are therefore microscopic algae (**phytoplankton**) that float near the water surface, where light availability is greatest. This contrasts with terrestrial plants, which require elaborate support structures to raise their leaves above neighbors. Plants are the major habitat-structuring feature in terrestrial ecosystems. Their physical structure governs the patterns of physical environment, organism activity, and ecosystem processes. In the open ocean, however, the environment is physically structured by vertical gradients in light, temperature, oxygen, and salinity. Ocean currents and deep circulation frequently remove phyto-

TABLE 10.1. Basic properties of water and air that influence ecosystem processes.

Property	Water	Air	Ratio of water to air
Oxygen concentration (ml L^{-1})	7.0	209.0	1:30
Diffusion coefficient (mm s^{-1})			
Oxygen	0.00025	1.98	1:8000
Carbon dioxide	0.00018	1.55	1:9000
Density (kg L^{-1})	1.000	0.0013	800:1
Viscosity (cP)	1.0	0.02	50:1
Heat capacity (cal L^{-1} (°C)$^{-1}$)	1000.0	0.31	3000:1

Data from Moss (1998).

plankton from the **euphotic zone**, the uppermost layer of water where there is enough light to support photosynthesis. Coping with this frequent disturbance requires rapid cell division, small size, and, for larger organisms, the capacity to swim. The small size of phytoplankton results in a high surface to volume ratio that maximizes their effectiveness in absorbing nutrients. Many marine plankton are particularly small (nanoplankton are 2 to 20 μm in diameter, and picoplankton are less than 2 μm in diameter) and have a competitive advantage where nutrients are extremely dilute. Picoplankton, for example, account for half of the plankton biomass of the highly nutrient-impoverished tropical oceans (Valiela 1995). The size and lifespan of marine organisms increases going up the food chain (Fig. 10.1).

The size of aquatic organisms determines their feeding strategy. Water is a polar molecule that sticks to the surface of organisms. The movement of small organisms and particles is impeded by these viscous forces. Large organisms, in contrast, can swim, and their speed is largely determined by inertia. The Reynolds number (*Re*) is the ratio of inertial to viscous forces and is a measure of the ease with which organisms can move through a viscous fluid like water.

$$Re = \frac{lv}{V_k} \tag{10.1}$$

The movement of organisms through water is not strongly impeded for organisms with a large length (l) and velocity (v) and under conditions of low kinematic viscosity (V_k) (Fig. 10.2). Small planktonic organisms must deal with life at a low Reynolds number, where viscous forces are much stronger than inertial forces. At these small sizes, swimming and filter feeding are

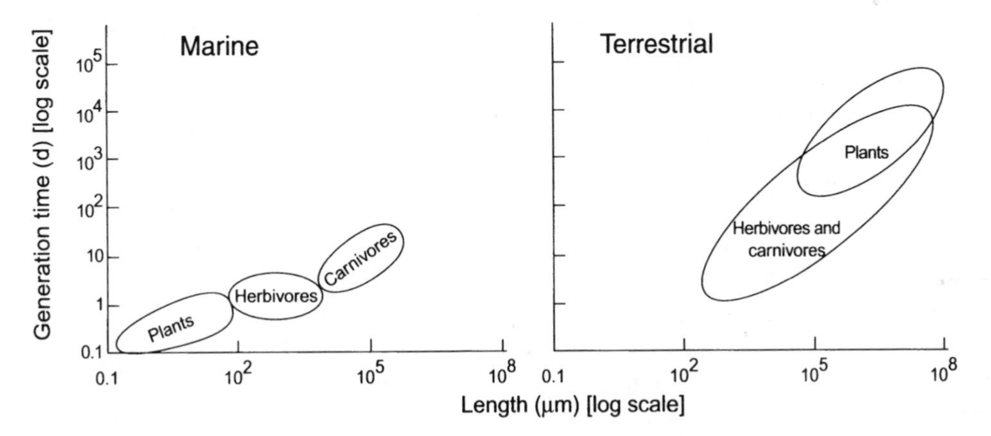

FIGURE 10.1. Body size and generation time for organisms in the ocean (Steele 1991) and on land of dominant plants, herbivores, and carnivores. In the ocean the dominant plants (picoplankton and nanoplankton) are generally smaller than the herbivores that feed on them, whereas on land, the dominant plants are often as large or larger than the herbivores that eat them.

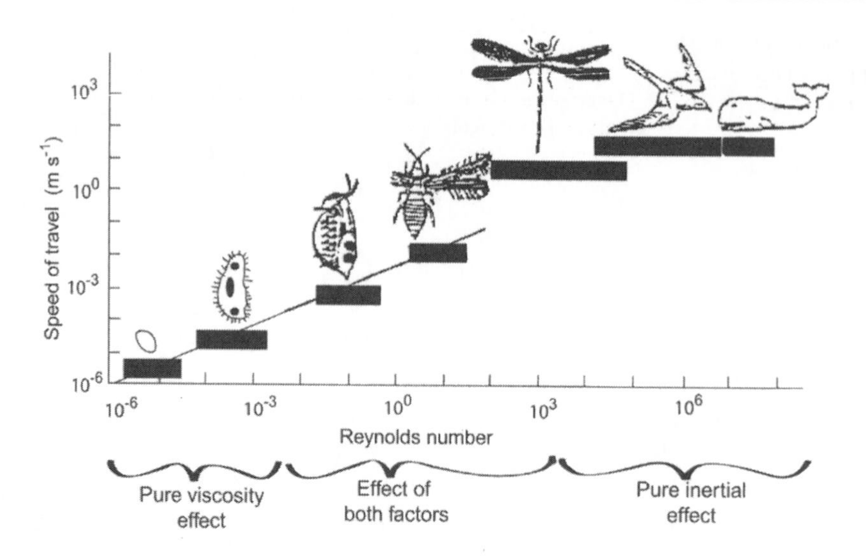

FIGURE 10.2. Range of Reynolds numbers for organisms of different lengths and speeds. Small organisms like phytoplankton have small Reynolds numbers and derive their nutrition by diffusion. As size and Reynolds number increase, nutrition based on movement (filter feeding and swimming) become progressively more important. (Redrawn with permission from Halsted Press; Schwoerbel 1987.)

energetically expensive, so diffusion is the major process that moves nutrients to the cell surface, just as with fine roots growing in the soil solution.

The small size and lack of nonphotosynthetic support structures in marine phytoplankton mean that marine primary producers require relatively little biomass to support a given photosynthetic capacity. The average primary producer biomass per unit area on land, for example, is 660-fold greater than in the ocean, although the average net primary production (NPP) per unit area on land is only 5-fold greater than in the ocean (Table 10.2) (Cohen 1994). Phytoplankton biomass of oceans and lakes turns over 20 to 40 times per year, or even daily under conditions that are favorable for growth, whereas turnover for terrestrial plant biomass often requires years to decades (Valiela 1995).

The air that surrounds terrestrial organisms delivers oxygen and other gases orders of magnitude more rapidly than occurs in water. The surface ocean water, for example, has an oxygen concentration 30-fold lower than in air (Table 10.1), and aquatic sediments are much more likely to be anaerobic than are terrestrial soils. Aquatic organisms therefore exhibit a variety of adaptations to acquire oxygen and withstand anaerobic conditions, whereas on land the acquisition of water and the avoidance or tolerance of desiccation are more common evolutionary themes.

The buoyancy of water produces an environment that is rich in small particles, including algal cells and suspended particles of detritus. **Filter feeders** are organisms that feed on suspended particles through use of a diverse array of tools, including hairy appendages on legs or mouth parts, sticky secretions, and silken nets. Filter feeders have no counterpart in terrestrial

TABLE 10.2. Characteristics of oceans and continents.

Unit	Oceans	Continents
Surface area (% of Earth)	71	29
Volume of life zone (% of Earth)	99.5	0.5
Living biomass (10^{15} g C)	2	560
Living biomass (g m^{-2})	5.6	3700
Dead organic matter (10^3 g m^{-2})	5.5	10
Net primary production (g C m^{-2} yr^{-1})	69	330
Residence time of C in living biomass (yr)	0.08	11.2

Data from Cohen (1994).

TABLE 10.3. Generic metazoan diversity of land and oceans.

Phyla	Ocean benthic	Ocean pelagic	Fresh-water	Symbiotic	Terrestrial
Total (33)	27	11	14	15	11
Endemic	10	1	0	4	1

Data from May (1994).

ecosystems (Gee 1991). Most filter feeders are about 100-fold larger than the suspended particles (**seston**) on which they feed. Filter feeders are often selective in the size of particles they ingest but process algae, bacteria, and suspended particles of organic and inorganic materials relatively indiscriminately. Much of this food may therefore be of relatively low quality, and substantial quantities of water must be processed to acquire sufficient energy and nutrients to support growth.

There are many more species on land than in the oceans but the broad phyletic diversity for coping with life is greater in the ocean. About 76% of all species occur on land, and most of these are insects. Of the 15% of the species that are marine, most are **benthic** animals of the ocean sediments. There are only about 20,000 photosynthetic species in aquatic ecosystems in contrast to the 300,000 species of terrestrial plants, and there are very few aquatic insects (Falkowski et al. 1999). At higher taxonomic levels, however, 80% the multicellular genera occur in the sea, and only 20% are on land (Table 10.3) (May 1994). The greater diversity of genera and phyla in the ocean than on land could reflect its longer evolutionary history, giving organisms more time to try out different fundamental body plans and functional types (May 1994). The larger number of species on land than in the ocean could reflect the greater heterogeneity and potential for spatial isolation in terrestrial habitats.

Life evolved in the sea. From there, both plants and multicellular animals moved to fresh waters and then to land (Moss 1998). Several groups of plants and animals subsequently followed the reverse path from land to fresh water to oceans. These transitions have occurred many times, indicating that the physiological adjustments required are not insurmountable on evolutionary time scales.

The marine environment is slightly more **saline** (salty) than the internal body fluids of marine organisms, so organisms must minimize loss of water and gain of salts to maintain ionic balance. Movement of organisms from marine to fresh water reverses this osmotic gradient and intensifies the costs of osmoregulation. Terrestrial plants and animals confront two contrasting problems. First, their source of water is usually fresh, requiring more energy to maintain osmotic gradients. Second, they are exposed to an aerial environment that promotes water loss and dehydration. One great advantage to life on land is greater availability of oxygen and the greater energy provided by aerobic metabolism. Disadvantages include greater dehydration and lower buoyancy of air than water (Table 10.1).

The benthic environment of marine sediments differs radically from terrestrial soils in its low oxygen availability. Oxygen concentration in deep waters is much lower than in air, and oxygen diffusion into water-saturated sediments is much slower than through the air-filled pores of terrestrial soils. Mixing of sediments by benthic organisms plays a critical role in promoting oxygen flux into sediments and therefore benthic decomposition. Redox reactions involving electron acceptors other than oxygen play a key role in the metabolism of benthic organisms and therefore in the patterns of carbon and nutrient cycling (see Chapter 3) (Valiela 1995).

The wide range in ecosystem structure among aquatic ecosystems reflects variations in physical environment. Most nonpelagic aquatic ecosystems are intermediate in structural properties between terrestrial and pelagic ecosystems. In the **littoral zone**, where the ocean and land meet, organisms can reduce the probability of being swept away by attaching to substrates. This gives rise to a diverse array of shallow-water ecosystems, including coral reefs, seagrass beds, and kelp forests. In ecosystems where physical attachment is possible, phytoplankton, benthic algal mats, multicellular algae such as sea lettuce (e.g., *Ulva*) and kelp (e.g., *Laminaria*), and vascular plants like eelgrass (*Zostera*) occur in various combinations. The relative abundance of different types of primary producers depends on many factors, including nutrient availability, water depth, the stability of substrates, and disturbance regime. Nutrient addition, for example, favors rapidly growing phytoplankton, which reduce the light available to multicellular algae at depth. Intermediate levels of disturbance and presence of stable substrates favor multicellular primary producers (Sousa 1985).

Oceans

The large area and low productivity per unit area of ocean cancel out, so the ocean contributes nearly half (about 40%) of global NPP. Although oceans cover 70% of Earth's surface, the average NPP per unit area is only 20% of that on land (Table 10.2). Aquatic productivity, however, is highly variable, just as on land. The most productive aquatic ecosystems, such as coral reefs, kelp forests, and eutrophic lakes, can be at least as productive as the most productive terrestrial ecosystems (Fig. 10.3). NPP in the open ocean, which accounts for 90% of the ocean area, however, is similar to that of terrestrial deserts and tundra. Because of its large area, the open ocean accounts for 60% of marine production, with picoplankton accounting for about 90% of this production (Valiela 1995).

Carbon and Light Availability

Photosynthesis is seldom carbon limited in the ocean. In marine pelagic ecosystems, for example, only 1% of the carbon in a given water volume is involved in primary production, whereas the nitrogen in this water may cycle through primary production 10 times a

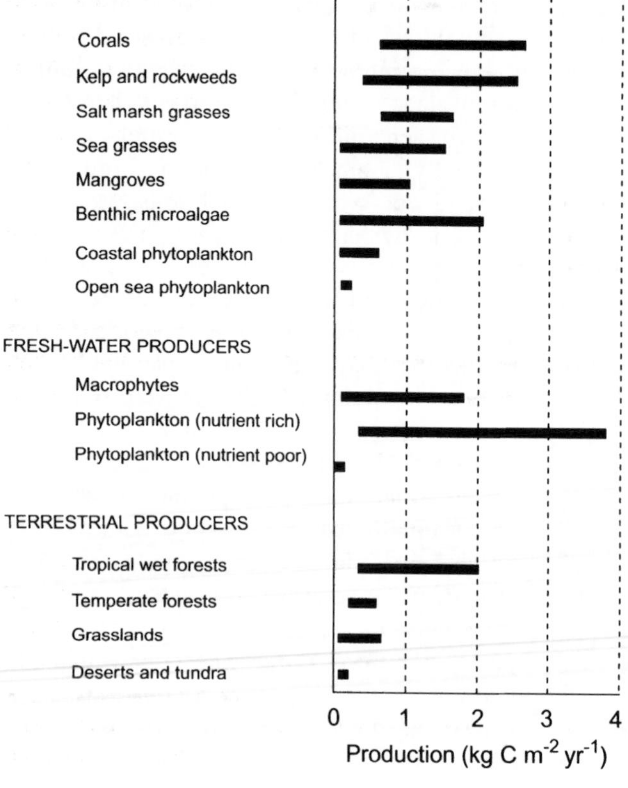

FIGURE 10.3. NPP of selected marine, fresh-water, and terrestrial ecosystems. Marine and fresh-water ecosystems exhibit the same range of NPP that occurs on land, but unproductive marine ecosystems (the open ocean) are much more widespread. (Redrawn with permission from Springer-Verlag; Valiela 1995.)

year (Thurman 1991). One reason for the lack of carbon limitation in the ocean is that inorganic carbon is available in several forms, including CO_2, bicarbonate (HCO_3^-), carbonate (CO_3^-), and carbonic acid (H_2CO_3). When CO_2 dissolves in water, a small part is transformed to carbonic acid, which in turn dissociates to bicarbonate, carbonate, and H^+ ions with a concomitant drop in pH.

$$H_2O + CO_2 \leftrightarrow H_2CO_3 \leftrightarrow H^+ + HCO_3^- \leftrightarrow 2H^+ + CO_3^{2-} \quad (10.2)$$

As expected from these equilibrium reactions, the predominant forms of inorganic carbon are free CO_2 and carbonic acid at low pH (the equation driven to the left), soluble bicarbonate at about pH 8 (typical of ocean waters), and carbonates at high pH (equation driven to the right). Bicarbonate accounts for 90% of the inorganic carbon in most marine waters. Phytoplankton in pelagic ecosystems use primarily CO_2 as their carbon source. CO_2 is then replenished from bicarbonate (Eq. 10.2). Some marine algae in the littoral zone, such as the macroalga *Ulva*, also use bicarbonate.

Water, algae, and suspended and dissolved materials absorb light that enters aquatic ecosystems, whereas light on land is absorbed primarily by the plant canopy. Light energy fuels photosynthesis in the same way on land and in the water. On land this energy is modified (absorbed or transmitted) primarily by vegetation as it moves down through the canopy (see Chapter 5). In water, however, the smaller biomass of plant cells and the significant absorption by water, dissolved organics, and suspended particles cause the medium itself to contribute substantially to the exponential decrease in light from the surface to depth. Light is depleted rapidly in coastal waters, where plankton, suspended sediments, and dissolved organic compounds are relatively abundant. The euphotic zone, where there is sufficient light to support phytoplankton growth, extends only to about 200 m in the clear waters of the open ocean (Fig. 10.4). Most of the ocean volume therefore cannot support the growth of primary producers whose carbon fixation depends on light.

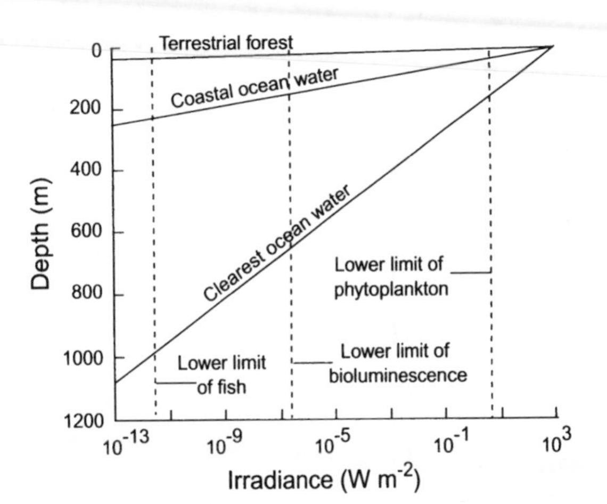

FIGURE 10.4. Light availability at different depths in forests and in coastal and open oceans (Chazdon and Fetcher 1984a). (Modified with permission from Springer-Verlag; Valiela 1995.)

In aquatic ecosystems blue light predominates at depth whereas red light predominates at depth in terrestrial canopies. In clear water there is an inverse relationship between wavelength and light transmission; blue light, which has a short wavelength and high energy, penetrates to greatest depth (Fig. 10.5). Turbid waters both absorb and scatter light, causing the longer wavelengths to become more rapidly depleted with depth. Humic compounds, for example, absorb ultraviolet (UV) and blue wavelengths. In contrast to aquatic systems, the high density of chlorophyll in terrestrial plant canopies selectively depletes the short-wavelength, high-energy blue light, so light at the bottom of a forest canopy is enriched in red compared to incoming solar radiation (Fig. 10.5). Green light, which is not absorbed efficiently by chlorophyll, also penetrates to the forest floor to a greater extent than does blue light. Eutrophic lakes with high chlorophyll concentrations have profiles of light quality more similar to those of forests than to the open ocean. Aquatic and terrestrial ecosystems thus differ in both the quantity and the quality of light that penetrates to depth.

Marine phytoplankton are like terrestrial shade plants. Photosynthesis saturates at rela-

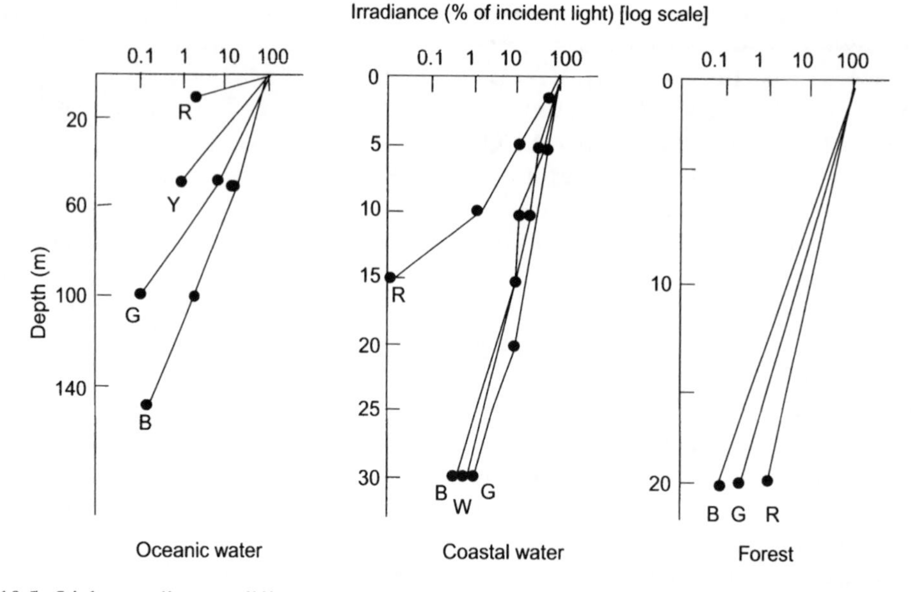

FIGURE 10.5. Light quality at different depths in ocean and coastal waters and in forests (Chazdon and Fetcher 1984a). R, red; Y, yellow; G, green; B, blue; W, white. (Modified with permission from Springer-Verlag; Valiela 1995.)

tively low light intensity, from 5 to 25% of full sun, depending on the algal group (Valiela 1995). Photosynthesis declines at higher light levels, so maximum photosynthesis usually occurs at about 10m depth on sunny days. One reason that phytoplankton function as shade plants is that they mix vertically through the water column, so an individual cell spends relatively little time near the surface. Photosynthetic acclimation to high light requires about 12h, which is probably longer than the time that most phytoplankton would be exposed to high light. In the ocean and clear lakes, UV-B radiation may also contribute to low photosynthetic rates in surface waters, raising questions about the consequences of ozone holes and increased UV-B at high latitudes. Light appears to limit ocean and lake production at the water surface, primarily beneath ice cover or during winter at high latitudes when the sun angle is low. At depth, light limits production in all aquatic habitats.

Primary producers exhibit a greater diversity of pigment types in the ocean than on land, presumably related to the complex light environment of water. Red algae, which are abundant at depth in tropical oceans, and brown algae like kelp, which are abundant at depth in cool temperate oceans, have pigments that absorb the blue light that is available to them. Surface algae tend to have green pigments. These differences in algal pigments may also be adaptations to light quantity per se.

Reduced sulfur compounds provide the energy for carbohydrate synthesis from CO_2 in some anaerobic aquatic habitats. Although most organisms depend on carbon fixed through light-dependent photosynthesis, some bacteria use the energy of reduced sulfur compounds to reduce CO_2 to form organic compounds. Entire ecosystems are built on such a base near **hydrothermal vents** in zones of seafloor spreading in midocean regions. Although hydrothermal vents account for only a tiny fraction of total ocean production, they support unique communities and complex food webs that are completely independent of energy input from the sun (Karl et al. 1980). Similar **chemosynthesis** occurs in anaerobic sediments, but these sulfur-dependent oxidation–reduction reactions usually account for only a small fraction of the total carbon budget of these environments.

Nutrient Availability

The euphotic zones of the ocean are frequently nutrient poor. In pelagic ecosystems of the open ocean, photosynthetic cells in the euphotic zone are spatially separated from the benthic supply of nutrients. This contrasts with terrestrial ecosystems in which transport tissues carry nutrients directly from the soil to photosynthetic cells in the canopy. The small size of phytoplankton causes diffusion to be the major process that moves nutrients to the cell surface, as described earlier. Production in these pelagic ecosystems is therefore generally nutrient limited, and algal uptake maintains low nutrient concentrations in the water of the euphotic zone (Fig. 10.6). Some phytoplankton swim (flagellates or ciliates) or sink (through changes in buoyancy) to reduce nutrient limitation by diffusion. Swimming can increase nutrient uptake in microplankton by 50 to 200%, but picoplankton cannot swim fast enough to overcome diffusion (Valiela 1995). Only large-celled algae can sink fast enough to overcome nutrient limitation by diffusion.

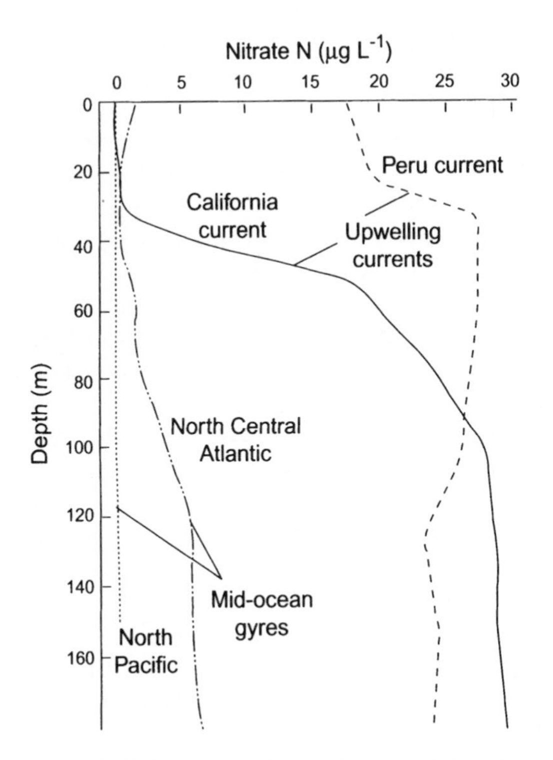

FIGURE 10.6. Depth profiles of nitrate and phosphate in midocean gyres and upwelling zones of the ocean. (Redrawn with permission from Saunders; Dugdale 1976.)

The nature of nutrient limitation in the open ocean is a complex consequence of element interactions. The open ocean is a nutritional desert, remote from sources of nutrient input. In the open ocean, phosphorus appears to be the master element that ultimately limits the productive capacity of the oceans (Tyrrell 1999, Sigman and Boyle 2000). Its supply to the open ocean depends on products of rock weathering that are transported to the ocean in rivers, deposited as dust from neighboring continents, or mixed upward from the deep ocean. Whenever phosphorus availability increases, nitrogen fixers such as cyanobacteria generally add nitrogen until phosphorus again limits their production. The open ocean, however, seldom builds up the high nitrate concentrations found in lakes, and phytoplankton production frequently responds more strongly to nitrogen than to phosphorus in short-term experiments (Fig. 10.7) (Valiela 1995, Tyrrell 1999). Ocean water converges strongly on a relatively constant N : P ratio of 14 to 16, suggesting that both nitrogen and phosphorus frequently limit production. This **Redfield ratio** reflects the relative requirement of the two elements by phytoplankton and most other organisms on Earth. Nitrogen limitation is widespread in coastal oceans, perhaps reflecting denitrification that occurs in anaerobic sediments (Falkowski et al. 1999).

Trace elements—which are cofactors for nitrogenase, the nitrogen-fixing enzyme, and which are also required by other phytoplankton—often limit ocean productivity. In the subequatorial gyres, the Subarctic Pacific, and the Southern Ocean surrounding Antarctica, surface nitrogen and phosphorus concentrations are relatively high, and about half of the available nitrogen and phosphorus are mixed to depth without being used to support primary production. In these regions, production fails to respond to addition of these nutrients, leading to a syndrome known as high-nitrogen, low-chlorophyll (HNLC) syndrome (Valiela 1995, Falkowski et al. 1999). Large-scale iron-addition experiments in these regions have

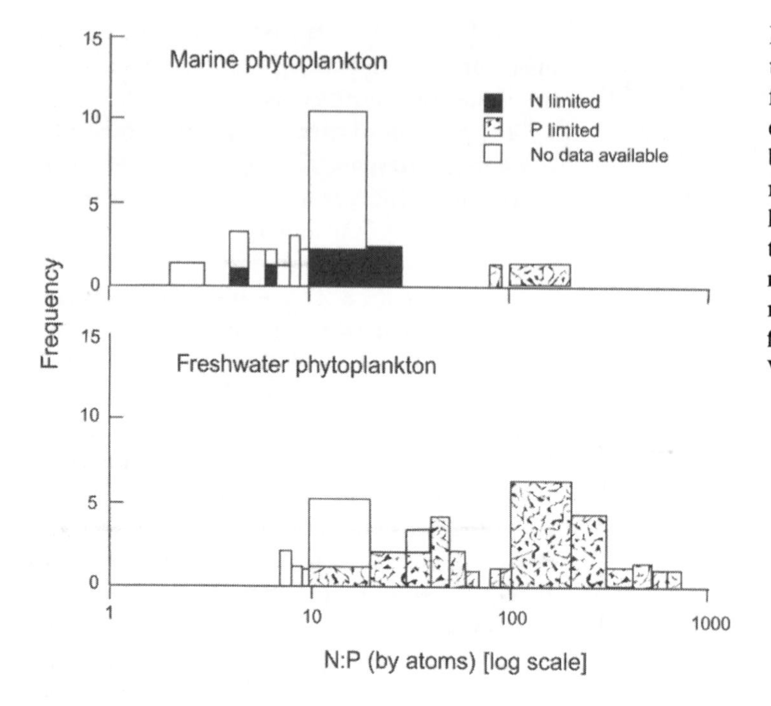

FIGURE 10.7. Frequency distribution of the N:P ratio in marine and fresh-water phytoplankton. Nutrient addition experiments (shaded bars) indicate that the high N:P ratios in lakes reflect phosphorus limitation of algal growth, whereas the generally low N:P ratios of marine phytoplankton commonly reflect nitrogen limitation. (Modified with permission from Springer-Verlag; Valiela 1995.)

caused phytoplankton blooms large enough to be seen from satellites, indicating that iron limits the capacity of phytoplankton to use nitrogen and phosphorus. During glacial periods there may have been 10-fold greater input of iron- and phosphorus-bearing dust to the oceans, thus stimulating ocean productivity and in turn lowering atmospheric CO_2 concentrations (Falkowski et al. 1999). The key role of iron in regulating production in some sectors of the open ocean has led to the suggestion that large-scale iron fertilization might stimulate ocean production sufficiently to scavange large amounts of CO_2 from the atmosphere and sequester it in deep oceans in the form of dead organic matter. The iron-addition experiments, however, show that this stimulation of production is relatively short-lived, presumably because other elements quickly become limiting to production, as soon as the iron demands of phytoplankton are met. Grazing is another factor that contributes to low phytoplankton biomass and productivity in portions of the open ocean. In some HNLC zones of the ocean there is simply not enough phytoplankton biomass to use the nutrients that are available (Valiela 1995).

The strong nutrient limitation and lack of CO_2 limitation of productivity in most of the world's oceans make it unlikely that marine productivity will respond directly to increasing atmospheric CO_2. Nutrient limitation of marine production, however, makes these ecosystems potentially vulnerable to anthropogenic nutrient inputs. Estuaries, for example, have been substantially altered by the large nutrient inputs from agricultural lands. Runoff and sewage effluents have substantially altered coastal ecosystems near heavily populated or agricultural areas (Howarth et al. 1996). Oxygen depletion by decomposers in the water column of the Mississippi Delta, for example, has created a large dead zone in the Gulf of Mexico. Although the impacts of nutrients on the open ocean may be more subtle and difficult to detect because most pollution sources are remote, they could, over the long term, be important because of the high degree of nutrient limitation of pelagic ecosystems and their large aerial extent.

Ocean productivity is ultimately limited by the rate of nutrient supply from the land or deep ocean waters. For this reason, productivity is greater in coastal waters than in the open ocean. Tidal mixing of sediment nutrients into the water column and oxygenation of the water column contribute to the high productivity of estuaries and intertidal and near-shore marine ecosystems (Nixon 1988). The energy input from waves, for example, exceeds that from the sun in some intertidal areas. Coral reefs are among the most productive ecosystems on Earth (Fig. 10.3). Frequent tidal flushing supplies nutrients to algae that grow on the surfaces of dead corals. These algae have high turnover rates because they are constantly grazed by fish. The biomass of algae in this ecosystem is therefore small, just like the biomass of phytoplankton in pelagic ecosystems.

In pelagic ecosystems, upwelling near the west coasts of continents provides the greatest rate of nutrient supply. Upwelling supports some of Earth's major fisheries off Peru, northwest Africa, eastern India, southwest Africa, and the western United States (Valiela 1995) (Fig. 10.6). In these areas, Coriolis forces cause winds and surface waters to move offshore (see Chapter 2). These surface waters are replaced by nutrient-rich waters from depth. Upwelling also occurs in the open ocean where major ocean currents diverge. This happens, for example, in the equatorial Pacific, where ocean currents diverge to the north and south and in the Southern Ocean, the North Atlantic, and the North Pacific (Valiela 1995). These regions have relatively high nutrient availability and productivity.

Vertical gradients in water density also influence the effectiveness of nutrient transport from subsurface to surface waters. In the central subtropical oceans, where upwelling does not occur, the strong vertical temperature gradient results in an extremely stable thermocline, in which low-density warm water is underlain by high-density cold water (see Chapter 2). This stable stratification of water minimizes the effectiveness of vertical mixing by waves and ocean currents, so nutrient availability and productivity of the subtropical ocean is extremely low. As latitude increases, however, surface temperature declines. This weakens the vertical density gradient, so storm waves and currents are more effective in mixing deep nutrient-rich waters to the surface. The strong westerly winds and storm tracks associated with the polar jet also contribute to effective mixing in high-latitude oceans. Temperate and polar ocean waters are therefore more nutrient-rich and productive than are tropical oceans.

The upward mixing of nutrients is greatest during winter, when surface waters are coldest, and the vertical stratification is least stable. Winter is also the time of year when strong equator-to-pole heating gradients generate the strongest winds (see Chapter 2). High-latitude oceans typically experience a **bloom** of phytoplankton in spring, after winter mixing has occurred and when light increases sufficiently to support high photosynthetic and growth rates. The bloom ends when nutrients are depleted by production, and most algae have been consumed by zooplankton grazers. The high productivity of high-latitude oceans supports rich fisheries, although many of these have been depleted by overfishing. The latitudinal variation in pelagic productivity also explains several other interesting ecological patterns, such as the annual migration of many whales and sea birds between the Antarctic and the Arctic Oceans to capitalize on spring blooms of high-latitude productivity. In addition, a high proportion of fish species at high latitudes have an **anadromous** life history, in which they exploit the productive marine environment to support growth during the adult phase and use the relatively predator-free freshwater environment to reproduce. This anadromous life history strategy is increasingly favored as latitude increases because marine productivity increases with increasing latitude, whereas terrestrial productivity declines with increasing latitude (Gross et al. 1988).

Carbon and Nutrient Cycling

Herbivory accounts for a threefold greater proportion of the carbon and nutrient transfer in pelagic than in terrestrial ecosystems (Fig. 10.8) (Cyr and Pace 1993). Although marine phytoplankton exhibit a range of structural and chemical defenses against herbivores, just like terrestrial plants, phytoplankton are relatively

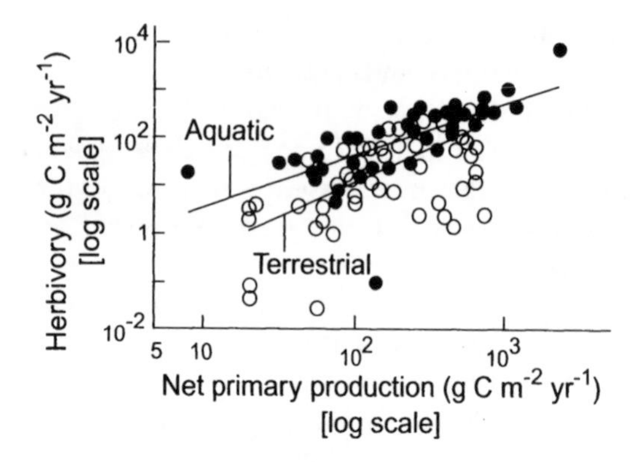

FIGURE 10.8. Comparative productivity and herbivory rates between aquatic and terrestrial ecosystems. (Redrawn with permission from *Nature*; Cyr and Pace 1993.)

digestible due to their lack of structural support tissue. The resulting high rate of herbivory by zooplankton in pelagic ecosystems transfers a large proportion of primary producer carbon from plants to animals. Herbivory is strongly correlated with NPP, so the secondary productivity of marine fisheries and other components of secondary production depend strongly on NPP (see Chapter 11). Food webs in the three-dimensional pelagic environment are frequently longer and more complex than those in the two-dimensional benthic environment (Thurman 1991). Because predation is strongly size dependent, the wide range of sizes of pelagic plankton (0.1 to 2000 μm) also contributes to long food chains and complex webs in pelagic ecosystems.

Decomposition within the euphotic zone recycles nutrients and contributes energy to higher trophic levels. Phytoplankton release about 10% (5 to 60%) of their production as exudates into the water column (Valiela 1995), a proportion of NPP similar to that which terrestrial plants transfer to the soil as root exudates and to support mycorrhizal fungi. Zooplankton spill phytoplankton cytoplasm into the water, as they eat, and excrete their own waste products. Pelagic bacteria break down the resulting organic compounds and mineralize the associated nutrients, which are then available to primary producers. This decomposition occurs relatively quickly because the carbon substrates are mostly labile organic compounds of low molecular weight with a low C:N ratio (Fenchel 1994). This contrasts with the structurally complex, carbon-rich compounds (cellulose, lignin, phenols, tannins) that dominate terrestrial detritus. Viruses play an important role in planktonic food webs, lysing bacteria and algae. Viral lysis may account for 5 to 25% of bacterial mortality in pelagic ecosystems (Valiela 1995). Pelagic bacteria and viruses are grazed by small (nanoplankton) flagellate protozoans, which in turn are eaten by larger zooplankton. The detritus-based food web (see Chapter 11) is therefore tightly integrated with the plant-based trophic system in pelagic food webs and contributes substantially to the energy and nutrients that support marine fisheries. This **microbial loop** in pelagic ecosystems recycles most of the carbon and nutrients within the euphotic zone, so nutrients are recycled through food webs multiple times before being lost to depth (Fig. 10.9).

Pelagic carbon cycling pumps carbon and nutrients from the ocean surface to depth (Fig. 10.9). Although most of the planktonic carbon acquired through photosynthesis returns to the environment in respiration, just as in terrestrial ecosystems, marine pelagic ecosystems also transport 5 to 20% of the carbon fixed in the euphotic zone into the deeper ocean (Valiela 1995). This process is called the **biological pump**. The carbon flux to depth correlates closely with primary production, so the environmental controls over NPP largely determine the rate of carbon export to the deep ocean. This carbon export consists of particulate dead organic matter (feces and dead cells) and the carbonate exoskeletons that provide structural rigidity to many marine organisms. Carbonate accounts for about 25% of the biotically fixed carbon that rains out of the euphotic zone (Houghton et al. 1996). The carbonates redissolve under pressure as they sink to depth. Over decades to centuries, some of this carbon in deep waters recirculates to the surface through upwelling and mixing. This long-term circulation pattern will cause the effects of the current increase in atmospheric CO_2 to influence marine biogeochemistry for centuries after its impacts are felt in terrestrial ecosys-

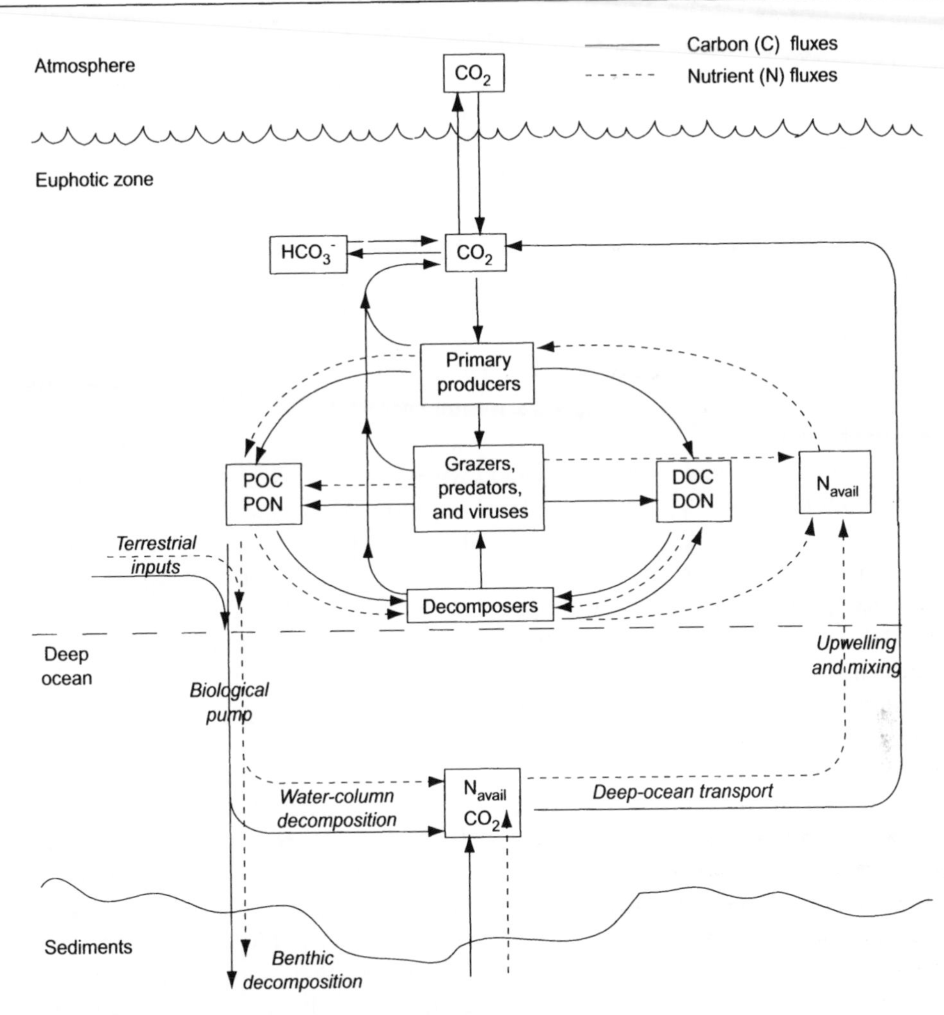

FIGURE 10.9. Major pools and net fluxes of carbon (C) and nitrogen (N) in the ocean. Phosphorus and other essential nutrients cycle in patterns similar to that shown for nitrogen. CO_2 in the euphotic zone equilibrates with bicarbonate in ocean water and with CO_2 in the atmosphere. CO_2 is depleted by photosynthesis by primary producers and is replenished by respiration of organisms and by upwelling and mixing from depth. Grazers consume primary producers and bacteria and are eaten by other animals and lysed by viruses. Each of these organisms releases dissolved and particulate forms of carbon and nitrogen (DOC, DON; POC, PON). Animals and decomposers also release available nitrogen (N_{avail}). DOC is consumed by bacteria, and available nutrients are absorbed by primary producers. Particulate carbon and nutrients produced by feces and dead organisms sink from the euphotic zone toward the sediments; as they sink, they decompose, releasing CO_2 and available nutrients. Benthic decomposition also releases CO_2 and available nutrients. Bottom waters, which are relatively rich in CO_2 and available nutrients, eventually return to the surface through mixing and upwelling; this augments the supply of available nutrients in the euphotic zone. DOC, dissolved organic carbon; DON, dissolved organic nitrogen; POC, particulate organic carbon; PON, particulate organic nitrogen.

tems. The net effect of the biological pump is to move carbon from the atmosphere to the deep waters and to ocean sediments. Carbon accumulation in midocean sediments is slow (about 0.01% of NPP) because most decomposition occurs in the water column before organic matter reaches the sediments and because these well-oxygenated sediments support decomposition of much of the remaining carbon (Valiela 1995).

The biological pump that transports carbon to depth carries with it the nutrients contained in dead organic matter. Decomposition continues as particles sink, so much of the decomposition occurs in the water column rather than in the sediments, particularly in the deep oceans. The rapid (about weekly) turnover of carbon and nutrients in phytoplankton in the euphotic zone (Falkowski et al. 1999) makes these nutrients vulnerable to loss from the ecosystem and contributes to the relatively open nutrient cycles of pelagic ecosystems. The longer-lived and larger primary producers on land can store and internally recycle nutrients for years. This reduces the proportion of nutrients that are annually cycled and contributes to the tightness of terrestrial nutrient cycles.

Benthic decomposition is more important on continental shelves than in the deep ocean because the coastal pelagic system is more productive, generating more detritus. In addition, the dead organic matter has less time to decompose before it reaches the sediments. Here oxygen consumption by decomposers depletes the oxygen enough that decomposition becomes oxygen limited, and organic matter accumulates or becomes a carbon source for methanogens and denitrifiers.

Lakes

Lakes consist of a range of ecosystem types, from pelagic systems to wetlands dominated by vascular plants (Wetzel 2001). The centers of deep lakes are structurally similar to marine ecosystems with discrete pelagic and benthic systems; phytoplankton are the major primary producers, and zooplankton are the major herbivores. The littoral zone of lakes generally experiences less disturbance from wave action and currents than in marine systems. This allows mats of algae to grow directly on lake sediments, even where light is only 0.1% of that present at the surface. The littoral zone of lakes often has rooted vascular plants, whose leaves extend above the water surface and shade the water, reducing the light available to phytoplankton and benthic algae. Floating aquatic plants like water lilies cause a similar reduction in light availability in the water column. Many lakeshore ecosystems and salt marshes, their marine equivalent, are structurally and functionally similar to terrestrial wetland ecosystems. There is therefore a continuum between the structural and functional properties of aquatic and terrestrial ecosystems.

The origin of lakes strongly affects their structure and functioning (Lodge 2001). Glacial lakes are abundant in young landscapes at high latitudes and altitudes. They are frequently interconnected with other lakes by short stream segments and have a low degree of endemism. Rivers create lakes in several ways, including isolation of former river channels (oxbow lakes) and periodically inundated swamps and floodplains such as the Pantanal of Bolivia and Brazil. These lakes are generally shallow and occasionally reconnect with adjacent rivers during floods. Tectonic lakes form along faults. They are often large, deep, and isolated from one another, providing an environment for substantial diversification, such as in the rift lakes in eastern Africa and Lake Baikal in Siberia. These large tectonically derived lakes harbor most of the endemic freshwater organisms. Over 80% of the open water animals of Lake Baikal, for example, are endemic (Burgis and Morris 1987). Other lakes form in volcanic craters, by damming of rivers, and other processes.

Controls over NPP

Photosynthesis in fresh-water ecosystems is seldom carbon limited, just as in the ocean. Groundwater entering fresh-water ecosystems is supersaturated with CO_2 derived from root and microbial respiration in soil (Kling et al. 1991). Most streams, rivers, and oligotrophic lakes are net sources of CO_2 to the atmosphere because the rates of water and CO_2 input from groundwater generally exceed the capacity of primary producers to use the CO_2 (Cole et al. 1994, Hope et al. 1994). Eutrophic lakes with their high algal biomass have a greater demand for CO_2 to support photosynthesis than do oligotrophic systems, but their organic accumulation and high decomposition rate in sediments provide a large CO_2 input to the water column.

This creates a strong vertical gradient in CO_2, with CO_2 concentration being drawn down at the surface, leading to CO_2 absorption from the atmosphere during the day (Carpenter et al. 2001). Some fresh-water vascular plants such as *Isoetes* use crassulacean acid metabolism (CAM) photosynthesis to acquire CO_2 at night and refix it by photosynthesis during the day (Keeley 1990). Other fresh-water vascular plants transport CO_2 from the roots to the canopy to supplement CO_2 supplied from the water column.

Vertical mixing is important in lakes, just as in marine pelagic ecosystems. Lake mixing occurs not only by wave action, as in the ocean, but also by lake **turnover** in most temperate and high-latitude lakes (Wetzel 2001). In autumn, the surface waters cool to 4°C, the temperature at which water is most dense. Once the surface waters cool to the point that water temperature is similar from top to bottom, the water column is readily mixed by wind. This causes surface waters to sink and brings nutrient-rich bottom waters to the surface. Turnover also occurs in spring, when surface waters warm to 4°C, leading to a spring bloom in production. When lakes do not turn over, oxygen becomes depleted at depth, leading to greater prevalence of anaerobic conditions. Warm-climate lakes do not experience this seasonal lake turnover if the surface waters remain much warmer and less dense than deep water throughout the year.

Nutrients, rather than light, water, or CO_2, are the resources that most consistently limit the productivity of aquatic ecosystems. Both N:P ratios in algae and experimental nutrient additions show that phosphorus limits algal production in the majority of unpolluted lakes, whereas nitrogen is the most common limiting element in coastal marine and salt marsh ecosystems (Fig. 10.7) (Schindler 1977, Valiela 1995). Why should nitrogen be the limiting element in temperate terrestrial ecosystems but phosphorus the limiting element in lakes that are embedded within this terrestrial matrix? At least two factors resolve this apparent paradox. The low mobility of phosphorus compared to nitrogen in soils retains phosphorus more effectively than nitrogen in terrestrial systems. In addition, lakes that receive large phosphorus inputs from pollution or other sources generally support the growth of nitrogen-fixing phytoplankton, such as cyanobacteria. These nitrogen fixers have a competitive advantage over nonfixers when nitrogen is scarce and phosphorus is available. Lakes therefore add their own nitrogen, whenever the phosphorus is sufficient to support nitrogen fixation. Nitrogen fixation in the surface water of lakes is seldom limited by light, as it may be in terrestrial and some stream ecosystems. For these reasons, lakes are seldom nitrogen limited. Nitrate concentrations are typically an order of magnitude higher in lake than in ocean water (Valiela 1995), again indicating the generally greater availability of nitrogen than of phosphorus in lakes.

Nutrient inputs to lakes from streams, groundwater, and atmospheric deposition strongly influence lake biogeochemistry. Lakes are generally small aquatic patches in a terrestrial matrix; they are therefore strongly influenced by inputs of macronutrients and base cations from groundwater and streams (Schindler 1978). The granitic bedrock of the Canadian Shield, from which soils were removed by continental glaciers during the Pleistocene, for example, have low rates of nutrient input from watersheds to lakes. The strong nutrient limitation of many of these lakes makes them vulnerable to change in response to nutrient inputs from agriculture or acid rain (Driscoll et al. 2001). Trout and other top predators in oligotrophic lakes may require decades to reach a large size, whereas this may occur in a few months or years in eutrophic lakes.

Anthropogenic addition of nutrients to lakes frequently causes **eutrophication**, a nutrient-induced increase in lake productivity. Eutrophication radically alters ecosystem structure and functioning. Increased algal biomass reduces water clarity, thereby reducing the depth of the euphotic zone. This in turn reduces the oxygen available at depth. The increased productivity also increases the demand for oxygen to support the decomposition of the large detrital inputs. If mixing is insufficient to provide oxygen at depth, the deeper waters no longer support fish and other oxygen-requiring heterotrophs. This situation is particularly severe in winter, when low temperature limits oxygen production from photosynthesis. In ice-covered lakes, ice and snow reduce light inputs that drive photosynthesis (providing oxygen) and prevent the surface mixing of oxygen into the lake. Lakes in which the entire water column becomes anaerobic during winter do not support fish. Even during summer, the accumulation of algal detritus at times of low surface mixing can deplete oxygen from the water column, leading to high fish mortality.

Carbon and Nutrient Cycling

Carbon and nutrient cycling processes in lakes are similar to those described for the ocean. Phytoplankton account for most primary production in large lakes. As in the ocean, most phytoplankton are grazed by zooplankton, so phytoplankton biomass is relatively low. Lakes without fish have large-bodied zooplankton like *Daphnia* that are efficient feeders on plankton and suspended organic matter. When fish are present, however, they reduce populations of large-bodied zooplankton and probably reduce the efficiency of energy transfer up the food chain (Brooks and Dodson 1965). As in the ocean, bacterial production is substantial in lakes, and the bacteria are an important component of the pelagic food web. Terrestrial dissolved and particulate organic matter is an important substrate for bacterial production in some lakes. Some of this terrestrial organic matter may be more recalcitrant than the bacterial substrates in the oceans and may be consumed more slowly by bacteria or accumulate in sediments. Decomposition in the sediments tends to be more important in lakes than in the ocean because more detritus reaches the bottom before it decomposes, so there is often a well-developed benthic food web similar to that in coastal sediments. Some of the nutrients released by decomposition return to the water column and are mixed to the surface by wave action and lake turnover.

The species composition of lakes strongly influences their physical properties and biogeochemistry. Inadvertent experiments in which fishermen or management agencies have introduced fish or benthic organisms to lakes have provided a wealth of evidence that species traits strongly affect the functioning of aquatic ecosystems (Spencer et al. 1991). In many lakes, the abundance of a top predator alters the abundance of their prey and indirectly the abundance of phytoplankton (see Chapter 11). Changes in benthic fauna can have equally large impacts. Introduction of the zebra mussel to the United States, for example, has displaced native mussels from many rivers and streams. The zebra mussel is a more effective filter feeder than their native counterparts, filtering from 10 to 100% of the water column per day (Strayer et al. 1999). The resulting decrease in density of phytoplankton and other edible particles reduced zooplankton abundance and shifted energy flow from the water column to the sediments.

Streams and Rivers

The structure of stream and river ecosystems depends on stream width and flow rate. The physical environment and therefore the biotic structure of stream ecosystems are dramatically different from those of lakes or the open ocean. Water is constantly moving downstream across the riverbed, bringing in new material from upstream and sweeping away anything that is not attached to the substrate or able to swim vigorously. Phytoplankton are therefore unimportant in streams, except in slow-moving or polluted rivers. The major primary producers of rapidly moving streams are **periphyton**, algae that attach to stable surfaces such as rocks and vascular plants. The slippery surfaces of rocks in a riverbed consist of periphyton and associated bacteria in a polysaccharide matrix. Submerged or emergent vascular plants and benthic mats become relatively more important in slow-moving stretches of the river. Within a given stretch of river, alternating pools and riffles differ in flow rate and ecosystem structure. Seasonal changes in discharge often radically alter the flow regime and therefore structure of these ecosystems. Desert streams, for example, have flash floods after intense rains but may have no surface flow during dry times of year (Fisher et al. 1998). Other streams have discharge peaks associated with snow melt. Large rivers may overflow their banks onto a floodplain during periods of high flow. Many tropical rivers flood annually, so floodplains alternate between being terrestrial and aquatic habitats.

The **river continuum concept** describes an idealized transition in ecosystem structure and functioning from narrow headwater streams to broad rivers (Fig. 10.10) (Vannote et al. 1980). Headwater streams are often shaded by terrestrial vegetation. These plants reduce light availability to aquatic primary producers and provide most of the organic matter input to the stream. Leaves and wood that fall into the stream are colonized by aquatic fungi and to a lesser extent by bacteria (Moss 1998). The resulting leaf packs that accumulate behind rocks, logs, and other obstructions are consumed by invertebrate **shredders** that break leaves and other detritus into pieces and digest the microbial jam on the surface of these particles, just as occurs in the soil (Wagener et al. 1998). This creates fresh surfaces for microbial attack and produces feces and other fine material that is carried downstream. Some of the fine particles are consumed in suspension by filter feeders like black fly larvae or from benthic sediments by **collectors** like oligochaete worms. The abundance of algae and their **grazers** is limited in headwater streams by low light avail-

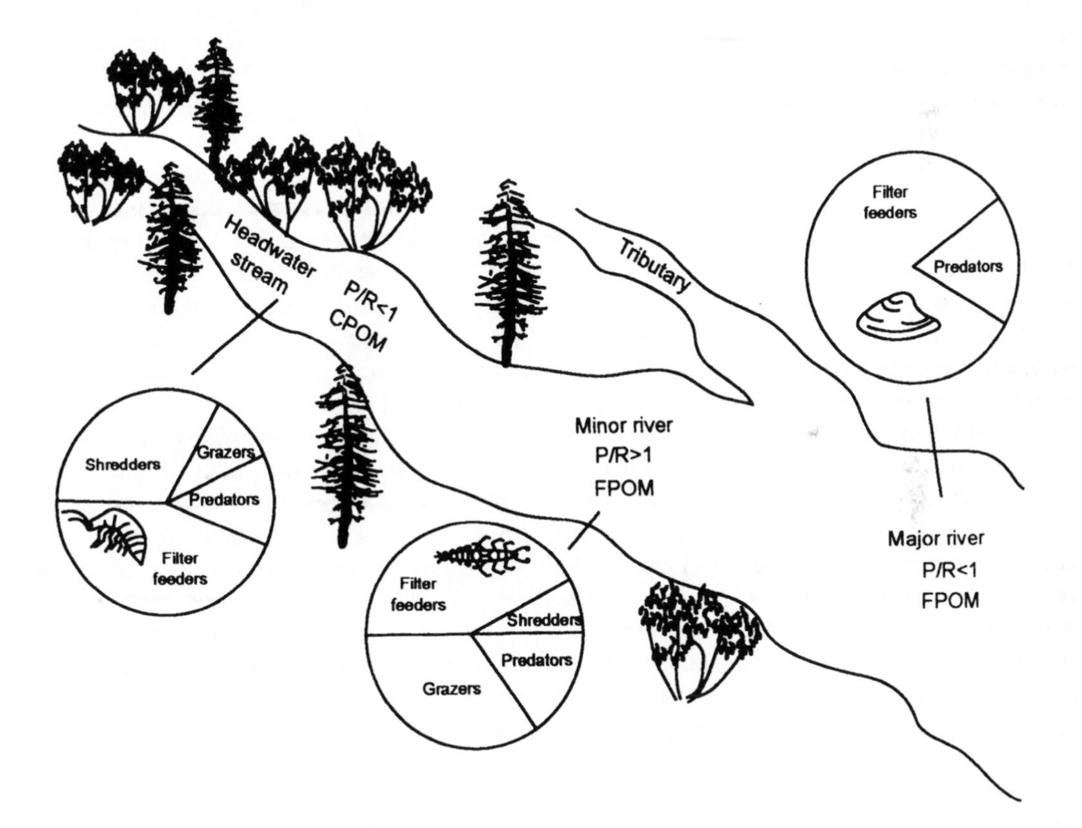

FIGURE 10.10. The river continuum concept (Vannote et al. 1980). Headwater streams have little in-stream production (P), so respiration (R) by decomposers and animals greatly exceeds production. Coarse particulate organic matter (CPOM) dominates the detrital pool. Shredders and collectors are the dominant invertebrates. In middle sections of rivers, more light is available, and in-stream production exceeds respiration. Fine particulate organic matter (FPOM) is the dominant form of organic matter, and collectors and grazers are the dominant organisms. Large rivers accumulate considerable organic-rich sediments dominated by collectors feeding on FPOM from upstream. Respiration by detrital organisms exceeds primary production.

ability. As headwater streams merge to form broader streams, the greater light availability supports more in-stream production, and the input of terrestrial detritus contributes proportionately less to stream energetics. This coincides with a change in the invertebrate community from one dominated by shredders to one dominated by collectors and grazers (Fig. 10.3). These middle reaches of rivers are typically less steep than headwaters and begin to accumulate sediments from upstream erosion. These sediments support rooted vascular plants and a benthic detrital community of collectors. The largest downstream reaches of rivers typically have a sediment bed and are dominated by collectors that live in the sediments. These large rivers may have submerged or emergent vascular plants, depending on the stability of the flow regime. There is a gradual increase in fish diversity from headwater streams to large rivers, whereas the diversity of benthic invertebrates is generally greatest in middle reaches of rivers (Poff et al. 2001).

There is massive variation among streams and rivers in their structure and functioning, just as in terrestrial and marine ecosystems. The river continuum concept provides a framework for predicting patterns of variation within a region but does not capture the large variation due to substrate and climate. Nutrient-poor regions of the tropics and boreal peatlands, for example, have large inputs of dissolved organic carbon (DOC) leading to black-water rivers. White-water rivers result from inputs of silt and other mineral particulates from glacial and agricultural erosion. Clear-water rivers lack these dissolved and suspended materials. Island streams are much shorter than the large rivers that drain the interiors of continents. Lakes and impoundments on rivers create abrupt shifts in habitats, food resources, and biota, punctuating the gradual changes of the river continuum (Ward and Stanford 1983).

Carbon and Nutrient Cycling

Stream productivity is governed by its interface with terrestrial ecosystems (Hynes 1975). Terrestrial ecosystems influence stream productivity directly through the input of detritus that fuels the detritus-based food chain and indirectly by determining the light environment that supports in-stream production. In forest headwater streams, the dominant energy input is terrestrial detritus that enters as **coarse particulate organic matter** (CPOM) (Fig. 10.10). This includes leaves, wood, and other material larger than 1 mm diameter. In forests, there is relatively little algal production in headwater streams because of low light availability. Algal production becomes proportionately more important in ecosystems with low canopy cover, such as in grasslands, tundras, and deserts. **Fine particulate organic matter** (FPOM) comes primarily from within the stream through the processing of CPOM by shredders, the abrasion of periphyton from rocks, and other processes. About a third of the leaf material consumed by shredders, for example, is released into the stream as FPOM (Giller and Malmqvist 1998). The third major organic carbon input to streams comes as dissolved organic carbon from terrestrial groundwater. DOC is the largest pool of organic carbon in most streams. In tropical black-water rivers and boreal peatlands, this carbon source to streams is particularly large and/or persistent. DOC inputs to streams can be an important energy source if the compounds are readily assimilated and metabolized by microbes. Tannins and other recalcitrant substances, however, are processed slowly in streams. Headwater streams are dominated by a detritus-based food chain, including fungi, shredders, and their predators. Heterotrophic respiration therefore considerably exceeds photosynthesis. Downstream, where rivers are wide enough to allow substantial light input, photosynthesis may be similar to or exceed heterotrophic respiration (Vannote et al. 1980). In these middle sections of rivers, heterotrophic respiration is supported by a mixture of FPOM imported from upstream, terrestrial inputs of CPOM (litter), DOC, and algal production. In large rivers with large sediment loads, water clarity may limit algal production, and detrital processing again dominates.

The frequency and magnitude of nutrient limitation to algal production in streams and rivers are more variable than in lakes (Newbold 1992). Many streams, particularly headwater streams, are not strongly nutrient limited, in part because turbulence reduces diffusion limitation. In addition, uptake of nutrients by stream organisms does not influence the supply of nutrients from upstream (Newbold 1992). The relative importance of nitrogen and phosphorus limitation varies among streams, depending on watershed parent material, landscape age, and land use. Phosphorus limitation of stream production, for example, is more common in the eastern United States, where the parent material is relatively old and weathered, than on younger parent materials, where phosphorus inputs are larger and nitrogen is more likely to limit production (Home and Goldman 1994).

There is a strong interaction between top–down and bottom–up controls over primary production in streams. Nutritional controls over the energy available to support higher trophic levels is generally the dominant control over stream productivity, but the types of predators present strongly influence the pathway of energy flow, just as in lakes (see Chapter 12).

Carbon and nutrients spiral down streams and rivers and the groundwater beneath them, rather than exchange vertically with the atmosphere and groundwater. Streams are not passive channels that carry materials from land to the ocean. The streams and their riparian zones process much of the material that enters them. The strong directional flow of water in streams and rivers carries the resulting products downstream, where they are repeatedly reprocessed in successive stream sections. Energy and nutrients therefore **spiral** down streams, rather than cycle vertically as they tend to do in most terrestrial ecosystems (Fisher et al. 1998). This leads to open patterns of nutrient cycling, in which the lateral transfers are much greater than the internal recycling (Giller and Malmqvist 1998). Stream productivity therefore depends highly on regular subsidies from the surrounding terrestrial matrix and is quite sensitive to changes in these inputs due to pollution or land use change. The **spiraling length** of a stream is the average horizontal distance between successive uptake events. It depends on the **turnover length** (the downstream distance moved while an element is in organisms) and the **uptake length** (the average distance that an atom moves from the time it is released until it is absorbed again). A representative spiraling length of a woodland stream is about 200 m. Of this distance, about 10% occurs as microorganisms flow downstream attached to CPOM and FPOM, 1% as consumers move downstream, and the remaining 89% after release of the nutrient by mineralization (Giller and Malmqvist 1998). A unit of nutrient therefore spends most of its time with relatively little movement, but moves rapidly once it is mineralized and soluble in the water. Spiraling is therefore not a gradual process but occurs in pulses. The patterns of **drift** of stream invertebrates is consistent with these generalizations. Invertebrates drift downstream when they are dislodged from substrates or disperse. Drift is a an important food source for fish but represents only about 0.01% of the invertebrate biomass of stream at any point in time. In other words, stream invertebrates are so effective in remaining attached to their substrates that carbon and nutrients spiral downstream primarily in the dissolved phase.

Headwater streams less than 10 m in width are particularly important in nutrient processing because they are the immediate recipient of most terrestrial inputs and account for up to 85% of the stream length within most drainage networks (Peterson et al. 2001). Small streams are particularly effective in cycling nitrogen (have shorter uptake lengths) because their shallow depths and high surface to volume ratios enhance nitrogen absorption by algae and bacteria that are attached to rocks and sediments. Uptake lengths for ammonium range from 10 to 1000 m and increase exponentially with increases in stream discharge (Peterson et al. 2001). Streams generally have much higher nitrate than ammonium concentrations, even when they occur in ammonium-dominated watersheds, because of preferential uptake of ammonium over nitrate by stream organisms and because nitrification rates are frequently high in riparian zones and in streams. For these reasons, the uptake length of nitrate is about 10-fold greater than that of ammonium. Thus nitrate is much more mobile than ammonium in streams, as on land, but for different reasons.

Because of high rates of nitrogen uptake and cycling by the streambed, most nitrogen that enters streams from terrestrial ecosystems is absorbed within minutes to hours and is processed multiple times before it reaches the ocean (Peterson et al. 2001). Nitrification and denitrification release to the atmosphere substantial proportions of the nitrogen that enters streams. Consequently, nitrogen inputs to oceans are much less than the quantity that enters the stream system.

The horizontal flow of carbon and nutrients in streams is similar to the vertical movement of elements through the soil on land but occurs over much larger distances. The basic steps in decomposition are identical on land and in aquatic ecosystems (Valiela 1995, Wagener et al. 1998). These steps include leaching of soluble materials from detritus, fragmentation of litter into small particles by invertebrates, and microbial decomposition of labile and recalcitrant substrates (see Chapter 7). On land, these processes begin at the soil surface, and small particles of organic matter move downward in the soil profile due to mixing by soil invertebrates, burial by new litter, and other processes. In stream ecosystems the same processes occur, but materials move horizontally tens of meters in the process.

Rivers and streams have a belowground component that is just as poorly understood as the soils of terrestrial ecosystems. The **hyporrheic zone** is the zone of groundwater that moves downstream within the streambed. Substantial decomposition occurs in the hyporrheic zone, releasing nutrients that support in-stream algal production. In intermittent streams, the hyporrheic zone is all that remains of the stream during dry periods. Water moves more slowly and therefore has a shorter processing length in the hyporrheic zone than in the stream channel, so the spiraling length is much shorter (Fisher et al. 1998).

Summary

The major differences between aquatic and terrestrial ecosystems result from the differences in the surrounding medium. The greater density of water than air supports photosynthetic organisms with minimal investment in structural support. Aquatic ecosystems therefore have negligible plant biomass but account for nearly half of Earth's NPP. Slow diffusion of gases in water cause oxygen to be much more strongly limiting in aquatic than terrestrial environments, particularly in sediments. Light and nutrients are the resources that most frequently limit aquatic production. NPP is largely restricted to the uppermost part of the water column, where there is sufficient light to drive photosynthesis. Within this zone, nutrients generally limit production. The magnitude and nature of nutrient limitation generally depends on nutrient inputs from terrestrial ecosystems. The open ocean, which receives least terrestrial inputs is strongly nutrient limited, especially in subtropical oceans where wind-driven mixing and upwelling are minimal.

Review Questions

1. How do water and air differ in density, viscosity, and rates of gas diffusion? How do these physical differences give rise to the large differences in structure and functioning that exist between terrestrial and aquatic ecosystems?
2. How do the controls over carbon and light availability differ between terrestrial and aquatic ecosystems?
3. How do the controls over nutrient availability and nutrient cycling differ between terrestrial and aquatic ecosystems?
4. What controls nutrient availability in the open ocean? How does this differ between the open ocean and the coastal zone? Between the open ocean, lakes and streams?
5. Describe the functioning of the microbial loop and biological pump in marine pelagic ecosystems. How does this differ from processes occurring in terrestrial soils?
6. How do ecosystem processes (community composition, patterns of production and decomposition, etc.) change from a headwater stream down to the mouth of a large river?

7. How do ecosystem structure and functioning change, when a stream is blocked by the construction of a dam?
8. How does the terrestrial matrix influence carbon and nutrient cycling in lakes and steams?
9. What determines the spiraling length of nutrients in a stream or river? How might human activities influence this spiraling length?

Additional Reading

Falkowski, P.G., R.T. Barber, and V. Smetacek. 1999. Biogeochemical controls and feedbacks on ocean primary production. Science 281:200–206.

Giller, P.S., and B. Malmqvist. 1998. *The Biology of Streams and Rivers*. Oxford University Press, Oxford, UK.

Moss, B. 1998. *Ecology of Fresh Waters: Man and Medium, Past to Future*. 3rd ed. Blackwell Scientific, Oxford, UK.

Schindler, D.W. 1977. Evolution of phosphorus limitation in lakes. *Science* 195:260–267.

Valiela, I. 1995. *Marine Ecological Processes*. Springer-Verlag, New York.

Vannote, R.I., G.W. Minshall, K.W. Cummings, J.R. Sedell, and C.E. Cushing. 1980. The river continuum concept. *Canadian Journal of Fisheries and Aquatic Sciences* 37:120–137.

Wagener, S.M., M.W. Oswood, and J.P. Schimel. 1998. Rivers and soils: Parallels in carbon and nutrient processing. *BioScience* 48:104–108.

Wetzel, R.G. 2001. *Limnology: Lake and River Ecosystems*. 3rd ed. Academic Press, San Diego, CA.

11
Trophic Dynamics

Trophic dynamics govern the movement of carbon, nutrients, and energy among organisms in an ecosystem. This chapter describes the controls over trophic dynamics of ecosystems.

Introduction

Although terrestrial animals consume a relatively small proportion of net primary production (NPP), they strongly affect energy flow and nutrient cycling. In earlier chapters we emphasized the interactions between plants and soil microbes, because these two groups directly account for about 90% of the energy transfers in most terrestrial ecosystems. Plants use solar energy to reduce CO_2 to organic matter, most of which senesces, dies, and directly enters the soil, where it is broken down by bacteria and fungi. Similarly, most nutrient transfers in ecosystems involve uptake by plants and return to the soil as dead organic matter, where nutrients are released and made available by microbial mineralization. In most terrestrial ecosystems the uncertainties in our estimates of primary production and decomposition exceed the total energy transfers from plants to animals. It is perhaps for this reason that terrestrial ecosystem ecologists have frequently ignored animals in classical studies of production and biogeochemical cycles. Aquatic ecologists, however, have been unable to ignore animals because most of the energy and nutrients are transferred from plants to animals rather than directly from plants to dead organic matter (see Chapter 10). Perhaps for this reason aquatic ecosystem ecologists have generally led the theoretical developments in this aspect of ecosystem ecology.

Understanding the factors governing energy and nutrient transfer to animals has societal implications. Most human populations depend heavily on high-protein foods derived from animals. The exponentially increasing human population requires more food in a world where many people already face an inadequate food supply. An ecologically viable strategy for efficiently providing food to feed the growing human population requires a good understanding of the ecological principles regulating the efficiency with which plants and animals support their growth and maintenance.

Overview

Energy transfers define the trophic structure of ecosystems. The simplest way to visualize the energetic interactions among organisms in an ecosystem is to trace the fate of a packet of energy from the time it enters the ecosystem until it leaves—without worrying about the identity of the organisms involved (Lindeman 1942). **Trophic transfers** involve the feeding of one organism on another or on dead organic matter. Plants are called **primary producers** or

autotrophs because they convert CO_2, water, and solar energy into biomass (see Chapter 5). **Heterotrophs** are organisms that derive their energy by eating live or dead organic matter. Heterotrophs function as part of two major trophic pathways, one that is based on live plants (the **plant-based trophic system**) and another that is based on dead organic matter (the **detritus-based trophic system**). The second trophic system is less immediately obvious to aboveground animals like us and is often overlooked, even though it usually accounts for most of the energy transfers through animals. **Consumers** (also termed **secondary producers**) are organisms that eat other live organisms. These include **herbivores**, which eat plants, **microbivores**, which eat bacteria and fungi, and **carnivores**, which eat animals. A group of organisms that are linked together by the linear transfer of energy and nutrients from one organism to another are referred to as a **food chain**. Grass, grasshoppers, and birds, for example, form a food chain. Those organisms that obtain their energy with the same number of transfers from plants or detritus belong to the same **trophic level**. Thus plants constitute the first trophic level; herbivores, the second; primary carnivores, the third; secondary carnivores that eat mainly primary carnivores, the fourth, etc., in a plant-based trophic system (Lindeman 1942, Odum 1959). Similarly, in the detritus-based trophic system, bacteria and fungi directly break down dead soil organic matter and absorb the breakdown products for their own growth and maintenance. These **primary detritivores** are the first trophic level in the detritus-based food chain and are fed on by animals in a series of trophic levels analogous to those in the plant-based trophic system (Fig. 11.1).

Although food chains are an easy way to visualize the trophic dynamics of an ecosystem, they are a gross oversimplification for organisms that eat more than one kind of food.

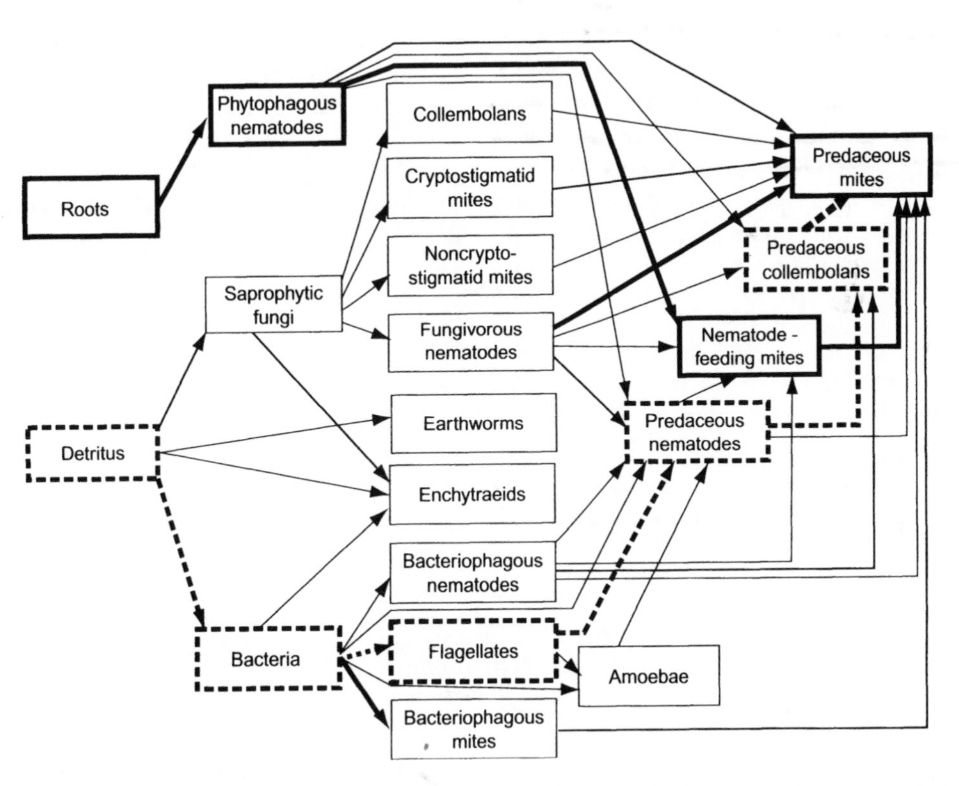

FIGURE 11.1. Pattern of energy flow through belowground portions of a grassland food web. Food webs consist of many interconnecting food chains, including plant-based (heavy solid lines) and detritus-based (dashed lines). (Modified with permission from *Biology and Fertility of Soils*; Hunt et al. 1987.)

People, for example, eat food from several trophic levels, including plants (first trophic level), cows (second trophic level), fish (second and higher trophic levels), and mushrooms (detritivores). Similarly, many birds and some mammals consume both herbivorous insects (plant-based trophic system) and worms (detritus-based trophic system). The actual energy transfers that occur in all ecosystems are therefore complex **food webs** (Fig. 11.1). We can trace the energy transfers through these food webs only by knowing the contribution of each trophic level to the diet of each animal in the ecosystem. Although the structure of food webs has been partially described for many ecosystems (Pimm 1984), the quantitative patterns of energy flow through food webs are generally poorly known, especially for detritus-based food webs.

The regulation of energy and nutrient flow through food webs is complicated and varies considerably among ecosystems. There are two idealized patterns, however, that bracket the range of possible controls. The availability of food at the base of the food chain (either plants or detritus) limits the production of upper trophic levels through **bottom–up controls**. Predators that regulate the abundance of their prey exert **top–down control** on food webs. Most trophic systems exhibit some combination of bottom–up and top–down controls, and the relative importance of these controls varies both temporally and spatially (Polis 1999).

Plant-Based Trophic Systems

Controls over Energy Flow Through Ecosystems

Plant production places an upper limit to the energy flow through plant-based food webs. The energy consumed by animals in the plant-based trophic system, on average, cannot exceed the energy that initially enters the ecosystem through primary production. This constitutes a fundamental constraint to the animal production that an ecosystem can support. When all terrestrial ecosystems are compared, herbivore production tends to increase with increasing primary production (Fig. 11.2). This relationship between primary and secondary production is particularly strong when comparisons are made among similar types of ecosystems. In the grasslands of Argentina, for example, the biomass of mammalian herbivores increases with increasing aboveground production along a gradient of water availability in both natural and managed grasslands (Fig. 11.3) (Osterheld et al. 1992). In the Serengetti grasslands of Africa, the large herds of ungulates also acquire most of their

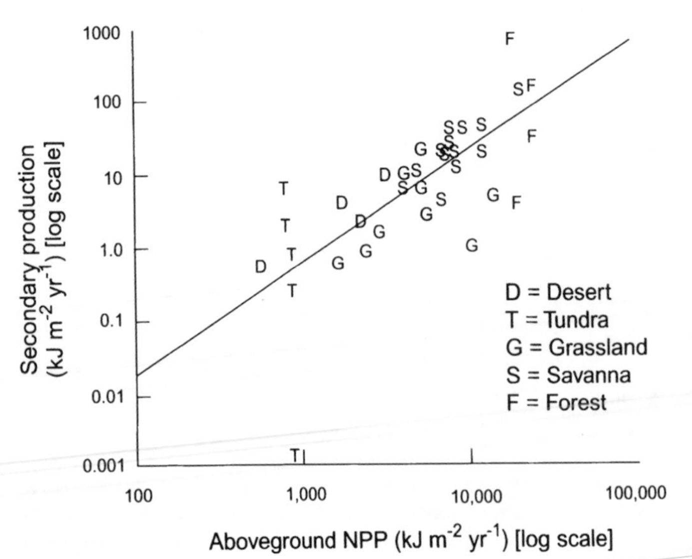

FIGURE 11.2. Log–log relationship between aboveground NPP and herbivore production. Note that 1 g of ash-free biomass is equivalent to 20 kJ of energy. Production of aboveground herbivores correlates with aboveground NPP across a wide range of ecosystems. (Redrawn with permission from *Nature*; McNaughton et al. 1989.)

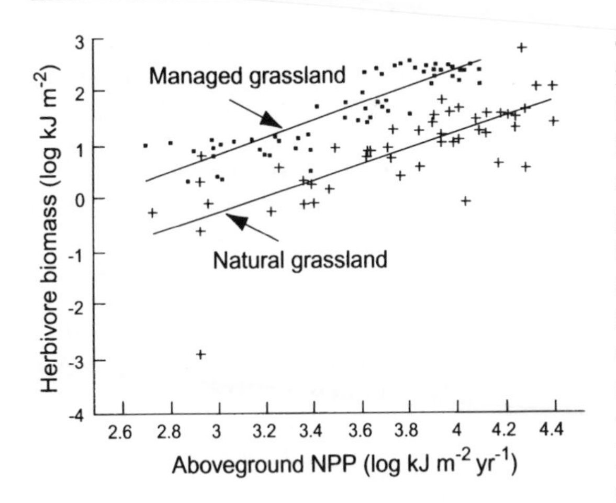

FIGURE 11.3. Log–log relationship between mammalian herbivore biomass and aboveground plant production in natural and managed grazing systems of South America. Herbivore biomass increases with increasing NPP. Animal biomass on the managed grassland is 10-fold greater than on the natural grassland at a given level of plant production, because managers control predation, parasitism, and disease and provide supplemental drinking water and minerals in managed systems. This difference in herbivore biomass between managed and unmanaged systems indicates that NPP is not the only constraint on animal production. (Redrawn with permission from *Nature*; Osterheld et al. 1992.)

food in the more productive grasslands (Sinclair 1979, McNaughton 1985). Similarly, productive forests generally have greater insect herbivory than do nonproductive forests. When forests are fertilized to increase their production, this usually increases the feeding by herbivores (Niemelä et al. 2001). The correlation between primary production and animal production within an ecosystem type is the basis of the world's large fisheries. Upwelling of nutrient-rich bottom waters supports a high production of algae, zooplankton, and fish (see Chapter 10). At the opposite extreme, **oligotrophic** (nutrient-poor) lakes on the Canadian Shield, an area whose soils were scraped away by continental glaciers during the Pleistocene, have low production of algae, zooplankton, and fish.

Subsidies are an important exception to the generalization that NPP within an ecosystem constrains secondary production. Most of the energetic base for headwater streams in forests, for example, comes from inputs of terrestrial litter. This **allochthonous** input (i.e., an input from outside the stream ecosystem) constitutes a subsidy that, together with **autochthonous production** (i.e., production occurring within the stream), provides the energy that supports aquatic food webs (see Chapter 10). Terrestrial food webs near oceans, rivers, and lakes are frequently subsidized by inputs of aquatic energy, for example, when birds or bears feed on fish or when spiders feed on marine detritus (Polis and Hurd 1996). Agricultural production is generally subsidized by inputs of nutrients, water, and/or fossil fuels (Schlesinger 1999).

Biome differences in herbivory reflect differences in NPP and plant allocation to structure. The most dramatic differences in herbivory among ecosystem types are consequences of variation in plant allocation to physical support. Lakes, oceans, and many rivers and streams are dominated by algae that allocate most of their energy to cytoplasm rather than to cellulose and structural support. Most algal cells are readily digested by zooplankton, so animals eat a large proportion of primary production and convert it into animal biomass. Even among algae, chlorophytes (naked green algae) are generally consumed more readily than algae that produce a protective outer coating, such as diatoms, dinoflagellates, and chrysophytes. At the opposite extreme, forests have a substantial proportion of production allocated to cellulose- and lignin-rich woody tissue that cannot be directly digested by animals. Some animals like ruminants (e.g., cows), caecal digesters (e.g., rabbits), and some insects (e.g., termites) support symbiotic gut microbes capable of cellulose breakdown; these animals assimilate some of the energy released by this microbial breakdown.

Among terrestrial ecosystems, there is a 1000-fold variation in the quantity of plant biomass consumed by herbivores (McNaughton et al. 1989). Herbivores consume the least biomass in unproductive ecosystems such as tundra (Fig. 11.4A). However, the energy consumed by herbivores is quite variable within and among other biomes. Consumption by herbivores shows a much stronger

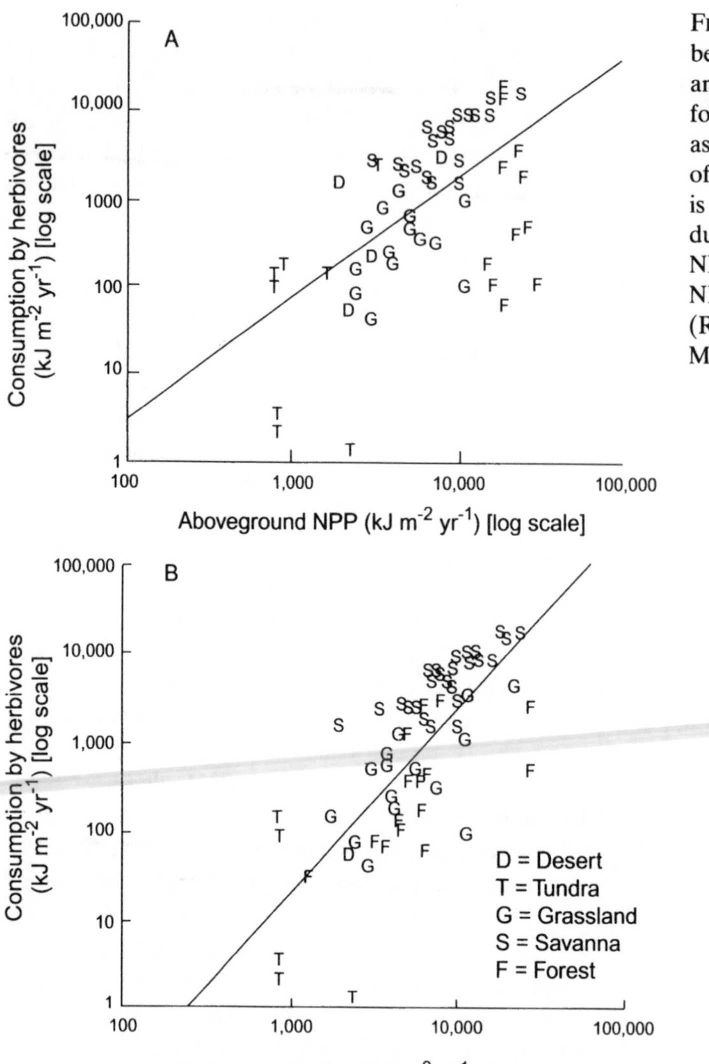

FIGURE 11.4. Log–log relationship between consumption by herbivores and (**A**) aboveground NPP and (**B**) foliage production. Note that 1 g of ash-free biomass is equivalent to 20 kJ of energy. Consumption by herbivores is more closely related to foliage production that to total aboveground NPP, because much of the aboveground NPP is inedible to most herbivores. (Redrawn with permission from *Nature*; McNaughton et al. 1989.)

relationship with production of edible tissue (e.g., leaves) (Fig. 11.4B) than with total aboveground NPP (Fig. 11.4A), because the woody support structures produced by plants contribute relatively little to herbivore consumption.

Plant chemical and physical defenses against herbivores reduce the proportion of energy transferred to herbivores in low-resource environments. It has been argued that predation rather than food availability must limit the abundance of herbivores because the world is covered by green biomass that has not been eaten by animals (Hairston et al. 1960). Not all green biomass, however, is sufficiently digestible to serve as food. In low-resource habitats, plants have a low protein content, high concentrations of chemical defenses against herbivores, and (frequently) physical defenses such as thorns. In African grasslands, for example, fertile "sweet veldt" grasslands support a higher diversity and production of herbivores than do the less fertile "sour veldt" grasslands. The same pattern is seen in tropical forests, where there are higher levels of chemical defense and lower levels of insect herbivory in infertile than in fertile forests (McKey et al. 1978).

Dry environments are also dominated by plants with low protein content and high levels of plant defense. Most herbivory in deserts is concentrated on seeds or on annual plants that germinate, grow, and reproduce within a few weeks after the onset of rains rather than on the dominant perennial plants with their higher concentrations of plant defenses. Thus, even in dry habitats, herbivory is concentrated on those segments of the community that lack well-developed physical and chemical defenses.

Three factors govern the allocation to defense in plants: (1) genetic potential, (2) the environment in which a plant grows, and (3) the seasonal program of allocation. Ecosystem differences in plant defense are determined most strongly by species composition. Different species in terrestrial and aquatic environments exhibit a wide range in both the type and quantity of defensive compounds produced. Terrestrial plants and marine kelps adapted to low-resource environments generally produce long-lived tissues with high concentrations of **carbon-based defense** compounds (i.e., organic compounds that contain no nitrogen, such as tannins, resins, and essential oils) (see Chapter 6). These compounds deter feeding by most herbivores (Coley et al. 1985, Hay and Fenical 1988). The tissue loss to herbivores is often similar (1 to 10%) to the annual allocation to reproduction (i.e., the allocation that most directly determines fitness). This suggests that there should be strong selection for effective chemical defenses against herbivores. When genotypes of a species are compared, for example, those individuals that allocate most strongly to defense grow most slowly (Fig. 11.5), suggesting a trade-off between allocation to growth vs. defense (Coley 1986). Plant species typical of high-nitrogen environments, particularly nitrogen-fixing species, often produce **nitrogen-based defenses** (i.e., organic compounds containing nitrogen, such as alkaloids) that are toxic to generalist herbivores. Nitrogen-based defenses are well developed, for example, in terrestrial legumes and freshwater cyanobacteria. Other species deter herbivores through production of sulfur-containing defenses, accumulation of selenium or silica, etc.

Any given genotype is usually less palatable when grown in a low-resource instead of a high-resource environment, due to a lower protein content and higher levels of plant defense. Under conditions of low nutrient or water availability, growth is constrained more strongly than is photosynthesis, so carbon tends to accumulate (Bryant et al. 1983) (see Chapter 6). Under these circumstances carbon may be allocated to chemical defense with modest negative impacts on growth rate. Finally, in a given environment, plants vary seasonally in their allocation to defense, with allocation to growth occurring when conditions are favorable and allocation to tissue differentiation and defense when conditions deteriorate (Lorio 1986, Herms and Mattson 1992).

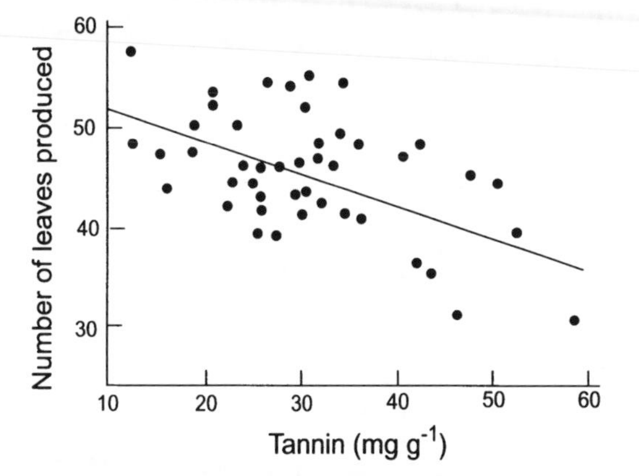

FIGURE 11.5. Relationship between rate of leaf production (an index of growth rate) and leaf tannin concentration in the tropical tree *Cecropia peltata*. Note the negative relationship between investment in defense and growth rate. (Redrawn with permission from *Oecologia*; Coley 1986.)

All three of these sources of variation in allocation to plant defense (genetics, environment, and seasonality) cause defensive measures to be most strongly expressed under conditions of low resource supply. This is the situation in which carbon is unlikely to be the resource that most directly limits growth (see Chapter 6). The most important effect of the plant secondary metabolites on herbivores is their toxicity to the animal, although toxicity to gut microbes and their tendency to bind with proteins (making protein less available to the animal) may also be important under some circumstances.

Herbivores magnify differences among ecosystems in production and energy flow

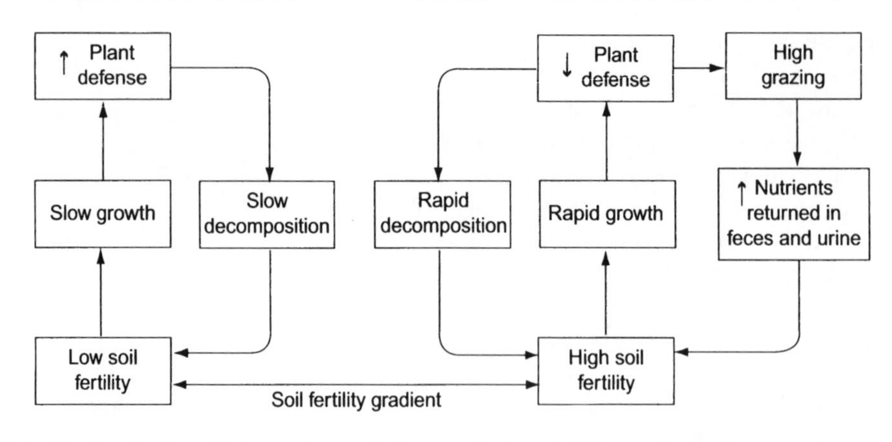

FIGURE 11.6. Feedbacks by which grazing and plant defense magnify differences among sites in soil fertility (Chapin 1991a). In infertile soils, herbivory selects for plant defenses, which reduce litter quality, decomposition, and nutrient supply rate. In fertile soils, herbivory speeds the return of available nutrients to the soil.

through food webs. Dominance by plants with well-developed plant defenses in low-resource environments tends to reduce the frequency of herbivory in these ecosystems because herbivores select patches in the landscape where food quality is relatively high (see Chapter 14). In addition, many plant defenses are toxic to soil microbes, which reduces decomposition rates (see Chapter 7) and further reduces soil fertility in low-resource environments (Northup et al. 1995) (Fig. 11.6).

In high-resource environments, however, where plants are more productive and more palatable, herbivores are more abundant. Their feeding results in a large input of available nutrients in feces and urine, which short-circuits decomposition and nitrogen mineralization and enhances the production of these ecosystems (Ruess and McNaughton 1987). Plants in these environments are frequently well adapted to herbivory. Grasslands with an evolutionary history of intensive grazing, for example, are often more productive when moderately grazed than in the absence of grazers. In the absence of grazers, species composition shifts to species that are less productive and have lower litter quality (McNaughton 1979, Milchunas and Lauenroth 1993, Hobbs 1996). Thus the net effect of herbivores in unmanaged terrestrial ecosystems is to magnify the differences in nutrient availability and production compared to regional patterns that simply reflect direct effects of parent material on nutrient supply and NPP (Chapin 1991a, Hobbs 1996). In managed ecosystems, there is often more grazing than would occur naturally (Fig. 11.3). Overgrazing can reduce production and plant cover and increase soil erosion, leading to a decline in soil resources and the productive potential of an ecosystem (Milchunas and Lauenroth 1993).

In contrast to terrestrial ecosystems, natural rates of herbivory in lakes sometimes reduce rates of nutrient cycling. Early in the season, high rates of herbivory often remove the small, edible, rapidly recycled phytoplankton in favor of more defended cyanobacteria. This reduces herbivory, causing the cycling of nutrients from primary producers to herbivores to decline.

Ecological Efficiencies

Energy losses at each trophic transfer limit the production of higher trophic levels. Not all of the biomass that is produced at one trophic level is consumed at the next level. Moreover, only some of the consumed biomass is digested and assimilated, and only some of the assimilated energy is converted into animal production (Fig. 11.7). Consequently, a relatively small fraction (generally less than 1 to 25%) of the energy available as food at one trophic level is

$$\text{Consumption efficiency } (E_{consump}) = \frac{I_n}{Prod_{n-1}}$$

$$\text{Assimilation efficiency } (E_{assim}) = \frac{A_n}{I_n}$$

$$\text{Production efficiency } (E_{prod}) = \frac{Prod_n}{A_n}$$

$$\text{Trophic efficiency } (E_{troph}) = (E_{consump}) \times (E_{assim}) \times (E_{prod}) = \frac{Prod_n}{Prod_{n-1}}$$

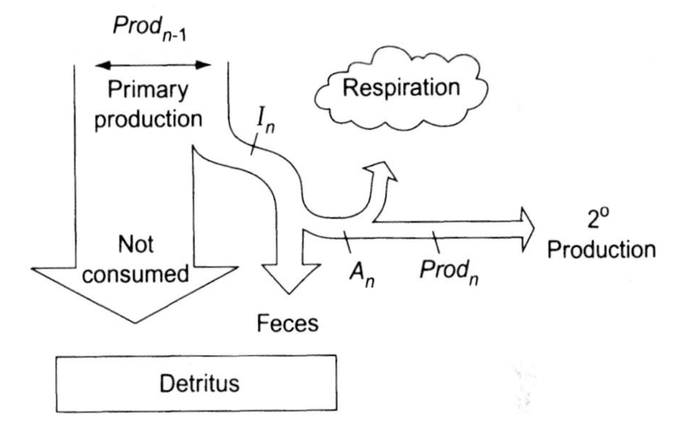

FIGURE 11.7. Components of trophic efficiency, which is the product of consumption efficiency, assimilation efficiency, and production efficiency. Production efficiency is the proportion of primary production that is ingested (I_n) by animals. Assimilation efficiency is the proportion of ingested food that is assimilated into the blood stream (A_n). Production efficiency is the proportion of assimilated energy that is converted to animal production. Most primary production is not consumed by animals and passes directly to the soil as detritus. Of the plant material consumed by herbivores, most is transferred to the soils as feces. Of the material assimilated by animals, most supports the energetic demands of growth and maintenance (respiration), and the remainder is converted to new animal biomass (secondary production).

converted into production at the next link in a food chain. This has profound consequences for the trophic structure of ecosystems because each link in the food chain has less energy available to it than did the preceding trophic link. In any plant-based trophic system, plants process the largest quantity of energy, with progressively less energy processed by herbivores, primary carnivores, secondary carnivores, etc. This leads to an inevitable **energy pyramid** (Elton 1927) and various ecological efficiencies (Lindeman 1942) that determine the quantity of energy transferred between successive trophic levels (Fig. 11.8). The production at each trophic link ($Prod_n$) depends on the production at the preceding trophic level ($Prod_{n-1}$) and the **trophic efficiency** (E_{troph}) with which the production of the prey ($Prod_{n-1}$) is converted into production of consumers ($Prod_n$).

$$Prod_n = Prod_{n-1} \times E_{troph} = Prod_{n-1} \times \left(\frac{Prod_n}{Prod_{n-1}} \right) \tag{11.1}$$

The trophic efficiency of each link in a food chain can be broken down into three ecological efficiencies (Fig. 11.7) related to the efficiencies of consumption ($E_{consump}$), assimilation (E_{assim}), and production (E_{prod}) (Lindeman 1942, Odum 1959, Kozlovsky 1968).

$$E_{troph} = E_{consump} \times E_{assim} \times E_{prod} \tag{11.2}$$

In terrestrial ecosystems, the distribution of biomass among trophic levels can be visualized as a **biomass pyramid** that is similar in structure to the energy pyramid, with greatest biomass in primary producers and progressively less biomass in higher trophic levels (Fig. 11.8). This occurs for at least two reasons: (1) As described earlier, the energy pyramid results in less energy available at each successive trophic link. (2) The large proportion of structural tissue in terrestrial plants minimizes the proportion of plant production that can be converted to secondary production. The decrease in biomass with successive links is most pronounced in forests, where the dominant plants are long lived and produce a large proportion of inedible biomass. Biomass pyramids are less broad in grasslands where plants have a lower allocation to woody structures, and there is a rela-

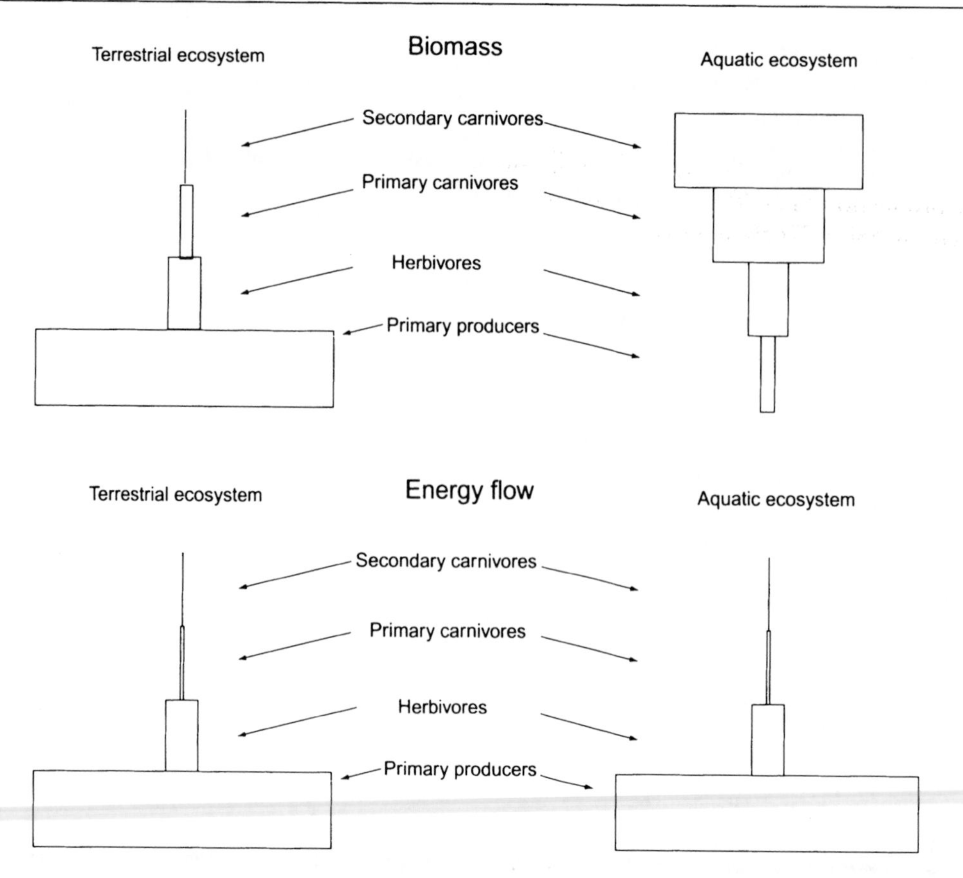

FIGURE 11.8. Pyramids of biomass and energy in a terrestrial and an aquatic food chain. The width of each box is proportional to its biomass or energy content. Pyramids of energy are structurally similar in terrestrial and aquatic food chains because energy is lost at each trophic transfer. Biomass pyramids differ between terrestrial and aquatic food chains because most plant biomass is not eaten on land, whereas most plant biomass (phytoplankton) is eaten and is short lived in aquatic ecosystems.

tively large biomass of herbivores and higher trophic levels.

In contrast to terrestrial ecosystems, freshwater and marine pelagic ecosystems have *less* biomass of primary producers than of higher trophic levels, leading to an **inverted biomass pyramid** (Fig. 11.8). This difference in trophic structure between terrestrial and pelagic ecosystems reflects fundamental differences in allocation and life history. The high buoyancy of water provides physical support for photosynthetic organisms, so algae do not require elaborate physical support structures and are much smaller and more edible than their terrestrial counterparts. Consequently, they are rapidly grazed, and their biomass does not accumulate. In summary, terrestrial ecosystems are characterized by large, long-lived plants, leading to a large plant biomass and relatively small biomass of higher trophic levels. Aquatic ecosystems, in contrast, are characterized by rapidly reproducing plants that are smaller and more short lived than organisms at higher trophic levels (see Fig. 10.1).

Regardless of the biomass distribution among trophic levels, there must always be more energy flow through the base of a trophic chain than at higher trophic levels. It is the energy pyramid rather than the biomass pyramid that describes the fundamental energetic relationships among trophic levels, because energy is lost at each trophic transfer; so there must always be a decline in energy available at each successive trophic level.

Consumption Efficiency

Consumption efficiency is determined primarily by food quality and secondarily by predation. Consumption efficiency is the proportion of the production at one trophic level that is ingested by the next trophic level (I_n) (Fig. 11.7).

$$E_{consump} = \frac{I_n}{Prod_{n-1}} \quad (11.3)$$

Unconsumed material eventually enters the detritus-based food chain as dead organic matter. On average, the quantity of food consumed by a given trophic level must be less than the production of the preceding trophic level, or the prey will be driven to extinction. In other words, the average consumption efficiency of a trophic link must be less than 100%. There may be times when the consumption by one trophic level exceeds that in the preceding level. Most vertebrate herbivores, for example, consume plants during winter, when there is no plant production. This is, however, typically offset by other times, such as summer, when plants usually produce much more biomass than animals can consume. Situations in which consumption efficiency is greater than 100% for prolonged periods lead to dramatic ecosystem changes. If predator control, for example, leads to a large deer population that consumes more plant biomass than is produced, the plant biomass will be reduced, altering plant species composition in ways that profoundly affect all ecosystem processes (see Chapter 12) (Pastor et al. 1988, Kielland and Bryant 1998, Paine 2000). Sometimes this occurs naturally. Some herbivores, such as beavers, typically overexploit their local food supply and move to new areas when their food is depleted.

The proportion of NPP consumed by herbivores varies at least 100-fold among ecosystems, from less than 1% to greater than 40% (Table 11.1). The major factor accounting for this wide range in herbivore consumption efficiency is differences in plant allocation to structure. Herbivore consumption efficiency is generally lowest in forests (less than 1 to 5%), where there is a large plant allocation to wood. Herbivore consumption efficiencies are higher in grasslands (10 to 60%), where most aboveground material is nonwoody, and highest (generally greater than 40%) in pelagic aquatic ecosystems, where most plant (i.e., algal) biomass is cell contents rather than cell walls. In these ecosystems, more algal biomass is often consumed by herbivores than dies and decomposes; this pattern contributes to inverted biomass pyramids (Fig. 11.8). In grasslands, consumption efficiencies are generally greater for ecosystems dominated by large mammals (25 to 50%) than those dominated by insects and small mammals (5 to 15%) (Detling 1988). The toxic nature of some plant tissues (due to presence of plant secondary metabolites) and inaccessibility of other tissues (e.g., roots to aboveground herbivores) constrain the herbivore consumption efficiency of terrestrial ecosystems. Nematodes, one of the major belowground herbivores, consume 5 to 15% of belowground NPP in grasslands (Detling 1988). Consumption efficiencies for belowground herbivores are not well documented, so whole-ecosystem estimates of consumption efficiencies almost always emphasize aboveground consumption. The highest consumption efficiencies in terrestrial ecosystems are on **grazing lawns**, such as those found in some African savannas (McNaughton 1985) and arctic wetlands (Jefferies 1988). These highly productive grasslands are maintained as a lawn by repeated herbivore grazing. Nutrient inputs in urine and feces from these herbivores

TABLE 11.1. Consumption efficiency of the herbivore trophic level in selected ecosystem types.

Ecosystem type	Consumption efficiency[a] (% of aboveground NPP)
Oceans	60–99
Managed rangelands	30–45
African grasslands	28–60
Herbaceous old fields (1–7 yr)	5–15
Herbaceous old fields (30 yr)	1.1
Mature deciduous forests	1.5–2.5

[a] Terrestrial estimates emphasize consumption by aboveground herbivores and may not accurately reflect the total ecosystem-scale consumption efficiency.
Data from Wiegert and Owen (1971) and Detling (1988).

promote rapid recycling of nutrients and therefore are a key factor supporting the grasslands' high production (Ruess et al. 1989) (Fig. 11.6).

Consumption efficiencies of carnivores are often higher than those of herbivores, ranging from 5 to 100%. Vertebrate predators that feed on vertebrate prey, for example, often have a consumption efficiency greater than 50%, indicating that more of their prey is eaten than enters the soil pool as detritus. Invertebrate carnivores often have a lower consumption efficiency (5 to 25%) than vertebrate carnivores. Consumption efficiency of a trophic level at the ecosystem scale must integrate vertebrate and invertebrate consumption, including animals that feed belowground, but these efficiencies are not well documented at the ecosystem scale. More frequently, consumption efficiency is documented for a single large herbivore for an ecosystem in which it is abundant.

The consumption efficiency of a trophic level depends on the biomass of consumers at that trophic level and factors governing their food intake. Over the long term, the quantity and quality of available food constitute the bottom–up controls over the population dynamics and biomass of consumers. In addition, predators exert top–down controls over consumer biomass. Bottom–up and top–down controls frequently interact. Insects feeding on low-quality foliage, for example, must eat more food over a longer time to meet their energetic and nutrient requirements for development. The longer development time required on low-quality food increases their vulnerability to predators and parasites. Rising atmospheric CO_2 concentration, which reduces leaf quality, for example, often increases the quantity of leaf material eaten by a caterpiller, because it must eat more food to meet its energetic requirements for development (Lindroth 1996). The resulting increase in development time, however, probably alters their interactions with higher trophic levels. Bottom–up controls related to NPP and food quality often explain differences among ecosystems in average consumer biomass and consumption, with greater consumer biomass in more productive ecosystems (Figs. 11.3 and 11.4). Predation, however, explains much of the interannual variation in consumer biomass and the quantity of food consumed.

People have substantially altered the trophic dynamics of ecosystems through their effects on consumer biomass. Stocking of lakes with salmonids, for example, increases predation on smaller fish, such as exotic alewife. Overfishing can have a variety of trophic effects, depending on the trophic level of the target fish. Overfishing of herbivorous fish in coral reefs, for example, allows macroalgae to escape grazing pressure and overgrow the corals, killing them in places. On land, stocking of cattle at densities higher than can be supported by primary production causes overgrazing and a decrease in plant biomass; this has led to the loss of productive capacity in many arid lands (Schlesinger et al. 1990). The consequences of human impacts on trophic systems are highly variable, but they often have profound effects on trophic levels up and down the food chain as well as on the target species (Pauly and Christensen 1995).

The bottom–up controls over consumption efficiency can be described in terms of the factors regulating food intake. Consumption by individual animals depends on the time available for eating, the time spent looking for food, the proportion of food that is eaten, and the rate at which food is consumed and digested. Each of these four determinants of consumption has important ecological, physiological, morphological, and behavioral controls that differ among animal species.

Animals do many things other than eating, including predator avoidance, digestion, reproduction, and sleeping. In addition, unfavorable conditions often restrict the time available for foraging, especially for **poikilothermic** animals such as insects, amphibians, and reptiles, whose body temperature depends on the environment. Because of these constraints, deer concentrate their feeding at dawn and dusk; desert rodents feed primarily at night; bears hibernate most of the winter; and mosquitoes feed most actively under conditions of low wind, moderate temperatures, and high humidity. **Activity budgets** describe the proportion of the time that an animal spends in various activities. Activity budgets differ among species, seasons, and habitats, but many animals spend a relatively small proportion of their time consuming food. Changes in climate or predator risk that influence activity budgets of an animal can profoundly alter food intake and therefore the energy available for animal production and maintenance.

Animals must find their food before they eat it. Most predators such as wolves spend more time looking for food than ingesting it. Other animals, including most herbivores, search for favorable habitats within a landscape, then spend most of their time ingesting food. Animals generally consume food faster than they can digest it, so some of the time spent in other activities simultaneously contributes to digestion of food.

Once an animal finds its food, it generally consumes only some of it. Many herbivores, for example, select only the youngest leaves of certain plant species and avoid other plant species, older leaves, stems, and roots. Similarly, carnivores may eat only certain parts of an animal and leave behind parts such as skin and bones. This selectivity places an upper limit on consumption efficiency, because there are components of production from one trophic level that are not consumed at the next level. Many animals become more selective as food availability increases. Lions, for example, eat less of their prey when food is abundant. Gypsy moths and snowshoe hares also preferentially feed on certain plant species, given the opportunity, but will feed on almost any plant during population outbreaks, after most palatable species have been consumed.

Selectivity also depends on the nutritional demands of an animal. Caribou and reindeer, for example, have a gut flora that is adapted to digest lichens, which are avoided by most other herbivores. These animals eat lichens in winter when low temperatures impose a high energy demand for **homeothermy** (maintenance of a constant body temperature). Lichens have a high energy content but little protein. In summer, however, when there is a high protein requirement for growth and lactation, these animals increase the proportion of nitrogen-rich vascular plant species in their diet (Klein 1982). Other herbivores may select plant species to minimize the accumulation of plant toxins. Moose and snowshoe hares in the boreal forest, for example, can consume only a certain amount of particular plant species before accumulation of plant toxins has detrimental physiological effects (Bryant and Kuropat 1980). They therefore tend to avoid plant species with high levels of toxic secondary metabolites. Selectivity by herbivores also depends on the community context. Mammalian generalist herbivores preferentially select plant species when they are uncommon because rare species are consumed too infrequently to reach a threshold of toxicity. Selectivity by these generalist browsers therefore tends to eliminate rare plant species and reduce plant diversity (Bryant and Chapin 1986).

Selectivity differs among animal species. Some grazers, like wildebeest in African savannas, are almost like lawn mowers. They follow the pulse of grass growth that occurs after rains and consume most plants that they encounter. Other animals, like impala, select leaves of relatively high nitrogen and low fiber content, especially in the dry season. Among mammals, there is a continuum from large-bodied **generalist herbivores**, which are relatively nonselective, to smaller-bodied **specialist herbivores**, which are highly specific in their food requirements. Similar patterns are seen among freshwater zooplankton; large-bodied cladocerans like *Daphnia* are generalist filter feeders, whereas same-size or smaller copepods are more selective (Thorp and Covich 2001). Specialization is even more pronounced among terrestrial insects. Some tropical insects, for example, eat only one part of a single plant species. The abundance of specialist insects could contribute to the high diversity of tropical forests, by preventing any one plant species from becoming extremely abundant.

Animals differ in their eating rate. This can be quantified as the bite rate times bite size for vertebrate herbivores or the rate at which soil moves through the gut of an earthworm. There is often an inverse relationship between feeding rate and selectivity, with selective herbivores spending more time looking for food and selecting among species and plant parts that they consume. Because gut capacity is limited, eating rate may be constrained by processing time. Animals that are more selective and have a lower feeding rate generally eat food that is more digestible, contributing to their shorter passage time.

Assimilation Efficiency

Assimilation efficiency depends on both the quality of the food and the physiology of the consumer. Assimilation efficiency is the proportion of ingested energy that is digested and assimilated (A_n) into the bloodstream (Fig. 11.7).

$$E_{\text{assim}} = \frac{A_n}{I_n} \tag{11.4}$$

Unassimilated material returns to the soil as feces, a component of the detrital input to ecosystems.

Assimilation efficiencies are often higher (5 to 80%) than consumption efficiencies (0.1 to 50%). Carnivores feeding on vertebrates tend to have higher assimilation efficiencies (about 80%) than do terrestrial herbivores (5 to 20%), because carnivores eat food that has less structural material than is present in terrestrial plants. Carnivores that kill large prey can avoid eating indigestible parts such as bones, whereas most terrestrial herbivores consume the indigestible cell wall structure in combination with cell contents. Among herbivores, species that feed on seeds, which have high concentrations of digestible, energy-rich storage reserves, have a higher assimilation efficiency than those feeding on leaves. Leaf-feeding herbivores, in turn, have higher assimilation efficiencies than those feeding on wood, which has higher concentrations of cellulose and lignin. Many aquatic herbivores have a particularly high assimilation efficiency (up to 80%) because of the low allocation to structure in many algae and other aquatic plants. Even in aquatic ecosystems, however, herbivores that feed on well-defended species have low assimilation efficiencies. Assimilation efficiencies of herbivores feeding on cyanobacteria, for example, can be as low as 20%.

The physiological properties of a consumer strongly influence assimilation efficiency. Ruminants, which carry a vat of cellulose-digesting microbes (the rumen) have a higher assimilation efficiency (about 50%) than do most nonruminant herbivores. One reason for the high assimilation efficiency of ruminants is the greater processing time than in nonruminants of similar size, giving more time for microbial breakdown of food. Homeotherms typically have higher assimilation efficiencies than do poikilotherms due to the warmer, more constant gut temperature, which promotes digestion and assimilation. Homeotherms therefore have an advantage over poikilotherms in both consumption and assimilation efficiency.

Production Efficiency

Production efficiency is determined primarily by animal metabolism. Production efficiency is the proportion of assimilated energy that is converted to animal production (Fig. 11.7). Production efficiency includes both growth of individuals and reproduction to produce new individuals.

$$E_{\text{prod}} = \frac{Prod_n}{A_n} \tag{11.5}$$

Assimilated energy that is not incorporated into production is lost to the environment as respiratory heat. Production efficiencies for individual animals vary 50-fold from less than 1 to greater than 50% (Table 11.2) and differ most dramatically between homeotherms (E_{prod} 1 to 3%) and poikilotherms (E_{prod} 10 to 50%). Homeotherms expend most of their assimilated energy maintaining a relatively constant body temperature. This high constant body temperature makes their activity less dependent on environmental temperature and increases their capacity to catch prey and avoid predation but makes homeotherms inefficient in producing new animal biomass. Among homeotherms, production efficiency decreases with decreasing body size because a small size results in a high surface to volume ratio and therefore a high rate of heat loss from the warm animal to the cold environment. In contrast, the production efficiency of poikilotherms is relatively high (about 25%) and tends to decrease with increasing body size. Some large-bodied

TABLE 11.2. Production efficiency of selected animals.

Animal type	Production efficiency (% of assimilation)
Homeotherms	
Birds	1.3
Small mammals	1.5
Large mammals	3.1
Poikilotherms	
Fish and social insects	9.8
Nonsocial insects	40.7
Herbivores	38.8
Carnivores	55.6
Detritus-based insects	47.0
Noninsect invertebrates	25.0
Herbivores	20.9
Carnivores	27.6
Detritus-based invertebrates	36.2

Data from Humphreys (1979).

animals, such as tuna, that belong to groups usually thought of as poikilotherms are partially homeothermic. Among poikilotherms, production efficiency is lowest in fish and social insects (about 10%), intermediate in noninsect invertebrates (about 25%), and highest in nonsocial insects (about 40%) (Table 11.2). Production efficiency often decreases with increasing age, because of changes in allocation to growth and reproduction.

Note that belowground NPP—including exudates and transfers to mycorrhizae—is large, poorly quantified, and frequently ignored in estimating trophic efficiencies. Our views of trophic efficiencies may change considerably as our understanding of belowground trophic dynamics improves. Fine roots, mycorrhizae, and exudates, for example, turn over quickly and may support high belowground consumption and assimilation efficiencies for herbivores such as nematodes that specialize on these carbon sources (Detling 1988).

Food Chain Length and Trophic Cascades

Production interacts with other factors to determine length of food chains and trophic structure of communities. Both the NPP and the inefficiencies of energy transfer at each trophic link constrain the amount of energy that is available at successive trophic levels and that therefore could influence the number of trophic levels that an ecosystem can support. The least productive ecosystems, for example, may have only plants and herbivores, whereas more productive habitats might also support multiple levels of carnivores (Fretwell 1977, Oksanen 1990). Detritus-based food chains also tend to be longer in more productive ecosystems (Moore and de Ruiter 2000). In some aquatic ecosystems, however, the trend can go in the opposite direction. Oligotrophic habitats can support inverted biomass pyramids in which large long-lived fish are more conspicuous than the algal and invertebrate populations that support them. Eutrophic lakes or rivers are often dominated by taxa at lower trophic levels that are less edible (Power et al. 1996a). The death and decay of these organisms may deplete dissolved oxygen, eventually making the habitat lethal for fish and other aquatic predators, shortening the length of food chains (Carpenter et al. 1998). When ecosystems are compared across broad productivity gradients, there is no simple relationship between NPP and the number of trophic levels (Pimm 1982, Post et al. 2000). Other factors such as environmental variability and the physical structure of the environment often have a greater effect on the number of trophic levels than does the energy available at the base of the food chain (Post et al. 2000).

The number of trophic levels influences the structure and dynamics of ecosystems through the action of **trophic cascades**, in which changes in the abundance at one trophic level alter the abundance of other trophic levels across more than one link in a food web (Pace et al. 1999). Trophic cascades result from strong predator–prey interactions between particular species (Paine 1980). Predation by one organism at one trophic level reduces the density of their prey, which releases its prey from consumer control (Carpenter et al. 1985, Pace et al. 1999). This trophic cascade (a top–down effect) causes an alternation among trophic levels in biomass of organisms (Power 1990). In streams, for example, if only algae are present, they grow until their biomass becomes nutrient limited,

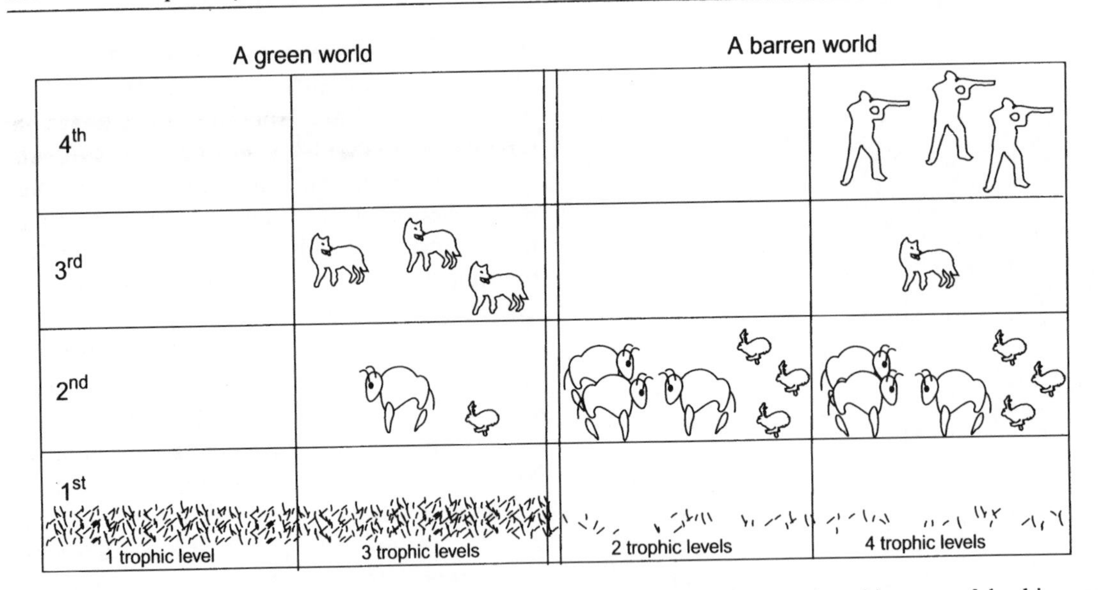

FIGURE 11.9. Effect of food chain length on primary producer biomass in situations in which trophic cascades operate. Plant biomass is abundant where there are odd numbers of trophic levels (1, 3, 5, etc.), because these have a low biomass of herbivores; plant biomass is reduced where there are even numbers of trophic levels (2, 4, 6, etc.), because these have a large biomass of herbivores.

producing a "green" surface (Fig. 11.9). If there are two trophic levels (plants and herbivores), the herbivores graze the plants to a low biomass level, leaving a barren surface. With three trophic levels, the secondary consumer reduces the biomass and grazing pressure of herbivores, which again allows algae to achieve a high biomass. Algal biomass is generally low when there is an even number (two, four, etc.) of trophic levels. An odd number of trophic levels in a trophic cascade reduces the biomass of herbivores and releases the algae, producing a "green" world (Fretwell 1977).

Trophic cascades have been demonstrated in a wide range of ecosystems, from the open ocean to the tropical rain forests and microbial food webs (Pace et al. 1999). Trophic cascades are best documented at the level of species rather than ecosystems, because they generally result from strong interactions between individual species (Paine 1980, Polis 1999). Trophic cascades are most important at the ecosystem scale, when a single species dominates a trophic level, for example when *Daphnia* is the dominant herbivore or a minnow-eating fish is the dominant carnivore in a lake (Polis 1999). Eutrophication often leads to strong species dominance, thereby providing conditions where trophic cascades can play an important role (Pace et al. 1999). Trophic cascades have important practical implications; introduction of minnow-eating fish, under the right circumstances, can release populations of zooplankton grazers, which graze down algal blooms and increase water clarity. Manipulation of trophic cascades to address management issues requires a sophisticated understanding of the ecology of the species involved and the factors governing their interactions. Interactions that were not anticipated frequently become important when trophic dynamics are altered, leading to unexpected responses to species introductions (Kitchell 1992). In most ecosystems there is a dynamic balance between bottom–up and top–down controls that is governed by a wide variety of ecological feedbacks (Power 1992b). In only some situations are ecosystem-scale trophic cascades a dominant feature of ecosystems (Polis 1999).

Seasonal Patterns

In terrestrial ecosystems, production by one trophic level seldom coincides with consump-

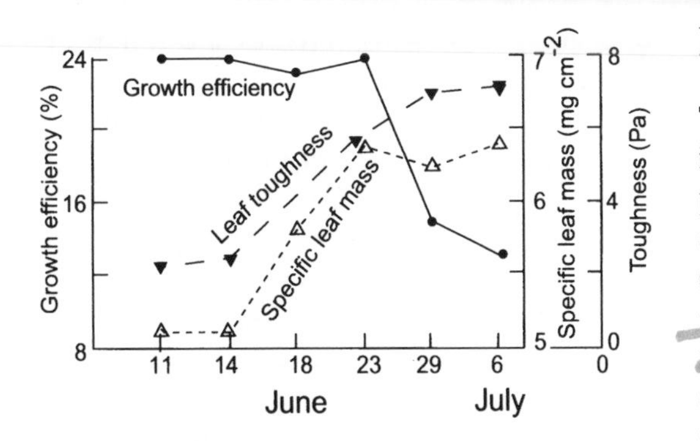

FIGURE 11.10. Seasonal pattern of specific leaf mass and leaf toughness of Finnish birch leaves and of growth efficiency of fourth instar larvae of birch moths. The herbivore grows at maximal efficiency until leaves become tough and mature. After this 2-week window of leaf development, the herbivore grows slowly. (Data from Ayres and MacLean 1987.)

tion by the next. The precise temporal relationship between predator and prey is highly variable, but some common patterns emerge. Plants and their insect predators often use similar temperature and photoperiodic cues to initiate spring growth. However, insects cannot afford to emerge before their food, so there is often a brief window in spring when plants are relatively free of invertebrate herbivory (Fig. 11.10). After insect emergence, there is often a brief window before leaves become too tough or toxic for insects to feed (Feeny 1970, Ayres and MacLean 1987). In contrast to insects, homeotherm herbivores continue to consume food during the cold season, when plants are dormant. These are, however, only two of many highly specific patterns of interactions between plants and their herbivores. Predation by higher trophic levels often focuses at times when prey are most vulnerable, such as when vertebrates are giving birth to young, when salmon are migrating, or when insects are moving actively in search of food. Again, the specific patterns are quite diverse and depend on the biology of predator and prey. The important point is that production by one trophic level and consumption by the next are seldom equal at any point in the annual cycle.

Nutrient Transfers

The *pathway* of nutrients through food chains is usually similar to that of energy. Nitrogen, phosphorus, and other nutrients in plants and animals are either organically bound or are dissolved in the cell contents. Nutrients contained in biomass eaten by animals therefore generally follow the same path through food chains as does energy, from plants to herbivores, to primary carnivores, to secondary carnivores, etc. At each link in the food chain, nutrients are digested and assimilated by animals, just as energy is digested and assimilated, although the efficiencies may differ substantially. As with energy, nutrient losses occur with each trophic transfer in the form of uneaten food, feces, and urine, so the quantity of nutrients transferred must decline with each successive trophic link. The pyramids of nutrient transfers are therefore similar in shape to those of energy flow. In terrestrial ecosystems, the pools of nutrients in organisms also decline with each successive trophic link, as is the case for pyramids of biomass. In pelagic aquatic ecosystems, however, the high trophic efficiencies of zooplankton and high turnover of primary producers generally result in inverted pyramids of nutrient pools, just as for aquatic biomass pyramids. In summary, the *general patterns* of nutrient transfer through ecosystems are similar to those of energy, although the quantitative dynamics may differ.

An important exception to this rule is sodium, which is required by animals for transmission of impulses in nerves and muscles. Most plants actively exclude sodium from roots and from leaves, so tissue concentrations are lower in plants than would be expected based on soil solution concentrations (see Chapter 8). Sodium is therefore more likely to be limiting to animals than to plants. Many terrestrial herbivores supplement the sodium acquired from food by ingesting soil or salts from **salt licks**, which are mineral-rich springs or outcrops. Sodium may therefore show a different pathway of trophic transfer than do other minerals.

A larger proportion of the nutrients contained in plant production pass through terrestrial herbivores than is the case for energy. Most terrestrial herbivores selectively feed on young tissues with a high concentration of digestible energy and low concentrations of cellulose and lignin. These young tissues have high concentrations of nitrogen and phosphorus. Because of selective herbivory on nutrient-rich tissues, a larger proportion of plant-derived nutrients cycle through plant-based trophic systems than is the case for carbon.

Terrestrial herbivores not only select nutrient-rich tissues but cycle nutrients more rapidly than do plants. Plants resorb about half the nitrogen and phosphorus from leaves during senescence, so plant litter generally has only half the nitrogen and phosphorus concentrations as does the live tissue eaten by herbivores (see Chapter 8). For this reason, herbivory is at least twice as important an avenue for nitrogen and phosphorus cycling in terrestrial ecosystems as it is for biomass and energy. The turnover time for nutrients in terrestrial herbivores is often shorter than in the plants on which they feed. Many terrestrial animals, particularly carnivores and homeotherms, eat more nutrients than they require for growth, due to the large energetic costs of movement, and in the case of homeotherms, for temperature regulation. Animals excrete excess nutrients in inorganic form or as urea, which is quickly hydrolyzed in soils (see Chapter 9). In summary, terrestrial herbivores speed nutrient cycling in at least two ways: (1) by removing plant tissues that are more nutrient-rich than would otherwise return to the soil in litterfall and (2) by returning nutrients to the soil in forms that can be directly used by plants (Fig. 11.6).

The ratio of elements required by plants and herbivores determines the nature of element limitation in organisms and the patterns of nutrient cycling in ecosystems. Both aquatic and terrestrial plants require nitrogen and phosphorus in a ratio of about 15:1 (the Redfield ratio; see Chapter 8) (Fig. 11.11). The N:P ratio in herbivores is generally less than in plants, particularly in lakes (Elser et al. 2000). Lake herbivores must therefore concentrate phosphorus more strongly than nitrogen to meet their nutritional demands and tend to excrete the excess nitrogen (Elser and Urabe 1999). In this way, herbivory speeds the recycling of nitrogen, relative to phosphorus, making nitrogen more available to phytoplankton and reinforcing the phosphorus limitation that characterizes many lakes. Differences in N:P ratios among grazers in lakes illustrate the importance of this effect. *Daphnia* is a rapidly growing cladoceran grazer that has a higher phosphorus concentration (lower N:P ratio) than the more slowly growing copepods. Under

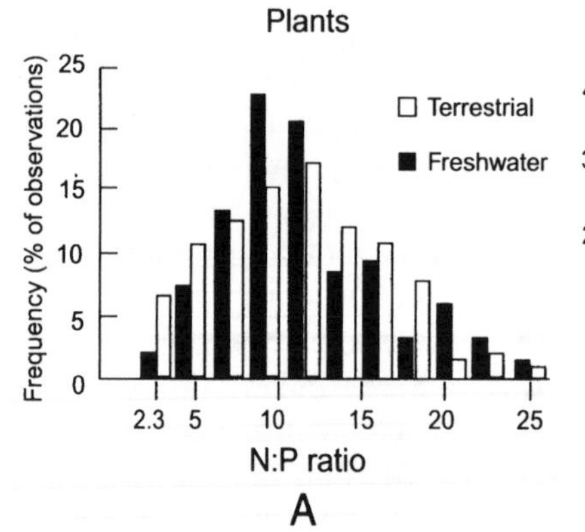

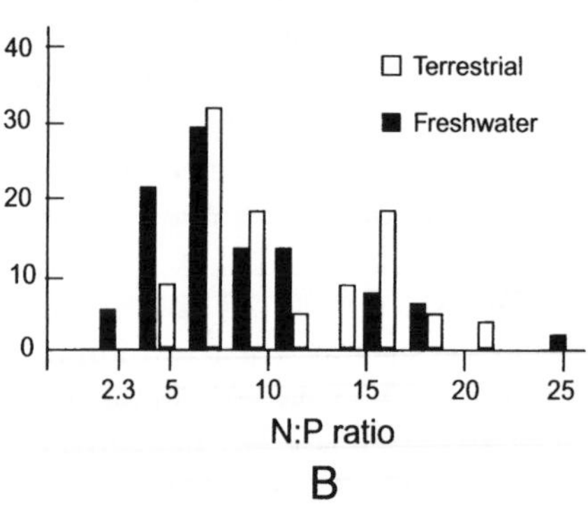

FIGURE 11.11. Frequency distribution of N:P ratios in terrestrial and freshwater ecosystems in (**A**) plants and (**B**) invertebrate herbivores. Ratios are mass of nitrogen relative to mass of phosphorus. The N:P ratio is lower in herbivores than in the plants on which they feed, particularly in fresh-water ecosystems. Herbivores therefore preferentially retain phosphorus and excrete nitrogen to the environment. (Redrawn with permission from *Nature*; Elser et al. 2000.)

conditions of *Daphnia* dominance, grazers concentrate more phosphorus and excrete more nitrogen than when copepods are the dominant grazer; this leads to phosphorus limitation of phytoplankton growth when *Daphnia* dominates and nitrogen limitation when copepods dominate. N:P ratios differ less between terrestrial plants and their herbivores than in aquatic ecosystems. Herbivory also accounts for less of the total nutrient return from plants to the environment in terrestrial than in aquatic ecosystems, so the impact of trophic shifts in N:P ratios on nutrient cycling in terrestrial ecosystems is uncertain. Nonetheless, the trend toward lower N:P ratios in terrestrial herbivores than in plants (Fig. 11.11) suggests that herbivores may be more phosphorus limited than the plants on which they feed and that phosphorus could be a more important nutritional constraint for animals than is generally recognized.

Detritus-Based Trophic Systems

Detritus-based trophic systems convert a much larger proportion of available energy into production than do plant-based trophic systems because they recycle unused organic matter. Plant and animal detritus is fed on by decomposer organisms (primarily bacteria and fungi), just as herbivores feed on live plants. As in the plant-based trophic system, there is a food chain of animals that feed on these decomposer organisms (Fig. 11.12). The principles governing this energy flow are similar to those in the plant-based food chain.

The quantity and quality of soil organic matter is the major determinant of the quantity of energy that flows through the detritus-based system. The detritus-based food chain exhibits losses of energy to growth and maintenance respiration and as feces, just as in plant-based food chains (Fig. 11.12). Moreover, each trophic transfer entails the excretion of inorganic N and P, which become available to plants, just as in the plant-based trophic system.

The major structural distinction between plant- and detritus-based systems is that the plant-based system involves a one-way flow of energy, as energy is either transferred up the food chain or is lost from the food chain as respiration, unconsumed production, or feces. In the detritus-based food chain, however, uneaten food, feces, and dead organisms again become substrate for decomposers at the base of the food chain (Fig. 11.12) (Heal and MacLean 1975). Energy flow in the detritus-based system therefore has a strong recycling component. Energy is conserved and is available to support detritus-based production until it is respired away or is converted to recalcitrant humic material. Due to the efficient use (and reuse) of energy that enters the base of the food chain, the detritus-based food web accounts for most of the energy flow and supports the greatest animal diversity in ecosystems (Heal and MacLean 1975).

The trophic efficiencies of the detritus-based trophic system are generally higher than in the plant-based trophic system. The consumption efficiency of detritus-based food chains is high (greater than 100%) because all of the potential "food" is consumed several times until it is eventually respired away. Assimilation efficiency is also high in decomposers (bacteria and fungi) because their digestion is extracellular, so, by definition, all the material that is consumed by decomposers is assimilated. Herbivores, the first link in the plant-based trophic system, in contrast, have assimilation efficiencies that are commonly 1 to 10%. Production efficiencies of decomposers (40 to 60%; see Chapter 7) and animals in detritus-based food chains (35 to 45%) are also higher than in plant-based trophic systems (Table 11.2). Together these high trophic efficiencies explain why the detritus-based trophic system accounts for most of the secondary production in ecosystems.

Integrated Food Webs

Mixing of Plant-Based and Detritus-Based Food Chains

Food webs blur the trophic position of each species in an ecosystem. In the real world, most animals feed on prey from more than one

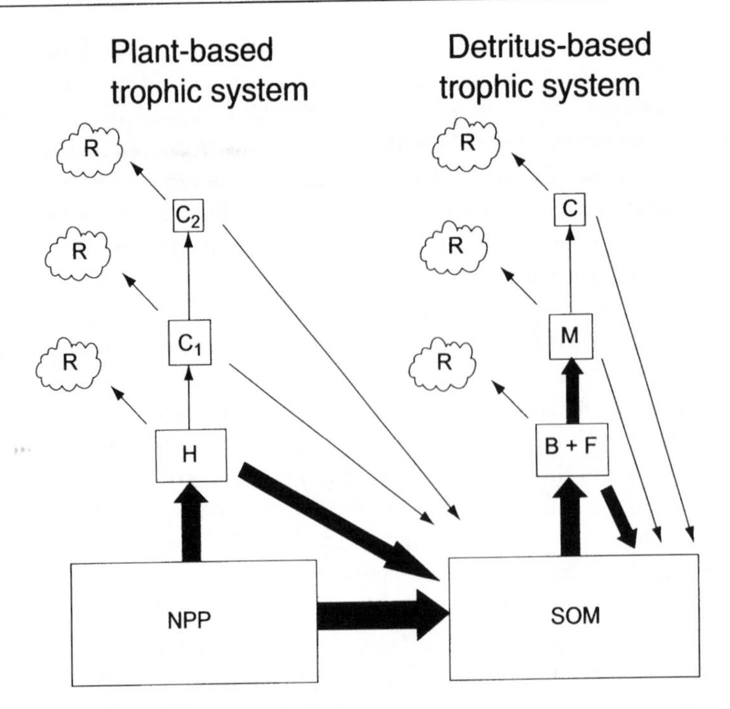

FIGURE 11.12. The two basic trophic systems in ecosystems (Heal and MacLean 1975). In the plant-based trophic system, some energy is transferred from live plants to herbivores (H), primary carnivores (C_1), secondary carnivores (C_2), etc. In the detritus-based trophic system, energy is transferred from dead soil organic matter (SOM) to bacteria (B) and fungi (F), microbivores (M), carnivores (C), etc. In both trophic systems, energy that is not assimilated at each trophic transfer passes to the detritus pool (as unconsumed organisms or as feces). The major difference between these two trophic systems is that energy passes in a unidirectional flow through the plant-based trophic system to herbivores and carnivores or to the detrital pool. In the detritus-based trophic system, however, material that is not consumed returns to the base of the food chain and can recycle multiple times through the food chain before it is respired (R) away or converted to recalcitrant humus.

trophic level, and many animals feed from both the plant-based and the detritus-based trophic systems and at different trophic levels within each system (Polis 1991). For this reason, it is difficult to assign most organisms to a single trophic level. Most fungivores, for example, feed on a mixture of mycorrhizal fungi that derive their energy from plants and saprophytic fungi that decompose dead organic matter. Bacteria also derive energy from root exudates (a component of NPP) and from dead organic matter. Soil animals that eat bacteria and fungi are therefore part of both the plant-based and the detritus-based trophic systems. Root-feeding mites and nematodes fall prey to animals that also eat detritus-based animals (Fig. 11.1). All soil food webs therefore process a mixture of plant and detrital energy and nutrients in ways that are difficult to unravel. Although food webs through aboveground animals have been studied more thoroughly, they also involve substantial detrital input from animals that feed on fungi or on soil animals. Robins, for example, feed on both earthworms and herbivorous insects. Bears eat plant roots and ants of terrestrial origin (plant- and largely detritus-based food chains, respectively) and fish from aquatic food webs. Many insects are detrital feeders at the larval stage but drink nectar or blood (plant-based trophic system), as adults. About 75% of food webs contain both plant- and detritus-based components (Moore and Hunt 1988), so mixed trophic systems are the rule rather than the exception.

The ecosystem consequence of this blurring of food webs is that each food web subsidizes, and is subsidized by, other food webs. Subsidies are therefore an important component of most ecosystems.

Scavengers such as vultures, hyenas, crabs, and many beetles are technically part of the detritus-based food web, although their consumption, assimilation, and production efficiencies are similar to those of carnivores. Scavengers often kill weakened animals, and many predators feed on prey that have been recently killed by other animals, further blurring the distinction between plant-based and detritus-based food chains.

Food Web Complexities

Parasites, pathogens, and diseases are trophically similar to predators. They derive their energy from host tissues and use the products of these cells for their own growth and reproduction, just like predators. It is difficult in practice, however, to separate the biomass of parasites, pathogens, and diseases from that of their hosts, so the concepts of consumption and assimilation efficiencies are seldom applied to these organisms. Parasites, pathogens, and diseases are therefore often treated as agents of mortality rather than as consumers.

Mutualists also confound the trophic picture. Mycorrhizal fungi can change from being mutualistic to parasitic, depending on environmental conditions and the nutritional status of the host plant (Koide 1991). Under mutualistic conditions, mycorrhizal fungi act as herbivores in transferring carbohydrates from plants to the fungus. However, nutrient transfer occurs in the opposite direction, so the trophic role of these two organisms depends on the constituent of interest. In summary, although the broad outlines of trophic dynamics have a clear conceptual basis, the complexities of nature and our poor understanding of belowground processes often make it difficult to describe these food webs quantitatively.

Summary

Resource supply and other factors controlling NPP constrain the energy that is available to higher trophic levels in plant-based trophic systems. These same factors govern the quantity and quality of litter input to the soil and therefore the energy available to the detritus-based trophic system. These factors constitute the bottom–up controls over trophic dynamics. The trophic efficiency with which energy is transferred from one trophic level to the next depends on the efficiencies of consumption, assimilation, and production. Consumption efficiency depends on the interaction of food quantity and quality with predation by higher trophic levels. Consumption efficiency of herbivores is lowest in unproductive habitats dominated by plants that are woody or well defended. Carnivores generally have higher consumption efficiency than do herbivores.

Assimilation efficiency is determined primarily by food quality. It is lower in unproductive than in productive habitats and lower for herbivores than for carnivores. In contrast to the other components of trophic efficiency, the production efficiency is determined primarily by animal physiology; poikilotherms have a higher production efficiency than do homeotherms. Most secondary production in terrestrial ecosystems occurs in the detritus-based trophic system. In this system, material that is not consumed or assimilated returns to the base of the food chain and continues to recycle through the food chain until it is respired or converted to recalcitrant humus. Most food webs contain both plant- and detritus-based components. Impacts, including those resulting from human activities, on any link in food webs frequently propagate to other links in food webs.

Review Questions

1. Describe the pathways of carbon flow in an herbivore-based food chain. How does the efficiency of conversion of food into consumer biomass differ between herbivores and carnivores? What determines the partitioning of assimilated energy between respiration and production?
2. What is the major structural difference between plant-based and detritus-based

food chains? Which food chain can support the greatest total production? Why?
3. What are the major structural differences between terrestrial and aquatic food chains? Why do these differences occur?
4. What plant traits determine the amount of herbivory that occurs? What ecological factors influence these plant traits?
5. What are the effects of herbivores on nitrogen cycling?
6. What are the mechanisms by which top predators influence abundance of primary producers in aquatic food chains? What determines the number of trophic links in an ecosystem? How does this affect ecosystem structure?

Additional Reading

Bryant, J.P., and P.J. Kuropat. 1980. Selection of winter forage by subarctic browsing vertebrates: The role of plant chemistry. *Annual Review of Ecology and Systematics* 11:261–285.

Carpenter, S.R., J.F. Kitchell, and J.R. Hodgson. 1985. Cascading trophic interactions and lake productivity. *BioScience* 35:634–649.

Coley, P.D., J.P. Bryant, and F.S. Chapin III. 1985. Resource availability and plant anti-herbivore defense. *Science* 230:895–899.

Heal, O.W., and J. MacLean, S.F. 1975. Comparative productivity in ecosystems-secondary productivity. Pages 89–108 *in* W.H. van Dobben, and R.H. Lowe-McConnell, editors. *Unifying Concepts in Ecology*. Junk, The Hague.

Herms, D.A., and W.J. Mattson. 1992. The dilemma of plants: To grow or defend. *Quarterly Review of Biology* 67:283–335.

Hobbs, N.T. 1996. Modification of ecosystems by ungulates. *Journal of Wildlife Management* 60: 695–713.

Lindeman, R.L. 1942. The trophic-dynamic aspects of ecology. *Ecology* 23:399–418.

Oksanen, L. 1990. Predation, herbivory, and plant strategies along gradients of primary productivity. Pages 445–474 *in* J.B. Grace, and D. Tilman, editors. *Perspectives on Plant Competition*. Academic Press, San Diego.

Paine, R.T. 2000. Phycology for the mammalogist: Marine rocky shores and mammal-dominated communities. How different are the structuring processes? *Journal of Mammalogy* 81:637–648.

Pastor, J., R.J. Naiman, B. Dewey, and P. McInnes. 1988. Moose, microbes, and the boreal forest. *BioScience* 38:770–777.

Polis, G.A. 1999. Why are parts of the world green? Multiple factors control productivity and the distribution of biomass. *Oikos* 86:3–15.

Power, M.E. 1992. Top-down and bottom-up forces in food webs: do plants have primacy? *Ecology* 73:733–746.

12 Community Effects on Ecosystem Processes

The traits and diversity or organisms and their interactions in communities strongly affect ecosystem processes. This chapter describes the patterns of community effects on ecosystem processes.

Introduction

Species traits interact with the physical environment to govern ecosystem processes. Up to this point, we have emphasized only the most general properties of organisms. We discussed primary producers, for example, as if they were a homogeneous group of organisms in an ecosystem and indicated that primary production can be broadly predicted from climate and parent material. Species differences in traits such as photosynthesis, root allocation, and litter quality, however, strongly affect the functioning of terrestrial ecosystems. Similarly, the phosphorus requirements and prey size preferences of zooplankton govern patterns of nutrient cycling in lakes. Under what circumstances must we know the traits of individual organisms within a trophic group to understand ecosystem processes? In this chapter, we explore this question through discussion of **functional types**—that is, groups of species that are ecologically similar in their effects on ecosystem processes (Chapin 1993a, Smith et al. 1997). Termites, homeotherm herbivores, nitrifying bacteria, and evergreen shrubs are examples of functional types that have predictable general effects on ecosystem processes. Predators, such as planktivorous fish, that feed on the same prey also constitute a functional type or feeding guild. However, no two species or individuals within a functional type are ecologically identical, so, as our understanding improves or our questions become more refined, we expect to recognize situations in which species diversity within functional types or genetic diversity within species has detectable ecosystem consequences.

Natural ecosystems are currently experiencing major changes in species diversity and the traits of dominant species. Earth is currently in the midst of the sixth major extinction event in the history of life (Pimm et al. 1995). Although the causes of earlier extinction events (e.g., the extinction of dinosaurs) are uncertain, they probably resulted from sudden changes in physical environment caused by factors such as asteroid impacts or pulses of volcanism. Current extinction rates are 100- to 1000-fold higher than prehuman extinction rates and could rise to 10,000-fold, if species that are currently threatened become extinct (Fig. 12.1). The current extinction event is unique in the history of life in that it is biotically driven, specifically by the effect of the human species on land use, species invasions, and atmospheric

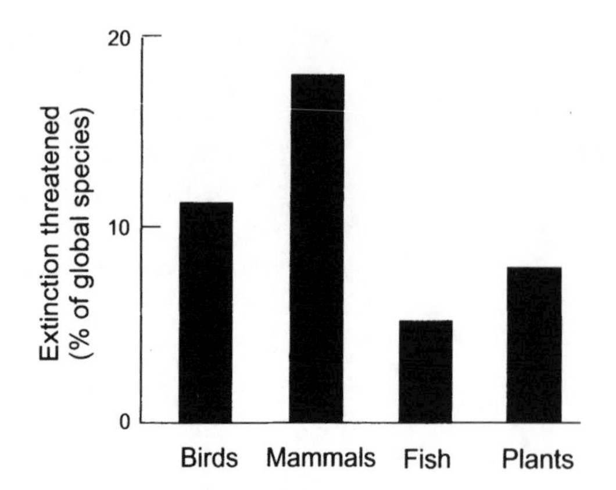

FIGURE 12.1. Percentage of major vertebrate and vascular plant species that are currently threatened with extinction. (Redrawn with permission from *Nature*; Chapin et al. 2000b.)

and environmental change. Although human impacts affect many processes at global scales (Vitousek 1994a) (see Chapter 15), the loss of species diversity is of particular concern because it is irreversible. For this reason, it is critical to understand the functional consequences of the current large losses in species diversity (Chapin et al. 2000b). Although global extinction of a species is a major conservation concern, localized extinctions or large changes in abundances happen more frequently and have the largest effects on functioning of ecosystems.

A second biotic change of global proportions is the frequent introduction of exotic species into ecosystems. People have intentionally and inadvertently moved thousands of species around the globe, leading toward a homogenization of the global biota. Exotic species often change the physical and biotic environment enough to alter the dominance or eliminate native species from an ecosystem. Although extinction and immigration of species are natural ecological processes, the dramatic increase in frequency of these events in recent decades is rapidly changing the types and numbers of species throughout the globe. There are many ethical, aesthetic, and economic reasons for concern about changes in biodiversity. The changes are occurring so quickly, however, that we must assess their functional consequences for population, community, and ecosystem processes. Changes in species composition often have a greater effect on ecosystem processes than do the direct impacts of global changes in atmospheric composition and climate. Understanding the nature of biotic impacts on ecosystem processes is therefore critical to predictions of future changes in ecosystems. In this chapter we focus on the ecosystem consequences of changes in communities and the associated changes in species traits.

Overview

The number, relative abundance, identity, and interactions of species all affect ecosystem processes. Species in a given trophic level almost always differ in some traits that affect ecosystem processes. Sun and shade species, for example, differ in the conditions under which they contribute to carbon inputs; presence of both types of species therefore increases the efficiency with which light is converted to net primary production (NPP) (see Chapter 5). Nitrogen-fixing cyanobacteria differ from nonfixing phytoplankton in their impacts on nitrogen cycling; presence of both types of species therefore influences the response of nitrogen cycling to variations in phosphorus inputs to lakes (see Chapter 10). No single species can perform all of the functional roles that are exhibited by a trophic level. A diversity of species is functionally important because it increases the range of organismic traits that are represented in an ecosystem and therefore the range of conditions under which ecosystem properties can be sustained. From an ecosystem perspective, species diversity is simply a summary variable that describes the total range of biological attributes of all the species in the ecosystem. The functional consequences of a change in diversity depend on the number of species present (**species richness**), their relative abundances (**species evenness**), the identity of species that are present (**species composition**), the interactions among species, and the temporal and spatial variation in these properties. Each of these components of diversity affects the functioning of ecosystems.

If all species were functionally different and contributed in unique ways to a given process, rates of ecosystem processes might change linearly as the number of species increased (Fig. 12.2A) (Vitousek and Hooper 1993, Sala et al. 1996). Nitrogen retention, for example, might increase as species with different rooting depths or preferred forms of nitrogen uptake are added to the ecosystem. In practice, however, the relationship between species number and any given ecosystem process tends to saturate with increasing numbers of species, because some species that are added are ecologically similar to species already present in the community (Tilman et al. 1996). Ecosystem processes probably differ in their sensitivity to the number of species in a given trophic level. Nutrient retention may be particularly sensitive to the number and functional diversity of plant species, whereas decomposition rate may depend more on the diversity of the microbial community.

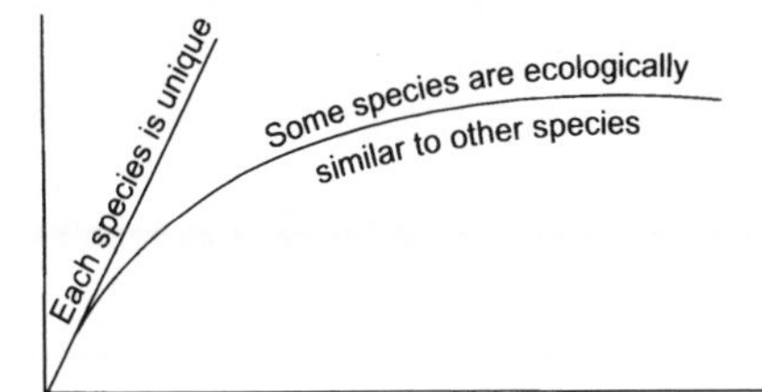

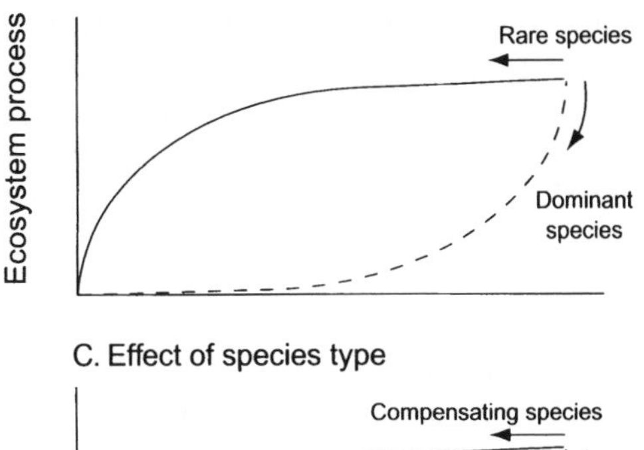

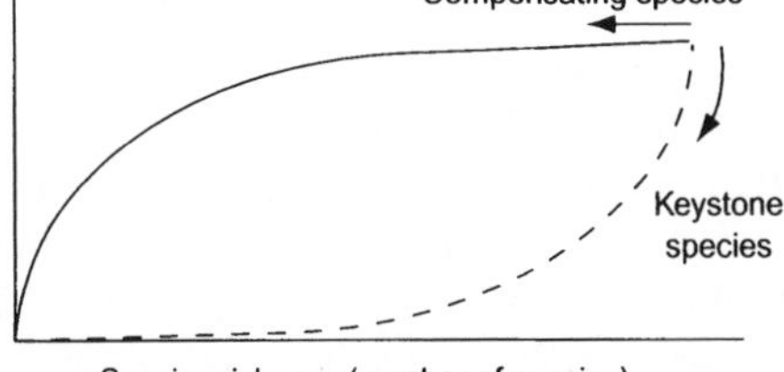

FIGURE 12.2. Expected relationship between ecosystem processes and the number of species, their relative abundance, and the type of species in an ecosystem (Vitousek and Hooper 1993, Sala et al. 1996). **A**, Some processes (or stocks) may increase linearly with increasing species number; others may show an assymptotic increase. **B**, Removal of dominant species from an ecosystem has greater impact on ecosystem processes than does removal of rare species. **C**, Similarly, the removal of keystone species has large ecosystem effects, whereas removal of one species of a functional type allows other species in that functional type to increase in abundance; this compensation would cause only a moderate impact on ecosystem processes, until most species from that functional type have been removed. The arrows show the expected change in ecosystem processes in response to species loss.

If all else were equal, a change in the abundance of a dominant species is more likely to have ecosystem effects than is a change in abundance of a rare species (Sala et al. 1996) (Fig. 12.2B), because dominant species account for most of the energy and nutrient flow through an ecosystem. Dominant species are also most likely to have strong effects on microenvironment. Loss of dominant conifers due to pathogen or insect outbreak, for example, alters microclimate and plant biomass so strongly that most ecosystem processes are affected (Matson and Waring 1984).

The functional attributes of species in a community are at least as important as the number of species present in determining the mechanisms by which species diversity influences ecosystem processes (Hooper et al., in press). Many species are ecologically more important than their abundance would suggest. **Keystone species** are an extreme example of species with strong effects. A keystone species is a functional type represented by only one species. It is ecologically distinct from other species in the ecosystem and has a much greater impact on ecosystem processes than would be expected from its biomass (Power et al. 1996b) (Fig. 12.2C). The tsetse fly in Africa, for example, has a large effect on ecosystem processes per unit of tsetse fly biomass, because it limits the density of many ecologically important mammals, including people. Loss of a keystone species has greater ecological impact than does the loss of one of a group of ecologically similar species because, in the latter case, the remaining species could continue to perform the relevant ecological functions of that functional type. The more species there are in a functional type, the less likely it is that a gain or loss of a single species from that functional type will have large ecosystem effects. Our challenge, as ecologists, is to identify the traits of organisms that have strong effects on ecosystems; species with these traits are likely to be strong interactors in ecosystems (Paine 2000).

Species interactions govern the traits that are expressed most clearly in ecosystems. The impact of a species on ecosystem processes depends on its interactions with other species. The impact of deer on terrestrial vegetation or the impact of *Daphnia* on algal biomass of lakes, for example, depends on the density of their predators. The mechanisms by which species diversity influences ecosystem processes often depend on species interactions such as competition, facilitation, and predation.

Species Effects on Ecosystem Processes

Species are most likely to have strong ecosystem effects when they alter interactive controls, which are the general factors that directly regulate ecosystem processes. These controls include the supply of resources that are essential for primary production, climate, functional types of organisms, disturbance regime, and human activities (see Chapter 1).

Species Effects on Resources

Resource Supply

Species traits that influence the supply of limiting resources have major impacts. The supply of resources required for growth of primary producers is one of the interactive controls to which ecosystem processes are most sensitive (see Chapter 1). These resources include light, nutrients, and, on land, water. For this reason, species traits that alter the supply of limiting resources will substantially alter ecosystem processes.

The introduction of a strong nitrogen fixer into a community that lacks such species can substantially alter nitrogen availability and cycling. Invasion by the exotic nitrogen-fixing tree *Myrica faya* in Hawaii, for example, increased nitrogen inputs, litter nitrogen concentration, and nitrogen availability (Vitousek et al. 1987) (Fig. 12.3). A nitrogen-fixing invader is most likely to be successful in ecosystems that are nitrogen limited; have no strong nitrogen fixers; and have adequate phosphorus, micronutrients, and light (Vitousek and Howarth 1991). Thus we expect large ecosystem changes from invasion of nitrogen-fixing species primarily in combinations of the following circumstances: (1) low nitrogen supply (early primary succession in the temperate zone and in other low-nitrogen environments), (2) distant from

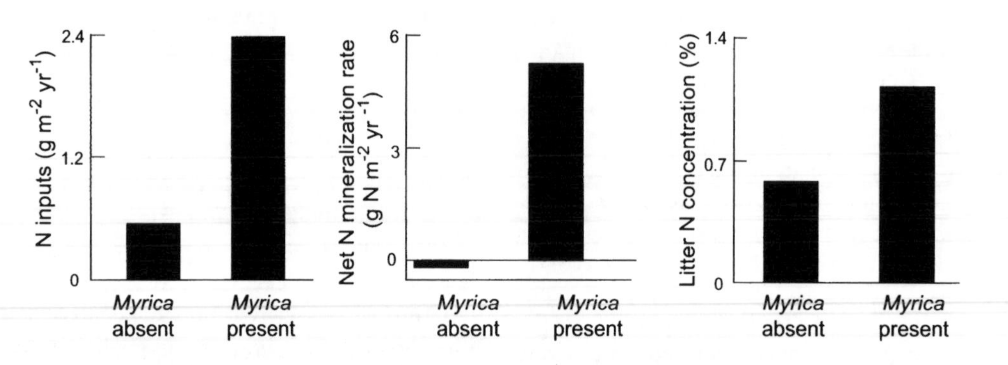

FIGURE 12.3. Impact of the nitrogen-fixing tree *Myrica faya* on nitrogen inputs, litter nitrogen concentration, and nitrogen mineralization rate in a Hawaiian montane forest (Vitousek et al. 1987).

parent populations of nitrogen-fixing species (e.g., islands), (3) low competition for light or phosphorus (e.g., early in succession and in phosphorus-enriched lakes or soils), and (4) reduction of shading by grazing (e.g., pastures). Exotic nitrogen-fixing species have particularly large effects when they escape the suite of species-specific herbivores or pathogens that often restrict nitrogen inputs from native nitrogen fixers.

Deep-rooted species can increase the volume of soil tapped by an ecosystem and therefore the pool of soil resources available to support production. The perennial bunch grasses that once dominated California grasslands, for example, have been largely replaced either by European annual grasses or by forests of Australian *Eucalyptus*. The deep-rooted *Eucalyptus* trees access a deeper soil profile than do annual grasses, so the forest absorbs more water and nutrients. In dry, nutrient-limited ecosystems, this substantially enhances ecosystem productivity and nutrient cycling (Fig. 12.4) but reduces species diversity. The introduction of deep-rooted phreatophytes in deserts also increases the productivity in watercourses but reduces diversity, because litter accumulation on the soil surface inhibits the growth of desert annuals (Berry 1970). Deep-rooted species can also tap nutrients that are available only at depth. A deep-rooted tundra sedge, for example, is the only species in arctic tussock tundra that accesses nutrients in the groundwater that flows over permafrost. By tapping nutrients at depth, the productivity of this sedge increases 10-fold in sites with abundant groundwater flow, whereas productivity of other species is unaffected by deep resources (Chapin et al. 1988). In the absence of this species, ecosystem productivity and nutrient cycling would be greatly reduced. In general, we expect large ecosystem impacts from invasion of deep-rooted species where growth-limiting resources are available at depth. Ecosystem differences in maximum rooting depth have implications for regional hydrology and climate.

Nitrification, denitrification and, consequently, gaseous nitrogen loss to the atmosphere are controlled by relatively few species of microorganisms. Nitrification rate also influences the susceptibility of nitrogen to leaching loss (see Chapter 7). Changes in the abundance of microorganisms controlling these processes might therefore alter nutrient availability through their effects on nutrient loss (Schimel 2001). Mycorrhizal fungi also influence the quantity of nutrients that are available to vegetation (see Chapter 8).

Animals can influence the resource base of the ecosystem by foraging in one area and depositing nutrients elsewhere in feces and urine (see Chapter 14). Sheep, for example,

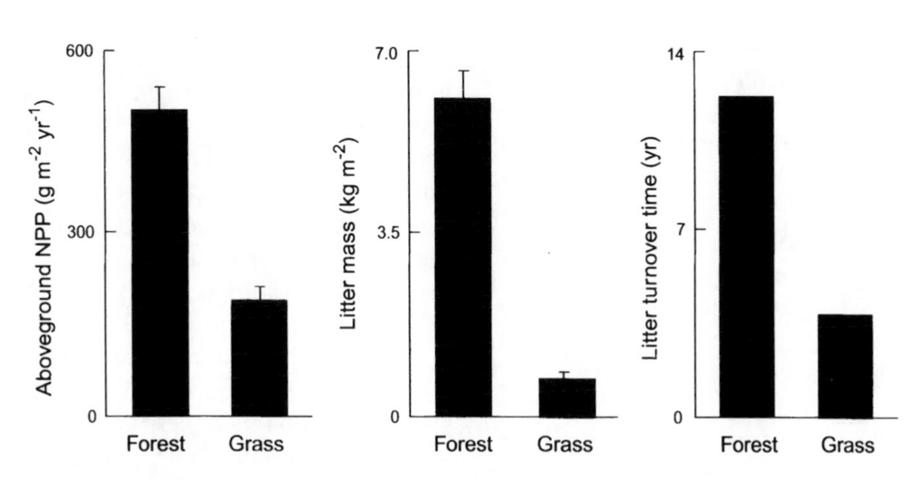

FIGURE 12.4. Comparison of ecosystem processes between two exotic communities that differ in rooting depth: annual grassland and a *Eucalyptus* forest in California (Robles and Chapin 1995). Data are means ± SE.

enrich soils on hilltops where they bed down at night. Migrating salmon perform a similar nutrient-transport role in streams. They feed primarily in the open ocean and then return to small streams where they spawn, die, and decompose. The nutrients carried by the salmon from the ocean can sustain a substantial proportion of the algal and insect productivity of small streams. These nutrient subsidies can be transferred to adjoining terrestrial habitats by bears and otters that feed on salmon or by predators of insects that emerge from streams.

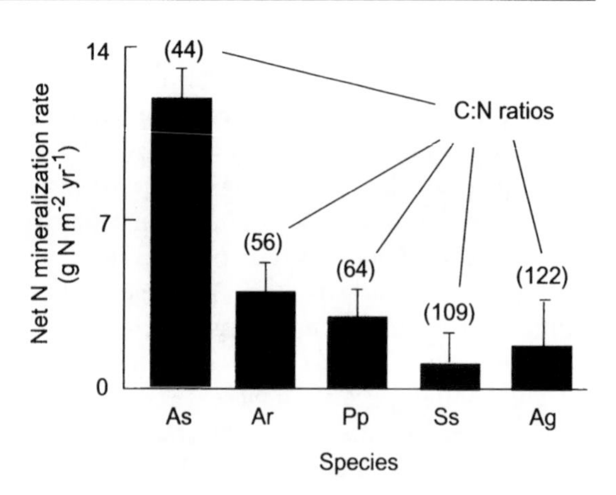

FIGURE 12.5. Effects of prairie grass species on nitrogen mineralization when grown on soils with containing $100\,g\,N\,m^{-2}$ (Wedin and Tilman 1990). Grasses range from early to late successional in the following order: *Agrostis scabra* (As), *Agropyron repens* (Ar), *Poa pratensis* (Pp), *Schizochyrium scoparium* (Ss), and *Andropogon gerardi* (Ag). Data are means ±95% confidence interval (CI). Numbers are the C:N ratios of aboveground biomass.

Nutrient Turnover

Species differences in litter quality magnify site differences in soil fertility. Differences among plant species in tissue quality strongly influence litter decomposition rates (see Chapter 7). Litter from low-nutrient-adapted species decomposes slowly because of the negative effects on soil microbes of low concentrations of nitrogen and phosphorus and high concentrations of lignin, tannins, waxes, and other recalcitrant or toxic compounds. This slow decomposition of litter from species characteristic of nutrient-poor sites reinforces the low nutrient availability of these sites (Hobbie 1992, Wilson and Agnew 1992). Species from high-resource sites, in contrast, produce rapidly decomposing litter due to its higher nitrogen and phosphorus content and fewer recalcitrant compounds, enhancing rates of nutrient turnover in nutrient-rich sites.

Experimental planting of species on a common soil shows that species differences in litter quality can alter soil fertility quite quickly. Early successional prairie grasses, whose litter has a low C:N ratio, for example, causes an increase in the nitrogen mineralization rate of soil within 3 years, compared to the same soil planted with late-successional species whose litter has a high C:N ratio (Wedin and Tilman 1990) (Fig. 12.5).

Seasonality of Resource Capture

Phenological specialization could increase resource capture. Phenological specialization in the timing of plant activity can increase the total time available for plants to acquire resources from their environment. This is most evident when coexisting species differ in the timing of their maximal activity. In mixed grasslands, for example, C_4 species are generally more active in the warmer, drier part of the growing season than are C_3 species. Consequently C_3 species account for most early season production, and C_4 species account for most late-season production. Similarly, in the Sonoran desert, there is a different suite of annuals that becomes active after winter than after summer rains. In both cases, phenological specialization probably enhances NPP and nitrogen cycling. In mixed-cropping agricultural ecosystems, phenological specialization is more effective in enhancing production than are species differences in rooting depth (Steiner 1982).

The ecosystem consequences of phenological specialization to exploit the extremes of the growing season are less clear. Evergreen forests, for example, have a longer photosynthetic season than deciduous forests, but most carbon gain occurs in midseason in both forest types, when conditions are most favorable

(Schulze et al. 1977). The early spring growth of spring ephemeral herbs in deciduous forests also has relatively little influence on nitrogen cycling because most nitrogen turnover occurs in midseason. Phenological specialization is an area in which species effects on ecosystem processes could be important, but these effects have been well documented primarily in agricultural ecosystems.

Species Effects on Climate

Species effects on microclimate influence ecosystem processes most strongly in extreme environments. This occurs because ecosystem processes are particularly sensitive to climate in extreme environments (Wilson and Agnew 1992, Hobbie 1995). Boreal mosses, for example, form thick mats that insulate the soil from warm summer air temperatures. The resulting low soil temperature retards decomposition, contributing to the slow rates of nutrient cycling that characterize these ecosystems (Van Cleve et al. 1991). Some mosses such as *Sphagnum* effectively retain water, as well as insulating the soil, leading to cold anaerobic soils that reduce decomposition rate and favor peat accumulation. The accumulation of nitrogen and phosphorus in undecomposed peat reduces growth of vascular plants. The shading of soil by plants is an important factor governing soil microclimate in hot environments. Establishment of many desert cactuses, for example, occurs primarily beneath the shade of "nurse plants."

Species effects on water and energy exchange can affect regional climate. Species differences in albedo or the partitioning between sensible and latent heat fluxes can have strong effects on local and regional climate. The lower transpiration rate of pasture grasses compared to deep-rooted tropical trees, for example, could lead to a significantly warmer, drier climate following widespread tropical deforestation because of the lower evapotranspiration and greater sensible heat flux of pastures (see Chapter 2). Changes in vegetation caused by overgrazing could alter regional climate. In the Middle East, for example, overgrazing reduced the cover of plant biomass. Model simulations suggest that the resulting increase in albedo reduced the total energy absorbed, the amount of sensible heat released to the atmosphere, and consequently the amount of convective uplift of the overlying air. Less moisture was therefore advected from the Mediterranean Sea, resulting in less precipitation and reinforcing the vegetation changes (Charney et al. 1977). These vegetation-induced climate feedbacks could have contributed to the desertification of the Fertile Crescent. Vegetation changes associated with fire in the boreal forest can have a cooling effect on climate. Late-successional conifers, which dominate the landscape in the absence of fire, have a low albedo and stomatal conductance and therefore transfer large amounts of sensible heat to the atmosphere. Postfire deciduous forests, in contrast, absorb less energy, due to their high albedo, and transmit more of this energy to the atmosphere as latent rather than sensible heat, resulting in less immediate warming of the atmosphere and more moisture available to support precipitation (Chapin et al. 2000a) (Fig. 12.6). If these vegetation changes were widespread, they could have a negative feedback to high-latitude warming and reduce the probability of fire. This

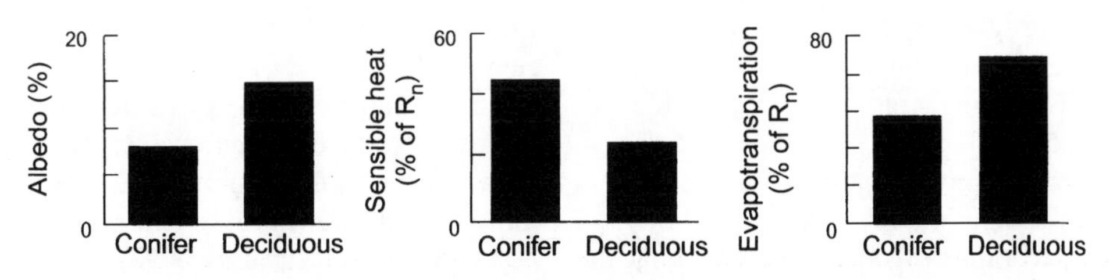

FIGURE 12.6. Sensible and latent heat fluxes from deciduous and conifer boreal forests (Baldocchi et al. 2000).

is one of the few negative feedbacks to regional warming that has been identified.

Species Effects on Disturbance Regime

Organisms that alter disturbance regime change the balance between equilibrium and nonequilibrium processes. Following disturbance, there are substantial changes in most ecological processes, including increased opportunities for colonization by new individuals and often an imbalance between inputs to and outputs from ecosystems (see Chapter 13). For this reason, animals or plants that enhance disturbance frequency or severity increase the importance of processes, such as colonization, that are particularly important under nonequilibrium conditions. The identity of plants that colonize following disturbance, in turn affects the capacity of the ecosystem to gain carbon and retain nutrients.

One of the major avenues by which animals affect ecosystem processes is through physical disturbance (Lawton and Jones 1995, Hobbs 1996). Gophers, pigs, and ants, for example, disturb the soil, creating sites for seedling establishment and favoring early successional species (Hobbs and Mooney 1991). Elephants have a similar effect, trampling vegetation and removing portions of trees (Owen-Smith 1988). By analogy, the Pleistocene megafauna may have promoted steppe grassland vegetation by trampling mosses and stimulating nutrient cycling (Zimov et al. 1995). The shift toward early successional or less woody vegetation generally leads to a lower biomass, a higher ratio of production to biomass, and a litter quality and microenvironment that favor decomposition (see Chapter 13). The associated enhancement of mineralization can either stimulate production (Zimov et al. 1995) or promote ecosystem nitrogen loss (Singer et al. 1984), depending on the magnitude of disturbance.

Beavers in North America are "ecosystem engineers" that modify the availability of resources to other organisms by changing the physical environment at a landscape scale (Jones et al. 1994, Lawton and Jones 1995). The associated flooding of organic-rich riparian soils produces anaerobic conditions that promote methanogenesis, so beaver ponds become hot spots of methane emissions (see Chapter 14) (Roulet et al. 1997). The recent recovery of beaver populations in North America after intensive trapping during the nineteenth and early twentieth centuries has substantially altered boreal landscapes, leading to a fourfold increase in methane emissions in regions where beaver are abundant (Bridgham et al. 1995).

The major ecosystem engineers in soils are earthworms in the temperate zone and termites in the tropics (Lavelle et al. 1997). Soil mixing by these animals alters soil development and most soil processes by disrupting the formation of distinct soil horizons, reducing soil compaction, and transporting organic matter to depth (see Chapter 7).

Plants also alter disturbance regime through effects on flammability. The introduction of grasses into a forest or shrubland, for example, can increase fire frequency and cause the replacement of forest by savanna (D'Antonio and Vitousek 1992). Similarly, boreal conifers are more flammable than deciduous trees because of their large leaf and twig surface area, canopies that extend to the ground surface (acting as ladders for fire to move into the canopy), low moisture content, and high resin content (Johnson 1992). For this reason, the invasion of the northern hardwood forests by hemlock in the early Holocene caused an increase in fire frequency (Davis et al. 1998). The resins in boreal conifers that promote fire also retard decomposition (Flanagan and Van Cleve 1983) and contribute to fuel accumulation.

Plants are often critical in stabilizing soils and reducing wind and soil erosion in early succession. This allows successional development and retains the soil resources that determine the structure and productivity of late-successional stages. Introduced dune grasses, for example, have altered soil accumulation patterns and dune morphology in the western United States (D'Antonio and Vitousek 1992), and early successional alpine vegetation stabilizes soils and reduces probability of landslides.

Species Interactions and Ecosystem Processes

Species interactions modify the impacts of individual species on ecosystem processes. Most ecosystem processes respond in complex ways to changes in the presence or absence of certain species, because *interactions* among species generally govern the extent to which species traits are expressed at the ecosystem level. Species interactions, including mutualism, trophic interactions (predation, parasitism, and herbivory), facilitation, and competition, may affect ecosystem processes directly by modifying pathways of energy and material flow or indirectly by modifying the abundances or traits of species with strong ecosystem effects (Wilson and Agnew 1992, Callaway 1995).

Mutualistic species interactions contribute directly to many essential ecosystem processes. Nitrogen inputs to terrestrial ecosystems, for example, are mediated primarily by mutualistic associations between plants and nitrogen-fixing microorganisms. Mycorrhizal associations between plant roots and fungi greatly aid plant nutrient uptake from soil, increase primary production, and speed succession. Decomposition is accelerated by the presence of highly integrated communities (consortia) of soil microorganisms in which each species contributes a distinct set of enzymes (see Chapter 7). Many mutualisms are highly specific, which increases the probability that loss of a single species will have cascading effects on the rest of the system.

Species that alter trophic dynamics can have large ecosystem impacts. When top predators are removed, prey populations sometimes explode and deplete their food resources, leading to a cascade of ecological effects (see Chapter 11). These **top–down controls** are particularly well developed in aquatic systems. The removal of sea otters by Russian fur traders, for example, allowed a population explosion of sea urchins that overgrazed kelp (Fig. 12.7) (Estes and Palmisano 1974). Recent overfishing in the North Pacific may have triggered similar sea urchin outbreaks, as killer whales moved closer to shore in search of food and switched to sea otters as an alternate prey (Estes et al. 1998). In the absence of dense sea urchin populations, kelp provides the physical structure for diverse subtidal communities and attenuates waves that otherwise augment coastal erosion and storm damage. The addition or removal of a fish species from lakes often has large keystone effects that cascade up or down the food chain (Carpenter et al. 1992, Power et al. 1996a). Many nonaquatic ecosystems also exhibit strong responses to changes in predator abundance (Hairston et al. 1960, Strong 1992, Hobbs 1996). Removal of wolves, for example, releases moose populations that graze down vegetation,

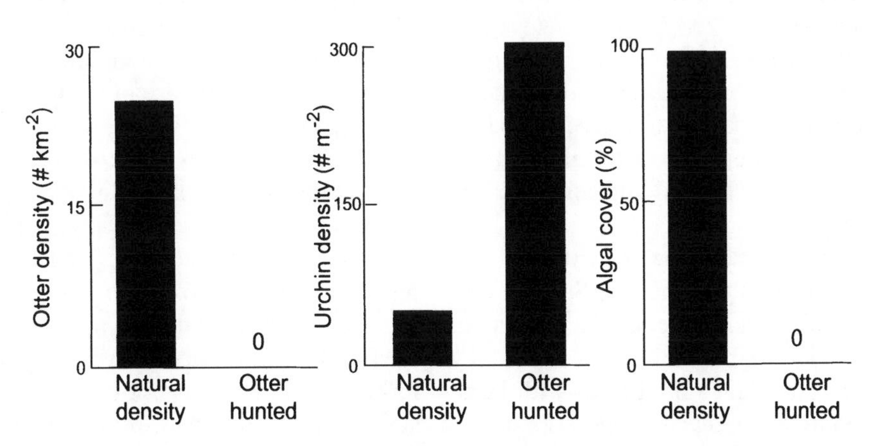

FIGURE 12.7. Density of sea otters and sea urchins, and percentage of macroalgal cover. The latter two parameters were measured at 3 and 9 m depth in the Aleutian Islands of Alaska (Estes and Palmisano 1974). Sites differed in otter density due to differential hunting pressure 300 yr earlier.

and removal of elephants or other keystone mammalian herbivores leads to encroachment of woody plants into savannas (Owen-Smith 1988). Disease organisms, such as rinderpest in Africa, can also act as a keystone species by greatly modifying competitive interactions and community structure (Bond 1993). Plant species that are introduced without their host-specific insect herbivores or pathogens often become aggressive invaders. The cactus *Opuntia*, for example, became surprisingly abundant when introduced to Australia, in part due to overgrazing; but it was reduced to manageable levels by a cactus-specific herbivore *Cactoblastis*. Other species that have become aggressive in the absence of their specialist herbivores include goldenrod (*Solidago* spp.) in Europe, wild rose (*Rosa* spp.) in Argentina, and star thistle (*Centaurea* spp.) in California.

Often these top–down controls by predators or pathogens have much greater effect on biomass and species composition of lower trophic levels than on the flow of energy or nutrients through the ecosystem (Carpenter et al. 1985), because declines in producer biomass are compensated by increased productivity and nutrient cycling rates by other trophic levels. Intensely grazed grassland systems such as the southern and southeastern Serengeti, for example, have a low plant biomass but rapid cycling of carbon and nutrients due to treading and excretion by large mammals. Grazing prevents the accumulation of standing dead litter, which return nutrients to soil in plant-available forms (McNaughton 1985, 1988). Keystone predators or grazers thus alter the *pathway* of energy and nutrient flow, modifying the balance between herbivore-based and detritus-based food chains, but we know less about their effects on total energy and nutrients cycling through ecosystems.

Many species effects on ecosystems are indirect and not easily predicted. Species that themselves have small effects on ecosystem processes can have large indirect effects if they influence the abundance of species with large direct ecosystem effects, as described for trophic interactions. Thus a seed disperser or pollinator that has little direct effect on ecosystem processes may be essential for persistence of a canopy species with a greater direct ecosystem impact. Stream predatory invertebrates alter the behavior of their prey, making them more vulnerable to fish predation, which leads to an increase in the weight gain of fish (Soluck and Richardson 1997). Thus all types of organisms—plants, animals, and microorganisms—must be considered in understanding the effects of biodiversity on ecosystem functioning. Although each of these examples is unique to a particular ecosystem, the ubiquitous nature of species interactions with strong ecosystem effects makes these interactions a general feature of ecosystem functioning (Chapin et al. 2000b). In many cases, changes in these interactions alter the traits that are expressed by species and therefore the effects of species on ecosystem processes. Consequently, simply knowing that a species is present or absent is insufficient to predict its impact on ecosystems. There is currently no clear theoretical framework to predict when these indirect effects are most important. Consequently, the introduction or loss of a species, such as a popular sport fish, often generates unanticipated surprises (Carpenter and Kitchell 1993).

Diversity Effects on Ecosystem Processes

Diversity within a functional type may enhance the efficiency of resource use and retention in ecosystems. Many species in a community appear functionally similar, for example, the nanoplankton in the ocean or the canopy trees in a tropical forest. What are the ecosystem consequences of changes in species diversity *within* a functional type? Evolutionary theory provides some clues. Ecologically similar species co-exist in a community in part because of niche partitioning. In other words, co-existing species differ slightly in their responses to environment, perhaps specializing to use different soil horizons, canopy heights, or times of season. They may also differ in the range of temperatures or water or nutrient availabilities that they exploit effectively (Tilman 1988). These subtle differences in environmental specialization might increase the efficiency of resource use by the community if some species use resources that would otherwise not be tapped by other species.

In experimental grassland communities, for example, plots that were planted with a larger number of species had greater plant cover and lower concentrations of inorganic soil nitrogen than did low-diversity plots (Fig. 12.8) (Tilman et al. 1996). The more diverse plots might use more resources because species have **complementary patterns of resource use**; in other words, species might differ in the types of resources, the location of their roots, or their timing of uptake. Alternatively, diverse plots might use resources more effectively because they are more likely to have a species that is highly effective in capturing resources (**sampling effect**) or are more likely to include species with complementary patterns of resource use (Hooper et al., in press). In other cases, low-diversity ecosystems are quite efficient in using soil resources. Crop or forest monocultures, for example, are often just as productive as mixed cropping systems (Vandermeer 1995) and mixed-species forest stands (Rodin and Bazilevich 1967). Although there are many examples of a positive relationship between species number and productivity or efficiency of resource use, this does not always occur. The effect of species richness frequently saturates at a much lower number of species (5 to 10) than characterize most natural communities. Determining the circumstances and mechanisms in which species number influences ecosystem processes is an active area of ecosystem research (Hooper et al., in press).

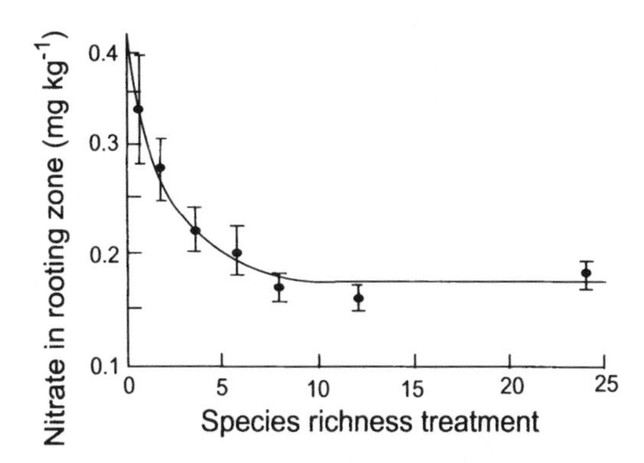

FIGURE 12.8. Effect of the number of plant species sown on a plot on the nitrate concentration in the rooting zone. Measurements were made 3 years after the plots were sown. Data are means ± SE. (Redrawn with permission from *Nature*; Tilman et al. 1996.)

Diversity of functionally similar species stabilizes ecosystem processes in the face of temporal variation in environment. In ecosystems in which functionally similar species differ in environmental response, this can buffer ecosystem processes from environmental fluctuations (McNaughton 1977, Chapin and Shaver 1985). Tropical tree species, for example, differ subtly in their growth response to nutrients (Fig. 12.9). Conditions that favor some species will likely reduce the competitive advantage of other functionally similar species, thus stabilizing the total biomass or activity by the entire community. In other words, in **compensation** for the reduced growth by some species, other species grow more. For example, in one study, annual variation in weather caused at least a twofold variation production by each of the major vascular plant species in arctic tussock tundra. Years that were favorable for some species, however, reduced the productivity of others, so there was no significant difference in productivity at the ecosystem scale among the 5 years examined (Chapin and Shaver 1985). Directional changes in environment can also cause less change in total biomass than in the biomass of individual species for similar reasons; some species respond positively to the change in environment, whereas other species respond negatively. This stabilization of biomass and production by diversity has been observed in many (but not all) studies (Cottingham et al. 2001), including grasslands, in response to the addition of water and nutrients (Lauenroth et al. 1978) and to grazing (McNaughton 1977); in tundra, in response to changes in temperature, light, and nutrients (Chapin and Shaver 1985); and in lakes, in response to acidification (Frost et al. 1995). This stability of processes provided by diversity has societal relevance. Many traditional farmers plant diverse crops, not to maximize productivity in a given year but to decrease the chances of crop failure in a bad year (Altieri 1990). Even the loss of rare species

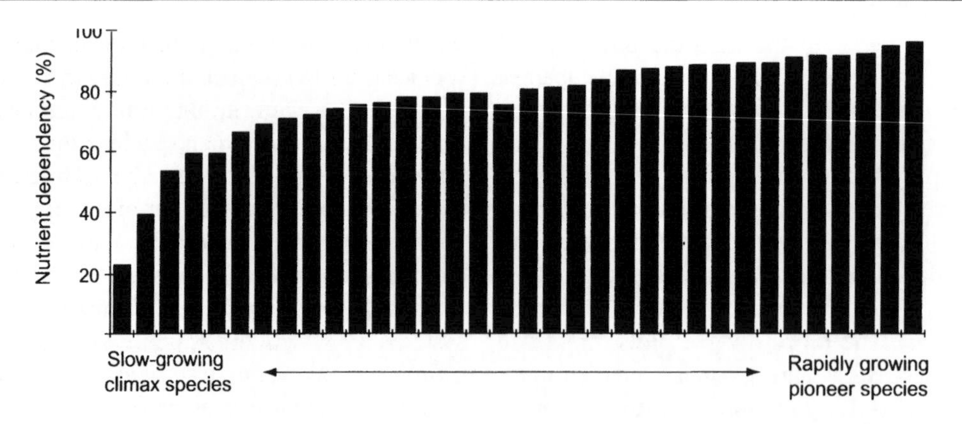

FIGURE 12.9. Graded environmental response of species within a functional type. Nutrient dependency is the growth rate of fertilized seedlings of canopy trees from the dry tropical forest relative to unfertilized seedlings. Groups of species that show a continuum in environmental response are more likely to compensate for the removal of an ecologically similar species than are species that are radically different in their ecological response. (Figure provided by permission of P. Huante; Huante et al. 1995.)

may jeopardize the resilience of ecosystems. Rare species that are functionally similar to abundant ones in rangelands, for example, become more common when their abundant counterparts are reduced by overgrazing. This compensation in response to release from competition minimizes the changes in ecosystem properties (Walker et al. 1999).

Species diversity also reduces the probability of outbreaks by pest species by diluting the availability of their hosts. Often the spread of outbreak species depends strongly on host density.

Diversity provides insurance against change in functioning under extreme or novel conditions. Species diversity not only stabilizes ecosystem processes in the face of annual variation in environment but also provides insurance against drastic change in ecosystem structure or processes in response to extreme events (Walker 1992, Chapin et al. 1997). Any change in climate or climatic extremes that is severe enough to cause extinction of one species is unlikely to eliminate all members from a functional type (Walker 1995). The more species there are in a functional type, the less likely it is that any extinction event or series of such events will have serious ecosystem consequences (Holling 1986). In a study of temperate grassland, for example, an extreme drought eliminated or reduced the growth of many plant species. The drought had least impact on productivity in those plots with highest species diversity (Fig. 12.10). Results from this field experiment are somewhat difficult to interpret because plots of low diversity were the result of long-term addition of nitrogen fertilizer; the nitrogen addition caused competitive elimina-

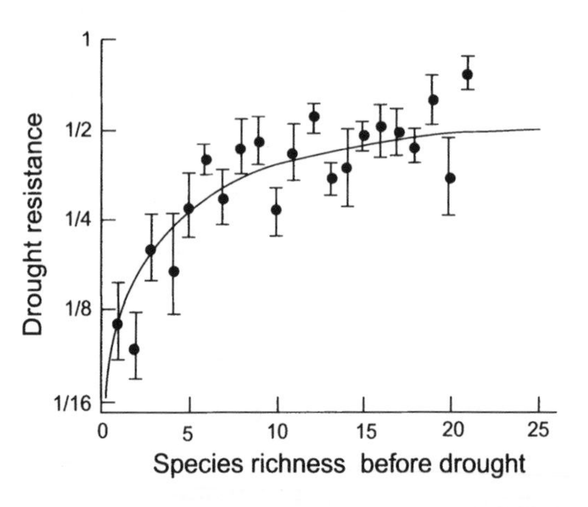

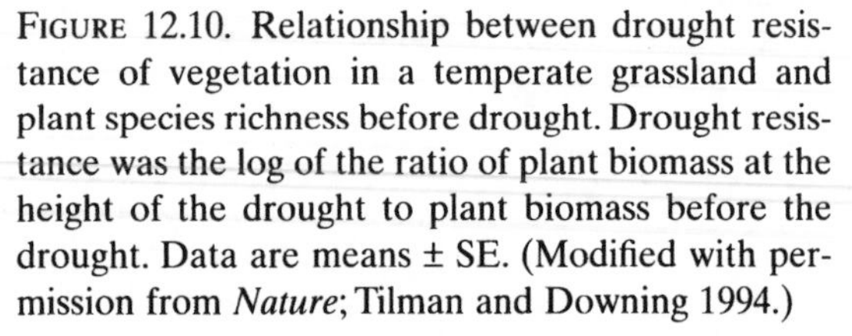
FIGURE 12.10. Relationship between drought resistance of vegetation in a temperate grassland and plant species richness before drought. Drought resistance was the log of the ratio of plant biomass at the height of the drought to plant biomass before the drought. Data are means ± SE. (Modified with permission from *Nature*; Tilman and Downing 1994.)

tion of the more drought-tolerant species. In a laboratory experiment that manipulated species diversity of mosses, communities with high species diversity maintained a higher biomass when exposed to drought than did less-diverse communities by facilitating the survival of tall dominant mosses (Mulder et al. 2001). Though theoretical studies also predict such buffering or insurance against extreme events, there is relatively little experimental evidence to test for this effect.

Differences in environmental response among functionally different species may accentuate ecosystem change. In contrast to the buffering provided by ecologically similar species, species that differ in their response to the environment *and* in their effects on ecosystem processes can make ecosystems vulnerable to change. Rising concentrations of atmospheric CO_2, for example, can reduce plant transpiration, resulting in increased magnitude or duration of soil moisture (Owensby et al. 1996). This, in turn, can shift the competitive balance from grasses to shrubs, promoting shrub encroachment into grasslands and savannas and causing replacement of one biome by another.

Global environmental change is causing many ecosystems to experience novel conditions of nitrogen deposition and atmospheric CO_2 concentrations. If the principles discussed in this chapter apply broadly, we expect that the diversity of natural ecosystems will be critical in determining the biotic properties of ecosystems (i.e., the diversity of functional types) and their vulnerability to change (the buffering provided by diversity within functional types).

Summary

The species diversity of Earth is changing rapidly due to frequent species extinctions (both locally and globally), introductions, and changes in abundance. We are, however, only beginning to understand the ecosystem consequences of these changes. Many species have traits that strongly affect ecosystem processes through their effects on the supply or turnover of limiting resources, microclimate, and disturbance regime. The impact of these species traits on ecosystem processes depends on the abundance of a species, its functional similarity to other species in the community, and species interactions that influence the expression of important traits at the ecosystem scale.

Diversity per se may be ecologically important if it leads to complementary use of resources by different species or increases the probability of including species with particular ecological effects. Because species belonging to the same functional type generally differ in their response to environment, diversity within a functional type may stabilize ecosystem processes in the face of temporal variation or directional changes in environment. Introduction of functionally different species to an ecosystem, in contrast, may accelerate the rate of ecosystem change. The effects of species traits on ecosystem processes are generally so strong that changes in the species composition or diversity of ecosystems are likely to alter their functioning, although the exact nature of these changes is frequently difficult to predict.

Review Questions

1. What are functional types? What is the usefulness of the functional-type concept if all species are ecologically distinct?
2. How is the expected ecosystem impact of the loss of a species affected by (a) the number of species in the ecosystem, (b) the abundance or dominance of the species that is eliminated, or (c) the type of species that is eliminated? Explain.
3. If a new species invades or is lost from an ecosystem, which species traits are most likely to cause large changes in productivity and nutrient cycling? Give examples that illustrate the mechanisms by which these species effects occur.
4. Which species traits have greatest effects on regional processes such as climate and hydrology?
5. How do species interactions influence the effect of a species on ecosystem processes?
6. How does the diversity of species *within a functional type* affect ecosystem processes?

What is the mechanism by which this occurs? Why is it important to distinguish between the effects of changes in species composition within vs. between functional types?

7. What are the mechanisms by which species diversity might affect nutrient uptake or loss in an ecosystem. Suggest an experiment to distinguish between these possible mechanisms. Design an agricultural ecosystem that maintains crop productivity but has tight nutrient cycles.

Additional Reading

Chapin, F.S. III, E.S. Zaveleta, V.T. Eviner, R.L. Naylor, P.M. Vitousek, S. Lavorel, H.L. Reynolds, D.U. Hooper, O.E. Sala, S.E. Hobbie, M.C. Mack, and S. Diaz. 2000. Consequences of changing biotic diversity. *Nature* 405:234–242.

Frost, T.M., S.R. Carpenter, A.R. Ives, and T.K. Kratz. 1995. Species compensation and complementarity in ecosystem function. Pages 224–239 *in* C.G. Jones, and J.H. Lawton, editors. *Linking Species and Ecosystems*. Chapman & Hall, New York.

Johnson, K.G., K.A. Vogt, H.J. Clark, O.J. Schmitz, and D.J. Vogt. 1996. Biodiversity and the productivity and stability of ecosystems. *Trends in Ecology and Evolution* 11:372–377.

Lawton, J.H., and C.G. Jones. 1995. Linking species and ecosystems: Organisms as ecosystem engineers. Pages 141–150 *in* C.G. Jones, and J.H. Lawton, editors. *Linking Species and Ecosystems*. Chapman & Hall, New York.

Power, M.E., D. Tilman, J.A. Estes, B.A. Menge, W.J. Bond, L.S. Mills, G. Daily, J.C. Castilla, J. Lubchenco, and R.T. Paine. 1996. Challenges in the quest for keystones. *BioScience* 46:609–620.

Vandermeer, J. 1995. The ecological basis of alternative agriculture. *Annual Review of Ecology and Systematics* 26:201–224.

Vitousek, P.M. 1990. Biological invasions and ecosystem processes: Towards an integration of population biology and ecosystem studies. *Oikos* 57:7–13.

Walker, B.H. 1995. Conserving biological diversity through ecosystem resilience. *Conservation Biology* 9:747–752.

Wilson, J.B., and D.Q. Agnew. 1992. Positive-feedback switches in plant communities. *Advances in Ecological Research* 23:263–336.

Part III
Patterns

13 Temporal Dynamics

Ecosystem processes constantly change in response to variation in environment over all time scales. This chapter describes the major patterns and controls over the temporal dynamics of ecosystems.

Introduction

Ecosystems are always recovering from past changes. In earlier chapters we emphasized ecosystem responses to the *current* environment. Ecosystems are, however, always responding to past changes that have occurred over all time scales (Holling 1973, Wu and Loucks 1995). These changes include relatively predictable daily and seasonal variations, less predictable changes in weather (e.g., passage of weather fronts, El Niño events, and glacial cycles), and occurrence of disturbances (e.g., tree falls, herbivore outbreaks, fires, and volcanic eruptions). Consequently, the behavior of an ecosystem is always influenced by both the current environment and many previous environmental fluctuations and disturbances.

The global environment is changing more rapidly than it has for millions of years. These changes result from an exponentially rising human population that shows an every-increasing technological capacity to alter Earth's environment and ecosystems. Perhaps the most urgent need in ecosystem ecology is to improve our understanding of factors governing the stability and change in ecological systems (Box 13.1). This understanding is critical to managing ecosystems so they sustain their diversity and other important ecological attributes and so ecosystems continue to produce the goods and services required by society. This chapter addresses the basis of the temporal dynamics of ecosystems.

Fluctuations in Ecosystem Processes

Interannual Variability

Ecosystem processes measured in 1 year are seldom representative of the long-term mean. Many ecosystem processes are sensitive to interannual variations in weather and to fluctuations in the internal dynamics of ecosystems, such as outbreaks of herbivores or pathogens. The same ecosystem, for example, can change from being a carbon source in one year to a carbon sink in the next (Goulden et al. 1998). The production at a particular trophic level can change from being limited by food to being limited by predation. Some of the interannual variability in ecosystem dynamics reflects processes that are potentially predictable, such as the cyclic variation in climate. El Niño events are a consequence of large-scale oscillations in the global ocean-atmosphere system that recur every 2 to 10 years (see Chapter 2). These are associated with relatively repeatable climatic

Box 13.1. Ecosystem Resilience and Change

An emerging challenge in ecosystem ecology is to improve our understanding of the properties and processes that allow ecosystems to persist in the face of environmental change. Ecological stability is an issue that has intrigued ecologists for decades. Discussions of stability initially focused on changes in natural populations of plants and animals but have since been broadened to include ecosystem processes. Although we may have an intuitive feel for what constitutes a stable ecosystem, stability is difficult to define. There are at least 160 different definitions of *stability* in the literature (Grimm and Wissel 1997). Despite the wide range of definitions, some fundamental generalizations emerge that provide a basis for discussing ecosystem responses to change.

The response of any system to a **perturbation** (i.e., a external force that displaces the system from equilibrium) can be described in terms of a few general properties (Holling 1986, Berkes and Folke 1998) (Fig. 13.1). The **response** of a system to perturbation describes the direction and magnitude of change in the system after a perturbation. The **resistance** of a system describes its tendency to remain in its reference state in the face of a perturbation; in ecological terms, resistance is the capacity of a system to maintain certain structures and functions despite disturbance. An ecosystem that shows little change in structure, productivity, or rate of nutrient cycling in response to a drought or fire, for example, is resistant to those perturbations. The resistance of an ecosystem to perturbation depends on the nature, magnitude, and duration of the perturbation as well as on the nature of the system. Ecosystems are often particularly vulnerable to new types of disturbances that they have not previously experienced. The **recovery** of a system describes the extent to which it returns to its original state after perturbation. An ecosystem recovers, if negative feedbacks that resist changes in ecosystem properties are stronger than positive feedbacks that push the ecosystem toward some new state (see Chapter 1). The recovery of an ecosystem depends on the magnitude of the response and the time since the perturbation.

Ecological **resilience** has been defined in many ways. One use of the term denotes elasticity, or the rate at which a system returns to a reference state following perturbation (May 1973, Pimm 1984). Systems with low resilience may never recover to their original state and are readily converted to a new state (Berkes and Folke 1998). The

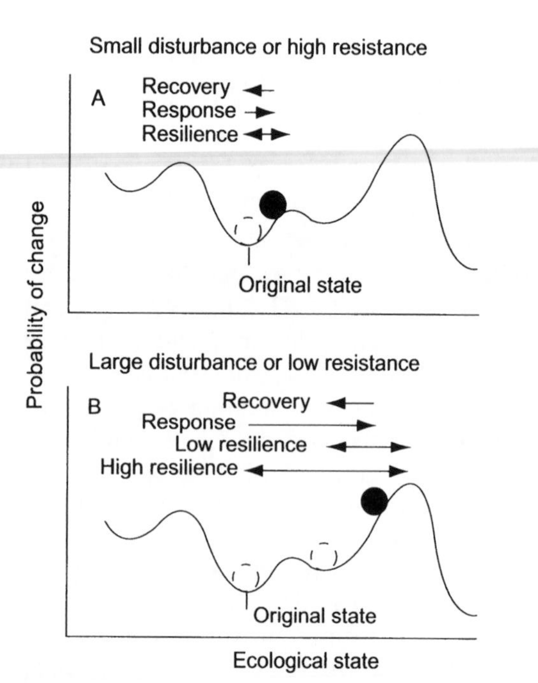

FIGURE 13.1. Properties of a system that influence its probability of changing state. The solid ball represents the state of the system after a perturbation. The open ball shows the most likely future states of the system. **A**, A system shows a small response to a perturbation if the perturbation is small or the system is resistant to change. **B**, After a perturbation, a system can assume many possible states; if it is highly resilient, it may return quickly to its original state; if it is less resilient, or if the perturbation is large, the system may move to a new state.

close interdependence of resilience and resistance and the many alternative definitions of these terms that have been proposed have contributed to the confusion in terminology related to resilience. The important point is that systems that maintain their properties despite disturbance (i.e., are resistant to change) and that return rapidly to their original state (i.e., are resilient) exhibit more stable and predictable ecosystem properties (Holling 1986).

These properties allow us to assess the sensitivity of ecosystem processes to change. They imply, however, that an ecosystem can be characterized by a reference state that is typical of the system. All ecosystems are highly resilient to regular daily and seasonal variations. Their properties may change, however, in response to directional changes in environment or to disturbances that are particularly novel or severe (Fig. 13.1). The changes that are most important to consider are those that have large effects and occur on time scales at least as long as those of the processes we are studying. The biomass of a forest stand, for example, is relatively resistant to rapid environmental variations, such as interannual variations in climate; biomass is, however, influenced by more long-lasting effects caused by wind storms, recent herbivore outbreaks, and successional development. Net primary production responds more rapidly to environment than does biomass and requires additional consideration of shorter-term changes, such as interannual variation in climate. Resilience must therefore be defined with respect to particular ecosystem properties and perturbations, rather than being considered an absolute property of an ecosystem.

patterns, such as drought in Southeast Asia and the continental interior of North America and rains in southwestern North America. Glacial cycles occur over thousands of years in response to variations in solar input and feedbacks from the biosphere to climate (see Chapter 2). An emerging challenge of climate and ecosystem modeling is to understand these climate oscillations well enough to predict or explain their effects on interannual variation in ecosystem processes.

The internal dynamics of ecosystems also generate large interannual fluctuations in ecosystem processes. The population density of herbivores, for example, can vary more than 100-fold over a few years, causing large fluctuations in plant biomass, nutrient cycling, and other processes (Fig. 13.2). The causes of these fluctuations and cycles are debated but probably reflect interactions among plants, herbivores, predators, and parasites. One important interaction, for example, may occur between plants and their herbivores. Herbivore populations often decline after a depletion of their food supply, due to insufficient food and/or buildup of predators. Theoretical studies suggest that, if the recovery of an herbivore population lags behind the vegetation recovery, this generates a population cycle, with a period that is four times the length of the time lag (May 1973). When snowshoe hares heavily browse their food, for example, plants produce new shoots that remain toxic for 2 to 3 years. This contributes to low densities of hare populations for 2 to 3 years after a population crash. This time lag may contribute to the 11-year period in the snowshoe hare cycle (Bryant 1981). An important challenge for ecosystem ecologists is to improve understanding of the interactions between external factors, such as climate, and internal ecosystem dynamics in causing the natural variability in ecosystem processes.

Long-Term Change

Today's ecosystem processes depend on both the current environment and past events. Legacies are the persistent effect of past events. Legacies affect ecosystem processes over a wide range of time scales. Individual redwood trees (*Sequoia sempervirens*) in coastal California, for example, can live for thousands

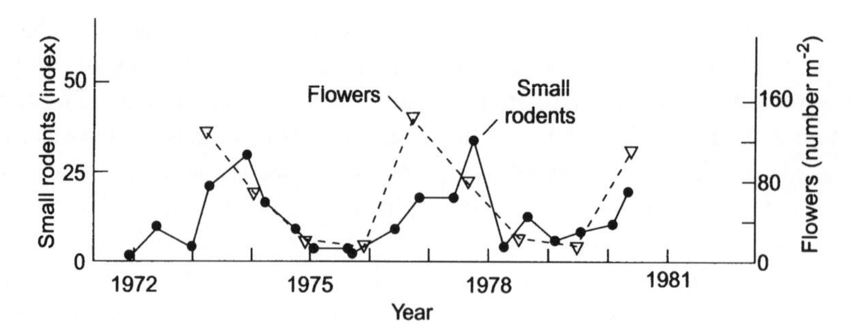

FIGURE 13.2. Interannual variation in flowering density of an understory shrub (*Vaccinium myrtillus*) and of small rodents in northern Finland (Laine and Henttonen 1983). These herbivores and their food plants show approximately 4-year cycles of abundance.

of years. During the Tertiary they occupied a warm moist environment throughout much of western North America. Their range is now restricted to valleys where coastal fog minimizes summer drought stress. Aspen clones in the Rocky Mountains range in age from a few years to as much as 10,000 years (Kemperman and Barnes 1976, Tuskan et al. 1996). The current distribution of redwoods and aspens is therefore a product of past population and community processes and cannot be fully understood with reference only to the present environment. Many species are still migrating poleward in response to the disappearance of continental ice sheets 10,000 years ago. The soils beneath these recent arrivals may still reflect the properties of earlier communities rather than being completely a function of current vegetation.

The current functioning of the East Siberian coastal plain also reflects processes that occurred thousands of years ago. During the Pleistocene, this region was a steppe grassland that accumulated highly organic loess soils and ice lenses that now occupy 50 to 70% of the soil volume (Zimov et al. 1995). As the ice melted in the warmer Holocene climate, the soils subsided, forming lakes with organic-rich sediments. Pleistocene-age organic matter is now the major carbon source for methane production by these lakes (Zimov et al. 1997). In other words, the current processing of carbon in these lakes is strongly influenced by processes that occurred 10,000 to 100,000 years ago.

Current ecosystem processes also respond to changes that have occurred more recently. Large areas of Europe and northeastern North America were deforested for agriculture in recent centuries and have more recently reverted to forests. Even forests older than 200 years still exhibit composition and dynamics that reflect their earlier history (Foster and Motzkin 1998) (Fig. 13.3). A plow layer is still evident in these forests, for example, resulting in a sharp vertical discontinuity in soil processes and nutrient supply. Net primary production (NPP) substantially exceeds the rates of heterotrophic respiration in these ecosystems, so vegetation and soils are actively accumulating carbon (Goulden et al. 1996). Any study of carbon dynamics that ignored these historical legacies would seriously misinterpret the relationship between carbon balance of

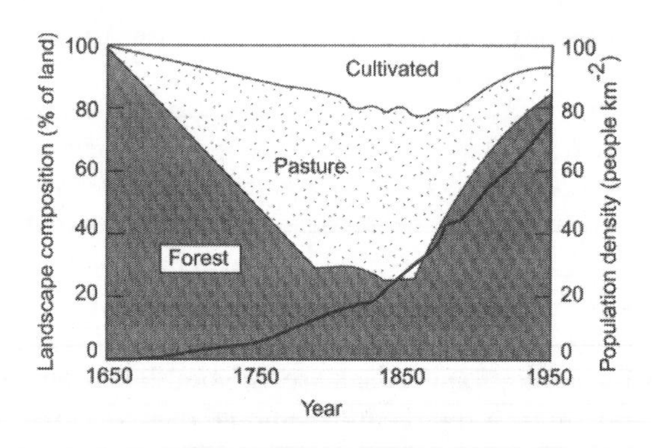

FIGURE 13.3. Changes in land use and population density in central Massachusetts (northeastern United States) since European colonization. Most forests in this region were previously croplands or pastures. (Redrawn with permission from *Ecosystems*; Foster and Motzkin 1998.)

ecosystems and current climate. Similarly, the legacies of the current environment will exert large effects on the future structure and functioning of ecosystems.

Disturbance

Conceptual Framework

Disturbance is a major cause of long-term fluctuations in the structure and functioning of ecosystems. We define **disturbance** as a relatively discrete event in time and space that alters the structure of populations, communities, and ecosystems and causes changes in resource availability or the physical environment (Pickett and White 1985, Pickett et al. 1999). Many natural disturbances, such as herbivore outbreaks, treefalls, fires, hurricanes, floods, glacial advances, and volcanic eruptions, exert these effects through reductions in live plant biomass or sudden changes in the pool of actively cycling soil organic matter. Disturbance is difficult to define unambiguously. Events such as intensive grazing and subzero temperatures that are normal features of some ecosystems seriously disrupt the functioning of others. Disturbance must therefore be defined in the context of the normal range of environmental variation that an ecosystem experiences. The dividing line between disturbance and normal function is somewhat arbitrary. Herbivory, for example, is often treated as part of the steady-state functioning of ecosystems, whereas stand-killing insect outbreaks are treated as disturbances. The processes are similar, however, and there is a continuum in size, severity, and frequency between these two extremes. Disturbance is clearly not an external event that "happens" to an ecosystem. Like other interactive controls, disturbance is an integral part of the functioning of all ecosystems that responds to and affects most ecosystem processes. Naturally occurring disturbances such as fires and hurricanes are therefore not "bad"; they are normal properties of ecosystems.

Human activities have altered the frequency and size of many natural disturbances, such as fires and floods, and have produced new types of disturbance, such as large-scale logging, mining, and wars. Many human disturbances have ecological effects that are similar to those of natural disturbances, so the study of either natural or human disturbances provides insights into the regulation of ecosystem processes and human impacts on these processes. Natural and anthropogenic disturbances interact with environmental gradients to create much of the spatial patterning in landscapes (see Chapter 14).

After disturbance, ecosystems undergo **succession**, a directional change in ecosystem structure and functioning resulting from biotically driven changes in resource supply. Disturbances that remove live or dead organic matter, for example, are colonized by plants that gradually reduce the availability of light at the soil surface and alter the availability of water and nutrients (Tilman 1985). If there were no further disturbance, succession would proceed toward a **climax**, the end point of succession (Clements 1916). At this point, the structure and rates of ecosystem processes approach a steady state in which resource demand by vegetation would be balanced by the rate of resource supply. In practice, however, new disturbances usually occur before succession reaches a climax, so individual stands of an ecosystem are seldom in steady state. Nonetheless, the concept of directional changes in vegetation after disturbance provides a useful framework for analyzing the role of disturbance in ecosystem processes.

Succession occurs in response to biotically driven changes in resource supply, which typically occur over time scales of years to centuries. Succession does not therefore include the seasonal fluctuations in ecosystem processes from summer to winter, which are driven more directly by climate than by the internal dynamics of ecosystems.

Disturbance Properties

The impact of disturbance on ecosystem processes depends on its severity, frequency, type, size, timing, and intensity. Together these attributes of disturbance constitute the **disturbance**

regime (Heinselman 1973). Disturbance regime is a filter that influences the types of organisms present and therefore the structure and functioning of ecosystems.

Disturbance severity is the magnitude of change in resource supply or environment caused by disturbance. For those disturbances that remove vegetation and/or soils, disturbance severity is the quantity of organic matter removed from plants or soil by the disturbance. **Primary succession** occurs after severe disturbances that remove or bury most products of ecosystem processes, leaving little or no organic matter or organisms. Disturbances leading to primary succession include volcanoes, glaciers, landslides, mining, flooding, coastal dune formation, and drainage of lakes. **Secondary succession** occurs on previously vegetated sites after disturbances such as fire, hurricanes, logging, and agricultural plowing. These disturbances remove or kill most live aboveground biomass but leave some soil organic matter and plants or plant propagules in place. Disturbance severity is probably the major factor determining the rate and trajectory of vegetation development after disturbance. A severe fire that kills all plants, for example, has a different effect on vegetation recovery than does a fire that burns only surface litter, allowing surviving vegetation to resprout. There is also a continuum in disturbance severity between large-scale defoliation events and the removal of a single leaf by a caterpillar or between landslides and the burial of surface litter by an earthworm. In other words, there is a continuum in disturbance severity between the day-to-day functioning of ecosystems and events that initiate primary succession (Fig. 13.4).

Many disturbances can be characterized by **intensity**, the energy released per unit area and time. Intensity of a disturbance often influences the severity of its effect. Intense marine storms, for example, dislodge many intertidal organisms, and intense hurricanes blow down many trees. Fire intensity is rate of heat production. It depends on the mass, consumption rate, and energy content of the fuel (Johnson 1992) and determines the temperatures that organisms experience and therefore their probability of surviving fire. Intense fires are not always severe, however. In fact, slow smoldering fires of low intensity frequently consume more fuel (are more severe) than are intense fires that move rapidly through a stand.

Disturbance frequency varies dramatically among ecosystems and among disturbance types. Herbivory occurs continuously in most ecosystems. At the opposite extreme, volcanoes

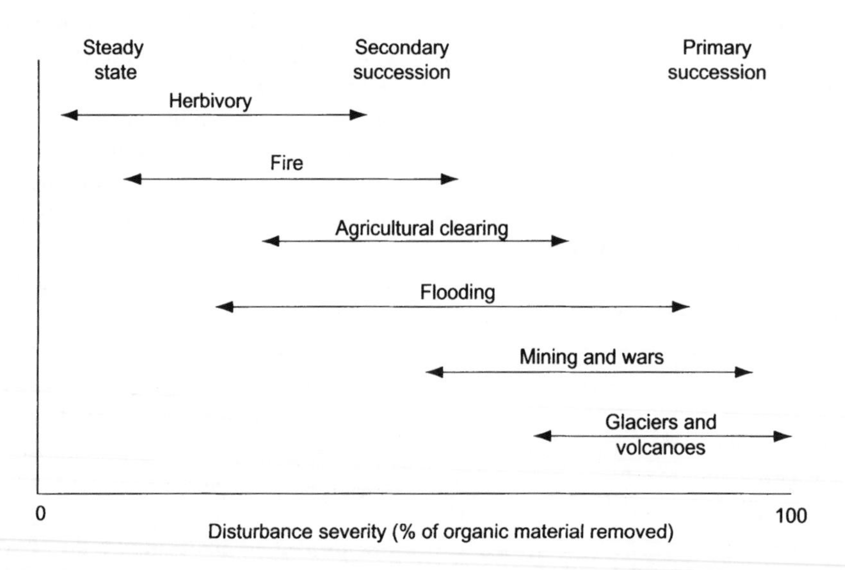

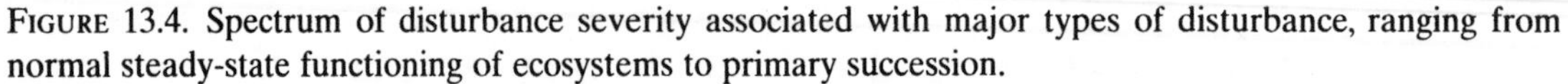
FIGURE 13.4. Spectrum of disturbance severity associated with major types of disturbance, ranging from normal steady-state functioning of ecosystems to primary succession.

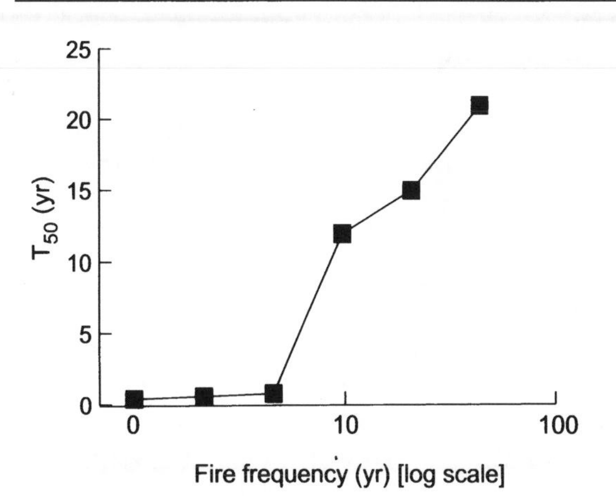

FIGURE 13.5. Relationship of fire frequency to the time (t_{50}) required for an ecosystem to accumulate 50% of its maximum biomass (Chapin and Van Cleve 1981). Ecosystems with frequent disturbance recover more quickly.

or floods may never have occurred in some locations. Average fire frequency ranges from once per year in some grasslands to once every several thousand years in some mesic forests. Ecosystems are usually most resilient to disturbances that occur frequently. Ecosystems that experience frequent fire, for example, support fire-adapted species that recover biomass more quickly than in ecosystems in which fire occurs infrequently (Fig. 13.5). Human activities often modify disturbance frequency through initiation or suppression of disturbance. Damming of streams can eliminate spring floods that scour sediments and detritus from channels, resulting in large changes in stream food webs and capacity to support fish (Power 1992a). Fire suppression in the giant sequoias (*Sequoiadendron gigantea*) of the Sierra Nevada mountains of California made this ecosystem more vulnerable to fire as a result of the growth of understory trees that formed a ladder for fire to reach from the ground to the canopy. Although the thick-barked sequoias are resistant to ground fires, they are vulnerable to fires that extend into the canopy. In this way, fire suppression has increased the risk of catastrophic fires that could eliminate giant sequoias.

Disturbance type influences ecosystem processes independent of frequency and severity. Organisms adapt to disturbances that occur relatively frequently in their current environment or in their evolutionary past. They are often vulnerable to novel disturbances. Benthic communities, for example, may recover slowly from bottom trawling that scrapes surface sediments. Many upland species are intolerant of flooding, whereas many trees from wet environments, such as tropical wet forests, have thin bark and are killed by fire. Some traits, such as heat-induced germination of chaparral postfire annuals, enable species to respond appropriately to specific types of disturbance. Other traits enable species to colonize many types of disturbances. Weedy species, for example, produce abundant small seeds that disperse long distances or remain dormant in the soil from one disturbance to the next. Their germination is often triggered by environmental conditions characteristic of most disturbed sites (Fenner 1985, Baskin and Baskin 1998), so they are relatively insensitive to disturbance type. Novel disturbances are more likely to lead to slow recovery or to trigger a new successional trajectory than are disturbances to which organisms are well adapted.

Disturbance size is highly variable. **Gap-phase succession**, for example, occurs in small gaps created by the death of one or a few plants. Many tropical wet forests or intertidal communities, for example, are mosaics of gaps of different ages. Other ecosystems develop after **stand-replacing disturbances** that can be hundreds of square kilometers in area. Disturbance size influences ecosystems primarily through its effects on landscape structure, which influences lateral flow of materials, organisms, and disturbance among patches in the landscape (see Chapter 14). Disturbance size, for example, affects the rate of seed input after fire. Small fires are readily colonized by seeds that blow in from surrounding unburned patches or are carried by mammals and birds, whereas regeneration in the middle of large fires, fields, or clearcuts may be limited be seed availability and be colonized primarily by light-seeded species that disperse long distances. Disturbance size also influences the spread of herbivores and pathogens that colonize early successional sites. **Disturbance pattern** on the landscape influences the effective size of a disturbance event. Disturbances often leave islands of undisturbed vegetation that act as propagule sources, causing the effective size of the disturbance to be much smaller than its area would suggest (Turner et al. 1997). A series of dams on a river creates a chain of interconnected lakes that can be colonized much more readily than isolated kettle lakes formed after glacial retreat or farm ponds formed by restricting groundwater flow.

The **timing of disturbance** often influences its impact. A strong freeze or fire that occurs during bud break has greater impact than one that occurs 2 weeks before bud break. Similarly, anaerobic conditions associated with flooding of the Mississippi River during the 1993 growing season caused more root and tree mortality than if the flood had occurred when roots were inactive. Hydroelectric dams may eliminate seasonal flooding associated with rain or snowmelt and alter flow based on electricity demand. Human activities often change the timing of disturbances such as grazing, fire, and flooding.

Disturbance is one of the key interactive controls that governs ecosystem processes (see Chapter 1) through its effects on other interactive controls (microclimate, soil resource supply, functional types of organisms, and probability of future disturbance). Postfire stands, for example, often have warm soils that have a high water content; this occurs due to the low albedo of the charred surface and the decrease in leaf area that transpires water and shades the soil. Fire both volatilizes nitrogen, which is lost from the site, and returns inorganic nitrogen and other nutrients to the soil in ash, thus altering soil resource supply. The net effect of fire is usually to enhance nutrient availability, although the magnitude of this effect depends on fire severity and intensity (Wan et al. 2001). Fire affects the functional types of plants in an ecosystem through its effects on differential survival and competitive balance in the postfire environment. Because of its sensitivity to, and effect on, other interactive controls, changes in disturbance regime alter the structure and functioning of the ecosystem.

Succession

Successional changes occurring over decades to centuries explain much of the local variation among ecosystems. Although climate, soils, and topography explain most of the broad global and regional patterns in ecosystem processes, disturbance regime and postdisturbance succession account for many of the local patterns of spatial variability (see Chapter 14). In this section, we describe common patterns of successional change in major ecosystem processes. These successional changes are most clearly delineated in primary succession, so we begin with a description of primary successional processes and then describe how the patterns differ between primary and secondary succession.

Ecosystem Structure and Composition

Primary Succession

Succession involves a change from a community governed by the dynamics of colonization to one governed by competition for resources. Vegetation development after disturbance is strongly influenced by the initial colonization events, which in turn depend on environment and the availability of propagules (Egler 1954, Connell and Slatyer 1977, Bazzaz 1996). Severe disturbances such as glaciers, volcanic eruptions, and mining eliminate most traces of previous vegetation and must be colonized from outside the disturbed site. Most initial colonizers of these primary successional sites have small seeds that can disperse long distances by wind. Fresh lava or land exposed by retreat of glaciers, for example, is first colonized by wind-dispersed spores of algae, cyanobacteria, and lichens, which form crusts that stabilize soils (Worley 1973). These are followed by small-seeded wind-dispersed vascular plants (primarily woody species), whose arrival rates depend largely on distance to seed source (Shiro and del Moral 1995). Late successional species with heavier seeds generally arrive more slowly (Fig. 13.6).

Many of the species that become abundant early in primary succession are free living or

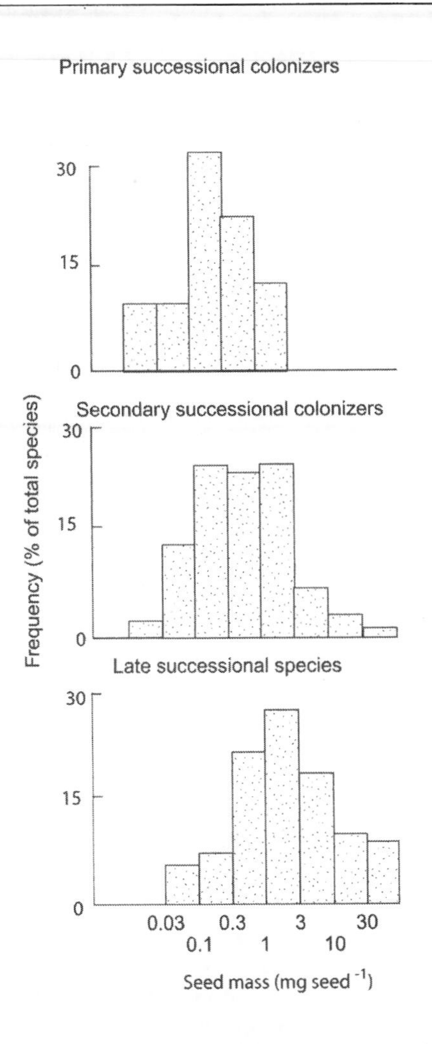

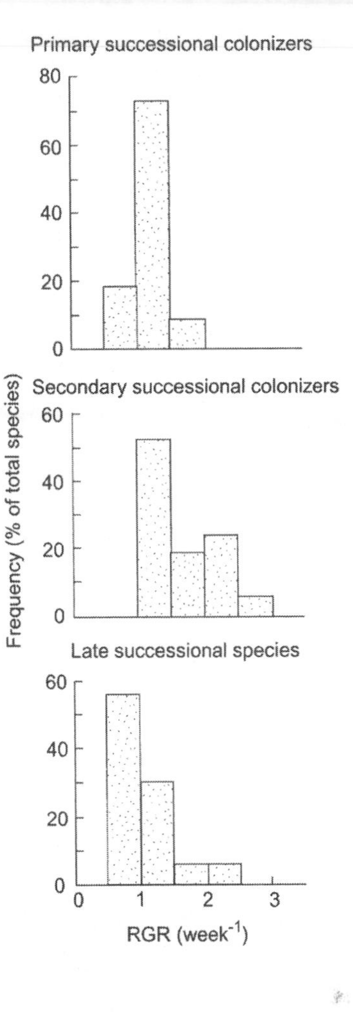

FIGURE 13.6. Frequency distribution of log seed mass and relative growth rate (RGR) for British species that are primary successional colonizers, secondary successional colonizers, and late successional species. (Data from Grime and Hunt 1975; Grime et al. 1981. Redrawn with permission from Blackwell Scientific; Chapin 1993b.)

symbiotic nitrogen fixers. Vascular plant species capable of symbiotic nitrogen fixation occur frequently (about 75% of sites studied) in early primary succession, although they dominate the vegetation only about 25% of the time (Walker 1993). These species are most common on glacial moraines and mudflows; intermediate on mine tailings, landslides, floodplains, and dunes; and least abundant on volcanoes and rock outcrops (Walker 1993). When early successional colonizers fix abundant nitrogen, their net effect is generally to promote the establishment and growth of later successional species (Walker 1993) (Fig. 13.7).

The long-term successional trajectory of vegetation is strongly influenced by the species composition of the initial colonizers because the opportunities for colonization decline as succession proceeds. In many forests, all tree species colonize in early succession, and the successional changes in dominance reflect species differences in size and growth rate (Egler 1954, Walker et al. 1986). In other cases, late successional species may establish more gradually. As succession proceeds, the soil becomes covered by leaf litter, making a less favorable seed bed, and there is increasing competition for light and other resources. The

Stage	Pioneer	*Dryas*	Alder	Spruce
Life history patterns	Dominance by light-seeded species	Dominance by rapidly growing species	Dominance by tall shrubs of intermediate longevity	Dominance by long-lived trees
Facilitative effects	↑ Survivorship	↑ N (weak) ↑ Growth (weak)	↑ SOM ↑ N ↑ Mycorrhizae ↑ Growth	↑ Germination
Inhibitory effects	↓ Germination (weak)	↓ Germination ↓ Survivorship ↑ Seed predation and mortality	↓ Germination ↓ Survivorship ↑ Seed predation and mortality Root competition Light competition	↓ Growth ↓ Survivorship ↑ Seed predation and mortality Root competition Light competition ↓ N
Impacts of herbivory	Minimal	Reduce growth of early successional species	Eliminate early successional species	Minimal

FIGURE 13.7. Interaction of life history patterns, facilitative and inhibitory effects, and herbivory in causing successional change after glacial retreat at Glacier Bay, Alaska. Life history patterns determine the type of species that dominate at each successional stage. The rate at which this dominance changes is determined by facilitative or inhibitory effects of the dominant species and by patterns of herbivory. In general, all four of these processes contribute simultaneously to successional change, with the most important processes being life history patterns in the pioneer stage, herbivory in mid-successional stages, facilitation in the alder stage, and inhibition in late succession. SOM, soil organic matter. (Modified with permission from *Ecological Monographs*; Chapin et al. 1994.)

effects of initial colonizers on their physical environment also influence the identity of species that can subsequently establish and compete effectively. After volcanic eruption in Hawaii, for example, there is usually a slow succession from short-statured vegetation dominated by algal crusts, herbaceous plants, and small shrubs to forests dominated by slowly growing tree-ferns and trees. An exotic nitrogen-fixing tree, *Myrica faya*, whose seeds are brought to early or mid-successional sites by birds, can, however, add sufficient nitrogen to alter substantially the nitrogen supply, production, and species composition of vegetation and therefore the successional trajectory (Vitousek et al. 1987) (Fig. 13.3).

In some cases, one or a few successional pathways predominate because there are only a few postdisturbance combinations of environment and potential colonizers. In other cases, multiple successional pathways are possible. After glacial retreat in 1800 at Glacier Bay, Alaska, for example, *Populus* (poplar) and *Picea* (spruce) were the major initial colonizers. Further retreat of the glacier, however, brought early successional habitat within dispersal distance of nitrogen-fixing alders, which then became an important early successional species

(Fastie 1995). Alders increased the nitrogen inputs and long-term productivity of later successional stages (Bormann and Sidle 1990). The late-successional forests on older sites at Glacier Bay therefore followed a different successional trajectory than forests on younger sites. Human activities strongly affect both the postdisturbance environment and availability of propagules, so future trajectories of succession will likely differ from those that currently predominate.

Secondary Succession

Secondary succession differs from primary succession in that many of the initial colonizers are already present on site immediately after disturbance. They may resprout from roots or stems that survived the disturbance or germinate from a soil **seed bank**—seeds produced after previous disturbance events and that remain dormant in the soil until postdisturbance conditions (light, wide temperature fluctuations, and/or high soil nitrate) trigger germination (Fenner 1985, Baskin and Baskin 1998). In many forests there is also a **seedling bank** of large-seeded species that show negligible growth beneath the dense shade of a forest canopy but grow rapidly in treefall gaps to become the next generation of canopy dominants. Other colonizers of secondary succession disperse into the disturbed site from adjacent areas, just as in primary succession. Those dispersing species include both small-seeded, wind-dispersed species and large-seeded, animal-dispersed species (Fig. 13.6). Initial colonizers grow rapidly to exploit the resources made available by disturbance. Gap-phase succession is seldom limited by propagule availability (Shugart 1980), whereas the successional trajectory of large disturbed sites may depend on the species that disperse to the site (Fastie 1995).

The changes in species composition that occur after the initial colonization of a site result from a combination of (1) the inherent life history traits of colonizers, (2) facilitation, (3) competitive (inhibitory) interactions, (4) herbivory, and (5) stochastic variation in the environment (Connell and Slatyer 1977, Pickett et al. 1987, Walker 1999). **Life history traits** include seed size and number, potential growth rate, maximum size, and longevity. These traits determine how quickly a species can get to a site, how quickly it grows, how tall it gets, and how long it survives. Most early secondary successional species arrive soon after a disturbance, grow quickly, are relatively short statured, and have a low maximum longevity, compared to late-successional species (Noble and Slatyer 1980) (Fig. 13.6; Table 13.1). Even if no species interactions occurred during succession, life history patterns alone would cause a shift in dominance from early to late successional species because of differences in arrival rate, size, and longevity.

Facilitation involves processes in which early successional species make the environment more favorable for the growth of later successional species. Facilitation is particularly important in severe physical environments, such as primary succession, where nitrogen fixation and addition of soil organic matter by early successional species ameliorates the environment and increases the probability that seedlings of other species will establish and grow (Callaway 1995, Brooker and Callaghan 1998). **Competition** is an interaction among two organisms or species

TABLE 13.1. Successional changes in life history patterns after glacial retreat in Glacier Bay, Alaska.

Genus	Successional stage	Seed mass (μg seed^{-1})	Maximum height (m)	Age at first reproduction (yr)	Maximum longevity (yr)
Epilobium	pioneer	72	0.3	1	20
Dryas	*Dryas*	97	0.1	7	50
Alnus	alder	494	4	8	100
Picea	spruce	2694	40	40	700

Data from Chapin et al. (1994).

that reduces the availability of resources to other individuals. Both competitive and facilitative interactions are widespread in plant communities (Callaway 1995, Bazzaz 1996); their relative importance in causing changes in species composition during succession probably depends on environmental severity (Connell and Slatyer 1977, Callaway 1995). **Herbivores and pathogens** account for much of the mortality of early successional plants. Selective browsing by mammals is particularly important in eliminating early successional species as succession proceeds (Bryant and Chapin 1986, Paine 2000). In intertidal communities, grazing by fish and invertebrates such as limpets exerts a similar effect.

In general, life history traits generally determine the *pattern* of species change through succession, whereas facilitation, competition, and herbivory determine the rate at which this occurs (Chapin et al. 1994). These processes interact with other less predictable events, such as storms or droughts, to cause the diversity of successional changes that occur in natural ecosystems (Pickett et al. 1987, Walker 1999) (Fig. 13.7).

Secondary succession can begin with soils that have either high or low nutrient availability. When initial nutrient availability is high, early successional species typically have high relative growth rates, supported by high rates of photosynthesis and nutrient uptake. These species reproduce at an early age and allocate a large proportion of NPP to reproduction (Table 13.1). Their strategy is to grow quickly under conditions of high resource supply, then disperse to new disturbed sites. These early successional species include many weeds that colonize sites disturbed by people. As succession proceeds, there is a gradual shift in dominance to species that have lower resource requirements and grow more slowly. In ecosystems with low initial availability of soil resources, succession proceeds more slowly and follows patterns similar to those described for primary succession. Because there is a continuum in disturbance characteristics between primary and secondary succession, the patterns of establishment and succession differ among ecosystem types with different disturbance regimes and even among different disturbance events in the same ecosystem type.

Carbon Balance

Primary Succession

In primary succession productivity and decomposition rates are often greatest in mid-succession. Primary succession begins with little live or dead organic matter, so NPP and decomposition are initially close to zero. NPP increases slowly at first because of low plant density, small plant size, and strong nitrogen limitation of growth. NPP and biomass generally increase most dramatically after nitrogen fixers colonize the site. The planting of nitrogen-fixing lupines on English mine wastes (Bradshaw 1983) and the natural establishment of nitrogen-fixing alders after retreat of Alaskan glaciers (Bormann and Sidle 1990), for example, cause sharp increases in plant biomass and NPP. In primary successional sequences that lack a strong nitrogen fixer, successional increases in biomass and NPP depend on other forms of nitrogen input, including atmospheric deposition, plant and animal detritus, and floods.

Long-term successional trajectories of biomass and NPP differ among ecosystems. A common pattern in forests is that NPP increases from early to mid-succession and then declines after the forest reaches its maximum leaf area index (LAI) (Fig. 13.8) (Ryan et al. 1997). Several processes are thought to contribute to these patterns. In some forests, hydraulic conductance declines in late succession, causing water to limit the leaf area that can be supported and therefore the NPP that the ecosystem can sustain (see Chapter 6). In other forests, nutrient supply declines in late succession, leading to a corresponding reduction in NPP (Van Cleve et al. 1991). The mortality of branches and trees often increases in late succession, as trees age. The combination of reduced NPP and increased mortality of plants and plant parts in late succession slows the rate of biomass accumulation, so biomass approaches a relatively constant value (steady state) (Fig. 13.9). There is little support for the earlier generalization (Odum 1969) that the decline in

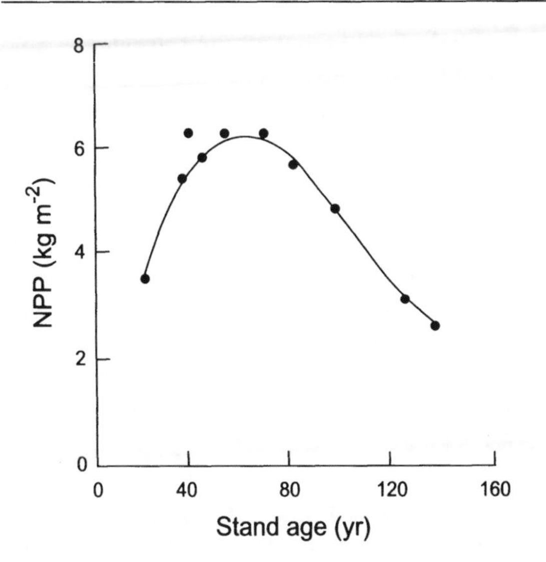

FIGURE 13.8. Successional changes in aboveground spruce production in eastern Russia. NPP declines after the forest reaches maximum LAI at about 60 yr of age. (Redrawn with permission from *Advances in Ecological Research*; Ryan 1997.)

production in late succession reflects increased maintenance respiration to support the increasing biomass. Most of the increased biomass of forests is wood, which consists mainly of dead cells that require no maintenance. In late succession interannual variation in climate and biotic processes such as pest outbreaks have a stronger influence on biomass and NPP than does successional change. Biomass can either increase through succession or decline in late succession, due to stand thinning. The long-term end points of successional trajectories in biomass and NPP are often uncertain because disturbance usually resets the successional clock before the ecosystem reaches steady state.

Over extremely long time scales, changes in rates of weathering and soil development lead to further changes in biomass and other ecosystem properties (see Chapter 3). Redwoods in California coastal forests, for example, are replaced by a pygmy forest of evergreen trees and shrubs after hundreds of thousands of years due to the formation of a hardpan that prevents drainage and creates anaerobic conditions that retard decomposition and root growth (Westman 1978). The slow-growing plants capable of surviving under these low-nutrient conditions produce litter with a high concentration of phenolics, which further reduces decomposition rate, resulting in a positive feedback that leads to progressively lower biomass, productivity, and nutrient turnover (Northup et al. 1995).

The decomposition rate at the start of primary succession is near zero, because there is little or no soil organic matter. The low organic content of these soils contributes to their low moisture-holding capacity and cation exchange capacity (CEC) (Fig. 13.10) (see Chapter 3). The *pattern* of change in decomposition through primary succession is similar to the pattern described for NPP. Decomposition, however, lags behind the changes in NPP, causing soil organic matter (SOM) to accumulate (Fig. 13.11). Initially decomposition is slow in primary succession because it is limited by the rate of litter input. Decomposition increases substantially in mid-succession in response to increases in the quantity and quality of litter. In forests, the late-successional decline in NPP reduces litter inputs to soils, causing decomposition (in grams per square meter) to decline. In ecosystems in which nutrient availability declines in late succession, this reduces litter quality and quantity, further reducing decomposition rate (Van Cleve et al. 1993).

Net ecosystem production (NEP) is the net rate of carbon accumulation by the ecosystem. It is determined primarily by the balance between NPP and carbon losses through

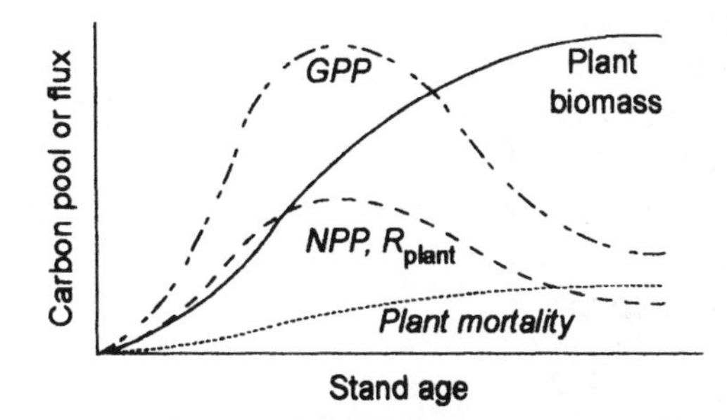

FIGURE 13.9. Idealized patterns of successional change in plant biomass, NPP, plant respiration (R_{plant}), and plant mortality of a forest. NPP often reaches a peak in mid-succession, and both production and respiration decline in late succession. GPP, gross primary production.

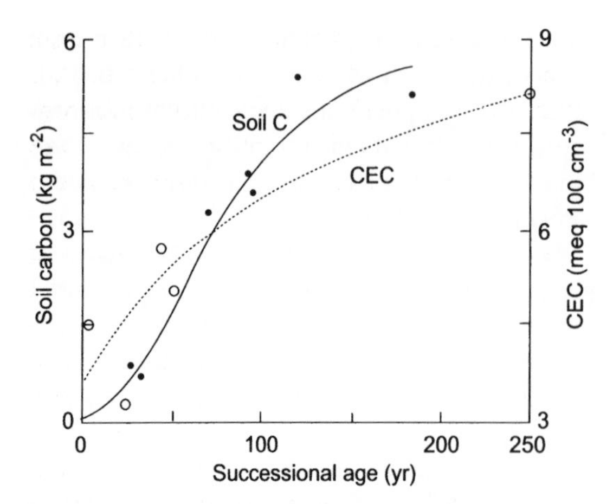

FIGURE 13.10. Accumulation during succession of soil organic carbon (Crocker and Major 1955) and associated change in cation exchange capacity (CEC) of mineral soil (Ugolini 1968) after deglaciation at Glacier Bay Alaska. Measurements were made to a depth of 45 cm in mineral soil. The accumulation of soil carbon contributes to the increased CEC, which retains nutrients to support plant growth.

heterotrophic respiration and leaching; heterotrophic respiration is typically the largest avenue of carbon loss. Because of the lag of heterotrophic respiration behind NPP, NEP is generally positive during early and mid-succession, leading to a net accumulation of carbon in ecosystems in both vegetation and soils (Fig. 13.11). This explains why midlatitude, north temperate forests that were established in abandoned agricultural lands one to two centuries ago are currently a net carbon sink (Goulden et al. 1996, Valentini et al. 2000). NEP should approach zero in late succession because NPP and heterotrophic respiration are approximately equal. At this point NEP is governed more by climate and pest outbreaks than by successional dynamics. In carbon-accumulating ecosystems such as peatlands, boreal forests, and arctic tundra, however, decomposition declines more strongly in late succession than does NPP, so NEP remains pos-

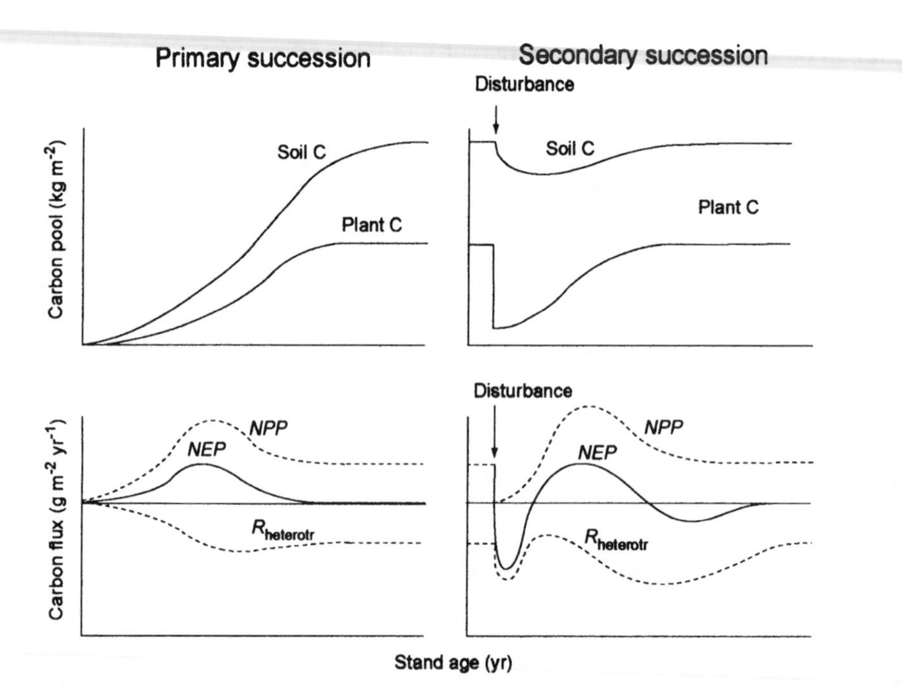

FIGURE 13.11. Idealized patterns of change in carbon pools (plants and soils) and fluxes (NPP, $R_{heterotr}$, and NEP) in primary and secondary succession. In early primary succession, plant and soil carbon accumulates slowly, because NPP is greater than heterotrophic respiration ($R_{heterotr}$)—that is, there is a positive net ecosystem production (NEP). In early secondary succession, soil carbon declines after disturbance because carbon losses from heterotrophic respiration exceed carbon gain from NPP, leading to a negative NEP. In late succession, plant and soil carbon approach the steady state (in an idealized situation), and NEP approaches zero. In both primary and secondary succession, NPP and NEP are maximal in mid-succession. The graphs assume negligible carbon loss to groundwater.

itive in late succession, and the ecosystem continues to accumulate carbon. This explains why some peatlands have larger carbon pools per unit area than most ecosystems, despite their low productivity. Because of ecosystem differences in disturbance frequency and environmental effects on carbon cycling processes, there is a wide range of potential successional patterns in NPP, heterotrophic respiration, and NEP among ecosystems.

Secondary Succession

The initial carbon pools and fluxes are much larger in secondary than in primary succession. Carbon dynamics are dramatically different in secondary succession from those in primary succession, because secondary succession begins with an initial stock of SOM. Immediately after disturbance, NPP is low in secondary succession because of low plant biomass, just as in primary succession (Fig. 13.11). NPP recovers more quickly in secondary than in primary succession, however, due to the generally rapid colonization and high growth rate of herbs, grasses, and resprouting perennial species. High availability of light, water, and nutrients supports the high growth potential of early successional vegetation in many secondary successional sequences. The herbaceous species that dominate most early secondary successional sites return most of their biomass to the soil each year. As perennial plants, particularly woody species, increase in abundance, biomass and NPP increase more rapidly, because woody species retain a larger proportion of their biomass than do herbs. Changes in biomass and NPP in middle and late secondary succession are similar to patterns described for primary succession (Fig. 13.11), because they are controlled by the same factors and processes—largely the soil resources available to support production and the growth potential of the species typical of the ecosystem.

In contrast to primary succession, decomposition is often more rapid early in secondary succession than at any other time (Fig. 13.11) because many disturbances transfer large amounts of labile carbon to soils and create an environment that is favorable for decomposition. The size of the initial carbon pool depends on the nature and severity of the disturbance. After a treefall, hurricane, or insect outbreak there are large inputs of new labile carbon from leaf and root death. Fire consumes some of the surface SOM but also adds new carbon to the soil through death of roots and unburned aboveground plant material. Disturbance also stimulates decomposition because the removal of vegetation allows more radiation to penetrate to the soil surface and reduces transpirational water loss. The resulting increases in soil temperature and soil water content generally enhance decomposition. The large quantity and high quality of litter of early secondary successional plants also promotes decomposition. In mid-succession the regrowing vegetation uses an increasing proportion of the available water and nutrients and reduces soil temperature by shading the soil surface. These changes in environment cause a decline in decomposition. Decomposition continues to decline in late succession because the decline in NPP reduces litter input, litter quality often declines, and the environment becomes less favorable than in early succession.

How do these contrasting patterns of NPP and decomposition affect NEP? In early secondary succession, ecosystem carbon pools decline (i.e., NEP is negative) because decomposition causes large carbon losses, and there is little NPP (Fig. 13.11). In early to mid-succession, before the peak in NPP, ecosystems begin accumulating carbon again, as soon as NPP outpaces decomposition. In late succession, ecosystems either approach a carbon balance of zero or continue to accumulate carbon at a slow rate, depending on the environmental limitations to NPP and decomposition. Other avenues of carbon loss from ecosystems, such as leaching of dissolved organic carbon, also influence NEP, but their patterns of successional change are not well documented.

Although the successional patterns of NPP, decomposition, and carbon stocks in plants and soils that we have described are frequently observed, the details and timing of these patterns differ substantially among ecosystems, depending on factors such as initial ecosystem carbon stocks, resource availability, disturbance severity, and successional pathway.

Nutrient Cycling

Primary Succession

Nutrient dynamics during succession are both a cause and a consequence of the dynamic interplay between NPP and decomposition. The most dramatic change in nutrient cycling during early primary succession is the accumulation of nitrogen in vegetation and soils. Most parent materials have extremely low nitrogen contents in the absence of biotic influences, so the initial nitrogen pools in the ecosystem are small and depend on atmospheric inputs. At this initial stage of primary succession, nitrogen is the element that most strongly limits plant growth and therefore the rates of accumulation of plant biomass and SOM (Vitousek et al. 1987, Chapin et al. 1994). The rate of nitrogen input, which frequently is associated with the establishment of nitrogen-fixing plants (both free-living cyanobacteria and symbiotic nitrogen fixers), therefore governs the initial dynamics of nutrient cycling in primary succession. Nitrogen typically accumulates at rates of 3 to $16\,g\,N\,m^{-2}\,yr^{-1}$ for 50 to 200 years, before approaching an asymptote of 200 to $500\,g\,N\,m^{-2}$ (Walker 1993). As leaves and roots of nitrogen-fixing plants senesce and are eaten by herbivores, the nitrogen is transferred from plants to the soil, where it is mineralized and absorbed by both nitrogen-fixing and non-nitrogen-fixing plants. Litter from non-nitrogen-fixing plants becomes an increasingly important source for nitrogen mineralization as primary succession proceeds. This causes the ecosystem to shift from an open nitrogen cycle, with substantial input from nitrogen fixation (see Chapter 9), to a more closed nitrogen cycle, in which plant growth depends on the mineralization of soil organic nitrogen. During mid-succession, plants and soil microbes are so efficient at accumulating nutrients that losses of nitrogen and other essential elements from ecosystems are often negligible (Fig. 13.12) (Vitousek and Reiners 1975). In late succession, nitrogen inputs to the ecosystem may largely balance nitrogen losses from leaching and denitrification, causing ecosystem nitrogen pools to approach a relatively stable size.

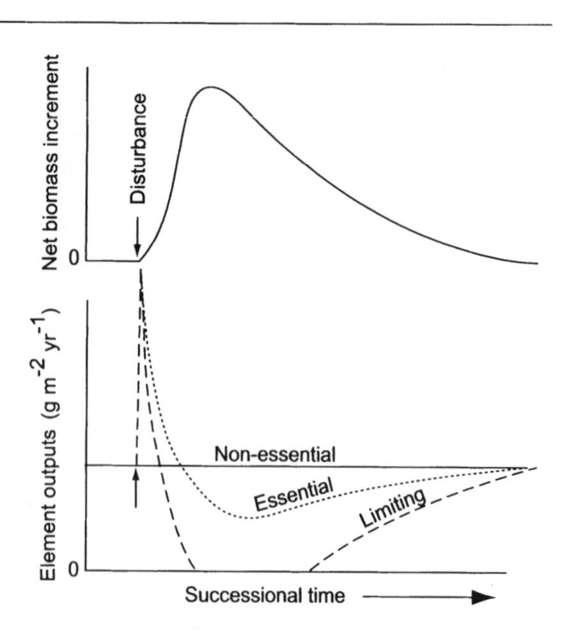

FIGURE 13.12. Changes through succession in net biomass increment in vegetation and in the losses of limiting, essential, and nonessential elements. In early succession, when there are large annual increments in biomass, elements that are required for this production (especially growth-limiting elements) accumulate in new plant and microbial biomass, so they are not lost from the ecosystem by leaching. In late succession, when the element requirements for new plant and microbial biomass are balanced by element release from the breakdown of dead organic matter, nutrient inputs to the ecosystem are approximately balanced by nutrient outputs, regardless of whether nutrients are required by vegetation or not. (Modified with permission from *BioScience*; Vitousek and Reiners 1975.)

Most evidence for these successional changes in nitrogen cycling comes from studies of **chronosequences**, series of sites that differ in age but are assumed to be similar with respect to other state factors (see Chapter 1). Since we seldom know whether sites in a chronosequence began their successional development under identical conditions, long-term studies of succession are critical in testing whether the patterns in nitrogen cycling observed in chronosequences truly reflect the actual successional changes that occur at a site.

The accumulation of nitrogen in the initial stages of primary succession governs the rates of internal cycling of other essential elements in ecosystems. Early in primary succession, inputs of most elements in precipitation and weathering may approximately equal outputs, because biological storage pools in vegetation and soils are small. Plants and soil microbes have a limited range of element ratios. Plants in nitrogen-limited ecosystems therefore accumulate most essential elements approximately in proportion to their rate of nitrogen accumulation, preventing the loss of these elements from the ecosystem in mid-succession (Fig. 13.12). If biomass and element pools stabilize in late succession, element losses increase until they approximately equal element inputs. Over time scales of soil development, there are additional changes in nutrient cycling that occur when the supply of weatherable minerals is depleted or becomes bound in unavailable forms. Availability of phosphorus and cations, for example, typically declines in old high-weathered sites, as they leach or become bound in unavailable forms (see Chapters 3 and 9). Under these circumstances, phosphorus or other elements may limit plant production (Chadwick et al. 1999), and cycling rates of these limiting elements regulate rates of cycling of nitrogen and other minerals.

Secondary Succession

Secondary succession after natural disturbances differs from primary succession because it generally begins with higher nitrogen availability. Natural disturbances that initiate secondary succession produce a pulse of nutrient availability because disturbance-induced changes in environment and litter inputs increase mineralization of dead organic matter and reduce plant biomass and nutrient uptake. Fires, which may volatilize large amounts of nitrogen, also return nutrients in ash, as described earlier, leading to high nutrient availability after fire (Wan et al. 2001). Plant growth is therefore generally not strongly nutrient-limited early in secondary succession after natural disturbances, and there is usually adequate nitrogen to support high rates of photosynthesis and growth (Scatena et al. 1996). The pulse of nutrient availability and the reduction in plant biomass and capacity for plant uptake after disturbance also increase the vulnerability of ecosystems to nutrient loss. High rates of nitrogen mineralization produce NH_4^+, which serves as a substrate for nitrification and its associated loss of nitrogen trace gases (see Chapter 9). The nitrate can then be denitrified, particularly if soils are wet, or leached below the rooting zone. The occurrence or extent of this nitrogen loss depends on the balance between nitrogen mineralization and uptake by plants and microbes. Rains that occur immediately after a fire, for example, can leach nitrate into groundwater and streams. The few studies of nutrient losses associated with natural disturbances show surprisingly small nitrogen losses to streams from wildfire (Stark and Steele 1977) or hurricanes (Schaefer et al. 2000). We know little, however, about other avenues of loss or whether these results are representative of natural disturbances.

The vulnerability of ecosystems to nutrient losses after disturbance has been illustrated in many forest harvest experiments, such as those at Coweta and Hubbard Brook in the United States. After stream discharge and chemistry had been monitored for several years, the forest was cut on an entire watershed, and regenerating vegetation was killed with herbicides (Bormann and Likens 1979). The combination of high decomposition and mineralization rates and absence of plant uptake after disturbance caused large losses of essential plant nutrients in stream water (see Fig. 9.7). When vegetation was allowed to regrow, the increased plant uptake caused nutrient losses in stream water to decline to preharvest levels. These studies show clearly that the dynamics of nutrient loss after disturbance are highly variable, with the extent of nutrient loss often depending on nutrient availability at the time of disturbance and the capacity of regenerating vegetation to absorb nutrients.

Anthropogenic disturbances create a wide range of initial nutrient availabilities. Some disturbances, such as mining, can produce an initial environment that is even less favorable than most natural primary successional habitats for initiation of succession. These habitats may have toxic by-products of mining or mineral material with a low capacity for water and nutrient retention. Some agricultural lands are abandoned to secondary succession after erosion or (in the tropics) formation of laterite soils (see Chapter 3), reducing the nutrient-supplying power of soils. Soils from some degraded lands also have concentrations of aluminum and other elements that are toxic to many plants. Secondary succession in degraded lands may be quite slow. At the opposite extreme, abandonment of rich agricultural lands or the logging of productive forests may create conditions of unusually high nutrient availability, leading to the potential loss of nutrients through leaching and denitrification. These nutrient losses are particularly dramatic in the tropics, where rapid mineralization and biomass burning associated with forest clearing release large amounts of nitrogen as trace gases (nitric oxide, NO_x, and nitrous oxide, N_2O) and as nitrate in groundwater (Matson et al. 1987). The impact of agricultural nutrient additions is particularly long lived for phosphorus because of its effective retention by soils. An understanding of the successional controls over nutrient cycling provides the basis for management strategies that minimize undesirable environmental impacts (see Chapter 16). The return of topsoil or planting of nitrogen-fixing plants on mine wastes, for example, greatly speeds successional development on these sites (Bradshaw 1983). Retention of some organic debris after logging may support microbial immobilization of nutrients that would reduce leaching loss.

Trophic Dynamics

The proportion of primary production consumed by herbivores is maximal in early to middle succession. In early primary and secondary succession, rates of herbivory may be low because of low food density, insufficient cover to hide vertebrate herbivores from their predators, and insufficient canopy to create a humid, nondesiccating environment for invertebrate herbivores. Herbivory is often greatest in early to middle secondary succession because the rapidly growing herbaceous and shrub species that dominates this stage have high nitrogen concentrations and a relatively low allocation to plant defense (see Chapter 11). This explains why abandoned agricultural fields, recent burn scars, or riparian areas are focal points for browsing mammals, insect herbivores, and their predators. In early successional boreal floodplains, for example, moose consume about 30% of aboveground NPP and account for a similar proportion of the nitrogen inputs to soil (Kielland and Bryant 1998). The abundant insect herbivores on these sites support a high diversity of neotropical migrant birds. Similarly, in temperate and tropical regions, early successional forests support large populations of deer and other browsers. In ecosystems in which nutrient availability declines from early to late succession, plants shift allocation from growth to defense (see Chapter 11). The resulting decline in forage quality reduces levels of consumption by most herbivores and higher trophic levels. Some insect outbreak species are an important exception to this successional pattern. They often attack late-successional trees that are weakened by environmental stress.

Vertebrate herbivores can either promote or retard succession, depending on their relative impact on early vs. late successional species. Vertebrate herbivores both respond to and contribute to successional change. The effects of herbivores on succession differ among ecosystems depending on the nature and specificity of herbivore–plant interactions. There are, however, several common patterns that emerge.

In forested regions, birds, rodents, and other vertebrates often enhance the dispersal of early successional species such as blackberries, junipers, and grasses into abandoned agricultural fields and other disturbed sites. Birds and squirrels also disperse the large seeds of late-successional species such as oak and hickory into early successional sites. These animal-mediated dispersal events are particularly important in secondary succession, where the rapid development of herbaceous vegetation makes it difficult for small-seeded woody species to compete and establish successfully.

The relatively low levels of plant defense in species that typically characterize early forest succession make these plants a nutritious target for generalist herbivores. Preferential feeding on these species reduces their height growth. Browsed plants respond to aboveground herbivory by reducing root allocation, making them less competitive for water and nutrients (Ruess et al. 1998). Many later successional species produce chemical defenses that deter generalist herbivores. Selective herbivory contributes to the competitive release of late successional species, enabling them to overtop and shade their early successional competitors. In this way, selective browsing by mammals frequently speeds successional change in forests (Pastor et al. 1988, Kielland and Bryant 1998, Paine 2000). In tropical rain forests mammalian herbivores maintain the diversity of understory seedlings that become the next generation of canopy dominants, because they feed preferentially on the "weedy" tree seedlings that are most common in the understory (Dirzo and Miranda 1991).

In contrast to forests, many grasslands and savannas are maintained by mammalian herbivores that prevent succession to forests. Elephants, for example, browse and uproot trees in African savannas. These savannas succeed to closed forests in areas where elephant populations have been reduced by overhunting. In North American prairies, grazers and fire restrict the invasion of trees. When these sources of disturbance are reduced, trees frequently invade and convert the grassland to forest. Similarly, at the end of the Pleistocene the decline in large mammals that occurred on many continents, in part from human hunting, contributed to the vegetation changes that occurred at that time (Flannery 1994, Zimov et al. 1995).

Herbivores have multiple effects on nutrient cycling in early succession. In the short term they enhance nutrient availability by returning available nutrients to the soil in feces and urine, which short-circuits the decomposition process (Kielland and Bryant 1998). Herbivory can also alter the temperature and moisture regime for decomposition at the soil surface by reducing leaf and root biomass. The quality of litter that a given plant produces is also enhanced by herbivory (Irons et al. 1991). Over the long term, however, herbivory accelerates plant succession by removing early successional species. This tends to reduce nutrient cycling rates and nutrient losses (Pastor et al. 1988, Kielland and Bryant 1998) (see Fig. 11.4).

Water and Energy Exchange

Disturbances that eliminate plant biomass increase runoff through a reduction in evapotranspiration. One of the most dramatic consequences of forest cutting or overgrazing is increased runoff to streams and rivers. As forests recover from disturbance, evapotranspiration increases, and runoff returns to predisturbance levels (Fig. 13.13). These watershed-scale observations provide a basis for understanding successional controls over water and energy exchange. The low evapotranspiration in early succession results from the small biomass of roots to absorb water from the soil and of leaves to transfer that water to the atmosphere. This low evapotranspiration therefore leads to high soil moisture (except at the surface where soil evaporation occurs) and runoff. In disturbances that remove the plant canopy and litter layer, impaction by rain drops on mineral soil reduces infiltration and increases overland flow and runoff. As root and

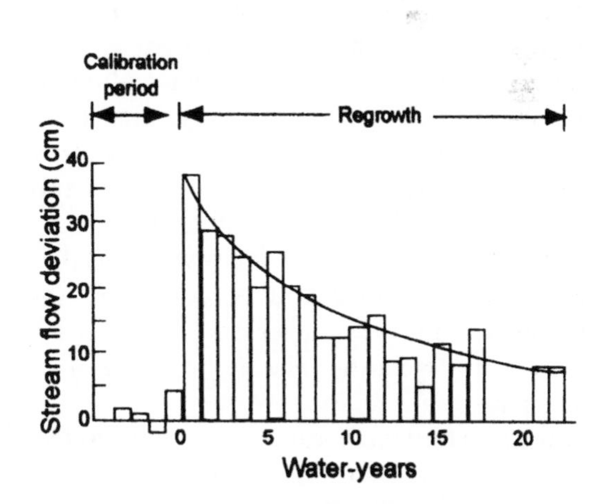

FIGURE 13.13. Runoff from a watershed in a North Carolina forest in the southeastern United States under natural conditions (the calibration period) and after forest harvest. Water yield from the watershed greatly increased in the absence of vegetation and approached preharvest levels within 20 years. (Hibbert 1967.)

leaf area increase through succession, there is a corresponding increase in evapotranspiration and decrease in runoff. The high nitrogen availability, high photosynthetic rate, and high leaf area early in secondary succession contribute to a high canopy conductance so evapotranspiration increases more rapidly than plant biomass as succession proceeds (Bormann and Likens 1979). As the canopy increases in height and complexity, solar energy is trapped more effectively, reducing albedo and increasing the energy available to drive evapotranspiration. The high surface roughness of tall complex canopies increases mechanical turbulence and mixing within the canopy. All of these factors contribute to high evapotranspiration in mid-succession.

Successional changes in albedo differ among ecosystems because of the wide range among ecosystems in albedo of bare soil (see Table 4.2). Many recently disturbed sites have a low albedo because of the dark color of moist exposed soils or of charcoal. Albedo increases when vegetation, with its generally higher albedo, begins to cover the soil surface (Fig. 13.14). Albedo probably declines again in late succession due to increased canopy complexity (see Chapter 4). In ecosystems that succeed from deciduous to conifer forest, this species shift causes a further reduction in albedo. The winter energy exchange of northern forests is influenced by snow, which has an albedo threefold to fivefold higher than vegetation (Betts and Ball 1997). Winter albedo of these forests declines through succession, first as vegetation grows above the snow, then as the canopy becomes more dense, and finally when (if) there is a switch from deciduous to evergreen vegetation. All of these changes increase the extent to which vegetation masks the snow from incoming solar radiation.

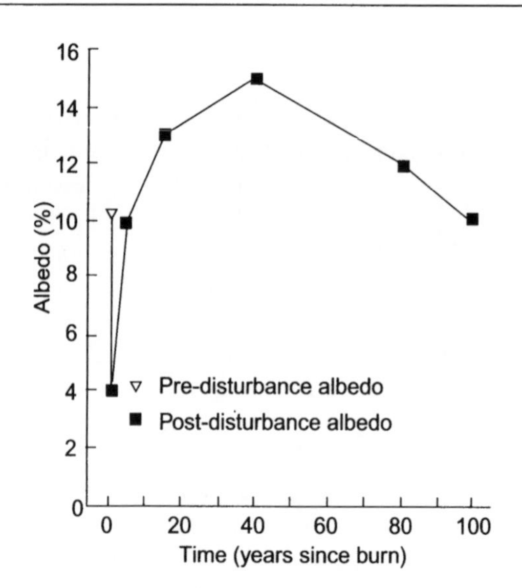

FIGURE 13.14. Successional changes in albedo after fire in Alaskan boreal forests (Chambers and Chapin, in press). The black postfire surface causes a decline in albedo. Albedo increases during the herbaceous and deciduous forest phases of succession and declines in late succession due to a switch to conifer vegetation. This successional change occurs more rapidly after moderate fires because of the more rapid replacement of deciduous species by conifers.

High surface temperatures that contribute to high emission of longwave radiation dominate energy budgets of early successional sites. Early successional sites often have a high surface temperature for several reasons. (1) The low albedo of recently disturbed sites maximizes radiation absorption and therefore the quantity of energy available at the surface. (2) The low leaf area, small root biomass, and low hydraulic conductance of dry surface soils limits the proportion of energy dissipated by evapotranspiration. (3) The relatively smooth surface of unvegetated or early successional sites minimizes mechanical turbulence that would otherwise transport the heat away from the surface. The resulting high surface temperature promotes emission of long-wave radiation (see Chapter 4).

The large longwave emission dissipates much of the absorbed radiation after disturbance, so *net* radiation (the net energy absorbed by the surface) is not as great as we might expect from the low albedo of these sites. For example, net radiation actually declines after fire in the boreal forest despite a reduction in albedo because of the large emission of longwave radiation (Chambers and Chapin, in press). The soil surface of unvegetated sites is prone to drying between rain events due to the combination of high surface temperatures and the low resupply of water from depth, due to the low hydraulic conductance of dry soils (see Chapter 4). Dry surface soils provide little moisture for surface evaporation and are good thermal insulators, so both evapotranspiration and average ground heat flux are often relatively low on unvegetated surfaces (Oke 1987). Consequently, sensible heat flux accounts for the largest proportion of energy that is dissipated from these sites to the atmosphere. The absolute magnitude of sensible heat flux from early successional sites differs among ecosystems and climate zones and depends on both net radiation (the energy available to be dissipated) and the energy partitioning among sensible, latent, and ground heat fluxes. As succession proceeds, latent heat fluxes become a more prominent component of energy transfer from the land to the atmosphere.

Temporal Scaling of Ecological Processes

Temporal extrapolation requires an understanding of the typical time scales of important ecological processes. Measurements of ecological processes are generally made over shorter time periods than the time scales over which we would like to make predictions. Few studies, for example, provide detailed information about the functioning of ecosystems over time scales of decades to centuries—the time scale over which ecosystems are likely to respond to global environmental change. **Temporal scaling** is the extrapolation of measurements made at one time interval to longer (or occasionally shorter) time intervals. Simply multiplying an instantaneous flux rate by 24h to get a daily rate or by 365 days to get an annual rate seldom gives a reasonable approximation because this ignores the temporal variation in driving variables and the time lags and thresholds in ecosystem responses to these drivers. Rates of photosynthesis, for example, differ between night and day and between summer and winter.

One approach to temporal scaling is to select measurements that are consistent with the time scale and question of interest. A second approach is to extrapolate results based on models that simulate processes accounting for important sources of variation over the time scale of interest. The key to temporal scaling is therefore to focus clearly on the processes that are important over the time scales of interest. Entire books have been written on temporal scaling based on isotopic measurements (Ehleringer et al. 1993), long-term measurements (Sala et al. 2000b), and modeling (Ehleringer and Field 1993, Waring and Running 1998). Here we provide a brief overview of these approaches.

Isotopic tracers provide an important tool for estimating long-term rates of net carbon exchange of plants and ecosystems because they integrate the net effect of carbon inputs and loses throughout the time period that carbon exchanges occur (see Boxes 5.1 and 7.1). The ^{13}C content of plants in dry environments, for example, provides an integrated measure of water use efficiency (WUE) during the time interval during which the plant material was produced. The ^{13}C content of soils in ecosystems that have changed in dominant vegetation from C_3 to C_4 plants provides an integrated measure of soil carbon turnover since the time that the vegetation change occurred. These measurements are appropriate for estimating long-term rates because they incorporate effects of processes that occur slowly or intermittently that might not be captured in short-term gas-exchange measurements. Seasonally integrated WUE measured with stable isotopes, for example, is affected by dry and wet periods that influence seasonal water and carbon exchange; whereas instantaneous measurements of gas exchange are unlikely to be representative of the entire annual cycle. Other examples of measurements that integrate over long time intervals include NPP, which integrates longer time periods than does photosynthesis or respiration, and changes in soil carbon stocks over succession, which integrate over longer time periods than do NPP and decomposition.

Process-based models are an important tool for temporal scaling because they make projections of the state of the ecosystem over longer time intervals (or at different times) than can be measured directly. The challenge in developing models for temporal extrapolation is the selection of the driving variables that account for the most important sources of temporal variation over the time scale of interest. The diurnal pattern of net photosynthesis can often be adequately simulated based on the relationship of net photosynthesis to light and temperature. Annual estimates of photosynthetic flux (gross primary production, GPP), however, also require information on seasonal variation in leaf biomass and photosynthetic capacity. In annual simulations, the diurnal variation in photosynthesis is less important to model explicitly because it is quite predictable, based on the empirical relationship between daily photosynthesis and mean daily temperature and light. **Slow variables**, such as successional changes in LAI or nitrogen availability, are often treated as constants in short-term ecological studies, but can become key controlling variables over longer time scales (Carpenter and Turner 2000). We must therefore think carefully about which critical driving variables are likely to change over the time scale of intended predictions and look for evidence of the relationship of ecological processes to these slow variables. Models of carbon flux based on the relationship of GPP and respiration to daily or monthly climate, for example, can be validated by comparing model output to patterns of carbon flux observed over longer time scales (e.g., interannual variation in carbon flux) (Clein et al., in press). There is a wide range of temporal scales over which important ecosystem controls vary (Fig. 13.15).

Spatial variation in driving variables sometimes gives hints as to which slow variables are important to include in long-term extrapolations. The spatial relationship between the distribution of biomes or plant functional types and climate, for example, has been used as a basis for predicting how vegetation might respond to future climatic warming (Prentice et al. 1992, VEMAP Members 1995). Spatial relationships with driving variables often reflect quasi-equilibrium relationships. Dry tropical forests, for example, occur where the average climate is warm and has a distinct dry season. Temporal extrapolations should also consider extreme events and time lags that may not be evident from an examination of spatial pattern. Ice storms, a spring freeze, intense droughts, 100-year floods, and other events with long-lasting effects strongly influence the structure and functioning of ecosystems long after they occur.

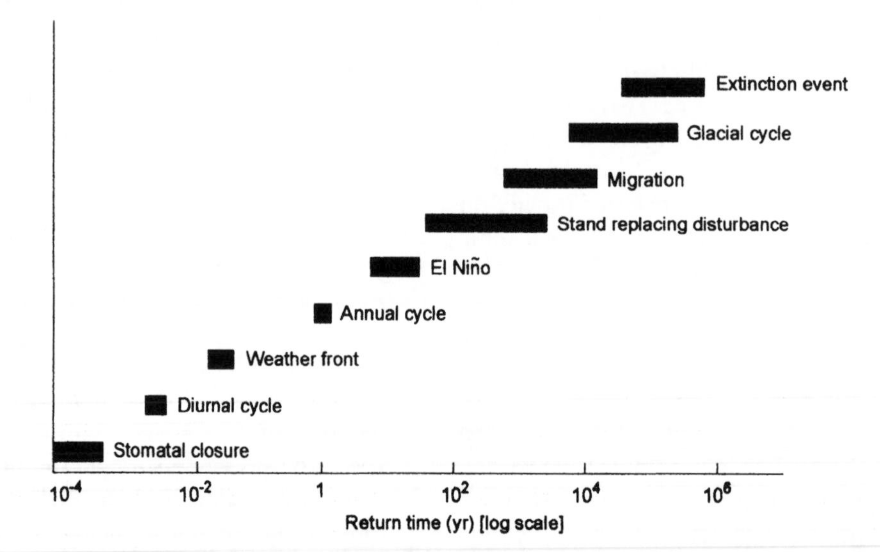

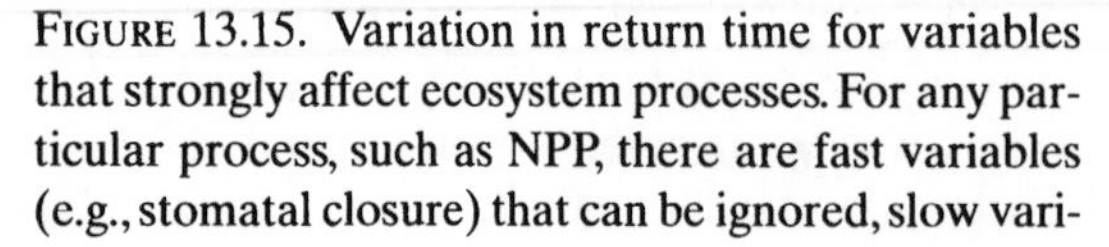

FIGURE 13.15. Variation in return time for variables that strongly affect ecosystem processes. For any particular process, such as NPP, there are fast variables (e.g., stomatal closure) that can be ignored, slow variables (e.g., El Niño or stand-replacing disturbance) that strongly affect the process, and extremely slow variables (e.g., glacial cycles) that can be treated as constants.

Summary

Rates of all ecosystem processes are constantly adjusting to past changes that have occurred over all time scales, ranging from sunflecks that last milliseconds to soil development that occurs over millions of years. Ecosystem processes that occur slowly, such as soil organic matter development, deviate most strongly from steady state and are most strongly affected by legacies of past events. Ecosystem processes are highly resistant and/or resilient to predictable changes in environment such as those that occur diurnally and seasonally and in response to disturbances to which the organisms are well adapted.

Because disturbance is a natural component of all ecosystems, the successional changes in ecosystem processes after disturbance are important for understanding regional patterns of ecosystem dynamics. These ecosystem changes are particularly sensitive to the severity, frequency, and type of disturbance. Carbon accumulates in vegetation and soils, leading to positive NEP through primary succession because changes in decomposition lag behind changes in NPP. NPP in forests is frequently greatest in mid-succession. Secondary succession begins with a large negative NEP due to low NPP and rapid decomposition, but carbon cycling in middle and late succession are similar to the patterns in primary succession.

Nutrient cycling changes through early primary succession as nitrogen fixers establish and add nitrogen to the ecosystem. Other elements cycle in proportion to the cycling of nitrogen. In secondary succession, however, nitrogen is generally most available in early succession. At this time, nitrogen and other elements are vulnerable to loss until the potential of plants and microbes to absorb nutrients exceeds the rate of net mineralization. This tightens the nitrogen cycle. Recycling within the ecosystem is strongest in mid-succession, when rates of nutrient mineralization constrain the rates of uptake by vegetation.

The role of herbivores in succession differs among ecosystem types and with successional stage. Mammals often accelerate the early successional changes in forests by eliminating palatable early successional species. In grasslands, however, herbivores prevent the establishment of woody species that might otherwise transform grasslands into shrublands and forests. Some insects have their greatest impact in late succession, particularly in forests, where they can be important agents of mortality.

Stand-replacing disturbances greatly reduce evapotranspiration and increase runoff. Evapotranspiration increases through succession more rapidly than might be expected from biomass recovery because early successional vegetation has high transpiration rates. Sensible heat flux tends to show the reverse successional pattern with high sensible heat flux (and/or longwave radiation) immediately after disturbance and lower sensible heat flux as rapidly growing mid-successional vegetation establishes, reflects radiation, and transfers water to the atmosphere.

Review Questions

1. Provide examples of ways in which the carbon and nitrogen cycling of an ecosystem might be influenced by the legacy of events that occurred 1 week ago, 5 years ago, 100 years ago, 2000 years ago.
2. What properties of disturbance regimes determine the ecological consequences of disturbance? How do these properties differ between treefalls in a tropical wet forest and fire in a dry conifer forest?
3. What are the major processes causing successional change in plant species? How do the relative importance of these processes differ between primary and secondary succession?
4. How do NPP, decomposition, and the carbon pools in plants and soils change through primary succession? At what successional stage does carbon accumulate most rapidly? Why? How do these patterns differ between primary and secondary succession? Why do these differences occur?
5. How does nitrogen cycling differ between primary and secondary succession? At what stages is this difference most pronounced?

6. How do trophic dynamics change through succession? Why?
7. How do water and energy exchange change through succession? What explains these patterns?
8. What are the major issues to consider in extrapolating information from one temporal scale to another? Describe ways in which this temporal extrapolation might be done.

Additional Reading

Bazzaz, F.A. 1996. *Plants in Changing Environments. Linking Physiological, Population, and Community Ecology*. Cambridge University Press, Cambridge, UK.

Bormann, F.H., and G.E. Likens. 1979. *Pattern and Process in a Forested Ecosystem*. Springer-Verlag, New York.

Chapin, F.S. III, L.R. Walker, C.L. Fastie, and L.C. Sharman. 1994. Mechanisms of primary succession following deglaciation at Glacier Bay, Alaska. *Ecological Monographs* 64:149–175.

Clements, F.E. 1916. *Plant Succession: An Analysis of the Development of Vegetation*. Publication 242. Carnegie Institution of Washington, Washington, DC.

Connell, J.H., and R.O. Slatyer. 1977. Mechanisms of succession in natural communities and their role in community stability and organization. *American Naturalist* 111:1119–1114.

Crocker, R.L., and J. Major. 1955. Soil development in relation to vegetation and surface age at Glacier Bay, Alaska. *Journal of Ecology* 43:427–448.

Fastie, C.L. 1995. Causes and ecosystem consequences of multiple pathways of primary succession at Glacier Bay, Alaska. *Ecology* 76:1899–1916.

Vitousek, P.M., and W.A. Reiners. 1975. Ecosystem succession and nutrient retention: A hypothesis. *BioScience* 25:376–381.

Vitousek, P.M., L.R. Walker, L.D. Whiteaker, D. Mueller-Dombois, and P.A. Matson. 1987. Biological invasion by *Myrica faya* alters ecosystem development in Hawaii. *Science* 238:802–804.

Zimov, S.A., V.I. Chuprynin, A.P. Oreshko, F.S. Chapin III, J.F. Reynolds, and M.C. Chapin. 1995. Steppe-tundra transition: An herbivore-driven biome shift at the end of the Pleistocene. *American Naturalist* 146:765–794.

14 Landscape Heterogeneity and Ecosystem Dynamics

Landscape heterogeneity determines the regional consequences of processes occurring in individual ecosystems. In this chapter we describe the major causes and consequences of landscape heterogeneity.

Introduction

Spatial heterogeneity within and among ecosystems is critical to the functioning of individual ecosystems and of entire regions. In previous chapters we emphasized the controls over ecosystem processes in relatively homogenous units or **patches** of an ecosystem. The spatial pattern of ecosystems in a region, however, also influences ecosystem processes. Riparian ecosystems between upland agricultural systems and streams or rivers, for example, may filter nitrate and other pollutants that would otherwise move into streams. Spatial patterns within ecosystems also influence ecosystem processes. The most rapid rates of nutrient cycling and greatest accumulations of organic matter in arid ecosystems, for example, occur beneath, rather than between, plants. The fragmentation of ecosystems into smaller units separated by other patch types influences the abundance and diversity of animals. All of the processes and mechanisms that operate in ecosystems (see Part II) have important spatial dimensions. In this chapter, we first discuss the concepts and characteristics of landscapes that aid in understanding and quantifying landscape interactions and then discuss sources of spatial heterogeneity within and among ecosystems and the consequences of that heterogeneity for interactions among ecosystems on a landscape.

Concepts of Landscape Heterogeneity

Spatial pattern exerts a critical control over ecological processes at all scales. Landscapes are mosaics of patches that differ in ecologically important properties. **Landscape ecology** addresses the causes and consequences of spatial heterogeneity (Urban et al. 1987, Forman 1995, Turner et al. 2001). This field focuses on both the interactions among patches on the landscape and the behavior and functioning of the landscape as a whole. Landscape processes can be studied at any scale, ranging from the mosaic of gopher mounds in a square meter of grassland to biomes that are patchily distributed across the globe. Landscape processes are frequently studied at the scale of stands of vegetation within a watershed or region.

Some landscape patches are **biogeochemical hot spots** with high process rates, causing them to be more important than their area would suggest. Beaver ponds, for example, are biogeochemical hot spots for methane emissions in boreal landscapes (Roulet et al. 1997), and recently cleared pastures in the central Amazon Basin are hotspots for nitrous oxide emissions (Matson et al. 1987). Hot spots are defined with respect to a particular process and occur at all spatial scales, from the rhizosphere surrounding a root to urine patches in a grazed pasture, to wetlands in a watershed, to tropical forests on the globe. The environmental controls over biogeochemical hot spots often differ radically from controls in the surrounding **matrix**—that is, the predominant patch type in the landscape. Only by studying processes in these hot spots can we understand these processes and extrapolate their consequences to larger scales. Landscape ecology therefore plays an essential role in understanding the Earth System because of the importance of estimating fluxes (and their controls) of energy and materials at regional and global scales.

The size, shape, and distribution of patches in the landscape govern interactions among patches. Patch size influences habitat heterogeneity. Large forest fragments in an agricultural landscape, for example, contain greater habitat heterogeneity and support more species and bird pairs than do small patches (Freemark and Merriam 1986, Wiens 1996). Patch size also influences the spread of propagules and disturbance from one patch to another. Seeds must travel farther to colonize large disturbed patches, such as fire scars or abandoned agricultural fields, than to colonize small disturbed patches. Patch size therefore affects recruitment and the capacity of regenerating vegetation to use the pulse of nutrients that accompanies disturbance (Rupp et al. 2000). **Patch shape** influences the effective size of patches by determining the average distance of each point in the patch to an edge. Patch size and shape together determine the ratio of edge to area of the patch. The edge-to-area ratio of lakes and streams, for example, is critical in determining the relative importance of aquatic and terrestrial production in supplying energy to aquatic food webs, which radically affects their functioning (see Chapter 10).

The population dynamics of many organisms depend on movement between patches, which is strongly influenced by their **connectivity** (Turner et al. 2001). Birds and small animals in an agricultural landscape, for example, use fence rows to travel among patches of suitable habitat. In a patchy environment, local populations may go extinct, and the dynamics of **metapopulations**—populations that consist of partially isolated subpopulations—depend on relative rates of local extinctions in patches and colonization from adjacent patches (Hanski 1999). Species conservation plans often encourage the use of corridors to facilitate movement among suitable habitat patches (Fahrig and Merriam 1985), although the effectiveness of corridors is debated (Rosenberg et al. 1997, Turner et al. 2001). Connectivity may be particularly critical at times of climatic change. Isolated nature reserves, for example, may contain species that cannot adapt or migrate in response to rapid environmental change. The effectiveness of corridors among patches depends on the size and mobility of organisms or the nature of disturbances that move among patches (Wu and Loucks 1995). A fence row, for example, may be a corridor for voles, a barrier for cattle, and invisible to birds.

Ecological **boundaries** are critical to the interactions among neighboring landscape elements (Gosz 1991). Animals like deer, for example, are edge specialists that forage in one patch type and seek protection from predation in another. The size of the patch and its edge-to-area ratio determine the total habitat available to edge specialists. Edges often experience a different physical environment than do the interiors of patches. Forest boundaries adjacent to clearcuts, for example, experience more wind and solar radiation and are drier than are patch interiors (Chen et al. 1995). In tropical rain forests the trees within 400 m of an edge experience more frequent blowdowns than do trees farther from an edge (Laurance and Bierregaard 1997). These differences in physical environment affect rates of disturbance and nutrient cycling, which translate into variations in recruitment, productivity, and competitive balance among species. The depths to which these edge effects penetrate differ among processes and ecosystems. Wind effects, for example, may penetrate more deeply from an edge than would availability of mycorrhizal propagules.

The abruptness of boundaries influences their role in the landscape (McCoy et al. 1986). Relatively broad gradients at the boundaries between biomes occur where there is a gradual shift in some controlling variable such as precipitation or temperature. Sharper boundaries tend to occur where there are steep gradients in physical variables that control the distribution of organisms and ecosystem processes or where an ecologically important functional type reaches its climatic limit. The boundary between a stream and its riparian zone reflects presence or absence of water above the ground surface. Similarly, the boundary between different parent materials can be quite sharp, with two sides of a boundary supporting different ecosystems. Climatically determined boundaries, such as treeline or the savanna–forest border, are useful places to study the effects of climatic change because species are at their physiologically determined range limits. Species in these situations may be sensitive to small changes in climate.

The **configuration**, or spatial arrangement, of patches in a landscape influences landscape properties because it determines which patches interact and the spatial extent of their interactions. Riparian areas are important because they are an interface between terrestrial and aquatic ecosystems. Their linear configuration and location make them much more important than their small aerial extent would suggest. An area of the same size that occurred elsewhere in the landscape would function quite differently.

Causes of Spatial Heterogeneity

Landscape heterogeneity stems from environmental variation, population and community processes, and disturbance. Spatial variation in state factors (e.g., topography and parent material) and interactive controls (e.g., disturbance and dominant plant species) determine the natural matrix of spatial variability in ecosystems (Holling 1992). Human activities are an increasing cause of changes in the spatial heterogeneity of ecosystems.

State Factors and Interactive Controls

Differences in abiotic characteristics and associated biotic processes account for the basic matrix of landscape variability. Temperature, precipitation, parent materials, and topography vary independently across Earth's surface. Some of these state factors, such as rock type, exhibit sharp boundaries and can therefore be classified into distinct patches. Others, including climate variables, vary more continuously and generate gradients in ecosystem structure and functioning. Analysis of these landscape classes and gradients shows that different factors control spatial patterns at different spatial scales. Regional-scale patterns of vegetation, net primary production (NPP), soil organic matter (SOM), litter quality, and nutrient availability in grasslands, for example, correlate with regional gradients in precipitation and temperature (Burke et al. 1989) (Fig. 14.1). In contrast, topography, soil texture, and land use history explain most variability at the scale of a few kilometers, and microsite variation accounts for variability within patches (Burke et al. 1999). Broad elevational patterns of ecosystem processes in tropical forests on the Hawaiian Islands are also governed largely by climate, with local variation reflecting the type and age of the parent material (Vitousek et al. 1997b, Chadwick et al. 1999). The resulting differences in soils give rise to consistent differences in nitrogen cycling (Pastor et al. 1984), phosphorus cycling (Lajtha and Klein 1988) and nitrous oxide emissions (Matson and Vitousek 1987). These comparative studies provide a basis for extrapolating ecosystem processes to regional scales based on the underlying spatial matrix of abiotic factors.

Community Processes and Legacies

Historical legacies, stochastic dispersal events, and other community processes can modify the underlying relationship between environment and the distribution of a species. Ecosystem processes depend not only on the current envi-

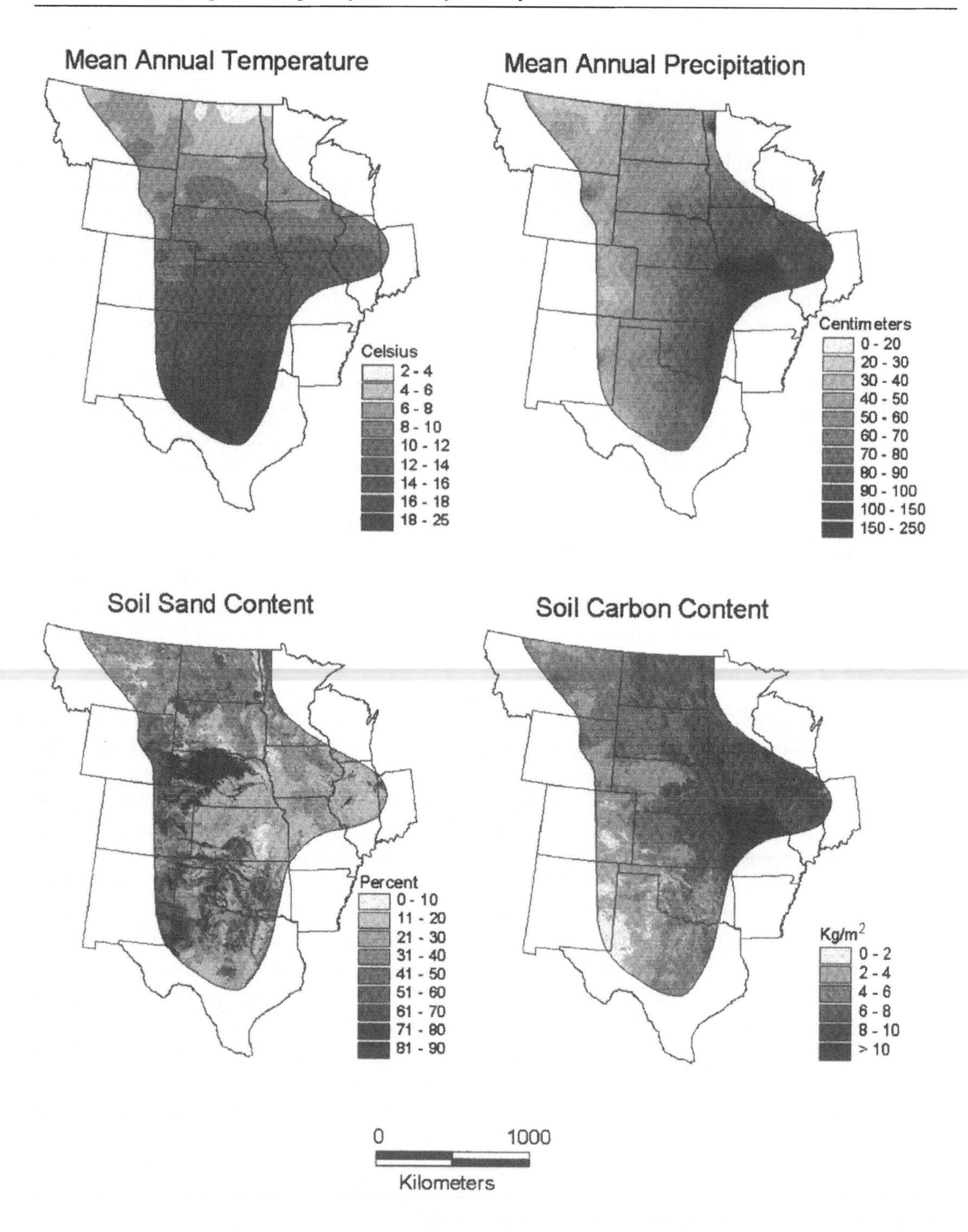

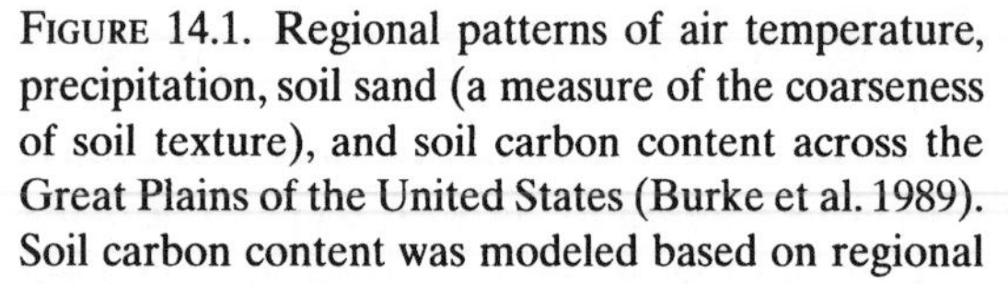

FIGURE 14.1. Regional patterns of air temperature, precipitation, soil sand (a measure of the coarseness of soil texture), and soil carbon content across the Great Plains of the United States (Burke et al. 1989). Soil carbon content was modeled based on regional databases of the environmental variables using the Century model. Soil carbon content varies regionally in ways that are predictable from climate and soil texture. (Figure provided by I. Burke.)

ronment but also on past events that influence the species present at a site (see Chapter 13). The patterns of local dominance resulting from past events can then affect the spatial pattern of processes, such as nutrient cycling (Frelich and Reich 1995, Pastor et al. 1998). White spruce and paper birch, for example, are common boreal trees that differ in both population parameters (e.g., age of first reproduction and seed dispersal range) and species effects (e.g., tissue nutrient concentration and decay rate) (Van Cleve et al. 1991, Pastor et al. 1998). The spatial dynamics of birch and spruce therefore translate into spatial patterns of ecosystem processes such as rates of nitrogen cycling. In semiarid ecosystems, soil processes are strongly influenced by the presence or absence of individual plants, resulting in "resources islands" beneath plant canopies (Burke and Lauenroth 1995). Herbivory also affects spatial patterning in ecosystems through its effects on SOM, nutrient availability, and NPP (Ruess and McNaughton 1987). The distribution of species on a landscape results from a combination of habitat requirements of a species, historical legacies (see Chapter 13), and stochastic events. Once these patterns are established, they can persist for a long time, if the species effects are strong. The fine-scale distribution of hemlock and sugar maple that developed in Minnesota several thousand years ago, for example, has been maintained because these two tree species each produce soil conditions that favor their own persistence (Davis et al. 1998).

Disturbance

Natural disturbances are ubiquitous in ecosystems and cause spatial patterning at many scales. The literature on **patch dynamics** views a landscape as a mosaic of patches of different ages generated by cycles of disturbance and postdisturbance succession (see Chapter 13) (Pickett and White 1985). In ecosystems characterized by gap-phase succession, the vegetation at any point in the landscape is always changing; but, averaged over a large enough area, the proportion of the landscape in each successional stage is relatively constant, forming a **shifting steady state mosaic**. Although every point in the landscape may be at different successional stages, the landscape as a whole may be close to steady state (Turner et al. 1993). This pattern is observed (1) in environmentally uniform areas, where disturbance is the main source of landscape variability; (2) when disturbances are small relative to the size of the landscape; and (3) when the rate of recovery is faster than the return time of the disturbance (Fig. 14.2). When disturbances are small and recovery is rapid, most of the landscape will be in middle to late successional stages. The formation of treefall gaps in gap-phase succession, for example, is the primary source of canopy turnover in ecosystems such as tropical rain forests, where large-scale disturbances are rare (Rollet 1983). Over time, gap-phase disturbance contributes to the maintenance of the productivity and nutrient dynamics of the entire forest. In the primary rain forests of Costa Rica, for example, the regular occurrence of treefalls results in maximum tree age of only 80 to 140 years (Hartshorn 1980). Light, and sometimes nutrient availability, increase in treefall gaps, providing resources that allow species with higher resource requirements to grow quickly and maintain themselves in the forest mosaic (Chazdon and Fetcher 1984b, Brokaw 1985). Disturbances by animals in grassland and shrublands can also generate a shifting steady state mosaic. Gophers, for example, disturb patches of California serpentine grasslands, causing patches to turn over every 3 to 5 years (Hobbs and Mooney 1991).

Large-scale infrequent disturbances alter the structure and processes of some ecosystems over large parts of a landscape. These disturbances result in large expanses of the landscape in the same successional stage and are termed **non–steady state mosaics**. After Puerto Rico's hurricane Hugo in 1989, for example, most of the trees in the hurricane path were broken off or blown over or lost a large proportion of their leaves, resulting in a massive transfer of carbon and nutrients from vegetation to the soils. The large pulse of high-quality litter increased decomposition rates substantially over large areas (Scatena et al. 1996).

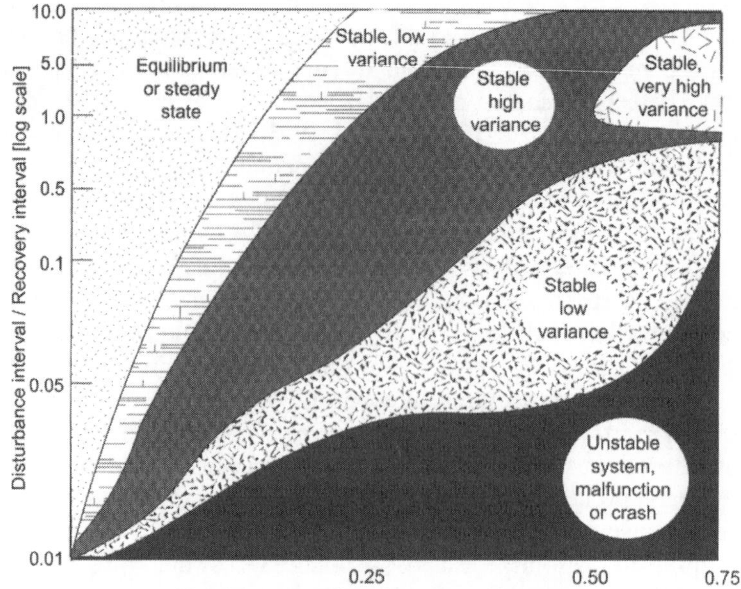

FIGURE 14.2. Effect of disturbance size (relative to the size of the landscape) and disturbance frequency (relative to the time required for the ecosystem to recover) on the stability of landscape processes. Landscapes are close to equilibrium when disturbances are small (relative to the size of the study area) and when they are infrequent (relative to the time required for ecosystem recovery). As disturbances become more frequent or larger, the landscape becomes more heterogeneous and there is increasing probability that the individual patches may undergo a different successional trajectory. (Redrawn with permission from *Landscape Ecology*, Vol. 8 © 1993 Kluwer Academic Publishers; Turner et al. 1993.)

Fire can also create large patches of a single successional stage on the landscape (Johnson 1992). The 1988 fires in Yellowstone National Park that burned 3200 km^2 of old pine forest altered large areas of the landscape. Fires of this magnitude and intensity recur every few centuries (Romme and Knight 1982). Long-term human fire suppression has increased the proportion of late-successional communities in many areas. This results in a more homogeneous and spatially continuous, fuel-rich environment in which fires can burn large areas. Even large disturbed areas, however, are often internally quite patchy. Fires, for example, generate islands of unburned vegetation and patches of varying burn severity. These unburned islands act as seed sources for post-fire succession and protective cover for wildlife, greatly reducing the effective size of the disturbance (Turner et al. 1997). In many cases, these patches become less distinct as succession proceeds, so spatial heterogeneity may decline with time in non–steady state mosaics.

Human-induced disturbances alter the natural patterns and magnitude of landscape heterogeneity. Half of the ice-free terrestrial surface has been transformed by human activities (Turner et al. 1990). We have cleared or selectively harvested forests, converted grasslands and savannas to pastures or agricultural systems, drained wetlands, flooded uplands, and irrigated dry lands. Isolated land use changes may augment landscape heterogeneity by creating small patches within a matrix of largely natural vegetation. As land use change becomes more extensive, however, the human-dominated patches become the matrix in which isolated fragments of natural ecosystems are embedded, causing a reduction in landscape heterogeneity. These contrasting impacts of human actions on landscape heterogeneity are illustrated by the practice of shifting agriculture.

Shifting agriculture is a source of landscape heterogeneity at low population densities but reduces landscape heterogeneity as human population increases. **Shifting agriculture**, also known as **slash-and-burn agriculture** or **swidden agriculture**, involves the clearing of forest for crops followed by a fallow period during which forests regrow, after which the cycle repeats. Shifting agriculture is practiced extensively in the tropics and in the past played an important role in clearing the forests of Europe and eastern North America. Small areas of forest are typically cleared of most trees and burned to release organically bound nutrients. The soil is left untilled, causing little loss of SOM. Crops are planted in species mixtures, with multiple plantings and harvests (Vandermeer 1990). As soil fertility drops, and insect and plant pests encroach, often within 3 to 5 years, the agricultural plots are abandoned and allowed to regrow to forest. The regrowing forests provide fuel and other products for 20 to 40 years until the cycle repeats. Shifting agriculture generates landscape heterogeneity at many scales, ranging from different aged patches within a forest to different crop species within a field.

With moderate human population densities that allowed sufficient fallow periods and judicious selection of land for cultivation, shifting agriculture existed sustainably for thousands of years and caused no directional change in biogeochemical cycles (Ramakrishnan 1992). As population density increases, land becomes scarcer, and the fallow periods are shortened or eliminated, leading to a more homogeneous agricultural landscape. Under these conditions, nutrient and organic matter losses during the agricultural phase cannot be recouped, and the system degrades, requiring larger areas to provide sufficient food. As the landscape becomes dominated by active cropland or early successional weedy species, the seed sources of mid-successional species are eliminated, preventing forest regrowth and further reducing the potential for landscape heterogeneity. In northeast India, for example, this shifting agriculture appears unsustainable when the rotation cycle declines below 10 years (Ramakrishnan 1992).

Interactions Among Sources of Heterogeneity

Landscape heterogeneity and disturbance history interact to influence further disturbance. Disturbance is more than a simple overlay on the spatial patterns governed by environment, because even slight variations in topography or edaphic factors can influence the frequency, type, or severity of natural disturbance or the probability that land will be cleared by people. Slope and aspect of a hillside affect solar irradiance, soil moisture, soil temperature, and evapotranspiration rate. These factors, in turn, contribute to variation in biomass accumulation, species composition, and fuel characteristics. Different parts of the landscape may therefore differ in susceptibility to fire. The resulting mosaic of patch types with different flammabilities can prevent a small, locally contained fire from moving across large areas. Slope and aspect can also directly influence the exposure of ecosystems to fire spread because fire generally moves uphill and tends to halt at ridgetops. Elevation and topographic position also influence the susceptibility of forest trees to wind-throw (Foster 1988).

Patchiness created by disturbance and other legacies influences the probability and spread of disturbance, thereby maintaining the mosaic structure of landscapes. The spread of fire, for example, creates patches of early successional vegetation in fire-prone ecosystems that are less flammable than late-successional vegetation (Starfield and Chapin 1996). In this way, past disturbances create a legacy that governs the probability and patch size of future disturbances. The effectiveness of these disturbance-generated early successional firebreaks depends on climate. At times of extreme fire weather, almost any vegetation will burn.

The past history of insect or pathogen outbreaks also generates a spatial pattern that determines the pattern of future outbreaks. In mountain hemlock ecosystems of the northwestern United States, low light and nutrient availability in old-growth stands makes trees

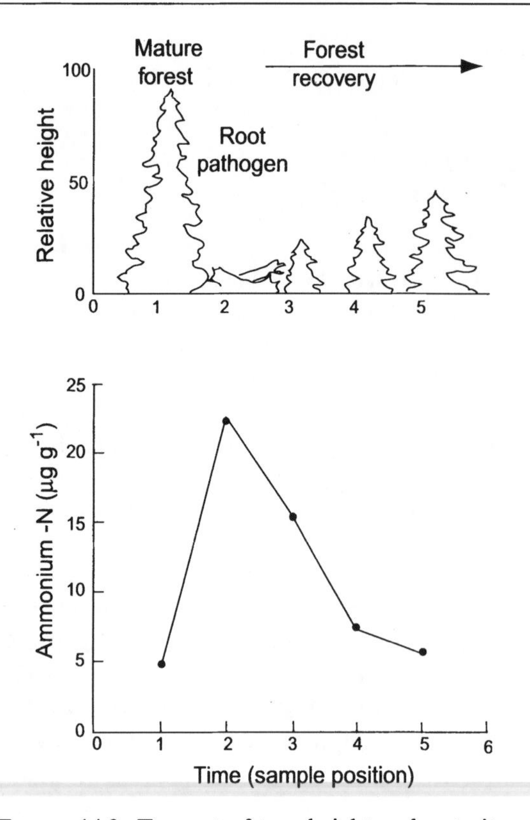

FIGURE 14.3. Transect of tree height and net nitrogen mineralization rate across a disturbance caused by root pathogens in a hemlock stand in the northwestern United States (Matson and Boone 1984). Nitrogen mineralization increases dramatically beneath trees recently killed by the pathogen (position 2). As trees recover (positions 3, 4, and 5), net nitrogen mineralization declines toward rates typical of undisturbed forest (position 1).

vulnerable to a root pathogen. The resulting tree death increases light, nitrogen mineralization, and nutrient availability, making the regrowing forest resistant to further attack (Matson and Boone 1984) (Fig. 14.3). The infections tend to move through stands in a wave-like pattern, attacking susceptible patches and creating resistant patches in their wake (Sprugel 1976), just as described for fire. Similarly, hurricanes that blow down large patches of trees generate early successional patches of short-statured trees that are less vulnerable to wind-throw. Even the fine-grained steady state mosaics that characterize gap-phase succession are self-sustaining because young trees that grow in a gap created by treefall are less likely to die than are older trees. In summary, disturbances that reduce the probability of future disturbance generate a negative feedback that tends to stabilize the disturbance regime of an ecosystem, resulting in a shifting steady state mosaic with a characteristic patch size and return interval. Any long-term trend in climate or soil resources that alters disturbance regime will probably alter the characteristic distribution of patch sizes on the landscape.

Disturbances that increase the probability of other disturbances complicate predictions of landscape pattern. Insect outbreaks that kill trees in a fine-scale mosaic, for example, can increase the overall flammability of the forest, increasing the probability of large fires (Holsten et al. 1995). The public concern about large fires after insect outbreaks then creates public pressure for salvage logging of insect-killed stands. This logging creates patches of clearcuts that are intermediate in size between those created by insects and those that might have been produced by a catastrophic fire. It is difficult to predict in advance which of these three alternative patch structures, or combination of them, will occur, because disturbances that increase the probability of other disturbances create a positive feedback that destabilizes the existing pattern of disturbance regime and landscape heterogeneity. Rule-based models that define conditions under which particular scenarios are likely to occur provide a framework for predictions in the face of multiple potential outcomes (Starfield 1991).

Human activities create positive and negative feedbacks to disturbances that alter the patch structure and functioning of landscapes. In principle, the effect of human-induced disturbances, such as land clearing, on landscape structure is no different from that of any other disturbance. However, the novel nature and the increasingly extensive occurrence of human disturbances are rapidly altering the structure of many landscapes. The construction of a road through the tropical wet forests of Rondonia, Brazil, for example, created a simple linear disturbance of negligible size. The sudden increase in human access, however, led to rapid clearing of forest patches that were much larger than the natural patches created by treefalls or the hand-

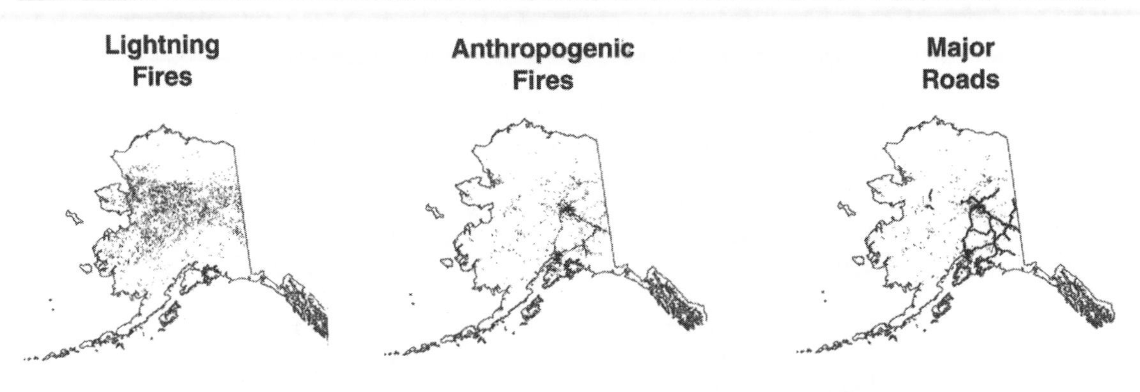

FIGURE 14.4. Locations of naturally and human-caused fires in Alaska (Gabriel and Tande 1983). The human-caused fires mirror the road and river transportation corridors, indicating the importance of human access in altering the regional fire regime.

cleared patches created by shifting agriculture. Similarly, road access is the major factor determining the distribution of fire ignitions in the boreal forest of interior Alaska (Fig. 14.4). In general, road access is one of the best predictors of the spread of human-induced disturbances in relatively natural landscapes (Dale et al. 2000).

Socioeconomic factors, such as farmer income, interact with site characteristics to influence human effects on the landscape pattern. Heterogeneous landscapes are often converted to fine-scale mosaics of agricultural and natural vegetation, whereas large areas suitable for mechanized agriculture are more likely to be deforested in large blocks. In northern Argentina, for example, patches of dry deciduous forests on the eastern slopes of the Andes were converted to small patches of cropland or modified by grazing into thorn-scrub grazing lands and secondary forests (Cabido and Zak 1999) (Fig. 14.5). On the adjacent plains, however, larger parcels were initially deforested for grazing and more recently converted to mechanized agriculture. Large holdings on the plains are owned by companies that make land use decisions based on the global economy. Small family producers in the mountains maintain a more traditional lifestyle that involves smaller, less frequent changes in land use.

Disturbance is increasingly used as a management tool to generate more natural stand and landscape structures. Forest harvest varies from 0 to 100% tree removal, and the sizes and shapes of clearcuts can be altered from the standard checkerboard pattern to mimic more natural disturbances (Franklin et al. 1997). Forest harvest regimes can also be designed to retain some of the functional attributes of late successional forests, such as the filtering function of riparian vegetation, the presence of large woody debris, and the retention of a few large trees as seed source and nesting habitat. Protection of these features can significantly reduce the ecological impact of forest harvest. Prescribed fire is increasingly used as a management tool, particularly in areas where a century of Smoky the Bear policy of complete fire suppression has led to unnaturally large fuel accumulations. Prescribed fires are typically lit when weather conditions are such that fire intensity and severity are low, so the fire can be readily controlled. In populated mediterranean regions, vegetation may be physically removed as a substitute for fire, because prescribed fires are considered unsafe. Natural fire, prescribed fire, and physical removal of vegetation are likely to differ in their impacts on ecosystem processes due to differences in the quantity of organic matter and nutrients removed; these differences affect subsequent regrowth. Ecologists are only beginning to understand the long-term consequences of different disturbance regimes for the structure and functioning of ecosystems and landscapes. As this understanding improves, more informed decisions can be made in using distur-

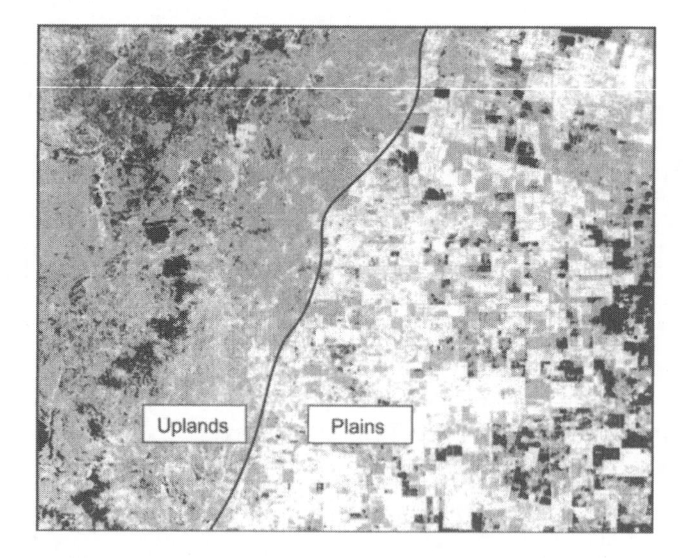

FIGURE 14.5. Seminatural vegetation (black), lands that have been modified by grazing (gray) and croplands (white) of the Cordoba region of northern Argentina in 1999 (Satellite based) (Cabido and Zak 1999). The plains to the east are more suitable for mechanized agriculture and are large land holdings with substantial areas converted to croplands. Lands to the west are more mountainous and less suitable for mechanized agriculture; they are owned by small farmers, each of whom maintains a heterogeneous mosaic of land use. The proportion of area converted to cropland is greater in large land holdings suitable for intensive agriculture. (Figure provided by M. Cabido and M. Zak.)

bance as a tool in ecosystem management (see Chapter 16). The goal of manipulating disturbance regime as a management tool is to mimic the ecological effects of disturbance under conditions in which the natural disturbance pattern has unacceptable societal consequences, for example in the rapidly expanding exurban-wildland interface where homes are being built in fuel-rich habitats that are the product of fire suppression.

Patch Interactions on the Landscape

Interactions among patches on the landscape influence the functioning of individual patches and the landscape as a whole. Landscape patches interact when things move across boundaries from one patch to another. This occurs through topographically controlled interactions, transfers through the atmosphere, biotic transfers, and the spread of disturbance. These transfers are critically important to the long-term sustainability of ecosystems because they represent losses from donor ecosystems and subsidies to recipient ecosystems. Large changes in these transfers constitute changes in inputs and/or outputs of resources and therefore substantially alter the functioning of ecosystems.

Topographic and Land-Water Interactions

Topographically controlled redistribution of materials is the predominant physical pathway by which materials move between ecosystems (Fig. 14.6). Gravity is a potent force for landscape interactions. It causes water to move downhill, carrying dissolved and particulate materials. Gravity is also the driving force for landslides, soil creep, and other forms of soil movement. These topographically controlled processes transfer materials from uplands to lowlands, from terrestrial to aquatic systems, and from fresh-water ecosystems to estuaries and oceans (Naiman 1996).

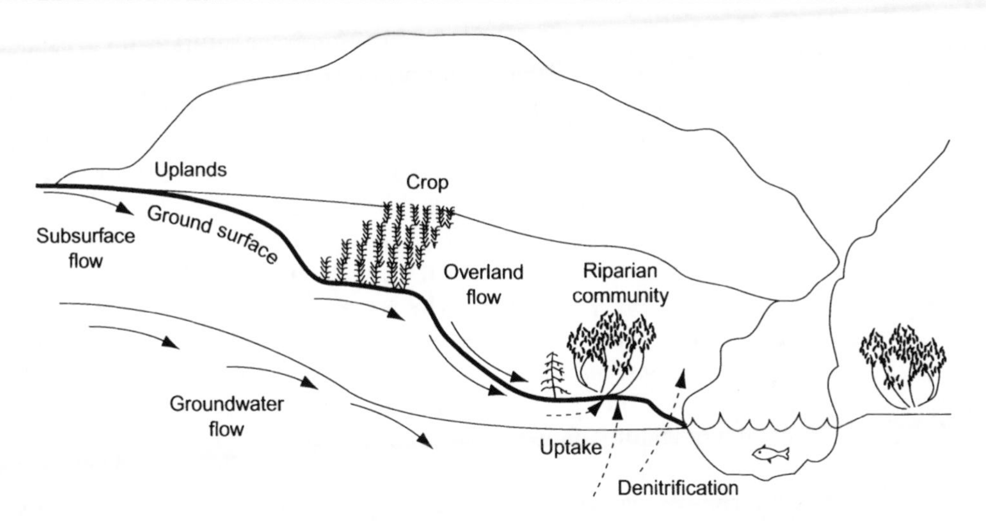

FIGURE 14.6. Topographically controlled interactions among ecosystems in a landscape via erosion and solution transfers in subsurface flow or groundwater. Riparian forest trees absorb nutrients primarily from well-aerated soils, whereas denitrification requires anoxic conditions, which generally occur below the water table. Nitrogen uptake and denitrification are the most important mechanisms by which riparian zones filter nitrogen from groundwater between upland ecosystems and streams.

The nature of donor ecosystems and their management govern the transfer of dissolved materials. Regions with intensive agriculture and those receiving substantial nitrogen deposition transfer substantial quantities of nitrate and phosphorus to rivers, lakes, and groundwater (Carpenter et al. 1998). There is, for example, a strong relationship between the total nitrogen input to the major watersheds of the world and nitrate loading in rivers (Fig. 14.7) (Howarth et al. 1996). At more local scales, the patterns of land use and urbanization influence the input of nutrients to lakes and streams. These increased fluxes of dissolved nitrogen have multiple environmental consequences, including health hazards, acidification, eutrophication, and reduced biodiversity of downstream fresh-water and marine ecosystems (Howarth et al. 1996, Nixon et al. 1996).

Erosion moves particulate material containing nutrients and organic matter from one ecosystem and deposits it in another. Erosion ranges in scale from silt suspended in flowing water to movement of whole mountainsides in landslides. The quantity of material moved depends on many physical factors, including slope position, slope gradient, the types of rocks and unconsolidated material underlying soils, and the types of erosional agents (e.g., amount and intensity of rainfall events) (see Chapter 3). The biological characteristics of ecosystems are also critical. Vegetation type, root strength, disturbance, management, and human development can be as important as the vertical gradient or parent material. Forest harvest on steep slopes in the northwestern United States, for example, has increased the

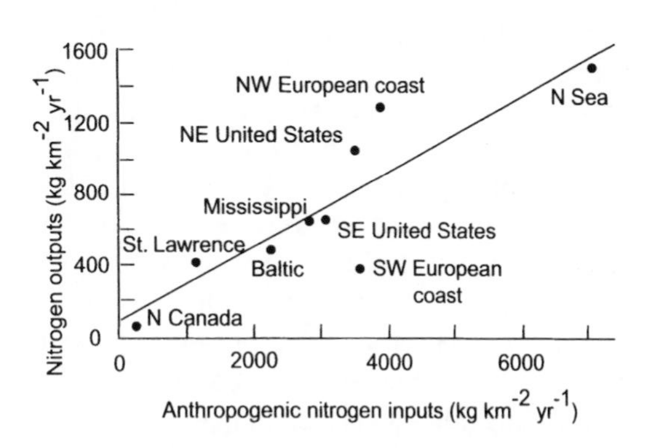

FIGURE 14.7. Relationship between the total anthropogenic nitrogen input to the major watersheds of the world and nitrate loading in rivers. (Redrawn with permission from *Biogeochemistry*, Vol. 35 © 1996 Kluwer Academic Publishers; Howarth et al. 1996.)

frequency of landslides. Similarly, upland agriculture often increases sedimentation and the associated transfer of nutrients and contaminants (Comeleo et al. 1996). Proper management of upslope systems through use of cover crops, reduced tillage, and other management practices can reduce erosional transfers of materials.

Landscape pattern influences the transfer of materials among ecosystems. Even in unmanaged landscapes, ecosystems interact with one another along topographic sequences, with nutrients leached from uplands providing a nutrient subsidy to midslope or lowland ecosystems (Shaver et al. 1991). The configuration of these ecosystems in the landscape determines the pattern of nutrient redistribution and their outputs to groundwater and streams. Riparian vegetation zones, including wetlands and floodplain forests, act as filters and sediment traps for the water and materials moving from uplands to streams (Fig. 14.6). Dominance of riparian zones by disturbance-adapted plants that tolerate soil deposition and have rapid growth rates contributes to the effectiveness of these landscape filters. Riparian zones play a particularly crucial role in agricultural watersheds, where they remove fertilizer-derived nitrogen as well as phosphorus and eroding sediments. The fine-textured, organic-rich soils and moist conditions characteristic of most riparian areas also promote denitrification of incoming nitrate. Plant uptake and denitrification together account for the decline in nitrate concentration as groundwater flows from agricultural fields through riparian forests to streams. Phosphorus is retained in riparian areas primarily by plant and microbial uptake and physical adsorption to soils because phosphorus has no pathway of gaseous loss.

The high productivity and nutrient status of riparian vegetation and the presence of water cause riparian areas to be intensively used by animals, including livestock in managed ecosystems. People also use riparian areas intensively for water, gravel, transportation corridors, and recreation. Long-term elevated inputs from heavily fertilized agricultural areas or from wetlands used for tertiary sewage treatment (i.e., to remove the products of microbial decomposition) can saturate their capacity to filter nutrients from groundwater. Any mode of overexploitation of riparian areas increases sediment and nutrient loading to streams and reduces shading, making fresh-water ecosystems more vulnerable to changes in land use within the watershed (Correll 1997, Lowrance et al. 1997, Naiman and Decamps 1997).

The properties of recipient ecosystems influence their sensitivity to landscape interactions. The vulnerability of ecosystems to inputs from other patches in the landscape depends largely on their capacity to sequester or transfer the inputs. Riparian areas, for example, may have a higher capacity to retain a pulse of nutrients or transfer them to the atmosphere by denitrification than do upland late-successional forests. Streams characterized by frequent floods are less likely to accumulate sediment inputs than are slow-moving streams and rivers, because floods flush sediments from river channels of steep stream reaches. Lakes on calcareous substrates or those that receive abundant groundwater input due to a location low in a watershed are better buffered against inputs of acidity and nutrients than are oligotrophic lakes on granitic substrates or lakes high in a watershed that receive less groundwater input (Webster et al. 1996).

Estuaries, the coastal ecosystems located where rivers mix with seawater, are a striking example of the way in which ecosystem properties influence their sensitivity to inputs from the landscape. They are among the most productive ecosystems on Earth (Howarth et al. 1996, Nixon et al. 1996). Their high productivity stems in part from the inputs they receive from land and from the physical structure of the ecosystem, which is stabilized by the presence of seagrasses and other rooted plants. This tends to dampen wave and tidal energy, reducing resuspension and increasing sedimentation. Salinity and other geochemical changes that occur as the waters mix lead to flocculation and settling of suspended particulates. Nutrient uptake by the rooted vegetation and phytoplankton, burial by sedimentation, and denitrification in anoxic sediments function as sinks for nutrients flowing from upstream watersheds, just as in riparian zones. Estuaries differ from one another in their capacity to act as sinks for incoming materials, due to variation in basin geometry, sediment input, and tidal interactions. The stability of the landscape on the Mississippi River Delta, for example, depends on regular delivery of sediments from upstream to replace soils removed by tidal erosion. Channels, levees, and other engineering solutions to flood control and water management may reduce the probability of flooding but greatly augment the land loss to coastal erosion (Costanza et al. 1990). Many estuaries, including the Gulf of Mexico near the entrance of the Mississippi River, are becoming saturated by nutrient enrichment within their watersheds, resulting in harmful algal blooms, loss of seagrass, and increasing frequency of anoxia or hypoxia and related fish kills (Mitsch et al. 2001).

Atmospheric Transfers

Atmospheric transport of gases and particles links ecosystems over large distances and coarse spatial scales. Gases emitted from managed or natural ecosystems are processed in the atmosphere and can be transported for distances ranging from kilometers to the globe. Particulates from biomass burning, wind-blown dust, sea spray, and anthropogenic sources can also be carried through the atmosphere from one ecosystem to another. Once deposited, they can alter the functioning of the recipient ecosystems (Fig. 14.8), just as with topographically controlled transfers.

In areas downwind of agriculture, ammonia gas (NH_3) and nitric oxides (NO_x) can represent a significant fraction of nitrogen deposition. Dutch heathlands, for example, receive 10-fold more nitrogen deposition than would occur naturally. The magnitude of these inputs is similar to the quantity of nitrogen that annually cycles through vegetation, greatly increasing the openness of the nitrogen cycle. Areas downwind of industry and fossil fuel combustion receive nitrogen largely as NO_x. Sulfur gases, including sulfur dioxide (SO_2), are also produced by fossil fuel combustion, although improved regulations have reduced these emissions and deposition relative to NO_x.

The large nitrogen inputs to ecosystems have important consequences for NPP, nutrient

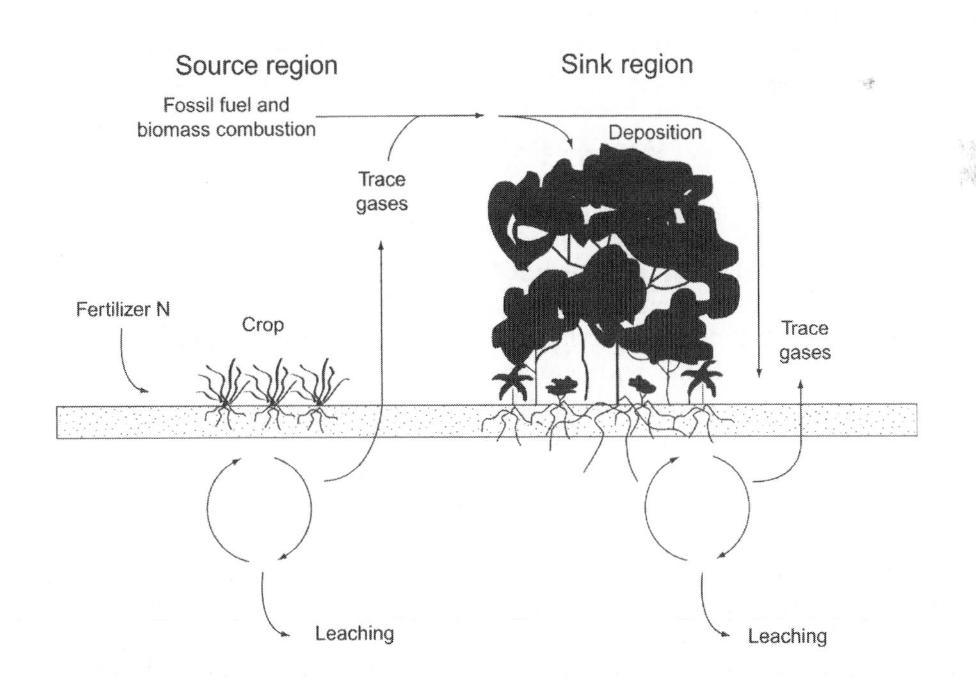

FIGURE 14.8. Atmospheric transfers of gases, solutions, and particulates among ecosystems. Inputs come from fossil fuel and biomass combustion and from trace gases originating from natural and managed ecosystems.

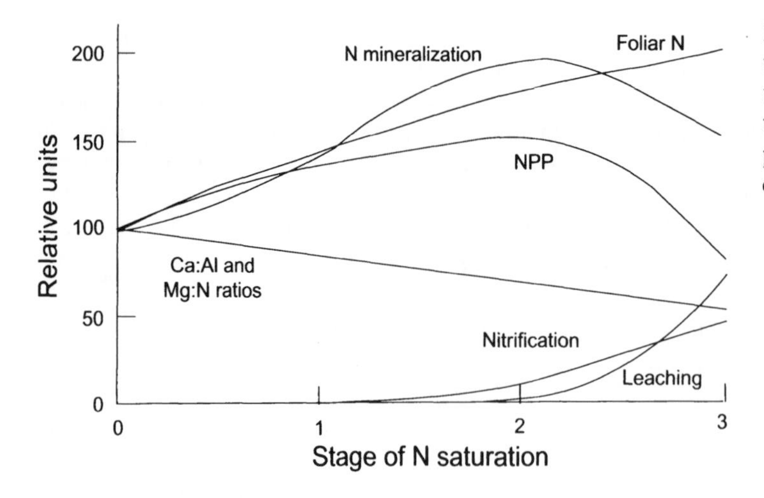

FIGURE 14.9. Changes hypothesized to occur as forests undergo long-term nitrogen deposition and nitrogen saturation. (Redrawn with permission from *BioScience*; Aber et al. 1998.)

cycling, trace gas fluxes, and carbon storage. Chronic nitrogen deposition initially reduces nitrogen limitation by increasing nitrogen cycling rates, foliar nitrogen concentrations, and NPP. Above some threshold, however, the ecosystem becomes saturated with nitrogen (Fig. 14.9) (Aber et al. 1998). As excess nitrate and sulfate leach from the soil, they carry with them cations to maintain charge balance, inducing calcium and magnesium deficiency in vegetation (Driscoll et al. 2001). In southern Sweden, for example, over half of the plant-available cations have been lost from the upper 70 cm of soil in the past half century, probably due to chronic exposure to acid precipitation (Hallbacken 1992). The exchange complex becomes more dominated by aluminum and hydrogen ions, increasing soil acidity and the likelihood of aluminum toxicity. Together this suite of soil changes often enhances frost susceptibility, impairs root development, and promotes herbivory, leading to forest decline in many areas of Europe and the northeastern United States (Schulze 1989, Aber et al. 1998). The major surprise, however, has been how resilient many forests have been to acid rain, often retaining most of the nitrogen inputs within the ecosystem for as much as two decades. The resilience of ecosystems depends in part on the magnitude of inputs (related to distance from pollution sources and amount of precipitation received) and initial soil acidity, which in turn depends on parent material and species composition. Many ecosystems, however, now show clear signs of nitrogen saturation, resulting in forest decline, loss of acid-neutralizing capacity in lakes, and increasing nitrogen inputs to streams (Aber et al. 1998, Carpenter et al. 1998, Driscoll et al. 2001).

Nearly all research on the transport, deposition, and ecosystem consequences of anthropogenic nitrogen has been conducted in the temperate zone. Further increases in nitrogen deposition will, however, likely occur primarily in the tropics and subtropics (Galloway et al. 1995), where plant and microbial growth are frequently limited by elements other than nitrogen. These ecosystems might therefore show more immediate nitrogen loss in trace gases or leaching in response to nitrogen deposition (Matson et al. 1998). On the other hand, soil properties such as high clay content or cation exchange capacity may allow tropical soils to sequester substantial quantities of nitrogen before they become leaky.

Biomass burning transfers nutrients directly from terrestrial pools to the atmosphere and then to down-wind ecosystems. Biomass combustion releases a suite of gases that reflect the elemental concentrations in vegetation and fire intensity. About half of dry biomass consists of carbon, so the predominant gases released are carbon compounds in various stages of oxidation, including carbon dioxide (CO_2), methane (CH_4), carbon monoxide (CO), and smaller quantities of nonmethane hydrocarbons. The atmospheric role of these gases varies. CO_2 and CH_4 are greenhouse gases, whereas carbon

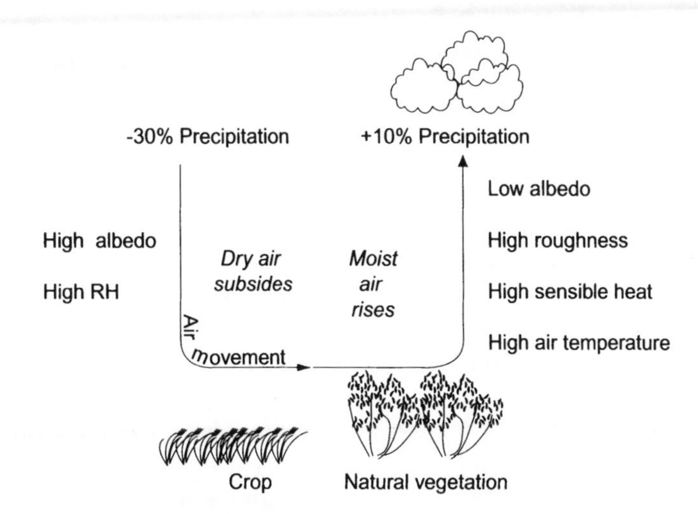

FIGURE 14.10. Effects on regional climate of conversion from heathland to barley croplands in southwestern Australia (Chambers 1998). The heathland absorbs more radiation (low albedo) and transmits a larger proportion of this energy to the atmosphere as sensible heat than does adjacent croplands. This causes air to rise over the heathland and draws in moist air laterally from the irrigated cropland; this causes subsidence of air over the cropland, just as with land–sea breezes. Rising moist air increases precipitation by 10% over heathland, whereas subsiding dry air reduces precipitation by 30% over the cropland. RH, relative humidity.

monoxide and nonmethane hydrocarbons react in the troposphere to produce ozone and other atmospheric pollutants that can affect downwind ecosystems (see Chapter 2). Nitrogen is also released in various oxidation states, including nitrogen oxides (NO and NO_2, together known as NO_x) and ammonia. The proportional release of these forms also depends on the intensity of the burn, with NO_x typically accounting for most of the emissions. Sulfur-containing gases; organic soot and other aerosol particles; elemental carbon; and many trace species of carbon, nitrogen, and sulfur also have important regional and global effects. Satellite and aircraft data show that these gases and aerosols in biomass-burning plumes can be transported long distances.

Windblown particles of natural and anthropogenic origins link ecosystems on a landscape. The role of the atmosphere as a transport pathway among ecosystems differs among elements. For some base cations (Ca^{2+}, Mg^{2+}, Na^+, and K^+) and for phosphorus, dust transport is the major atmospheric link among ecosystems. At the local to regional scales, dust from roads or rivers can alter soil pH and other soil properties that account for regional zonation of vegetation and land–atmosphere exchange (Walker and Everett 1991, Walker et al. 1998). At the global scale, Saharan dust is transported across the Atlantic Ocean and deposited on the Amazon by tropical easterlies. Although the annual input of dust is small, it contributes substantially to soil development over the long term (Graustein and Armstrong 1983). Similarly, dust from the Gobi desert is deposited in the Hawaiian Islands at the rate of 0.1 $g\,m^{-2}$ per century. In old soils (those more than 2 million years old), dust input can be the largest source of base cations (Chadwick et al. 1999).

Land–atmosphere exchange of water and energy in one location influences downwind climate. Oceans and large lakes moderate the climate of adjacent land areas by reducing temperature extremes and increasing precipitation (see Chapter 4). Human alteration of the land surface is now occurring so extensively that it also has significant effects on downwind ecosystems. Conversion of Australian heathlands to agriculture has, for example, increased precipitation over heathlands and reduced it by 30% over agricultural areas (Fig. 14.10) (see Chapter

4). At a global scale, the clearing of land for agriculture has reduced regional albedo and evapotranspiration, leading to greater sensible heat flux (Chase et al. 2000). At all spatial scales, the atmospheric transfer of heat and water vapor from one ecosystem to another strongly affects ecosystem processes in downwind ecosystems. The climatic impacts on downwind ecosystems of reservoirs, irrigation of arid lands, and land use change are seldom included in assessments of the potential effects of these management projects.

Movement of Plants and Animals on the Landscape

The movement and dispersal of plants and animals link ecosystems on a landscape. Large animals typically consume forage from high-quality patches and deposit it where they rest or sleep. Sheep in New Zealand, for example, often camp on ridges at night, moving nutrients upward and counteracting the downward nutrient transport by gravity. Marine birds transfer so much phosphorus from marine foods to the land that the **guano** deposited in their traditional nesting areas has served as a major source of phosphorus for fertilizer. **Anadromous fish**—that is, marine fish that enter fresh water to breed, also transport marine-derived nutrients to terrestrial ecosystems. These fish carry the nutrients up rivers and streams, where they become an important food item for terrestrial predators, which transport the marine-derived nutrients to riparian and upland terrestrial ecosystems (Willson et al. 1998, Helfield and Naiman 2001). These nutrient subsidies by animals contribute to the spatial patterning in ecosystem processes. Insects that feed on seaweed and other marine detritus are an important food source for spiders on islands, merging marine and terrestrial food webs (Polis and Hurd 1996).

Animals also transfer plants, especially as seeds, on fur and in feces. Many plants have evolved life history strategies to take advantage of this efficient form of dispersal. This dispersal mechanism has contributed to the spread of invasive plants. Feral pigs, a nonnative herbivore in Hawaiian rain forests, for example, transfer seeds of invasive plants such as the passion vine, which alters patterns of nutrient cycling. Similarly, the alien bird white eye spreads the alien nitrogen fixer *Myrica faya* (Woodward et al. 1990), which alters the nitrogen status of native ecosystems (Vitousek et al. 1987). Thus invasions of both plants and animals from one ecosystem to another can contribute to a variety of ecosystem changes.

Animals that move among patches can have effects that differ among patch types. Edge specialists such as deer, for example, may concentrate their browsing in one habitat type but seek protection from predators and deposit nutrients in another. At a larger scale, migratory birds move seasonally among different ecosystem types. Lesser snow geese, for example, overwinter in the southern United States and breed in the Canadian Arctic. Populations of this species have increased by more than an order of magnitude as a result of increased use of agricultural crops (rice, corn, and wheat) on the wintering grounds and reduced hunting pressure. This species now exceeds the carrying capacity of its summer breeding grounds and has converted productive arctic salt marshes into unvegetated barrens (Jefferies and Bryant 1995).

People are an increasing cause of lateral transfers of materials among ecosystems, through addition of fertilizers, pesticides, etc., and removal of crops and forest products, and diversion of water. The nutrient transfers in food from rural to urban areas are substantial. The resulting nutrient inputs to aquatic systems occur in locations where riparian zones and other ecological filters are often degraded or absent. Water diversion by people has substantially altered rates and patterns of land use change in arid areas at the expense of rivers and wetlands (see Chapter 4). As water becomes increasingly scarce in the coming decades, pressures for water diversion are likely to increase.

Disturbance Spread

Patch size and arrangement determine the spread of disturbance across a landscape. Disturbance is a critical interactive control over ecosystem processes that is strongly influenced by horizontal spread from one patch to another. Fire and many pests and pathogens move most readily across continuous stretches of disturbance-prone vegetation. Fire breaks of nonflammable vegetation, for example, are an effective mechanism of reducing fire risk at the urban-wildland interface. Fires create their own fire breaks because postfire vegetation is generally less flammable than that which precedes a fire. Theoretical models suggest that, when less than half of the landscape is disturbance prone, frequency is less important than severity in determining the impacts of disturbance. When large proportions of the landscape are susceptible to disturbance, however, the frequency of disturbance becomes increasingly important (Gardner et al. 1987, Turner et al. 1989). The size of patches also influences the spread of disturbance. Landscapes dominated by large patches tend to have a low frequency of large fires. Landscapes with small patches, have greater edge-to-area ratio, so fires tend to spread more frequently into less flammable vegetation (Rupp et al. 2000).

Patchy agricultural landscapes are less prone to spread of pests and pathogens than are large continuous monocultures. Intensive agriculture has reduced landscape patchiness in several respects. The average size of individual fields and the proportion of the total area devoted to agriculture has generally increased, as has the use of genetically uniform varieties. This can lead to rapid spread of pests across the landscape.

Human Land Use Change and Landscape Heterogeneity

Human modification of landscapes has fundamentally altered the role of ecosystems in regional and global processes. Much of the land use change has occurred within the last two to three centuries, a relatively short time in the context of evolution or landscape development. Since 1700, for example, the land area devoted to crop production has increased 466% to a current 15 million km^2 worldwide, an area almost twice the size of the conterminous United States. Many areas of the world are therefore dominated by a patchwork of agricultural fields, pastures, and remnant unmanaged ecosystems. Similar patchworks of cut and regrowing forest interspersed with small areas of old-growth forest are common on every continent. Human-dominated landscapes supply large amounts of food, fiber, and other ecosystem services to society. Two general patterns of land use change emerge: (1) **extensification**, or the increase in area affected by human activities, and (2) **intensification**, or the increased inputs applied to a given area of land or water.

Extensification

Land use changes include both conversions and modifications (Meyer and Turner 1992). **Land use conversion** involves a human-induced change in ecosystem type to one dominated by different physical environment or plant functional type, for example, the change from forest to pasture or from stream to reservoir. **Land use modification** is the human alteration of an ecosystem in ways that significantly affect ecosystem processes, community structure, and population dynamics without radically changing the physical environment or dominant plant functional type. Examples include alteration of natural forest to managed forest, savanna management as grazing lands, and alteration of traditional low-input agriculture to high-intensity agriculture. In aquatic ecosystems this includes the alteration of flood frequency by dams and levees or the stocking of lakes for sport fishing. Both types of land use change alter the functioning of ecosystems, the interaction of patches on the landscape, and the functioning of landscapes as a whole.

Deforestation is an important conversion in terms of spatial extent and ecosystem and global consequences. Forests cover about 30% of the terrestrial surface, about three times the total agricultural land area. Globally, forest area has decreased about 15% (i.e., by 9 million km^2) since preagricultural times. Much of the European and the Indian subcontinents, for example, were prehistorically blanketed by forests but over the last 5 to 10 centuries have supported extensive areas of agriculture. Similarly, North America was once contiguously wooded from the Atlantic seaboard to the Mississippi River, but large areas of this forest were cleared by European settlers at rates similar to those that now characterize tropical forests (Dale et al. 2000).

Today, conversion of forests to pasture or agriculture is one of the dominant land use changes in the humid tropics. The magnitude of this land use change is uncertain, but one intermediate estimate suggests that 75,000 km^2 of **primary tropical forest**—that is, forest that had never been cleared—are now cleared annually, either permanently or through the expansion of shifting cultivation (Melillo et al. 1996). Another 37,000 km^2 of primary forest are logged annually for commercial use. Finally, approximately 145,000 km^2 of **secondary forest**—that is, forests that have regrown after earlier clearing—are cleared each year. At this rate, most primary tropical forests may disappear in the next several decades. The rates and patterns of forest clearing vary regionally, so primary forests are likely to persist longer in some regions such as Amazonia and Central Africa than elsewhere.

The trajectory of landscape change caused by deforestation depends on both the nature of the original forests and the land use that follows. The permanent or long-term conversion of forests to managed ecosystems involves burning or removal of most of the biomass and often leads to large losses of carbon, nitrogen, and other nutrient elements from the system. Logging, in contrast, removes only the commercially valuable trees and may cause less carbon and nutrient loss from soils. The nutrient losses that accompany deforestation can cause nutrient limitations to plant and microbial growth. They can also alter adjacent ecosystems, particularly aquatic ecosystems, and influence the atmosphere and climate through changes in trace gas fluxes (see Chapter 9) and water and energy exchange (see Chapter 4).

Reforestation of abandoned agricultural land through natural succession or active tree planting is also changing landscapes, particularly in the eastern United States, Europe, China, and Russia (see Fig. 13.2). In the eastern United States, for example, much of the land that was originally cleared reverted to forest dominated by native species. In Chile, however, primary forests are being replaced by plantations of rapidly growing exotic trees such as *Pinus radiata* (Armesto et al. 2001). These plantations have low diversity and a quite different litter chemistry and pattern of nutrient cycling than do the primary forests that they replace. The characteristics of the regrowing forests also depend on the previous types of land use (Foster et al. 1996, Foster and Motzkin 1998). Long-term and intensive agricultural practices can compact the soil, alter soil structure and drainage capability, deplete the soil organic matter, reduce soil water-holding capacity, reduce nutrient availability, deplete the seed bank of native species, and introduce new weedy species. The forests that regenerate on such land may therefore differ substantially from the original forest and from those regrow on less-intensively managed lands (Motzkin et al. 1996). Grazing intensity and accompanying land management practices also influence potential revegetation, with more intensively grazed systems often taking longer to regain forest biomass. Natural reforestation under these conditions may proceed slowly or not at all.

Use of native grasslands, savannas, and shrublands for cattle grazing is the most extensive modification of natural ecosystems occurring today. Globally, thousands of square kilometers of savanna are burned annually to maintain productivity for cattle grazing. Although both fire and grazing are natural components of most grasslands, changes in the frequency and/or severity of burning and grazing alters ecosystem processes. Burning releases nutrients and stimulates the production of new leaves that have a higher protein content and are more palatable to grazers. Conversely, grazers reduce fire probability by reducing the accumulation of grass biomass and leaf litter. Fire and grazers both prevent establishment of most trees, which might otherwise convert savannas to woodlands or forests. When fire frequency increases substantially, however, the loss of carbon and nitrogen from the system can reduce soil fertility and water retention (and therefore productivity). It can also affect regional trace gas budgets and deposition in downwind ecosystems and the transfer of nutrients and sediments to aquatic ecosystems.

Vegetation in many grassland ecosystems co-evolved with large herbivores. The steppe ecosystems of the central United States, eastern Africa, and Argentina, for example, supported high densities of animals, so grazing by domestic cattle does not necessarily degrade them (Milchunas et al. 1988, Owen-Smith 1988, Osterheld et al. 1992). Grazing is, in fact, an essential component of nutrient cycling and the maintenance of productivity of these grasslands (McNaughton 1979, 1985). In these ecosystems, grazers maintain the competitive advantage of native grazing-tolerant grasses. Overgrazing is more likely to occur in environments with low water or nutrient availability, where intensive grazing may exceed the productive capacity of the ecosystem. Under these circumstances, the cover of palatable grasses may decline, making the soil more prone to erosion, which feeds back to further loss of productivity (Schlesinger et al. 1990). Overgrazing can contribute to the encroachment of shrubs or succulents into grasslands. In other cases, cattle contribute to shrub encroachment through dispersal of seeds in dung (Brown and Archer 1999). Some of the most productive grasslands, such as the pampas of Argentina or the tallgrass prairies of Ukraine or the midwestern United States have been largely converted to mechanized agriculture.

Expansion of marine fishing has altered marine food webs globally, with cascading effects on most ecosystem processes. The area of the world's oceans that are actively fished has increased substantially, in part because technological advances allow fish and benthic invertebrates to be harvested more efficiently and stored for longer times before returning to markets. Most of Earth's continental shelves, the most productive marine ecosystems, are now actively fished, as are productive high-latitude open oceans. Removal of fish has cascading effects on pelagic ecosystems because fish predation has large top–down effects on the biomass and on species composition of zooplankton, which in turn affect primary productivity by phytoplankton and the recycling of nutrients within the water column (see Chapter 10). Harvesting of benthic invertebrates, such as clams, crabs, and oysters, also has large ecosystem effects because of direct habitat disturbance and the effects of these organisms on detrital food webs and benthic decomposition. The globalization of marine fisheries has a broader impact than we might expect, because many large fish are highly mobile and migrate for thousands of kilometers. Large changes in their populations therefore have ecological effects that diffuse widely thoughout the ocean and even into fresh-water ecosystems in the case of anadromous fish.

Intensification

Intensification of agriculture frequently reduces landscape heterogeneity and increases the transfer of nutrients and other pollutants to adjacent ecosystems. Agricultural intensification involves the intensive use of high-yield crop varieties combined with tillage; irrigation; and industrially produced fertilizers, pesticides, and herbicides. Intensification has allowed food production to keep pace with the rapid human population growth (see Fig. 8.1) (Evans 1980). Although this practice has minimized the aerial extent of land required for agriculture, it has nearly eliminated some ecosystem types that would naturally occupy areas of high soil fertility. Intensive agriculture is most developed on relatively flat areas such as floodplains and prairies that are suitable for irrigation and use of large farm machinery. The high cost of this equipment requires that large areas be cultivated, largely eliminating natural patterns of landscape heterogeneity.

Agricultural intensification generates biogeochemical hot spots that alter ecosystem processes in ways that impact the local, regional, and global environment (Matson et al. 1997). Tillage increases decomposition rate by reducing the physical protection of SOM and altering soil microclimate (see Chapters 3 and 7). In this way 30 to 50% of the original soil carbon is lost from permanently cultivated agri-

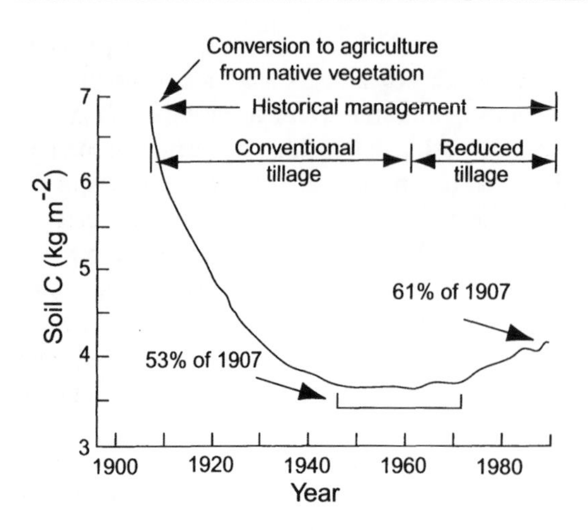

FIGURE 14.11. Simulation of loss of SOM after conversion of grassland to agriculture, followed by a small increase with conversion to low-till agriculture. Losses of soil carbon reduce the productive potential of the soil and transfer carbon in the form of CO_2 from the soil to the atmosphere. New techniques of low-till and no-till agriculture can reduce the magnitude of soil carbon loss or, under some circumstances, lead to a small net carbon accumulation. (Redrawn with permission from *Science*, Vol. 277 © 1997 American Association for the Advancement of Science; Matson et al. 1997.)

cultural systems (Fig. 14.11). These soil carbon losses often occur within 20 years in temperate ecosystems or within 5 years in tropical systems, depending on soil temperature and moisture. The large regular inputs of nutrients required to sustain intensive agriculture (see Fig. 8.1) increases the emissions of nitrogen trace gases that play a significant role in the global nitrogen cycle and link these ecosystems with downwind ecosystems (see Chapters 9 and 15).

Nutrient loading on land increases nonpoint sources of pollution for neighboring aquatic ecosystems (Carpenter et al. 1998). Phosphorus additions on land have particular large effects for at least two reasons. First, primary production of most lakes is phosphorus limited and therefore responds sensitively to even small phosphorus additions. Second, much of the phosphorus added to agricultural fields is chemically fixed, so orders of magnitude more phosphorus are generally added to fields than is absorbed by crops. These large additions represent a massive reservoir of phosphorus that will continue to enter aquatic ecosystems long after farmers stop adding fertilizer. Phosphorus inputs from human and livestock sewage have similarly long-lasting effects.

Intensive agriculture for rice production in flooded ecosystems produces a different type of biogeochemical hot spot. Rice feeds half of the world's population, with 90% of the production occurring in Asia. The most intensive rice cultivation takes place in periodically flooded fields, where reduced oxygen supply and decomposition lead to greater accumulation of SOM than in upland agricultural systems. Flooding creates an ideal environment for methanogens, which produce methane during decomposition of organic matter under anaerobic conditions (see Chapter 7). Paddy agricultural systems are therefore important sources of atmospheric methane and now produce about half as much methane as all natural wetlands. Together with cattle production, rice cultivation accounts for much of the increase in atmospheric methane (see Chapter 15).

Land use change caused greater ecological impact during the twentieth century than any other global change. Understanding and projecting future changes in land use are therefore critical to predicting and managing future changes in the Earth System. Development of plausible scenarios for the future requires close collaboration among climatologists, ecologists, agronomists, and social scientists. Optimistic scenarios that assume that the growing human population will be fed rather than die from famines, wars, or disease epidemics project continued large changes in land use, particularly in developing countries (Alcamo 1994). What actually occurs in the future is, of course, uncertain, but these and other scenarios suggest that land use change will continue to be the major cause of global environmental change in the coming decades. Ecologists working together with policy makers, planners, and managers have the opportunity to develop approaches that will minimize the impact of future landscape changes (see Chapter 16). This vision must recognize the large effects of land use change on landscape processes and their consequences on local to global scales.

Spatial Heterogeneity and Scaling

Extrapolation of ecosystem processes to large spatial scales requires an understanding of the role of spatial heterogeneity in ecosystem processes. Efforts to estimate the cumulative effect of ecosystem processes at regional and global scales has contributed to the increased recognition of the importance of landscape processes in ecosystem dynamics. Estimates of global productivity or annual carbon sequestration, for example, require that rates measured (or modeled) in a few locations be extrapolated over large areas. Many approaches to spatial extrapolation have been used, each with its advantages and disadvantages. A paint-by-numbers approach estimates the flux or pool for a large area by multiplying the average value for each patch type (e.g., the yield of major types of crops or the carbon stocks of different forest types) times the aerial extent of that patch type. This provides a rough approximation that can guide process-based research. This approach requires the selection of representative values of processes and accurate estimates of the area of each patch type. Satellite imagery now provides improved estimates of the aerial extent of many patch types, but spatial and temporal variation in processes makes it difficult to find good representative sites from which data can be extrapolated. This extrapolation approach can be combined with empirical regression relationships (rather than a single representative value) to estimate process rates for each patch type. Carbon pools in forests, for example, might be estimated as a function of temperature or normalized difference vegetation index (NDVI) rather than assuming that a single value could represent the carbon stocks of all forests.

Process-based models make up another approach to estimating fluxes or pools over large areas. These estimates are based on maps of input variables for an area (e.g., maps of climate, elevation, soils, and satellite-based indices of leaf area) and a model that relates input variables to the ecosystem property that is simulated by the model (Potter et al. 1993, VEMAP Members 1995) (Box 14.1). Regional evapotranspiration, for example, can be estimated from satellite data on vegetation composition and maps of temperature and pre-

---

Box 14.1. Spatial Scaling Through Ecological Modeling

The complexity of ecological controls over all the processes that influence ecosystem carbon balance makes long-term projections of terrestrial carbon storage a daunting task. Making these projections is, however, critical to improving our understanding of the relative role of terrestrial ecosystems in the global carbon balance. Experiments that test the multiple combinations of environmental conditions influencing terrestrial carbon storage are difficult to design. Modeling allows a limited amount of empirical information to be greatly extended through simulation of complex combinations of environmental–biotic interactions. One important use of ecosystem models has been to identify the key controls that govern long-term changes in terrestrial carbon storage (net ecosystem production, NEP). This application of ecosystem understanding is central to societal issues and policy formulation.

Many of key processes regulating NEP involve changes that occur over decades to centuries. The temporal resolution of the models must therefore be coarse, with **time steps** (the shortest unit of time simulated by the model) of a day, month, or year. Use of relatively long time steps such as weeks or months reduces the level of detail that can be considered. Temperature and moisture controls over decomposition, for example, can still be observed with an annual time step. The short-term pulses of decomposition

associated with drying and wetting cycles or grazing by soil fauna, however, are subsumed in the shape of the annual temperature and moisture response curves of decomposition and in the decomposition coefficients. The processes that are unique to the rhizosphere and bulk soil must also be ignored, when these environments are lumped into a single soil pool in a model. In this case, only the more general controls such as temperature, moisture, and chemistry can be included.

The basic structure of a model of NEP must include the pools of carbon in the soils and vegetation. It must also include the fluxes of carbon from the atmosphere to plants (gross primary production [GPP] or NPP), from plants to the atmosphere (plant respiration, harvest, and combustion), from plants to soil (litterfall), and from soil to the atmosphere (decomposition and disturbance). Models differ in the detail with which these and other pools and fluxes are represented. Plants, for example, might be considered a single pool or might be separated into different plant parts (leaves, stems, and roots), functional types of plants (e.g., trees and grasses in a savanna), or chemical fractions such as cell wall and cell contents. Under some circumstances, certain fluxes (e.g., fire and leaching) are ignored. There is no single "best" model of NEP. Each model has a unique set of objectives, and the model structure must be designed to meet these objectives. We briefly describe how three models incorporate information about controls over NEP, emphasizing how the differences in model structure make each model appropriate to particular questions or ecosystems.

Perhaps the biggest challenge in model development is deciding which processes to include. One approach is to use a hierarchical series of models to address different questions at different scales (Reynolds et al. 1993). Models of leaf-level photosynthesis and of microclimate within a canopy have been developed and extensively tested for agricultural crops, based on the basic principles of leaf biochemistry and the physics of radiation transfer within canopies. One output of these models is a regression relationship between environment at the top of the canopy and net photosynthesis by the canopy. This environment–photosynthesis regression relationship can then be incorporated into models operating at larger temporal and spatial scales to simulate NPP, without explicitly including all the details of biochemistry and radiation transfer. This hierarchical approach to modeling provides an opportunity to **validate** the model output (i.e., compare the model predictions with data obtained from field observations or experimental manipulations) at several scales of temporal and spatial resolution, providing confidence that the model captures the important underlying processes at each level of resolution.

The Terrestrial Ecosystem Model (TEM) (Fig. 14.12) was designed to simulate the carbon budget of ecosystems for all locations on Earth at 0.5° longitude by 0.5° latitude resolution (60,000 grid cells) for time periods of a century or more (McGuire et al. 2001). TEM has a relatively simple structure and a monthly time step, so it can run efficiently in large numbers of grid cells for long periods of time. Soil, for example, consists of a single carbon pool. The model assumes simple universal relationships between environment and ecosystem processes based on general principles that have been established in ecosystem studies. The model assumes, for example, that decomposition rate of the soil carbon pool depends on the size of this pool and is influenced by the temperature, moisture, and C:N ratio of the soil. TEM incorporates feedbacks that constrain the possible model outcomes. The nitrogen released by decomposition, for example, determines the nitrogen available for NPP, which in turn governs carbon inputs to the soil and therefore the pool of soil carbon available for decomposition. This simplified representation of ecosystem carbon dynamics is sufficient to capture global patterns of carbon storage (McGuire et al. 2001), making the model useful in simulating regional and global patterns of soil carbon storage.

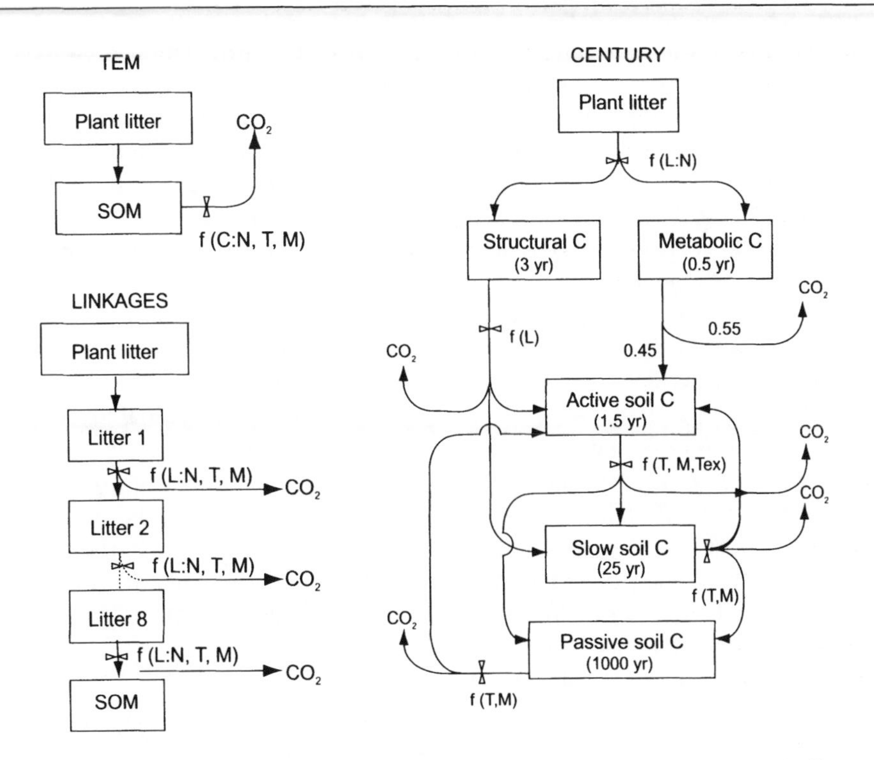

FIGURE 14.12. The decomposition portion of three terrestrial ecosystem models: TEM (McGuire et al. 1995), LINKAGES (Pastor and Post 1986), and CENTURY (Parton et al. 1987). Inputs from the vegetation component of these models is shown as plant litter. Arrows indicate the fluxes of carbon from litter to other pools and eventually to CO_2. The bow ties indicate controls over these fluxes (or the partitioning of the flux to two pools) as functions (*f*) of C:N ratio (C:N), lignin (L), lignin:N ratio (L:N), temperature (*T*), and moisture (*M*). In CENTURY we show representative residence times of different carbon pools in grassland soils.

The CENTURY model was originally developed to simulate changes in soil carbon storage in grasslands in response to variation in climate, soils, and tillage (Parton et al. 1987, 1993) (Fig. 14.12). It has since been adapted to most global ecosystem types. In CENTURY, the soil is subdivided into three compartments (active, slow, and passive soil carbon pools) that are defined empirically by turnover rates observed in soils. The active pool represents microbial biomass and labile carbon in the soil that has a turnover time of days to years. The slow pool consists of more recalcitrant materials, with a turnover time of years to decades. The passive pool is humified carbon that is stabilized on mineral surfaces. It has turnover times of hundreds to thousands of years. The detailed representation of soil pools in CENTURY enables it to estimate changes in decomposition under situations in which a change in disturbance regime or climate alters the decomposition of some soil pools more than other pools. A change in climate, for example, primarily affects the active and slow pools, with the passive pool remaining protected by clay minerals; tillage, however, enhances the decomposition of all soil pools.

The litter layer is much better developed in forests than in grasslands, so much of the forest decomposition occurs in the forest floor above mineral soil. Soil texture therefore has less influence on decomposition in forests than in grasslands. The LINKAGES model follows the decomposition of each litter cohort (i.e., each year's litterfall) separately for 8 years, based on the temperature, moisture, and lignin:N ratio of that litter cohort (Pastor and Post 1986) (Fig. 14.12). After 8 years, the remaining organic matter is transferred to an SOM pool, which decomposes as a function of the size of the pool, temperature, and moisture, as in the other ecosystem carbon models.

How do we know whether the patterns of NEP estimated by global-scale models are realistic? A comparison of model results with field data for the few locations where NEP has been measured provides one reality check. At these sites, measurements of NEP over several years spanning a range of weather conditions provides a measure of how that ecosystem responds to variation in climate. This allows a test of the model's ability to capture the effects of ecosystem structure and climate on NEP.

The seasonal and interannual patterns of atmospheric CO_2 provide a second reality check for global models of NEP. The temporal and spatial patterns of atmospheric CO_2 are the direct consequence of net ecosystem exchange by the terrestrial biosphere (including human activities) and the oceans. Atmospheric transport models describe the patterns of redistribution of water, energy, and CO_2 through Earth's atmosphere. These transport models can be run in inverse mode to estimate the spatial and temporal patterns of CO_2 uptake and release from the land and oceans that are required to produce the observed patterns of CO_2 concentration in the atmosphere (Fung et al. 1987, Tans et al. 1990). The global patterns of CO_2 sources and sinks estimated from the atmospheric transport models can then be compared with the patterns estimated from ecosystem models. Any large discrepancy between these two modeling approaches provide hints about processes and/or locations where either the ecosystem or the atmospheric transport models may have not adequately captured the important controls over carbon exchange and transport.

cipitation that are used as inputs to an ecosystem model (Running et al. 1989). Estimates from ecosystem models are sensitive to the quality of the input data (which will always be inadequate at large spatial scales) and to the degree of generality of the relationships simulated by the models. Rapid improvements in the technology and availability of remotely sensed data from satellites are providing new sources of spatially explicit data on ecosystem variables such as leaf area index (LAI), soil moisture, and surface temperature. Field research can then relate these remote-sensing signatures to properties and processes that are important in ecosystems. The relationship of ecosystem carbon exchange to light, temperature, and LAI can be determined in field studies and incorporated into the model structure (Clein et al., in press). The generality of relationships used in ecosystem models can then be tested through comparisons of model output with field data and through intercomparisons of models that differ in their structure but use the same input data (VEMAP Members 1995, Cramer et al. 2000).

Any extrapolation exercise requires consideration of biogeochemical hot spots with high process rates. Regional extrapolation of methane flux at high latitudes, for example, should consider beaver ponds (Roulet et al. 1997) and thermokarst lakes (Zimov et al. 1997) because they have high fluxes relative to their area. Estimates of NEP, however, require differentiation between young and old forests, because forest age is the major determinant of NEP (see Chapter 13).

Extrapolation requires a careful consideration of characteristic length scales of key processes (Fig. 14.13). General circulation models that simulate climate at the global scale, for example, often have a spatial resolution of 2° by 2° of longitude and latitude, which is sufficient for incorporating the differential heating by land and ocean but inadequately represents fine-scale processes, such as cloud formation. Successional models of ecosystems characterized by gap dynamics might simulate processes at the scale of individual trees, whereas ecosystems characterized by stand-replacing fires might be adequately represented by patches several hectares or square kilometers in size. Results from models that simulate processes at large spatial and temporal scales often focus on slow variables that strongly influence long-

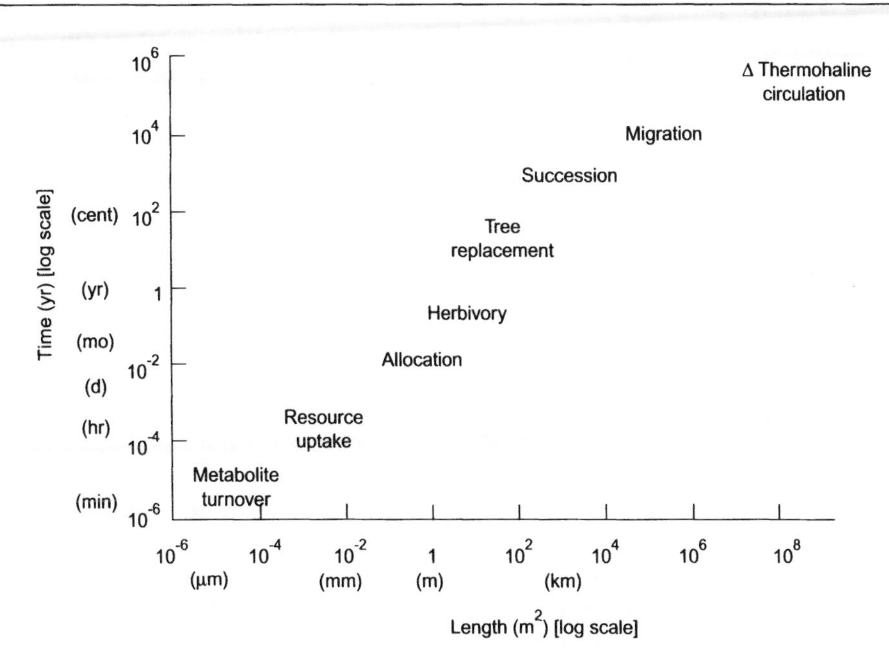

FIGURE 14.13. Temporal and spatial scales at which selected ecosystem processes occur. The study of any ecosystem process requires understanding at least one level below (to provide mechanistic understanding) and one level above (to provide context with respect to patterns of temporal and spatial variability).

term processes in ecosystems (Carpenter and Turner 2000). The availability of data and the computing power required for model simulations provide additional pragmatic constraints to the spatial resolution of regional models. Satellite data are readily available at 1-km resolution for the globe, but data with finer resolution are more expensive and less continuously available.

Some processes can be extrapolated to large scales without explicitly considering landscape interactions. The extrapolation of carbon flux, for example, may be adequately represented in the short term from an understanding of its response to climate, vegetation, and stand age. Over the long term, erosion, leaching, and fire transfer significant amounts of carbon among patches. Other processes, such as vegetation change in response to changing climate, are sensitive to rates of species migration and disturbance spread. Spatially explicit models that incorporate the spread of disturbance among patches on a landscape are critical for projections of long-term changes in vegetation and disturbance regime (Gardner et al. 1987, Rupp et al. 2000).

Development of new measurement techniques that quantify processes at coarse spatial scales provides an additional basis for scaling. Measurements of carbon and water exchange that were traditionally measured on individual leaves or patches of soil can now be measured by eddy correlation techniques over entire ecosystems, using either micrometeorological towers or aircraft. These approaches measure ecosystem fluxes at scales of tens of meters to kilometers and therefore integrate fluxes across both hot and cold spots in the landscape.

A global monitoring network that measures atmospheric CO_2 and CH_4 concentrations provides an additional scaling tool. Atmospheric transport models can be run in **inverse mode** to estimate the regional and global patterns of CO_2 or CH_4 fluxes. Inverse modeling asks what seasonal and temporal patterns of fluxes would be required to produce the observed patterns of CO_2 concentration in Earth's atmosphere. Each of the currently available approaches to scaling has limitations, so the development of new approaches and comparisons of results from different approaches are active areas of research on global change.

Summary

Spatial heterogeneity within and among ecosystems is critical to the functioning of individual ecosystems and of entire regions. Landscapes are mosaics of patches that differ in ecologically important properties. Some patches, for example, are biogeochemical hot spots that are much more important than their area would suggest. The size, shape, connectivity, and configuration of patches on a landscape influence their interactions. Large patches, for example, may contain greater heterogeneity of resources and environment and have a smaller proportion of edge habitat. The shape and connectivity of patches influences their effective size and heterogeneity in ways that differ among organisms and processes. The distribution of patches on a landscape is important because it determines the nature of transfers of materials and disturbance among adjacent patches. The boundaries between patches have unique properties that are important to edge specialists. Boundaries also have physical and biotic properties that differ from the centers of patches, so differences among patches in edge-to-area ratios, due to patch size and shape, influence the average rates of processes in a patch.

State factors, such as topography and parent material, govern the underlying matrix of spatial variability in landscapes. This physically determined pattern of variability is modified by biotic processes and legacies in situations where species strongly affect their environment. These landscape patterns and processes in turn influence disturbance regime, which further modifies the landscape pattern. Humans are exerting increasing control over landscape patterns and change. Land use decisions that convert one land-surface type to anther (e.g., deforestation, reforestation, shifting agriculture) or that modify its functioning (e.g., cattle grazing on rangelands) influence both the sites where those activities are being carried out and the functioning of neighboring ecosystems and the landscape as a whole. Human effects on ecosystems are becoming both more extensive (i.e., affecting more area) and more intensive (i.e., having greater impact per unit area).

Ecosystems do not exist as isolated units on the landscape. They interact through the movement of water, air, materials, organisms, and disturbance from one patch to another. Topographically controlled movement of water and materials to downslope patches depends on the arrangement of patches on the landscape and the properties of those patches. Riparian areas, for example, are critical filters that reduce the transfer of nutrients and sediments from upland ecosystems to streams, lakes, estuaries, and oceans. Aerial transport of nutrients, water, and heat strongly influences the nutrient inputs and climate of downwind ecosystems. These aerial transfers among ecosystems are now so large and pervasive as to have strong effects on the functioning of the entire biosphere. Animals transport nutrients and plants at a more local scale and influence patterns of colonization and ecosystem change. The spread of disturbance among patches influences both the temporal dynamics and the average properties of patches on a landscape. The connectivity of ecosystems on the landscape is rarely incorporated into management and planning activities. The increasing human impacts on landscape interactions must be considered in any long-term planning for the sustainability of managed and natural ecosystems.

Review Questions

1. What is a landscape? What properties of patches determine their interactions in a landscape?
2. How do fragmentation and connectivity influence the functioning of a landscape?
3. Give examples of spatial heterogeneity in ecosystem structure at scales of 1 m, 10 m, 1 km, 100 km, and 1000 km. How does spatial heterogeneity at each of these scales affect the way in which these ecosystems func-

tion? In other words, if heterogeneity at each scale disappeared, what would be the differences in the way in which these ecosystems function?

4. What are the major natural and anthropogenic sources of spatial heterogeneity in a landscape? How do these sources of heterogeneity influence the way in which these landscapes function? How do interactions among these sources of heterogeneity affect landscape dynamics?
5. What is the difference between a shifting steady state mosaic and a non–steady state mosaic? Give examples of each.
6. What is the difference between intensification and extensification? What has been the role of each in ecosystem and global processes?
7. Which ecosystem processes are most strongly affected by landscape pattern? Why?
8. What properties of boundaries influence the types of interactions that occur between patches within a landscape?
9. Describe how patches within a landscape interact through (1) the flow of water, (2) transfers of materials through the atmosphere, (3) movement of animals, and (4) the movement of disturbance. What properties of landscapes and patches influence the relative importance of these mechanisms of patch interaction?
10. What issues must be considered in extrapolating processes measured at one scale to larger areas? How does the occurrence of hot spots influence approaches to spatial scaling?

Additional Reading

Forman, R.T.T. 1995. *Land Mosaics: The Ecology of Landscapes and Regions*. Cambridge University Press, Cambridge, UK.

Foster, D.R., and G. Motzkin. 1998. Ecology and conservation in the cultural landscape of New England: Regional forest dynamics in central New England. *Ecosystems* 1:96–119.

Holling, C.S. 1992. Cross-scale morphology, geometry, and dynamics of ecosystems. *Ecological Monographs* 62:447–502.

Matson, P.A., W.J. Parton, A.G. Power, and M.J. Swift. 1997. Agricultural intensification and ecosystem properties. *Science* 227:504–509.

Meyer, W.B., and B.L. Turner III. 1992. Human population growth and global land-use/cover change. *Annual Review of Ecology and Systematics* 23:39–61.

Naiman, R.J., and H. Decamps. 1997. The ecology of interfaces: Riparian zones. *Annual Review of Ecology and Systematics* 28:621–658.

O'Neill, R.V., D.L. DeAngelis, J.B. Waide, and T.F.H. Allen. 1986. *A Hierarchical Concept of Ecosystems*. Princeton University Press, Princeton, NJ.

Pickett, S.T.A., and M.L. Cadenasso. 1995. Landscape ecology: Spatial heterogeneity in ecological systems. *Science* 269:331–334.

Turner, M.G. 1989. Landscape ecology: The effect of pattern on process. *Annual Review of Ecology and Systematics* 20:171–198.

Turner, M.G., R.H. Gardner, and R.V. O'Neill. 2001. *Landscape Ecology in Theory and Practice: Pattern and Process*. Springer-Verlag, New York.

Urban, D.L., R.V. O'Neill, and H.H. Shugart. 1987. Landscape ecology. *BioScience* 37:119–127.

Waring, R.H., and S.W. Running. 1998. *Forest Ecosystems: Analysis at Multiple Scales*. Academic Press, New York.

Part IV
Integration

15
Global Biogeochemical Cycles

The magnitude of biotic and human impacts on ecosystem processes becomes clear when summed at the global scale. This chapter describes the global cycles of water and several biotically important elements.

Introduction

A thorough understanding of the cycling of water, carbon, nitrogen, phosphorus, and sulfur is key to understanding ecosystems, the biosphere, and the entire Earth System. Human activities have dramatically altered element cycles since the beginning of the Industrial Revolution. Burning of fossil fuels in particular has increased emissions of CO_2, nitric oxides, and several sulfur gases. Mining and agriculture have also altered the availability and mobility of carbon, nitrogen, phosphorus, and sulfur. Changes in these biogeochemical cycles have altered Earth's climate, speeding up the global hydrologic cycle, which in turn feeds back to other biogeochemical cycles. These changes affect ecosystems at all scales, ranging from individual organisms to the entire biosphere. In this chapter, we focus on the global cycles of carbon, nitrogen, phosphorus, sulfur, and water, summarizing at the global scale the natural pools and fluxes in the cycles and the factors responsible for change.

The Global Carbon Cycle

Photosynthetic uptake of carbon from the atmosphere and oceans provides the fuel for most biotic processes. This reduced carbon makes up about half of the mass of Earth's organic matter. Biological systems, in turn, respire CO_2 when they use organic carbon for growth and metabolism. The controls over the carbon cycle depend on time scale, ranging from millions of years, by which cycling is controlled by movements of Earth's crust, to seconds, by which cycling is controlled by photosynthetic rate and surface–air exchange (see Chapters 5 to 7).

Carbon is distributed among four major pools: the atmosphere, oceans, land (soils and vegetation), and sediments and rocks (Fig. 15.1) (Reeburgh 1997, McCarthy et al. 2001). Atmospheric carbon, which consists primarily of CO_2, is the smallest but most dynamic of these pools. It **turns over**—that is, is completely replenished—every 3 to 4 years, primarily through its removal by photosynthesis and return by respiration. The metabolic processes of organisms therefore constitute the engine that drives the global carbon cycle on time scales of seconds to centuries.

Carbon is present in the oceans as dissolved organic carbon (DOC), dissolved inorganic carbon (DIC), and particulate organic carbon (POC), which consists of both live organisms and dead materials. Most (98%) of this carbon is in inorganic form, primarily as bicarbonate (90%), with most of the rest as carbonate. Free

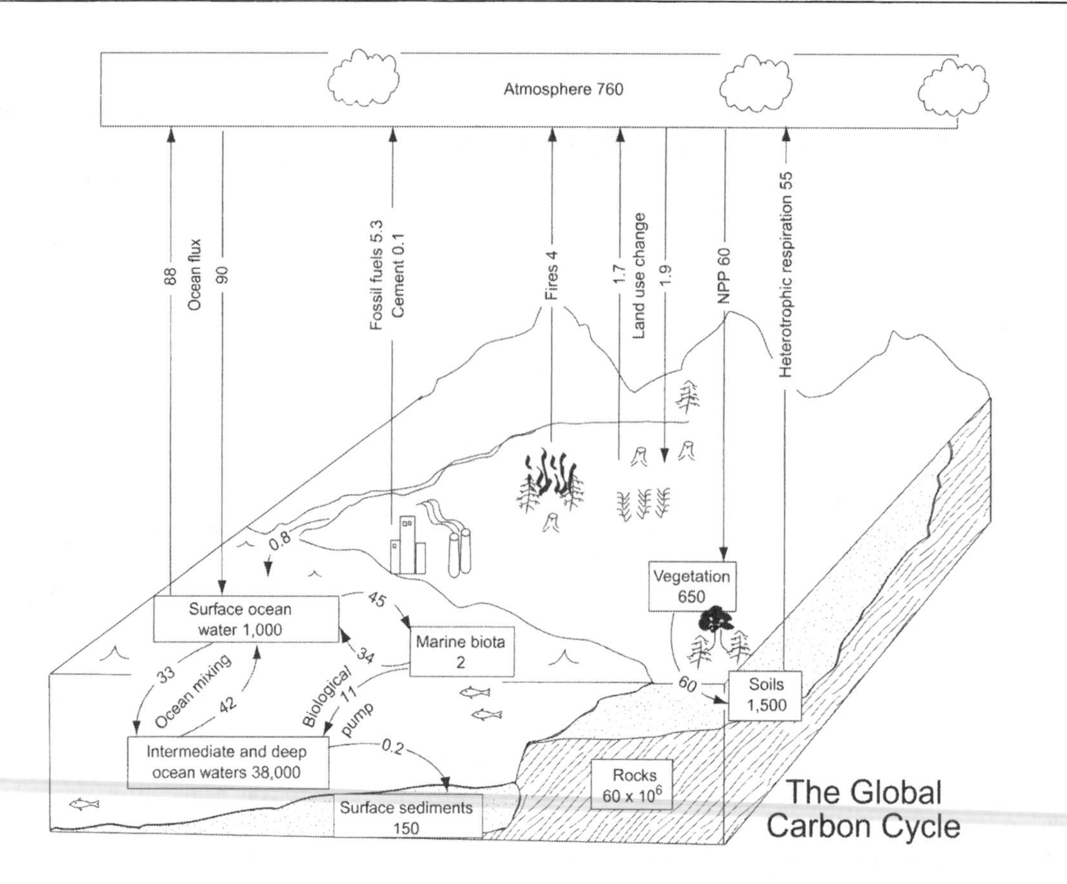

FIGURE 15.1. The global carbon cycle showing approximate magnitudes of the major pools (boxes) and fluxes (arrows) in units of pedagrams per year (1 pedagram = 10^{15} g). The carbon pools that contribute to carbon cycling over decades to centuries are the atmosphere, land (vegetation and soils), and oceans. On land, the carbon gain by vegetation is slightly greater than the carbon loss in respiration, leading to net carbon storage on land. The net carbon input to the oceans is also slightly greater than the net carbon return to the atmosphere. The terrestrial biosphere accounts for 50 to 60% of global net primary production (NPP). Most (80%) of the marine NPP is released to the environment by heterotrophic respiration, and the remaining 20% goes to the deep oceans by the biological pump. Ocean upwelling returns most of this carbon to the surface ocean waters. Human activities cause a net carbon flux to the atmosphere through combustion of fossil fuels, cement production, and land use change. The carbon content of the atmosphere was calculated from the 1999 CO_2 concentration of 367 ppmv (Prentice et al. 2001). Data for pools of carbon in vegetation are from Saugier et al. (2001); for the surface ocean from Houghton et al. (1996); and for the deep ocean, sediments, and rocks from Reeburgh 1997). Remaining data are from Prentice et al. (2001).

CO_2, the form that is directly used by most marine primary producers, accounts for less than 1% of this inorganic pool. These three forms of DIC are in a pH-dependent equilibrium (see Chapter 10). The marine biota account for only 2 Pg (2×10^{15} g) carbon, although they cycle as much carbon annually as does terrestrial vegetation. The carbon in marine biota turns over every 2 to 3 weeks.

The ocean's surface waters that interact with the atmosphere contain about 1000 Pg carbon, similar to the quantity in the atmosphere (Fig. 15.1). The capacity of the ocean to take up carbon is constrained by three categories of processes that operate at different time scales (Schlesinger 1997). In the short term, the surface exchange rate depends on wind speed, surface temperature, and the CO_2 concentration of surface waters. On daily to monthly time scales, the CO_2 concentration in the surface water depends on photosynthesis and pH-dependent buffering reactions. Finally, the surface waters are a relatively small pool (only 75 to 200 m deep) of water that exchanges relatively slowly with deeper layers because the warm, low-salinity surface water is less dense than deeper layers (see Fig. 2.8). Carbon that enters surface waters is transported slowly to depth by two major mechanisms. Organic detritus and its calcium carbonate ($CaCO_3$) skeletal content, which form in the euphotic zone, sink to deeper waters, a process termed the **biological pump** (see Chapter 10). Bottom-water formation in the polar seas transports dissolved carbon to depth, a process termed the **solubility pump** (see Chapter 2). Once carbon reaches intermediate and deep waters, it is stored for hundreds to thousands of years before returning to the surface through upwelling. Most (97%) of the ocean carbon is in the intermediate and deep waters (Fig. 15.1).

The terrestrial biosphere contains the largest biological reservoir of carbon. There is nearly as much carbon in terrestrial vegetation as in the atmosphere, and there is at least twice as much carbon in soils as in the atmosphere (Fig. 15.1) (Jobbágy and Jackson 2000). Terrestrial net primary production (NPP) is slightly greater than that in the ocean, but due to the much larger plant biomass on land, terrestrial plant carbon has a turnover time of about 11 years, compared to 2 to 3 weeks in the ocean. NPP is about half of gross primary production (GPP; i.e., photosynthetic carbon gain) on land ($60\,Pg\,yr^{-1}$ out of $120\,Pg\,yr^{-1}$) and in the ocean ($45\,Pg\,yr^{-1}$ out of $103\,Pg\,yr^{-1}$) (Prentice et al. 2001). Soil carbon turns over on average every 25 years. These average turnover times mask large differences in turnover time among components of the terrestrial carbon cycle. Photosynthetically fixed carbon in chloroplasts turns over on time scales of seconds through photorespiration (see Chapter 5). Leaves and roots are replaced over weeks to years, and wood is replaced over decades to centuries. Components of soil organic matter (SOM) also have quite different turnover times, with labile forms turning over in minutes and humus having turnover times of decades to thousands of years (see Chapter 7).

Carbon in rocks and surface sediments accounts for well over 99% of Earth's carbon ($10^7\,Pg$) (Reeburgh 1997, Schlesinger 1997). This carbon pool cycles extremely slowly, with turnover times of millions of years. Factors governing the turnover of these pools are geologic processes associated with the rock cycle, including the movement of continental plates, volcanism, uplift, and weathering (see Chapter 3).

Human activities make a significant contribution to the global carbon cycle. Combustion of fossil fuels releases CO_2 from petroleum products. Cement production releases CO_2 from carbonate rocks. Land use conversion releases carbon by biomass burning and enhanced decomposition. Together these fluxes are about 15% of the carbon cycled by terrestrial or by marine production, making human activities the third largest biotically controlled flux of carbon to the atmosphere.

Long-Term Change in Atmospheric CO_2

Critical processes in the carbon cycle occur on all time scales. The important processes are photosynthesis and respiration on time scales of seconds to years; NPP, SOM turnover, and disturbance on time scales of decades to centuries; and uplift, volcanism, weathering, and ocean sedimentation over thousands to millions of years. Atmospheric CO_2 concentration has changed dramatically through Earth's history, with concentrations 10-fold higher than today (greater than 3000 ppmv) likely to have occurred several times in the last hundred millions years. CO_2 concentrations have also been below 300 ppmv, first about 20 million years ago, in part due to reduced volcanism (Pagani et al. 1999, Pearson and Palmer 2000) and most recently just before the beginning of the Industrial Revolution 150 years ago (Webb and Bartlein 1992). Geochemical processes determine variation in CO_2 on geological time scales. These include the weathering of silicate rocks (which consumes CO_2 and releases bicarbonate), burial of organic carbon in sediments, and volcanism (which release CO_2) (Berner 1997). Biological processes influence geochemical cycling in many ways, for example by increasing weathering rates (see Chapter 3). Although critical on long time scales, the rates of these geochemical processes are so slow compared to anthropogenic changes that they do not influence current trajectories of change in atmospheric CO_2.

Over the last 400,000 years, changes in solar input associated with variations in Earth's orbit (see Fig. 2.15) have caused cyclic variation in atmospheric CO_2 concentrations associated with glacial–interglacial cycles (Fig. 15.2) (Petit et al. 1999, Sigman and Boyle 2000). CO_2 concentration declined during glacial periods and increased during interglacials. These changes in CO_2 concentration are much larger than can be explained simply by changes in light intensity and temperature in response to altered solar input. The large biospheric changes must result from amplification by biogeochemical feedbacks in the Earth System. Several feedbacks could contribute to these atmospheric changes (Sigman and Boyle 2000). (1) Increased transport of dust off the less-vegetated continents during glacial periods may have increased iron, phosphorus, and silica transport and enhanced NPP in high-latitude oceans, leading to increased CO_2 uptake and transport to depth via the biological pump (see Chapter 10). (2) Extensive winter sea ice around Antarctica may have reduced out-gassing of CO_2 in locations of upwelling of CO_2-rich deep waters (Stephens and Keeling 2000). (3) Some additional carbon may have been stored on land during glacial periods on continental shelves exposed by the drop in sea level or in permafrost at high latitudes (Zimov et al. 1999). However, terrestrial systems were probably a net carbon source during glacial periods, due to the replacement of forests by grasslands, deserts, tundra, and ice sheets. Marine foraminiferan fossils acquired a more terrestrial ^{13}C signature during glacial periods; this could reflect the movement of terrestrial carbon from the land to the atmosphere and subsequently to the ocean (Bird et al. 1994,

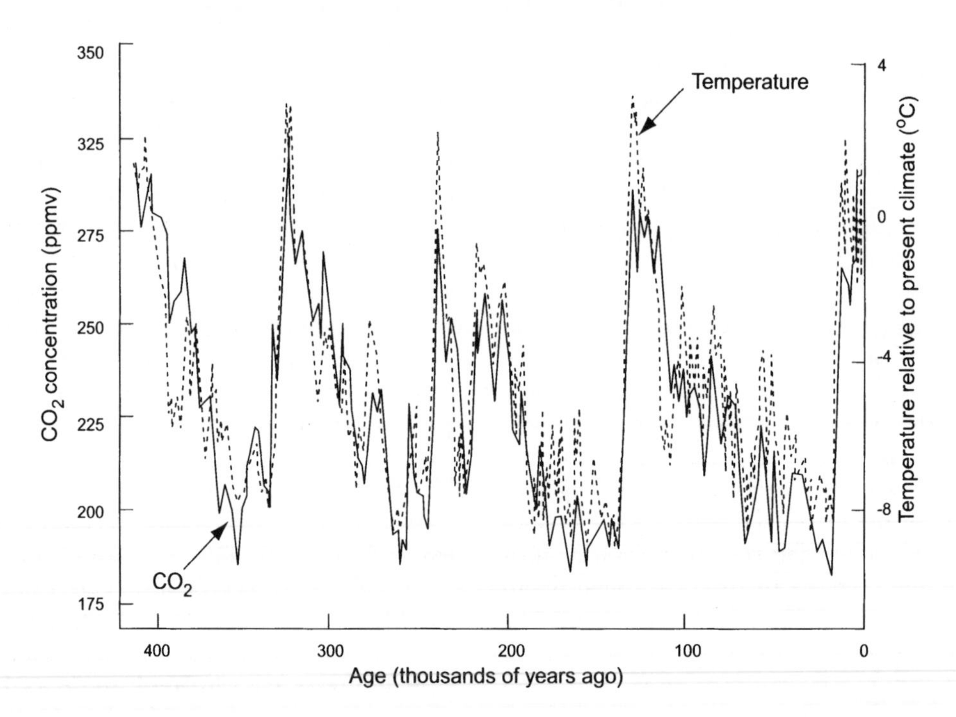

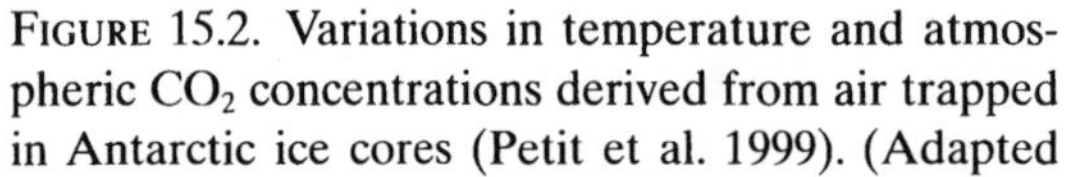
FIGURE 15.2. Variations in temperature and atmospheric CO_2 concentrations derived from air trapped in Antarctic ice cores (Petit et al. 1999). (Adapted from IPCC Assessment Report 2001; Folland et al. 2001.)

Crowley 1995). An improved understanding of controls over these potential feedback mechanisms is important because it could indicate how the Earth System will respond to current trends of increasing temperature and atmospheric CO_2.

Atmospheric CO_2 concentration has been relatively stable over the last 12,000 years, ranging from about 260 to 280 ppmv in preindustrial times. Several high-resolution ice-core records demonstrate that human-induced increases in CO_2 over the past century are an order of magnitude more rapid than any that occurred over the previous 20,000 years and probably the last 400,000 years (Petit et al. 1999). This sudden change in the global cycles of carbon and other elements induced by human activities are so large as to suggest that Earth has entered a new geologic epoch, the **Anthropocene** (see Fig. 2.12).

Anthropogenic Changes in the Carbon Cycle

The burning of fossil fuels is the primary cause of current increases in atmospheric CO_2. About 5.3 $Pg\,yr^{-1}$ of carbon are emitted from fossil fuel combustion and 0.1 $Pg\,yr^{-1}$ from cement production (Prentice et al. 2001). We are less certain about the estimated $1.7 \pm 0.8\,Pg\,yr^{-1}$ of carbon emitted due to deforestation, because the aerial extent and carbon dynamics associated with this land use change are less well documented. CO_2 flux from land use change must include both the net sources of carbon to the atmosphere (e.g., deforestation and agricultural conversion) and the net sinks (e.g., forest regrowth, forest plantations, fire suppression, woody encroachment, and changes in agricultural management) (Houghton et al. 1999).

Anthropogenic emissions have caused atmospheric CO_2 concentration to increase exponentially since the beginning of the Industrial Revolution (Fig. 15.3). A comparison of the annual increment in carbon content of the atmosphere with known emissions shows that only about half the anthropogenic carbon that is emitted to the atmosphere remains there. The remainder is taken up on land or in the oceans. Measurements of $\delta^{13}C$ in atmospheric CO_2 and measurements of atmospheric O_2 (Keeling et al. 1993, Bender et al. 1996) help identify more precisely the location of these missing sinks of CO_2 (Box 15.1).

Estimates of carbon sources and sinks can also be developed using biomass inventories (Dixon et al. 1994), inverse atmospheric modeling (Tans et al. 1990), and carbon cycle modeling. Inventory studies measure changes in regional or global carbon pools over time by repeatedly measuring the same forest plots. These records suggest that 0.5 $Pg\,yr^{-1}$ of carbon is accumulating in northern forests due to increasing rates of tree growth (FCCC 2000). Data from the tropics are less certain but atmospheric sampling suggests that tropical regions are approximately in balance with the atmosphere, so there must be a net sink in unmanaged forests that approximately balances the carbon losses from deforestation and biomass burning (Schimel et al. 2001). The large quantity and spatial variability of soil carbon pools makes it difficult to use an inventory approach to detect changes in soil carbon pools. Even the quantity of carbon in terrestrial vegetation is uncertain; two recent global syntheses differed by 25% (500 Pg vs. 650 Pg) in their estimate of the size of this pool (Prentice et al. 2001, Saugier et al. 2001).

Inverse modeling can be used to estimate the global pattern of net sources and sinks of CO_2 that are required to produce the observed latitudinal and seasonal patterns of atmospheric CO_2 concentration. Atmospheric transport models and global networks of observations of CO_2 concentrations suggest that the gradient in CO_2 concentration from the Northern Hemisphere to the Southern Hemisphere is not as large as might be expected from the geographical distribution of anthropogenic CO_2 emission, which is largely a Northern Hemisphere source. These models therefore support the existence of a Northern Hemisphere sink for CO_2. Latitudinal gradients in the ^{13}C and O_2 suggest that about half of the sink is on land; about 1.9 $Pg\,yr^{-1}$ of anthropogenic carbon is absorbed by both land and ocean (Prentice et al. 2001). Much of the terrestrial sink appears to be in north temperate forests; North America and Eurasia contribute about equally

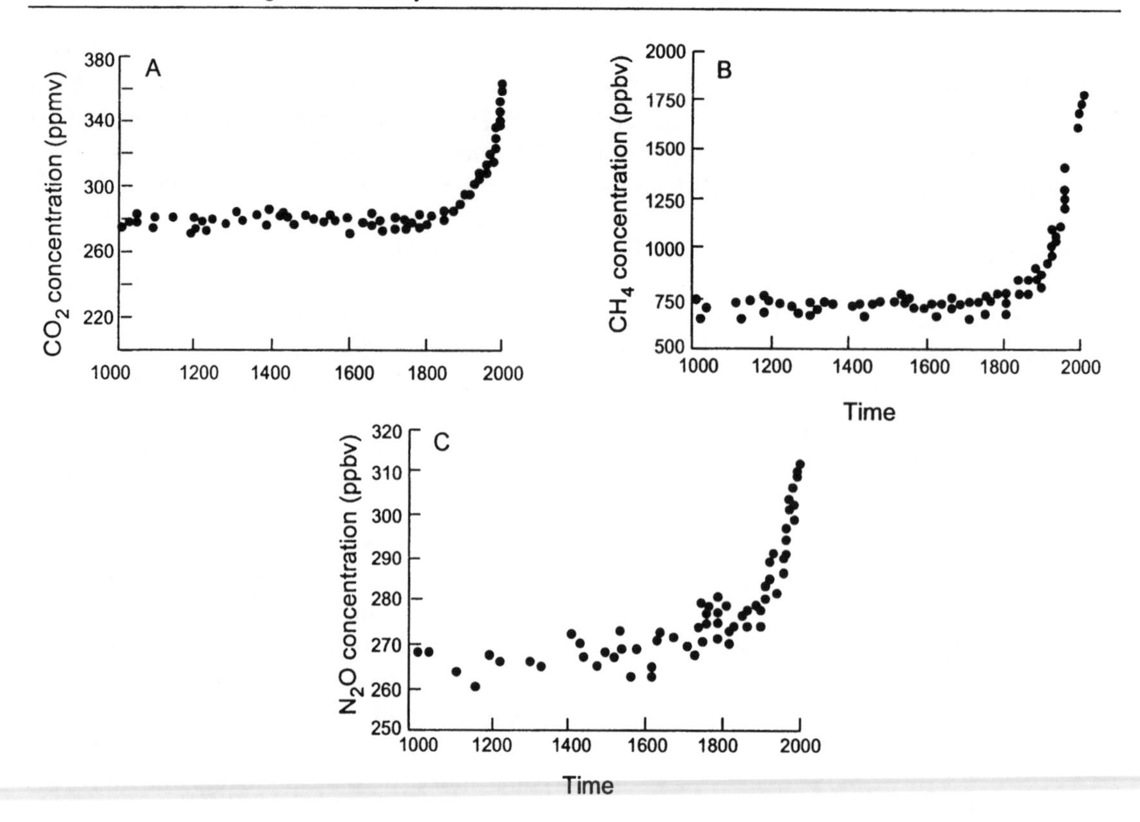

FIGURE 15.3. Changes over the last thousand years in the atmospheric concentrations of three radiatively active gases that are influenced by human activities (Prather et al. 2001, Prentice et al. 2001). Data shown are a composite of time series from air trapped in Antarctic ice cores and from direct atmospheric measurements. (Adapted from IPCC Assessement Report 2001; Prentice et al. 2001.)

to this sink per unit area of land, despite substantial differences in climate and human impact (Schimel et al. 2001). Boreal forests may now be a source of carbon, in part due to increased wildfire. Tropical forests appear to be in approximate balance with the atmosphere, due to similar magnitudes of net carbon uptake from unmanaged forests and carbon loss from deforestation.

Terrestrial Sinks for CO_2

Four potentially important mechanisms contribute to the Northern Hemisphere terrestrial sink for CO_2: land use change, CO_2 fertilization, inadvertent nitrogen fertilization, and climate effects (Schimel 1995). The conversion of forests to agricultural lands dominated land use change in the middle and high latitudes until the mid-twentieth century. Today, forest regrowth in previously harvested areas or in abandoned agricultural lands has enhanced carbon storage. The widespread suppression of wildfire also enhances the mid-latitude carbon sink, because it reduces fire emissions and allows woody plants to encroach into grasslands (Houghton et al. 2000). These are probably the most important causes of the north-temperate terrestrial carbon sink (Schimel et al. 2001).

CO_2 fertilization contributes to carbon storage. Photosynthesis typically increases 20 to 40% under doubled CO_2 in short-term studies. Carbon storage by ecosystems is, however, much less responsive to CO_2 than is the short-term response of individual plants because plant growth becomes nutrient limited as nutrients become sequestered in live and dead organic matter (Shaver et al. 1992, Schimel 1995) (see Chapter 5). Nutrient availability

Box 15.1. Partitioning of Carbon Uptake Between the Land and the Oceans

Only about half of the anthropogenic CO_2 that enters the atmosphere remains there. The land or oceans take up the remainder. Changes in the oxygen content of the atmosphere provide a measure of relative importance of land and ocean uptake. Net terrestrial uptake of CO_2 is accompanied by a net release of oxygen, with a 1:1 ratio of moles of CO_2 absorbed to moles of O_2 released. When CO_2 dissolves in ocean water, however, this causes no net release of oxygen. This difference in exchange processes can be used to partition the total CO_2 uptake between terrestrial and ocean components (Keeling et al. 1996b).

The relative abundance of the two stable isotopes of carbon (^{13}C and ^{12}C) in the atmosphere provides a measure of the relative activity of the terrestrial and oceanic components of the global carbon cycle (Ciais et al. 1995). Fractionation during photosynthesis by C_3 plants discriminates against ^{13}C, causing biospheric carbon to be depleted in ^{13}C by about 18‰ relative to the atmosphere. Exchanges with the ocean, however, involve relatively small fractionation effects. Changes in the $^{13}C:^{12}C$ ratio of atmospheric CO_2 therefore indicate the relative magnitude of terrestrial and oceanic CO_2 uptake.

Measurement of the global pattern and temporal changes in oxygen concentration and the $^{13}C:^{12}C$ ratio of atmospheric CO_2 suggest that the land and ocean contribute about equally to the removal of anthropogenic CO_2 from the atmosphere (Prentice et al. 2001). There are, however, many assumptions and complications in using either of these approaches to estimate the relative magnitudes of terrestrial and oceanic carbon uptake. The advantage of atmospheric measurements is that they give an integrated estimate of all uptake processes on Earth, because of the relatively rapid rate at which the atmosphere mixes.

therefore limits the long-term rate of terrestrial carbon sequestration, and the overall effect of CO_2 fertilization on terrestrial carbon storage appears to be much smaller than that due to reforestation.

Nitrogen typically limits the growth of plants in nontropical terrestrial ecosystems (see Chapter 8). Nitrogen additions through fertilization or atmospheric deposition of air pollutants like NO_x and nitric acid from fossil-fuel burning might therefore augment carbon storage, although this effect may be relatively small (Nadelhoffer et al. 1999). Increases in carbon storage due to nitrogen fertilization may be limited in scope because nitrogen deposition causes forest degradation above some threshold (Schulze 1989, Aber et al. 1998). Further nitrogen deposition might cause carbon loss from forests. The net effect of anthropogenic nitrogen deposition on carbon cycling and sinks is regionally variable and highly uncertain (Holland et al. 1997).

Finally, climate changes (including changes in temperature, moisture, and radiation) affect carbon storage through their effects on carbon inputs (photosynthesis) and outputs (respiration). In the short term, warming is expected to reduce carbon storage in soils by increasing heterotrophic respiration. The associated increase in rates of mineralization of nitrogen and other nutrients from organic matter could, however, increase nutrient uptake and production by vegetation (Shaver et al. 2000). Vegetation generally has a much higher C:N ratio (160:1) than does soil organic matter (15:1) (Schlesinger 1997), so the transfer of a given quantity of nitrogen from the soil to plants enhances carbon storage (Vukicevic et al. 2001). In addition, ecosystem respiration acclimatizes to temperature, so it increases less in response to warming than might be expected from short-term measurements (Luo et al. 2001). The response of Alaskan arctic tundra to regional warming is consistent with these ideas.

These ecosystems initially responded to warming with increased carbon loss but subsequently returned to near-zero carbon balance with the atmosphere (Oechel et al. 2000). Increases in precipitation can also cause variable responses in ecosystem carbon balance. Improved soil moisture can increase plant growth, but in some ecosystems this carbon storage term may be offset by increased decomposition. Although climate effects can be substantial, their net effect on terrestrial carbon storage is uncertain.

Terrestrial carbon exchange with the atmosphere varies substantially among years, due to interannual variation in climate. Terrestrial ecosystems tend to be a net carbon source in warm years and a net carbon sink in cool years, because fires and respiration increase more strongly with temperature than does NPP (Vukicevic et al. 2001, Schimel et al. 2001). Some of this interannual variation may reflect El Niño southern oscillation (ENSO) events, perhaps explaining why the terrestrial biosphere was a particularly strong sink for carbon in the early 1990s.

The relative importance of the various mechanisms of enhanced carbon storage in the terrestrial biosphere is uncertain (Schimel 1995, Schimel et al. 2001). Nonetheless, the multiple and potentially interacting sinks appear sufficient to account for the observed movement of anthropogenic CO_2 from the atmosphere to land (Table 15.1). These sink mechanisms (forest regrowth, CO_2 fertilization, and land use change) are likely to saturate and become less effective in the future (Schimel et al. 2001). The most effective mechanism of stabilizing atmospheric CO_2 concentration is therefore to reduce emissions.

Even if anthropogenic carbon emissions to the atmosphere were stopped immediately, the elevated atmospheric CO_2 concentration caused by past emissions will persist for decades to centuries. The lifetime of anthropogenic effects on atmospheric CO_2 depends primarily on the turnover time of terrestrial and oceanic carbon pools that interact with the atmosphere (Braswell and Moore 1994). Many of these key pools turn over slowly. Soil, for example, has an average turnover time of 25 years (Fig. 15.1), with some soil pools turning over even more slowly (see Chapter 7). The reequilibration time of the atmosphere is usually 1.5 to 3 times longer than the turnover time, explaining the persistence of anthropogenic effects on the atmosphere. Delays in efforts to reduce fossil-fuel emissions may therefore have unexpectedly long-lasting effects.

TABLE 15.1. Average (1980–1989) annual emissions and fate of anthropogenic carbon.[a]

Sources and sinks of anthropogenic carbon	Annual net flux ($Pg\,C\,yr^{-1}$)
Anthropogenic carbon sources	7.1 ± 1.1
Fossil fuel and cement production	5.5 ± 0.5
Net emissions from tropical land use change	1.6 ± 1.0
Carbon sinks	7.1
Storage in the atmosphere	3.2 ± 0.2
Oceanic uptake	1.6 ± 1.0
Terrestrial uptake	2.1
CO_2 fertilization	1.0 ± 0.5
Forest regrowth in the Northern Hemisphere	0.5 ± 0.5
Nitrogen deposition	0.6 ± 0.3
Other	0.2 ± 2.0

[a] About half of the anthropogenic carbon emissions remain in the atmosphere. The oceans and terrestrial biosphere absorb the remaining anthropogenic emissions.
Data from Schimel (1995).

The Global Methane Budget

Human activities are responsible for increasing methane concentrations in the atmosphere. Although the methane (CH_4) concentration of the atmosphere (1.8 ppmv) is much less than that of CO_2 (370 ppmv), CH_4 is about 20 times more effective per molecule as a greenhouse gas than is CO_2. Like CO_2, the CH_4 concentration of the atmosphere has increased exponentially since the beginning of the Industrial Revolution (Fig. 15.3). The CH_4 increase accounts for 20% of the increased greenhouse warming potential of the atmosphere (Cicerone and Oremland 1988, Khalil and Rasmussen 1990). Documenting the major global sources and sinks of atmospheric CH_4 is therefore important for understanding the recent increases in global temperature and the potential for future climate warming.

Methane is produced only under anaerobic conditions (see Chapter 7). Wetlands account for 70% of the naturally produced CH_4, with the remainder coming primarily from freshwater sediments, fermentation in the guts of animals (e.g., termites and ruminants), and various geological sources (Table 15.2). Anthropogenic methane sources are about twice as large as the natural sources, which explains why CH_4 accumulates in the atmosphere, despite its high reactivity and rapid turnover (9 years). Fossil-fuel extraction and refining; waste management (landfills, animal wastes, and domestic sewage treatment); and agricultural sources (rice paddies, biomass burning, and fermentation in guts of domestic ruminants like cattle) are each an important CH_4 source.

Estimates of the magnitude of most of the natural and anthropogenic CH_4 sources are highly uncertain (Table 15.2) (McCarthy et al. 2001). Important new sources are still being identified, including high-latitude lakes and reservoirs on organic-rich substrates (Zimov et al. 1997, St. Louis et al. 2000).

CH_4 reacts readily with OH radicals in the atmosphere in the presence of sunlight. This photochemical process is the major sink for atmospheric CH_4, accounting for 85% of the CH_4 consumption (Table 15.2). Additional CH_4 mixes into the stratosphere, where it reacts with ozone (see Chapter 2) or is removed by methanotrophs in soils (see Chapter 7). The annual atmospheric accumulation of CH_4 is about 10% of the annual anthropogenic flux.

TABLE 15.2. Global sources and sinks of methane.

Methane sources and sinks	Annual flux ($Tg\,CH_4\,yr^{-1}$)	
Natural sources	160	
Wetlands		115
Termites and ruminants		20
Ocean sediments		10
Fresh-water sediments		5
Geological sources		10
Anthropogenic sources[a]	375	
Fossil-fuel use		100
Waste management		90
Fermentation by cattle		85
Biomass burning		40
Rice paddies		60
Total sources	535	
Sinks	515	
Reaction with OH		445
Removal in stratosphere		40
Removal by soils		30
Atmospheric increase	30	

[a] Reservoirs are estimated as an additional $70\,Tg\,yr^{-1}$ anthropogenic source (St. Louis et al. 2000).
Data from Schlesinger (1997).

The Global Nitrogen Cycle

The productivity of many unmanaged ecosystems on both land and sea and of most managed agricultural and forestry ecosystems is limited by the supply of available nitrogen. In contrast to carbon, almost all of the nitrogen that is relevant to biogeochemistry is in a single pool (the atmosphere) with comparatively small quantities in the oceans, rocks, and sediments (Reeburgh 1997) (Fig. 15.4). Organic nitrogen pools are minuscule relative to the atmospheric pool and occur primarily in soils and terrestrial vegetation. Although nitrogen makes up 78% of the atmosphere, it is nearly all N_2 and is unavailable to most organisms. The major pathway by which N_2 is transformed to biologically available forms is via nitrogen fixation by bacteria in soils and aquatic systems or living in association with plants. The global quantity of nitrogen fixed annually by natural ecosystems is quite uncertain, ranging between 90 and $190\,Tg\,yr^{-1}$ for terrestrial ecosystems and between 40 and $200\,Tg\,yr^{-1}$ for marine ecosystems. Lightning strikes probably add an additional 3 to $10\,Tg\,yr^{-1}$ of nitrogen to the available pool. Before human alteration, the amount of nitrogen coming into the biosphere via nitrogen fixation was approximately balanced by return to the unavailable pools via denitrification and burial in sediments). In contrast to carbon, nitrogen cycles quite tightly within terrestrial ecosystems, with the annual throughput being about 10-fold greater than inputs and losses.

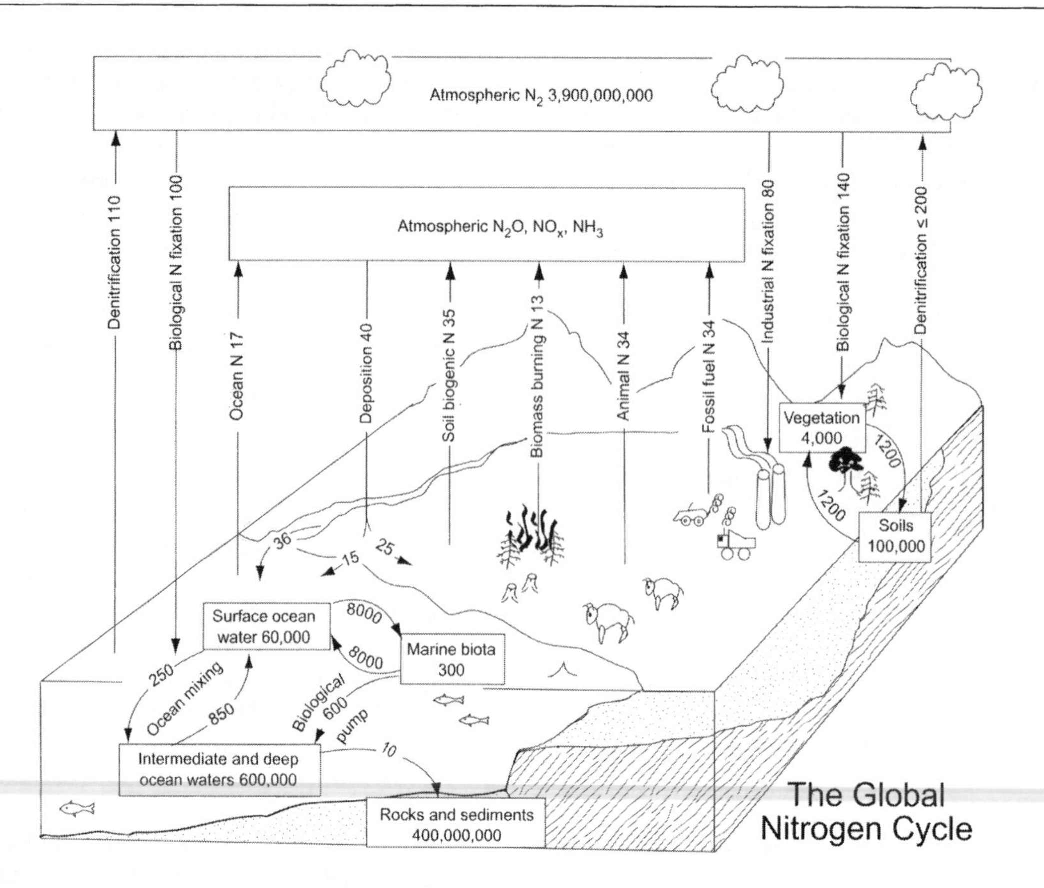

FIGURE 15.4. The global nitrogen cycle showing approximate magnitudes of the major pools (boxes) and fluxes (arrows) in units of teragrams per year (1 teragram = 10^{12} g). The atmosphere contains the vast majority of Earth's nitrogen. The amount of nitrogen that annually cycles through terrestrial vegetation is 9-fold greater than inputs by nitrogen fixation. In the ocean the annual cycling of nitrogen through the biota is 80-fold greater than inputs by nitrogen fixation. Denitrification is the major output of nitrogen to the atmosphere. Human activities increase nitrogen inputs through fertilizer production, planting of nitrogen-fixing crops, and combustion of fossil fuels. Nitrogen fluxes in the biological pump and ocean mixing were calculated from Fig. 15.6, assuming an N:P ratio of 15. Data for marine nitrogen fixation are from Karl et al. (in press) for nitrogen deposition from Holland et al. (1997), for nitrogen in vegetation and soils (calculated from Fig. 15.1, assuming a C:N ratio of 160 for vegetation and 15 for soils) from Schlesinger (1997), and for nitrogen in marine biota from Galloway (1996) and Reeburgh (1997). Remaining data are from Table 15.3 and Schlesinger (1997).

Anthropogenic Changes in the Nitrogen Cycle

In the past century human activities have approximately doubled the quantity of nitrogen cycling between terrestrial ecosystems and the atmosphere. Globally, human activities now convert N_2 to reactive forms at about the same rate as natural processes, through industrial fixation of nitrogen and the planting of nitrogen-fixing crops. The Haber process, which uses energy from fossil fuels to convert N_2 to ammonia gas (NH_3) to produce fertilizers, fixes more nitrogen than any other anthropogenic process. Industrial fixation of nitrogen by the Haber process increased substantially in the 1940s, reaching 30 Tg yr^{-1} by 1970 and 80 Tg yr^{-1} by 2000 (Fig. 15.5); it is projected to increase to 120 Tg yr^{-1} by 2025 (Galloway et al. 1995). Initially, most nitrogen fertilizer was applied in developed nations, but by 2000 almost two thirds of the fertilizer nitrogen was applied in

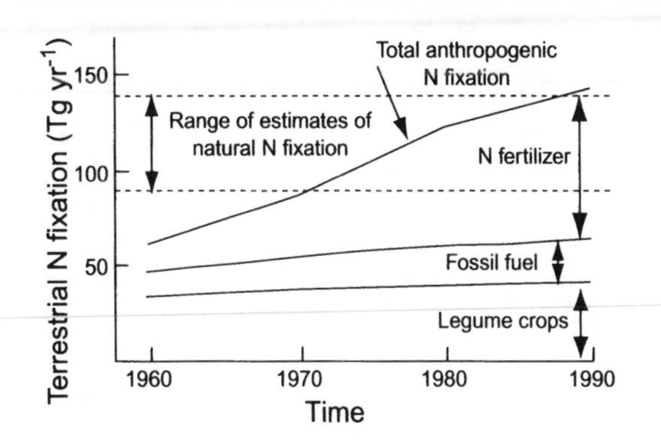

FIGURE 15.5. Anthropogenic fixation of nitrogen in terrestrial ecosystems over time compared with the range of estimates of natural biological nitrogen fixation on land. (Redrawn with permission from *Globol Biogeochemical Cycles*; Galloway et al. 1995).

developing nations, with 40% of this total being applied in the tropics and subtropics. Much of the projected increase in fertilizer use is expected to occur in the less-developed nations of the world.

Cultivation of nitrogen-fixing crops such as soybeans, alfalfa, and peas adds fixed nitrogen over and above that which is added via biological fixation in natural ecosystems. Some nitrogen fixation is also carried out by free-living and associative nitrogen fixers like *Azolla* that commonly occur in rice paddies. Annual fixation rates in crop systems are about 32 to 53 Tg yr^{-1}, 20 to 40% of total biotic fixation that occurs on land.

Human activities account for most of the nitrogen trace gases transferred from Earth to the atmosphere (Table 15.3). In addition to the large pool of relatively unreactive N_2, the atmosphere contains several nitrogen trace gases, including nitric oxides (NO and NO_2), nitrous oxide (N_2O), and NH_3. Although the pools and fluxes of these nitrogen trace gases are much smaller than those of N_2 (Fig. 15.4), they play a much more active role in atmospheric chemistry and have been more strongly affected by human activities (see Chapter 9).

N_2O, which is increasing at the rate of 0.2 to 0.3% yr^{-1} (Fig. 15.3), is an inert gas that is 200-fold more effective than CO_2 as a greenhouse gas and contributes about 6% of the greenhouse warming (Ramaswamy et al. 2001). Nitrification and denitrification in the oceans and in tropical soils are the major natural sources of N_2O (Schlesinger 1997). Human activities have nearly doubled N_2O flux from Earth to the atmosphere, primarily through agricultural fertilization (Fig. 15.3). Other anthropogenic N_2O sources include cattle and feedlots, biomass burning, and various industrial sources.

TABLE 15.3. Global sources to and sinks from the atmosphere of nitrogen trace gases.

Sources and sinks	NO_x-N[a]	N_2O-N[a]	NH_3-N[b]	Total N
Natural biogenic sources	7.8	10	23	40.8
Oceans	0	4[c]	13	17
Soils	2.8[d]	6	10	18.8
Lightning	5	0	0	5
Anthropogenic sources	44.2	8.1	52.2	104.5
Cultivated soils	2.8[d]	4.2	9	16
Biomass burning	7.1	0.5	5	12.6
Domestic animals	0	2.1	32	34.1
Fossil fuels/industrial	33	1.3	2.2	36.5
Other	0.7	0	4	4.7
Sinks				
Atmospheric destruction	??	12.3	1	
Deposition on land	40[e]??	0	57	
Annual accumulation	0	3.9	0	

[a] Data from Prather et al. (2001).
[b] Data from Schlesinger (1997).
[c] Data from Karl et al. (in press).
[d] Soil NO_x flux uncertain; assumed to be 50% natural, 50% anthropogenic.
[e] Data from Holland et al. (1997).

N_2O is broken down in the stratosphere, where it catalyzes the destruction of stratospheric ozone.

Human activities have tripled the flux of NH_3 from land to the atmosphere (Table 15.3). Domestic animals are now the single largest global source of ammonia; agricultural fertilization, biomass burning, and human sewage are important additional sources. Cultivated soils, which account for only 10% of the ice-free land area (see Table 6.6), account for about half of the ammonia flux from soils to the atmosphere. In summary, activities associated with agriculture (animal husbandry, fertilizer addition, and biomass burning) are the major cause for increased ammonia transport to the atmosphere and account for 60% of the global flux. Ammonia is a reactant in many atmospheric reactions that form aerosols and generate air pollution. Ammonia is also the main acid-neutralizing agent in the atmosphere, raising the pH of rainfall, cloud water, and aerosols. Most of the ammonia emitted to the atmosphere returns to Earth in precipitation.

Human activities have increased NO_x flux to the atmosphere sixfold to sevenfold, primarily through the combustion of fossil fuels (Table 15.3). Nitrification is the largest natural terrestrial source of NO (see Chapter 9). Fertilizer addition has increased the magnitude of this source, with additional NO coming from biomass burning. Preindustrial NO_x fluxes were greater in tropical than temperate ecosystems, due to frequent burning of tropical savannas, soil emissions, and production by lightning (Holland et al. 1999). Most NO_x deposition now occurs in the temperate zone, where deposition rates have increased fourfold since preindustrial times. Unlike N_2O, NO is highly reactive and alters atmospheric chemistry rather than accumulating in the atmosphere. Nitric oxide is a precursor to the photochemical production of tropospheric ozone (O_3), a major component of smog, and is often the rate-limiting reactant in ozone formation. When its concentrations are high, O_3 can be produced via the oxidation of carbon monoxide, nonmethane hydrocarbons, and methane. It also affects the concentration of the hydroxyl radical, the main oxidizing (or cleansing) chemical in the atmosphere and thus indirectly affects the concentrations of many other gases. The oxidation of NO leads to the formation of nitric acid, a component of acid precipitation and an increasingly large source of nitrogen inputs to ecosystems.

Nitrogen deposition affects many ecosystem processes. The widespread nitrogen limitation of plant production in nontropical ecosystems results in an effective retention of a large proportion of anthropogenic nitrogen that is deposited in ecosystems, particularly in young, actively growing forests that are accumulating nutrients in vegetation (see Fig. 13.11). In some cases nitrogen deposition may stimulate carbon uptake and storage, although the global magnitude of carbon storage accounted for by nitrogen deposition is uncertain (Table 15.1) (Holland et al. 1997).

Nitrogen accumulation in production and organic matter storage cannot increase indefinitely. After long-term chronic nitrogen inputs, nitrogen supply may exceed plant and microbial demands, resulting in **nitrogen saturation** (Agren and Bosata 1988, Aber et al. 1998). When ecosystems become nitrogen saturated, nitrogen losses to stream water, groundwater, and the atmosphere should increase and eventually approach nitrogen inputs. Nitrogen saturation is often associated with declines in forest productivity and increased tree mortality in coniferous forests in Europe (Schulze 1989) and the United States (Aber et al. 1995).

Temperate forests vary regionally in the rate at which they approach nitrogen saturation, depending on rates of nitrogen inputs and the capacity of soils to buffer these inputs (Berendse et al. 1993, Vitousek et al. 1997a, Aber et al. 1998). In tropical forests, where nitrogen availability is typically high relative to plant and microbial demands, anthropogenic nitrogen deposition may lead to immediate nitrogen losses (Hall and Matson 1998); this could have potentially negative effects on plant and soil processes (Matson et al. 1999). In general, the capacity of a forest ecosystem to retain nitrogen is linked to its productive potential and to its current degree of nitrogen limitation (Aber et al. 1995, Magill et al. 1997).

The addition of limiting nutrients can alter species dominance and reduce the diversity of ecosystems. Nitrogen addition to grasslands or heathlands, for example, increases the dominance of nitrogen-demanding grasses, which then suppress other plant species (Berendse et al. 1993). These species changes can convert nutrient-poor, diverse heathlands to species-poor forests and grasslands (Aerts and Berendse 1988). Loss of species diversity with nitrogen addition therefore occurs at both patch and landscape scales.

Human activities increase the nitrogen losses from terrestrial ecosystems and nitrogen transfer to aquatic ecosystems. The massive nitrogen additions to terrestrial ecosystems, in the form of deposition, fertilization, food imports, and growth of nitrogen-fixing crops, have led to a dramatic increase in nitrogen concentrations in surface and groundwaters over the past century. Nitrate concentrations in the Mississippi River have more than doubled since the 1960s (Turner and Rabalais 1991), and nitrate concentrations in other major rivers of the United States have increased 3- to 10-fold in the past century (see Fig. 14.10) (Howarth et al. 1996). Nitrate concentrations in many lakes, streams, and rivers of Europe have likewise increased, as have concentrations in most aquifers (Vitousek et al. 1997a).

The Global Phosphorus Cycle

Phosphorus is unique among the major biogeochemical cycles because it has only a tiny gaseous component and has no biotic pathway that brings new phosphorus into ecosystems. Therefore, until recently, ecosystems derived most available phosphorus from organic forms, and phosphorus cycled quite tightly within terrestrial ecosystems. Like nitrogen, phosphorus is an essential nutrient that is frequently in short supply. Marine and fresh-water sediments and terrestrial soils account for most phosphorus on Earth's surface (Fig. 15.6). Most of this store is not directly accessible to the biota. Most phosphorus in soils, for example, occurs primarily in insoluble forms such as calcium or iron phosphate. Most organic phosphorus is in plant or microbial biomass, and the recycling of that organic matter when it dies is the major source of available phosphorus to organisms.

The physical transfers of phosphorus around the global system are constrained by the lack of a major atmospheric gaseous component. Leaching losses in natural ecosystems are also low due to the low solubility of phosphorus. Instead, phosphorus moves around the global system primarily through wind erosion and runoff of particulates in rivers and streams to the oceans. The major flux in the global phosphorus cycle (excluding human activities) is via hydrologic transport from land to the oceans. In the oceans, some of those phosphorus-containing particulates are recycled by marine biota, but a much larger portion (90%) is buried in sediments. Because there is no atmospheric link from oceans to land, the flow is one-way on short time scales (Smil 2000). On geological time scales (tens to hundreds of millions of years), phosphorus-containing sedimentary rocks are exposed and weathered, resupplying phosphorus to the terrestrial biosphere.

Anthropogenic Changes in the Phosphorus Cycle

Human activities have enhanced the mobility of phosphorus and altered its natural cycling by accelerating erosion and wind- and water-borne transport. Inorganic phosphorus fertilizers have been produced since the mid-1800s, but the amount produced and applied has increased dramatically since the mid-twentieth century (Fig. 15.7), coincident with the intensification of agriculture that accompanied the Green Revolution (Smil 2000). Between 1850 and 2000, agricultural systems received about 550 Tg of new phosphorus. The annual application of phosphorus to agricultural ecosystems (10 to 15 $Tg\,yr^{-1}$) is 20 to 30% of that which cycles naturally through all terrestrial ecosystems (Fig. 15.6).

Human land use change has also increased phosphorus losses from ecosystems. Water and wind erosion cause a 15 $Tg\,yr^{-1}$ phosphorus loss

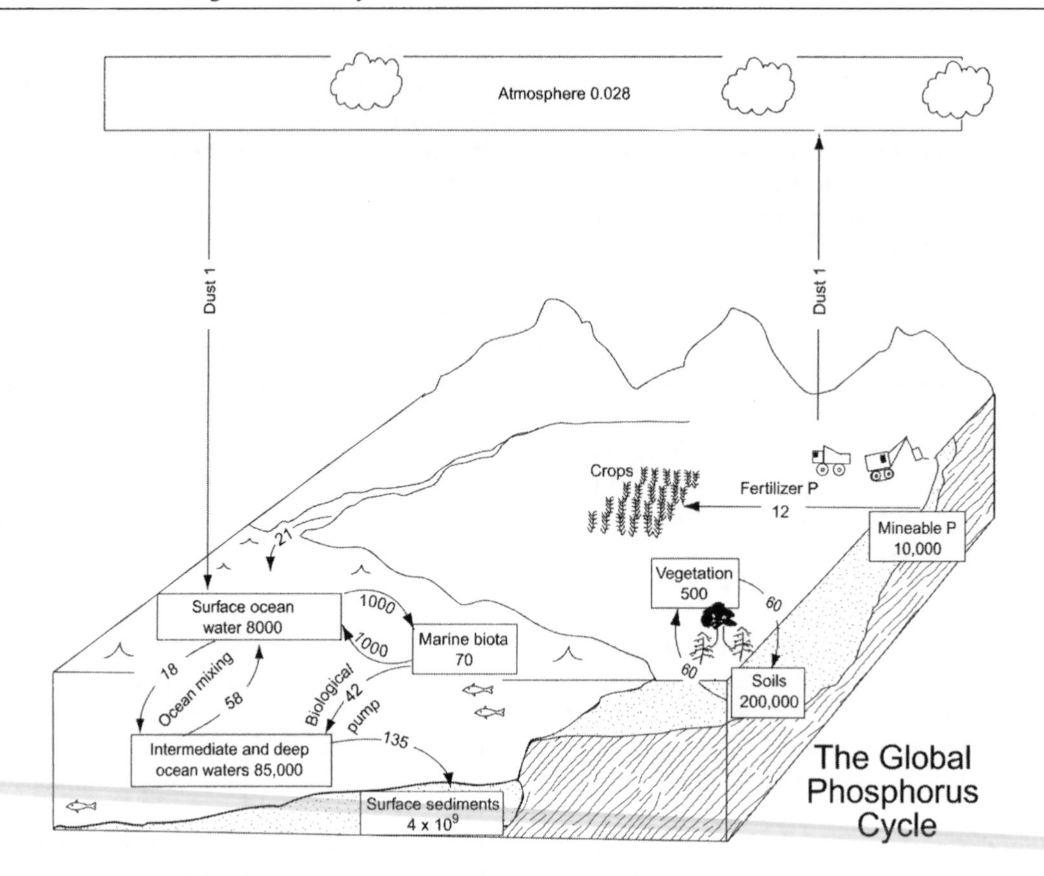

FIGURE 15.6. The global phosphorus cycle showing approximate magnitudes of the major pools (boxes) and fluxes (arrows) in units of teragrams per year. Most phosphorus that participates in biogeochemical cycles over decades to centuries is present in soils, sediments, and the ocean. Phosphorus cycles tightly between vegetation and soils on land and between marine biota and surface ocean water in the ocean. The major human effects on the global phosphorus cycle have been application of fertilizers (about 20% of that which naturally cycles through vegetation) and the erosional loss from crop and grazing lands. Data for phosphorus pools in vegetation, marine biota, and ocean water are from Smil (2000), for the atmospheric pool and for ocean mixing and the biological pump from Reeburgh (1997). Remaining data are from Schlesinger (1997).

from the world's croplands, an amount similar to the annual fertilizer inputs. Overgrazing has also increased erosional losses, mobilizing about 13 $Tg\,yr^{-1}$ of phosphorus from grazing lands (Smil 2000). The production of human and animal wastes have led to point and non-point sources of phosphorus. The total phosphorus losses from terrestrial ecosystems due to human activities are about twice the annual fertilizer inputs.

Together, these changes have increased the transport of phosphorus around the world (Howarth et al. 1995). Because phosphorus commonly limits production in lakes, the inadvertent phosphorus fertilization of fresh-water ecosystems can lead to eutrophication and associated negative consequences for aquatic organisms and society (see Chapter 14). Phosphorus transport by wind-blown dust can also affect down-wind ecosystems, such as the Southern Ocean.

The Global Sulfur Cycle

The global cycle of sulfur shares characteristics with the global cycles of nitrogen and phosphorus. The sulfur cycle, like the nitrogen cycle, has a significant atmospheric component. The gaseous forms in the atmosphere have low concentrations but play important roles. Like phosphorus, sulfur is primarily rock derived. Sea water, sediments, and rocks are the largest reservoirs of sulfur (Fig. 15.8). The atmosphere contains little sulfur. Before human activities of the past several centuries, sulfur became available to the biosphere primarily through the weathering of sedimentary pyrite. Once weathered, sulfur moves through the global system by hydrologic transport or emission to the atmosphere as a reduced sulfur gas or sulfur-containing particles. About $100\,Tg\,yr^{-1}$ of sulfur, moving mostly as dissolved sulfate, was transported through rivers to the coastal margins or open oceans in the preindustrial world (Galloway 1996).

Sulfur can be reduced to sulfide or to other trace sulfur gases in anaerobic environments such as wetlands and coastal sediments. The emission of sulfate from sea water and sulfur trace gases from oceans ($160\,Tg\,yr^{-1}$) is about 10-fold greater than that from continents (Fig. 15.8). Marine biogenic emission include dimethylsulfide (DMS), one of the primary sources of atmospheric sulfate; emissions of sulfur dioxide (SO_2) from volcanic eruptions are the other major source.

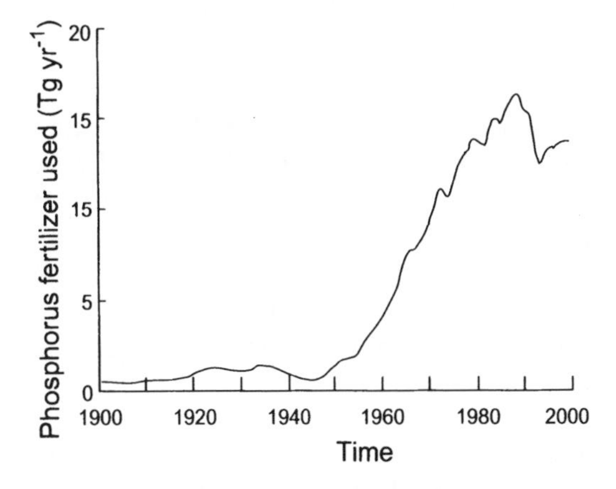

FIGURE 15.7. Changes in the global use of inorganic phosphorus fertilizers during the twentieth century. (Redrawn with permission from *Annual Review of Energy in the Environment*, Volume 25 ©2000 by Annual Reviews www.AnnualReviews.org; Smil 2000.)

Sulfur emitted to the atmosphere typically has a short residence time. It is oxidized to sulfate by reaction with OH radicals. Sulfate rains out downwind within a few days, generally as sulfuric acid. Sulfuric acid has low equilibrium vapor pressure, so it quickly condenses to form sulfate in cloud droplets, which readily evaporate to form sulfate aerosols. These aerosols have both direct and indirect effects on Earth's energy budget. Their direct effect is to backscatter (reflect) incoming shortwave radiation, thus reducing solar inputs and tending to reduce global temperature. Their indirect effects are more complicated and difficult to predict. As particulates, they act as cloud condensation nuclei by providing a surface on which water can condense, thereby influencing cloud formation, cloud lifetimes, and cloud droplet size. The density of cloud condensation nuclei and droplet size in turn govern cloud albedo. The uncertainty of the direction and magnitude of these effects of sulfate aerosols on cloud albedo is a key reason for concern about the anthropogenic changes in the global sulfur cycle.

Human activities now transfer about $100\,Tg\,yr^{-1}$ of sulfur to the atmosphere and oceans, increasing the natural cycling rate by about 50% (Fig. 15.8). Half of this sulfur arises from fossil-fuel combustion and ore refining; and the rest comes from mobilization of sulfur in dust from farming, animal husbandry, erosion of exposed sediments, and other sources. Much of the anthropogenic sulfur moves through the atmosphere and is deposited on land, where it can accumulate in soils or biota, or is discharged to the oceans in solution.

Reconstruction of global temperature records from ice cores shows that sulfur dioxide from volcanic emissions is a major cause of interannual climate variation over long time scales. Consequently, the dramatic increase in sulfur aerosols due to anthropogenic emissions will undoubtedly play an important role in future climate changes. The negative forcing due to sulfur emissions and their associated direct and indirect effects could range from 0 to $1.5\,W\,m^2$, a negative forcing that partially offsets the warming due to greenhouse gases (Penner et al. 2001).

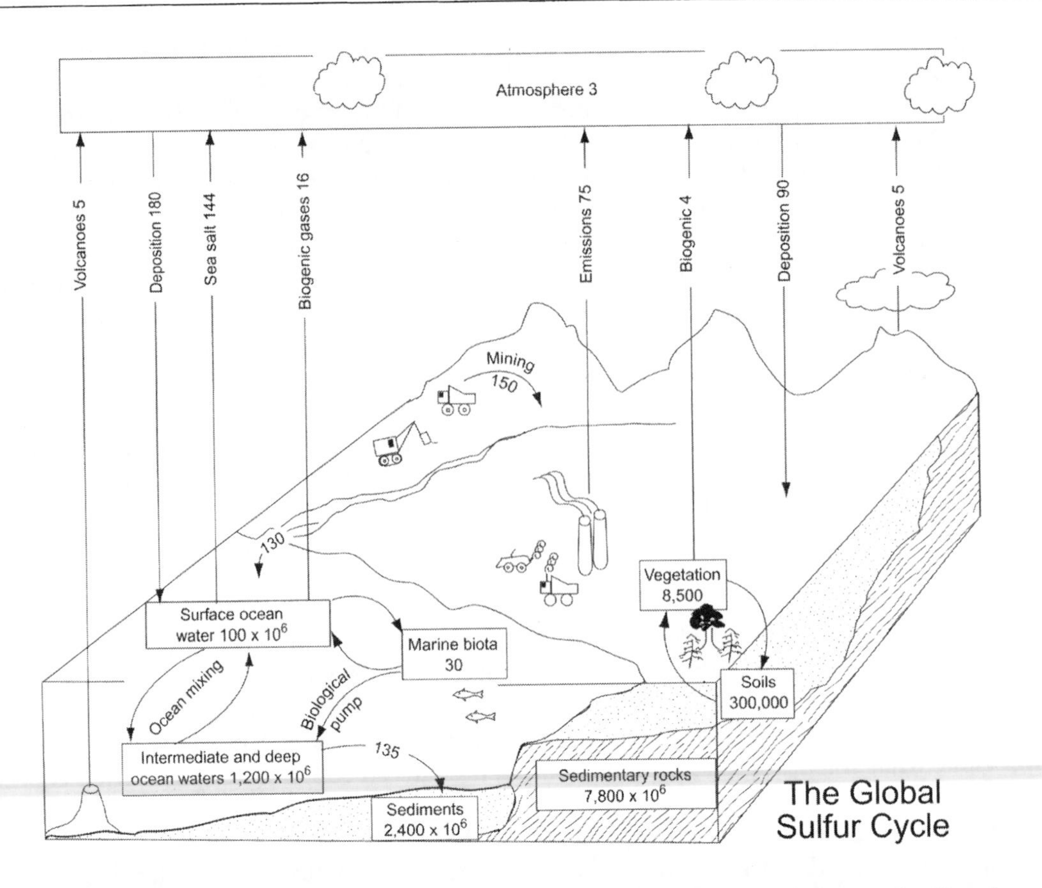

FIGURE 15.8. The global sulfur cycle showing approximate magnitudes of the major pools (boxes) and fluxes (arrows) in units of teragrams per year. Most sulfur is in rocks, sediments, and ocean waters. The major fluxes in the sulfur cycle are through the biota and various trace gas fluxes. Human activities have doubled the global fluxes of sulfur through mining and increased gas emissions. Data for anthropogenic emissions are from Penner et al. (2001) and for the sulfur pools in marine biota and soils from Reeburgh (1997). Remaining data are from Galloway (1996) and Schlesinger (1997).

The Global Water Cycle

Only a tiny fraction of Earth's water (0.01%) is in soils, where it is accessible to plants and available to support the activities of terrestrial organisms. Most of Earth's water is in the oceans (96.5%), ice caps and glaciers (2.4%), and groundwater (1%) (Fig. 15.9) (Schlesinger 1997). About 91% of the water that evaporates from oceans returns there as precipitation. Another 9% of ocean evaporation (40,000 km^3) moves over the land, where it falls as precipitation and returns to the oceans as river runoff. The total evaporation from land (71,000 km^3) is about 15% of total global evaporation, although land occupies about 30% of Earth's surface; this indicates that average evapotranspiration rates are about half as great on land as over the oceans. In regions of adequate soil moisture, vegetation enhances evaporation from land compared to a free water surface (see Chapter 4). There are large regional variations in evaporation rate over both land and ocean related to climate and, in the case of land, to water availability and transpiration rates of vegetation. Of the terrestrial precipitation (111,000 km^3), about one third comes from the oceans (40,000 km^3), and two thirds (71,000 km^3) is evaporated from land and recycled. Evaporation and precipitation are both highly variable, both regionally and seasonally.

The quantity of water in the atmosphere is only 2.6% of that which annually cycles through the atmosphere in evaporation and transpiration, giving a turnover time of about 10 days. Precipitation is therefore tightly linked to evapotranspiration from upwind ecosystems over time scales of hours to weeks. Soil water, in contrast, has a turnover time of about a year, with inputs from precipitation and outputs to evapotranspiration and runoff. This turnover time makes soil moisture sensitive to seasonal and interannual variations in precipitation and evapotranspiration.

Anthropogenic Changes in the Water Cycle

Human activities have altered the global hydrologic cycle primarily through changes in climate and Earth's energy budget. The twentieth-century rise in global mean air temperature of 0.6°C has increased rates of evaporation, which in turn increased precipitation. Precipitation in the contiguous United States, for example, increased by 10% between 1910 and 2000, with projections of a 3 to 15% increase in response to the projected temperature increase of 1.5 to 3.5°C during the twenty-first century. Past changes in precipitation have varied spa-

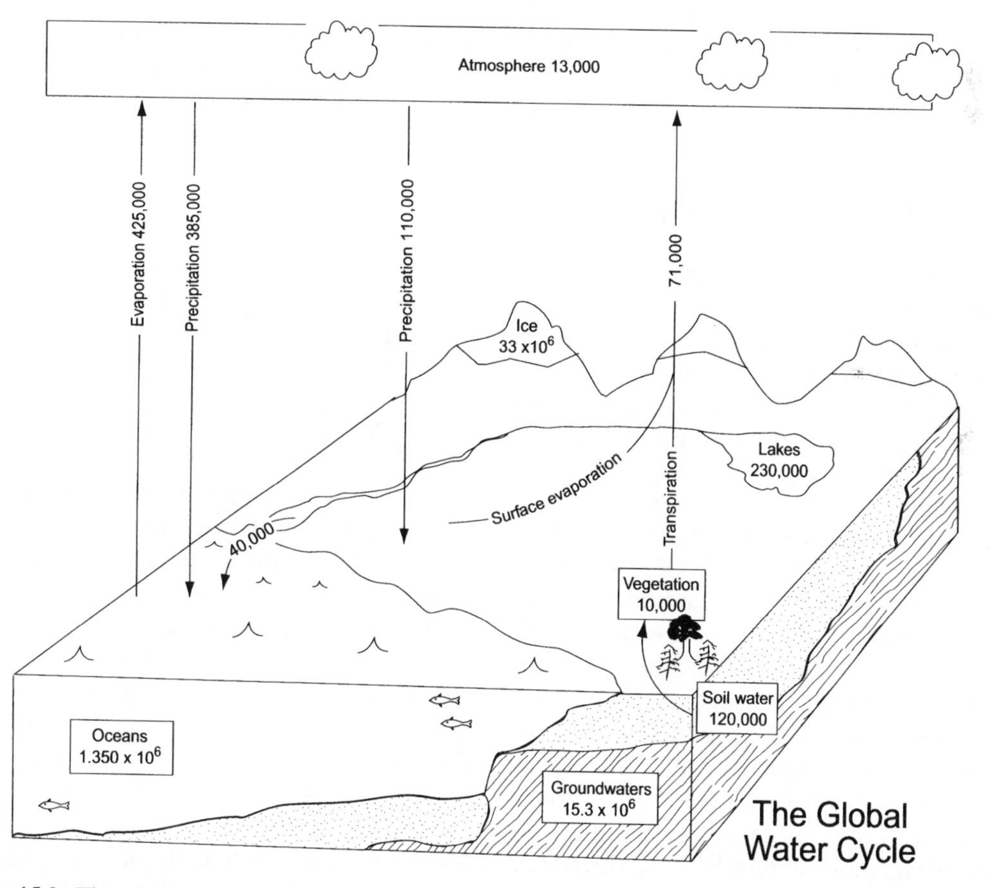

FIGURE 15.9. The global water cycle showing approximate magnitudes of the major pools (boxes) and fluxes (arrows) in units of cubic kilometers per year (Schlesinger 1997). Most water is in the oceans, ice, and groundwater, where it is not directly accessible to terrestrial organisms. The major water fluxes are evapotranspiration, precipitation, and runoff.

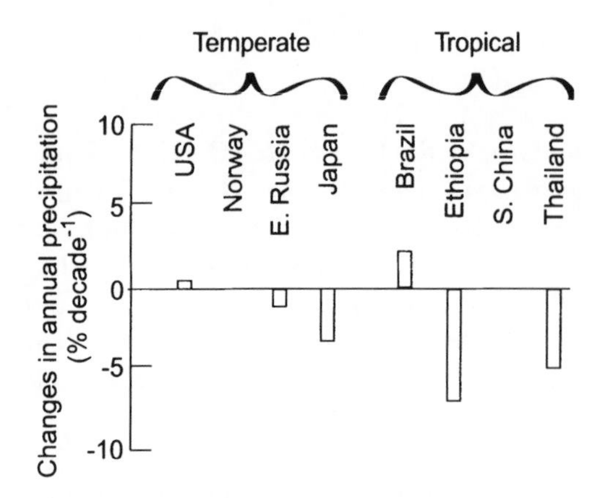

FIGURE 15.10. Regional changes in precipitation between 1950 and 2000 (Folland et al. 2001). There is substantial regional variation in the direction of precipitation changes.

tially, as will certainly be the case in the future (Fig. 15.10). The ecological impact of a given change in precipitation also varies among ecosystems. A small change in precipitation in arid regions, for example, could have much greater ecological impact than larger changes in areas that currently receive abundant rainfall. Projected future changes in precipitation will affect river flows, groundwater recharge, the water relations of natural ecosystems, and the water available to managed ecosystems.

Land use changes alter the hydrologic cycle by altering (1) the quantity of energy absorbed, (2) the pathway of energy loss, and (3) the moisture content and temperature of the atmosphere. Conversion from tropical rain forest to pasture, for example, leads to less energy absorption because of increased albedo and a larger proportion of energy dissipated to the atmosphere as sensible rather than latent heat (Gash and Nobre 1997). The warmer, drier atmosphere allows less precipitation, favoring the persistence of pastures rather than return to rain forests (see Fig. 2.11). Conversion of heathland to agriculture in western Australia also reduced precipitation by 30% (Chambers 1998) (see Fig. 14.10). When land use changes are extensive, they can have continental-scale effects on temperature and precipitation, often at locations remote from the region of land-cover change, as a result of large-scale adjustments in atmospheric circulation (Chase et al. 2000). Land-cover changes in Southeast Asia, for example, have particularly large effects on global-scale climate through atmospheric teleconnections.

Ecosystems are generally more sensitive to soil moisture (terrestrial ecosystems) or runoff (aquatic ecosystems) than to precipitation. The projected increases in both evaporation and precipitation complicate projections of changes in water availability. Decreases in precipitation will probably reduce soil moisture, but regions with increased precipitation may still experience reduced soil moisture, if evaporation increases more than precipitation. Models generally project increased soil moisture at high latitudes, large-scale soil drying in continental portions of northern mid-latitudes in summer due to higher temperatures and insufficient increases (or reductions) of rainfall. Many areas that are currently important for agriculture, such as Ukraine or the midwestern United States, may be particularly prone to future drought; and grain-producing areas may migrate northward to areas that are currently too cold to support intensive agriculture. These changes in location of soil moisture suitable for agriculture will have major regional and national economic and social impacts.

Consequences of Changes in the Water Cycle

Society depends most directly on some of the smallest and most vulnerable pools in the global hydrologic cycle. Agriculture, for example, relies on soil water derived from precipitation, a relatively small pool that would be altered quickly in response to major changes in the balance between precipitation and evapotranspiration. In some areas soil moisture derived from precipitation is supplemented by irrigation, which withdraws water from lakes, rivers, and groundwater. Lands under irrigation have increased fivefold during the twentieth century (Fig. 15.11) (Gleick 1998). During this period there was an eightfold increase in the water used to support human activities, which paralleled a fourfold increase in human population and a 50% increase in per capita water

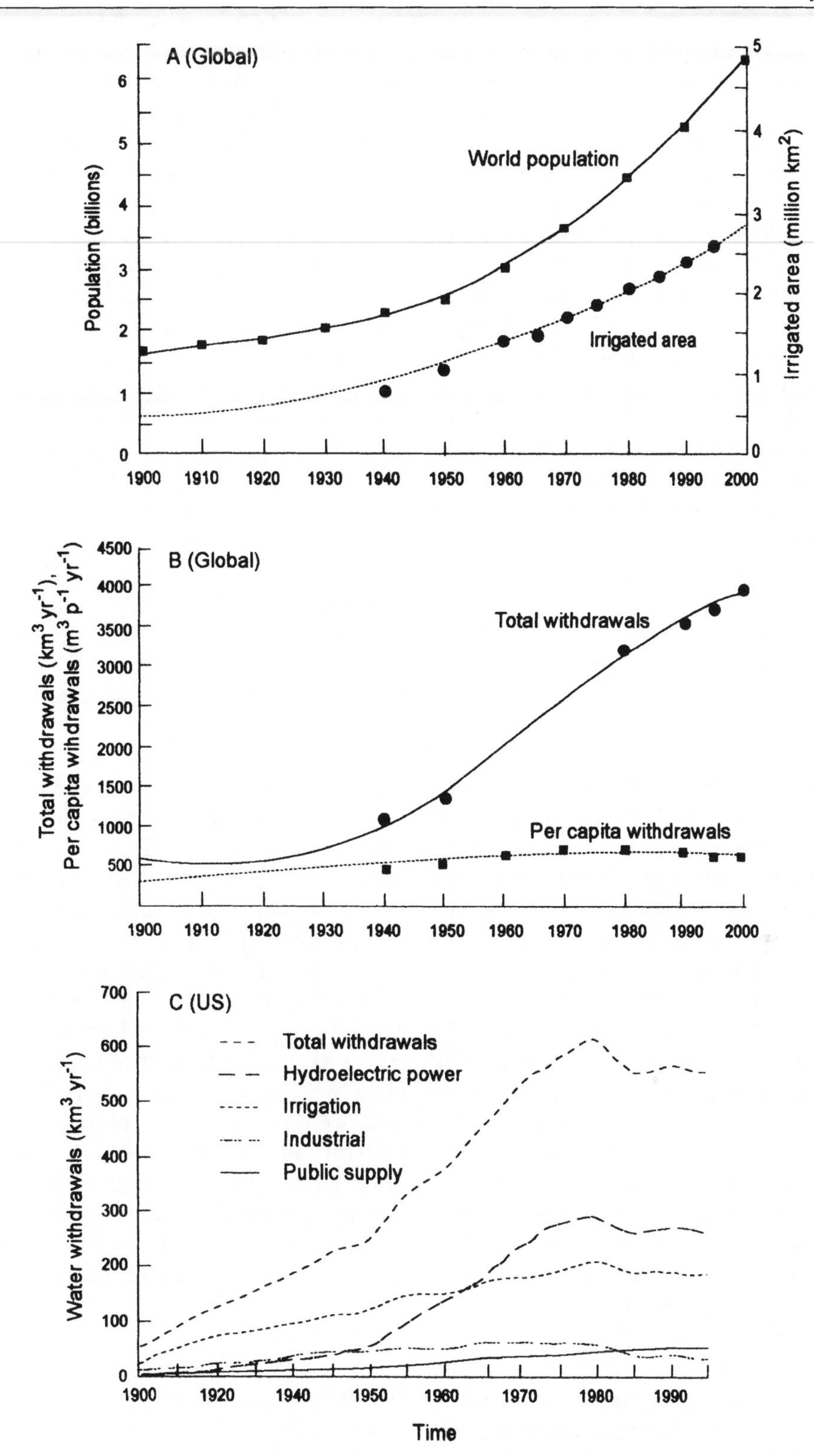

FIGURE 15.11. Trends in (**A**) world population and global land area under irrigation; (**B**) water withdrawals to support human activities, expressed as a global total and on a per capita basis; and (**C**) water withdrawal in the United States, separated by economic sector. (Redrawn with permission from Island Press; Gleick 1998.)

consumption. Humans now use 54% of the accessible runoff (see Chapter 4). Most of this water is used for hydroelectric power and irrigation (Fig. 15.11).

The scarcity of water is only part of the hydrologic problems facing society. Approximately 40% of the world's population had no access to adequate sanitation in 1990 (Gleick 2000), and 20% had no clean drinking water. The shortage of clean water is particularly severe in the developing nations, where future population growth and water requirements are likely to be greatest. The projected increases in human demands for fresh water will certainly have strong impacts on aquatic ecosystems, through eutrophication and pollution, diversion of fresh water for irrigation, and modification of flow regimes by dams and reservoirs (see Chapter 10).

Summary

Ecological processes and human activities play major roles in most biogeochemical cycles. The magnitude of biotic and human impacts on ecosystem processes becomes clear when summed at the global scale. Biotic processes (photosynthesis and respiration) constitute the engine that drives the global carbon cycle. The four major carbon pools that contribute to carbon cycling over decades to centuries are the atmosphere, land, oceans, and surface sediments. On land the carbon gain by vegetation is slightly greater than the carbon loss in respiration, leading to net carbon storage on land. The net carbon input to the oceans is also slightly greater than the net carbon return to the atmosphere. Marine primary production is about the same as that on land. Most (80%) of this marine NPP is released to the environment by respiration, with the remaining 20% going to the deep oceans by the biological pump. Ocean upwelling returns most of this carbon to the surface ocean waters; only small quantities are deposited in sediments. Human activities cause a net carbon flux to the atmosphere through combustion of fossil fuels, cement production, and land use change. This flux is equivalent to 14% of terrestrial heterotrophic respiration.

The atmosphere contains the vast majority of Earth's nitrogen. The amount of nitrogen that annually cycles through terrestrial vegetation is 9-fold greater than inputs by nitrogen fixation. In the ocean the annual cycling of nitrogen through the biota is 500-fold greater than inputs by nitrogen fixation. Denitrification is the major output of nitrogen to the atmosphere. Human activities have doubled the quantity of nitrogen fixed by the terrestrial biosphere through fertilizer production, planting of nitrogen-fixing crops, and combustion of fossil fuels.

Most phosphorus that participates in biogeochemical cycles over decades to centuries is present in soils, sediments, and the ocean. Phosphorus cycles tightly between vegetation and soils on land and between marine biota and surface ocean water in the ocean. The major human impact on the global phosphorus cycle has been application of fertilizers (about 20% of that which naturally cycles through vegetation) and erosional loss from crop and grazing lands (about half of that which annually cycles through vegetation). Most sulfur is in rocks, sediments, and ocean waters. The major fluxes in the sulfur cycle are through the biota and various trace gas fluxes. Human activities have substantially increased global fluxes of sulfur through mining and increased gas emissions.

Most water is in the oceans, ice, and groundwater, where it is not directly accessible to terrestrial organisms. The major water fluxes are evapotranspiration, precipitation, and runoff. Human activities have speeded up the global hydrologic cycle by increasing global temperature, which enhances evapotranspiration and therefore precipitation, and by diverting more than half of the accessible fresh water for human use. Availability of adequate fresh water will be an increasingly scarce resource for society, if current human population trends continue.

Review Questions

1. How do the major global cycles (carbon, nitrogen, phosphorus, sulfur, and water) differ from one another in terms of (a) the major pools and (b) the major fluxes? In

which cycles are soil pools and fluxes largest? In which cycles are atmospheric pools and fluxes largest?

2. How do the controls over the global carbon cycle differ between time scales of months, decades, and millennia? How has atmospheric CO_2 varied on each of these time scales, and what has caused this variation?
3. How have human activities altered the global carbon cycle? What are the mechanisms that explain why some of the CO_2 generated by human activities becomes sequestered on land?
4. What are the major causes and the climatic consequences of increased atmospheric concentrations of CO_2, CH_4, and N_2O? What changes in human activities would be required to reduce the rate of increase of these gases? What policies would be most effective in reducing atmospheric concentrations of these gases, and what would be the societal consequences of these policy changes?
5. What are the major natural sources and sinks of atmospheric methane? How might these be changed by recent changes in climate and atmospheric composition?
6. What are the major natural sources and sinks of atmospheric N_2O? How might these be changed by recent changes in climate and land use?
7. How have human activities changed the global nitrogen cycle? How have these changes affected the nitrogen cycle in unmanaged ecosystems?
8. How do changes in the nitrogen cycle affect the global carbon cycle? In what types of ecosystems would you expect these nitrogen effects on the carbon cycle to be strongest? Why?
9. How have human activities changed the global phosphorus and sulfur cycles? How do changes in these cycles affect the global cycles of other elements?
10. How have human activities changed the global water cycle? If the world has so much water, and this water is replenished so frequently by precipitation, why are people concerned about changes in the global water cycle? In what regions of the world will changes in the quantity and quality of water have greatest societal impact? Why?

Additional Reading

Aber, J., W. McDowell, K. Nadelhoffer, A. Magill, G. Bernstson, M. Kamakea, S. McNulty, W. Currie, L. Rustad, and I. Fernandez. 1998. Nitrogen saturation in temperate forest ecosystems. *BioScience* 48:921–934.

Cicerone, R.J., and R.S. Oremland. 1988. Biogeochemical aspects of atmospheric methane. *Global Biogeochemical Cycles* 2:299–327.

Galloway, J.N. 1996. Anthropogenic mobilization of sulfur and nitrogen: Immediate and delayed consequences. *Annual Review of Energy in the Environment* 21:261–292.

Houghton, J.T., Y. Ding, D.J. Griggs, M. Noguer, P.J. van der Linden, X. Dai, K. Maskell, and C.A. Johnson, editors. 2001. *Climate Change 2001: The Scientific Basis*. Cambridge University Press, Cambridge, UK.

Matson, P.A., W.H. McDowell, A.R. Townsend, and P.M. Vitousek. 1999. The globalization of N deposition: Ecosystem consequences in tropical environments. *Biogeochemistry* 46:67–83.

Reeburgh, W.S. 1997. Figures summarizing the global cycles of biogeochemically important elements. *Bulletin of the Ecological Society of America* 78: 260–267.

Schimel, D.S., et al. 2001. Recent patterns and mechanisms of carbon exchange by terrestrial ecosystems. *Nature* 414:169–172.

Schlesinger, W.H. 1997. *Biogeochemistry: An Analysis of Global Change*. Academic Press, San Diego, CA.

Smil, V. 2000. Phosphorus in the environment: Natural flows and human interferences. *Annual Review of Energy in the Environment* 25:53–88.

Vitousek, P.M., and P.A. Matson. 1993. Agriculture, the global nitrogen cycle, and trace gas flux. Pages 193–208 *in* R.S. Oremland, editor. *The Biogeochemistry of Global Change: Radiative Trace Gases*. Chapman & Hall, New York.

16
Managing and Sustaining Ecosystems

Human activities influence all of Earth's ecosystems. This chapter summarizes the nature of these impacts, the principles by which important ecological properties can be sustained, and the management approaches that have been developed to maximize sustainability.

Introduction

Humans depend on Earth's ecosystems for food, shelter, and other essential goods and services. Ecosystems provide well-recognized **goods**, including timber, forage, fuels, medicines, and precursors to industrial products. Ecosystems also provide underrecognized **services**, such as recycling of water and chemicals, mitigation of floods, pollination of crops, and cleansing of the atmosphere (Daily 1997). The harvest and management of these resources are a major component of the global economy. Our purposeful use and misuse of these resources have endangered them, and many apparently unrelated activities have had indirect and unintended negative effects on them.

The overuse or misuse of resources can alter the functioning of ecosystems and the services they provide (see Chapter 1). Land use change, for example, can degrade the capacity of watersheds to purify water, leading to large treatment costs to cities. Degradation and loss of wetlands can expose communities to increased damage from floods and storm surges. Decimation of populations of insect pollinators has reduced yields of many crops (Daily 1997). Introductions and invasions of nonnative species such as killer bees, fire ants, and zebra mussels, through the actions of humans, cause enormous damage to living resources and threaten human health. Human activities also indirectly affect ecosystem goods and services through changes in the atmosphere, hydrologic systems and climate (see Chapter 15).

The growing scale of human activities suggests that all ecosystems are influenced, directly or indirectly, by our activities. No ecosystem functions in isolation, and all are influenced by human activities taking place in adjacent communities and around the world. Human activities are leading to global changes in most major interactive controls over ecosystem processes: climate (global warming), soil and water resources (nitrogen deposition, erosion, diversions), disturbance regime (land use change, fire control), and functional types of organisms (species introductions and extinctions). In many cases, at the scale of regions, these global changes interact with each other and with local changes. All ecosystems are therefore experiencing directional changes in ecosystem controls, creating novel conditions and, in many cases, positive feedbacks that lead to new types of ecosystems. These changes in interactive controls will inevitably change the properties of ecosystems; some of these changes are detrimental to society.

Our use, mismanagement, and unintended effects on ecosystems have put some of them at risk, imperiling our own well-being. In this chapter, we describe some general ecological principles that may contribute to formulating management approaches and minimizing impacts. We conclude that maintaining Earth's ecosystems, even the "wild" ones, in the face of anthropogenic changes will require new management approaches. We review some of these that draw on ecosystem ecology and other sciences to manage and sustain ecosystems and the benefits we derive from them.

Ecosystem Concepts in Management

Given the strong directional changes in most interactive controls, all ecosystems on Earth must be managed to sustain their important properties. Ecosystems that are managed for food and fiber must be managed to maintain their productive potential. Ecosystems that have been degraded by human activities should be managed to restore their original properties and ecosystem services. In ecosystems that have been less influenced by human activities, the management challenge is to protect ecosystem functions and conserve biological diversity, by both reducing rates of land conversion and planning for conservation in the face of continued human development, climate change, and other global changes. Management of all of these conditions requires an application of solid scientific information and principles. In this section, we discuss several of the emerging concepts and principles of ecosystem ecology that are important for sustainable management.

Natural Variability

Observations about natural temporal and spatial variability in ecosystem functioning provide hints about the potential responses of ecosystems to human-caused changes. Natural variability has allowed us to understand something about the driving forces that determine the structure and functioning of ecosystems, and how changes in those forces lead to ecological change. Hypotheses about the mechanisms of ecosystem response to change can be developed and tested with spatial and temporal historical data (Swetnam et al. 1999).

Historical and regional information on variability can be useful in many management decisions. Records of hydrologic flows and ecosystem characteristics, for example, allowed predictions of the impacts of altered hydrologic flows in the Everglades (Harwell 1997). Fire records and information on associated ecosystem processes provide insights into consequences of fire management. In cases in which ecosystem conditions desired by managers or stakeholders are not within the bounds of natural ecosystem variability, the mismatch indicates that the desired conditions require reassessment (Landres et al. 1999).

Past patterns of ecosystem variability are not always a good predictor of current and future changes. The multiple and interacting nature of current changes (see Chapter 1) differ from those that ecosystems experienced in the past. Moreover, time lags and nonlinearities in response of different parts of ecosystems to change make the responses of whole ecosystems difficult to predict and manage based on past experience. Management therefore also benefits from experimentation and hypothesis testing.

Resilience and Stability

Understanding the basis for ecosystem resilience provides the basis for sustaining ecosystem properties in the face of human-induced change. Ecosystems that maintain their properties despite disturbance (i.e., are resistant to change) and that return rapidly to their original state (i.e., are resilient) exhibit more stable and predictable ecosystem properties (Holling 1986) (see Box 13.1). Lakes provide some of the most useful tests of hypotheses about resistance and resilience. Phosphorus inputs to lakes can push a system from one stable state (clear-water, oligotrophic system) to another (turbid-water, eutrophic system). The relationship between phosphorus turnover rate and food chain length provides an index of resilience that is useful for comparing the response of different lakes to a given disturbance (Cottingham and Carpenter 1994). Lakes with high phosphorus turnover remove phosphorus from actively cycling pools more quickly than do lakes with slow turnover. Short (planktivore-dominated) food chains process phosphorus more rapidly than do long (piscivore) food chains, except under high phosphorus levels. This type of study provides the basis for predictions about vulnerability of lakes to pulses of phosphorus inputs from agriculture or livestock in the watershed. Lakes also differ in their resistance to changes in pH resulting from acid rain. Lakes on granitic parent material with low acid-neutralizing capacity are more likely to acidify in response to acid rain than are lakes on limestone (Driscoll et al. 2001). The resulting changes in acidity alter trophic structure and abundance of fish.

Sustainable management of ecosystems benefits from a landscape perspective that considers interactions among ecosystems. A lake cannot be managed sustainably, for example, without considering the nutrient inputs from the surrounding landscape, and forest production can be managed most sustainably as a landscape mosaic by taking account of disturbances such as hurricanes, fire, and logging. The resilience and sustainability of lakes depends on a range of control processes that function at different scales to mitigate the effects of disturbance (Carpenter and Cottingham 1997). These control processes include the filtration effects of riparian vegetation and wetlands, the role of game fish in trophic dynamics, and the absorption of nutrients by macrophytes. When these components are intact, landscapes containing lakes can withstand perturbations such as droughts, floods, forest fires, and some land use change. Management of landscapes at coarse spatial scales requires different information from management of individual lakes, fields, or forest stands. At coarse spatial scales, monitoring of food webs in lakes is not feasible, so land use records, remote sensing, and surveys of fishing activity and success provide more useful input to models. An important implication of a landscape focus is that it requires the recognition of ecosystem response to multiple forcings.

State Factors and Interactive Controls

Directional changes in state factors or interactive controls limit the sustainability of ecosystems. State factors provide a useful framework within which to examine regulation of ecosystem processes (see Chapter 1). State factors and interactive controls exert such strong control over ecosystem processes that changes in these controlling factors as a result of human activities inevitably alter ecosystems and reduce the extent to which their current properties can be sustained. Management practices can, however, strongly influence the degree of sustainability. In particular, management that focuses on negative feedbacks, which tend to maintain the ecosystem in its current state, is the key to ecosystem sustainability (see Chapter 1). These negative feedbacks counterbalance positive feedbacks that tend to push the ecosystem toward some new state. Ecosystems will be sustainable when the net effect of negative feedbacks exceeds the effects of positive feedbacks.

At least three considerations influence the degree to which ecosystems can be managed sustainably. First, the sustainability of productivity and other ecosystem characteristics requires that state factors and interactive controls be conserved as much as possible. The interactive controls most readily managed are resources (e.g., through fertilizer and irrigation), disturbance regime, and functional types of organisms. Maintenance of long-term agricultural productivity requires management practices that retain soil organic matter (SOM), which in turn provides a buffered supply of water and nutrients (see Chapter 7).

Second, negative feedbacks among important interactive controls increase the sustainability of ecosystem processes. Biocontrol in agriculture uses negative feedbacks between predators and prey to limit the impact of pest insects on crops (Huffaker 1957). When these negative feedbacks are weakened, management must be intensified. Positive feedbacks such as mycorrhizal or nitrogen-fixing mutualisms may push an ecosystem toward some new state. These can also be constructive management tools in promoting the recovery of degraded ecosystems (Perry et al. 1989).

Finally, linkages among ecosystems in a landscape are most likely to enhance sustainability when they generate negative feedbacks among processes in these ecosystems. Aquatic ecosystems, for example, are recipients of nutrient runoff from land. They are vulnerable to ecological changes that occur on land but have only modest direct effects on their donor ecosystems. Land use practices that consider only terrestrial ecosystems therefore unavoidably affect aquatic ecosystems, unless laws or regulations constrain fertilizer inputs to land. Laws and regulations can create either positive or negative feedbacks between unmanaged ecosystems and human society. Hunting regulations, for example, that limit human harvest when game populations are reduced (a negative feedback) provide a more stable population regulation than regulations that provide subsidies or price supports when game or fish populations decline (a positive feedback).

Application of Ecosystem Knowledge in Management

Forest Management

The challenge for sustainable forestry is to define the attributes of forested ecosystems that are ecologically and societally important and to maximize these ecosystem services in the face of change. Management for sustainable timber production is one of many possible objectives for forest ecosystem management and provides a good example of the need for ecosystem ecology in management. Several issues are addressed by sustainable forestry. Nutrient supply rates, for example, must be sufficient to support rapid growth yet not so high that they lead to large nutrient losses. The rate at which stands are harvested must be balanced with their rate of regeneration following logging. Species diversity typical of natural mosaics of forest stands should be maintained. The sizes and arrangement of logged patches should provide a seminatural landscape mosaic with dependable seed sources and patterns of forest edges that allow natural use and movement of animal populations.

Addressing these issues requires attention to and management of interactive controls; understanding disturbance regime, plant functional types, and soil resources is critical. In the northwestern United States, for example, old-growth Douglas fir forests attain an age of more than 500 years (Wills and Stuart 1994). Fire and wind-throw of individual trees are the major natural disturbances, creating mosaics of tree ages at multiple spatial scales. Logging is now the most widespread disturbance in this region. Logging differs from the natural disturbance regime by affecting larger areas, occurring more frequently, removing the nitrogen bound in biomass, and increasing the probability of soil erosion. On some sites, planting of nitrogen-fixing alders in association with regenerating Douglas fir can compensate for nitrogen losses during logging (Binkley et al. 1992) and could also reduce erosion. On the other hand, management with alder could have undesirable effects in nitrogen-rich sites, where nitrogen is not limiting, and competition from alders during early succession could reduce the productivity of tree seedlings and potentially lead to higher nutrient losses. Strategies for forest management that embrace ecological principles will recognize inherent variability in ecosystem state factors and interactive controls and will select management practices in a broad environmental context.

Fisheries Management

Formulation of management options for fisheries requires an understanding of ecosystem resilience. Management options include marine reserves, quota systems, new approaches for setting fishing limits based on population sizes, and economic incentives for long-term population maintenance. Unrestricted fish harvest can reduce sustainability by replacing the natural negative feedbacks to population changes with positive feedback responses that drive harvested populations to low levels. Supply-and-demand economics and government subsidies, for example, often maintain or *increase* fishing intensity when fish populations decline (Ludwig et al. 1993, Hilborn et al. 1995, Pauly and Christensen 1995). This contrasts with the *decreasing* predation pressure that would accompany a decline in prey population in an unmanaged ecosystem (Francis 1990).

Management of the North Pacific salmon fishery has instituted a negative feedback on fishing pressure through tight regulation of fishing activity. Commercial and subsistence fishing are allowed only after enough fish have moved into spawning streams to ensure adequate recruitment. This negative feedback to fishing pressure may contribute to the record-high salmon catches from this fishery after 30 years of management (Ludwig et al. 1993). Sustaining the fishery also requires protection of spawning streams from changes in other interactive controls. These include dams that prevent winter floods (disturbance regime), warming of streams by removal of riparian vegetation of logged sites (microclimate), species introductions (functional types), and inputs of silt and nutrients in runoff or municipal sewage (nutrient resources).

A common approach to sustainable management is to harvest only the production in excess of that which would occur when density-dependent mortality limits fisheries stocks (termed **surplus production**) (Rosenberg et al. 1993, Hilborn et al. 1995). The existence and magnitude of surplus production may depend on the stability of interactive controls (e.g., physical environment, nutrients, and predation pressure) and the extent to which these interactive controls respond to changes in fisheries stocks. The major challenge in fisheries management is to estimate surplus production in the face of fluctuating interactive controls and uncertainty in the relationship between these controls and the fish population size. It has been sharply debated whether any ecosystem is sustainable when subjected to continuous human harvest (Ludwig et al. 1993, Rosenberg et al. 1993).

Ecosystem Restoration

Ecosystem restoration often benefits from introduction of positive feedbacks that push the ecosystem to a new, more desirable state. Many ecosystems become degraded through a combination of human impacts, including soil loss, air and water pollution, habitat fragmentation, water diversion, fire suppression, and introduction of exotic species. In degraded agricultural systems and grazing lands, the challenge is to restore them to a sufficiently productive state to provide goods and services to humans. In other cases, the goal is to restore the natural composition, structure, processes, and dynamics of the original ecosystem (Christensen et al. 1996). Advances in restoration practices involve identifying the impediments to recovery of ecosystem structure and function and overcoming these impediments with artificial interventions that often use or mimic natural processes and interactive controls.

Interventions can be applied to any component of ecosystems, but hydrology and soil and plant community characteristics are commonly the focus of effort (Dobson et al. 1997) (Box 16.1). Soil organic matter loss and low soil fertility are common problems in heavily managed agricultural and pasture systems and in forests or grasslands reestablishing on mine wastes. Fertilizers and nitrogen-fixing trees can restore soil nutrients and organic inputs (Bradshaw 1983). Reduced tillage often slows or eliminates losses of SOM. Once soil characteristics are appropriate, plant species can be reintroduced by seeding, planting, or natural immigration (Dobson et al. 1997). The scientific basis for restoration ecology is actively developing. Remaining questions include the steps required to regain the full range of species in restored sites, the importance of the initial composition of species in determining long-term characteristics, and the importance of soil organisms in ecosystem recovery.

Management for Endangered Species

Management for endangered species requires a landscape perspective. The focus of endangered-species protection has generally been

Box 16.1. Everglades Restoration Study

Major human impacts on the natural hydrology of the Everglades ecosystem in the southeastern United States began in the early twentieth century. In response to hurricanes, flooding, and the resulting loss of human life and property, the U.S. Army Corps of Engineers built levees, canals, pumping stations, and water-control structures that separated the remaining Everglades from growing urban areas and divided them into basins (Davis and Ogden 1994). The northern Everglades were partitioned into a series of water conservation areas (WCAs) and the Everglades Agricultural Area south of Lake Okeechobee, which included 1900 km^2 of sugar cane plantations (DeAngelis et al. 1998). The water flow to the remaining "natural" Everglades declined sharply and occurred as pulsed releases by water-control structures. These hydrologic changes caused pronounced fluctuations in water levels and increased the frequency of major drying events (DeAngelis et al. 1998). The survival of many species, including birds, alligators, and crocodiles, depends on reasonable regularity in the rise and fall of water level throughout the year. Since the 1940s, the nesting populations of wading birds declined by 90% (Davis and Ogden 1994). Land use change such as agricultural drainage destroyed many high-elevation, short-hydroperiod wetlands. Other human-induced effects on the system include storm water runoff from the phosphorus-enriched Everglades Agricultural Area, mercury pollution, and invasive species (DeAngelis et al. 1998).

The large-scale changes in the Florida Everglades and associated ecosystems required a landscape approach to restore the natural hydrology on which so much of the flora and fauna depend. The goals of the south Florida ecosystem restoration include the maintenance of ecological processes (e.g., disturbance regimes, hydrologic processes, and nutrient cycles) and maintenance of viable populations of all native species. The U.S. Army Corps of Engineers was charged with both improving protection of Everglades National Park and providing sufficient water to meet the demands of a large urban and agricultural economy. Planned construction projects include the creation of storm water treatment areas to remove phosphorus from the water and to allow increased water diversion into the Everglades (DeAngelis et al. 1998). Additional land is being purchased to provide areas of water storage and a buffer zone between natural areas and the expanding urban zone.

An ecosystem model was developed to evaluate the potential effectiveness of various rehabilitation and management options. This spatially explicit landscape model was linked with individual-based modeling of 10 higher trophic-level indicator species to provide quantitative predictions relevant to the goals of the Everglades Restoration (DeAngelis et al. 1998). These indicator species—including the Florida panther, white ibis, and American crocodile—all differ in their use of the landscape and resources and span a range of habitat needs and trophic interactions (Davis and Ogden 1994). The simultaneous success of all of these species in a restored Everglades would imply health of the overall ecosystem (DeAngelis et al. 1998). In this ecosystem approach, models of higher trophic-level indicator species use information from models at intermediate trophic levels (fish, aquatic macroinvertebrates such as crayfish, and several reptile and amphibian functional types), simulated as size-structured populations, and lower trophic levels (periphyton, aggregated mesofauna, and macrophytes), using process models. These species-specific models are then layered on a landscape geographic information system (GIS) model that includes hydrological and abiotic factors such as surface elevations, vegetation types, soil types, road locations, and water levels (DeAngelis et al. 1998). South Florida provides an example of the incorporation of scientific knowledge of ecosystem processes into long-term state and national ecosystem management efforts.

the establishment of protected areas containing populations of the target species and vegetation associated with those species. Establishment of parks is, however, insufficient protection for species when humans continue to influence important state factors and interactive controls, such as climate, fire regime, water flows, or species introductions (Jensen et al. 1993). If the climate changes, for example, animals may be trapped inside a park that no longer has a suitable climate or vegetation. Selection of parks that have a range of elevations provides an opportunity for organisms to migrate vertically to higher elevations in response to climate warming. Habitat fragmentation and land use change also alter the natural linkages among ecosystems inside and outside of parks. Nearly all parks therefore require management to compensate for human impact. The boundaries of Yellowstone National Park, for example, block migration of elk to traditional wintering areas, so winter food supplements must be provided. These winter food supplements in combination with the extirpation of natural predators release the elk population from their natural population controls. Managers must therefore allow hunting or relocation of elk as an alternative mechanism of population regulation. Using management to replace interactive controls, rather than to protect the interactive controls, is a complex and difficult task, especially when management has multiple, often conflicting goals.

Integrative Approaches to Ecosystem Management

Managing and sustaining ecosystems in a rapidly changing world requires new management approaches that consider ecosystems as interacting components of social and biophysical landscapes and a broader ecological perspective than management focusing on a single species or product. Ecological sustainability cannot be divorced from economic and cultural sustainability. A policy that promotes ecological sustainability at the expense of its human residents cannot be effectively implemented. Conversely, programs of economic development that sacrifice long-term ecological or cultural sustainability cannot be sustained over the long term. An emerging challenge is to address regional sustainability in ways that simultaneously consider the ecological, economic, and cultural costs and benefits of particular policies. Development of a scientific basis for ecosystem sustainability requires close collaboration among ecologists, resource managers, economists, sociologists, anthropologists, and others. Development and implementation of the resulting policies requires involvement of scientists, resource managers, policy makers, land owners, industrial and recreational users, and other stakeholders. This comprehensive **ecosystem approach** (Bengston 1994) uses both the ecological principles outlined in this book and the principles and understanding developed in many fields of social science. Its objectives, scale, and roles for science and management differ significantly from more traditional management approaches (Table 16.1).

In general, an ecosystem approach considers the range of goods and services provided by an ecosystem and manages them in light of their interactions and trade-offs. Management must focus on the scale and pattern of the system, rather than being constrained by jurisdictional boundaries. An ecosystem approach is place based, designed around the traits of the ecosystem and its political and social landscape. Its goal is to sustain or increase the capacity of a system to provide desired goods and services over the long term. Finally, the ecosystem approach integrates social and economic information with biophysical information and explicitly considers the provision of human needs.

The use of an ecosystem approach is evident in the concepts and implementation of ecosystem management plans, integrated conservation and development plans, and approaches that assign value to ecosystem functions.

Ecosystem Management

Ecosystem management is the application of ecological science to resource management to

TABLE 16.1. Differences between traditional forest management and an ecosystem approach to forest management.

Characteristic	Traditional forest management	Forest ecosystem management
Objectives	• maximizes commodity production	• maintains ecosystem integrity, while allowing for sustainable commodity production
	• maximizes net present value	• maintains future options
	• maintains forest harvest at levels less than or equal to their growth or renewal	• aims to sustain ecosystem productivity over time, with short-term considerations of factors such as aesthetics and social acceptability of harvest practices
Scale	• stand scale within political or ownership boundaries	• ecosystem and landscape scale
Role of science	• views forest management as an applied science	• views forest management as combining science and social factors
Role of management	• focuses on outputs demanded by people (e.g., timber, recreation, wildlife)	• focuses on inputs and processes (e.g., diversity and ecological processes) that give rise to outputs
	• strives for management that fits industrial production	• strives for management that mimics natural processes and productivity
	• considers timber the primary output	• considers all species important and considers that services (protecting watersheds, recreation, etc.) are of equal importance with goods timber
	• strives to avoid impending timber shortages	• strives to avoid biodiversity loss and soil degradation
	• views forests as a crop production system	• views forests as a natural ecosystem
	• values economic efficiency	• values cost effectiveness and social acceptability

Adapted from WRI (2000), after Bengston (1994).

promote long-term sustainability of ecosystems and the delivery of essential ecosystem goods and services to society. The concept was adopted by the U.S. Forest Service in 1992 and has since been developing in theory and application. Although there are virtually hundreds of definitions of ecosystem management (Table 16.2), the concept includes a set of common principles: (1) long-term sustainability as a fundamental value; (2) clear, operational goals; (3) sound ecological understanding; (4) understanding of connectedness and complexity; (5)

TABLE 16.2. Selected definitions of ecosystem management.

- Regulating internal ecosystem structure and function, plus inputs and outputs to achieve socially desirable conditions (Agee and Johnson 1987).
- The careful and skillful use of ecological, economic, social, and managerial principles in managing ecosystems to produce, restore, or sustain ecosystem integrity and desired conditions, uses, product, values, and services over the long term (Overbay 1992).
- The strategy by which, in aggregate, the full array of forest values and functions is maintained at the landscape level. Coordinated management at the landscape level, including across ownerships, is an essential component (Society of American Foresters 1993).
- Integration of ecological, economic, and social principles to manage biological and physical systems in a manner that safeguards the ecological sustainability, natural diversity, and productivity of the landscape (Wood 1994).

Modified from Christensen et al. (1996).

recognition of the dynamic character of ecosystems; (6) attention to scale and context; (7) inclusion of humans as a component of ecosystems; and (8) incorporation of adaptive approaches (Christensen et al. 1996). Most integrated ecosystem management programs explicitly facilitate public participation and collaborative decision making.

Long-term sustainability is the fundamental objective of ecosystem management. It is achievable in part via the inclusion of sound ecological models and understanding that incorporate the complex and dynamic character of ecosystems and acknowledge humans as inherent components of ecosystems. Change and uncertainty are intrinsic characteristics of most ecosystems, and ecosystem management is an approach that acknowledges the occurrence of stochastic events as well as predictable variability (Holling 1993). Ecosystem management must therefore be flexible enough to learn from scientific analysis and advances and to adapt to institutional and environmental change.

An ecosystem management approach is especially critical for the management of complex systems such as watersheds and marine fisheries, where management must consider multiple changes and the linkage among ecosystems through the movement of water, air, animals, and plants. The ocean ecosystem that is relevant to salmon fisheries combines freshwater rivers and streams, coastal ecosystems, intermediate continental shelf waters, and the deep ocean, all of which are characterized by complex dynamics that vary in space and time in ways that are poorly understood. The complexity and scales of change in marine ecosystems are only partially understood, including seasonal variations in productivity, regional-scale El Niño climatic events, and long-term changes in salinity and ocean temperature. Large natural fluctuations in the abundances of marine fish are the rule more than the exception. Our limited powers of direct observation also result in a fragmented knowledge of the diversity, abundance, and interactions of marine organisms. One challenge of ecosystem management is to reconcile the disparity between the spatial and temporal scales at which humans make resource management decisions and those at which ecosystem properties operate (Christensen et al. 1996).

Ecosystem management goes beyond a single focus on commodity resources and harvesting limits. Instead, it embraces sustainability as the criterion for commodity provision and/or other uses. Ecosystem management is therefore concerned with multiple functions, thresholds in processes, and trade-offs among different management consequences. It frequently considers, for example, both productivity and biodiversity (Johnson et al. 1996). All ecosystem management projects strive for an integrated understanding and management of the ecological, social, economic, and political aspects of resource use to maximize long-term sustainability (Box 16.2).

Adaptive management, involving experimentation in the design and implementation of policies, is central to effective management of ecosystems. An **adaptive policy** is one that is designed from the outset to test hypotheses about the ways in which ecosystem processes respond to human actions. In this way, if the policy fails, learning occurs, so better policies can be applied in the future. Perhaps as a result of frequent management failures and gaps in scientific knowledge, the concept of adaptive management has become central to the implementation of ecosystem management. One advantage of adaptive management stems from the high degree of uncertainty in real-life complex systems. Instead of delaying timely action due to the lack of certainty, adaptive management promotes the opportunity to learn from management experience. The lack of action in the face of uncertainty can have ecosystem and societal consequences that are at least as great as actions based on reasonable hypotheses about how the ecosystem functions. Hypotheses that underlie adaptive management may consider the probabilities of both desired outcomes and ecological disasters (Starfield et al. 1995). The optimal policy, for example, may be one that has a moderate probability of desirable outcomes and a low probability of causing an ecological disaster.

While hundreds of projects have embraced an ecosystem management perspective, most

Box 16.2. Great Barrier Reef, Australia

Coral reef ecosystems have recently come under increasing consumptive pressure and damage from polluting activities on land. The Great Barrier Reef Marine Park Authority (GBRMPA) in Australia developed an ecosystem management plan to protect this valuable and diverse coral reef based on the principles of ecosystem management (Christensen et al. 1996). The park is a protected area managed for multiple uses with oversight by the GBRMPA. Prohibited activities include oil and gas exploration, mining, littering, spear fishing, and harvesting of large fish. The park is divided into three zones of use intensity: preservation zones, which allow only strictly controlled scientific research; marine national park zones, which allow scientific, educational, and recreational use; and general-use zones, which permit various uses, including recreational and commercial fishing under guidelines established to maintain ecosystem integrity. The preservation zones provide a valuable baseline for understanding and evaluating patterns of change in overall ecosystem behavior. Additional special-use zones protect critical breeding or nesting sites and provide important protection for natural areas or research. An ecotourism strategy has been developed that allows floating structures in special areas for viewing the natural reef free from fishing impacts. Public involvement and education are an essential component to this ecosystem management plan that incorporates public accountability, operational efficiency, minimal regulation, adaptability to changing circumstances, and scientific credibility.

still fall short in implementation. Impediments to success include jurisdictional debates; lack of political will or foresight; conflicting desires for, and trade-offs among, goods and services; and the limitations of scientific information. Nonetheless, adoption of this approach has improved the extent to which management meets its goals and gives optimism that future development of the theory and practice of ecosystem management will be ecologically and societally beneficial (Yaffee et al. 1996, Peine 1999, WRI 2000).

Integrated Conservation and Development Projects

Integrated conservation and development projects (ICDPs) represent a new approach to conservation in the developing world. ICDPs focus equally on biological conservation and human development, typically through externally funded, locally based projects (Wells and Brandon 1993, Kremen et al. 1994). In the past, conservation and development projects typically were considered separately, by different organizations, sometimes with conflicting goals and conflicting consequences (Sutherland 2000). It is now widely accepted that the two directives are more likely to be successful if considered together; the main goal of ICDPs is to link these previously opposing goals. In response to the failure of conservation and development projects to succeed separately, ICDPs emerged in the 1980s and established formal partnerships between conservation organizations and development agencies in an effort to create environmentally sound, economically sustainable alternatives to destructive land use change (Kremen et al. 1994, Alpert 1996).

An important objective of ICDPs is to determine the types, intensities, and distribution of resource use that are compatible with the conservation of biodiversity and the maintenance of ecological processes (Alpert 1996). Most ICDPs therefore have the following characteristics: (1) They link conservation of natural habitats with the improvement of living conditions in the local communities. (2) They are site based and tailored to specific problems such as impending loss of exceptional habitat. (3) They attract international expertise, local support, and external sources of income, and (4) they adapt to conditions in the developing world such as heavy dependence on natural resources, high population growth, and high opportunity costs of protected areas (Alpert 1996). ICDPs often team a nongovernmental organization; a foreign donor agency; a national agency in charge of forestry, wildlife, or parks; and, if possible, local traditional and official leaders. Projects incorporate biological information and scientific knowledge of ecosystem processes, as well as the interests of managers and local communities in their design and implementation.

One major challenge of successful ICDPs is to develop an appropriate research mechanism to collect the scientific data needed to guide the dual objective of conservation and development (Kremen et al. 1994). It is critical to monitor biodiversity and ecosystem processes across space and time and at multiple levels of ecological organization (species, communities, ecosystems, and landscape) and their responses to management (Noss 1990, Kremen et al. 1994). Ecological and socioeconomic indicators can identify the causes and consequences of habitat loss, monitor changes in resource use and harvesting impacts, and evaluate the success of various management programs (Kremen et al. 1994, 1998). A successful monitoring program is essential to test the hypothesis that economic development linked to conservation promotes conservation.

In the 1990s, more than 100 ICDP projects were initiated, including over 50 in at least 20 countries in sub-Saharan Africa (Alpert 1996). A review of African projects concluded that ICDPs do not provide a definitive solution to habitat loss, but they can offer medium-term solutions to local conflicts between biological conservation and natural resource use in economically poor, remote areas of exceptional ecological importance (Alpert 1996). An earlier review of 36 ICDPs worldwide concluded that only 5 of the projects had contributed positively to wildlife conservation (Kremen et al. 1994). Limited tourist revenue potential, lack of local management capacity, political unrest, large human populations, customary rights to land or resources enclosed by reserves, or the absence of an official protected area can pose significant impediments to the success of a project (Alpert 1996). The ICDPs most successful in promoting conservation contain significant community participation, which fosters improved community attitudes toward conservation. As with other kinds of ecosystem management, getting the science right is an essential, but insufficient, step. Over the long run, as we learn from successes and failures, the approaches employed in ICDPs will evolve to address remaining impediments and challenges.

Valuation of Ecosystem Goods and Services

The concept of ecosystem goods and services takes a pragmatic view in asking how changes to ecosystems will affect the products and services that humans derive from ecological systems. Recognizing and estimating the economic value of the broad range of services provided by ecosystems is an important aspect of the ecosystem approach to management (Costanza et al. 1997, Daily 1997). The concept of **goods** and **services** has been particularly important in reminding the public and policy makers that they depend on ecosystems for food, fiber, fuel, pharmaceuticals, and industrial products as well as many services like water purification, mitigation of floods, pollination of crop and wild plants, pest control, composition of the atmosphere, and aesthetic beauty.

The recognition that the world's ecosystems are capital assets, and that they can yield ecosystem goods and services under proper management, has led to a need and an opportunity for ecosystem valuation. Human societies must often choose between alternative uses of the environment. Should a wetland be preserved, used for sewage treatment, or drained and converted to agriculture? Which services should fresh-water systems be managed for (Table 16.3)? Individuals and societies are constantly making decisions about how to use ecosystem goods and services, but these

TABLE 16.3. Examples of services provided by rivers, lakes, aquifers, and wetlands.

Water supply
Drinking, cooking, washing, and other household uses
Manufacturing, thermoelectric power, and other industrial uses
Irrigation of crops, lawns, etc.
Aquaculture
Supply of goods other than water
Fish
Waterfowl
Clams and mussels
Nonextractive or in-stream benefits
Flood control
Transportation
Recreational swimming, boating, etc.
Pollution dilution and water-quality protection
Hydroelectric generation
Bird and wildlife habitat
Enhanced property values
Nonuser values

Data from Postel and Carpenter (1997).

decisions often ignore the value of the resource and assume that the ecosystem service was "free" (Daily et al. 2000).

Valuation of ecosystem services is a way to organize information to help make such decisions (Daily et al. 2000). Valuation of ecosystem services requires sound ecological information and a clear understanding of alternatives and impacts. Ecological understanding is critical, for example, to characterize the services provided by ecosystems and the processes by which they are generated. This information is frequently site specific, so local ecological knowledge is needed. Ecological and economic information must then be integrated to make sound decisions. Several approaches have been used to assign economic value to ecosystem goods and services. These include direct valuation approaches, by which economic worth is assigned according to avoided costs or to market value, and indirect approaches, which estimate the worth of a good or service through surveys and contingent valuation (Goulder and Kennedy 1997) (Table 16.4). Although the methodology is still developing, economic valuations of the provision of goods or services have successfully contributed to decision making about ecosystem management at a variety of scales (Daily et al. 2000) (Box 16.3).

Valuation of goods and services requires both their identification and an assessment of their temporal and spatial variability, their vulnerability to stresses, and the extent to which they can be predicted using simulation models or indicators. Ecologists often have a sense for what allows an ecosystem or landscape to provide a given service, but we have a limited ability to quantify that service and to predict it in the future. When, for example, and under what conditions, will a watershed provide clean water? Do all mangroves protect shorelines?

Valuation of ecosystem services is often applied to a single variable or resource (e.g.,

TABLE 16.4. Examples of ecosystem services and valuation methods.

Service	Valuation method
Inputs that support production	
Pest control	avoided cost
Flood control	avoided cost
Soil fertility	avoided cost
Water filtration	avoided cost
Sustenance of plants and animals	
Consumptive uses	direct valuations based on market prices
Nonconsumptive uses	indirect valuations (travel cost or contingent aluation methods)
Provision of existence values	indirect valuations (contingent valuation method)
Provision of option values	empirical assessments of individual risk aversion

Data from Goulder and Kennedy (1997).

Box 16.3. Water Purification for New York City

New York City has a long tradition of clean water. This water, which originates in the Catskill Mountains, was once bottled and sold because of its high purity. In recent years, the Catskills natural ecological purification system has been overwhelmed by sewage and agricultural runoff, causing the water to drop below accepted health standards. The cost of a filtration plant to purify this water was estimated at $6 to $8 billion in capital costs, plus annual operating costs of $300 million, a high price to pay for what once could be obtained for free.

This high cost prompted investigation of the cost of restoring the integrity of the watershed's natural purification services. The cost of this environmental solution was approximately $1 billion to purchase and halt development on critical lands within the watershed, to compensate land owners for restrictions on private development, and to subsidize the improvement of septic systems. The great cost savings provided by ecosystem services was selected by the city as the preferred alternative. This choice provided additional valuable services including flood control and sequestration of carbon in plants and soils.

Excerpted from the President's Committee of Advisors on Science and Technology (1998).

fresh-water supply) and does not integrate across the numerous goods and services that occur in an ecosystem. A more comprehensive approach is essential to estimate the trade-offs among management strategies that maximize different ecosystem goods and services. This problem is clearly illustrated in efforts to limit the environmental costs of fertilization in agriculture. In the midwestern United States, wetlands are often used to reduce fertilizer loss from agriculture, due to their capacity to filter nutrients and sediments from laterally moving groundwater and surface water. A singular focus on wetlands and the provision of clean water in rivers, however, prevents a complete analysis of the environmental effects of fertilization and the steps needed to reduce them. Fertilizer nitrogen is also lost to the atmosphere, causing global and regional air pollution and is lost to aquifers, where it can affect the quality of drinking water. Researchers are only beginning to ask which good or service is being valued and protected, and at what cost to others (Naylor and Drew 1998).

Despite these and other limitations, the valuation of ecosystem services is becoming an important tool that contributes to sustaining ecosystems. The protection of highly valued and well-understood services (such as clean water) through the protection of ecosystems is increasingly viewed as a wise alternative to expensive construction and engineering projects (Box 16.3). With increasing knowledge, the benefits of protecting the less-known ecosystem services will become more widely recognized (Daily and Ellison 2002).

Summary

Human activities influence all ecosystems on Earth. Ecosystems are directly impacted by activities such as resource harvests, land conversion, and management and are indirectly influenced by human-caused changes in atmospheric chemistry, hydrology, and climate. Because human activities strongly influence most of Earth's ecosystems, it follows that we should also take responsibility for their care and protection. Part of that responsibility must be to slow the rate and extent of global changes in climate, biogeochemical cycles, and land use. In addition, active management of all ecosystems is required to maintain populations, species, and ecosystem functions in the face of anthropogenic change and to sustain the provision of goods and services that humans receive from them.

State factors and interactive controls exert such strong control over ecosystem processes that changes in these controlling factors inevitably alter ecosystems and reduce the extent to which their current properties can be sustained. Management practices can, however, strongly influence the degree of sustainability. If the goal of management is to enhance sustainability of managed and unmanaged ecosystems, then state factors and interactive controls must be conserved as much as possible and negative feedbacks, which contribute to maintaining these controls, must be strengthened within and among ecosystems. Directional changes in many of these ecosystem controls heighten the challenge of sustainably managing natural resources and threaten the sustainability of natural ecosystems everywhere.

The ecosystem approach to management applies ecological understanding to resource management to promote long-term sustainability of ecosystems and the delivery of essential ecosystem goods and services to society. This requires a landscape or regional perspective to account for interactions among ecosystems and explicitly includes humans as components of this regional system. Ecosystem management acknowledges the importance of stochastic events and our inability to predict future conditions with certainty. Adaptive management is a commonly used approach to ecosystem management that takes actions based on hypotheses of how management will affect the ecosystem. According to the results of these experiments, management policies are modified to improve sustainability.

ICDPs apply adaptive management to conservation in the developing world. ICDPs focus equally on biological conservation and human development. The main goal of ICDPs is to link these often previously opposing goals, based on the assumption that local human populations will place immediate socioeconomic security before conservation concerns. Fundamental principles underlying ecosystem management in general, and ICDPs in particular, are that people are an integral component of regional systems and that planning for a sustainable future requires solutions that are ecologically, economically, and culturally sustainable.

Review Questions

1. What are the major direct and indirect effects of human activities on ecosystems? Give examples of the magnitude of human impacts on ecosystems.
2. How does the resilience of an ecosystem influence its sustainability in the face of human-induced environmental change? What ecological properties of ecosystems influence their sustainability?
3. Describe a management approach that would maximize ecosystem sustainability. What factors or events are most likely to cause this management approach to fail?
4. What are ecosystem goods and services? How can an understanding of ecosystem services be used in management decisions?
5. What is ecosystem management? What is the role of humans in ecosystems in the context of ecosystem management?
6. What are the advantages and disadvantages of adaptive management as an approach to managing ecosystems?
7. What have ICDPs taught us about the advisability of including humans as components of ecosystems?

Additional Reading

Alpert, P. 1996. Integrated conservation and development projects: Examples from Africa. *BioScience* 46:845–855.

Carpenter, S. R., and J.F. Kitchell, editors. 1993. *The Trophic Cascade in Lakes*. Cambridge University Press, Cambridge, UK.

Chapin, F.S. III, M.S. Torn, and M. Tateno. 1996. Principles of ecosystem sustainability. *American Naturalist* 148:1016–1037.

Christensen, N.L., A.M. Bartuska, J.H. Brown, S. Carpenter, C. D'Antonio, R. Francis, J.F. Franklin, J.A. MacMahon, R.F. Noss, D.J. Parsons, C.H. Peterson, M.G. Turner, and R.G. Woodmansee. 1996. The report of the Ecological Society of America committee on the scientific basis for ecosystem management. *Ecological Applications* 6:665–691.

Daily, G.C. 1997. *Nature's Services: Societal Dependence on Natural Ecosystems*. Island Press, Washington, DC.

Holling, C.S. 1986. Resilience of ecosystems: Local surprise and global change. Pages 292–317 *in* W.C. Clark, and R.E. Munn, editors. *Sustainable Development and the Biosphere*. Cambridge University Press, Cambridge, UK.

Lubchenco, J., A.M. Olson, L.B. Brubaker, S.R. Carpenter, M.M. Holland, S.P. Hubbell, S.A. Levin, J.A. MacMahon, P.A. Matson, J.M. Melillo, H.A. Mooney, C.H. Peterson, H.R. Pulliam, L.A. Real, P.J. Regal, and P.G. Risser. 1991. The sustainable biosphere initiative: An ecological research agenda. *Ecology* 72:371–412.

Matson, P.A., W.J. Parton, A.G. Power, and M.J. Swift. 1997. Agricultural intensification and ecosystem properties. *Science* 227:504–509.

Sutherland, W.J. 2000. *The Conservation Handbook: Research, Management and Policy*. Blackwell Scientific, Oxford, UK.

Abbreviations

a_n	nutrient productivity
A_n	quantity of energy or material assimilated by a trophic level
ABA	abscisic acid
ADP	adenosine diphosphate
AM	arbuscular mycorrhizae
ATP	adenosine triphosphate
B	bacteria
b	buffer capacity of soil
B_{animal}	biomass of animals
B_{microb}	biomass of soil microbes
B_{plant}	biomass of plants
BS	bundle sheath cell
C	degrees Celsuis; carnivore
c	bulk soil phosphorus concentration
C_1	primary carnivore
C_2	secondary carnivore
C_3	related to the photosynthetic pathway whose initial carboxylation products are three-carbon sugars
C_4	related to the photosynthetic pathway whose initial carboxylation products are four-carbon acids
C_p	specific heat at constant pressure
$\%C_{S1}$	percentage of soil carbon derived from the initial vegetation type
C_{S2}	^{13}C content of second soil type
$^{13}C_{std}$	^{13}C content of a standard
C_{V1}	^{13}C content of vegetation from the initial vegetation type
C_{V2}	^{13}C content of vegetation from the second vegetation type
CAM	crassulacean acid metabolism
CEC	cation exchange capacity
CFC	chlorofluorocarbon
cm	centimeter (10^{-2} m)
C:N	carbon:nitrogen ratio
CPOM	coarse particulate organic matter
d	day
D	deuterium
D	diffusion coefficient of phosphorus
DDT	an insecticide
DIC	dissolved inorganic carbon
DMS	dimethylsulfoxide
DNA	deoxyribonucleic acid
DOC	dissolved organic carbon
DON	dissolved organic nitrogen
E	rate of evapotranspiration of an ecosystem
e	root elongation rate
E_{assim}	assimilation efficiency
$E_{consump}$	consumption efficiency
E_{prod}	production efficiency
E_{troph}	trophic efficiency
ENSO	El Niño southern oscillation
F	fungi
$F_{anim\text{-}soil}$	flux of carbon from animals to the soil in feces and dead animals
F_{CH_4}	methane emission from the ecosystem to the atmosphere
$F_{disturb}$	flux of carbon from an ecosystem to the atmosphere due to disturbance
F_{emiss}	emission to the atmosphere of volatile organic compounds by plants
F_{harv}	flux of carbon from an ecosystem to the atmosphere due to human harvest

F_{herbiv}	consumption of plants by animals
$F_{lateral}$	lateral flux of carbon into or out of an ecosystem
F_{leach}	flux of carbon from an ecosystem to groundwater by leaching
$F_{micro\text{-}anim}$	consumption of microbial biomass by animals
F_n	component of the gravitational force that is normal to the slope and therefore contributes to friction that resists erosion
F_p	component of the gravitational force that is parallel to the slope and therefore drives erosion
$F_{pl\text{-}fire}$	flux from plants to the atmosphere due to combustion during fires
$F_{pl\text{-}soil}$	flux from plants to soil
$F_{soil\text{-}fire}$	flux of carbon from dead organic matter to the atmosphere due to combustion during fires
F_t	total gravitational force
FPAR	fraction of photosynthetically active radiation
FPOM	fine particulate organic matter
g	gram
g	gravitational acceleration
G	ground heat flux
GBRMPA	Great Barrier Reef Marine Park Authority
GIS	geographic information system
GPP	gross primary production
h	hour
h	height (m)
H	sensible heat flux
H	herbivore
HNLC	high-nitrogen, low-chlorophyll
I	irradiance at any point in the canopy
I_0	irradiance at the top of the canopy
I_{max}	maximum uptake rate
I_n	quantity of energy or material ingested by trophic level n
ICDP	integrated conservation and development project
ITCZ	intertropical convergence zone
J	joule
J_p	rate of water flow through plants
J_s	rate of water flow through the soil
k	extinction coefficient; decomposition constant
K	degrees Kelvin
K	equilibrium constant; shortwave radiation
K_{in}	incoming shortwave radiation
K_m	affinity of roots for phosphorus
K_{out}	outgoing shortwave radiation
kg	kilogram (10^3 g)
kJ	kilojoule (10^3 J)
km	kilometer (10^3 m)
kPa	kilopascal (10^3 Pa)
L	longwave radiation; latent heat of vaporization
L	liter
l	length
L_{in}	incoming longwave radiation
L_{out}	outgoing longwave radiation
L_p	hydraulic conductivity of plant xylem
L_s	hydraulic conductivity of soil
L_t	litter mass at time t
L_0	litter mass at time zero
LAI	leaf area index
LE	latent heat flux
LUE	light use efficiency
m	meter
m	mass
M	microbivore
M_a	angular momentum
Mes	mesophyll
Mg	megagram (10^6 g)
mg	milligram (10^{-3} g)
MJ	megajoule (10^6 J)
mL	milliliter (10^{-3} L)
mm	millimeter (10^{-3} m)
MPa	megapascal (10^6 Pa)
MRT	mean residence time
N	north
n	sample size
N_{avail}	available nitrogen; available nutrients
NADP	nicotinamide adenine dinucleotide phosphate (in its oxidized form)
NADPH	nicotinamide adenine dinucleotide phosphate (in its reduced form)
NAO	North Atlantic oscillation
NBP	net biome production
NDVI	normalized difference vegetation index

NEE	net ecosystem exchange of CO_2 between the ecosystem and the atmosphere
NEE_{dark}	net ecosystem exchange measured in the dark
NEE_{light}	net ecosystem exchange measured in the light
NEP	net ecosystem production
NIR	near infrared radiation
nm	nanometer (10^{-9} m)
nmol	nanomole (10^{-9} mole)
NO_x	nitric oxides in general (includes NO and NO_2)
N:P	nitrogen to phosphorus ratio
NPP	net primary production
NUE	nitrogen use efficiency
O	organic
O_a	highly decomposed organic horizon of soil
O_e	moderately decomposed organic horizon of soil
O_i	slightly decomposed organic horizon of soil
P	pressure; precipitation
P	production
P_{area}	photosynthetic rate (per unit leaf area)
P_{mass}	photosynthetic rate (per unit leaf mass)
Pa	pascal
PAR	photosynthetically active radiation
PBL	planetary boundary layer
PCB	polychlorinated biphenyl (an industrial class of compounds containing chlorine)
PEP	phosphoenolpyruvate
Pg	pedagram (10^{15} g)
pH	negative log of H^+ activity
PNA	Pacific North America pattern
POC	particulate organic carbon
PON	particulate organic nitrogen
ppmv	parts per million by volume
ppt	parts per thousand
$Prod_n$	production by trophic level n
$Prod_{n-1}$	Production at the preceeding tropic level
Q_{10}	proportional increase in the rate of a process with a 10°C increase in temperature
r	radius
R	runoff; radiation; respiration
R	regolith; universal gas constant
R^*	phosporus uptake threshold
R_{animal}	animal respiration
$R_{ecosyst}$	ecosystem respiration
R_{growth}	growth respiration
$R_{heterotr}$	heterotrophic respiration
R_{ion}	respiration associated with ion uptake
R_{mi}	mitochondrial respiration
R_{microb}	microbial respiration
R_{maint}	maintenance respiration
R_{net}	net radiation
R_{plant}	plant respiration
R_{sam}	isotope ratio of a sample
R_{std}	isotope ratio of a standard
Re	Reynolds number
RH	relative humidity
Rubisco	ribulose-bis-phosphate carboxylase
RuBP	ribulose-bis-phosphate
s	second
S	heat storage by a surface
S	water storage by an ecosystem; south
SE	standard error
SLA	specific leaf area
SOM	soil organic matter
SRL	specific root length
t	time
t_r	residence time
T	temperature; transpiration rate; absolute temperature
TEM	Terrestrial Ecosystem model
Tg	teragram (10^{12} g)
UV	ultraviolet
v	velocity
V_k	kinematic viscosity
VAM	vesicular arbuscular mycorrhizae
VIS	visible radiation
VPD	vapor pressure deficit
W	watt
WCA	water conservation area
WUE	water use efficiency
yr	year
α	albedo
δ	del; difference in isotope concentration relative to a standard
Δ	change in a quantity

ε	emissivity
μg	microgram (10^{-6} g)
μL	microliter (10^{-6} L)
μm	micrometer (10^{-6} m)
μmol	micromole (10^{-6} moles)
ρ	density
σ	Stefan-Boltzman constant
Ψ_m	matric potential
Ψ_o	osmotic potential
Ψ_p	pressure potential
Ψ_{plant}	total plant water potential
Ψ_{soil}	total soil water potential
Ψ_t	total water potential

Glossary

A horizon. Uppermost mineral horizon of soils.

Abiotic. Not directly caused or induced by organisms.

Abiotic condensation. Nonenzymatic reaction of quinones with other organic materials in soil.

Abscisic acid. Plant hormone that is transported from roots to leaves and causes a reduction in stomatal conductance.

Absolute humidity. Vapor density.

Absorbence. Fraction of the global solar irradiance incident on a surface that is absorbed.

Absorbed photosynthetically active radiation. Visible light (400 to 700 nm) absorbed by plants.

Acclimation. Morphological or physiological adjustment by an individual plant to compensate for the change in performance caused by a change in one environmental factor (e.g., temperature).

Accumulation. Buildup of storage products resulting from an excess supply over demand.

Acid rain. Rain that has low pH, due to high concentrations of sulfuric and nitric acid released from combustion of fossil fuels.

Active transport. Energy-requiring transport of ions or molecules across a membrane against an electrochemical gradient.

Activity budget. Proportion of time that an animal spends in various activities.

Actual evapotranspiration. Annual evapotranspiration at a site; a climate index that integrates temperature and moisture availability.

Actual vegetation. Vegetation that actually occurs on a site.

Adaptation. Genetic adjustment by a population to maximize performance in a particular environment.

Adaptive management. Management involving experimentation in the design and implementation of policies so subsequent management can be modified based on learning from these experiments.

Adiabatic lapse rate. Change in temperature experienced by a parcel of air as it moves vertically in the atmosphere due to a change in atmospheric pressure. The dry adiabatic lapse rate is the change in temperature that occurs if the air does not exchange energy with the surrounding air (about $9.8°C\,km^{-1}$). The moist adiabatic lapse rate also includes temperature changes due to release of latent heat as water vapor condenses. The observed lapse rate varies regionally, depending on surface heating and atmospheric moisture but averages about $6.5°C\,km^{-1}$.

Advection. Net horizontal transfer of gases or water.

Aerobic. Occurring in the presence of oxygen.

Aerodynamic conductance. Boundary layer conductance of a canopy.

Aerosol. Small (0.1 to $10\,\mu m$) particles suspended in air.

Aggregate. Clumps of soil particles bound together by polysaccharides, fungal hyphae, or minerals.

Albedo. Fraction of the incident shortwave radiation reflected from a surface.

Alfisol. Soil order that develops beneath temperate and subtropical forests, characterized by less leaching than spodosol.

Allocation. Proportional distribution of photosynthetic products or newly acquired nutrients among different organs or functions in a plant.

Allochthonous input. Input of energy and nutrients from outside the ecosystem; synonymous with *subsidy*.

Allometric relationship. Regression relationship that describes the biomass of some part of an organism as a function of some easily measured parameter (e.g., plant biomass as a function of stem diameter and height).

Ammonification. Conversion of organic nitrogen to ammonium due to the breakdown of litter and soil organic matter; synonymous with *nitrogen mineralization*.

Amorphous minerals. Minerals with no regular arrangements of atoms.

Anadromous. Life cycle in which reproduction occurs in lakes, streams, or rivers while the adult phase occurs primarily in the ocean.

Anaerobic. Occurring in the absence of oxygen.

Andisol. Soil order characterized by young soils on volcanic substrates.

Angular momentum. Force possessed by a rotating body.

Anion. Negatively charged ion.

Anion exchange capacity. Capacity of a soil to hold exchangeable anions on positively charged sites at the surface of soil minerals and organic matter.

Anthropocene. Geologic epoch characterized by human impacts, initiated with the Industrial Revolution.

Anthropogenic. Resulting from or caused by people.

Arbuscular mycorrhizae. Mycorrhizae that exchange carbohydrates between plant roots and fungal hyphae via arbuscules; also termed vesicular arbuscular mycorrhizae or endomycorrhizae.

Arbuscules. Exchange organs between plant and mycorrhizal fungus that occur within plant cells.

Aridisol. Soil order that develops in arid climates.

Aspect. Compass direction in which a slope faces.

Assimilation. Incorporation of an inorganic resource (e.g., CO_2 or NH_4^+) into organic compounds; transfer of digested food from the intestine to the bloodstream of an animal.

Assimilation efficiency. Proportion of ingested energy that is assimilated into the bloodstream of an animal.

Assimilatory nitrate reduction. Conversion of nitrate to amino acids by soil microbes.

Autochthonous production. Production occurring within the ecosystem.

Autotroph. Organism that produces organic matter from CO_2 and environmental energy rather than by consuming organic matter produced by other organisms. Most produce organic matter by photosynthesis; synonymous with *primary producer*.

Available energy. Absorbed energy that is not stored or conducted into the ground; it is the energy available for turbulent exchange with the atmosphere.

B horizon. Soil horizon with maximum accumulation of iron and aluminum oxides and clays.

Backscatter. Reflection from small particles.

Base cations. Nonhydrogen, nonaluminum cations.

Base flow. Background stream flow from groundwater input in the absence of recent storm events.

Base saturation. Percentage of the total exchangeable cation pool that is accounted for by base cations.

Benthic. Associated with aquatic sediments.

Biofilm. Microbial community embedded in a matrix of polysaccharides secreted by bacteria.

Biogenic. Biologically produced.

Biogeochemical cycling. Biologically mediated cycling of materials in ecosystems.

Biogeochemistry. Biologically influenced chemical processes in ecosystems.

Biological pump. Flux of carbon and nutrients in feces and dead organisms from the euphotic zone to deeper waters and the sediments of the ocean.

Biomass. Quantity of living material (e.g., plant biomass).

Biomass burning. Combustion of plants and soil organic matter following forest clearing.

Biomass pyramid. Quantity of biomass in different trophic levels of an ecosystem.

Biome. General class of ecosystems (e.g., tropical rain forest, arctic tundra).

Biosphere. Biotic component of Earth, including all ecosystems and living organisms.

Biotic. Caused or induced by organisms.

Bloom. Rapid increase in phytoplankton biomass.

Bottom–up controls. Regulation of consumer populations by quantity and quality of food.

Bottom water. Deep ocean water below about 1000 m depth.

Boundary layer. Thin layer around a leaf or root in which the conditions differ from those in the bulk atmosphere or soil, respectively.

Boundary layer conductance. Conductance of water vapor across the boundary of a leaf or canopy; canopy boundary layer conductance is also termed *aerodynamic conductance*.

Bowen ratio. Ratio of sensible to latent heat flux.

Brine rejection. Exclusion of salt during formation of ice crystals in sea ice.

Bulk density. Mass of dry soil per unit volume.

Buffering capacity. Capacity of the soil to release cations to replace ions lost by uptake or leaching.

Bulk density. Mass of soil per unit volume.

Bulk soil. Soil outside the rhizosphere.

Bundle sheath cells. Cells surrounding the vascular bundle of a leaf; site of C_3 photosynthesis in C_4 plants.

C horizon. Soil horizon that is relatively unaffected by the soil forming processes.

C_3 photosynthesis. Photosynthetic pathway in which CO_2 is initially fixed by Rubisco, producing three-carbon sugars.

C_4 photosynthesis. Photosynthetic pathway in which CO_2 is initially fixed by PEP carboxylase during the day, producing four-carbon organic acids.

Calcic horizon. Hard calcium (or magnesium) carbonate-rich horizon formed in deserts; formerly termed *caliche*.

Caliche. Calcic horizon.

Canopy closure. Time during succession at which crowns of adjacent trees overlap to produce a relatively uniform canopy.

Canopy conductance. Measure of the physiological controls over water vapor transfer from the ecosystem to the atmosphere. It equals the average stomatal conductance of individual leaves times LAI.

Canopy interception. Fraction of precipitation that does not reach the ground.

Carbon-based defense. Organic compounds that contain no nitrogen and defend plants against pathogens and herbivores.

Carbon-fixation reactions. Those reactions in photosynthesis that use the products of the light-harvesting reactions to reduce CO_2 to sugars.

Carboxylase. Enzyme that catalyzes the reaction of a substrate with CO_2.

Carboxylation. Attachment of CO_2 to an acceptor molecule.

Carnivore. Organism that eats live animals.

Carrier. Protein involved in ion transport across a membrane.

Catalyst. Molecule that speeds the conversion of substrates to products.

Catena. Sequence of soils or ecosystems between crests of hills and floors of adjacent drainages, whose characteristics change from point to point, depending on drainage and other land-surface processes.

Cation. Positively charged ion.

Cation exchange capacity. Capacity of a soil to hold exchangeable cations on negatively charged sites at the surface of soil minerals and organic matter.

Cavitation. Breakage of water columns under tension in the xylem.

Cellobiase. Enzyme that breaks down cellobiose to form glucose.

Cellobiose. Organic compound composed of two glucose units formed by cellulose breakdown.

Charge density. Charge per unit hydrated volume of the ion.

Chelation. Reversible combination, usually with high affinity, with a metal ion (e.g., iron, copper).

Chemical alteration. Chemical changes in dead organic matter during decomposition.

Chemical weathering. Changes due to chemical reactions between the materials and the atmosphere or water.

Chemodenitrification. Abiotic conversion of nitrite to nitric oxide (NO).

Chemosynthesis. Synthesis of organic matter fueled by oxidation–reduction reactions unrelated to photosynthesis.

Chlorofluorocarbon. Organic chemicals containing chlorine and/or fluorine; gases that destroy stratospheric ozone.

Chlorophyll. Green pigment involved in light capture by photosynthesis.

Chloroplast. Organelles that carry out photosynthesis.

Chronosequence. Sites that are similar to one another with respect to all state factors except time since disturbance.

Circadian rhythms. Innate physiological cycles in organisms that have a period of about 24h.

Clay. Soil particles less than 0.002mm diameter.

Climate modes. Relatively stable configurations of global atmospheric circulation.

Climate system. Interactive system made up of the atmosphere, hydrosphere, biosphere, cryosphere, and land surface.

Climatic climax. End point of succession that is determined only by climate.

Climax. End point of succession where the structure and rates of ecosystem processes reach steady state and where resource consumption by vegetation is balanced by the rate of resource supply.

Closed system. System in which the internal transfers of substances are much greater than inputs and outputs.

Cloud condensation nuclei. Aerosols around which water vapor condenses to form clouds.

C:N ratio. Ratio of carbon mass to nitrogen mass.

CO_2 compensation point. CO_2 concentration at which net photosynthesis equals zero.

Coarse particulate organic matter. Organic matter in aquatic ecosystems, including leaves and wood, that is larger than 1mm diameter.

Collector. Benthic macroinvertebrate that feeds on fine organic particles; includes filtering collectors that consume suspended particles and gathering collectors that consume deposited particles.

Co-metabolism. Breakdown of a substrate by a series of enzymes that are produced by different microbes.

Community. Group of co-existing organisms in an ecosystem.

Compensation. Increased growth of some species in a community, due to release of resources, in response to reduced growth by other species.

Compensation point. Temperature, CO_2 concentration or light level at which net carbon exchange by a leaf is zero (i.e., photosynthesis equals respiration).

Competition. Interactions among organisms that use the same limiting resources (resource competition) or that harm one another in the process of seeking a resource (interference competition).

Complementary resource use. Use of resources that differ in type, depth, or timing by co-occurring species.

Conductance. Flux per unit driving force (e.g., concentration gradient); inverse of resistance.

Configuration. Spatial arrangement of patches in a landscape.

Connectivity. Degree of connectedness among patches in a landscape.

Consortium. Group of genetically unrelated bacteria, each of which produce only some of the enzymes required to break down complex molecules.

Consumer. Organism that meets its energetic and nutritional needs by eating other living organisms.

Consumption efficiency. Proportion of the production at one trophic level that is ingested by the next tropic level.

Convection. Heat transfer by turbulent movement of a fluid (e.g., air or water).

Coriolis effect. Tendency, due to Earth's rotation, of objects to be deflected to the right in the Northern Hemisphere and to the left in the Southern Hemisphere.

Cortex. Layers of root cells outside the endodermis involved in nutrient uptake.

Coupling. Effectiveness of atmospheric mixing between the canopy and the atmosphere.

Crassulacean acid metabolism. Photosynthetic pathway in which stomates open and carbon is fixed at night into four-carbon acids. During the day stomates close, C_4 acids are decarboxylated, and CO_2 is fixed by C_3 photosynthesis.

Crystalline minerals. Minerals with highly regular arrangements of atoms.

Cytoplasm. Contents of a cell that are contained within its plasma membrane but outside the vacuole and the nucleus.

Deciduous. Shedding leaves in response to specific environmental cues.

Decomposer. Organism that breaks down dead organic matter and consumes the resulting energy and nutrients for its own production.

Decomposition. Breakdown of dead organic matter through fragmentation, chemical alteration, and leaching.

Decomposition constant. Constant that describes the exponential breakdown of a tissue.

Decoupling coefficient. Measure of the extent to which the canopy is decoupled from the atmosphere.

Deep water. Ocean water greater than 1000 m depth.

Deforestation. Conversion of forest to a nonforest ecosystem type.

Demand. Requirement; used in the context of the control of the rate of a process (e.g., nutrient uptake) by the amount needed.

Denitrification. Conversion of nitrate to gaseous forms (N_2, NO, and N_2O).

Deposition. Atmospheric input of materials to an ecosystem.

Detritivore. Organism that derives energy from breakdown of dead organic matter.

Detritus. Dead plant and animal material, including leaves, stems, roots, dead animals, and animal feces.

Detritus-based trophic system. Organisms that consume detritus or energy derived from detritus.

Diffuse radiation. Radiation that is scattered by particles and gases in the atmosphere.

Diffusion. Net movement of molecules or ions along a concentration gradient due to their random kinetic activity.

Diffusion shell. Zone of nutrient depletion around individual roots caused by active nutrient uptake at the root surface and diffusion to the root from the surrounding soil.

Direct radiation. Radiation that comes directly from the sun without scattering or reradiation by the atmosphere or objects in the environment.

Discrimination. Preferential reaction with the lighter isotope of an element or compound containing that element.

Dissolved organic carbon. Water-soluble organic carbon compounds.

Dissolved organic nitrogen. Water-soluble organic nitrogen compounds.

Disturbance. Relatively discrete event in time and space that alters the structure of populations, communities, and ecosystems and causes changes resource availability or the physical environment.

Disturbance regime. The range of severity, frequency, type, size, timing, and intensity of disturbances characteristic of an ecosystem or region.

Disturbance severity. Magnitude of change in resource supply or environment caused by a disturbance.

Doldrums. Region near the equator with light winds and high humidity.

Down regulation. Decrease in capacity to carry out a reaction; for example down regulation of CO_2 uptake in response to elevated CO_2.

Downwelling. Downward movement of surface ocean water, due to high density associated with high salinity and low temperature.

Drift. Invertebrates that move downstream in flowing water.

E horizon. Heavily leached horizon beneath the A horizon; formed in humid climates.

Eccentricity. Degree of ellipticity of Earth's orbit around the sun.

Ecosystem. Ecological system consisting of all the organisms in an area and the physical environment with which they interact.

Ecosystem approach. Management of ecosystem goods and services provided by ecosystems in light of their interactions and trade-offs.

Ecosystem ecology. Study of the interactions between organisms and their environment as an integrated system.

Ecosystem engineer. Organisms that alter resource availability by modifying the physical properties of soils and litter.

Ecosystem good. Substance produced by an ecosystem and used by people (e.g., oxygen, food, or fiber).

Ecosystem management. Application of ecological science to resource management to promote long-term sustainability of ecosystems and the delivery of essential ecosystem goods and services to society.

Ecosystem processes. Inputs or losses of materials and energy to and from the ecosystem and the transfers of these substances among components of the system.

Ecosystem respiration. Sum of plant respiration and heterotrophic respiration.

Ecosystem service. Societally important consequences of ecosystem processes (e.g., water purification, mitigation of floods, pollination of crops).

Ectomycorrhizae. Mycorrhizal association in some woody plants in which a large part of the fungal tissue is found outside the root.

Efficiency. Rate of a process per unit plant resource.

El Niño. Warming of surface water throughout the central and eastern tropical Pacific Ocean.

Electron-transport chain. Series of membrane-bound enzymes that produce ATP and NADPH as a result of passing electrons down an electropotential gradient.

Emergent properties. Properties of organisms, communities, or ecosystems that are not immediately obvious from study of processes at finer levels of organization.

Emissivity. Coefficient that describes the maximum rate at which a body emits radiation, relative to a perfect (black body) radiator, which has a value of 1.0.

Endocellulase. Enzyme that breaks down the internal bonds to disrupt the crystalline structure of cellulose.

Endodermis. Layer of suberin-coated cells between the cortex and xylem of roots; water penetrates this layer only by moving through the cytoplasm of these cells.

Endomycorrhizae. Mycorrhizal association in many herbaceous species and some trees in which a large part of the fungal tissue is found inside the root; also termed *arbuscular mycorrhizae*.

Energy pyramid. Quantity of energy transferred between successive trophic levels.

Entisols. Soil order characterized by minimal soil development.

Enzyme. Organic molecule produced by an organism that catalyzes a chemical reaction.

Epidermis. Layer of cells on the surface of a leaf or root.

Equilibrium. State of balance between opposing forces.

Estuary. Coastal ecosystem where a river mixes with seawater.

Euphotic zone. Uppermost layer of water in aquatic ecosystems where there is enough light to support photosynthesis.

Eutrophic. Nutrient rich.

Eutrophication. Nutrient-induced increase in productivity.

Evapotranspiration. Water loss from an ecosystem by transpiration and surface evaporation.

Evergreen. Retention of green leaves throughout the year.

Exocellulase. Enzyme that cleaves off disaccharide units from the ends of cellulose chains, forming cellobiose.

Exodermis. Layer of suberin-coated cells just beneath the epidermis of roots of some species.

Exoenzyme. Enzyme that is secreted by an organism into the environment.

Extensification. Expansion of the aerial extent of land cover change due to human activities.

Extinction coefficient. Constant that describes the exponential decrease in irradiance through a canopy.

Exudation. Secretion of soluble organic compounds by roots into the soil.

Facilitation. Processes by which some species make the environment more favorable for the growth of other species.

Fast variable. Variable that changes rapidly.

Feedback. Response in which the product of one of the final steps in a chain of events affects one of the first steps in this chain; fluctuations in rate or concentration are minimized with negative feedbacks or amplified with positive feedbacks.

Fermentation. Anaerobic process that breaks down labile organic matter to produce organic acids and CO_2.

Ferrell cell. Atmospheric circulation cell between 30° and 60° N or S latitude.

Field capacity. Water held by a soil after gravitational water has drained.

Filter feeder. Aquatic animal that feeds on suspended particles.

Fine particulate organic matter. Particulate organic matter in aquatic ecosystems that is smaller than 1 mm diameter.

Fire intensity. Rate of heat production.

Fixation. Covalent binding of an ion to a mineral surface.

Flux. Flow of energy or materials from one pool to another.

Food chain. Group of organisms that are linked together by the linear transfer of energy and nutrients from one organism to another.

Food web. Group of organisms that are linked together by the transfer of energy and nutrients that originates from the same source.

Forward modeling. Modeling that estimates the outputs of a simulation model based on the temporal and spatial patterns of inputs.

Fractionation. Preferential incorporation of a light isotope (e.g., ^{12}C vs. ^{13}C).

Fragmentation. Breaking up of intact litter into small pieces.

Fulvic acids. Humic compounds that are relatively water soluble due to their extensive side chains and many charged groups.

Functional type. A group of species that is similar with respect to their impacts on community or ecosystem processes (effects functional type); functional types have also been defined with respect to their similarity of response to a given environmental change, such as elevated CO_2 (response functional types).

Gap-phase succession. Succession that occurs in small patches within a stand due to death of individual plants or plant parts.

Gelisol. Soil order characterized by presence of permafrost.

Generalist herbivore. Herbivore that is relatively nonselective in its choice of plant species.

Geotropism. Growth response of plant organs with respect to gravity.

Gley soil. Blue-gray soil due to loss of ferric iron; formed under anaerobic conditions.

Graminoid. Grasslike plant (grasses, sedges, and rushes).

Grazer. Herbivore that consumes herbaceous plants (terrestrial ecosystems) or periphyton (aquatic ecosystems).

Grazing lawn. Productive grassland or wetland ecosystem in which plants are heavily grazed but supported by large nutrient inputs from grazers.

Greenhouse effect. Warming of the atmosphere due to atmospheric absorption of infrared radiation.

Greenhouse gas. Atmospheric gas that absorbs infrared radiation.

Gross primary production. Net carbon input to ecosystems—that is, net photosynthesis expressed at the ecosystem scale ($g\,C\,m^{-2}\,yr^{-1}$).

Ground heat flux. Heat transferred from the surface into the soil.

Groundwater. Water in soil and rocks beneath the rooting zone.

Growth. Production of new biomass.

Growth respiration. Respiration to support biosynthesis (the production of new biomass).

Guano. Large accumulations of seabird feces.

Gyre. Large circulation systems in surface ocean waters.

Hadley cell. Atmospheric circulation cell between the equator and 30° N or S latitude, driven by rising air where the sun's rays are perpendicular to Earth's surface.

Halocline. Relatively sharp vertical gradient in salinity in a lake or ocean.

Halophyte. Plant species that typically grows on saline soils.

Hard pan. Soil horizon with low hydraulic conductivity.

Hartig net. Hyphae that penetrate cell walls of root cortical cells in ectomycorrhizae.

Heat capacity. Amount of energy required to raise the temperature of unit volume of a body by 1°C.

Heat of fusion. Energy required to change a substance from a solid to a liquid without a change in temperature.

Heat of vaporization. Energy required to change a gram of a substance from a liquid to a vapor without change in temperature.

Heat storage. Energy stored by an object due to an increase in temperature.

Herbivore. Organism that eats live plants.

Herbivory. Consumption of plants by animals.

Heterocyst. Specialized nonphotosynthetic cells of phototrophs that protect nitrogenase from denaturation by oxygen.

Heterotrophic respiration. Respiration by nonautotrophic organisms.

Heterotroph. Organism that consumes organic matter produced by other organisms rather than producing organic matter from CO_2 and environmental energy; includes decomposers, consumers, and parasites.

Histosol. Soil order characterized by highly organic soils due to poor drainage and low oxygen.

Homeothermy. Maintenance of a constant body temperature.

Horizon. Layer in a soil profile. The horizons, from top to bottom, are the O horizon, which consists of organic matter above mineral soil; the A horizon, a dark layer with substantial organic matter; the E horizon, which is heavily leached; a B horizon, where iron and aluminum oxides and clays accumulate; and a C horizon, which is relatively unaffected by soil-forming processes.

Horse latitudes. Latitudes 30° N and S, characterized by weak winds and high temperatures.

Hot spot. Zone of high rates of biogeochemical processes in a soil or landscape.

Humic acid. Relatively insoluble humic compounds with extensive networks of aromatic rings and few side chains.

Humification. Nonenzymatic process by which recalcitrant breakdown products of decomposition are complexed to form humus.

Humin. Relatively insoluble humic compounds with extensive networks of long-chain, nonpolar groups.

Humus. Amorphous soil organic matter that is the final product of decomposition.

Hydraulic conductivity. Capacity of a given volume of a substance (such as soil) to conduct water; this defines the relationship between discharge and the hydraulic gradient causing it.

Hydraulic lift. Vertical movement of water through roots from moist to dry soils along a gradient in water potential.

Hydrothermal vent. Vent that emits reduced gases such as H_2S in zones of sea-floor spreading.

Hyphae. Filamentous structures that make up the vegetative body of fungi.

Hyporrheic zone. Zone of flowing groundwater within the streambed or riverbed.

Ice-albedo feedback. Atmospheric warming caused by warming-induced decrease in albedo due to earlier melting of sea ice.

Igneous rocks. Rocks formed when magma from Earth's core cools near the surface.

Immobilization. Removal of inorganic nutrients from the available pool by microbial uptake and chemical fixation.

Inceptisol. Soil order characterized by weak soil development.

Infiltration. Movement of water into the soil.

Integrated conservation and development project. Project in developing nation that focuses simultaneously on biological conservation and human development.

Intensification. Intensive application of water, energy, and fertilizers to agricultural ecosystems to enhance their productivity.

Intensity. Energy released by a disturbance per unit area and time.

Interactive controls. Factors that control and respond to ecosystem characteristics, including resource supply, modulators, major functional types of organisms, disturbance regime, and human activities.

Interception. Contact of nutrients with roots due to the growth of roots to the nutrients; fraction of precipitation that does not reach the ground (canopy interception).

Intermediate water. Middle layer of ocean water between about 200 and 1000 m depth.

Intertropical convergence zone. Region of low pressure and rising air where surface air from the Northern and Southern Hemispheres converge.

Inverse modeling. Modeling that estimates the temporal and spatial patterns of inputs required to produce the observed temporal and spatial patterns of model outputs.

Inversion. Increase in atmospheric temperature with height.

Inverted biomass pyramid. Biomass pyramid in which there is a smaller biomass of primary producers than of upper trophic levels; typical of pelagic ecosystems of lakes, streams, and oceans.

Ionic binding. Electrostatic attraction between oppositely charged ions or surfaces.

Irradiance. Radiant energy flux density received at a surface—that is, the quantity of radiant energy received at a surface per unit time.

Jet stream. Strong winds over a broad height range in the upper troposphere.

Katabatic winds. Downslope winds that occur at night when air cools, becomes more dense, and flows downhill.

Kelvin waves. Large-scale ocean waves that travel back and forth across the ocean.

Keystone species. Species that has a much greater impact on ecosystem processes than would be expected from its biomass; functional type represented by a single species.

La Niña. Sea surface temperatures in the equatorial Pacific Ocean associated with strong upwelling of cold water off South America and warm currents in the western Pacific.

Labile. Easily decomposed.

Land breeze. Night breeze from the land to the ocean caused by the higher surface temperature over the ocean at night.

Landscape. Mosaic of patches that differ in ecologically important properties.

Land use conversion. Human-induced change of an ecosystem to one that is dominated by a different physical environment or different plant functional types.

Land use modification. Human alteration of an ecosystem in ways that significantly affect ecosystem processes, community structure and population dynamics without changing the physical environment or the dominant plant functional type of the ecosystem.

Latent heat flux. Energy transferred between a surface and the atmosphere by the evaporation of water or the condensation of water vapor.

Latent heat of vaporization. Heat absorbed by evaporation or released by condensation of water (or of other substances) when the phase changes.

Laterite. Iron-rich layer in tropical soils that have hardened irreversibly on exposure to repeated saturation and drying cycles; also termed *plinthite layers*.

Law of the minimum. Plant growth is limited by a single resource at any one time; another resource becomes limiting only when the supply of the first resource is increased above the point of limitation.

Leaching. Downward movement of materials in solution. This can occur from the canopy to the soil, from soil organic matter to the soil solution, from one soil horizon to another, or from the ecosystem to groundwater or aquatic ecosystems.

Leaf area index. Leaf area per unit ground area. Projected LAI is the leaf area projected onto a horizontal plane. Total LAI is the total surface area of leaves, including the upper and lower surface of flat leaves and the cylindrical surface of conifer needles; it is approximately twice the value of projected LAI, except in the case of conifer needles, where the projected leaf area is multiplied by π (3.1416) to get total leaf area.

Leaf mass ratio. Ratio of leaf mass to total plant mass.

Legacy. Effect of past events on the current functioning of an ecosystem.

Life history traits. Traits (e.g., seed size and number, potential growth rate, maximum size, and longevity) of an organism that determine how quickly a species can get to a site, how quickly it grows, how tall it gets, and how long it survives.

Light compensation point. Light intensity at which net photosynthesis equals zero.

Light-harvesting reactions. Reactions of photosynthesis that transform light energy into chemical energy.

Light saturation. Range of light intensities above which the rate of photosynthesis is insensitive to light intensity.

Light use efficiency. Ratio of gross primary production to absorbed photosynthetically active radiation at the leaf or ecosystem scale.

Limitation. Reduced rate of a process (e.g., net primary production, growth or photosynthesis) due to inadequate supply of a resource (e.g., light) or low temperature.

Lithosphere. Hard outermost shell of Earth.

Litter. Dead plant material that is sufficiently intact to be recognizable.

Litterbag. Mesh bag used to measure decomposition rate of detritus.

Litterfall. Shedding of aboveground plant parts and death of plants.

Littoral zone. Shore of a lake or ocean.

Loam. Soil with substantial proportions of at least two size classes of soil particles.

Loess. Soil derived from wind-blown silt particles.

Longwave radiation. Radiation with wavelengths 3000 to 30,000 nm.

Macrofauna. Soil animals larger than 10 mm in length.

Macronutrients. Nutrients that are required in large quantities by organisms.

Macropores. Large pores between soil aggregates that allow rapid movement of water, roots, and soil animals.

Maintenance respiration. Respiration used to support maintenance of live biomass.

Mantle. Fungal hyphae that surround the root in ectomycorrhizae; also termed *sheath*.

Mass flow. Bulk transport of solutes due to the movement of soil solution.

Mass wasting. Downslope movement of soil or rock material under the influence of gravity without the direct aid of other media such as water, air, or ice.

Matric potential. Component of water potential caused by adsorption of water to surfaces; it is considered a component of pressure potential in some treatments.

Matrix. Predominant patch type in a landscape.

Mean residence time. Mass divided by the flux into or out of the pool over a given time period; synonymous with *turnover time*.

Mechanical weathering. Physical fragmentation of the rock without chemical change.

Mesofauna. Soil animals 0.2 to 10 mm in length.

Mesopause. Boundary between the mesosphere and thermosphere.

Mesophyll cells. Photosynthetic cells in a leaf.

Mesosphere. Atmospheric layer between the stratosphere and the thermosphere, which is characterized by a decrease in temperature with height.

Metamorphic rocks. Sedimentary or igneous rocks that are modified by exposure to heat or pressure.

Metapopulations. Populations of a species that consist of partially isolated subpopulations.

Methanogen. Methane-producing bacteria.

Methanotroph. Methane-consuming bacteria.

Microbial loop. Microbial food web (including both plant- and detritus-based organic material) that recycles carbon and nutrients within the euphotic zone.

Microbial transformation. Transformation of plant-derived substrates into microbial-derived substrates as a result of microbial turnover.

Microbivore. Organism that eats microbes.

Microfauna. Soil animals less than 0.2mm in length.

Micronutrients. Nutrients that are required in small quantities by organisms.

Milankovitch cycles. Cycles of solar input to Earth caused by regular variations in Earth's orbit (eccentricity, tilt, and precession).

Mineralization. Conversion of carbon and nutrients from organic to inorganic forms due to the breakdown of litter and soil organic matter. Gross mineralization is the total amount of nutrients released via mineralization (regardless of whether it is subsequently immobilized or not). Net mineralization is the *net* accumulation of inorganic nutrients in the soil solution over a given time interval.

Modulator. Factor that influences growth rate but is not consumed in the growth process (e.g., temperature, ozone).

Mollisol. Soil order characterized by an organic-rich, fertile A horizon that grades into a B horizon.

Monsoon. Tropical or subtropical system of air flow characterized by a seasonal shift between prevailing onshore and offshore winds.

Mutualism. Symbiotic relationship between two species that benefits both partners.

Mycorrhizae. Symbiotic relationship between plant roots and fungal hyphae, in which the plant acquires nutrients from the fungus in return for carbohydrates that constitute the major carbon source for the fungus.

Mycorrhizosphere. Zone of soil that is directly influenced by mycorrhizal hyphae.

Negative feedback. Interaction in which two components of a system have opposite effects on one another; this reduces the rate of change in the system.

Net biome production. Net ecosystem production at the regional scale; includes patches that have accumulated carbon and those that have lost carbon through disturbance and other processes during the time period of measurement.

Net ecosystem exchange. Net carbon exchange between the land or ocean and the atmosphere; equals net ecosystem production minus transport of carbon to groundwater or to deep ocean water.

Net ecosystem production. Net annual carbon accumulation by the ecosystem.

Net primary production. Quantity of new plant material produced annually (gross primary production minus plant respiration); includes new biomass, hydrocarbon emissions, root exudates, and transfers to mycorrhizae.

Net radiation. Balance between the inputs and outputs of shortwave and longwave radiation.

Niche. Ecological role of an organism in an ecosystem.

Nitrification. Conversion of ammonium to nitrate in the soil. Autotrophic nitrifiers use the energy yield from NH_4^+ oxidation to fix carbon used in growth and maintenance, analogous to the way plants use solar energy to fix carbon via photosynthesis. Heterotrophic nitrifiers gain their energy from breakdown of organic matter.

Nitrogenase. Enzyme that converts dinitrogen to ammonium.

Nitrogen-based defense. Plant defensive compound containing nitrogen.

Nitrogen fixation. Conversion of dinitrogen gas to ammonium.

Nonoccluded phosphorus. Exchangeable phosphate that is loosely adsorbed to surfaces of iron and aluminum oxides or calcium carbonate.

Non–steady state mosaic. Landscape that is not in equilibrium with the current environment because large-scale disturbances cause large proportions of the landscape to be in one or a few successional stages.

Normalized difference vegetation index. Index of vegetation greenness.

Nutrient cycling. Mineralization and uptake of nutrients within an ecosystem patch.

Nutrient productivity. Instantaneous rate of carbon gain per unit nutrient.

Nutrient spiraling. Mineralization and uptake of nutrients that occurs as dead organic matter, dissolved nutrients, and organisms move along a section of a stream or river.

Nutrient uptake. Nutrient absorption by plant roots.

Nutrient use efficiency. Growth per unit of plant nutrient; ratio of nutrients to biomass lost in litterfall; also calculated as nutrient productivity times residence time.

O horizon. Organic horizon above mineral soil.

Occluded phosphorus. Unavailable phosphate that is most tightly bound to oxides of iron and aluminum.

Oligotrophic. Nutrient poor.

Omnivore. Organism that eats food from several trophic levels.

Orographic effects. Effects due to presence of mountains.

Osmotic potential. Component of water potential due to the presence of substances dissolved in water.

Overland flow. Movement of water over the soil surface.

Oxidation. Loss of electrons by an electron donor in oxidation–reduction reactions.

Oxisol. Soil order found in the wet tropics characterized by highly weathered, leached soils.

Oxygenase. Enzyme that catalyzes a reaction with oxygen.

Ozone hole. Zone of destruction of stratospheric ozone at high southern and high northern latitudes. This hole allows increased penetration of UV radiation to Earth's surface.

Parent material. Rocks or other substrates that generate soils through weathering.

Patch. Relatively homogeneous stand of an ecosystem in a landscape.

PEP carboxylase. Initial carboxylating enzyme in C_4 photosynthesis.

Pelagic. Open water.

Percolation. Saturated flow of water through a soil.

Periphyton. Algae that attach to rocks, vascular plants, and any other stable surfaces.

Permafrost. Permanently frozen ground—that is, soil that remains frozen for at least 2 years.

Permanent wilting point. Water held by a soil that cannot be extracted by plant uptake.

Perturbation. An external force that displaces a system from equilibrium.

pH. Negative log of the hydrogen ion concentration; denotes the activity of H^+ ions and thus the acidity of the system.

Phagocytosis. Consumption of material by a cell by enclosing it in a membrane-bound structure that enters the cell.

Phenology. Time course of periodic events in organisms that are correlated with climate (e.g., budbreak).

Phloem. Long-distance transport system in plants for flow of carbohydrates and other solutes.

Phosphatase. Enzyme that hydrolyzes phosphate from a phosphate-containing organic compound.

Photo-oxidation. Oxidation of compounds by light energy; photosynthetic enzymes can be photo-oxidized under conditions of high light.

Photoperiod. Daylength.

Photoprotection. Protection of photosynthetic pigments from destruction by high light.

Photorespiration. Production of CO_2 due to the oxygenation reaction catalyzed by Rubisco.

Photosynthesis. Biochemical process that uses light energy to convert CO_2 to sugars. Net photosynthesis is the net carbon input to ecosystems; synonymous at the ecosystem level with gross primary production.

Photosynthetic capacity. Photosynthetic rate per unit leaf mass measured under favorable conditions of light, moisture, and temperature.

Photosynthetic light use efficiency. Rate of photosynthesis per unit light.

Photosynthetic nitrogen use efficiency. Rate of photosynthesis per unit nitrogen.

Photosynthetically active radiation. Visible light; radiation with wavelengths between 400 and 700 nm.

Phototroph. Nitrogen-fixing microorganism that produces its own organic carbon through photosynthesis.

Phreatophyte. Deep-rooted plant that taps groundwater.

Phyllosphere decomposition. Decomposition that occurs on leaves before leaf fall.

Phytoplankton. Microscopic algae suspended in the surface water of aquatic ecosystems.

Pixel. Individual cell of a satellite image that provides a generalized spectral response for that area.

Planetary boundary layer. The layer of the atmosphere that is directly affected by the fluxes and friction of Earth's surface.

Planetary wave. Large (greater than 1500 km length) wave in the atmosphere.

Plankton. Microscopic organisms suspended in the surface water of aquatic ecosystems.

Plant-based trophic system. Plants, herbivores, and organisms that consume herbivores and their predators.

Plant defense. Chemical or physical property of plants that deters herbivores.

Plasmodesmata. Cytoplasmic connections between adjacent cortical cells.

Plinthite layers. Laterite layers in tropical soils.

Podzol. Spodosol.

Poikilothermic. Organism whose body temperature depends on the environment.

Polar cell. Atmospheric circulation cell between 60° and the pole driven by subsidence at the poles.

Polar front. Boundary between the polar and subtropical air masses characterized by rising air and frequent storms.

Polyphenol. Soluble organic compound with multiple phenolic groups.

Pool. Quantity of energy or material in an ecosystem compartment such as plants or soil.

Positive feedback. Interaction in which two components of a system have a positive effect on the other or in which both have a negative effect on one another; this amplifies the rate of change in the system.

Potential biota. Organisms that are present in a region and could potentially occupy the site.

Potential vegetation. Vegetation that would occur in the absence of human disturbance.

Precession. A "wobbling" in Earth's axis of rotation with respect to the stars, determining the date during the year when solstices and equinoxes occur.

Precipitation. Water input to an ecosystem as rain and snow.

Pressure potential. Component of water potential generated by gravitational forces and by physiological processes of organisms.

Prevailing wind. Most frequent wind direction.

Primary forest. Forest that has never been cleared.

Primary minerals. Minerals present in the rock or unconsolidated parent material before chemical changes have taken place.

Primary producers. Organisms that convert CO_2, water, and solar energy into biomass (i.e., plants); synonymous with *autotroph*.

Primary production. Conversion of CO_2, water, and solar energy into biomass. Gross primary production is the net carbon input to ecosystems, or the net photosynthesis expressed at the ecosystem scale ($g C m^{-2} yr^{-1}$). Net primary production is the net carbon accumulation by vegetation (GPP minus plant respiration).

Primary succession. Succession following severe disturbances that remove or bury most products of ecosystem processes, leaving little or no organic matter or organisms.

Production efficiency. Proportion of assimilated energy that is converted to animal production, including both growth and reproduction.

Profile. Vertical cross-section of soil.

Protease. Protein-hydrolyzing enzyme.

Proteoid roots. Dense clusters of fine roots produced by certain families such as the Proteaceae.

Protozoan. Single-celled animal.

Quality. Chemical nature of live or dead organic matter that determines the ease with which it is broken down by herbivores or decomposers, respectively.

Quantum yield. Moles of CO_2 fixed per mole of light quanta absorbed; the initial slope of the light-response curve.

Quinone. Highly reactive class of compounds produced from polyphenols.

R horizon. Unweathered bedrock at the base of a soil profile.

Rain shadow. Zone of low precipitation downwind of a mountain range.

Recalcitrant. Not readily decomposed.

Recovery. Extent to which a system returns to its original state following perturbation.

Redfield ratio. Ratio of nitrogen to phosphorus (approximately 14) giving optimal growth of algae.

Radiatively active gases. Gases that absorb infrared radiation.

Redox potential. Electrical potential of a system due to the tendency of substances in it to lose or accept electrons.

Reduction. The gain of electrons by an electron acceptor in oxidation–reduction reactions.

Regolith. Unweathered bedrock layer.

Relative accumulation rate. Nutrient uptake per unit plant nutrient.

Relative growth rate. Growth per unit plant biomass.

Relative humidity. Ratio of the actual amount of water held in the atmosphere compared to maximum that could be held at that temperature.

Release. Sudden increase in growth, when resource availability increases in response to death or reduced growth of neighboring individuals.

Residence time. Average time that an element or tissue remains in a system, calculated as the pool size divided by the input; synonymous with *turnover time*.

Resilience. Rate at with which a system returns to its reference state after a perturbation.

Resistance. Tendency of a system to remain in its reference state in the face of a perturbation.

Resorption. Withdrawal of nutrients from tissues during their senescence.

Resorption efficiency. Proportion of the maximum leaf nutrient pool that is resorbed before leaf fall.

Resource. Substance that is taken up from the environment and consumed in growth (e.g., light, CO_2, water, nutrients).

Respiration. Biochemical process that converts carbohydrates into CO_2 and water, releasing energy that can be used for growth and maintenance. Respiration can be associated with trophic groups (plant respiration, animal respiration, microbial respiration) or combinations of groups (heterotrophic respiration: animal plus microbial respiration; ecosystem respiration: heterotrophic plus plant respiration). Alternatively, can be defined by the way in which the resultant energy is used (maintenance respiration, growth respiration, respiration to support ion uptake).

Response. Direction and magnitude of change in the system following a perturbation.

Rhizosphere. Zone of soil that is directly influenced by roots.

River continuum concept. Idealized transition in ecosystem structure and function from narrow headwater streams to broad rivers.

Rock cycle. Formation, transformation, and weathering of rocks.

Root cap. Cells at the tips of roots that produce mucilaginous carbohydrates that lubricate the movement of roots through soil.

Root exudation. Diffusion and secretion of organic compounds from roots into the soil.

Root hair. Elongate epidermal cell of the root that extends out into the soil.

Root:shoot ratio. Ratio of root biomass to shoot biomass.

Roughness element. Obstacle to air flow (e.g., a tree) that creates mechanical turbulence.

Rubisco. Ribulose bisphosphate carboxylase; photosynthetic enzyme that catalyzes the initial carboxylation in C_3 photosynthesis.

Runoff. Water loss from an ecosystem in streams and rivers.

Saline. Salty.

Salinization. Salt accumulation due to evaporation of surface water.

Salt flat. Depression in an arid area that receives runoff but has no outlet.

Salt lick. Mineral-rich springs or outcrops that are used by animals as a source of minerals.

Sampling effect. Increased probability of encountering a species with particular traits in a species-rich community due simply to the greater number of species present.

Sand. Soil particles 0.05 to 2 mm diameter.

Saprovore. Organism that eats other live organisms in a detritus-based food chain.

Sapwood. Total quantity of functional conducting tissue of the xylem.

Saturated flow. Drainage of water under the influence of gravity.

Savanna. Grassland with scattered trees or shrubs.

Sea breeze. Daytime onshore breeze that occurs on coastlines due to differential heating of the land and water.

Secondary forest. Forest that has regrown after earlier clearing.

Secondary metabolites. Compounds produced by plants that are not essential for normal growth and development.

Secondary minerals. Crystalline and amorphous products that are formed through the reaction of materials released during weathering.

Secondary producers. Herbivores and carnivores.

Secondary succession. Succession that occurs on previously vegetated sites after a disturbance in which there are residual effects of organisms and organic matter from organisms present before the disturbance.

Sedimentary rocks. Rocks formed from sediments.

Seed bank. Seeds produced after previous disturbances that remain dormant in the soil until postdisturbance conditions (light, wide temperature fluctuations, and/or high soil nitrate) trigger germination.

Seedling bank. Seedlings beneath a canopy that show negligible growth beneath the dense shade of a forest canopy but grow rapidly in treefall gaps.

Selective preservation. Increase in concentration of recalcitrant material as a result of decomposition of labile substrates.

Senescence. Programmed breakdown of plant tissues.

Sensible heat. Heat energy that can be sensed (e.g., by a thermometer) and involves no change in state.

Sensible heat flux. Energy transferred between a surface and the near-surface atmosphere by conduction and movement to the bulk atmosphere by convection.

Seston. Particles suspended in the water column, including algae, bacteria, detritus, and mineral particles.

Severity. Proportion of the organic matter lost from the vegetation and surface soils due to disturbance.

Shade leaf. Leaf that is acclimated to shade or is produced by a plant adapted to shade.

Shifting agriculture. Clearing of forest for crops followed by a fallow period during which forests regrow, after which the cycle repeats; synonymous with *slash-and-burn* or *swidden agriculture*.

Shifting steady-state mosaic. Landscape in which patches differ in successional stage, but the landscape as a whole is at steady state (i.e., there is no directional change in the relative proportions of different successional stages).

Shortwave radiation. Radiation with wavelengths 300 to 3000 nm.

Shredder. Invertebrate that breaks leaves and other detritus into pieces and digests the microbial jam on the surface of these particles.

Siderophore. Organic chelate produced by plant roots.

Silt. Soil particles 0.002 to 0.05 mm diameter.

Sink. Part of the plant that shows a net import of a compound.

Sink strength. Demand of a plant organ or process for carbohydrates.

Slash-and-burn agriculture. Shifting agriculture.

Slow variable. Variable that changes slowly.

Snow–albedo feedback. Atmospheric warming caused by warming-induced decrease in albedo due to earlier snowmelt.

Soil creep. Downhill movement of soil; dubious character covered with soil.

Soil order. Major soil groupings in the U.S. soil taxonomic classification.

Soil organic matter. Dead organic matter in the soil that has decomposed to the point that its original identity is uncertain.

Soil phase. Soils belonging to the same soil type that differ in landscape position, stoniness, or other soil properties.

Soil resources. Water and nutrients available in the soil.

Soil series. Soils belonging to the same order that differ in profile characteristics, such as number and types of horizons, thickness, and horizon properties.

Soil structure. Binding together of soil particles to form aggregates.

Soil types. Soils belonging to the same soil series but having different textures of the A horizon.

Solifluction. Downslope flow of saturated soils above a frozen layer.

Solubility pump. Downward flux of carbon from surface to deep waters due to the downwelling of CO_2-rich North Atlantic or Antarctic waters.

Sorption. Binding of an ion to a mineral surface, ranging from electrostatic attraction to covalent binding.

Source. Part of a plant that shows a net export of a compound.

Southern oscillation. Atmospheric pressure changes over the southeastern Pacific and Indian Ocean.

Specialist herbivore. Herbivore that specializes on consumption of one or a few plant species or tissues.

Species composition. Identity of species in an ecosystem.

Species diversity. Number, evenness, and composition of species in an ecosystem; the total range of biological attributes of all species present in an ecosystem.

Species evenness. Relative abundances of species in an ecosystem.

Species richness. Number of species in an ecosystem.

Specific heat. Energy required to warm a gram of a substance by 1°C.

Specific leaf area. Ratio of leaf area to leaf mass.

Specific root length. Root length per unit root mass.

Spiraling length. Average horizontal distance that a nutrient moves between successive uptake events.

Spodosol. Soil order characterized by highly leached soils that develop in cold climates; Formerly termed *podzols*.

Stand-replacing disturbance. Large disturbances that affect entire stands of vegetation.

State factors. Independent variables that control the characteristics of soils and ecosystems (climate, parent material, topography, potential biota, and time).

Steady state. State of a system in which increments are approximately equal to losses, when averaged over a long time (e.g., the turnover time of the system); there are no directional changes in the major pools in a system at steady state.

Stem flow. Water that flows down stems to the ground.

Stomata. Pores in the leaf surface through which water and CO_2 are exchanged between the leaf and the atmosphere.

Stomatal conductance. Flux of water vapor or CO_2 per unit driving force between the leaf and the atmosphere.

Stratopause. Boundary between the stratosphere and the mesosphere.

Stratosphere. Atmospheric layer above the troposphere, which is characterized by an increase in temperature with height.

Strength of soil. Amount of force required to initiate slope failure.

Stress. Environmental factor that reduces plant performance; physical force that promotes mass wasting of soils.

Stroma. Gel matrix within the chloroplast in which the carbon-fixation reactions occur.

Subduction. Downward movement of a plate margin beneath another plate.

Suberin. Hydrophobic waxy substance that occurs in the cell walls of the endodermis and exodermis of plant roots.

Sublimation. Vaporization of a solid.

Subsidy. Energy or nutrient transfers from one ecosystem to another; synonymous with *allochthonous input*.

Succession. Directional change in ecosystem structure and functioning resulting from biotically driven changes in resource supply.

Sunfleck. Short period of high irradiance that interrupts a general background of low diffuse radiation.

Sun leaf. Leaf that is acclimated to high light or is produced by a plant adapted to high light.

Supply rate. Rate of input of a resource (e.g., nitrate supply rate).

Surface conductance. Potential of the leaf and soil surfaces in the ecosystem to lose water.

Surface water. Surface layer of the ocean heated by the sun and mixed by winds, typically 75 to 200 m deep.

Swidden agriculture. Shifting agriculture.

Systems ecology. Study of the ecosystem as a group of components linked by fluxes of materials or energy.

Taiga. Boreal forest.

Teleconnections. Dynamic interactions that interconnect distant regions of the atmosphere.

Temporal scaling. Extrapolation of measurements made at one time interval to longer (or occasionally shorter) time intervals.

Texture. Particle size distribution of soils.

Thermocline. Relatively sharp vertical temperature gradient in a lake or ocean.

Thermohaline circulation. Global circulation of deep and intermediate ocean waters driven by downwelling of cold saline surface water off of Greenland and Antarctica.

Thermosphere. Outermost layer of the atmosphere, which is characterized by an increase in temperature with height.

Throughfall. Water that drops from the canopy to the ground.

Thylakoids. Membrane-bound vesicles in chloroplasts in which the light-harvesting reactions of photosynthesis occur.

Tilt. Angle of Earth's axis of rotation and the plane of its orbit around the sun.

Time step. Shortest unit of time simulated by a model.

Top–down controls. Regulation of population dynamics by predation.

Toposequence. Series of ecosystems that are similar except with respect to their topographic position.

Trade winds. Easterly winds between 30° N and S latitudes.

Transformation. Conversion of the organic compounds contained in litter to recalcitrant organic compounds in soil humus.

Transpiration. Water movement through stomates from plants to the atmosphere.

Transporter. Membrane-bound protein that transports ions across cell membranes.

Trophic cascade. Top–down effect of predators on the biomass of organisms at lower trophic levels; results in alternation of high and low biomass of organisms in successive trophic levels.

Trophic efficiency. Proportion of production of prey that is converted to production of consumers at the next trophic level.

Trophic interactions. Feeding relationships among organisms.

Trophic level. Organisms that obtain their energy with the same number of steps removed from plants or detritus.

Trophic transfer. Flux of energy or materials due to consumption of one organism by another.

Tropopause. Boundary between the troposphere and the stratosphere.

Troposphere. Lowest layer of the atmosphere, which is continually mixed by weather systems and is characterized by a decrease in temperature with height.

Tundra. Ecosystem type that is too cold to support growth of trees.

Turbulence. State of air or water movement in which velocities exhibit irregular fluctuations capable of transporting heat and materials much more rapidly than by diffusion. Mechanical turbulence is caused by the uneven slowing of air by a rough surface. Convective turbulence is caused by the increased buoyancy of surface air caused by heat transfer from the surface.

Turnover. Replacement of a pool; ratio of the flux to the pool size; lake mixing that occurs when surface waters become more dense than deep waters.

Turnover length. Downstream distance moved in a stream while an element is in organisms.

Turnover time. Average time that an element spends in a system (pool/input); synonymous with *residence time*.

Ultisol. Soil order characterized by substantial leaching in a warm, humid environment.

Unsaturated flow. Water movement through soils with a water content less than field capacity.

Uplift. Upward movement of Earth's surface.

Uptake. Absorption of water or mineral by an organism or tissue.

Uptake length. Average distance that an atom moves down stream from the time it is released by mineralization until it is absorbed again.

Upwelling. Upward movement of deep and intermediate ocean water, usually driven by offshore winds near coasts.

Validation. Comparison of model predictions with data.

Vapor density. Mass of water per volume of air; absolute humidity.

Vapor pressure. Partial pressure exerted by water molecules in the air.

Vapor pressure deficit. Difference in actual vapor pressure and the vapor pressure in air of the same temperature and pressure that is saturated with water vapor; loosely used to describe the difference in vapor pressure in air immediately adjacent to an evaporating surface and the bulk atmosphere, although strictly speaking the air masses are at different temperatures.

Vertisol. Soil order characterized by swelling and shrinking clays.

Vesicular arbuscular mycorrhizae. Synonymous with *arbuscular mycorrhizae*.

Water-holding capacity. Difference in soil water content between field capacity and permanent wilting point.

Water potential. Potential energy of water relative to pure water at the soil surface.

Water saturated. All soil pores filled with water.

Watershed. Drainage area of a stream, river, or lake leading to a single outlet for its runoff; synonymous with *catchment*. In England, the term refers to a ridge that separates two drainages.

Water use efficiency. Ratio of gross primary production to water loss; also sometimes calculated as the ratio of net primary production to cumulative transpiration (growth water use efficiency).

Water vapor feedback. Additional greenhouse effect provided by water vapor, when the atmosphere warms and increases its water vapor content.

Weathering. Processes by which parent rocks and minerals are altered to more stable forms. Physical weathering breaks rocks into smaller fragments with greater surface area. Chemical weathering results from chemical reactions between rock minerals and the atmosphere or water.

Westerlies. Winds that blow from the west.

Xanthophyll cycle. Transfer of absorbed energy to xanthophyll and eventually to heat at times when electron acceptors are not available to transfer electrons to carbon-fixation reactions.

Xeric. Characterized by plants that are tolerant of dry conditions.

Xylem. Water-conducting tissue of plants.

Zooplankton. Microscopic animals suspended in the surface water of aquatic ecosystems.

References

Aber, J., W. McDowell, K. Nadelhoffer, A. Magill, G. Bernstson, M. Kamakea, S. McNulty, W. Currie, L. Rustad, and I. Fernandez. 1998. Nitrogen saturation in temperate forest ecosystems. *BioScience* 48:921–934.

Aber, J.D., A. Magill, S.G. McNulty, R.D. Boone, K.J. Nadelhoffer, M. Downs, and R. Hallett. 1995. Forest biogeochemistry and primary production altered by nitrogen saturation. *Water Air and Soil Pollution* 85:1665–1670.

Aber, J.D., and J.M. Melillo. 1991. *Terrestrial ecosystems*. Saunders College, Orlando, FL.

Adams, P.C. 1999. *The dynamics of white spruce populations on a boreal river floodplain*. Ph.D. Dissertation. Duke University. Durham, NC.

Aerts, R. 1995. Nutrient resorption from senescing leaves of perennials: Are there general patterns? *Journal of Ecology* 84:597–608.

Aerts, R. 1997. Climate, leaf litter chemistry and leaf litter decomposition in terrestrial ecosystems: A triangular relationship. *Oikos* 79:439–449.

Aerts, R., and F. Berendse. 1988. The effect of increased nutrient availability on vegetation dynamics in wet heathlands. *Vegetatio* 76:63–69.

Aerts, R., and F.S. Chapin III. 2000. The mineral nutrition of wild plants revisited: A re-evaluation of processes and patterns. *Advances in Ecological Research* 30:1–67.

Agee, J., and D. Johnson 1988. *Ecosystem Management for Parks and Wilderness*. University of Washington Press, Seattle, WA.

Agren, G., and E. Bosata. 1988. Nitrogen saturation of terrestrial ecosystems. *Environmental Pollution* 54:185–197.

Ahrens, C.D. 1998. *Essentials of Meteorology: An Invitation to the Atmosphere*. Wadsworth, Belmont, CA.

Alcamo, J., editor. 1994. *IMAGE 2.0: Integrated Modeling of Global Climate Change*. Kluwer Academic, Dordrecht, the Netherlands.

Allen, M.F. 1991. *The Ecology of Mycorrhizae*. Cambridge University Press, Cambridge, UK.

Alpert, P. 1996. Integrated conservation and development projects: Examples from Africa. *BioScience* 46:845–855.

Altieri, M.A. 1990. Why study traditional agriculture? Pages 551–564 *in* C.R. Carrol, J.H. Vandermeer, and P.M. Rosset, editors. *Agroecology*. McGraw-Hill, New York.

Amundson, R., and H. Jenny. 1997. On a state factor model of ecosystems. *BioScience* 47:536–543.

Anderson, R.V., D.C. Coleman, and C.V. Cole. 1981. Effects of saprotrophic grazing on net mineralisation. Pages 201–216 *in* F.E. Clark and T. Rosswall, editors. *Terrestrial Nitrogen Cycles: Processes, Ecosystem Strategies and Management Impacts*. Ecological Bulletins, Stockholm.

Andersson, T. 1991. Influence of stemflow and throughfall from common oak (*Quercus robur*) on soil chemistry and vegetation patterns. *Canadian Journal of Forest Research* 21:917–924.

Armesto, J.J., R. Rozzi, and J. Caspersen. 2001. Temperate forests of North and South America. Pages 223–249 *in* F.S. Chapin III, O.E. Sala, and E. Huber-Sannwald, editors. *Global Biodiversity in a Changing Environment: Scenarios for the 21st Century*. Springer-Verlag, New York.

Aston, A.R. 1979. Rainfall interception by eight small trees. *Journal of Hydrology* 42:383–396.

Ayres, M.P., and S.F. MacLean Jr. 1987. Development of birch leaves and the growth energetics of *Epirrita autumnata* (Geometridae). *Ecology* 68: 558–568.

Baede, A.P.M., E. Ahlonsou, Y. Ding, and D. Schimel. 2001. The climate system: An overview. Pages

85–98 *in* J.T. Houghton, Y. Ding, D.J. Griggs, M. Noguer, P.J. van der Linden, X. Dai, K. Maskell, and C.A. Johnson, editors. *Climate Change 2001: The Scientific Basis*. Cambridge University Press, Cambridge, UK.

Bailey, R.G. 1998. *Ecoregions: The Ecosystem Geography of the Oceans and Continents*. Springer-Verlag, New York.

Baldocchi, D., F.M. Kelliher, T.A. Black, and P.G. Jarvis. 2000. Climate and vegetation controls on boreal zone energy exchange. *Global Change Biology* 6(Suppl. 1):69–83.

Barber, S.A. 1984. *Soil Nutrient Bioavailability*. Wiley, New York.

Barry, R.G., and R.J. Chorley. 1970. *Atmosphere, Weather, and Climate*. Holt, Rinehart & Winston, New York.

Baskin, C.C., and J.M. Baskin. 1998. *Seeds: Ecology, Biogeography, and Evolution of Dormancy and Germination*. Academic Press, San Diego, CA.

Bates, T.R., and J.P. Lynch. 1996. Stimulation of root hair elongation in *Arabidopsis thaliana* by low phosphorus availability. *Plant Cell and Environment* 19:529–538.

Bazzaz, F.A. 1996. *Plants in Changing Environments. Linking Physiological, Population, and Community Ecology*. Cambridge University Press, Cambridge, UK.

Beare, M.H., R.W. Parmelee, P.F. Hendrix, W. Cheng, D.C. Coleman, and D.A. Crossley Jr. 1992. Microbial and faunal interactions and effects on litter nitrogen and decomposition in agroecosystems. *Ecological Monographs* 62:569–591.

Bender, M., T. Ellis, P. Tans, R. Francey, and D. Lowe. 1996. Variability in the O_2/N_2 ratio of Southern Hemisphere air, 1991–1994: Implications for the carbon cycle. *Global Biogeochemical Cycles* 10: 9–21.

Bengston, D.N. 1994. Changing forest values and ecosystem management. *Society & Natural Resources* 7:515–533.

Berendse, F., and R. Aerts. 1987. Nitrogen-use efficiency: A biologically meaningful definition? *Functional Ecology* 1:293–296.

Berendse, F., R. Aerts, and R. Bobbink. 1993. Atmospheric nitrogen deposition and its impact on terrestrial ecosystems. Pages 104–121 *in* C.C. Vos and P. Opdam, editors. *Landscape Ecology of a Stressed Environment*. Chapman & Hall, London.

Berg, B., and G. Ekbohm. 1991. Litter mass-loss rates and decomposition patterns in some needle and leaf litter types. Long-term decomposition in a Scots pine forest. VII. *Canadian Journal of Botany* 69:1449–1456.

Berg, B., M.-B. Johansson, V. Meentemeyer, and W. Kratz. 1998. Decomposition of tree root litter in a climatic transect of coniferous forests in northern Europe: A synthesis. *Scandinavian Journal of Forest Research* 13:202–212.

Berg, B., and H. Staaf. 1980. Decomposition rate and chemical changes of Scots pine needle litter. II. Influence of chemical composition. Pages 373–390 *in* T. Persson, editor. *Structure and Function of Northern Coniferous Forests: An Ecosystem Study*. Ecological Bulletins, Stockholm.

Bergh, J., and S. Linder. 1999. Effects of soil warming during spring on photosynthetic recovery in boreal Norway spruce stands. *Global Change Biology* 5:245–253.

Berkes, F., and C. Folke. 1998. Linking social and ecological systems for resilience and sustainability. Pages 1–25 *in* F. Berkes and C. Folke, editors. *Linking Social and Ecological Systems: Management Practices and Social Mechanisms for Building Resilience*. Cambridge University Press, Cambridge, UK.

Berner, R.A. 1997. The rise of plants and their effect on weathering and atmosphere CO_2. *Science* 276: 544–546.

Berry, J., and O. Björkman. 1980. Photosynthetic response and adaptation to temperature in higher plants. *Annual Review of Plant Physiology* 31: 491–543.

Berry, W.L. 1970. Characteristics of salts secreted by *Tamarix aphylla*. *American Journal of Botany* 57:1226–1230.

Betts, A.K., and J.H. Ball. 1997. Albedo over the boreal forest. *Journal of Geophysical Research-Atmospheres* 102:28901–28909.

Billings, W.D. 1952. The environmental complex in relation to plant growth and distribution. *Quarterly Review of Biology* 27:251–265.

Billings, W.D., and H.A. Mooney. 1968. The ecology of arctic and alpine plants. *Biological Review* 43:481–529.

Binkley, D., P. Sollins, R. Bell, D. Sachs, and D. Myrold. 1992. Biogeochemistry of adjacent conifer and alder-conifer stands. *Ecology* 73:2022–2033.

Bird, M.I., J. Lloyd, and G.D. Farquhar. 1994. Terrestrial carbon storage at the LGM. *Nature* 371:585.

Birkeland, P.W. 1999. *Soils and Geomorphology*. Oxford University Press, New York.

Bloom, A.J., F.S. Chapin III, and H.A. Mooney. 1985. Resource limitation in plants—An economic analogy. *Annual Review of Ecology and Systematics* 16:363–392.

Bond, W.J. 1993. Keystone species. Pages 237–253 *in* E.-D. Schulze and H.A. Mooney, editors. *Ecosys-*

tem Function and Biodiversity. Springer-Verlag, Berlin.

Borchert, R. 1994. Soil and stem water storage determine phenology and distribution of tropical dry forest trees. *Ecology* 75:1437–1449.

Bormann, B.T., and R.C. Sidle. 1990. Changes in productivity and distribution of nutrients in a chronosequence at Glacier Bay National Park, Alaska. *Journal of Ecology* 78:561–578.

Bormann, F.H., and G.E. Likens. 1979. *Pattern and Process in a Forested Ecosystem*. Springer-Verlag, New York.

Bormann, F.H., G.E. Likens, T.G. Siccama, R.S. Pierce, and J.S. Eaton. 1974. The export of nutrients and recovery of stable conditions following deforestation at Hubbard Brook. *Ecological Monographs* 44:255–277.

Bradley, R.L., and J.W. Fyles. 1996. Interactions between tree seedling roots and humus forms in the control of soil C and N cycling. *Biology and Fertility of Soils* 23:70–79.

Bradshaw, A.D. 1983. The reconstruction of ecosystems. *Journal of Ecology* 20:1–17.

Brady, N.C., and R.R. Weil. 2001. *The Nature and Properties of Soils*. 13th Edition. Prentice Hall, Upper Saddle River, NJ.

Braswell, B.H., and B. Moore III. 1994. The lifetime of excess atmospheric carbon-dioxide. *Global Biogeochemical Cycles* 8:23–28.

Bridgham, S.D., C.A. Johnston, J. Pastor, and K. Updegraff. 1995. Potential feedbacks of northern wetlands on climate change. *BioScience* 45:262–274.

Brokaw, N.V.L. 1985. Gap-phase regeneration in a tropical forest. *Ecology* 66:682–687.

Brooker, R.W., and T.V. Callaghan. 1998. The balance between positive and negative plant interactions and its relationship to environmental gradients: A model. *Oikos* 81:196–207.

Brooks, J.L., and S.I. Dodson. 1965. Predation, body size and composition of plankton. *Science* 150:28–35.

Brown, J.R., and S. Archer. 1999. Shrub invasion of grassland: Recruitment is continuous and not regulated by herbaceous biomass or density. *Ecology* 80:2385–2396.

Bryant, J.P. 1981. Hare trigger. *Natural History* 90:46–53.

Bryant, J.P., and F.S. Chapin III. 1986. Browsing-woody plant interactions during boreal forest plant succession. Pages 213–225 *in* K. Van Cleve, F.S. Chapin III, P.W. Flanagan, L.A. Viereck, and C.T. Dyrness, editors. *Forest Ecosystems in the Alaskan Taiga: A Synthesis of Structure and Function*. Springer-Verlag, New York.

Bryant, J.P., and P.J. Kuropat. 1980. Selection of winter forage by subarctic browsing vertebrates: The role of plant chemistry. *Annual Review of Ecology and Systematics* 11:261–285.

Bryant, J.P., F.S. Chapin III, and D.R. Klein. 1983. Carbon/nutrient balance of boreal plants in relation to vertebrate herbivory. *Oikos* 40:357–368.

Burgess, S.S.O., M.A. Adams, N.C. Turner, and C.K. Ong. 1998. The redistribution of soil water by tree root systems. *Oecologia* 115:306–311.

Burgis, M.J., and P. Morris. 1987. *The Natural History of Lakes*. Cambridge University Press, Cambridge, UK.

Burke, I.C., and W.K. Lauenroth. 1995. Biodiversity at landscape to regional scales. Pages 304–311 *in* V.H. Heywood, editor. *Global Biodiversity Assessment*. Cambridge University Press, Cambridge, UK.

Burke, I.C., W.K. Lauenroth, R. Riggle, P. Brannen, B. Madigan, and S. Beard. 1999. Spatial variability of soil properties in the shortgrass steppe: The relative importance of topography, grazing, microsite, and plant species in controlling spatial patterns. *Ecosystems* 2:422–438.

Burke, I.C., D.S. Schimel, C.M. Yonker, W.J. Parton, L.A. Joyce, and W.K. Lauenroth. 1990. Regional modeling of grassland biogeochemistry using GIS. *Landscape Ecology* 4:45–54.

Burke, I.C., C.M. Yonker, W.J. Parton, C.V. Cole, K. Flach, and D.S. Schimel. 1989. Texture, climate, and cultivation effects on soil organic matter content in U.S. grassland soils. *Soil Science Society of America Journal* 53:800–805.

Cabido, M.R., and M.R. Zak. 1999. *Vegetación del Norte de Córdoba*. Imprenta Nico, Córdoba, Argentina.

Caldwell, M.M., T.E. Dawson, and J.H. Richards. 1998. Hydraulic lift: Consequences of water efflux from the roots of plants. *Oecologia* 113:151–161.

Callaway, R.M. 1995. Positive interactions among plants. *Botanical Review* 61:306–349.

Canadell, J., R.B. Jackson, J.R. Ehleringer, H.A. Mooney, O.E. Sala, and E.-D. Schulze. 1996. Maximum rooting depth of vegetation types at the global scale. *Oecologia* 108:585–595.

Carpenter, S., and K. Cottingham. 1997. Resilience and restoration of lakes. *Conservation Ecology* [online] 1:2.

Carpenter, S.R., and J.F. Kitchell, editors. 1993. *The Trophic Cascade in Lakes*. Cambridge University Press, Cambridge, UK.

Carpenter, S.R., and M.G. Turner. 2000. Hares and tortoises: Interactions of fast and slow variables in ecosystems. *Ecosystems* 3:495–497.

Carpenter, S.R., N.F. Caraco, D.L. Correll, R.W. Howarth, A.N. Sharpley, and V.H. Smith. 1998. Nonpoint pollution of surface waters with phosphorus and nitrogen. *Ecological Applications* 9: 559–568.

Carpenter, S.R., S.G. Fisher, N.B. Grimm, and J.F. Kitchell. 1992. Global change and freshwater ecosystems. *Annual Review of Ecology and Systematics* 23:119–139.

Carpenter, S.R., J.J. Hodgson, J.F. Kitchell, M.L. Pace, D. Bade, K.L. Cottingham, T.E. Essington, J.N. Houser, and D.E. Schindler. 2001. Trophic cascades, nutrients, and lake productivity: Whole-lake experiments. *Ecological Monographs* 71:163–186.

Carpenter, S.R., J.F. Kitchell, and J.R. Hodgson. 1985. Cascading trophic interactions and lake productivity. *BioScience* 35:634–639.

Cerling, T.E. 1999. Paleorecords of C_4 plants and ecosystems. Pages 445–469 *in* R.F. Sage and R.K. Monson, editors. *C_4 Plant Biology*. Academic Press, San Diego, CA.

Chabot, B.F., and D.J. Hicks. 1982. The ecology of leaf life spans. *Annual Review of Ecology and Systematics* 13:229–259.

Chadwick, O.A., L.A. Derry, P.M. Vitousek, B.J. Huebert, and L.O. Hedin. 1999. Changing sources of nutrients during 4 million years of soil and ecosystem development. *Nature* 397:491–497.

Chambers, S. 1998. *Short- and Long-Term Effects of Clearing Native Vegetation for Agricultural Purposes*. Ph.D. dissertation. Flinders University of South Australia, Flinders.

Chambers, S.D., and F.S. Chapin III. In press. Fire effects on surface-atmosphere energy exchange in Alaskan black spruce ecosystems. *Journal of Geophysical Research*.

Chapin, F.S. III. 1989. The cost of tundra plant structures: Evaluation of concepts and currencies. *American Naturalist* 133:1–19.

Chapin, F.S. III. 1983. Direct and indirect effects of temperature on arctic plants. *Polar Biology* 2: 47–52.

Chapin, F.S. III. 1991a. Effects of multiple environmental stresses on nutrient availability and use. Pages 67–88 *in* H.A. Mooney, W.E. Winner, and E.J. Pell, editors. *Response of Plants to Multiple Stresses*. Academic Press, San Diego, CA.

Chapin, F.S. III. 1993a. Functional role of growth forms in ecosystem and global processes. Pages 287–312 *in* J.R. Ehleringer and C.B. Field, editors. *Scaling Physiological Processes: Leaf to Globe*. Academic Press, San Diego, CA.

Chapin, F.S. III. 1991b. Integrated responses of plants to stress. *BioScience* 41:29–36.

Chapin, F.S. III. 1980. The mineral nutrition of wild plants. *Annual Review of Ecology and Systematics* 11:233–260.

Chapin, F.S. III. 1993b. Physiological controls over plant establishment in primary succession. Pages 161–178 *in* J. Miles and D.W.H. Walton, editors. *Primary Succession*. Blackwell Scientific, Oxford, UK.

Chapin, F.S. III, and R.A. Kedrowski. 1983. Seasonal changes in nitrogen and phosphorus fractions and autumn retranslocation in evergreen and deciduous taiga trees. *Ecology* 64:376–391.

Chapin, F.S. III, and G.R. Shaver. 1985. Individualistic growth response of tundra plant species to environmental manipulations in the field. *Ecology* 66:564–576.

Chapin, F.S. III, and K. Van Cleve. 1981. Plant nutrient absorption and retention under differing fire regimes. Pages 301–321 *in* H.A. Mooney, T.M. Bonnickson, N.L. Christensen, J.E. Lotan, and W.A. Reiners, editors. *Fire Regimes and Ecosystem Processes*. USDA Forest Service General Technical Report Wo-26. USDA Forest Service, Washington, DC.

Chapin, F.S. III, W. Eugster, J.P. McFadden, A.H. Lynch, and D.A. Walker. 2000a. Summer differences among arctic ecosystems in regional climate forcing. *Journal of Climate* 13:2002–2010.

Chapin, F.S. III, N. Fetcher, K. Kielland, K.R. Everett, and A.E. Linkins. 1988. Productivity and nutrient cycling of Alaskan tundra: Enhancement by flowing soil water. *Ecology* 69:693–702.

Chapin, F.S. III, J. Follett, and K.F. O'Connor. 1982. Growth, phosphate absorption, and phosphorus chemical fractions in two *Chionochloa* species. *Journal of Ecology* 70:305–321.

Chapin, F.S. III, D.A. Johnson, and J.D. McKendrick. 1980a. Seasonal movement of nutrients in plants of differing growth form in an Alaskan tundra ecosystem: Implications for herbivory. *Journal of Ecology* 68:189–209.

Chapin, F.S. III, A.D. McGuire, J. Randerson, R. Pielke Sr., D. Baldocchi, S.E. Hobbie, N. Roulet, W. Eugster, E. Kasischke, E.B. Rastetter, S.A. Zimov, and S.W. Running. 2000b. Arctic and boreal ecosystems of western North America as components of the climate system. *Global Change Biology* 6(Suppl. 1):1–13.

Chapin, F.S. III, P.C. Miller, W.D. Billings, and P.I. Coyne. 1980b. Carbon and nutrient budgets and their control in coastal tundra. Pages 458–482 *in* J. Brown, P.C. Miller, L.L. Tieszen, and F.L. Bunnell, editors. *An Arctic Ecosystem: The Coastal Tundra at Barrow, Alaska*. Dowden, Hutchinson, & Ross, Stroudsburg, PA.

Chapin, F.S. III, L. Moilanen, and K. Kielland. 1993. Preferential use of organic nitrogen for growth by a non-mycorrhizal arctic sedge. *Nature* 361:150–153.

Chapin, F.S. III, E.-D. Schulze, and H.A. Mooney. 1990. The ecology and economics of storage in plants. *Annual Review of Ecology and Systematics* 21:423–448.

Chapin, F.S. III, G.R. Shaver, A.E. Giblin, K.G. Nadelhoffer, and J.A. Laundre. 1995. Response of arctic tundra to experimental and observed changes in climate. *Ecology* 76:694–711.

Chapin, F.S. III, G.R. Shaver, and R.A. Kedrowski. 1986a. Environmental controls over carbon, nitrogen, and phosphorus chemical fractions in *Eriophorum vaginatum* L. in Alaskan tussock tundra. *Journal of Ecology* 74:167–195.

Chapin, F.S. III, M.S. Torn, and M. Tateno. 1996. Principles of ecosystem sustainability. *American Naturalist* 148:1016–1037.

Chapin, F.S. III, K. Van Cleve, and P.R. Tryon. 1986b. Relationship of ion absorption to growth rate in taiga trees. *Oecologia* 69:238–242.

Chapin, F.S. III, B.H. Walker, R.J. Hobbs, D.U. Hooper, J.H. Lawton, O.E. Sala, and D. Tilman. 1997. Biotic control over the functioning of ecosystems. *Science* 277:500–504.

Chapin, F.S. III, L.R. Walker, C.L. Fastie, and L.C. Sharman. 1994. Mechanisms of primary succession following deglaciation at Glacier Bay, Alaska. *Ecological Monographs* 64:149–175.

Chapin, F.S. III, E.S. Zaveleta, V.T. Eviner, R.L. Naylor, P.M. Vitousek, S. Lavorel, H.L. Reynolds, D.U. Hooper, O.E. Sala, S.E. Hobbie, M.C. Mack, and S. Diaz. 2000c. Consequences of changing biotic diversity. *Nature* 405:234–242.

Charney, J.G., W.J. Quirk, S.-H. Chow, and J. Kornfield. 1977. A comparative study of effects of albedo change on drought in semiarid regions. *Journal of Atmospheric Sciences* 34:1366–1385.

Chase, T.N., R.A. Pielke Sr., T.G.F. Kittel, R.R. Nemani, and S.W. Running. 2000. Simulated impacts of historical land cover changes on global climate in northern winter. *Climate Dynamics* 16:93–105.

Chazdon, R.L., and N. Fetcher. 1984a. Light environments of tropical forests. Pages 27–36 *in* E. Medina, H.A. Mooney, and C. Vazquez-Yanes, editors. *Physiological Ecology of Plants of the Wet Tropics*. W. Junk, The Hague.

Chazdon, R.L., and N. Fetcher. 1984b. Photosynthetic light environments in a lowland rain forest in Costa Rica. *Journal of Ecology* 72:553–564.

Chazdon, R.L., and R.W. Pearcy. 1991. The importance of sunflecks for forest understory plants. *BioScience* 41:760–766.

Chen, J., J.F. Franklin, and T.A. Spies. 1995. Growing-season microclimatic gradients from clearcut edges into old-growth Douglas-fir forests. *Ecological Applications* 5:74–86.

Cheng, W., Q. Zhange, D.C. Coleman, C.R. Carroll, and C.A. Hoffman. 1996. Is available carbon limiting microbial respiration in the rhizosphere? *Soil Biology and Biochemistry* 28:1283–1288.

Christensen, N.L., A.M. Bartuska, J.H. Brown, S. Carpenter, C. D'Antonio, R. Francis, J.F. Franklin, J.A. MacMahon, R.F. Noss, D.J. Parsons, C.H. Peterson, M.G. Turner, and R.G. Woodmansee. 1996. The report of the Ecological Society of America committee on the scientific basis for ecosystem management. *Ecological Applications* 6:665–691.

Ciais, P., P.P. Tans, J.W.C. White, M. Trolier, R.J. Francey, J.A. Berry, D.R. Randall, P.J. Sellers, J.G. Collatz, and D.S. Schimel. 1995. Partitioning of ocean and land uptake of CO_2 as inferred by ($d^{13}C$) measurements from the NOAA Climate Monitoring and Diagnostics Laboratory Global Air Sampling Network. *Journal of Geophysical Research-Atmospheres* 100:5051–5070.

Cicerone, R.J., and R.S. Oremland. 1988. Biogeochemical aspects of atmospheric methane. *Global Biogeochemical Cycles* 2:299–327.

Clarholm, M. 1985. Interactions of bacteria, protozoa and plants leading to mineralization of soil nitrogen. *Soil Biology and Biochemistry* 17:181–187.

Clark, D.A., S. Brown, D.W. Kicklighter, J.Q. Chambers, J.R. Thomlinson, and J. Ni. 2001a. Measuring net primary production in forests: Concepts and field methods. *Ecological Applications* 11:356–370.

Clark, D.A., S. Brown, D.W. Kicklighter, J.Q. Chambers, J.R. Thomlinson, J. Ni, and E.A. Holland. 2001b. Net primary production in tropical forests: An evaluation and synthesis of existing data. *Ecological Applications* 11:371–384.

Clarkson, D.T. 1985. Factors affecting mineral nutrient acquisition by plants. *Annual Review of Plant Physiology* 36:77–115.

Clein, J.S., and J.P. Schimel. 1994. Reduction in microbial activity in birch litter due to drying and rewetting events. *Soil Biology and Biochemistry* 26:403–406.

Clein, J.S., B.L. Kwiatkowski, A.D. McGuire, J.E. Hobbie, E.B. Rastetter, J.M. Melillo, and D.W. Kicklighter. 2000. Modeling carbon responses of tundra ecosystems to historical and projected climate: A comparison of a plot- and a global-scale ecosystem model to identify process-based uncertainties. *Global Change Biology* 6(Suppl. 1): 127–140.

Clein, J.S., A.D. McGuire, X. Zhang, D.W. Kicklighter, J.M. Melillo, S.C. Wofsy, P.G. Jarvis, and J.M. Massheder. In Press. Historical and projected carbon balances of mature black spruce ecosystems across North America: The role of carbon-nitrogen interactions. *Plant and Soil.*

Clements, F.E. 1916. *Plant Succession: An Analysis of the Development of Vegetation.* Carnegie Institution of Washington Publication 242. Carnegie Institution of Washington, Washington, DC.

Cohen, J.E. 1994. Marine and continental food webs: Three paradoxes. *Philosophical Transactions of the Royal Society of London, Series B* 343:57–69.

COHMAP. 1988. Climatic changes of the last 18,000 years: Observations and model simulations. *Science* 241:1043–1052.

Cole, J.J., N.F. Caracao, G.W. Kling, and T.K. Kratz. 1994. Carbon dioxide supersaturation in the surface waters of lakes. *Science* 265:1568–1570.

Coleman, D.C. 1994. The microbial loop concept as used in terrestrial soil ecology studies. *Microbial Ecology* 28:245–250.

Coley, P.D. 1986. Costs and benefits of defense by tannins in a neotropical tree. *Oecologia* 70:238–241.

Coley, P.D., J.P. Bryant, and F.S. Chapin III. 1985. Resource availability and plant anti-herbivore defense. *Science* 230:895–899.

Comeleo, R.L., J.F. Paul, P.V. August, J. Copeland, C. Baker, S.S. Hale, and R.W. Latimer. 1996. Relationships between watershed stressors and sediment contamination in Chesapeake Bay estuaries. *Landscape Ecology* 11:307–319.

Connell, J.H., and R.O. Slatyer. 1977. Mechanisms of succession in natural communities and their role in community stability and organization. *American Naturalist* 111:1114–1119.

Cornelissen, J.H.C. 1996. An experimental comparison of leaf decomposition rates in a wide range of temperate plant species and types. *Journal of Ecology* 84:573–582.

Correll, D.L. 1997. Buffer zones and water quality protection: General principles. Pages 7–20 *in* N.E. Haycock, T.P. Burt, K.W.T. Goulding, and G. Pinay, editors. *Buffer Zones: Their Processes and Potential in Water Protection.* Quest Environmental, Harpenden.

Costa, M.H., and J.A. Foley. 1999. Trends in the hydrological cycle of the Amazon basin. *Journal of Geophysical Research* 104:14189–14198.

Costanza, R., R. d'Arge, R. de Groot, S. Farber, M. Grasso, B. Hannon, K. Limburg, S. Naeem, R.V. O'Neill, J. Paruelo, R.G. Raskin, P. Sutton, and M. van den Belt. 1997. The value of the world's ecosystem services and natural capital. *Nature* 387:253–260.

Costanza, R., F.H. Sklar, and M.L. White. 1990. Modeling coastal landscape dynamics. *BioScience* 40:91–107.

Cottingham, K.L., and S.R. Carpenter. 1994. Predictive indices of ecosystem resilience in models of north temperate lakes. *Ecology* 75:2127–2138.

Cottingham, K.L., B.L. Brown, and J.T. Lennon. 2001. Biodiversity may regulate the temporal variability of ecological processes. *Ecology Letters* 4:72–85.

Coûteaux, M.-M., P. Bottner, and B. Berg. 1995. Litter decomposition, climate and litter quality. *Trends in Ecology and Evolution* 10:63–66.

Cowles, H.C. 1899. The ecological relations of the vegetation on the sand dunes of Lake Michigan. *Botanical Gazette* 27:95–117.

Craine, J.M., F.S. Chapin III, D.A. Wedin, and P.B. Reich. In press. Development of grassland root systems and their effects on ecosystem properties. *Plant and Soil.*

Craine, J.M., J. Froehle, D.G. Tilman, D.A. Wedin, and F.S. Chapin III. 2001. The relationships among root and leaf traits of 76 grassland species and relative abundance along fertility and disturbance gradients. *Oikos* 93:274–285.

Craine, J.M., D.A. Wedin, and F.S. Chapin III. 1999. Predominance of ecophysiological over environmental controls over CO_2 flux in a Minnesota grassland. *Plant and Soil* 207:77–86.

Cramer, W., A. Bondeau, F.I. Woodward, I.C. Prentice, R.A. Betts, V. Brovkin, P.M. Cox, V. Fisher, J.A. Foley, A.D. Friend, C. Kucharik, M.R. Lomas, N. Ramankutty, S. Sitch, B. Smith, A. White, and C. Young-Molling. 2001. Global response of terrestrial ecosystem structure and function to CO_2 and climate change: Results from six dynamic global vegetation models. *Global Change Biology.* 7:357–373.

Crews, T.E., K. Kitayama, J.H. Fownes, R.H. Riley, D.A. Herbert, D. Mueller-Dombois, and P.M. Vitousek. 1995. Changes in soil phosphorus fractions and ecosystem dynamics across a long chronosequence in Hawaii. *Ecology* 76:1407–1424.

Crocker, R.L., and J. Major. 1955. Soil development in relation to vegetation and surface age at Glacier Bay, Alaska. *Journal of Ecology* 43:427–448.

Crowley, T.J. 1995. Ice-age terrestrial carbon changes revisited. *Global Biogeochemical Cycles* 9:377–389.

Curtis, P.S., and X. Wang. 1998. A meta-analysis of elevated CO_2 effects on woody plant mass, form, and physiology. *Oecologia* 113:299–313.

Cyr, H., and M.L. Pace. 1993. Magnitude and patterns of herbivory in aquatic and terrestrial ecosystems. *Nature* 343:148–150.

D'Antonio, C.M., and P.M. Vitousek. 1992. Biological invasions by exotic grasses, the grass-fire cycle, and global change. *Annual Review of Ecology and Systematics* 23:63–87.

Daily, G.C. 1997. *Nature's Services: Societal Dependence on Natural Ecosystems*. Island Press, Washington, DC.

Daily, G.C. and K. Ellison. 2002. *The New Economy of Nature* Island Press, Washington, DC.

Daily, G.C., T. Soderqvist, S. Aniyar, K. Arrow, P. Dasgupta, P.R. Ehrlich, C. Folke, A.-M. Jansson, B.-O. Jansson, N. Kautsky, S. Levin, J. Lubchenco, K.-G. Maler, D. Simpson, D. Starrett, D. Tilman, and B. Walker. 2000. Ecology: The value of nature and the nature of value. *Science* 289:395–396.

Dale, V.H., S. Brown, R. Haeuber, N.T. Hobbs, N. Huntly, R.J. Naiman, W.E. Riebsame, M.G. Turner, and T. Valone. 2000. Ecological principles and guidelines for managing the use of land. *Ecological Applications* 10:639–670.

Davidson, E.A., P.A. Matson, P.M. Vitousek, R. Riley, K. Dunkin, G. Garcia-Mendez, and J.M. Maass. 1993. Process regulation of soil emissions of NO and N_2O in a seasonally dry tropical forest. *Ecology* 74:130–139.

Davies, W.J., and J. Zhang. 1991. Root signals and the regulation of growth and development of plants in drying soil. *Annual Review of Plant Physiology and Molecular Biology* 42:55–76.

Davis, M.B., R.R. Calcote, S. Sugita, and H. Takahara. 1998. Patchy invasion and the origin of a hem-lock-hardwood forest mosaic. *Ecology* 79:2641–2659.

Davis, S.D., and H.A. Mooney. 1986. Tissue water relations of four co-occurring chaparral shrubs. *Oecologia* 70:527–535.

Davis, S.M., and J.C. Ogden, editors. 1994. *Everglades: The Ecosystem and Its Restoration*. St Lucie, Delray Beach, FL.

Dawson, T.E. 1993. Water sources of plants as determined from xylem-water isotopic composition: Perspectives on plant competition, distribution, and water relations. Pages 465–496 *in* J.R. Ehleringer, A.E. Hall, and G.D. Farquhar, editors. *Stable Isotopes and Plant Carbon-Water Relations*. Academic Press, San Diego, CA.

DeAngelis, D.L., and W.M. Post. 1991. Positive feedback and ecosystem organization. Pages 155–178 *in* M. Higashi and T.P. Burns, editors. *Theoretical Studies of Ecosystems: The Network Perspective*. Cambridge University Press, Cambridge, UK.

DeAngelis, D.L., L.J. Gross, M.A. Huston, W.F. Wolff, D.M. Fleming, E.J. Comiskey, and S.M. Sylvester. 1998. Landscape modeling for Everglades ecosystem restoration. *Ecosystems* 1:64–75.

Del Grosso, S.J., W.J. Parton, A.R. Mosier, D.S. Ojima, A.E. Kulmala, and S. Phongpan. 2000. General model for N_2O and N_2 gas emissions from soils due to denitrification. *Global Biogeochemical Cycles* 14:1045–1060.

Delmas, R., C. Jambert, and Serga. 1997. Global inventory of NO_x sources. *Nutrient Cycling in Agroecosystems* 48:51–60.

Demming-Adams, B., and W.W. Adams. 1996. The role of xanthophyll cycle carotenoids in the protection of photosynthesis. *Trends in Plant Sciences* 1:21–26.

Detling, J.K. 1988. Grasslands and savannas: Regulation of energy flow and nutrient cycling by herbivores. Pages 131–148 *in* L.R. Pomeroy and J.J. Alberts, editors. *Concepts of Ecosystem Ecology*. Springer-Verlag, New York.

Detling, J.K., D.T. Winn, C. Procter-Gregg, and E.L. Painter. 1980. Effects of simulated grazing by below-ground herbivores on growth, CO_2 exchange, and carbon allocation patterns of *Bouteloua gracilis*. *Journal of Applied Ecology* 17:771–778.

Dingman, S.L. 2001. *Physical Hydrology*. Prentice-Hall, Upper Saddle River, NJ.

Dirzo, R., and A. Miranda. 1991. Altered patterns of herbivory and diversity in the forest understory: A case study of the possible consequences of contemporary defaunation. Pages 273–287 *in* P.W. Price, T.M. Lewinsohn, G.W. Fernandes, and W.W. Benson, editors. *Plant-Animal Interactions: Evolutionary Ecology in Tropical and Temperate Regions*. Wiley, New York.

Dixon, R.K., S. Brown, R.A. Houghton, A.M. Solomon, M.C. Trexler, and J. Wisniewski. 1994. Carbon pools and flux of global forest ecosystems. *Science* 263:185–190.

Dobson, A.P., A.D. Bradshaw, and A.J.M. Baker. 1997. Hopes for the future: Restoration ecology and conservation biology. *Science* 277:515–522.

Dokuchaev, V.V. 1879. Abridged historical account and critical examination of the principal soil classifications existing. *Transactions of the Petersburg Society of Naturalists* 1:64–67.

Drake, B.G., G. Peresta, E. Beugeling, and R. Matamala. 1996. Long-term elevated CO_2 exposure in a Chesapeake Bay wetland: Ecosystem gas exchange, primary production, and tissue nitrogen. Pages 197–214 *in* G.W. Koch and H.A. Mooney, editors. *Carbon Dioxide and Terrestrial Ecosystems*. Academic Press, San Diego, CA.

Driscoll, C.T., G.B. Lawrence, A.J. Bulger, T.J. Butler, C.S. Cronan, C. Eagar, K.F. Lambert, G.E. Likens, J.L. Stoddard, and K.C. Weathers. 2001. Acidic deposition in the northeastern United States: Sources and inputs, ecosystem effects and management strategies. *BioScience* 51:180–198.

Dugdale, R.C. 1976. Nutrient cycles. Pages 141–172 *in* D.H. Cushing and J.J. Walsh, editors. *The Ecology of the Seas*. Saunders, Philadelphia.

Egler, F.E. 1954. Vegetation science concepts. I. Initial floristic composition, a factor in old-field vegetation development. *Vegetatio* 4:414–417.

Ehleringer, J.R. 1993. Carbon and water relations in desert plants: An isotopic perspective. Pages 155–172 *in* J.R. Ehleringer, A.E. Hall, and G.D. Farquhar, editors. *Stable Isotopes and Plant Carbon-Water Relations*. Academic Press, San Diego, CA.

Ehleringer, J.R., N. Buchmann, and L.B. Flanagan. 2000. Carbon isotope ratios in belowground carbon cycle processes. *Ecological Applications* 10:412–422.

Ehleringer, J.R., and C.B. Field, editors. 1993. *Scaling Physiological Processes: Leaf to Globe*. Academic Press, San Diego, CA.

Ehleringer, J.R., and H.A. Mooney. 1978. Leaf hairs: Effects on physiological activity and adaptive value to a desert shrub. *Oecologia* 37:183–200.

Ehleringer, J.R., and C.B. Osmond. 1989. Stable isotopes. Pages 281–300 *in* R.W. Pearcy, J. Ehleringer, H.A. Mooney, and P.W. Rundel, editors. *Plant Physiological Ecology: Field Methods and Instrumentation*. Chapman & Hall, London.

Ehleringer, J.R., A.E. Hall, and G.D. Farquhar, editors. 1993. *Stable Isotopes and Plant Carbon-Water Relations*. Academic Press, San Diego, CA.

Ellenberg, H. 1979. Man's influence on tropical mountain ecosystems in South-America: 2nd Tansley Lecture. *Journal of Ecology* 67:401–416.

Ellenberg, H. 1978. *Vegetation von Mittleuropa*. Eugen Ulmer, Stuttgart.

Ellsworth, D.S. 1999. CO_2 enrichment in a maturing pine forest: Are CO_2 exchange and water status in the canopy affected? *Plant Cell and Environment* 22:461–472.

Elser, J.J., and J. Urabe. 1999. The stoichiometry of consumer-driven nutrient recycling: Theory, observations, and consequences. *Ecology* 80:735–751.

Elser, J.J., W.F. Fagan, R.F. Denno, D.R. Dobberfuhl, A. Folarin, A. Huberty, S. Interlandl, S.S. Kilham, E. McCauley, K.L. Schulz, E.H. Siemann, and R.W. Sterner. 2000. Nutritional constraints in terrestrial and freshwater food webs. *Nature* 408:578–580.

Elton, C.S. 1927. *Animal Ecology*. Macmillan, New York.

Enquist, B.J., and K.J. Niklas. 2001. Invariant scaling relations across tree-dominated communities. *Nature* 410:655–660.

Enríquez, S., C.M. Duarte, and K. Sand-Jensen. 1993. Patterns in decomposition rates among photosynthetic organisms: The importance of detritus C:N:P content. *Oecologia* 94:457–471.

Estes, J.A., and J.F. Palmisano. 1974. Sea otters: Their role in structuring nearshore communities. *Science* 185:1058–1060.

Estes, J.A., M.T. Tinker, T.M. Williams, and D.F. Doak. 1998. Killer whale predation on sea otters linking oceanic and nearshore ecosystems. *Science* 282:473–476.

Eugster, W., W.R. Rouse, R.A. Pielke, J.P. McFadden, D.D. Baldocchi, T.G.F. Kittel, F.S. Chapin III, G.E. Liston, P.L. Vidale, E. Vaganov, and S. Chambers. 2000. Land-atmosphere energy exchange in arctic tundra and boreal forest: available data and feedbacks to climate. *Global Change Biology* 6(Suppl. 1):84–115.

Evans, L.T. 1980. The natural history of crop yield. *American Scientist* 68:388–397.

Eviner, V.T., and F.S. Chapin III. 1997. Plant-microbial interactions. *Nature* 385:26–27.

Eviner, V.T., F.S. Chapin III, and C.E. Vaughn. 2000. Nutrient manipulations in terrestrial ecosystems. Pages 291–307 *in* O.E. Sala, R.B. Jackson, H.A. Mooney, and R.W. Howarth, editors. *Methods in Ecosystem Science*. Springer-Verlag, New York.

Fahey, T., C. Bledsoe, R. Day, R. Ruess, and A. Smucker. 1998. *Fine Root Production and Demography*. CRC Press, Boca Raton, FL.

Fahrig, L., and G. Merriam. 1985. Habitat patch connectivity and population survival. *Ecology* 66:1762–1768.

Falkowski, P.G., R.T. Barber, and V. Smetacek. 1999. Biogeochemical controls and feedbacks on ocean primary production. *Science* 281:200–206.

Farquhar, G.D., and T.D. Sharkey. 1982. Stomatal conductance and photosynthesis. *Annual Review of Plant Physiology* 33:317–345.

Fastie, C.L. 1995. Causes and ecosystem consquences of multiple pathways of primary succession at Glacier Bay, Alaska. *Ecology* 76:1899–1916.

FCCC. 2000. *Methodological Issue. Land-use, land-use change and forestry. Synthesis report on national greenhouse gas information reported by Annex I Parties for the land-use change and forestry sector and agricultural soils category.* Paper presented at the Framework Convention on Climate Change. Subsidiary Body for Scientific and Technological Advice, Bonn.

Federov, A.V., and S.G. Philander. 2000. Is El Niño changing? *Science* 288:1997–2002.

Feeny, P.P. 1970. Seasonal changes in oak leaf tannins and nutrients as cause of spring feeding by winter moth caterpillars. *Ecology* 51:565–581.

Fenchel, T. 1994. Microbial ecology on land and sea. *Philosophical Transactions of the Royal Society of London, Series B* 343:51–56.

Fenn, M.E., M.A. Poth, J.D. Aber, J.S. Baron, B.T. Bormann, D.W. Johnson, A.D. Lemly, S.G. McNulty, D.F. Ryan, and R. Stottlemeyer. 1998. Nitrogen excess in North American ecosystems: Predisposing factors, ecosystem responses and management strategies. *Ecological Applications* 8:706–733.

Fenner, M. 1985. *Seed Ecology*. Chapman & Hall, London.

Field, C. 1983. Allocating leaf nitrogen for the maximization of carbon gain: Leaf age as a control on the allocation program. *Oecologia* 56:341–347.

Field, C.B. 1991. Ecological scaling of carbon gain to stress and resource availability. Pages 35–65 *in* H.A. Mooney, W.E. Winner, and E.J. Pell, editors. *Integrated Responses of Plants to Stress*. Academic Press, San Diego.

Field, C., and H.A. Mooney. 1986. The photosynthesis-nitrogen relationship in wild plants. Pages 25–55 *in* T.J. Givnish, editor. *On the Economy of Plant Form and Function*. Cambridge University Press, Cambridge, UK.

Field, C., F.S. Chapin III, P.A. Matson, and H.A. Mooney. 1992. Responses of terrestrial ecosystems to the changing atmosphere: A resource-based approach. *Annual Review of Ecology and Systematics* 23:201–235.

Firestone, M.K., and E.A. Davidson. 1989. Microbiological basis of NO and N_2O production and consumption in soil. Pages 7–21 *in* M.O. Andreae, and D.S. Schimel, editors. *Exchange of Trace Gases Between Terrestrial Ecosystems and the Atmosphere*. J Wiley, New York.

Fisher, R.F., and D. Binkley. 2000. *Ecology and Management of Forest Soils*. J Wiley, New York.

Fisher, S.G., N.B. Grimm, E. Martí, R.M. Holmes, and J.B. Jones Jr. 1998. Material spiraling in stream corridors: A telescoping ecosystem model. *Ecosystems* 1:19–34.

Flanagan, P.W., and K. Van Cleve. 1983. Nutrient cycling in relation to decomposition and organic matter quality in taiga ecosystems. *Canadian Journal of Forest Research* 13:795–817.

Flanagan, P.W., and A.K. Veum. 1974. Relationships between respiration, weight loss, temperature, and moisture in organic residues on tundra. Pages 249–277 *in* A.J. Holding, O.W. Heal, S.F. Maclean Jr., and P.W. Flanagan, editors. *Soil Organisms and Decomposition in Tundra*. Tundra Biome Steering Committee, Stockholm.

Flannery, T.F. 1994. *The Future Eaters*. Reed Books, Victoria, BC, Canada.

Fog, K. 1988. The effect of added nitrogen on the rate of decomposition of organic matter. *Biological Review* 63:433–462.

Foley, J.A., J.E. Kutzbach, M.T. Coe, and S. Levis. 1994. Feedbacks between climate and boreal forests during the Holocene epoch. *Nature* 371: 52–54.

Foley, J.A., I.C. Prentice, N. Ramankutty, S. Levis, D. Pollard, S. Sitch, and A. Haxeltine. 1996. An integrated biosphere model of land surface processes, terrestrial carbon balance, and vegetation dynamics. *Global Biogeochemical Cycles* 10:603–628.

Folland, C.K., T.R. Karl, J.R. Christy, R.A. Clarke, G.V. Gruza, J. Jouzel, M.E. Mann, J. Oerlemans, M.J. Salinger, and S.-W. Wang. 2001. Observed climate variability and change. Pages 99–181 *in* J.T. Houghton, Y. Ding, D.J. Griggs, M. Noguer, P.J. van der Linden, X. Dai, K. Maskell, and C.A. Johnson, editors. *Climate Change 2001: The Scientific Basis*. Cambridge University Press, Cambridge, UK.

Forman, R.T.T. 1995. *Land Mosaics: The Ecology of Landscapes and Regions*. Cambridge University Press, Cambridge, UK.

Foster, D.R. 1988. Disturbance history, community organization and vegetation dynamics of the old-growth Pisgah Forest, southwestern New Hampshire. *Journal of Ecology* 76:105–134.

Foster, D.R., and G. Motzkin. 1998. Ecology and conservation in the cultural landscape of New England: Regional forest dynamics in central New England. *Ecosystems* 1:96–119.

Foster, D.R., D.A. Orwig, and J.S. McLachlan. 1996. Ecological and conservation insights from reconstructive studies of temperate old-growth forests. *Trends in Ecology and Evolution* 11:419–424.

Francis, R.C. 1990. Fisheries science and modeling: A look to the future. *Natural Resource Modeling* 4:1–10.

Franklin, J.F., D.R. Berg, D.A. Thornburgh, and J.C. Tappeiner. 1997. Alternative silvicultural approaches to timber harvesting: Variable retention harvest systems. Pages 111–140 *in* K.A. Kohm and J.F. Franklin, editors. *Creating a Forestry for the 21st Century: The Science of Ecosystem Management*. Island Press, Washington, DC.

Freemark, K.E., and H.G. Merriam. 1986. Importance of area and habitat heterogeneity to bird assemblages in temperate forest fragments. *Biological Conservation* 31:95–105.

Frelich, L.E., and P.B. Reich. 1995. Spatial patterns and succession in a Minnesota southern-boreal forest. *Ecological Monographs* 65:325–346.

Fretwell, S.D. 1977. The regulation of plant communities by food chains exploiting them. *Perspectives in Biology and Medicine* 20:169–185.

Frost, T.M., S.R. Carpenter, A.R. Ives, and T.K. Kratz. 1995. Species compensation and complementarity in ecosystem function. Pages 224–239 *in* C.G. Jones and J.H. Lawton, editors. *Linking Species and Ecosystems*. Chapman & Hall, New York.

Fung, I.Y., C.J. Tucker, and K.C. Prentice. 1987. Application of advanced very high resolution radiometer vegetation index to study atmosphere-biosphere exchange of CO_2. *Journal of Geophysical Research* 92D:2999–3015.

Gabriel, H.W., and G.F. Tande. 1983. A Regional Approach to Fire History in Alaska. USDI Bureau of Land Management BLM-Alaska Technical Report 9. USDI Bureau of Land Management, Anchorage, AK.

Galloway, J.N. 1996. Anthropogenic mobilization of sulfur and nitrogen: Immediate and delayed consequences. *Annual Review of Energy in the Environment* 21:261–292.

Galloway, J.N., W.H. Schlesinger, H. Levy II, A. Michaels, and J.L. Schnoor. 1995. Nitrogen fixation: Anthropogenic enhancement-environmental response. *Global Biogeochemical Cycles* 9:235–252.

Gardner, R.H., B.T. Milne, M.G. Turner, and R.V. O'Neil. 1987. Neutral models for the analysis of broad-scale landscape pattern. *Landscape Ecology* 1:19–28.

Gardner, W.R. 1983. Soil properties and efficient water use: An overview. Pages 45–64 *in* H.M. Taylor, W.R. Jordan, and T.R. Sinclair, editors. *Limitations to Efficient Water Use in Crop Production*. American Society of Agronomy, Madison, WI.

Garnier, E. 1991. Resource capture, biomass allocation and growth in herbaceous plants. *Trends in Ecology and Evolution* 6:126–131.

Gash, J.H.C., and C.A. Nobre. 1997. Climatic effects of Amazonian deforestation: Some results from ABRACOS. *Bulletin of the American Meteorological Society* 78:823–830.

Gee, J.H.R. 1991. Specialist aquatic feeding mechanisms. Pages 186–209 *in* R.S.K. Barnes and K.H. Mann, editors. *Fundamentals of Aquatic Ecology*. Blackwell Scientific, Oxford, UK.

Gholz, H.L., D.A. Wedin, S.M. Smitherman, M.E. Harmon, and W.J. Parton. 2000. Long-term dynamics of pine and hardwood litter in contrasting environments: Toward a global model of decomposition. *Global Change Biology* 6:751–765.

Gill, R.A., and R.B. Jackson. 2000. Global patterns of root turnover for terrestrial ecosystems. *New Phytologist* 147:13–31.

Giller, P.S., and B. Malmqvist. 1998. *The Biology of Streams and Rivers*. Oxford University Press, Oxford, UK.

Gleason, H.A. 1926. The individualistic concept of the plant association. *Bulletin of the Torrey Botanical Club* 53:7–26.

Gleick, P.H. 1998. *The World's Water 1998–1999. The Biennial Report on Freshwater Resources*. Island Press, Washington, DC.

Gleick, P.H. 2000. *The World's Water 2000–2001. The Biennial Report on Freshwater Resources*. Island Press, Washington, DC.

Gollan, T., N.C. Turner, and E.-D. Schulze. 1985. The responses of stomata and leaf gas exchange to vapor pressure deficits and soil water content. III. In the sclerophyllous woody species *Nerium oleander*. *Oecologia* 65:356–362.

Golley, F. 1961. Energy values of ecological materials. *Ecology* 42:581–584.

Golley, F.B. 1993. *A History of the Ecosystem Concept in Ecology: More Than the Sum of the Parts*. Yale University Press, New Haven, CT.

Goode, J.G., R.J. Yokelson, D.E. Ward, R.A. Susott, R.E. Babbitt, M.A. Davies, and W.M. Hao. 2000. Measurements of excess O_3, CO_2, CO, CH_4, C_2H_4, C_2H_2, HCN, NO, NH_3, HCOOH, CH_3COOH, HCHO, and CH_3OH in 1997 Alaskan biomass burning plumes by airborne Fourier transform infrared spectroscopy (AFTIR). *Journal of Geophysical Research* 105:22147–22166.

Gorham, E. 1991. Biogeochemistry: Its origins and development. *Biogeochemistry* 13:199–239.

Gosz, J.R. 1991. Fundamental ecological characteristics of landscape boundaries. Pages 8–30 *in* M.M. Holland, P.G. Risser, and R.J. Naiman, editors. *Ecotones: The Role of Landscape Boundaries in the Management and Restoration of Changing Environments*. Chapman & Hall, New York.

Goulden, M.L., J.W. Munger, S.-M. Fan, B.C. Daube, and S.C. Wofsy. 1996. CO_2 exchange by a deciduous forest: Response to interannual climate variability. *Science* 271:1576–1578.

Goulden, M.L., S.C. Wofsy, J.W. Harden, S.E. Trumbore, P.M. Crill, S.T. Gower, T. Fries, B.C. Daube, S.M. Fan, D.J. Sutton, A. Bazzaz, and J.W. Munger. 1998. Sensitivity of boreal forest carbon balance to warming. *Science* 279:214–217.

Goulder, L.H., and D. Kennedy. 1997. Valuing ecosystem services: Philosophical bases and empirical methods. Pages 23–48 *in* G.C. Daily, editor. *Nature's Services: Societal Dependence on Natural Ecosystems*. Island Press, Washington, DC.

Gower, S.T. 2002. Productivity of terrestrial ecosystems. Pages 516–521 *in* H.A. Mooney, and J. Canadell, editors. *Encyclopedia of Global Change*. Vol. 2. Blackwell Scientific, Oxford, UK.

Gower, S.T., C.J. Kucharik, and J.M. Norman. 1999. Direct and indirect estimation of leaf area index, F_{APAR}, and net primary production of terrestrial ecosystems. *Remote Sensing of the Environment* 70:29–51.

Graedel, T.E., and P.J. Crutzen. 1995. *Atmosphere, Climate, and Change*. Scientific American Library, New York.

Graetz, R.D. 1991. The nature and significance of the feedback of change in terrestrial vegetation on global atmospheric and climatic change. *Climatic Change* 18:147–173.

Graustein, W.C., and R.L. Armstrong. 1983. The use of strontium-87/Strontium-86 ratios to measure atmospheric transport into forested watersheds. *Science* 219:289–292.

Grime, J.P., and R. Hunt. 1975. Relative growth rate: Its range and adaptive significance in a local flora. *Journal of Ecology* 63:393–422.

Grime, J.P., G. Mason, A.V. Curtis, J. Rodman, S.R. Band, M.A.G. Mowforth, A.M. Neal, and S. Shaw. 1981. A comparative study of germination characteristics in a local flora. *Journal of Ecology* 69: 1017–1059.

Grimm, V., and C. Wissel. 1997. Babel, or the ecological stability discussions: An inventory and analysis of terminology and guide for avoiding confusion. *Oecologia* 109:323–334.

Grogan, P., and F.S. Chapin III. 2000. Nitrogen limitation of production in a Californian annual grassland: The contribution of arbuscular mycorrhizae. *Biogeochemistry* 49:37–51.

Gross, M.R., R.M. Coleman, and R.M. McDowell. 1988. Aquatic productivity and the evolution of anadromous fish migration. *Science* 239:1291–1293.

Guenther, A., C. Hewitt, D. Erickson, R. Fall, C. Geron, T. Graedel, P. Harley, L. Klinter, M. Lerdau, W. McKay, T. Pierce, B. Scholes, R. Steinbrecher, R. Tallamraju, R. Taylor, and P. Zimmerman. 1995. A global model of natural volatile organic compound emissions. *Journal of Geophysical Research* 100D: 8873–8892.

Gulledge, J., A. Doyle, and J. Schimel. 1997. Different NH_4^+-inhibition patterns of soil CH_4 consumption: A result of distinct CH_4 oxidizer populations across sites? Soil Biology and Biochemistry 29:13–21.

Gulmon, S.L., and H.A. Mooney. 1986. Costs of defense on plant productivity. Pages 681–698 *in* T.J. Givnish, editor. *On the Economy of Plant Form and Function*. Cambridge University Press, Cambridge, UK.

Gutierrez, J.R., and W.G. Whitford. 1987. Chihuahuan desert annuals: Importance of water and nitrogen. *Ecology* 68:2032–2045.

Hagen, J.B. 1992. *An Entangled Bank: The Origins of Ecosystem Ecology*. Rutgers University Press, New Brunswick, NJ.

Hairston, N.G., F.E. Smith, and L.B. Slobodkin. 1960. Community structure, population control and competition. *American Naturalist* 94:421–425.

Hall, S.J., and P.A. Matson. 1998. Nitrogen oxide emissions after nitrogen additions in tropical forests. *Nature* 400:152–155.

Hallbacken, L. 1992. *The Nature and Importance of Long-Term Soil Acidification in Swedish Forest Ecosystems*. Swedish University of Agricultural Sciences, Department of Ecology and Environmental Research, Uppsala.

Hanski, I. 1999. *Metapopulation Ecology*. Oxford University Press, Oxford, UK.

Harden, J.W., S.E. Trumbore, B.J. Stocks, A. Hirsch, S.T. Gower, K.P. O'Neill, and E.S. Kasischke. 2000. The role of fire in the boreal carbon budget. *Global Change Biology* 6(Suppl. 1):174–184.

Harte, J., and A.P. Kinzig. 1993. Mutualism and competition between plants and decomposers: Implications for nutrient allocation in ecosystems. *American Naturalist* 141:829–846.

Hartshorn, G.S. 1980. Neotropical forest dynamics. Biotropica 12:23–30.

Harwell, M.A. 1997. Ecosystem management of south Florida. *BioScience* 47:499–512.

Hay, M.E., and W. Fenical. 1988. Marine plant-herbivore interactions: The ecology of chemical defense. *Annual Review of Ecology and Systematics* 19:111–145.

Haynes, R.J. 1986. The decomposition process: Mineralization, immobilization, humus formation, and degradation. Pages 52–126 *in* R.J. Haynes, editor. *Mineral Nitrogen in the Plant-Soil System*. Academic Press, Orlando, FL.

Heal, O.W., and S.F. MacLean. 1975. Comparative productivity in ecosystems: Secondary productivity. Pages 89–108 *in* W.H. van Dobben and R.H. Lowe-McConnell, editors. *Unifying Concepts in Ecology*. Junk, The Hague.

Hedin, L.O., J.J. Armesto, and A.H. Johnson. 1995. Patterns of nutrient loss from unpolluted, old-growth temperate forests: Evaluation of biogeochemical theory. *Ecology* 76:493–509.

Heinselman, M.L. 1973. Fire in the virgin forests of the Boundary Waters Canoe Area, Minnesota. *Quaternary Research* 3:329–382.

Helfield, J.M., and R.J. Naiman. 2001. Effects of salmon-derived nitrogen on riparian forest growth and implications for stream productivity. *Ecology* 82:2403–2409.

Herms, D.A., and W.J. Mattson. 1992. The dilemma of plants: To grow or defend. *Quarterly Review of Biology* 67:283–335.

Heywood, V.H., and R.T. Watson, editors. 1995. *Global Biodiversity Assessment*. Cambridge University Press, Cambridge, UK.

Hibbert, A.R. 1967. Forest treatment effects on water yield. Pages 527–543 *in* W.E. Sopper and H.W. Lull, editors. *International Symposium on Forest Hydrology*. Pergamon Press, New York.

Hikosaka, K., and T. Hirose. 2001. Nitrogen uptake and use by competing individuals in a *Xanthium canadense* stand. Oecologia 126:174–181.

Hilborn, R., C.J. Walters, and D. Ludwig. 1995. Sustainable exploitation of renewable resources. *Annual Review of Ecology and Systematics* 26:45–67.

Hirose, T., and M.J.A. Werger. 1987. Maximizing daily canopy photosynthesis with respect to the leaf nitrogen allocation pattern in the canopy. *Oecologia* 72:520–526.

Hobbie, S.E. 1995. Direct and indirect effects of plant species on biogeochemical processes in arctic ecosystems. Pages 213–224 *in* F.S. Chapin III and C. Körner, editors. *Arctic and Alpine Biodiversity: Patterns, Causes and Ecosystem Consequences*. Springer-Verlag, Berlin.

Hobbie, S.E. 1992. Effects of plant species on nutrient cycling. *Trends in Ecology and Evolution* 7:336–339.

Hobbie, S.E. 2000. Interactions between litter lignin and soil nitrogen availability during leaf litter decomposition in a Hawaiian montane forest. *Ecosystems* 3:484–494.

Hobbie, S.E., and F.S. Chapin III. 1996. Winter regulation of tundra litter carbon and nitrogen dynamics. *Biogeochemistry* 35:327–338.

Hobbie, S.E., and P.M. Vitousek. 2000. Nutrient regulation of decomposition in Hawaiian montane forests: Do the same nutrients limit production and decomposition? *Ecology* 81:1867–1877.

Hobbs, N.T. 1996. Modification of ecosystems by ungulates. *Journal of Wildlife Management* 60: 695–713.

Hobbs, R.J., and H.A. Mooney. 1991. Effects of rainfall variability and gopher disturbance on serpentine annual grassland dynamics. *Ecology* 72:59–68.

Hodge, A., D. Robinson, B. Griffiths, and A. Fitter. 1999. Why plants bother: Root proliferation results in increased nitrogen capture from an organic patch when two grasses compete. *Plant, Cell and Environment* 22:811–820.

Högberg, P., and I.J. Alexander. 1995. Roles of root symbioses in African woodland and forest: Evidence from ^{15}N abundance and foliar nutrient concentrations. *Journal of Ecology* 83:217–224.

Holdridge, L.R. 1947. Determination of world plant formations from simple climatic data. *Science* 105:367–368.

Holland, E.A., B.H. Braswell, J.F. Lamarque, A. Townsend, J. Sulzman, J.F. Muller, F. Dentener, G. Grasseur, H. Levy, J.E. Penner, and G.J. Roelofs. 1997. Variations in the predicted spatial distribution of atmospheric nitrogen deposition and their impact on carbon uptake by terrestrial ecosystems. *Journal of Geophysical Research* 102:15849–15866.

Holland, E.A., F.J. Dentener, B.H. Braswell, and J.M. Sulzman. 1999. Contemporary and pre-industrial global reactive nitrogen budgets. *Biogeochemistry* 46:1–37.

Holling, C.S. 1992. Cross-scale morphology, geometry, and dynamics of ecosystems. *Ecological Monographs* 62:447–502.

Holling, C.S. 1993. Investing in research for sustainability. *Ecological Applications* 3:552–555.

Holling, C.S. 1973. Resilience and stability of ecological systems. *Annual Review of Ecology and Systematics* 4:1–23.

Holling, C.S. 1986. Resilience of ecosystems: Local surprise and global change. Pages 292–317 *in* W.C. Clark and R.E. Munn, editors. *Sustainable Development and the Biosphere*. Cambridge University Press, Cambridge, UK.

Holloway, J.M., R.A. Dahlgren, B. Hansen, and W.H. Casey. 1998. Contribution of bedrock nitrogen to high nitrate concentrations in stream water. *Nature* 395:785–788.

Holsten, E.H., R.A. Werner, and R.L. Develice. 1995. Effects of a spruce beetle (*Coleoptera: Scolytidae*) outbreak and fire on Lutz spruce in Alaska. *Environmental Entomology* 24:1539–1547.

Home, A.J., and C.R. Goldman. 1994. *Limnology*. McGraw-Hill, New York.

Hooper, D.U., and P.M. Vitousek. 1998. Effects of plant composition and diversity on nutrient cycling. *Ecological Monographs* 68:121–149.

Hooper, D.U., F.S. Chapin III, J.J. Ewel, J.P. Grime, A. Hector, P. Inchausti, S. Lavorel, Lawton, D. Lodge, M. Loreau, S. Naeem, B. Schmid, H. Setälä, A.J. Symstad, J. Vandermeer, P.M. Vitousek, and D.A. Wardle. In press. Effects of biodiversity on ecosystem functioning: A consensus of current knowledge and needs for future research. *Ecological Applications*.

Hope, D., M.F. Billett, and M.S. Cresser. 1994. A review of the export of carbon in river water: Fluxes and processes. *Environmental Pollution* 84:301–324.

Houghton, J.T., L.G. Meira Filho, B.A. Callander, N. Harris, A. Kattenberg, and K. Maskell, editors. 1996. *Climate Change 1995. The Science of Climate Change*. Cambridge University Press, Cambridge, UK.

Houghton, R.A., J.L. Hackler, and K.T. Lawrence. 1999. The US carbon budget: Contributions from land-use change. *Science* 285:574–578.

Houghton, R.A., D.L. Skole, C.A. Nobre, J.L. Hackler, K.T. Lawrence, and W.H. Chomentowski. 2000. Annual fluxes of carbon from deforestation and regrowth in the Brazilian Amazon. *Nature* 403:301–304.

Howarth, R.W., G. Billen, D. Swaney, A. Townsend, N. Jaworski, K. Lajtha, J.A. Downing, R. Elmgren, N. Caraco, T. Jordon, F. Berendse, J. Freney, V. Kudeyarov, P. Murdoch, and Z. Zhaoliang. 1996. Regional nitrogen budgets and N and P fluxes for the drainages to the North Atlantic Ocean: Natural and human influences. *Biogeochemistry* 35:75–139.

Howarth, R.W., H. Jensen, R. Marino, and H. Postma. 1995. Transport and processing of phosphorus in near-shore and oceanic waters. Pages 323–345 *in* H. Tiessen, editor. *Phosphorus in the Global Environment: Transfers, Cycles, and Management*. J Wiley, Chichester, UK.

Hu, S., F.S. Chapin III, M.K. Firestone, C.B. Field, and N.R. Chiariello. 2001. Nitrogen limitation of microbial decomposition in a grassland under elevated CO_2. *Nature* 409:188–191.

Huante, P., E. Rincón, and I. Acosta. 1995. Nutrient availability and growth rate of 34 woody species from a tropical deciduous forest in Mexico. *Functional Ecology* 9:849–858.

Huante, P., E. Rincón, and F.S. Chapin III. 1998. Effect of changing light availability on nutrient foraging in tropical deciduous tree-seedlings. *Oikos* 82:449–458.

Huffaker, C.B. 1957. Fundamentals of biological control of weeds. *Hilgardia* 27:101–157.

Humphreys, W.F. 1979. Production and respiration in animal populations. *Journal of Animal Ecology* 48:427–454.

Hunt, H.W., D.C. Coleman, E.R. Ingham, E.T. Elliott, J.C. Moore, S.L. Rose, C.P.P. Reid, and C.R. Morley. 1987. The detrital food web in a shortgrass prairie. *Biology and Fertility of Soils* 3:57–68.

Hutley, L.B., D. Doley, D.J. Yates, and A. Boonsaner. 1997. Water-balance of an Australian subtropical rain-forest at altitude: The ecological and physiological significance of intercepted cloud and fog. *Australian Journal of Botany* 45:311–329.

Hynes, H.B.N. 1975. The stream and its valley. *Verhandlungen der Internationalen Vereinigung für Theoretische und Angewandte Limnologie* 19:1–15.

Ingestad, T., and G.I. Ågren. 1988. Nutrient uptake and allocation at steady-state nutrition. *Physiologia Plantarum* 72:450–459.

Insam, H. 1990. Are the soil microbial biomass and basal respiration governed by the climatic regime? *Soil Biology and Biochemistry* 22:525–532.

Irons, J.G. III, J.P. Bryant, and M.W. Oswood. 1991. Effects of moose browsing on decomposition rates of birch leaf litter in a subarctic stream. *Canadian Journal of Fisheries and Aquatic Science* 48: 442–444.

Jackson, R.B., J. Canadell, J.R. Ehleringer, H.A. Mooney, O.E. Sala, and E.-D. Schulze. 1996. A global analysis of root distributions for terrestrial biomes. *Oecologia* 108:389–411.

Jackson, R.B., J.S. Sperry, and T.E. Dawson. 2000. Root water uptake and transport: Using physiological processes in global predictions. Trends in Plant Science 5:482–488.

Jaeger, C.H. III, S.E. Lindow, W. Miller, E. Clark, and M.K. Firestone. 1999a. Mapping sugar and amino acid availability in soil around roots with bacterial sensors of sucrose and tryptophan. *Applied and Environmental Microbiology* 65:2685–2690.

Jaeger, C.H., R.K. Monson, M.C. Fisk, and S.K. Schmidt. 1999b. Seasonal partitioning of nitrogen and soil microorganisms in an alpine ecosystem. *Ecology* 80:1883–1891.

Jarvis, P.G. 1976. The interpretation of the variations in leaf water potential and stomatal conductance found in canopies in the field. *Philosophical Transactions of the Royal Society of London, Series B* 273:593–610.

Jarvis, P.G., and J.W. Leverenz. 1983. Productivity of temperate, deciduous and evergreen forests. Pages 233–280 *in* O.L. Lange, P.S. Nobel, C.B. Osmond, and H. Ziegler, editors. *Encyclodedia of Plant Physiology, new series*. Springer-Verlag, Berlin.

Jarvis, P.G., and K.G. McNaughton. 1986. Stomatal control of transpiration: Scaling up from leaf to region. *Advances in Ecological Research* 15:1–49.

Jefferies, R.L. 1988. Pattern and process in arctic coastal vegetation in response to foraging by lesser snow geese. Pages 281–300 *in* M.J.A. Werger, P.J.M. Van der Aart, H.J. During, and J.T.A. Verhoeven, editors. *Plant Form and Vegetation Structure. Adaptation, Plasticity and Relation to Herbivory*. SPB Academic, The Hague.

Jefferies, R.L., and J.P. Bryant. 1995. The plant-vertebrate herbivore interface in arctic ecosystems. Pages 271–281 *in* F.S. Chapin III and C. Körner, editors. *Arctic and Alpine Biodiversity: Patterns, Causes, and Ecosystem Consequences*. Springer-Verlag, Berlin.

Jenny, H. 1941. *Factors of Soil Formation*. McGraw-Hill, New York.

Jenny, H., R.J. Arkley, and A.M. Schultz. 1969. The pigmy forest-podsol ecosystem and its dune associates of the Mendocino Coast. *Madroño* 20:60–74.

Jensen, D.B., M.S. Torn, and J. Harte. 1993. *In Our Own Hands: A Strategy for Conserving California's Diversity*. University of California Press, Berkeley.

Jobbágy, E.G., and R.B. Jackson. 2000. The vertical distribution of soil organic carbon and its relation to climate and vegetation. *Ecological Applications* 10:423–436.

Johnson, E.A. 1992. *Fire and Vegetation Dynamics. Studies from the North American Boreal Forest*. Cambridge University Press, Cambridge, UK.

Johnson, K.G., K.A. Vogt, H.J. Clark, O.J. Schmitz, and D.J. Vogt. 1996. Biodiversity and the productivity and stability of ecosystems. *Trends in Ecology and Evolution* 11:372–377.

Johnson, N.M. 1971. Mineral equilibria in ecosystem geochemistry. *Ecology* 52:529–531.

Jonasson, S., and F.S. Chapin III. 1985. Significance of sequential leaf development for nutrient balance of the cotton sedge, *Eriophorum vaginatum* L. *Oecologia* 67:511–518.

Jonasson, S., A. Michelsen, and I.K. Schmidt. 1999. Coupling of nutrient cycling and carbon dynamics in the Arctic, integration of soil microbial and plant processes. *Applied Soil Ecology* 11:135–146.

Jones, C.G., J.H. Lawton, and M. Shachak. 1994. Organisms as ecosystem engineers. *Oikos* 69: 373–386.

Jones, H.G. 1992. *Plants and Microclimate: A Quantitative Approach to Environmental Plant Physiology*. Cambridge University Press, Cambridge, UK.

Karl, D., A. Michaels, B. Bergman, D. Capone, E. Carpenter, R. Letelier, F. Lipschultz, H. Paerl, D. Sigman, and L. Stal. In press. Di-nitrogen fixation in the world's oceans. *Biogeochemistry*.

Karl, D.M., C.O. Wirsen, and H.W. Jannasch. 1980. Deep sea primary production at the Galapagos hydrothermal vents. *Science* 207:1345–1347.

Kasischke, E.S., N.L. Christensen, and B.J. Stocks. 1995. Fire, global warming, and the carbon balance of boreal forests. *Ecological Applications* 5: 437–451.

Kates, R.W., B.L. Turner, and W.C. Clark. 1990. The great transformation. Pages 1–17 *in* B.L. Turner, W.C. Clark, R.W. Kates, J.F. Richards, J.T. Mathews, and W.B. Meyer, editors. *The Earth as Transformed by Human Action*. Cambridge University Press, Cambridge, UK.

Keeley, J.E. 1990. Photosynthetic pathways in freshwater aquatic plants. *Trends in Ecology and Evolution* 5:330–333.

Keeling, C.D., J.F.S. Chin, and T.P. Whorf. 1996a. Increased activity of northern vegetation inferred from atmospheric CO_2 measurements. *Nature* 382:146–149.

Keeling, R.F., R.P. Najjar, and M.L. Bender. 1993. What atmospheric oxygen measurements can tell us about the global carbon-cycle. *Global Biogeochemical Cycles* 7:37–67.

Keeling, R.F., S.C. Piper, and M. Heimann. 1996b. Global and hemispheric CO_2 sinks deduced from changes in atmospheric O_2 concentration. *Nature* 381:218–221.

Kelliher, F.M., and R. Jackson. 2001. Evaporation and the water balance. Pages 206–217 *in* A. Sturman and R. Spronken-Smith, editors. *The Physical Environment: A New Zealand Perspective*. Oxford University Press, Melbourne, Australia.

Kelliher, F.M., R. Leuning, M.R. Raupach, and E.-D. Schulze. 1995. Maximum conductances for evaporation from global vegetation types. *Agricultural and Forest Meteorology* 73:1–16.

Kellogg, E.A. 1999. Phylogenetic aspects of the evolution of C_4 photosynthesis. Pages 411–444 *in* R.F. Sage and R.K. Monson, editors. *C_4 Plant Biology*. Academic Press, San Diego, CA.

Kemperman, J.A., and B.V. Barnes. 1976. Clone size in American aspens. *Canadian Journal of Botany* 54:2603–2607.

Khalil, M.A.K., and R.A. Rasmussen. 1990. Constraints on the global sources of methane and analysis of recent budgets. *Tellus* 42B:229–236.

Kielland, K. 1994. Amino acid absorption by arctic plants: Implications for plant nutrition and nitrogen cycling. *Ecology* 75:2373–2383.

Kielland, K. 1997. Role of free amino acids in the nitrogen economy of arctic cryptogams. *Ecoscience* 4:75–79.

Kielland, K. 1999. Short-circuiting the nitrogen cycle: Ecophysiological strategies of nitrogen uptake in plants from marginal environments. Pages 376–398 *in* N. Ae, J. Arihara, K. Okada, and A. Srinivasan, editors. *Plant Nutrient Acquisition: New Perspectives*. Springer-Verlag, Tokyo.

Kielland, K., and J. Bryant. 1998. Moose herbivory in taiga: Effects on biogeochemistry and vegetation dynamics in primary succession. *Oikos* 82:377–383.

Kitchell, J.F., editor. 1992. *Food Web Management: A Case Study of Lake Mendota*. Springer-Verlag, New York.

Kleiden, A., and H.A. Mooney. 2000. A global distribution of biodiversity inferred from climatic constraints: Results from a process-based modelling study. *Global Change Biology* 6:507–523.

Klein, D.R. 1982. Fire, lichens, and caribou. *Journal of Range Management* 35:390–395.

Kling, G.W., G.W. Kipphut, and M.C. Miller. 1991. Arctic lakes and streams as gas conduits to the atmosphere: Implications for tundra carbon budgets. *Science* 251:298–301.

Koerselman, W., and A.F.M. Mueleman. 1996. The vegetation N:P ratio: A new tool to detect the nature of nutrient limitation. *Journal of Applied Ecology* 33:1441–1450.

Koide, R.T. 1991. Nutrient supply, nutrient demand and plant response to mycorrhizal infection. *New Phytologist* 117:365–386.

Körner, C. 1999. *Alpine Plant Life*. Springer-Verlag, Berlin.

Körner, C. 1994. Leaf diffusive conductances in the major vegetation types of the globe. Pages 463–490 *in* E.-D. Schulze and M.M. Caldwell, editors. *Ecophysiology of Photosynthesis*. Springer-Verlag, Berlin.

Körner, C., and W. Larcher. 1988. Plant life in cold climates. *Symposium of the Society of Experimental Biology* 42:25–57.

Körner, C., J.A. Scheel, and H. Bauer. 1979. Maximum leaf diffusive conductance in vascular plants. *Photosynthetica* 13:45–82.

Kozlovsky, D.G. 1968. A critical evaluation of the trophic level concept.I. Ecological efficiencies. *Ecology* 49:48–60.

Kozlowski, T.T., P.J. Kramer, and S.G. Pallardy. 1991. *The Physiological Ecology of Woody Plants*. Academic Press, San Diego, CA.

Kramer, P.J., and J.S. Boyer. 1995. *Water Relations of Plants and Soils*. Academic Press, San Diego, CA.

Kremen, C.K., A.M. Merenlender, and D.D. Murphy. 1994. Ecological monitoring: A vital need for integrated conservation and development programs in the tropics. *Conservation Biology* 8:388–397.

Kremen, C., I. Raymond, and K. Lance. 1998. An interdisciplinary tool for monitoring conservation impacts in Madagascar. *Conservation Biology* 12: 549–563.

Kroehler, C.J., and A.E. Linkins. 1991. The absorption of inorganic phosphate from ^{32}P-labeled inositol hexaphosphate by *Eriophorum vaginatum*. *Oecologia* 85:424–428.

Kronzucker, H.J., M.Y. Siddiqi, and A.M. Glass. 1997. Conifer root discrimination against soil nitrate and the ecology of forest succession. *Nature* 385:59–61.

Kucharik, C.J., J.A. Foley, C. Delire, V.A. Fisher, M.T. Coe, J. Lenters, C. Young-Molling, N. Ramankutty, J.M. Norman, and S.T. Gower. 2000. Testing the performance of a dynamic global ecosystem model: Water balance, carbon balance and vegetation structure. *Global Biogeochemical Cycles* 14: 795–825.

Kummerow, J., B.A. Ellis, S. Kummerow, and F.S. Chapin III. 1983. Spring growth of shoots and roots in shrubs of an Alaskan muskeg. *American Journal of Botany* 70:1509–1515.

Laine, K., and H. Henttonen. 1983. The role of plant production in microtine cycles in northern Fennoscandia. *Oikos* 40:407–418.

Lajtha, K., and M. Klein. 1988. The effect of varying phosphorus availability on nutrient use by *Larrea tridentata*, a desert evergreen shrub. *Oecologia* 75:348–353.

Lambers, H., and H. Poorter. 1992. Inherent variation in growth rate between higher plants: A search for physiological causes and ecological consequences. *Advances in Ecological Research* 23:187–261.

Lambers, H., O.K. Atkin, and I. Scheurwater. 1996. Respiratory patterns in roots in relation to their functioning. Pages 323–362 *in* Y. Waisel, A. Eshel, and U. Kafkaki, editors. *Plant Roots: The Hidden Half*. Marcel Dekker, New York.

Lambers, H., F.S. Chapin III, and T. Pons. 1998. *Plant Physiological Ecology*. Springer-Verlag, Berlin.

Landres, P.B., P. Morgan, and F.J. Swanson. 1999. Overview of the use of natural variability concepts in managing ecological systems. *Ecological Applications* 9:1179–1188.

Landsberg, J.J., and S.T. Gower. 1997. *Applications of Physiological Ecology to Forest Management*. Academic Press, San Diego, CA.

Larcher, W. 1995. *Physiological Plant Ecology*. Springer-Verlag, Berlin.

Lauenroth, W.K., and O.E. Sala. 1992. Long-term forage production of North American shortgrass steppe. *Ecological Applications* 2:397–403.

Lauenroth, W.K., J.L. Dodd, and P.L. Simms. 1978. The effects of water- and nitrogen-induced stresses on plant community structure in a semiarid grassland. *Oecologia* 36:211–222.

Laurance, W.F., and R.O. Bierregaard, editors. 1997. *Tropical Forest Remnants: Ecology, Management and Conservation of Fragmented Forests*. University of Chicago Press, Chicago.

Lavelle, P., D. Bignell, and M. Lepage. 1997. Soil function in a changing world: The role of invertebrate ecosystem engineers. *European Journal of Soil Biology* 33:159–193.

Lawton, J.H., and C.G. Jones. 1995. Linking species and ecosystems: Organisms as ecosystem engineers. Pages 141–150 *in* C.G. Jones and J.H. Lawton, editors. *Linking species and ecosystems*. Chapman & Hall, New York.

Lee, R.B. 1982. Selectivity and kinetics of ion uptake by barley plant following nutrient deficiency. *Annals of Botany* 50:429–449.

Lee, R.B., and K.A. Rudge. 1987. Effects of nitrogen deficiency on the absorption of nitrate and ammonium by barley plants. *Annals of Botany* 57: 471–486.

Lieth, H. 1975. Modeling the primary productivity of the world. Pages 237–263 *in* H. Lieth and R.H. Whittaker, editors. *Primary Productivity of the Biosphere*. Springer-Verlag, Berlin.

Likens, G.E., F.H. Bormann, R.S. Pierce, J.S. Eaton, and N.M. Johnson. 1977. *Biogeochemistry of a Forested Ecosystem*. Springer-Verlag, New York.

Lindeman, R.L. 1942. The trophic-dynamic aspects of ecology. *Ecology* 23:399–418.

Lindroth, R.L. 1996. CO_2-mediated changes in tree chemistry and tree-Lepidopteran interactions. Pages 105–120 *in* G.W. Koch and H.A. Mooney, editors. *Carbon Dioxide and Terrestrial Ecosystems*. Academic Press, San Diego, CA.

Lindsay, W.L. 1979. *Chemical Equilibria in Soils*. J Wiley, New York.

Lipson, D.A., S.K. Schmidt, and R.K. Monson. 1999. Links between microbial population dynamics and nitrogen availability in an alpine ecosystem. *Ecology* 80:1623–1631.

Lipson, D.A., T.K. Raab, S.K. Schmidt, and R.K. Monson. 2001. An empirical model of amino acid transformations in an alpine soil. *Soil Biology and Biochemistry* 33:189–198.

Liston, G.E. 1999. Interrelationships among snow distribution, snowmelt, and snow cover depletion: Implications for atmospheric, hydrologic and ecologic modeling. *Journal of Applied Meteorology* 38:1474–1487.

Liston, G.E., and M. Sturm. 1998. A snow-transport model for complex terrain. *Journal of Glaciology* 44:498–516.

Livingston, G.P., and G.L. Hutchinson. 1995. Enclosure-based measurement of trace gas exchange: Applications and sources of error. Pages 14–51 *in* P.A. Matson and R.C. Harriss, editors. *Biogenic Trace Gases: Measuring Emissions from Soil and Water*. Blackwell Scientific, Oxford, UK.

Lloyd, J., and J.A. Taylor. 1994. On the temperature dependence of soil respiration. *Functional Ecology* 8:315–323.

Lodge, D.M. 2001. Lakes. Pages 277–313 *in* F.S. Chapin III, O.E. Sala, and E. Huber-Sannwald, editors. *Global Biodiversity in a Changing Environment: Scenarios for the 21st Century*. Springer-Verlag, New York.

Lorio, P.L. Jr. 1986. Growth-differentiation balance: A basis for understanding southern pine beetle-tree interactions. *Forest Ecology and Management* 14:259–273.

Los, S.O., G.J. Collatz, P.J. Sellers, C.M. Malmström, N.H. Pollack, R.S. DeFries, L. Bounoua, M.T. Parris, C.J. Tucker, and D.A. Dazlich. 2000. A global 9-yr biophysical land surface dataset from NOAA AVHRR data. *Journal of Hydrometeorology* 1:183–199.

Lousier, J.D., and S.S. Bamforth. 1990. Soil protozoa. Pages 97–136 *in* D.L. Dindal, editor. *Soil Biology Guide*. J Wiley, New York.

Lovett, G.M. 1994. Atmospheric deposition of nutrients and pollutants in North America: An ecological perspective. *Ecological Applications* 4:629–650.

Lowrance, R., L.S. Altier, J.D. Newbold, R.R. Schnabel, P.M. Groffman, J.M. Denver, D.L. Correll, J.W. Gilliam, J.L. Robinson, R.B. Brinsfield, K.W. Staver, W. Lucas, and A.H. Todd. 1997. Water quality functions of riparian forest buffer systems in the Chesapeake Bay watershed. *Environmental Management* 21:687–712.

Ludwig, D., R. Hilborn, and C. Walters. 1993. Uncertainty, resource exploitation, and conservation: Lessons from history. *Science* 260:17, 36.

Luo, Y., S. Wan, D. Hui, and L.L. Wallace. 2001. Acclimatization of soil respiration to warming in a tall grass prairie. *Nature* 413:622–625.

MacArthur, R.H., and E.O. Wilson. 1967. *The Theory of Island Biogeography*. Princeton University Press, Princeton, NJ.

MacLean, D.A., and R.W. Wein. 1978. Weight loss and nutrient changes in decomposing litter and forest floor material in New Brunswick forest stands. *Canadian Journal of Botany* 56:2730–2749.

Magill, A.H., J.D. Aber, J.J. Hendricks, R.D. Bowden, J.M. Melillo, and P. Steudler. 1997. Biogeochemical response of forest ecosystems to simulated chronic nitrogen deposition. *Ecological Applications* 7: 402–415.

Mann, M.E., R.S. Bradley, and M.K. Hughes. 1999. Northern hemisphere temperatures during the past millennium: Inferences, uncertainties and limitations. *Geophysical Research Letters* 26:759–762.

Margalef, R. 1968. *Perspectives in Ecological Theory*. University of Chicago Press, Chicago.

Margolis, H., R. Oren, D. Whitehead, and M.R. Kaufmann. 1995. Leaf area dynamics of conifer forests. Pages 181–223 *in* W.K. Smith and T.M. Hinckley, editors. *Ecophysiology of Coniferous Forests*. Academic Press, San Diego, CA.

Marschner, H. 1995. *Mineral Nutrition in Higher Plants*. Academic Press, London.

Mary, B., S. Recous, D. Darwis, and D. Robin. 1996. Interactions between decomposition of plant residues and nitrogen cycling in soil. *Plant and Soil* 181:71–82.

Matson, P.A., and R.D. Boone. 1984. Natural disturbance and nitrogen mineralization: Wave-form dieback of mountain hemlock in the Oregon Cascades. *Ecology* 65:1511–1516.

Matson, P.A., and R.C. Harriss. 1988. Prospects for aircraft-based gas exchange measurements in ecosystem studies. *Ecology* 69:1318–1325.

Matson, P.A., and P.M. Vitousek. 1987. Cross-system comparisons of soil nitrogen transformations and nitrous oxide flux in tropical forest ecosystems. *Global Biogeochemical Cycles* 1:163–170.

Matson, P.A., and P.M. Vitousek. 1981. Nitrogen mineralization and nitrification potentials following clearcutting in the Hoosier National Forest, Indiana. *Forest Science* 27:781–791.

Matson, P.A., and R.H. Waring. 1984. Effects of nutrient and light limitation on mountain hemlock: susceptibility to laminated root rot. *Ecology* 65:1517–1524.

Matson, P.A., W.H. McDowell, A.R. Townsend, and P.M. Vitousek. 1999. The globalization of N deposition: Ecosystem consequences in tropical environments. *Biogeochemistry* 46:67–83.

Matson, P.A., R.L. Naylor, and I. Ortiz-Monasterio. 1998. The integration of environmental, agronomic, and economic aspect of fertilizer management. *Science* 280:112–115.

Matson, P.A., W.J. Parton, A.G. Power, and M.J. Swift. 1997. Agricultural intensification and ecosystem properties. *Science* 227:504–509.

Matson, P.A., P.M. Vitousek, J. Ewel, M. Mazzarino, and G. Robertson. 1987. Nitrogen transformations following tropical forest felling and burning on a volcanic soil. *Ecology* 68:491–502.

Matson, P.A., C. Volkmann, K. Coppinger, and W.A. Reiners. 1991. Annual nitrous oxide flux and soil nitrogen characteristics in sagebrush steppe ecosystems. *Biogeochemistry* 14:1–12.

May, R.M. 1994. Biological diversity: Differences between land and sea. *Philosophical Transactions of the Royal Society of London, Series B* 343: 105–111.

May, R.M. 1973. *Stability and Complexity in Model Ecosystems*. Princeton University Press, Princeton, NJ.

McAndrews, J.H. 1966. Postglacial history of prairie, savanna, and forest in northwestern Minnesota. *Torrey Botanical Club Memoir* 22:1–72.

McCarthy, J.J., O.F. Canziani, N.A. Leary, D.J. Dokken, and K.S. White, editors. 2001. *Climate Change 2001: Impacts, Adaptation, and Vulnerability*. Cambridge University Press, Cambridge, UK.

McCoy, E.D., S.S. Bell, and K. Walters. 1986. Identifying biotic boundaries along environmental gradients. *Ecology* 67:749–759.

McGuire, A.D., J.W. Melillo, D.W. Kicklighter, and L.A. Joyce. 1995. Equilibrium responses of soil carbon to climate change: Empirical and process-based estimates. *Journal of Biogeography* 22:785–796.

McGuire, A.D., S. Sitch, J.S. Clein, R. Dargaville, G. Esser, J. Foley, M. Heimann, F. Joos, J. Kaplan, D.W. Kicklighter, R.A. Meier, J.M. Melillo, B. Moore II, I.C. Prentice, N. Ramankutty, T. Reichenau, A. Schloss, H. Tian, L.J. Williams, and U. Wittenberg. 2001. Carbon balance of the terrestrial biosphere in the twentieth century: Analyses of CO_2, climate and land-use effects with four process-based models. *Global Biogeochemical Cycles* 15:183–206.

McKane, R.B., E.B. Rastetter, G.R. Shaver, K.J. Nadelhoffer, A.E. Giblin, J.A. Laundre, and F.S. Chapin III. 1997. Climatic effects on tundra carbon storage inferred from experimental data and a model. *Ecology* 78:1170–1187.

McKey, D., P.G. Waterman, C.N. Mbi, J.S. Gartlan, and T.T. Struhsaker. 1978. Phenolic content of vegetation in two African rain forests: Ecological implications. *Science* 202:61–63.

McNaughton, K.G. 1976. Evaporation and advection I: Evaporation from extensive homogeneous surfaces. *Quarterly Journal of the Royal Meteorological Society* 102:181–191.

McNaughton, K.G., and P.G. Jarvis. 1991. Effects of spatial scale on stomatal control of transpiration. *Agricultural and Forest Meteorology* 54:279–302.

McNaughton, S.J. 1977. Diversity and stability of ecological communities: A comment on the role of empiricism in ecology. *American Naturalist* 111:515–525.

McNaughton, S.J. 1985. Ecology of a grazing ecosystem: The Serengeti. *Ecological Monographs* 53:259–294.

McNaughton, S.J. 1979. Grazing as an optimization process: Grass-ungulate relationships in the Serengeti. *American Naturalist* 113:691–703.

McNaughton, S.J. 1988. Mineral nutrition and spatial concentrations of African ungulates. *Nature* 334: 343–345.

McNaughton, S.J., M. Oesterheld, D.A. Frank, and K.J. Williams. 1989. Ecosystem-level patterns of primary productivity and herbivory in terrestrial habitats. *Nature* 341:142–144.

McNulty, S.G., J.D. Aber, T.M. McLellan, and S.M. Katt. 1990. Nitrogen cycling in high elevation forests of the northeastern U.S. in relation to nitrogen deposition. *Ambio* 19:38–40.

Melillo, J.M., J.D. Aber, and J.F. Muratore. 1982. Nitrogen and lignin control of hardwood leaf litter decomposition dynamics. *Ecology* 63:621–626.

Melillo, J.M., I.C. Prentice, G.D. Farquhar, E.-D. Schulze, and O.E. Sala. 1996. Terrestrial biotic responses to environmental change and feedbacks to climate. Pages 445–481 *in* J.T. Houghton, L.G. Meira Filho, B.A. Callander, N. Harris, A. Kattenberg, and K. Maskell, editors. *Climate Change 1995. The Science of Climate Change.* Cambridge University Press, Cambridge, UK.

Meyer, W.B., and B.L. Turner III. 1992. Human population growth and global land-use/cover change. *Annual Review of Ecology and Systematics* 23: 39–61.

Milchunas, D.G., and W.K. Lauenroth. 1993. Quantitative effects of grazing on vegetation and soils over a global range of environments. *Ecological Monographs* 63:327–366.

Milchunas, D.G., O.E. Sala, and W.K. Lauenroth. 1988. A generalized model of the effects of grazing by large herbivores on grassland community structure. *American Naturalist* 132:87–106.

Miller, R.W., and R.L. Donahue. 1990. *Soils. An Introduction to Soils and Plant Growth.* Prentice-Hall, Englewood Cliffs, NJ.

Milliman, J.D., and R.H. Meade. 1983. World-wide delivery of river sediment to the oceans. *Journal of Geology* 91:1–12.

Mitsch, W.J., J.W. Day, J.W. Gilliam, P.M. Groffman, D.L. Hey, G.W. Randall, and N. Wang. 2001. Reducing nitrogen loading to the Gulf of Mexico from the Mississippi River Basin: Strategies to counter a persistent ecological problem. *BioScience* 51: 373–388.

Monserud, R.A., and J.D. Marshall. 1999. Allometric crown relations in three northern Idaho conifer species. *Canadian Journal of Forest Research* 29: 521–535.

Monteith, J.L., and M.H. Unsworth. 1990. *Principles of Environmental Physics.* Hodder & Stoughton, London.

Mooney, H.A. 1972. The carbon balance of plants. *Annual Review of Ecology and Systematics* 3:315–346.

Mooney, H.A. 1986. Photosynthesis. Pages 345–373 *in* M.J. Crawley, editor. *Plant Ecology.* Blackwell, Oxford, UK.

Mooney, H.A., and E.L. Dunn. 1970. Convergent evolution of mediterranean-climate evergreen sclerophyll shrubs. *Evolution* 24:292–303.

Mooney, H.A., J. Canadell, F.S. Chapin III, J.R. Ehleringer, C. Körner, R.E. McMurtrie, W.J. Parton, L.F. Pitelka, and E.-D. Schulze. 1999. Ecosystem physiology responses to global change. Pages 141–189 *in* B. Walker, W. Steffen, J. Canadell, and J. Ingram, editors. *The Terrestrial Biosphere and Global Change: Implications for Natural and Managed Ecosystems.* Cambridge University Press, Cambridge, UK.

Moore, J.C., and P.C. de Ruiter. 2000. Invertebrates in detrital food webs along gradients of productivity. Pages 161–183 *in* D.C. Coleman and P.F. Hendrix, editors. *Invertebrates as Webmasters in Ecosystems.* CABI, Oxford, UK.

Moore, J.C., and H.W. Hunt. 1988. Resource compartmentation and the stability of real ecosystems. *Nature* 333:261–263.

Morris, J.T. 1980. The nitrogen uptake kinetics of *Spartina alterniflora* in culture. *Ecology* 61:1114–1121.

Moss, B. 1998. *Ecology of Fresh Waters: Man and Medium, Past to Future.* Blackwell Scientific, Oxford, UK.

Motzkin, G., D. Foster, A. Allen, J. Harrod, and R. Boone. 1996. Controlling site to evaluate history: Vegetation patterns of a New England sand plain. *Ecological Monographs* 66:345–365.

Mulder, C.P.H., D.D. Uliassi, and D.F. Doak. 2001. Physical stress and diversity-productivity relationships: The role of positive interactions. *Proceedings of the National Academy of Sciences, U.S.A.* 98:6704–6708.

Nadelhoffer, K.J., B.A. Emmett, P. Gundersen, O.J. Kjonaas, C.J. Koopmans, P. Schleppi, A. Tietema, and R.F. Wright. 1999. Nitrogen deposition makes a minor contribution to carbon sequestration in temperate forests. *Nature* 398:145–148.

Nadelhoffer, K.J., A.E. Giblin, G.R. Shaver, and A.E. Linkins. 1992. Microbial processes and plant nutrient availability in arctic soils. Pages 281–300 *in* F.S. Chapin III, R.L. Jefferies, J.F. Reynolds, G.R. Shaver, and J. Svoboda, editors. *Arctic Ecosystems in a Changing Climate: An Ecophysiological Perspective*. Academic Press, San Diego, CA.

Nadelhoffer, K., G. Shaver, B. Fry, A. Giblin, L. Johnson, and R. McKane. 1996. ^{15}N natural abundances and N use by tundra plants. *Oecologia* 107:386–394.

Nadkarni, N. 1981. Canopy roots: Convergent evolution in rainforest nutrient cycles. *Science* 214: 1023–1024.

Naiman, R.J. 1996. Water, society and landscape ecology. *Landscape Ecology* 11:193–196.

Naiman, R.J., and H. Decamps. 1997. The ecology of interfaces: Riparian zones. *Annual Review of Ecology and Systematics* 28:621–658.

Näsholm, T., A. Ekblad, A. Nordin, R. Giesler, M. Högberg, and P. Högberg. 1998. Boreal forest plants take up organic nitrogen. *Nature* 392: 914–916.

Näsholm, T., K. Huss-Danell, and P. Högberg. 2000. Uptake of organic nitrogen in the field by four agriculturally important plant species. *Ecology* 81:1155–1161.

Naylor, R.L., and W.M. Drew. 1998. Valuing mangrove ecosystem services in Kosrae, Micronesia. *Environment and Development Economics* 3:471–490.

Neff, J.C., E.A. Holland, F.J. Dentener, W.H. McDowell, and K.M. Russell. 2002. The origin, composition and rates of organic nitrogen deposition: A missing piece of the nitrogen cycle? *Biogeochemistry* 57/58:99–136.

Nepstad, D.C., C.R. deCarvalho, E.A. Davidson, P.H. Jipp, P.A. Lefebvre, G.H. Negreiros, E.D. da Silva, T.A. Stone, S.E. Trumbore, and S. Vieira. 1994. The role of deep roots in the hydrological and carbon cycles of Amazonian forests and pastures. *Nature* 372:666–669.

New, M.G., M. Hulme, and P.D. Jones. 1999. Representing 20th century space-time climate variability, I: Development of a 1961–1990 mean monthly terrestrial climatology. *Journal of Climate* 12: 829–856.

Newbold, J.D. 1992. Cycles and spirals of nutrients. Pages 379–408 *in* P. Calow and G.E. Petts, editors. *The Rivers Handbook: Hydrological and Ecological Principles*. Blackwell Scientific, Oxford, UK.

Newman, E.I. 1985. The rhizosphere: Carbon sources and microbial populations. Pages 107–121 *in* A.H. Fitter, D. Atkinson, D.J. Read, and M. Busher, editors. *Ecological Interactions in Soil*. Blackwell, Oxford, UK.

Niemelä, P., F.S. Chapin III, K. Danell, and J.P. Bryant. 2001. Animal-mediated responses of boreal forest to climatic change. *Climatic Change* 48:427–440.

Nixon, S.W. 1988. Physical energy inputs and the comparative ecology of lake and marine ecosystems. *Limnology and Oceanography* 33:1005–1025.

Nixon, S.W., J.W. Ammerman, L.P. Atkinson, V.M. Berounsky, G. Billen, W.C. Boicourt, W.R. Boynton, T.M. Church, D.M. Ditoro, R. Elmgren, J.H. Garber, A.E. Giblin, R.A. Jahnke, N.J.P. Owens, M.E.Q. Pilson, and S.P. Seitzinger. 1996. The fate of nitrogen and phosphorus at the land-sea margin of the North Atlantic Ocean. *Biogeochemistry* 35:141–180.

Noble, I.R., and R.O. Slatyer. 1980. The use of vital attributes to predict successional changes in plant communities subject to recurrent disturbances. *Vegetatio* 43:5–21.

Northup, R.R., Z. Yu, R.A. Dahlgren, and K.A. Vogt. 1995. Polyphenol control of nitrogen release from pine litter. *Nature* 377:227–229.

Norton, J.M., and M.K. Firestone. 1991. Metabolic status of bacteria and fungi in the rhizosphere of ponderosa pine seedlings. *Applied and Environmental Microbiology* 57:1161–1167.

Norton, J.M., J.L. Smith, and M.K. Firestone. 1990. Carbon flow in the rhizosphere of Ponderosa pine seedlings. *Soil Biology and Biochemistry* 22:445–449.

Noss, R.F. 1990. Indicators for monitoring biodiversity: A hierarchical approach. *Conservation Biology* 4:355–364.

Nulsen, R.A., K.J. Bligh, I.N. Baxter, E.J. Solin, and D.H. Imrie. 1986. The fate of rainfall in a mallee and heath vegetated catchment in southern

Western Australia. *Australian Journal of Ecology* 11:361–371.

O'Leary, M.H. 1988. Carbon isotopes in photosynthesis. *BioScience* 38:325–336.

O'Neill, R.V., D.L. DeAngelis, J.B. Waide, and T.F.H. Allen. 1986. *A Hierarchical Concept of Ecosystems*. Princeton University Press, Princeton, NJ.

Oades, J.M. 1989. An introduction to organic matter in mineral soils. Pages 89–160 *in* J.B. Dixon and S.B. Weed, editors. *Minerals in Soil Environments*. Soil Science Society of America, Madison, WI.

Odum, E.P. 1959. *Fundamentals of Ecology*. Saunders, Philadelphia.

Odum, E.P. 1969. The strategy of ecosystem development. *Science* 164:262–270.

Oechel, W.C., G.L. Vourlitis, S.J. Hastings, R.C. Zulueta, L. Hinzman and D. Kane. 2000. Acclimation of ecosystem CO_2 exchange in the Alaskan Arctic in response to decadal climate warming. *Nature* 406:978–981.

Oke, T.R. 1987. *Boundary Layer Climates*. Methuen, London.

Oksanen, L. 1990. Predation, herbivory, and plant strategies along gradients of primary productivity. Pages 445–474 *in* J.B. Grace and D. Tilman, editors. *Perspectives on Plant Competition*. Academic Press, San Diego, CA.

Olsen, J.S. 1963. Energy storage and the balance of producers and decomposers in ecological systems. *Ecology* 44:322–331.

Osterheld, M., O.E. Sala, and S.J. McNaughton. 1992. Effect of animal husbandry on herbivore-carrying capacity at a regional scale. *Nature* 356:234–236.

Overbay, J.C. 1992. Ecosystem management. Pages 3–15 *in Taking an Ecological Approach to Management*. USDA Forest Service General Technical Report Wo-WSA-3. USDA Forest Service, Washington, DC.

Ovington, J.D. 1962. Quantitative ecology and the woodland ecosystem concept. *Advances in Ecological Research* 1:103–192.

Owen-Smith, R.N. 1988. *Megaherbivores: The Influence of Very Large Body Size on Ecology*. Cambridge University Press, Cambridge, UK.

Owensby, C.E., J.M. Ham, A. Knapp, C.W. Rice, P.I. Coyne, and L.M. Auen. 1996. Ecosystem-level responses of tallgrass prairie to elevated CO_2. Pages 147–162 *in* G.W. Koch and H.A. Mooney, editors. *Carbon Dioxide and Terrestrial Ecosystems*. Academic Press, San Diego, CA.

Pace, M.L., J.J. Cole, S.R. Carpenter, and J.F. Kitchell. 1999. Trophic cascades revealed in diverse ecosystems. *Trends in Ecology and Evolution* 14:483–488.

Pagani, M., M.A. Arthur, and K.H. Freeman. 1999. Miocene evolution of atmosphere carbon dioxide. *Paleoceanography* 14:273–292.

Paine, R.T. 1980. Food webs: Linkage, interaction strength and community infrastructure. *Journal of Animal Ecology* 49:667–685.

Paine, R.T. 2000. Phycology for the mammalogist: Marine rocky shores and mammal-dominated communities. How different are the structuring processes? *Journal of Mammalogy* 81:637–648.

Parton, W.J., D.S. Schimel, C.V. Cole, and D.S. Ojima. 1987. Analysis of factors controlling soil organic matter levels in Great Plains grasslands. *Soil Science Society of America Journal* 51:1173–1179.

Parton, W.J., J.M.O. Scurlock, D.S. Ojima, T.G. Gilmanov, R.J. Scholes, D.S. Schimel, T. Kirchner, J.-C. Menaud, T. Seastedt, E. Garcia Moya, A. Kamnalrut, and J.I. Kinyamario. 1993. Observations and modeling of biomass and soil organic matter dynamics for the grassland biome worldwide. *Global Biogeochemical Cycles* 7:785–809.

Passioura, J.B. 1988. Response to Dr. P.J. Kramer's article, "Changing concepts regarding plant water relations." *Plant, Cell and Environment* 11:569–571.

Pastor, J., and W.M. Post. 1986. Influence of climate, soil moisture, and succession on forest carbon and nitrogen cycles. *Biogeochemistry* 2:3–27.

Pastor, J., J.D. Aber, C.A. McClaugherty, and J.M. Melillo. 1984. Aboveground primary production and N and P cycling along a nitrogen mineralization gradient on Blackhawk Island, Wisconsin. *Ecology* 65:256–268.

Pastor, J., B. Dewey, R. Moen, D.J. Mladenoff, M. White, and Y. Cohen. 1998. Spatial patterns in the moose-forest-soil ecosystem on Isle Royale, Michigan, USA. *Ecological Applications* 8: 411–424.

Pastor, J., R.J. Naiman, B. Dewey, and P. McInnes. 1988. Moose, microbes, and the boreal forest. *BioScience* 38:770–777.

Paton, T.R., G.S. Humphreys, and P.B. Mitchell. 1995. *Soils: A New Global View*. Yale University Press, New Haven, CT.

Paul, E.A., and F.E. Clark. 1996. *Soil Microbiology and Biochemistry*. 2nd Edition Academic Press, San Diego, CA.

Pauly, D., and V. Christensen. 1995. Primary production required to sustain global fisheries. *Nature* 374:255–257.

Payette, S., and L. Filion. 1985. White spruce expansion at the tree line and recent climatic change. *Canadian Journal of Forest Research* 15:241–251.

Pearcy, R.W. 1990. Sunflecks and photosynthesis in plant canopies. *Annual Review of Plant Physiology* 41:421–453.

Pearson, P.N., and M.R. Palmer. 2000. Atmosphere carbon dioxide concentrations over the past 60 million years. *Nature* 406:695–699.

Peine, J.D. 1999. *Ecosystem Management for Sustainability*. Lewis Publishers, Boca Raton, LA.

Penner, J.E., M. Andreae, H. Annegarn, L. Barrie, J. Feichter, D. Hegg, A. Jayaraman, R. Leaitch, D. Murphy, J. Nganga, and G. Pitari. 2001. Aerosols, their direct and indirect effects. Pages 289–348 *in* J.T. Houghton, Y. Ding, D.J. Griggs, M. Noguer, P.J. van der Linden, X. Dai, K. Maskell, and C.A. Johnson, editors. *Climate Change 2001: The Scientific Basis*. Cambridge University Press, Cambridge, UK.

Penning de Vries, F.W.T. 1975. The cost of maintenance processes in plant cells. *Annals of Botany* 39:77–92.

Penning de Vries, F.W.T., A.H.M. Brunsting, and H.H. van Laar. 1974. Products, requirements, and efficiency of biosynthesis: A quantitative approach. *Journal of Theoretical Biology* 45:339–377.

Perez-Harguindeguy, N., S. Diaz, J.H.C. Cornelissen, F. Vendramini, M. Cabido, and A. Castellanos. 2000. Chemistry and toughness predict leaf litter decomposition rates over a wide spectrum of functional types and taxa in central Argentina. *Plant and Soil* 218:21–30.

Perry, D.A., M.P. Amaranthus, J.G. Borchers, S.L. Borchers, and R.E. Brainerd. 1989. Bootstrapping in ecosystems. *BioScience* 39:230–237.

Peterson, B.J., W.M. Wolheim, P.J. Mujlholland, J.R. Webster, J.L. Meyer, J.L. Tank, E. Marti, W.B. Bowdon, H.M. Valett, A.E. Hershey, W.H. McDowell, W.K. Dodds, S.K. Hamilton, S. Gregory, and D.D. Morrall. 2001. Control of nitrogen export from watersheds by headwater streams. *Science* 292:86–90.

Petit, J.R., J. Jouzel, D. Raynaud, N.I. Barkov, J.M. Barnola, I. Basile, M. Bender, J. Chappellza, M. Davis, G. Delaygue, M. Delmotte, V.M. Kotlyakov, M. Legrand, V.Y. Lipenkov, C. Lorius, L. Pepin, C. Ritz, E. Saltzman, and M. Stievenard. 1999. Climate and atmospheric history of the past 420,000 years from the Vostok ice core, Antarctica. *Nature* 399:429–436.

Pickett, S.T.A., and P.S. White. 1985. *The Ecology of Natural Disturbance as Patch Dynamics*. Academic Press, New York.

Pickett, S.T.A., S.L. Collins, and J.J. Armesto. 1987. A hierarchical consideration of causes and mechanisms of succession. *Vegetatio* 69:109–114.

Pickett, S.T.A., J. Kolasa, and C.G. Jones. 1994. *Ecological Understanding. The Nature of Theory and the Theory of Nature*. Academic Press, New York.

Pickett, S.T.A., J. Wu, and M.L. Cadenasso. 1999. Patch dynamics and the ecology of disturbed ground: A framework for synthesis. Pages 707–722 *in* L.R. Walker, editor. *Ecosystems of Disturbed Ground*. Elsevier, Amsterdam.

Pielke, R.A. Sr., and R. Avisar. 1990. Influence of landscape structure on local and regional climate. *Landscape Ecology* 4:133–156.

Pimm, S.L. 1984. The complexity and stability of ecosystems. *Nature* 307:321–326.

Pimm, S.L. 1982. *Food Webs*. Chapman & Hall, New York.

Pimm, S.L., G.J. Russell, J.L. Gittleman, and T.M. Brooks. 1995. The future of biodiversity. *Science* 269:347–350.

Poff, N.L., P.L. Angermeier, S.D. Cooper, P.S. Lake, K.D. Fausch, K.O. Winemiller, L.A.K. Mertes, M.W. Oswood, J. Reynolds, and F.J. Rahel. 2001. Fish diversity in streams and rivers. Pages 315–349 *in* F.S. Chapin III, O.E. Sala, and E. Huber-Sannwald, editors. *Global Biodiversity in a Changing Environment: Scenarios for the 21st Century*. Springer-Verlag, New York.

Polis, G.A. 1991. Complex trophic interactions in deserts: An empirical critique of food-web theory. *American Naturalist* 138:123–155.

Polis, G.A. 1999. Why are parts of the world green? Multiple factors control productivity and the distribution of biomass. *Oikos* 86:3–15.

Polis, G.A., and S.D. Hurd. 1996. Linking marine and terrestrial food webs: Allochthonous input from the ocean supports high secondary productivity on small islands and coastal land communities. *American Naturalist* 147:396–423.

Pomeroy, J., N. Hedstrom, and J. Parviainen. 1999. The snow mass balance of Wolf Creek, Yukon: Effects of snow sublimation and redistribution. Pages 15–30 *in* J.W. Pomeroy and R.J. Granger, editors. *Wolf Creek Research Basin: Hydrology, Ecology, Environment*. National Water Research Institute, Environment Canada, Saskatoon, Canada.

Poorter, H. 1994. Construction costs and payback time of biomass: A whole-plant perspective. Pages 111–127 *in* J. Roy and E. Garnier, editors. *A Whole-Plant Perspective on Carbon-Nitrogen Interactions*. SPB Academic Publishing, The Hague.

Post, D.M., M.L. Pace, and N.G. Hairston. 2000. Ecosystem size determines food-chain length in lakes. *Nature* 405:1047–1049.

Post, W.M., W.R. Emanuel, P.J. Zinke, and A.G. Stangenberger. 1982. Soil carbon pools and world life zones. *Nature* 298:156–159.

Postel, S., and S. Carpenter. 1997. Freshwater ecosystem services. Pages 195–214 *in* G.C. Daily, editor. *Nature's Services: Societal Dependence on Natural Ecosystems*. Island Press, Washington, DC.

Postel, S.L., G.C. Daily, and P.R. Ehrlich. 1996. Human appropriation of renewable fresh water. *Science* 271:785–788.

Potter, C.S., J.T. Randerson, C.B. Field, P.A. Matson, P.M. Vitousek, H.A. Mooney, and S.A. Klooster. 1993. Terrestrial ecosystem production: A process model based on global satellite and surface data. *Global Biogeochemical Cycles* 7:811–841.

Power, M.E. 1990. Effects of fish in river food webs. *Science* 250:411–415.

Power, M.E. 1992a. Hydrologic and trophic controls of seasonal algal blooms in northern California rivers. *Archivs fur Hydrobiologie* 125:385–410.

Power, M.E. 1992b. Top-down and bottom-up forces in food webs: Do plants have primacy? *Ecology* 73:733–746.

Power, M.E., M.S. Parker, and J.T. Wootton. 1996a. Disturbance and food chain length in rivers. Pages 286–297 *in* G.A. Polis and K.O. Winemiller, editors. *Food Webs: Integration of Patterns and Dynamics*. Chapman & Hall, New York.

Power, M.E., D. Tilman, J.A. Estes, B.A. Menge, W.J. Bond, L.S. Mills, G. Daily, J.C. Castilla, J. Lubchenco, and R.T. Paine. 1996b. Challenges in the quest for keystones. *BioScience* 46:609–620.

Prather, M., D. Ehhalt, F. Dentener, R. Derwent, E. Dlugokencky, E. Holland, I. Isaksen, J. Katima, V. Kirchhoff, P. Matson, P. Midgley, and M. Wang. 2001. Atmospheric chemistry and greenhouse gases. Pages 239–287 *in* J.T. Houghton, Y. Ding, D.J. Griggs, M. Noguer, P.J. van der Linden, X. Dai, K. Maskell, and C.A. Johnson, editors. *Climate Change 2001: The Scientific Basis*. Cambridge University Press, Cambridge, UK.

Prentice, I.C., W. Cramer, S.P. Harrison, R. Leemans, R.A. Monserud, and A.M. Solomon. 1992. A global biome model based on plant physiology and dominance, soil properties and climate. *Journal of Biogeography* 19:117–134.

Prentice, I.C., G.D. Farquhar, M.J.R. Fasham, M.L. Goulden, M. Heimann, V.J. Jaramillo, H.S. Kheshgi, C. Le Quéré, R.J. Scholes, and D.W.R. Wallace. 2001. The carbon cycle and atmospheric carbon dioxide. Pages 183–237 *in* J.T. Houghton, Y. Ding, D.J. Griggs, M. Noguer, P.J. van der Linden, X. Dai, K. Maskell, and C.A. Johnson, editors. *Climate Change 2001: The Scientific Basis*. Cambridge University Press, Cambridge, UK.

Prescott, C.E. 1995. Does nitrogen availability control rates of litter decomposition in forests? *Plant and Soil* 168–169:83–88.

Prescott, C.E., R. Kabzems, and L.M. Zabek. 1999. Effects of fertilization on decomposition rate of *Populus tremuloides* foliar litter in a boreal forest. *Canadian Journal of Forest Research* 29: 393–397.

President's Committee of Advisors on Science and Technology Panel on Biodiversity and Ecosystems. March 1988. *Teaming with Life: Investing in Science to Understand and Use America's Living Capital.* PCAST Executive Secretariat, Washington, DC.

Press, F., and R. Siever. 1986. *Earth*. Freeman, New York.

Pugnaire, F.I., and F.S. Chapin III. 1992. Environmental and physiological factors governing nutrient resorption efficiency in barley. *Oecologia* 90:120–126.

Raab, T.K., D.A. Lipson, and R.K. Monson. 1999. Soil amino acid utilization among species of the Cyperaceae: Plant and soil processes. *Ecology* 80:2408–2419.

Raich, J.W., and W.H. Schlesinger. 1992. The global carbon dioxide flux in soil respiration and its relationship to climate. *Tellus* 44B:81–99.

Ramakrishnan, P.S. 1992. *Shifting Agriculture and Sustainable Development: An Interdisciplinary Study from North-Eastern India*. Parthenon Publishing Group, Park Ridge, NJ.

Ramaswamy, V., O. Boucher, J. Haigh, D. Hauglustaine, J. Haywood, G. Myhre, T. Nakajima, G.Y. Shi, and S. Solomon. 2001. Radiative forcing of climate change. Pages 349–416 *in* J.T. Houghton, Y. Ding, D.J. Griggs, M. Noguer, P.J. van der Linden, X. Dai, K. Maskell, and C.A. Johnson, editors. *Climate Change 2001: The Scientific Basis*. Cambridge University Press, Cambridge, UK.

Randerson, J., F.S. Chapin III, J. Harden, M. Harmon, and J. Neff. In press. Scaling terrestrial net carbon fluxes: A definition of net ecosystem production that includes disturbance and non-CO_2 fluxes. *Ecological Applications.*

Raper, C.D. Jr., D.L. Osmond, M. Wann, and W.W. Weeks. 1978. Interdependence of root and shoot activities in determining nitrogen uptake rate of roots. *Botanical Gazette* 139:289–294.

Rastetter, E.B., and G.R. Shaver. 1992. A model of multiple-element limitation for acclimating vegetation. *Ecology* 73:1157–1174.

Read, D.J. 1991. Mycorrhizas in ecosystems. *Experientia* 47:376–391.

Read, D.J., and R. Bajwa. 1985. Some nutritional aspects of the biology of ericaceous mycorrhizas.

Proceedings of the Royal Society of Edinburgh 85B:317–332.

Redfield, A.C. 1958. The biological control of chemical factors in the environment. *American Scientist* 46:205–221.

Reeburgh, W.S. 1997. Figures summarizing the global cycles of biogeochemically important elements. *Bulletin of the Ecological Society of America* 78:260–267.

Reich, P.B., D.S. Ellsworth, M.B. Walters, J.M. Vose, C. Gresham, K.J.C. Volin, and W.D. Bowman. 1999. Generality of leaf trait relationships: A test across six biomes. *Ecology* 80:1955–1969.

Reich, P.B., M.B. Walters, and D.S. Ellsworth. 1997. From tropics to tundra: Global convergence in plant functioning. *Proceedings of the National Academy of Sciences U. S. A.* 94:13730–13734.

Reynolds, J.F., D.W. Hilbert, and P.R. Kemp. 1993. Scaling ecophysiology from the plant to the ecosystem: A conceptual framework. Pages 127–140 *in* J.R. Ehleringer and C.B. Field, editors. *Scaling Physiological Processes: Leaf to Globe.* Academic Press, San Diego, CA.

Reynolds, J.F., and J.H.M. Thornley. 1982. A shoot:root partitioning model. *Annals of Botany* 49:585–597.

Rice, E.L. 1979. Allelopathy: An update. *Botanical Review* 45:15–109.

Ritchie, M.E., D. Tilman, and J.M.H. Knops. 1998. Herbivore effects on plant and nitrogen dynamics in oak savanna. *Ecology* 79:165–177.

Robertson, G.P. 1989. Nitrification and denitrification in humid tropical ecosystems: Potential controls on nitrogen retention. Pages 55–69 *in* J. Proctor, editors. *Mineral Nutrients in Tropical Forest and Savanna Ecosystems*. Blackwell Scientific, Oxford, UK.

Robertson, G.P., and E.A. Paul. 2000. Decomposition and soil organic matter dynamics. Pages 104–116 *in* O.E. Sala, R.B. Jackson, H.A. Mooney, and R.W. Howarth, editors. *Methods in Ecosystem Science*. Springer-Verlag, New York.

Robertson, G.P., K.M. Klingensmith, M.J. Klug, E.A. Paul, J.R. Crum, and B.G. Ellis. 1997. Soil resources, microbial activity, and primary production across an agricultural ecosystem. *Ecological Applications* 7:158–170.

Robinson, D. 1994. The responses of plants to non-uniform supplies of nutrients. *New Phytologist* 127:635–674.

Robles, M., and F.S. Chapin III. 1995. Comparison of the influence of two exotic species on ecosystem processes in the Berkeley Hills. *Madroño* 42: 349–357.

Rodin, L.E., and N.I. Bazilevich. 1967. *Production and Mineral Cycling in Terrestrial Vegetation.* Oliver & Boyd, Edinburgh, UK.

Rollet, B. 1983. La regeneration naturelle dans las trouees. *Revue Bois et Florets des Tropiques* 202:19–34.

Romme, W.H., and D.H. Knight. 1982. Landscape diversity: The concept applied to Yellowstone Park. *BioScience* 32:644–670.

Rosenberg, A.A., M.J. Fogarty, M.P. Sissenwine, J.R. Beddington, and J.G. Shepherd. 1993. Achieving sustainable use of renewable resources. *Science* 262:828–829.

Rosenberg, D.K., B.R. Noon, and E.C. Meslow. 1997. Biological corridors: Form, function, and efficacy. *BioScience* 47:677–687.

Roulet, N.T., P.M. Crill, N.T. Comer, A. Dove, and R.A. Boubonniere. 1997. CO_2 and CH_4 flux between a boreal beaver pond and the atmosphere. *Journal of Geophysical Research* 102:29313–29319.

Rovira, A.D. 1969. Plant root exudates. *Botanical Review* 35:35–56.

Ruess, R.W., and S.J. McNaughton. 1987. Grazing and the dynamics of nutrient and energy regulated microbial processes in the Serengeti grasslands. *Oikos* 49:101–110.

Ruess, R.W., R.L. Hendrick, and J.P. Bryant. 1998. Regulation of fine root dynamics by mammalian browsers in early successional Alaskan taiga forests. *Ecology* 79:2706–2720.

Ruess, R.W., D.S. Hik, and R.L. Jefferies. 1989. The role of lesser snow geese as nitrogen processors in a sub-arctic salt marsh. *Oecologia* 89:23–29.

Ruess, R.W., K. Van Cleve, J. Yarie, and L.A. Viereck. 1996. Contributions of fine root production and turnover to the carbon and nitrogen cycling in taiga forests of the Alaskan interior taiga forests on the Alaskan interior. *Canadian Journal of Forest Research* 26:1326–1336.

Ruimy, A., P.G. Jarvis, D.D. Baldocchi, and B. Saugier. 1996. CO_2 fluxes over plant canopies and solar radiation: A review. *Advances in Ecological Research* 26:1–68.

Running, S.W., R.R. Nemani, D.L. Peterson, L.E. Band, D.F. Potts, L.L. Pierce, and M.A. Spanner. 1989. Mapping regional forest evapotranspiration and photosynthesis by coupling satellite data with ecosystem simulation. *Ecology* 70:1090–1101.

Rupp, T.S., A.M. Starfield, and F.S. Chapin III. 2000. A frame-based spatially explicit model of subarctic vegetation response to climatic change: Comparison with a point model. *Landscape Ecology* 15:383–400.

Ryan, M.G., D. Binkley, and J.H. Fownes. 1997. Age-related decline in forest productivity: Pattern and process. *Advances in Ecological Research* 27:213–262.

Ryan, M.G., S. Linder, J.M. Vose, and R.M. Hubbard. 1994. Respiration of pine forests. *Ecological Bulletin* 43:50–63.

Saint Louis, V.L., C.A. Kelly, E. Duchemin, J.W.M. Rudd, and D.M. Rosenberg. 2000. Reservoir surfaces as sources of greenhouse gases to the atmosphere: A global estimate. *BioScience* 50:766–775.

Sala, O.E., F.S. Chapin III, J.J. Armesto, E. Berlow, J. Bloomfield, R. Dirzo, E. Huber-Sanwald, L.F. Huenneke, R. Jackson, A. Kinzig, R. Leemans, D. Lodge, H.A. Mooney, M. Osterheld, N.L. Poff, M.T. Sykes, B.H. Walker, M. Walker, and D.H. Wall. 2000a. Global biodiversity scenarios for the year 2100. *Science* 287:1770–1776.

Sala, O.E., R.B. Jackson, H.A. Mooney, and R.W. Howarth, editors. 2000b. *Methods in Ecosystem Science*. Springer-Verlag, New York.

Sala, O.E., W.K. Lauenroth, S.J. McNaughton, G. Rusch, and X. Zhang. 1996. Biodiversity and ecosystem functioning in grasslands. Pages 129–149 *in* H.A. Mooney, J.H. Cushman, E. Medina, O.E. Sala, and E.-D. Schulze, editors. *Functional Role of Biodiversity: A Global Perspective*. Wiley, Chichester, UK.

Sala, O.E., W.J. Parton, L.A. Joyce, and W.K. Lauenroth. 1988. Primary production of the central grassland region of the United States. *Ecology* 69:40–45.

Saugier, B., J. Roy, and H.A. Mooney. 2001. Estimations of global terrestrial productivity: Converging toward a single number? Pages 543–557 *in* J. Roy, B. Saugier, and H.A. Mooney, editors. *Terrestrial Global Productivity*. Academic Press, San Diego, CA.

Scatena, F.N., S. Moya, C. Estrada, and J.D. Chinea. 1996. The first five years in the reorganization of aboveground biomass and nutrient use following Hurricane Hugo in the Bisley Experimental Watersheds, Luquillo Experimental Forest, Puerto Rico. *Biotropica* 28:424–440.

Schaefer, D.A., W.H. McDEowell, F.N. Scatena, and C.E. Asbury. 2000. Effects of hurricane disturbance on stream water concentrations and fluxes in eight tropical forest watersheds of the Luquillo Experimental Forest, Puerto Rico. *Journal of Tropical Ecology* 16:189–207.

Schimel, D.S. 1995. Terrestrial ecosystems and the carbon cycle. *Global Change Biology* 1:77–91.

Schimel, D.S., J.I. House, K.A. Hibbard, P. Bousquet, P. Ciais, P. Peylin, B.H. Braswell, M.J. Apps, D. Baker, A. Bondeau, J. Canadell, G. Churkina, W. Cramer, A.S. Denning, C.B. Field, P. Friedlingstein, C. Goodale, M. Heimann, R.A. Houghton, J.M. Melillo, B. Moore, D. Murdiyarso, I. Noble, S.W. Pacala, I.C. Prentice, M.R. Raupach, P.J. Rayner, R.J. Scholes, W.L. Steffen, and C. Wirth. 2001. Recent patterns and mechanisms of carbon exchange by terrestrial ecosystems. *Nature* 414: 169–172.

Schimel, J.P. 2001. Biogeochemical models: Implicit vs. explicit microbiology. Pages 177–183 *in* E.-D. Schulze, S.P. Harrison, M. Heimann, E.A. Holland, J.J. Lloyd, I.C. Prentice, and D. Schimel, editors. *Global Biogeochemical Cycles in the Climate System*. Academic Press, San Diego, CA.

Schimel, J.P., and F.S. Chapin III. 1996. Tundra plants compete effectively with soil microbes for amino-acid nitrogen. *Ecology* 77:2142–2147.

Schimel, J.P., K. Van Cleve, R.G. Cates, T.P. Clausen, and P.B. Reichardt. 1996. Effects of balsam poplar (*Populus balsamifera*) tannins and low molecular weight phenolics on microbial activity in taiga floodplain soil: Implications for changes in N cycling during succession. *Canadian Journal of Botany* 74:84–90.

Schimper, A.F.W. 1898. *Pflanzengeographie auf Physiologischer Grundlage*. Fisher, Jena, Germany.

Schindler, D.W. 1985. The coupling of elemental cycles by organisms: Evidence from whole-lake chemical perturbations. Pages 225–250 *in* W. Stumm, editor. *Chemical Processes in Lakes*. Wiley, New York.

Schindler, D.W. 1977. Evolution of phosphorus limitation in lakes. *Science* 195:260–267.

Schindler, D.W. 1978. Factors regulating phytoplankton production and standing crop in the world's lakes. *Limnology and Oceanography* 23:478–486.

Schlesinger, W.H. 1997. *Biogeochemistry: An Analysis of Global Change*. Academic Press, San Diego, CA.

Schlesinger, W.H. 1977. Carbon balance in terrestrial detritus. *Annual Review of Ecology and Systematics* 8:51–81.

Schlesinger, W.H. 1999. Carbon sequestration in soils. *Science* 284:2095–2097.

Schlesinger, W.H., J.F. Reynolds, G.L. Cunningham, L.F. Huenneke, W.M. Jarrell, R.A. Virginia, and W.G. Whitford. 1990. Biological feedbacks in global desertification. *Science* 247:1043–1048.

Schulze, E.-D. 1989. Air pollution and forest decline in a spruce (*Picea abies*) forest. *Science* 244: 776–783.

Schulze, E.-D., and F.S. Chapin III. 1987. Plant specialization to environments of different resource availability. Pages 120–148 *in* E.D. Schulze and

H. Zwolfer, editors. *Potentials and Limitations in Ecosystem Analysis*. Springer-Verlag, Berlin.

Schulze, E.-D., M. Fuchs, and M.I. Fuchs. 1977. Spatial distribution of photosynthetic capacity and performance in a mountain spruce forest of northern Germany. III. The significance of the evergreen habit. *Oecologia* 30:239–248.

Schulze, E.-D., F.M. Kelliher, C. Körner, J. Lloyd, and R. Leuning. 1994. Relationship among maximum stomatal conductance, ecosystem surface conductance, carbon assimilation rate, and plant nitrogen nutrition: A global ecology scaling exercise. *Annual Review of Ecology and Systematics* 25:629–660.

Schulze, E.-D., R.H. Robichaux, J. Grace, P.W. Rundel, and J.R. Ehleringer. 1987. Plant water balance. *BioScience* 37:30–37.

Schulze, E.-D., C. Wirth, and M. Heimann. 2000. Climate change: Managing forests after Kyoto. *Science* 289:2058–2059.

Schuur, E.A.G. Unpublished. The sensitivity of tropical forest growth to climate: Revisiting the relationship between net primary productivity and global climate. *Ecology*.

Schuur, E.A.G., O.A. Chadwick, and P.A. Matson. 2001. Carbon cycling and soil carbon storage in mesic to wet Hawaiian montane forests. *Ecology* 82:3182–3196.

Schwoerbel, J. 1987. *Handbook of Limnology*. Halsted Press, New York.

Selby, M.J. 1993. *Hillslope Materials and Processes*. Oxford University Press, Oxford, UK.

Semikhatova, O.A. 2000. Ecological physiology of plant dark respiration: Its past, present and future. *Botanishcheskii Zhurnal* 85:15–32.

Shaver, G.R., W.D. Billings, F.S. Chapin III, A.E. Giblin, K.J. Nadelhoffer, W.C. Oechel, and E.B. Rastetter. 1992. Global change and the carbon balance of arctic ecosystems. *BioScience* 61:415–435.

Shaver, G.R., J. Canadell, F.S. Chapin III, J. Gurevitch, J. Harte, G. Henry, P. Ineson, S. Jonasson, J. Melillo, L. Pitelka, and L. Rustad. 2000. Global warming and terrestrial ecosystems: A conceptual framework for analysis. *BioScience* 50: 871–882.

Shaver, G.R., K.J. Nadelhoffer, and A.E. Giblin. 1991. Biogeochemical diversity and element transport in a heterogeneous landscape, the North Slope of Alaska. Pages 105–126 *in* M.G. Turner and R.H. Gardner, editors. *Quantitative Methods in Landscape Ecology*. Springer-Verlag, New York.

Shiro, T., and R. del Moral. 1995. Species attributes in early primary succession on volcanoes. *Journal of Vegetation Science* 6:517–522.

Shugart, H.H., and D.C. West. 1980. Forest succession models. *BioScience* 30:308–313.

Shukla, J., C. Nobre, and P. Sellers. 1990. Amazon deforestation and climate change. *Science* 247: 1322–1325.

Sigman, D.M., and E.A. Boyle. 2000. Glacial/interglacial variations in atmospheric carbon dioxide. *Nature* 407:859–869.

Simard, S.W., D.A. Perry, M.D. Jones, D.D. Myrold, D.M. Durall, and R. Molina. 1997. Net transfer of carbon between ectomycorrhizal tree species in the field. *Nature* 388:579–582.

Sinclair, A.R.E. 1979. The eruption of the ruminants. Pages 82–103 *in* A.R.E. Sinclair and M. Norton-Griffiths, editors. *Serengeti: Dynamics of an Ecosystem*. University of Chicago Press, Chicago.

Singer, F., W. Swank, and E. Clebsch. 1984. Effects of wild pig rooting in a deciduous forest. *Journal of Wildlife Management* 48:464–473.

Smart, D.R., and A.J. Bloom. 1988. Kinetics of ammonium and nitrate uptake among wild and cultivated tomatoes. *Oecologia* 76:336–340.

Smil, V. 2000. Phosphorus in the environment: Natural flows and human interferences. *Annual Review of Energy in the Environment* 25:53–88.

Smith, J.L., and E.A. Paul. 1990. The significance of soil microbial biomass estimations. Pages 357–396 *in* J. Bollag and G. Stotsky, editors. *Soil Biochemistry*. Marcel Dekker, New York.

Smith, S.E., and D.J. Read. 1997. *Mycorrhizal Symbiosis*. Academic Press, London.

Smith, T., H.H. Shugart, and F.I. Woodward. 1997. *Plant Functional Types*. Cambridge University Press, Cambridge, UK.

Society of American Foresters. 1993. *Sustaining Long-Term Forest Health and Productivity*. Society of American Foresters, Bethesda, MD.

Soluck, D.A., and J.S. Richardson. 1997. The role of stoneflies in enhancing growth of trout: A test of the importance of predator-predator facilitation within a stream community. *Oikos* 80:214–219.

Sousa, W.P. 1985. The role of disturbance in natural communities. *Annual Review of Ecology and Systematics* 15:353–391.

Spencer, C.N., B.R. McClelland, and J.A. Stanford. 1991. Shrimp stocking, salmon collapse, and eagle displacement. *BioScience* 41:14–21.

Sperry, J.S. 1995. Limitations on stem water transport and their consequences. Pages 105–124 *in* B.L. Gartner, editor. *Plant Stems: Physiology and Functional Morphology*. Academic Press, San Diego, CA.

Sprugel, D.G. 1976. Dynamic structure of wave-generated *Abies balsamea* forests in the northeastern United States. *Journal of Ecology* 64:880–911.

Starfield, A.M. 1991. Qualitative rule-based modeling. *BioScience* 40:601–604.

Starfield, A.M., and F.S. Chapin III. 1996. Model of transient changes in arctic and boreal vegetation in response to climate and land use change. *Ecological Applications* 6:842–864.

Starfield, A.M., J.D. Roth, and K. Ralls. 1995. "Mobbing" in Hawaiian monk seals: The value of simulation modeling in the absence of apparently crucial data. *Conservation Biology* 9:166–174.

Stark, J.M. 2000. Nutrient transformations. Pages 215–234 *in* O.E. Sala, R.B. Jackson, H.A. Mooney, and R.W. Howarth, editors. *Methods in Ecosystem Ecology*. Springer-Verlag, New York.

Stark, J.M., and M.K. Firestone. 1995. Mechanisms for soil moisture effects on nitrifying bacteria. *Applied Environmental Microbiology* 61:218–221.

Stark, J.M., and S.C. Hart. 1997. High rates of nitrification and nitrate turnover in undisturbed coniferous forests. *Nature* 385:61–64.

Stark, N.M., and R. Steele. 1977. Nutrient content of forest shrubs following burning. *American Journal of Botany* 64:1218–1224.

Steele, J.H. 1991. Can ecological theory cross the land-sea boundary? *Journal of Theoretical Biology* 153:425–436.

Steiner, K. 1982. *Intercropping in Tropical Smallholder Agriculture with Special Reference to West Africa*. German Agency for Technical Cooperation (GTZ), Eschborn, Germany.

Stephens, B.B., and R.F. Keeling. 2000. The influence of Antarctic sea ice on glacial-interglacial CO_2 variations. *Nature* 404:171–174.

Stevenson, F.J. 1994. *Humus Chemistry: Genesis, Composition, Reactions*. Wiley, New York.

Stock, W.D., and O.A.M. Lewis. 1984. Uptake and assimilation of nitrate and ammonium by an evergreen Fynbos shrub species *Protea repens* L. (Proteaceae). *New Phytologist* 97:261–268.

Strayer, D.L., N.F. Caraco, J.J. Cole, S. Findlay, and M.L. Pace. 1999. Transformation of freshwater ecosystems by bivalves. *BioScience* 49:19–27.

Strong, D.R. 1992. Are trophic cascades all wet? Differentiation and donor-control in speciose ecosystems. *Ecology* 73:747–754.

Sturm, M., J.P. McFadden, G.E. Liston, F.S. Chapin III, J. Holmgren, and M. Walker. 2001. Snow-shrub interactions in arctic tundra: A feedback loop with climatic implications. *Journal of Climate* 14:336–344.

Sturman, A.P., and N.J. Tapper. 1996. *The Weather and Climate of Australia and New Zealand*. Oxford University Press, Oxford, UK.

Sucoff, E. 1972. Water potential in red pine: Soil moisture, evapotranspiration, crown position. *Ecology* 52:681–686.

Sutherland, W.J. 2000. *The Conservation Handbook: Research, Management and Policy*. Blackwell Scientific, Oxford, UK.

Swank, W.T., and J.E. Douglass. 1974. Streamflow greatly reduced by converting deciduous hardwood stands to pine. *Science* 185:857–859.

Swetnam, T.W., C.D. Allen, and J.L. Betancourt. 1999. Applied historical ecology: Using the past to manage for the future. *Ecological Applications* 9:1189–1206.

Swift, M.J., O.W. Heal, and J.M. Anderson. 1979. *Decomposition in Terrestrial Ecosystems*. Blackwell Scientific, Oxford, UK.

Tans, P.P., I.Y. Fung, and T. Takahashi. 1990. Observational constraints on the global CO_2 budget. *Science* 247:1431–1438.

Tansley, A.G. 1935. The use and abuse of vegetational concepts and terms. *Ecology* 16:284–307.

Taylor, B.R., D. Parkinson, and W.F.J. Parsons. 1989. Nitrogen and lignin as predictors of litter decay rates: A microcosm test. *Ecology* 70:97–104.

Terashima, I., and K. Hikosaka. 1995. Comparative ecophysiology of leaf and canopy photosynthesis. *Plant, Cell and Environment* 18:1111–1128.

Thorp, J.H., and A.P. Covich, editors. 2001. *Ecology and Classification of North American Freshwater Invertebrates*. Academic Press, San Diego, CA.

Thurman, H.V. 1991. *Introductory Oceanography*. Macmillan, New York.

Tiessen, H. 1995. Introduction and synthesis. Pages 1–6 *in* H. Tiessen, editor. *Phosphorus in the Global Environment: Transfers, Cycles and Management*. J Wiley, Chichester, UK.

Tietema, A., and C. Beier. 1995. A correlative evaluation of nitrogen cycling in the forest ecosystems of the EC projects NITREX and EXMAN. *Forest Ecology and Management* 71:143–151.

Tilman, D. 1988. *Plant Strategies and the Dynamics and Function of Plant Communities*. Princeton University Press, Princeton, NJ.

Tilman, D. 1985. The resource-ratio hypothesis of plant succession. *American Naturalist* 125:827–852.

Tilman, D., and J.A. Downing. 1994. Biodiversity and stability in grasslands. *Nature* 367:363–365.

Tilman, D., D. Wedin, and J. Knops. 1996. Productivity and sustainability influenced by biodiversity in grassland ecosystems. *Nature* 379:718–720.

Townsand, A.R., P.M. Vitousek, and S.E. Trumbore. 1995. Soil organic matter dynamics along gradients in temperature and land use on the island of Hawaii. *Ecology* 76:721–733.

Trenberth, K., and T.J. Haar. 1996. The 1990–1995 El Niño-southern oscillation event: Longest on record. *Geophysical Research Letters* 23:57–60.

Trimble, S.W., F.H. Weirich, and B.L. Hoag. 1987. Reforestation reduces stream flow in the southeastern United States. *Water Resource Research* 23:425–437.

Trumbore, S.E. 1993. Comparison of carbon dynamics in tropical and temperate soils using radiocarbon measurements. *Global Biogeochemical Cycles* 7:275–290.

Trumbore, S.E., and J.W. Harden. 1997. Accumulation and turnover of carbon in organic and mineral soils of the BOREAS northern study area. *Journal of Geophysical Research* 102:28817–28830.

Turner, B.L. II, W.C. Clark, R.W. Kates, J.F. Richards, J.T. Mathews, and W.B. Meyer, editors. 1990. *The Earth as Transformed by Human Action*. Cambridge University Press, Cambridge, UK.

Turner, M.G., R.H. Gardner, V.H. Dale, and R.V. O'Neill. 1989. Predicting the spread of disturbance across heterogeneous landscapes. *Oikos* 55:121–129.

Turner, M.G., R.H. Gardner, and R.V. O'Neill. 2001. *Landscape Ecology in Theory and Practice: Pattern and Process*. Springer-Verlag, New York.

Turner, M.G., W.H. Romme, R.H. Gardner, and W.W. Hargrove. 1997. Effects of patch size and fire pattern on early post-fire succession on the Yellowstone Plateau. *Ecological Monographs* 67:422–433.

Turner, M.G., W.H. Romme, R.H. Gardner, R.V. O'Neill, and T.K. Kratz. 1993. A revised concept of landscape equilibrium: Disturbance and stability on scaled landscapes. *Landscape Ecology* 8:213–227.

Turner, R.E., and N.N. Rabalais. 1991. Changes in Mississippi River water quality this century. *BioScience* 41:140–147.

Tuskan, G.A., K.E. Francis, S.L. Russ, W.H. Romme, and M.G. Turner. 1996. RAPDs demonstrate genetic diversity within and among aspen populations in Yellowstone National Park, USA. *Canadian Journal of Forest Research* 26:2088–2098.

Tyrrell, T. 1999. The relative influences of nitrogen and phosphorus on oceanic primary production. *Nature* 400:525–531.

Uehara, G., and G. Gillman. 1981. *The Minerology, Chemistry, and Physics of Tropical Soils with Variable Charge Clays*. Westview Press, Boulder, CO.

Ugolini, F.C. 1968. Soil development and alder invasion in a recently deglaciated area of Glacier Bay, Alaska. Pages 115–148 *in* J.M. Trappe, F.F. Franklin, R.F. Tarrant, and G.M. Hansen, editors. *Biology of Alder*. USDA Forest Service, Pacific Northwest Forest and Range Experiment Station, Portland, OR.

Ugolini, F.C., and H. Spaltenstein. 1992. Pedosphere. Pages 123–153 *in* S.S. Butcher, R.J. Charlson, G.H. Orians, and G.V. Wolfe, editors. *Global Biogeochemical Cycles*. Academic Press, London.

Urban, D.L., R.V. O'Neill, and H.H. Shugart. 1987. Landscape ecology. *BioScience* 37:119–127.

Valentini, R., G. Matteucci, A.J. Dolman, E.-D. Schulze, C. Rebmann, E.J. Moors, A. Granier, P. Gross, N.O. Jensen, K. Pilegaard, A. Lindroth, A. Grelle, C. Bernhofer, A. Grünwald, M.R. Ceulemans, A.S. Kowalski, T. Vesala, Ü. Rannik, P. Berbigler, D. Loustau, J. Guömundsson, H. Thorgeirsson, A. Ibrom, K. Morgenstern, R. Clement, J. Moncrieff, L. Montagnani, S. Minerbi, and P.G. Jarvis. 2000. Respiration as the main determinant of carbon balance in European forests. *Nature* 404:861–864.

Valiela, I. 1995. *Marine Ecological Processes*. Springer-Verlag, New York.

Van Breemen, N., and A.C. Finzi. 1998. Plant-soil interactions: Ecological aspects and evolutionary implications. *Biogeochemistry* 42:1–19.

Van Cleve, K., F.S. Chapin III, C.T. Dryness, and L.A. Viereck. 1991. Element cycling in taiga forest: State-factor control. *BioScience* 41:78–88.

Van Cleve, K., W.C. Oechel, and J.L. Hom. 1990. Response of black spruce (*Picea mariana*) ecosystems to soil temperature modification in interior Alaska. *Canadian Journal of Forest Research* 20:1530–1535.

Van Cleve, K., L. Oliver, R. Schlentner, L.A. Viereck, and C.T. Dyrness. 1983. Productivity and nutrient cycling in taiga forest ecosystems. *Canadian Journal of Forest Research* 13:747–766.

Van Cleve, K., J. Yarie, R. Erickson, and C.T. Dyrness. 1993. Nitrogen mineralization and nitrification in successional ecosystems on the Tanana River floodplain, interior Alaska. *Canadian Journal of Forest Research* 23:970–978.

Vance, E.D., and F.S. Chapin III. 2001. Substrate-environment interactions: Multiple limitations to microbial activity in taiga forest floors. *Soil Biology and Biochemistry* 33:173–188.

Vandermeer, J. 1995. The ecological basis of alternative agriculture. *Annual Review of Ecology and Systematics* 26:201–224.

Vandermeer, J.H. 1990. Intercropping. Pages 481–516 *in* C.R. Carrol, J.H. Vandermeer, and P.M. Rosset, editors. *Agroecology*. McGraw-Hill, New York.

Vannote, R.I., G.W. Minshall, K.W. Cummings, J.R. Sedell, and C.E. Cushing. 1980. The river continuum concept. *Canadian Journal of Fisheries and Aquatic Sciences* 37:120–137.

VEMAP Members. 1995. Vegetation/ecosystem modeling and analysis project: Comparing biogeography and biogeochemistry models in a continental-scale study of terrestrial ecosystem responses to climate change and CO_2 doubling. *Global Biogeochemical Cycles* 9:407–437.

Verbyla, D.L. 1995. *Satellite Remote Sensing of Natural Resources*. CRC Press, Boca Raton, FL.

Verhoef, H.A., and L. Brussaard. 1990. Decomposition and nitrogen mineralization in natural and agro-ecosystems: The contribution of soil animals. *Biogeochemistry* 11:175–211.

Vitousek, P.M. 1994a. Beyond global warming: Ecology and global change. *Ecology* 75:1861–1876.

Vitousek, P.M. 1994b. Factors controlling ecosystem structure and function. Pages 87–97 *in* R. Amandson, J. Harden, and M. Singer, editors. *Factors of Soil Formation: A 50th Anniversary Retrospective*. Soil Science Society of America, Madison, WI.

Vitousek, P.M. 1984. Litterfall, nutrient cycling, and nutrient limitation in tropical forests. *Ecology* 65:285–298.

Vitousek, P.M. 1982. Nutrient cycling and nutrient use efficiency. *American Naturalist* 119:553–572.

Vitousek, P.M., and H. Farrington. 1997. Nitrogen limitation and soil development: Experimental test of a biogeochemical theory. *Biogeochemistry* 37:63–75.

Vitousek, P.M., and D.U. Hooper. 1993. Biological diversity and terrestrial ecosystem biogeochemistry. Pages 3–14 *in* E.-D. Schulze and H.A. Mooney, editors. *Biodiversity and Ecosystem Function*. Springer-Verlag, Berlin.

Vitousek, P.M., and R.W. Howarth. 1991. Nitrogen limitation on land and in the sea: How can it occur? *Biogeochemistry* 13:87–115.

Vitousek, P.M., and P.A. Matson. 1984. Mechanisms of nitrogen retention in forest ecosystems: A field experiment. *Science* 225:51–52.

Vitousek, P.M., and P.A. Matson. 1988. Nitrogen transformations in a range of tropical forest soils. *Soil Biology and Biochemistry* 20:361–367.

Vitousek, P.M., and W.A. Reiners. 1975. Ecosystem succession and nutrient retention: A hypothesis. *BioScience* 25:376–381.

Vitousek, P.M., J.D. Aber, R.W. Howarth, G.E. Likens, P.A. Matson, D.W. Schindler, W.H. Schlesinger, and G.D. Tilman. 1997a. Human alteration of the global nitrogen cycle: Sources and consequences. *Ecological Applications* 7:737–750.

Vitousek, P.M., O.A. Chadwick, T. Crews, J. Fownes, D. Hendricks, and D. Herbert. 1997b. Soil and ecosystem development across the Hawaiian Islands. *GSA Today* 7:1–8.

Vitousek, P.M., J.R. Gosz, C.C. Grier, J.M. Melillo, and W.A. Reiners. 1982. A comparative analysis of potential nitrification and nitrate mobility in forest ecosystems. *Ecological Monographs* 52:155–177.

Vitousek, P.M., P.A. Matson, and D.R. Turner. 1988. Elevational and age gradients in Hawai'ian montane rainforest: Foliar and soil nutrients. *Oecologia* 77:565–570.

Vitousek, P.M., H.A. Mooney, J. Lubchenco, and J.M. Melillo. 1997c. Human domination of Earth's ecosystems. *Science* 277:494–499.

Vitousek, P.M., L.R. Walker, L.D. Whiteaker, and P.A. Matson. 1993. Nutrient limitation to plant growth during primary succession in Hawaii. *Biogeochemistry* 23:197–215.

Vitousek, P.M., L.R. Walker, L.D. Whiteaker, D. Mueller-Dombois, and P.A. Matson. 1987. Biological invasion by *Myrica faya* alters ecosystem development in Hawaii. *Science* 238:802–804.

Vogt, K.A., C.C. Grier, and D.J. Vogt. 1986. Production, turnover, and nutrient dynamics in above- and belowground detritus of world forests. *Advances in Ecological Research* 15:303–377.

Vrba, E.S., and S.J. Gould. 1986. The hierarchical expansion of sorting and selection: Sorting and selection cannot be equated. *Paleobiology* 12: 217–228.

Vukicevic, T., B.H. Braswell, and D. Schimel. 2001. A diagnostic study of temperature controls on global terrestrial carbon exchange. *Tellus* 53B:150–170.

Wagener, S.M., M.W. Oswood, and J.P. Schimel. 1998. Rivers and soils: Parallels in carbon and nutrient processing. *BioScience* 48:104–108.

Walker, B.H. 1992. Biodiversity and ecological redundancy. *Conservation Biology* 6:18–23.

Walker, B.H. 1995. Conserving biological diversity through ecosystem resilience. *Conservation Biology* 9:747–752.

Walker, B.H., A. Kinzig, and J. Langridge. 1999. Plant attribute diversity, resilience, and ecosystem function: The nature and significance of dominant and minor species. *Ecosystems* 2:95–113.

Walker, D.A., and K.R. Everett. 1991. Loess ecosystems of northern Alaska: Regional gradient and toposequence at Prudhoe Bay. *Ecological Monographs* 61:437–464.

Walker, D.A., J.G. Bockheim, F.S. Chapin III, W. Eugster, J.Y. King, J.P. McFadden, G.J. Michaelson, F.E. Nelson, W.C. Oechel, C.-L. Ping, W.S.

Reeburgh, S. Regli, N.I. Shiklomanov, and G.L. Vourlitis. 1998. A major arctic soil pH boundary: Implications for energy and trace-gas fluxes. *Nature* 394:469–472.

Walker, L.R. 1993. Nitrogen fixers and species replacements in primary succession. Pages 249–272 *in* J. Miles and D.W.H. Walton, editors. *Primary succession on land*. Blackwell, Oxford, UK.

Walker, L.R. 1999. Patterns and processes in primary succession. Pages 585–610 *in* L.R. Walker, editor. *Ecosystems of Disturbed Ground*. Elsevier, Amsterdam.

Walker, L.R., J.C. Zasada, and F.S. Chapin III. 1986. The role of life history processes in primary succession on an Alaskan floodplain. *Ecology* 67:1243–1253.

Walker, T.W., and J.K. Syers. 1976. The fate of phosphorus during pedogenesis. *Geoderma* 1:1–19.

Wall, D.H., G. Adams, and A.N. Parsons. 2001. Soil biodiversity. Pages 47–82 *in* F.S. Chapin III, O.E. Sala, and E. Huber-Sannwald, editors. *Global Biodiversity in a Changing Environment: Scenarios for the 21st century*. Springer-Verlag, New York.

Wallwork, J.A. 1976. *The Distribution and Diversity of Soil Fauna*. Academic Press, New York.

Walters, M.B., and C.B. Field. 1987. Photosynthetic light acclimation in two rainforest *Piper* species with different ecological amplitudes. *Oecologia* 72:449–456.

Walters, M.B., and P.B. Reich. 1999. Low-light carbon balance and shade tolerance in the seedlings of woody plants: Do winter deciduous and broad-leaved evergreen species differ? *New Phytologist* 143:143–154.

Wan, S., D. Hui, and Y. Luo. 2001. Fire effects on nitrogen pools and dynamics in terrestrial ecosystems: A meta-analysis. *Ecological Applications* 11:1349–1365.

Ward, J.V., and J.A. Stanford. 1983. The serial discontinuity concept of running waters. Pages 29–42 *in* T.D. Fontaine and S.M. Bartell, editors. *Dynamics of Lotic Ecosystems*. Ann Arbor Science, Ann Arbor, MI.

Waring, R.H., and S.W. Running. 1998. *Forest Ecosystems: Analysis at Multiple Scales*. Academic Press, New York.

Weaver, C.P., and R. Avissar. 2001. Atmospheric disturbances caused by human modification of the landscape. *Bulletin of the American Meteorological Society* 82:269–281.

Webb, T., and P.J. Bartlein. 1992. Global changes during the last three million years: Climatic controls and biotic responses. *Annual Review of Ecology and Systematics* 23:141–173.

Webster, K.E., T.K. Kratz, C.J. Bowser, and J.J. Magnuson. 1996. The influence of landscape position on lake chemical responses to drought in northern Wisconsin. *Limnology and Oceanography* 41:977–984.

Webster, P.J., and T.N. Palmer. 1997. The past and future of El Niño. *Nature* 390:562–564.

Wedin, D.A., and D. Tilman. 1990. Species effects on nitrogen cycling: A test with perennial grasses. *Oecologia* 84:433–441.

Wells, M.P., and K.E. Brandon. 1993. The principles and practice of buffer zones and local participation in biodiversity conservation. *Ambio* 22:157–162.

Westman, W.E. 1978. Patterns of nutrient flow in the pygmy forest region of northern California. *Vegetatio* 36:1–15.

Wetzel, R.G. 2001. *Limnology: Lake and River Ecosystems*. Academic Press, San Diego, CA.

Whalen, S.C., and J.C. Cornwell. 1985. Nitrogen, phosphorus, and organic carbon cycling in an arctic lake. *Canadian Journal of Fisheries and Aquatic Sciences* 42:797–808.

Whittaker, R.H. 1975. *Communities and Ecosystems*. Macmillan, New York.

Whittaker, R.H., and W.A. Niering. 1965. Vegetation of the Santa Catalina Mountains, Arizona. (II) A gradient analysis of the south slope. *Ecology* 46:429–452.

Whittaker, R.H., G.E. Likens, F.H. Bormann, J.S. Eaton, and T.H. Siccama. 1979. The Hubbard Brook ecosystem study: Forest nutrient cycling and element behavior. *Ecology* 60:203–220.

Wiegert, R.G., and D.F. Owen. 1971. Trophic structure, available resources and population density in terrestrial versus aquatic ecosystems. *Journal of Theoretical Biology* 30:69–81.

Wiens, J.A. 1996. Wildlife in patchy environments: Metapopulations, mosaics, and management. Pages 53–84 *in* D.R. McCullough, editor. *Metapopulations and Wildlife Conservation*. Island Press, Washington, DC.

Wilcox, H.E. 1991. Mycorrhizae. Pages 731–765 *in* Y. Waisel, A. Eshel, and U. Kafkaki, editors. *Plant Roots: The Hidden Half*. Marcel Dekker, New York.

Williams, M., E.B. Rastetter, D.N. Fernandes, M.L. Goulden, G.R. Shaver, and L.C. Johnson. 1997. Predicting gross primary productivity in terrestrial ecosystems. *Ecological Applications* 7:882–894.

Wills, R.D., and J.D. Stuart. 1994. Fire history and stand development of a Douglas-fir/hardwood forest in northern California. *Northwest Science* 68:205–212.

Willson, M.F., S.M. Gende, and B.H. Marston. 1998. Fishes and the forest. *BioScience* 48:455–462.

Wilson, J.B., and D.Q. Agnew. 1992. Positive-feedback switches in plant communities. *Advances in Ecological Research* 23:263–336.

Winner, W.E., H.A. Mooney, K. Williams, and S. von Caemmerer. 1985. Measuring and assessing SO_2 effects on photosynthesis and plant growth. Pages 118–132 *in* W.E. Winner and H.A. Mooney, editors. *Sulfur Dioxide and Vegetation*. Stanford University Press, Stanford, CA.

Wood, C.A. 1994. Ecosystem management: Achieving the new land ethic. *Renewable Natural Resources Journal* 12:6–12.

Woodward, F.I. 1987. *Climate and Plant Distribution*. Cambridge University Press, Cambridge, CA.

Woodward, S., P.M. Vitousek, K. Benvenuto, P.A. Matson. 1990. Use of the exotic tree *Myrica Faya* by native and exotic birds in Hawaii Volcanoes National Park. *Pacific Science* 44:88–93.

Worley, I.A. 1973. The "black crust" phenomenon in upper Glacier Bay, Alaska. *Northwest Science* 47:19–34.

WRI. 2000. *World Resources 2000–2001. People and Ecosystems: The Fraying Web of Life*. World Resources Institute, Washington, DC.

Wright, I.J., P.B. Reich, and M. Westoby. 2001. Strategy shifts in leaf physiology, structure and nutrient content between species of high- and low-rainfall and high- and low-nutrient habitats. *Functional Ecology* 15:423–434.

Wu, J., and O.L. Loucks. 1995. From balance of nature to hierarchical patch dynamics: A paradigm shift in ecology. *Quarterly Review of Biology* 70:439–466.

Yaffee, S.L., A.F. Phillips, I.C. Frentz, P.W. Hardy, S.M. Maleki, and B.E. Thorpe. 1996. *Ecosystem Management in the United States*. Island Press, Washington, DC.

Zak, D.R., D. Tilman, R.R. Parmenter, C.W. Rice, F.M. Fisher, J. Vose, D. Milchunas, and C.W. Martin. 1994. Plant production and soil microorganisms in late-successional ecosystems: a continental-scale study. *Ecology* 75:2333–2347.

Zech, W., and I. Kogel-Knabner. 1994. Patterns and regulation of organic matter transformation in soils: Litter decomposition and humification. Pages 303–335 *in* E.-D. Schulze, editor. *Flux Control in Biological Systems: From Enzymes to Populations and Ecosystems*. Academic Press, San Diego, CA.

Zimmermann, M.H. 1983. *Xylem Structure and the Ascent of Sap*. Springer, New York.

Zimov, S.A., V.I. Chuprynin, A.P. Oreshko, F.S. Chapin III, J.F. Reynolds, and M.C. Chapin. 1995. Steppe-tundra transition: An herbivore-driven biome shift at the end of the Pleistocene. *American Naturalist* 146:765–794.

Zimov, S.A., S.P. Davidov, G.M. Zimova, A.I. Davidova, F.S. Chapin III, and M.C. Chapin. 1999. Contribution of disturbance to high-latitude amplification of atmospheric CO_2. *Science* 284: 1973–1976.

Zimov, S.A., Y.V. Voropaev, I.P. Semiletov, S.P. Davidov, S.F. Prosiannikov, F.S. Chapin III, M.C. Chapin, S. Trumbore, and S. Tyler. 1997. North Siberian lakes: A methane source fueled by Pleistocene carbon. *Science* 277:800–802.

Index